T0142256

Studies in Computational Intelligence

Volume 667

About this Series

The series "Studies in Computational Intelligence" (SCI) publishes new developments and advances in the various areas of computational intelligence—quickly and with a high quality. The intent is to cover the theory, applications, and design methods of computational intelligence, as embedded in the fields of engineering, computer science, physics and life sciences, as well as the methodologies behind them. The series contains monographs, lecture notes and edited volumes in computational intelligence spanning the areas of neural networks, connectionist systems, genetic algorithms, evolutionary computation, artificial intelligence, cellular automata, self-organizing systems, soft computing, fuzzy systems, and hybrid intelligent systems. Of particular value to both the contributors and the readership are the short publication timeframe and the worldwide distribution, which enable both wide and rapid dissemination of research output.

More information about this series at http://www.springer.com/series/7092

Patricia Melin · Oscar Castillo
Janusz Kacprzyk
Editors

Nature-Inspired Design of Hybrid Intelligent Systems

Editors
Patricia Melin
Division of Graduate Studies and Research
Tijuana Institute of Technology
Tijuana, BC
Mexico

Janusz Kacprzyk
Polish Academy of Sciences
Systems Research Institute
Warsaw
Poland

Oscar Castillo
Division of Graduate Studies and Research
Tijuana Institute of Technology
Tijuana, BC
Mexico

ISSN 1860-949X ISSN 1860-9503 (electronic)
Studies in Computational Intelligence
ISBN 978-3-319-83650-8 ISBN 978-3-319-47054-2 (eBook)
DOI 10.1007/978-3-319-47054-2

Printed on acid-free paper

This Springer imprint is published by Springer Nature
The registered company is Springer International Publishing AG
The registered company address is: Gewerbestrasse 11, 6330 Cham, Switzerland

Preface

We describe in this book, recent advances on the design of hybrid intelligent systems based on nature-inspired optimization and their application in areas such as intelligent control and robotics, pattern recognition, medical diagnosis, time series prediction and optimization of complex problems. The book is organized into seven main parts, with each containing a group of chapters around a similar subject. The first part consists of chapters related to theoretical aspects of type-2 and intuitionistic fuzzy logic, i.e., the chapters that propose new concepts and algorithms based on type-2 and intuitionistic fuzzy systems. The second part contains chapters on neural networks theory, which are basically chapters dealing with new concepts and algorithms in neural networks. The second part also contains chapters describing applications of neural networks in diverse areas, such as time series prediction and pattern recognition. The third part contains chapters that present enhancements to metaheuristics based on fuzzy logic techniques describing new nature-inspired optimization algorithms that use fuzzy dynamic adaptation of parameters. The fourth part presents diverse applications of nature-inspired optimization algorithms. The fifth part contains chapters describing applications of fuzzy logic in diverse areas, such as time series prediction and pattern recognition. The sixth part contains chapters describing new optimization algorithms and their applications in different areas. Finally, the seventh part contains chapters that present the design and application of different hybrid intelligent systems.

In the first part of theoretical aspects of type-2 and intuitionistic fuzzy logic, there are eight chapters that describe different contributions that propose new models, concepts and algorithms centered on type-2 and intuitionistic fuzzy systems. The aim of using fuzzy logic is to provide uncertainty management in modeling complex problems.

In the second part of neural networks theory and applications, there are eight chapters that describe different contributions that propose new models, concepts and algorithms centered on neural networks. The aim of using neural networks is to provide learning and adaptive capabilities to intelligent systems. There are also chapters that describe different contributions on the application of these kinds of neural models to solve complex real-world problems, such as time series prediction, medical diagnosis, and pattern recognition.

In the third part of fuzzy logic for the augmentation of nature-inspired optimization metaheuristics, there are ten chapters that describe different contributions that propose new models and concepts, which can be considered as the basis for enhancing nature-inspired algorithms with fuzzy logic. The aim of using fuzzy logic is to provide dynamic adaptation capabilities to the optimization algorithms, and this is illustrated with the cases of the bat algorithm, harmony search and other methods. The nature-inspired methods include variations of ant colony optimization, particle swarm optimization, bat algorithm, as well as new nature-inspired paradigms.

In the fourth part of nature-inspired optimization applications, there are seven chapters that describe different contributions on the application of these kinds of metaheuristic algorithms to solve complex real-world optimization problems, such as time series prediction, medical diagnosis, robotics, and pattern recognition.

In the fifth part of fuzzy logic applications there are six chapters that describe different contributions on the application of these kinds of fuzzy logic models to solve complex real-world problems, such as time series prediction, medical diagnosis, fuzzy control, and pattern recognition.

In the sixth part of optimization, there are nine chapters that describe different contributions that propose new models, concepts and algorithms for optimization inspired in different paradigms. The aim of using these algorithms is to provide general optimization methods and solution to some real-world problem in areas, such as scheduling, planning and project portfolios.

In the seventh part, there are eight chapters that present nature-inspired design and applications of different hybrid intelligent systems. There are also chapters that describe different contributions on the application of these kinds of hybrid intelligent systems to solve complex real-world problems, such as time series prediction, medical diagnosis, and pattern recognition.

In conclusion, the edited book comprises chapters on diverse aspects of fuzzy logic, neural networks, and nature-inspired optimization metaheuristics and their application in areas such as such as intelligent control and robotics, pattern recognition, time series prediction and optimization of complex problems. There are theoretical aspects as well as application chapters.

Tijuana, Mexico — Patricia Melin
Tijuana, Mexico — Oscar Castillo
Warsaw, Poland — Janusz Kacprzyk
June 2016

Contents

Part I Type-2 and Intuitionistic Fuzzy Logic

General Type-2 Fuzzy Edge Detection in the Preprocessing of a Face Recognition System 3
Claudia I. Gonzalez, Patricia Melin, Juan R. Castro, Olivia Mendoza and Oscar Castillo

An Overview of Granular Computing Using Fuzzy Logic Systems 19
Mauricio A. Sanchez, Oscar Castillo and Juan R. Castro

Optimization of Type-2 and Type-1 Fuzzy Integrator to Ensemble Neural Network with Fuzzy Weights Adjustment 39
Fernando Gaxiola, Patricia Melin, Fevrier Valdez and Juan R. Castro

Interval Type-2 Fuzzy Possibilistic C-Means Optimization Using Particle Swarm Optimization 63
Elid Rubio and Oscar Castillo

Choquet Integral and Interval Type-2 Fuzzy Choquet Integral for Edge Detection 79
Gabriela E. Martínez, D. Olivia Mendoza, Juan R. Castro, Patricia Melin and Oscar Castillo

Bidding Strategies Based on Type-1 and Interval Type-2 Fuzzy Systems for Google AdWords Advertising Campaigns 99
Quetzali Madera, Oscar Castillo, Mario Garcia and Alejandra Mancilla

On the Graphical Representation of Intuitionistic Membership Functions for Its Use in Intuitionistic Fuzzy Inference Systems 115
Amaury Hernandez-Aguila, Mario Garcia-Valdez and Oscar Castillo

A Gravitational Search Algorithm Using Type-2 Fuzzy Logic for Parameter Adaptation 127
Beatriz González, Fevrier Valdez and Patricia Melin

Part II Neural Networks Theory and Applications

Particle Swarm Optimization of the Fuzzy Integrators for Time Series Prediction Using Ensemble of IT2FNN Architectures 141
Jesus Soto, Patricia Melin and Oscar Castillo

Long-Term Prediction of a Sine Function Using a LSTM Neural Network 159
Magdiel Jiménez-Guarneros, Pilar Gómez-Gil, Rigoberto Fonseca-Delgado, Manuel Ramírez-Cortés and Vicente Alarcón-Aquino

UAV Image Segmentation Using a Pulse-Coupled Neural Network for Land Analysis 175
Mario I. Chacon-Murguia, Luis E. Guerra-Fernandez and Hector Erives

Classification of Arrhythmias Using Modular Architecture of LVQ Neural Network and Type 2 Fuzzy Logic 187
Jonathan Amezcua and Patricia Melin

A New Method Based on Modular Neural Network for Arterial Hypertension Diagnosis 195
Martha Pulido, Patricia Melin and German Prado-Arechiga

Spectral Characterization of Content Level Based on Acoustic Resonance: Neural Network and Feedforward Fuzzy Net Approaches 207
Juan Carlos Sanchez-Diaz, Manuel Ramirez-Cortes, Pilar Gomez-Gil, Jose Rangel-Magdaleno, Israel Cruz-Vega and Hayde Peregrina-Barreto

Comparison of Optimization Techniques for Modular Neural Networks Applied to Human Recognition 225
Daniela Sánchez, Patricia Melin, Juan Carpio and Hector Puga

A Competitive Modular Neural Network for Long-Term Time Series Forecasting 243
Eduardo Méndez, Omar Lugo and Patricia Melin

Part III Fuzzy Metaheuristics

Differential Evolution Using Fuzzy Logic and a Comparative Study with Other Metaheuristics 257
Patricia Ochoa, Oscar Castillo and José Soria

An Adaptive Fuzzy Control Based on Harmony Search and Its Application to Optimization 269
Cinthia Peraza, Fevrier Valdez and Oscar Castillo

A Review of Dynamic Parameter Adaptation Methods for the Firefly Algorithm 285
Carlos Soto, Fevrier Valdez and Oscar Castillo

Fuzzy Dynamic Adaptation of Parameters in the Water Cycle Algorithm 297
Eduardo Méndez, Oscar Castillo, José Soria and Ali Sadollah

Fireworks Algorithm (FWA) with Adaptation of Parameters Using Fuzzy Logic 313
Juan Barraza, Patricia Melin, Fevrier Valdez and Claudia González

Imperialist Competitive Algorithm with Dynamic Parameter Adaptation Applied to the Optimization of Mathematical Functions . . . 329
Emer Bernal, Oscar Castillo and José Soria

Modification of the Bat Algorithm Using Type-2 Fuzzy Logic for Dynamical Parameter Adaptation 343
Jonathan Pérez, Fevrier Valdez and Oscar Castillo

Flower Pollination Algorithm with Fuzzy Approach for Solving Optimization Problems 357
Luis Valenzuela, Fevrier Valdez and Patricia Melin

A Study of Parameters of the Grey Wolf Optimizer Algorithm for Dynamic Adaptation with Fuzzy Logic 371
Luis Rodríguez, Oscar Castillo and José Soria

Gravitational Search Algorithm with Parameter Adaptation Through a Fuzzy Logic System 391
Frumen Olivas, Fevrier Valdez and Oscar Castillo

Part IV Metaheuristic Applications

Particle Swarm Optimization of Ensemble Neural Networks with Type-1 and Type-2 Fuzzy Integration for the Taiwan Stock Exchange 409
Martha Pulido, Patricia Melin and Olivia Mendoza

A New Hybrid PSO Method Applied to Benchmark Functions 423
Alfonso Uriarte, Patricia Melin and Fevrier Valdez

On the Use of Parallel Genetic Algorithms for Improving the Efficiency of a Monte Carlo-Digital Image Based Approximation of Eelgrass Leaf Area I: Comparing the Performances of Simple and Master-Slaves Structures 431
Cecilia Leal-Ramírez, Héctor Echavarría-Heras, Oscar Castillo and Elia Montiel-Arzate

Social Spider Algorithm to Improve Intelligent Drones Used in Humanitarian Disasters Related to Floods 457
Alberto Ochoa, Karina Juárez-Casimiro, Tannya Olivier, Raymundo Camarena and Irving Vázquez

An Optimized GPU Implementation for a Path Planning Algorithm Based on Parallel Pseudo-bacterial Potential Field 477
Ulises Orozco-Rosas, Oscar Montiel and Roberto Sepúlveda

Estimation of Population Pharmacokinetic Parameters Using a Genetic Algorithm 493
Carlos Sepúlveda, Oscar Montiel, José. M. Cornejo Bravo and Roberto Sepúlveda

Optimization of Reactive Control for Mobile Robots Based on the CRA Using Type-2 Fuzzy Logic 505
David de la O, Oscar Castillo and Jose Soria

Part V Fuzzy Logic Applications

A FPGA-Based Hardware Architecture Approach for Real-Time Fuzzy Edge Detection 519
Emanuel Ontiveros-Robles, José González Vázquez, Juan R. Castro and Oscar Castillo

A Hybrid Intelligent System Model for Hypertension Diagnosis 541
Ivette Miramontes, Gabriela Martínez, Patricia Melin and German Prado-Arechiga

Comparative Analysis of Designing Differents Types of Membership Functions Using Bee Colony Optimization in the Stabilization of Fuzzy Controllers 551
Leticia Amador-Angulo and Oscar Castillo

Neuro-Fuzzy Hybrid Model for the Diagnosis of Blood Pressure 573
Juan Carlos Guzmán, Patricia Melin and German Prado-Arechiga

Microcalcification Detection in Mammograms Based on Fuzzy Logic and Cellular Automata 583
Yoshio Rubio, Oscar Montiel and Roberto Sepúlveda

Sensor Less Fuzzy Logic Tracking Control for a Servo System with Friction and Backlash 603
Nataly Duarte, Luis T. Aguilar and Oscar Castillo

Part VI Optimization: Theory and Applications

Differential Evolution with Self-adaptive Gaussian Perturbation 617
M.A. Sotelo-Figueroa, Arturo Hernández-Aguirre, Andrés Espinal and J.A. Soria-Alcaraz

Optimization Mathematical Functions for Multiple Variables Using the Algorithm of Self-defense of the Plants 631
Camilo Caraveo, Fevrier Valdez and Oscar Castillo

Evaluation of the Evolutionary Algorithms Performance in Many-Objective Optimization Problems Using Quality Indicators 641
Daniel Martínez-Vega, Patricia Sanchez, Guadalupe Castilla, Eduardo Fernandez, Laura Cruz-Reyes, Claudia Gomez and Enith Martinez

Generating Bin Packing Heuristic Through Grammatical Evolution Based on Bee Swarm Optimization 655
Marco Aurelio Sotelo-Figueroa, Héctor José Puga Soberanes, Juan Martín Carpio, Héctor J. Fraire Huacuja, Laura Cruz Reyes, Jorge Alberto Soria Alcaraz and Andrés Espinal

Integer Linear Programming Formulation and Exact Algorithm for Computing Pathwidth 673
Héctor J. Fraire-Huacuja, Norberto Castillo-García, Mario C. López-Locés, José A. Martínez Flores, Rodolfo A. Pazos R., Juan Javier González Barbosa and Juan M. Carpio Valadez

Iterated VND Versus Hyper-heuristics: Effective and General Approaches to Course Timetabling 687
Jorge A. Soria-Alcaraz, Gabriela Ochoa, Marco A. Sotelo-Figueroa, Martín Carpio and Hector Puga

AMOSA with Analytical Tuning Parameters for Heterogeneous Computing Scheduling Problem 701
Héctor Joaquín Fraire Huacuja, Juan Frausto-Solís, J. David Terán-Villanueva, José Carlos Soto-Monterrubio, J. Javier González Barbosa and Guadalupe Castilla-Valdez

Increase Methodology of Design of Course Timetabling Problem for Students, Classrooms, and Teachers 713
Lucero de M. Ortiz-Aguilar, Martín Carpio, Héctor Puga, Jorge A. Soria-Alcaraz, Manuel Ornelas-Rodríguez and Carlos Lino

Solving the Cut Width Optimization Problem with a Genetic Algorithm Approach 729
Hector Joaquín Fraire-Huacuja, Mario César López-Locés, Norberto Castillo García, Johnatan E. Pecero and Rodolfo Pazos Rangel

Part VII Hybrid Intelligent Systems

A Dialogue Interaction Module for a Decision Support System Based on Argumentation Schemes to Public Project Portfolio 741
Laura Cruz-Reyes, César Medina-Trejo, María Lucila Morales-Rodríguez, Claudia Guadalupe Gómez-Santillan, Teodoro Eduardo Macias-Escobar, César Alejandro Guerrero-Nava and Mercedes Pérez-Villafuerte

Implementation of an Information Retrieval System Using the Soft Cosine Measure 757
Juan Javier González Barbosa, Juan Frausto Solís, J. David Terán-Villanueva, Guadalupe Castilla Valdés, Rogelio Florencia-Juárez, Lucía Janeth Hernández González and Martha B. Mojica Mata

TOPSIS-Grey Method Applied to Project Portfolio Problem 767
Fausto Balderas, Eduardo Fernandez, Claudia Gomez, Laura Cruz-Reyes and Nelson Rangel V

Comparing Grammatical Evolution's Mapping Processes on Feature Generation for Pattern Recognition Problems 775
Valentín Calzada-Ledesma, Héctor José Puga-Soberanes, Alfonso Rojas-Domínguez, Manuel Ornelas-Rodríguez, Juan Martín Carpio-Valadez and Claudia Guadalupe Gómez-Santillán

Hyper-Parameter Tuning for Support Vector Machines by Estimation of Distribution Algorithms 787
Luis Carlos Padierna, Martín Carpio, Alfonso Rojas, Héctor Puga, Rosario Baltazar and Héctor Fraire

Viral Analysis on Virtual Communities: A Comparative of Tweet Measurement Systems 801
Daniel Azpeitia, Alberto Ochoa-Zezzatti and Judith Cavazos

Improving Decision-Making in a Business Simulator Using TOPSIS Methodology for the Establishment of Reactive Stratagems 809
Alberto Ochoa, Saúl González, Emmanuel Moriel, Julio Arreola and Fernando García

Non-singleton Interval Type-2 Fuzzy Systems as Integration Methods in Modular Neural Networks Used Genetic Algorithms to Design 821
Denisse Hidalgo, Patricia Melin and Juan R. Castro

Part I
Type-2 and Intuitionistic Fuzzy Logic

General Type-2 Fuzzy Edge Detection in the Preprocessing of a Face Recognition System

Claudia I. Gonzalez, Patricia Melin, Juan R. Castro, Olivia Mendoza and Oscar Castillo

Abstract In this paper, we present the advantage of using a general type-2 fuzzy edge detector method in the preprocessing phase of a face recognition system. The Sobel and Prewitt edge detectors combined with GT2 FSs are considered in this work. In our approach, the main idea is to apply a general type-2 fuzzy edge detector on two image databases to reduce the size of the dataset to be processed in a face recognition system. The recognition rate is compared using different edge detectors including the fuzzy edge detectors (type-1 and interval type-2 FS) and the traditional Prewitt and Sobel operators.

Keywords General type-2 fuzzy logic · Alpha planes · Fuzzy edge detection · Pattern recognition systems · Neural networks

1 Introduction

Edge detection is one of the most common approaches to detect discontinuities in gray-scale images. Edge detection can be considered an essential method used in the image processing area and can be applied in image segmentation, object recognition systems, feature extraction, and target tracking [2].

There are several edge detection methods, which include the traditional ones, such as Sobel [29], Prewitt [27], Canny [2], Kirsch [9], and those based in type-1 [8, 25, 30, 31], interval type-2 [1, 20, 23, 24] and general fuzzy systems [6, 14]. In Melin et al. [14] and Gonzalez et al. [6], some edge detectors based on GT2 FSs have been proposed. In these works the results achieved by the GT2 FS are compared with others based on a T1 FS and with an IT2 FS. According with the

C.I. Gonzalez · J.R. Castro · O. Mendoza
Autonomous University of Baja California, Tijuana, Mexico

P. Melin · O. Castillo (✉)
Tijuana Institute of Technology, Tijuana, Mexico
e-mail: ocastillo@tectijuana.mx

P. Melin et al. (eds.), *Nature-Inspired Design of Hybrid Intelligent Systems*,
Studies in Computational Intelligence 667, DOI 10.1007/978-3-319-47054-2_1

results obtained in these papers, the conclusion is that the edge detector based on GT2 FS is better than an IT2 FS and a T1 FS.

In other works, like in [21], an edge detector based on T1 FS and other IT2 FS are implemented in the preprocessing phase of a face recognition system. According to the recognition rates achieved in this paper, the authors conclude that the recognition system has better performance when the IT2 fuzzy edge detector is applied.

In this paper, we present a face recognition system, which is performed with a monolithic neural network. In the methodology, two GT2 fuzzy edge detectors are applied over two face databases. In the first edge detector, a GT2 FS is combined with the Prewitt operator and the second with the Sobel operator. The edge datasets achieved by these GT2 fuzzy edge detectors are used as the inputs of the neural network in a face recognition system.

The aim of this work is to show the advantage of using a GT2 fuzzy edge detector in pattern recognition applications. Additionally, make a comparative analysis with the recognition rates obtained by the GT2 against the results achieved in [21] by T1 and T2 fuzzy edge detectors.

The remainder of this paper is organized as follows. Section 2 gives a review of the background on GT2 FS. The basic concepts about Prewitt, Sobel operator, low-pass filter and high-pass filter are described in Sect. 3. The methodology used to develop the GT2 fuzzy edge detector is explained in Sect. 4. The design of the recognition system based on monolithic neural network is presented in Sect. 5. The recognition rates achieved by the face recognition system and the comparative results are show in Sect. 6. Finally, Sect. 7 offers some conclusions about the results.

2 Overview of General Type-2 Fuzzy Sets

The GT2 FSs have attracted attention from the research community, and have been applied in different areas, like pattern recognition, control systems, image processing, robotics, and decision-making to name a few [3, 5, 13, 16, 28, 34]. It has been demonstrated that a GT2 FS can have the ability to handle great uncertainties.

In the following, we present a brief description about GT2 FS theory, which are used in the methodology proposed in this paper.

2.1 *Definition of General Type-2 Fuzzy Sets*

A general type-2 fuzzy set $(\tilde{A})$ consists of the primary variable x having domain *X*, the secondary variable u with domain in J_x^u at each $x \in X$. The secondary membership grade $\mu_{\tilde{A}}(x, u)$ is a 3D membership function where $0 \leq \mu_{\underset{A}{\sim}}(x, u) \leq 1$ [18, 37, 38]. It can be expressed by (1)

$$\tilde{A} = \left\{ \left((x, u), \mu_{\tilde{A}}(x, u) \right) | \forall x \in X, \ \ \forall u \in J_x^u \subseteq [0, 1] \right\}. \tag{1}$$

The footprint of uncertainty (FOU) of $\left(\tilde{A}\right)$ is the two-dimensional support of $\mu_{\tilde{A}}(x, u)$ and can be expressed by (2)

$$FOU\left(\tilde{A}\right) = \{(x, u) \in X \times [0,1] | \mu_{\tilde{A}}(x, u) > 0\}. \tag{2}$$

2.2 General Type-2 Fuzzy Systems

The general structure of a GT2 FLS is shown in Fig. 1 and this consists of five main blocks, which are the input fuzzification, fuzzy inference engine, fuzzy rule base, type-reducer, and defuzzifier [16].

In a GT2 FLS first the fuzzifier process maps a crisp input vector into other GT2 input FSs. In the inference engine, the fuzzy rules are combined and provide a mapping from GT2 FSs input to GT2 FSs output. This GT2 FSs output is reduced to a T1 FSs by the type-reduction process [35, 36].

There are different type-reduction methods, the most commonly used are the Centroid, Height and Center-of-sets type reduction. In this paper, we applied Centroid type-reduction. The Centroid definition $C_{\tilde{A}}$ of a GT2 FLS [11, 12, 17] is expressed in (3)

$$\begin{aligned} C_{\tilde{A}} = \{(z_i, \mu(z_i)) | z_i \in \frac{\sum_{i=1}^{N} x_i \theta_i}{\sum_{i=1}^{N} \theta_i}, \\ \mu(z_i) \in f_{x_1}(\theta_1) \times \cdots \times f_{x_N}(\theta_N), \quad \theta_i \in J_{x_1} \times \cdots \times J_{x_N}\} \end{aligned} \tag{3}$$

where θ_i is a combination associated to the secondary degree $f_{x_1}(\theta_1) * \cdots * f_{x_N}(\theta_N)$.

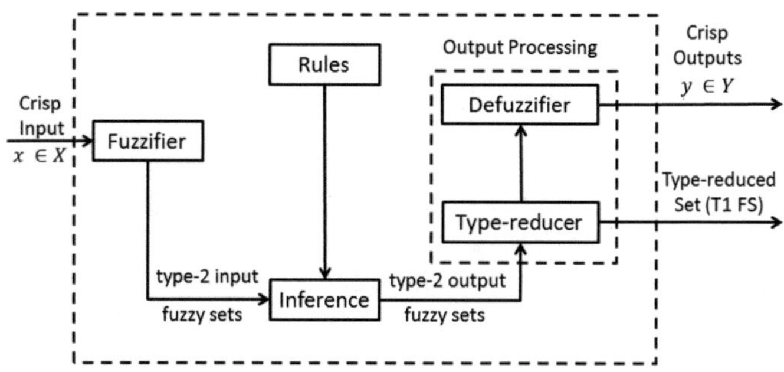

Fig. 1 General type-2 fuzzy logic system

2.3 General Type-2 Fuzzy Systems Approximations

Due to the fact that a GT2 FSs defuzzification process is computationally more complex than T1 and IT2 FSs; several approximation techniques have been proposed, some of them are the zSlices [33, 34] and the $\alpha-$plane representation [15, 19]. In these two approaches, the 3D GT2 membership functions are decomposed using different cuts to achieve a collection of IT2 FSs.

In this paper, the defuzzifier process is performed using $\alpha-$plane approximation, which is defined as follows.

An α-plane for a GT2 FS $\tilde{A}$, is denoted by $\tilde{A}\alpha$, and it is the union of all primary membership functions of $\tilde{A}$, which secondary membership degrees are higher or equal than α $(0 \leq \alpha \geq 1)$ [15, 19]. The $\alpha-$plane is expressed in (4)

$$\tilde{A}_\alpha = \{(x,u)|\mu_{\tilde{A}}(x,u) \geq \alpha,\ \ \forall x \in X,\ \ \forall u \in [0,1]\} \tag{4}$$

3 Edge Detection and Filters

In this section, we introduce some concepts about filters and edge detectors (Prewitt and Sobel) using in image processing areas; since, these are used as a basis in our investigation.

3.1 Prewitt Operator

The Prewitt operator is used for edge detection in digital images. This consists of a pair of 3 × 3 convolution kernels which are defined in (5) and (6) [7].

$$\text{Prewittx} = \begin{bmatrix} -1 & -1 & -1 \\ 0 & 0 & 0 \\ 1 & 1 & 1 \end{bmatrix} \tag{5}$$

$$\text{Prewitty} = \begin{bmatrix} -1 & 0 & 1 \\ -1 & 0 & 1 \\ -1 & 0 & 1 \end{bmatrix} \tag{6}$$

The kernels in (5) and (6) can be applied separately to the input image (I), to produce separate measurements of the gradient component (7), (8) in horizontal (gx) and vertical orientation (gy), respectively [7].

$$gx = \text{Prewittx} * I \tag{7}$$

$$gy = \text{Prewitty} * I \tag{8}$$

The gradient components (7) and (8) can be combined together to find the magnitude of the gradient at each point and the orientation of that gradient [22, 29]. The gradient magnitude (G) is given by (9).

$$G = \sqrt{gx^2 + gy^2} \tag{9}$$

3.2 Sobel Operator

Sobel operator is similar to the Prewitt operator. The only difference is that the Sobel operator use the kernels expressed in (10) and (11) to detect the vertical and horizontal edges.

$$\text{Sobelx} = \begin{bmatrix} -1 & -2 & -1 \\ 0 & 0 & 0 \\ 1 & 2 & 1 \end{bmatrix} \tag{10}$$

$$\text{Sobely} = \begin{bmatrix} -1 & 0 & 1 \\ -2 & 0 & 2 \\ -1 & 0 & 1 \end{bmatrix} \tag{11}$$

3.3 Low-Pass Filter

Low-pass filters are used for image smoothing and noise reduction; this allows only passing the low frequencies of the image [21]. Also is employed to remove high spatial frequency noise from a digital image. This filter can be implemented using (12) and the mask (highM) used to obtained the highPF is expressed in (13).

$$\text{lowPF} = \text{lowM} * I \tag{12}$$

$$\text{lowM} = \frac{1}{25} * \begin{bmatrix} 1 & 0 & 0 & 0 & 0 \\ 0 & 1 & 0 & 0 & 0 \\ 0 & 0 & 1 & 0 & 0 \\ 0 & 0 & 0 & 1 & 0 \\ 0 & 0 & 0 & 0 & 1 \end{bmatrix} \tag{13}$$

3.4 High-Pass Filter

High-pass filter only allows the high frequency of the image to pass through the filter and that all of the other frequency are blocked. This filter will highlight regions with intensity variations, such as an edge (will allow to pass the high frequencies) [21]. The high-pass (highPF) filter is implemented using (14)

$$\text{highPF} = \text{highM} * I \tag{14}$$

where highM in (14) represents the mask used to obtained the highPF and this is defined by (15)

$$\text{highM} = \begin{bmatrix} -1/16 & -1/8 & -1/16 \\ -1/8 & 3/4 & -1/8 \\ -1/16 & -1/8 & -1/16 \end{bmatrix} \tag{15}$$

4 Edge Detection Improved with a General Type-2 Fuzzy System

In our approach, two edge detectors are improved, in the first a GT2 FS is combined with Prewitt operator and the second with the Sobel operator. The general structure used to obtain the first GT2 fuzzy edge detector is shown in Fig. 2. The second fuzzy edge detector has a similar structure; we only change the kernel using the Sobel operators in (10) and (11), which are described in Sect. 3.

The GT2 fuzzy edge detector is calculated as follows. To start, we select an input image (I) of the images database; after that, the horizontal gx (7) and vertical gy (8)

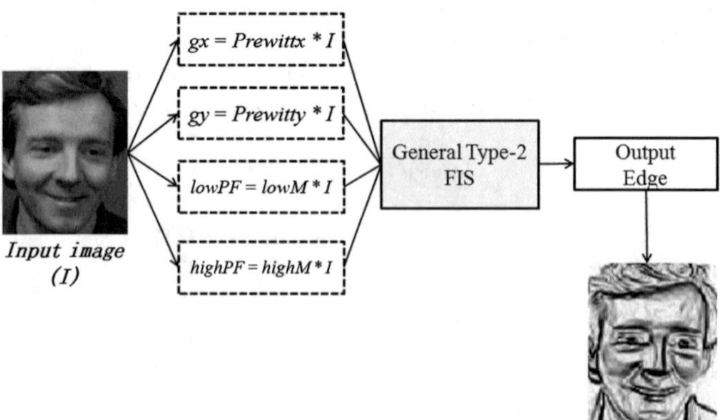

Fig. 2 Edge detector improved with Prewitt operator and GT2 FSs

image gradients are obtained; moreover, the low-pass (12) and high-pass (14) filters are also applied over (I).

The GT2 FIS was built using four inputs and one output. The inputs are the values obtained by gx (7), gy (8), lowPF (12) and highPF (14); otherwise, the output inferred represents the fuzzy gradient magnitude which is labeled as Output Edge.

An example of the input and output membership functions used in the GT2 FIS is shown in Figs. 3 and 4 respectively.

In order to objectively compare the performance of the proposed edge detectors against the results achieved in Mendoza [21], we use a similar knowledge base of fuzzy rules; these rules were designed as follows:

1. If (dx is LOW) and (dy is LOW) then (OutputEdge is HIGH)
2. If (dx is MIDDLE) and (dy is MIDDLE) then (OutputEdge is LOW)
3. If (dx is HIGH) and (dy is HIGH) then (OutputEdge is LOW)
4. If (dx is MIDDLE) and (highPF is LOW) then (OutputEdge is LOW)
5. If (dy is MIDDLE) and (highPF is LOW) then (OutputEdge is LOW)
6. If (lowPF is LOW) and (dy is MIDDLE) then (OutputEdge is HIGH)
7. If (lowPF is LOW) and (dx is MIDDLE) then (OutputEdge is HIGH).

5 Face Recognition System Using Monolithic Neural Network and a GT2 Fuzzy Edge Detector

The aim of this work is to apply a GT2 fuzzy edge detector in a preprocessing phase in a face recognition system. In our study case, the recognition system is performed using a Monolithic Neural Networks. As already mentioned in Sect. 5, the edge detectors were designed using GT2 fuzzy combined with Prewitt and Sobel operator.

In Fig. 5 as an illustration, the general structure used in the proposed face recognition system is shown. The methodology used in the process is summarized in the following steps:

A. **Select the input images database**

In the simulation results, two benchmark face databases were selected; in which are included the ORL [32] and the Cropped Yale [4, 10, 26].

B. **Applied the edge detection in the input images**

In this preprocessing phase, the two GT2 fuzzy edge detectors described in Sect. 5 were applied on the ORL and Cropped Yale database.

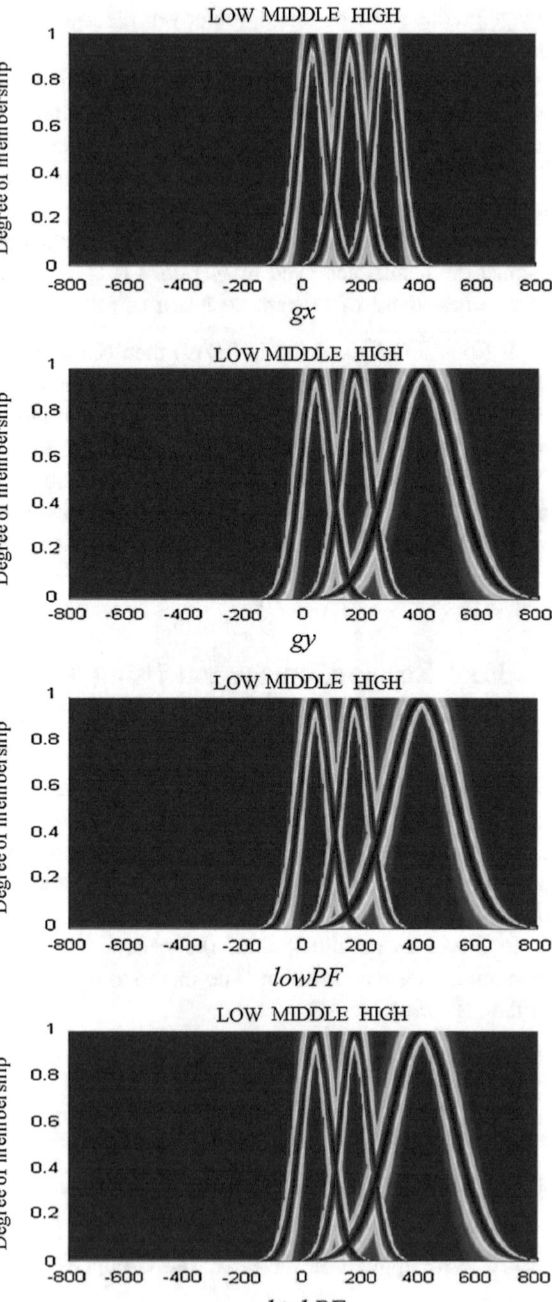

Fig. 3 Input membership functions using in the GT2 FIS

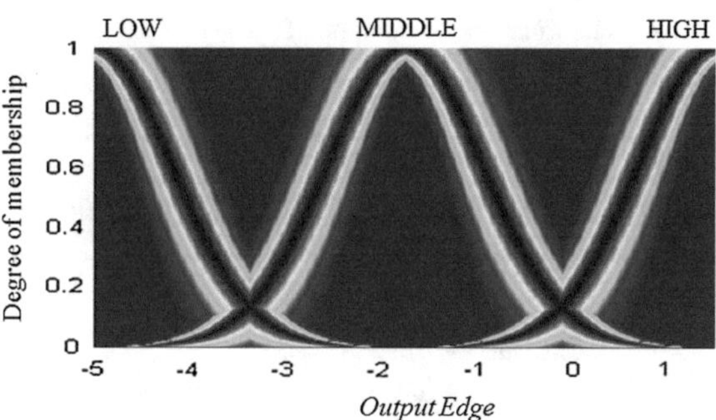

Fig. 4 Output membership function using in the GT2 FIS

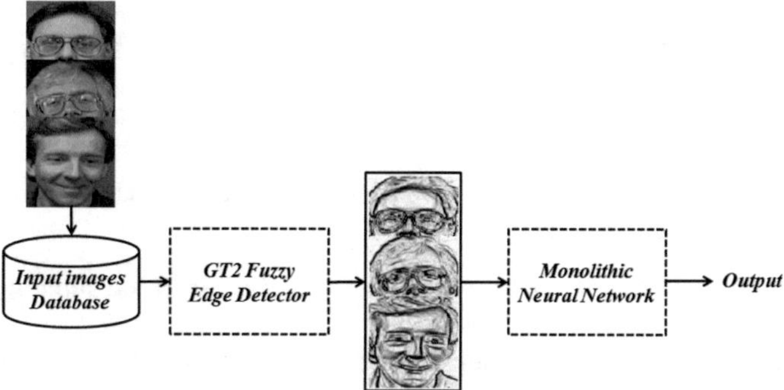

Fig. 5 General structure for the face recognition system

C. **Training the monolithic neural network**

The images obtained in the edge detection phase are used as the inputs of the neural network. In order to evaluate more objectively the recognition rate, the *k*-fold cross validation method was used. The training process is defined as follows:

1. Define the parameters for the monolithic neural network [21].
   - Layers hidden: two
   - Neuron number in each layer: 200
   - Learning algorithm: Gradient descent with momentum and adaptive learning
   - Error goal: 1e-4.
2. The indices for training and test k-folds were calculated as follows:
   - Define the people number (p).
   - Define the sample number for each person (s).

Table 1 Information for the tested database of faces

Database	People number (p)	Samples number (s)	Fold size (m)	Training set size (i)	Test size (t)
ORL	40	10	80	320	80
Cropped Yale	38	10	76	304	76

- Define the k-folds $(k = 5)$.
- Calculate the number of samples (m) in each fold using (16)

$$m = (s/k) \cdot p \tag{16}$$

- The train data set size (i) is calculated in (17)

$$i = m(k - 1) \tag{17}$$

- Finally, the test data set size (18) are the number of samples in only one fold.

$$t = m \tag{18}$$

- The train set and test set obtained for the three face database used in this work are shown in Table 1.

3. The neural network was training $k - 1$ times, one for each training fold calculated previously.
4. The neural network was testing k times, one for each fold test set calculated previously.

Finally, the mean of the rates of all the k-folds are calculated to obtain the recognition rate.

6 Experimental Results

This section provides a comparison of the recognition rates achieved by the face recognition system when different fuzzy edge detectors were applied.

In the experimental results several edge detectors were analyzed in which are included the Sobel operator, Sobel combined with T1 FLS, IT2 FLS, and GT2 FLS. Besides these, the Prewitt operator, Prewitt based on T1 FLS, IT2 FLS, and GT2 FLS are also considered. Additional to this, the experiments were also validated without using any edge detector.

The tests were executed using the ORL and the Cropped Yale database; an example of these faces database is shown in Table 2. The parameters used in the monolithic neural network are described in Sect. 5. Otherwise, the training set and

Table 2 Faces database

Database	Examples
ORL	
Cropped Yale	

Table 3 Recognition rate for ORL database using GT2 fuzzy edge detector

Fuzzy system edge detector	Mean rate (%)	Standard deviation	Max rate (%)
Sobel + GT2 FLS	87.97	0.0519	96.50
Prewitt + GT2 FLS	87.68	0.0470	96.25

testing set that we considered in the tests are presented in Table 1, and these values depend on the database size used.

It is important to mention that all values presented below are the results of the average of 30 simulations achieved by the monolithic neural network. For this reason, the results presented in this section cannot be compared directly with the results achieved in [21]; because, in [21] only are presented the best solutions.

In the first test, the face recognition system was performed using the Prewitt and Sobel GT2 fuzzy edge detectors. This test was applied over the ORL data set. The mean rate, standard deviation, and max rate values achieved by the system are shown in Table 3. In this Table, we can note that better results were obtained when the Sobel GT2 fuzzy edge detector was applied; with a mean rate of 87.97, an standard deviation of 0.0519, and max rate of 96.50.

As a part of this test in Table 4, the results of 30 simulations are shown; these results were achieved by the system when the Prewitt GT2 fuzzy edge detector is applied.

In another test, the system was considered using the Cropped Yale database. The numeric results for this experiment are presented in Table 5. In this Table, we can notice that both edge detectors achieved the same max rate value; but, the mean rate was better with the Sobel + GT2 FLS.

As part of the goals of this work, the recognition rate values achieved by the system when the GT2 fuzzy edge detector is used were compared with the results obtained when the neural network is training without edge detection, the Prewitt operator, the Prewitt combined with T1 and IT2 FSs; also, the Sobel operator, the Sobel edge detector combined with T1 and IT2 FSs. The results achieved after to apply these different edge detection methods are shown in Tables 6 and 7.

Table 4 Recognition rate for ORL database using Prewitt GT2 fuzzy edge detector

Simulation number	Mean rate (%)	Standard deviation	Max rate (%)
1	83.50	0.0408	88.75
2	87.75	0.0357	92.50
3	89.75	0.0323	93.75
4	88.50	0.0427	95.00
5	70.25	0.3728	91.25
6	86.75	0.0420	91.25
7	89.00	0.0162	91.25
8	87.25	0.0205	88.75
9	86.75	0.0227	90.00
10	90.00	0.0441	**96.25**
11	89.50	0.0381	95.00
12	90.00	0.0265	92.50
13	89.75	0.0205	92.50
14	89.25	0.0360	92.50
15	92.00	0.0189	93.75
16	88.00	0.0447	93.75
17	86.25	0.0605	**96.25**
18	90.25	0.0323	95.00
19	90.50	0.0189	92.50
20	88.25	0.0227	91.25
21	89.00	0.0323	92.50
22	89.25	0.0167	91.25
23	89.00	0.0503	95.00
24	87.00	0.0447	91.25
25	89.25	0.0189	92.50
26	85.00	0.0776	93.75
27	87.00	0.0512	93.75
28	89.75	0.0399	95.00
29	86.00	0.0408	90.00
30	86.00	0.0503	93.75
Average	87.68	0.0470	92.75

Table 5 Recognition rate for Cropped Yale database using GT2 fuzzy edge detector

Fuzzy system edge detector	Mean rate (%)	Standard deviation	Max rate
Sobel + GT2 FLS	93.16	0.0328	100
Prewitt + GT2 FLS	97.58	0.0328	100

Table 6 Recognition rate for ORL database

Fuzzy system edge detector	Mean rate (%)	Standard deviation	Max rate
None	2.59	0.0022	5.00
Sobel operator	2.70	0.0037	5.00
Sobel + T1FLS	86.16	0.0486	93.75
Sobel + IT2FLS	87.35	0.0373	95.00
Sobel + GT2 FLS	**87.97**	**0.0519**	**96.50**
Prewitt operator	2.70	0.0036	5.00
Prewitt + T1FLS	87.03	0.0386	93.75
Prewitt + IT2FLS	87.54	0.0394	95.00
Prewitt + GT2 FLS	87.68	0.0470	96.25

Table 7 Recognition rate for Cropped Yale database

Fuzzy system edge detector	Mean rate (%)	Standard deviation	Max rate
None	2.83	0.0042	6.57
Sobel operator	2.63	0.0025	2.63
Sobel + T1FLS	97.52	0.0293	100
Sobel + IT2FLS	97.70	0.0314	100
Sobel + GT2 FLS	**98.11**	**0.0314**	**100**
Prewitt operator	2.80	0.0050	5.26
Prewitt + T1FLS	94.28	0.0348	100
Prewitt + IT2FLS	94.35	0.0304	100
Prewitt + GT2 FLS	97.58	0.0328	100

The results obtained over the ORL database are presented in Table 6; so, in this Table we can notice that the mean rate value is better when the Sobel GT2 fuzzy edge detector is applied with a value of 87.97. In these results, we can also observe that the mean rate and max rate values obtained with the Prewitt + GT2 FLS were better than the Prewitt + IT2 FLS and Prewitt + T1 FLS.

Otherwise, the results achieved when the Cropped Yale database is used are shown in Table 7. In this Table, we observed that the best performance (mean rate) of the neural network is obtained when the Sobel + GT2 FLS was applied; nevertheless, we can notice than the max rate values obtained by all the fuzzy edge detectors was of 100 %.

7 Conclusions

In summary, in this paper we have presented two edge detector methods based on GT2 FS. The edge detection was applied in two image databases before the training phase of the monolithic neural network.

Based on the simulation results presented in Tables 6 and 7, we can conclude that the edge detection based on GT2 FS represent a good way to improve the performance in a face recognition system.

In general, the results achieved in the simulations were better when the fuzzy edge detection was applied; since the results were very low when the monolithic neural network was performed without edge detection; even so, when the traditional Prewitt and Sobel edge detectors were applied.

Acknowledgment We thank the MyDCI program of the Division of Graduate Studies and Research, UABC, and the financial support provided by our sponsor CONACYT contract grant number: 44524.

References

1. Biswas, R. and Sil, J., "An Improved Canny Edge Detection Algorithm Based on Type-2 Fuzzy Sets," Procedia Technology, vol. 4, pp. 820–824, Jan. 2012.
2. Canny, J. "A Computational Approach to Edge Detection", in IEEE Transactions on Pattern Analysis and Machine Intelligence, vol. PAMI-8, no. 6, pp. 679–698, 1986.
3. Doostparast Torshizi, A. and Fazel Zarandi, M. H., "Alpha-plane based automatic general type-2 fuzzy clustering based on simulated annealing meta-heuristic algorithm for analyzing gene expression data.," Comput. Biol. Med., vol. 64, pp. 347–59, Sep. 2015.
4. Georghiades, A. S., Belhumeur, P. N., Kriegman, D. J., "From Few to Many: Illumination Cone Models for Face Recognition under Variable Lighting and Pose," in IEEE Transactions on Pattern Analysis and Machine Intelligence, vol. 23, no. 6, pp. 643-660, 2001.
5. Golsefid, S. M. M., Zarandi, F. and Turksen, I. B., "Multi-central general type-2 fuzzy clustering approach for pattern recognitions," Inf. Sci. (Ny)., vol. 328, pp. 172–188, Jan. 2016.
6. Gonzalez, C. I., Melin, P., Castro, J. R., Mendoza, O. and Castillo O., "An improved sobel edge detection method based on generalized type-2 fuzzy logic", Soft Computing, vol. 20, no. 2, pp. 773-784, 2014.
7. Gonzalez, R. C., Woods, R. E. and Eddins, S. L., "Digital Image Processing using Matlab," in Prentice-Hall, 2004.
8. Hu, L., Cheng, H. D. and Zhang, M., "A high performance edge detector based on fuzzy inference rules," Information Sciences, vol. 177, no. 21, pp. 4768–4784, Nov. 2007.
9. Kirsch, R., "Computer determination of the constituent structure of biological images," Computers and Biomedical Research, vol. 4, pp. 315–328, 1971.
10. Lee, K. C., Ho, J. and Kriegman, D., "Acquiring Linear Subspaces for Face Recognition under Variable Lighting," in IEEE Transactions on Pattern Analysis and Machine Intelligence, vol. 27, no. 5, pp. 684-698, 2005.
11. Liu, F., "An efficient centroid type-reduction strategy for general type-2 fuzzy logic system," Information Sciences, vol. 178, no. 9, pp. 2224–2236, 2008.
12. Liu, X., Mendel, J. M. and Wu, D., "Study on enhanced Karnik–Mendel algorithms: Initialization explanations and computation improvements," Information Sciences, vol. 184, no. 1, pp. 75–91, 2012.
13. Martínez, G. E., Mendoza, O., Castro, J. R., Melin, P. and Castillo, O., "Generalized type-2 fuzzy logic in response integration of modular neural networks," IFSA World Congress and NAFIPS Annual Meeting (IFSA/NAFIPS), pp. 1331-1336, 2013.

14. Melin, P., Gonzalez, C. I., Castro, J. R., Mendoza, O. and Castillo O., "Edge-Detection Method for Image Processing Based on Generalized Type-2 Fuzzy Logic," in IEEE Transactions on Fuzzy Systems, vol. 22, no. 6, pp. 1515-1525, 2014.
15. Mendel, J. M., "Comments on α -Plane Representation for Type-2 Fuzzy Sets: Theory and Applications," in IEEE Transactions on Fuzzy Systems, vol.18, no.1, pp. 229-230, 2010.
16. Mendel, J. M., "General Type-2 Fuzzy Logic Systems Made Simple: A Tutorial," in IEEE Transactions on Fuzzy Systems, vol. 22, no. 5, pp.1162-1182, 2014.
17. Mendel, J. M., "On KM Algorithms for Solving Type-2 Fuzzy Set Problems," in IEEE Transactions on Fuzzy Systems, vol. 21, no. 3, pp. 426–446, 2013.
18. Mendel, J. M. and John, R. I. B., "Type-2 fuzzy sets made simple," in IEEE Transactions on Fuzzy Systems, vol. 10, no. 2, pp. 117–127, 2002.
19. Mendel, J. M., Liu, F. and Zhai, D., "α-Plane Representation for Type-2 Fuzzy Sets: Theory and Applications," in IEEE Transactions on Fuzzy Systems, vol.17, no.5, pp. 1189-1207, 2009.
20. Mendoza, O., Melin, P. and Castillo, O., "An improved method for edge detection based on interval type-2 fuzzy logic," Expert Systems with Applications, vol. 37, no. 12, pp. 8527–8535, Dec. 2010.
21. Mendoza, O., Melin, P. and Castillo, O., "Neural networks recognition rate as index to compare the performance of fuzzy edge detectors," in Neural Networks (IJCNN), The 2010 International Joint Conference on, pp. 1–6, 2010.
22. Mendoza, O., Melin, P. and Licea, G., "A hybrid approach for image recognition combining type-2 fuzzy logic, modular neural networks and the Sugeno integral," Information Sciences, vol. 179, no. 13, pp. 2078–2101, 2009.
23. Mendoza, O., Melin, P. and Licea, G., "A New Method for Edge Detection in Image Processing Using Interval Type-2 Fuzzy Logic," 2007 IEEE International Conference on Granular Computing (GRC 2007), pp. 151–151, Nov. 2007.
24. Mendoza, O., Melin, P. and Licea, G., "Interval type-2 fuzzy logic for edges detection in digital images," International Journal of Intelligent Systems (IJIS), vol. 24, no. 11, pp. 1115–1133, 2009.
25. Perez-Ornelas, F., Mendoza, O., Melin, P., Castro, J. R., Rodriguez-Diaz, A., "Fuzzy Index to Evaluate Edge Detection in Digital Images," PLOS ONE, vol. 10, no. 6, pp. 1-19, 2015.
26. Phillips, P. J., Moon, H., Rizvi, S. A. and Rauss, P. J, "The FERET Evaluation Methodology for Face-Recognition Algorithms," in IEEE Transactions on Pattern Analysis and Machine Intelligence, vol. 22, no.10, pp. 1090–1104, 2000.
27. Prewitt, J. M. S., "Object enhancement and extraction"," B.S. Lipkin, A. Rosenfeld (Eds.), Picture Analysis and Psychopictorics, Academic Press, New York, NY, pp. 75–149, 1970.
28. Sanchez, M. A., Castillo, O. and Castro, J. R., "Generalized Type-2 Fuzzy Systems for controlling a mobile robot and a performance comparison with Interval Type-2 and Type-1 Fuzzy Systems," Expert Syst. Appl., vol. 42, no. 14, pp. 5904–5914, Aug. 2015.
29. Sobel, I., "Camera Models and Perception", Ph.D. thesis, Stanford University, Stanford, CA, 1970.
30. Talai, Z. and Talai, A., "A fast edge detection using fuzzy rules," 2011 International Conference on Communications, Computing and Control Applications (CCCA), pp. 1–5, Mar. 2011.
31. Tao, C., Thompson, W. and Taur, J., "A fuzzy if-then approach to edge detection," Fuzzy Systems, pp. 1356–1360, 1993.
32. The USC-SIPI Image Database. Available 00 http://www.sipi.usc.edu/database/.
33. Wagner, C., Hagras, H., "Employing zSlices based general type-2 fuzzy sets to model multi level agreement", 2011 IEEE Symposium on Advances in Type-2 Fuzzy Logic Systems (T2FUZZ), pp. 50–57, 2011.
34. Wagner, C., Hagras, H., "Toward general type-2 fuzzy logic systems based on zSlices", in IEEE Transactions on Fuzzy Systems, vol. 18, no. 4, pp. 637–660, 2010.
35. Zadeh, L. A., Fuzzy Sets, vol. 8, Academic Press Inc., USA, 1965.

36. Zadeh, L. A., "Outline of a New Approach to the Analysis of Complex Systems and Decision Processes," in IEEE Transactions on Systems, Man, and Cybernetics, vol. SMC-3, no. 1, pp. 28–44, 1973.
37. Zhai, D. and Mendel, J. M., "Centroid of a general type-2 fuzzy set computed by means of the centroid-flow algorithm," Fuzzy Systems (FUZZ), 2010 IEEE International Conference on, pp. 1–8, 2010.
38. Zhai, D. and Mendel, J. M., "Uncertainty measures for general Type-2 fuzzy sets," Information Sciences, vol. 181, no. 3, pp. 503–518, 2011.

An Overview of Granular Computing Using Fuzzy Logic Systems

Mauricio A. Sanchez, Oscar Castillo and Juan R. Castro

Abstract As Granular Computing has gained interest, more research has lead into using different representations for Information Granules, i.e., rough sets, intervals, quotient space, fuzzy sets; where each representation offers different approaches to information granulation. These different representations have given more flexibility to what information granulation can achieve. In this overview paper, the focus is only on journal papers where Granular Computing is studied when fuzzy logic systems are used, covering research done with Type-1 Fuzzy Logic Systems, Interval Type-2 Fuzzy Logic Systems, as well as the usage of general concepts of Fuzzy Systems.

Keywords Granular · Fuzzy logic · Fuzzy systems · Type-2 fuzzy sets

1 Introduction

With an inspiration on the human cognitive process, where personal experience adds to choosing the best level of abstraction which can achieve a better decision; Granular Computing (GrC) [1] uses numerical evidence to set the best possible level of resolution for achieving a good decision model. GrC works on existing models, or modeling methods, by adding an abstraction layer based on its inspiration of human cognition and with such modifications obtains better models which can lead to improved model performance as well an improved interpretation of its parts, viz. Information Granules.

M.A. Sanchez (✉) · O. Castillo
Tijuana Institute of Technology, Tijuana, Mexico
e-mail: mauricio.sanchez@tectijuana.mx

O. Castillo
e-mail: ocastillo@tectijuana.mx

J.R. Castro
Autonomous University of Baja California, Tijuana, Mexico
e-mail: jrcastror@uabc.edu.mx

P. Melin et al. (eds.), *Nature-Inspired Design of Hybrid Intelligent Systems*,
Studies in Computational Intelligence 667, DOI 10.1007/978-3-319-47054-2_2

For a GrC model to represent anything a medium is necessary. This medium is called an Information Granule, which is a grouping of similar information taken from some type of previous experimental evidence. These Information Granules can have any type of representation, e.g. intervals, rough sets [2], fuzzy sets [3], fuzzy rough sets [4], quotient space [5], intuitionistic fuzzy sets [6], neutrosophic sets [7], among others. Although each one has their own research surrounding their GrC-based implementations, as well as having their own strengths and weaknesses, in the end, choosing a representation is a matter of choice. Information Granules in the form of fuzzy sets (FS) have been chosen for the survey in this paper, since they have a high maturity level where they can be expertly designed or be created by an algorithm based on experimental evidence, they can be inferred via a fuzzy logic system (FLS) where a group of FS which form different rules in the IF…THEN format process inputs into outputs which ultimately represent a given model of information, knowledge, or behavior. Also, fuzzy sets have two basic trends in complexity which have been used with GrC: Type-1 Fuzzy Sets (T1 FS), and Interval Type-2 Fuzzy Sets (IT2 FS). Although a third trend exists, which are Generalized Type-2 Fuzzy Sets (GT2 FS), but no known research exists in journal form where they are used with GrC, hence its omittance.

In this paper, a survey was made which shortly examines all journal publications where GrC is used when Information Granules are represented by FS. For this case, the following combinations of keywords were searched for granular computing, information granule, fuzzy, type-1, and type-2.

This paper is divided into four main sections, first, a quick background look into T1 FLS and IT2 FLS; then, the general overview is presented, separated into research publication done with T1 FLS, IT2 FLS, and other fuzzy concepts; afterwards, some trends in current publications are shown; finally, some concluding remarks are added.

2 Fuzzy Logic Systems

With traditional logic, available values are {1,0}; with three-valued logic, available values are {1,0,1}, or other variations, e.g., {0,1,2}; whereas in the case of Fuzzy Logic (FL), available values are between [0,1]. This interval leads to the possibility expressing ambiguities, or imprecision, when dealing with decision making situations as no longer only did two or three choices exist, but an infinite of possible choices between the said intervals.

A FLS transforms inputs into outputs via its inference. This fuzzy inference system comprises of rules, which are defined by various membership functions. This inference gives the possibility of applying an FLS in any area where a decision might be required. an FLS can be manually or automatically designed; manual design [8–10] is usually done by experts where they choose the best rule description and distribution of individual membership functions which best describe the problem to be solved, on the other hand, automatic design [11–13] is based on assessing the best rule description and membership functions based on previously obtained experimental data.

As previously stated, information granules can be represented by Fuzzy Sets. These fuzzy information granules can make use of the fuzzy inference's capabilities and transform inputs into outputs. This precedence has given Fuzzy Information Granules much attention, such that this paper is solely focused on implementations of Information Granules using FLSs.

2.1 Type-1 Fuzzy Logic Systems

A Type-1 Fuzzy Logic System (T1 FLS) has membership functions which only represent a type of imprecision, or ambiguity. Shown in Fig. 1, is a sample generic Gaussian membership function. A Type-1 Fuzzy Set A, denoted by $\mu_A(x)$, where $x \in X$ and X is the Universe Of Discourse. Represented by $A = \{x, \mu_A(x) | x \in X\}$, this being an FS which takes on values in the interval $[0,1]$.

The rules for a T1 FLS take the form of Eq. (1), where a mapping exists from input space into output space. Where R^l is a rule, x_p is an input p, F_p^l is an input membership function on the lth rule and pth input; y is the output on membership function G^l n the lth rule. Here, F and G take on the form of $\mu_F(x)$ and $\mu_G(x)$.

$$R^l\text{:IF}\, x_1 \,\text{is}\, F_1^l \,\text{and} \ldots \text{and}\, x_p \,\text{is}\, F_p^l,\ \text{THEN}\, y \,\text{is}\, G^l; \quad l = 1, \ldots, M \tag{1}$$

The inference is first performed rule-wise by Eq. (2) when done with t-norm connectors $(\tilde{*})$. Where μ_{B^l} is the resulting membership function in the consequents per each rule's inference, and Y is the space belonging to the consequents.

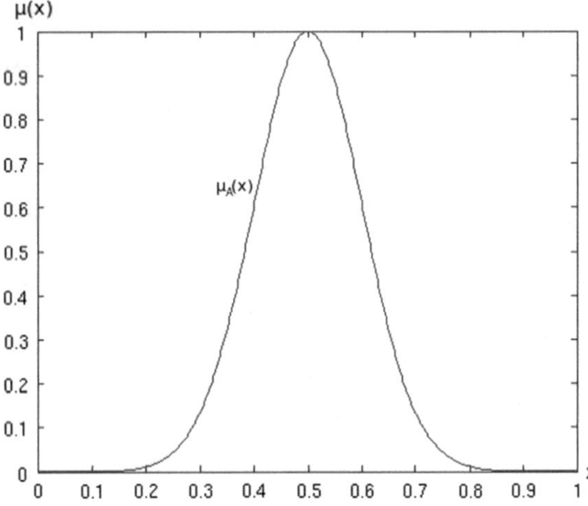

Fig. 1 Sample type-1 fuzzy membership function

$$\mu_{B^l}(y) = \mu_{G^l}(y) \tilde{*} \left\{ \left[\sup_{x_1 \in X_1} \mu_{x_1}(x_1) \tilde{*} \mu_{F_1^l}(x_1) \right] \right.$$
$$\left. \tilde{*} \cdots \tilde{*} \left[\sup_{x_p \in X_p} \mu_{x_p}(x_p) \tilde{*} \mu_{F_p^l}(x_p) \right] \right\}, \quad y \in Y \tag{2}$$

The defuzzification process can be done in various ways, all achieving a very similar result, for examples: centroid, center-of-sums, or height, described by Eqs. (3), (4), and (5), respectively. Where y_i is a discrete position from Y, $y_i \in Y$, $\mu_B(y)$ is a FS which has been mapped from the inputs, c_{B^l} denotes the centroid on the lth output, a_{B^l} is the area of the set, and $\bar{y}^l$ is the point which has the maximum membership value in the lth output set.

$$y_c(x) = \frac{\sum_{i=1}^{N} y_i \mu_B(y_i)}{\sum_{i=1}^{N} \mu_B(y_i)} \tag{3}$$

$$y_a(x) = \frac{\sum_{l=1}^{M} c_{B^l} a_{B^l}}{\sum_{l=1}^{M} a_{B^l}} \tag{4}$$

$$y_h(x) = \frac{\sum_{l=1}^{M} \bar{y}^l \mu_{B^l}(\bar{y}^l)}{\sum_{l=1}^{M} \mu_{B^l}(\bar{y}^l)} \tag{5}$$

2.2 *Interval Type-2 Fuzzy Logic Systems*

An Interval Type-2 Fuzzy Logic System (IT2 FLS) has membership functions which can represent varying degrees of uncertainty. Shown in Fig. 2, is a sample generic Gaussian membership function with uncertain standard deviation. An Interval Type-2 Fuzzy Set $\tilde{A}$ is denoted by $\mu_{\tilde{A}}(x) = \left[\underline{\mu}_{\tilde{A}}(x), \bar{\mu}_{\tilde{A}}(x) \right]$, where $\underline{\mu}_{\tilde{A}}(x)$ and $\bar{\mu}_{\tilde{A}}(x)$ are the lower and upper membership function which conform the IT2 FS. Represented by $\tilde{A} = \left\{ \left((x, u), \mu_{\tilde{A}}(x, u) = 1 \right) \middle| \forall x \in X, \forall u \in [0,1] \right\}$, where x is a subset of the Universe Of Discourse X, and u is a mapping of X into [0,1].

The rule's format for an IT2 FLS is the same as that of a T1 FLS, as shown in Eq. (6). When inferred upon, it maps from an input space into an output space.

$$R^l\text{:IF}\, x_1 \text{ is } \tilde{F}_1^l \text{ and} \ldots \text{and}\, x_p \text{ is } \tilde{F}_1^l,\ \text{THEN}\, y \text{ is } \tilde{G}^l; \quad l = 1, \ldots, M \tag{6}$$

The Fuzzy Inference from an IT2 FLS is best represented by Eq. (7). Where μ_{B^l} is the resulting membership function in the consequents per each rule's inference, and Y is the space belonging to the consequents.

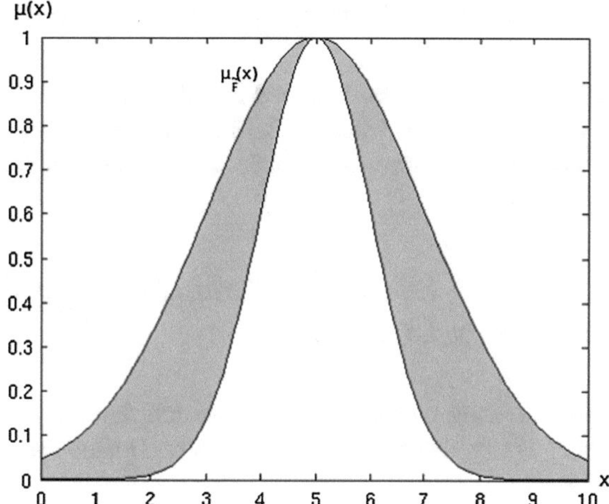

Fig. 2 Sample interval type-2 fuzzy membership function

$$\mu_{\tilde{B}}(y) = \left\{ \left[\sup_{x_1 \in X_1} \mu_{X_1}(x_1) \mu_{F_1^l}(x_1) \right] \tilde{*} \cdots \tilde{*} \left[\sup_{x_p \in X_p} \mu_{X_p}(x_p) \mu_{F_p^l}(x_p) \right] \right\} \tilde{*} \mu_{G^l}(y), \quad y \in Y \tag{7}$$

The type reducer of an IT2 FLS can be done in various ways, one of these techniques is that of center-of-sets (cos), as shown in Eq. (8), where $Y_{\cos}$ is an interval set defined by two points $[y_l^i, y_r^i]$, which are obtained from Eqs. (9) to (10).

$$Y_{\cos}(x) = \int_{y^1 \in [y_l^1, y_r^1]} \cdots \int_{y^M \in [y_l^M, y_r^M]} \int_{f^1 \in [\underline{f}^1, \bar{f}^1]} \cdots \int_{f^M \in [\underline{f}^M, \bar{f}^M]} 1 \Big/ \frac{\sum_{i=1}^M f^i y^i}{\sum_{i=1}^M f^i} \tag{8}$$

$$y_l = \frac{\sum_{i=1}^M f_l^i y_l^i}{\sum_{i=1}^M f_l^i} \tag{9}$$

$$y_r = \frac{\sum_{i=1}^M f_r^i y_r^i}{\sum_{i=1}^M f_r^i} \tag{10}$$

Finally, the defuzzification process is done by Eq. (11), which obtains a crisp output value.

$$y(x) = \frac{y_l + y_r}{2} \tag{11}$$

3 Overview of Fuzzy Information Granule Research

Overviews have been separated into three categories: where T1 FLSs are used with granular concepts, where IT2 FLSs are used with granular concepts, and where general Fuzzy concepts are used with granular concepts.

3.1 Fuzzy Information Granules Using Type-1 Fuzzy Logic Systems

The following overviews show research done in the area of Granular Computing using Fuzzy Information Granules constructed from T1 FSs and inferred via an FLS.

In Pedrycz et al. [14] multiple fuzzy models are constructed by separate locally available data, in which later on a global model is formed at a higher level of hierarchy, therefore obtaining a granular fuzzy model. The principle of justifiable granularity is used throughout to show how granular parameters of the models are formed.

In Pedrycz and Homenda [15] the principle of justifiable granularity is introduced, which supports the creation of information granules by optimizing a design process where experimental data is balanced with granule specificity.

In Pedrycz [16] multiple items are revised. First, a goal of design concerned with revealing and representing structure in a data set is oriented toward the underlying relational aspects of experimental data; second, a goal that deals with their formation of quality mappings which is directly affected by the information granules over which is operates. A fuzzy clustering is used in order to obtain fuzzy models, in which a generalized version of the FCM is proposed that has an additive form of the objective function with a modifiable component of collaborative activities.

In Pedrycz et al. [17] an augmented FCM is proposed which models with the purpose of a balance between the structural content of the inputs and output spaces, with Mamdani and Takagi-Sugeno fuzzy rules. A direct link between the development of fuzzy models and the idea of granulation-degranulation is also studied.

In Pedrycz et al. [18] a fuzzy clustering scheme is proposed which uses viewpoints to represent domain knowledge. Where the user introduced a point of view of the data, treated as prototypes, the clustering takes place using these viewpoints and constructs its fuzzy model.

In Bargiela et al. [19], they define a granular mapping over information granules and are mapped into a collection of granules expressed in an output space. This is done by first defining an interaction between information granules and experimental data, and then using these measures of interaction in the expression of a granular mapping.

In Balamash et al. [20] they formulate information granules as an optimization task where selected information granules are refined into a family of more detailed information granules, that way the partition requirement becomes satisfied. The Conditional FCM is used as a clustering method for information granulation.

In Wang et al. [21] a modified FCM is used in conjunction with information granulation in order to solve the problem of time series long-term prediction, obtaining more meaningful and semantically sound entities in the form of information granules that is interpretable and easily comprehended by humans.

In Lu et al. [22] a numeric time series is segmented into a collection of time windows for which each one is built into a fuzzy granule exploiting the principle of justifiable granularity. Subsequently, a granular model is reached by mining logical relationships of adjacent granules.

In Pedrycz et al. [23] a general study is made on several algorithmic schemes associated with Granular Computing. A study of the implications of using the principle of justifiable granularity and its benefits is made. The overall quality of granular models is also evaluated via the determination of the area under curve, where this curve is formed in the coverage-specificity coordinates. Various numerical results are discussed which display the essential features of each algorithmic scheme.

In Pedrycz et al. [24] a granular data description is proposed which characterizes numeric data via a collection of information granules, such that the key structure of the data, its topology and essential relationships are described in the form of a family of fuzzy information granules. An FCM clustering algorithm is used as a granulation method followed by the implementation of the principle of justifiable granularity as to obtain meaningful information granules.

In Rubio et al. [25] through the use of Neutrosophic Logic, which extends the concepts of Fuzzy Logic and uses a three-valued logic that uses an indeterminacy value, granular neural-fuzzy models are created. This leads to more meaningful and simpler granular models with better generalization performance when compared to other recent modeling techniques.

In Yu et al. [26] a fundamental issue of the design of meaningful information granules is covered. This issue is that of the tradeoff between the specificity of an information granule and its experimental evidence. For this, an optimization task is proposed for which it discusses its solution and properties.

In Pedrycz et al. [27] a concept of granular representation of numeric membership functions of fuzzy sets is introduced. It gives a synthetic and qualitative view of fuzzy sets and their processing. An optimization task is also introduced alongside, which refers to a notion of consistency of a granular representation. This consistency is also related to several operations, such as *and* and *or* consistency of granular representations, implemented by t-norms and t-conorms.

In Panoutsos et al. [28] a systematic modeling approach using granular computing and neuro-fuzzy modeling is presented. A granular algorithm extracts relational information and data characteristics from an initial set of information, which are then converted into a linguistic rule-base of a fuzzy system. Afterwards it is elicited and optimized.

In Zadeh [29] a theoretic proposal of fuzzy information granulation based on human reasoning is shown. The idea is centered on the logic of fuzzy boundaries between granules when their sum is tightly packed, such that crisp information granules cannot depict a better description, but fuzzy granules can. Later on, various principal types of granules are proposed: possibilitic, veristic and probabilistic.

In Pedrycz et al. [30] a mechanism for fuzzy feature selection is introduced, which considers features to be granular rather than numeric. By varying the level of granularity, the contribution level of each feature is modified with respect to the overall feature space. This is done by transforming feature values into intervals, they consider broader intervals to be less essential to the full feature space. The quantification of the importance of each feature is acquired via an FCM model.

In Kang et al. [31] granular computing is introduced into formal concept analysis by providing a unified model for concept lattice building and rule extraction on a fuzzy granularity base. They present hat maximal rules are complete and nonredundant, and users who want to obtain decision rules should generate maximal rules.

In Park et al. [32] a fuzzy inference system based on information granulation and genetic optimization is proposed. It focuses on capturing essential relationships between information granules rather than concentrating on plain numeric data. It uses an FCM to build an initial fuzzy model which then using a genetic algorithm optimizes the apex of the fuzzy sets in respect to its structure and parameters.

In Choi [33] a methodology for information granulation-based genetically optimized fuzzy inference system is proposed which has a tuning method with a variant identification ratio for structural and parametric optimization of the reasoning system. This tuning is done via a mechanism of information granulation. The structural optimization of the granular model is done via an FCM algorithm, afterward the genetic algorithm takes over in order to optimize the granular model.

In Novák [34] the concept that a granule can be modeled in formal fuzzy logic is introduced. It proposes the concepts of intension and extension of granules, internal and external operations, various kinds of relations, motion of granules, among other concepts. And proposes a more complex concept of association which is formed from granules.

In Bargiela et al. [35] a model of granular data emerging through a summarization and processing of numeric data is proposed, where the model supports data analysis and contributes to further interpretation activities. An FCM is used to reveal the structure in the data, where a Tchebyschev metric is used to promote the development of easily interpretable information granules. A deformation effect of the hyperbox geometry of granules is discussed; where they show that this deformation can be quantified. Finally, a two-level topology of information granules is also shown where the core of the topology is in the form of hyperbox information granules.

In Pedrycz et al. [36] a concept of granular mappings is introduced which offer a design framework, which is focused on forming conceptually meaningful information granules. The directional nature of the mapping is also considered during the granulation process. An FCM algorithm is used to facilitate the granulation of the input space, and the principle of justifiable granularity is used to construct the information granules in the output space.

In Pedrycz et al. [37] the concept of granular prototypes which generalize the numeric representation of clusters is shown, where more details about the data structure are captured through this concept. The FCM algorithm is used as a granulation method. An optimization of the fuzzy information granules is guided by the coverage criterion. Various optimal allocations of information granularity schemes are researched.

In Pedrycz [38] an idea of granular models is introduced, where generalizations of numeric models that are formed as a result of an optimal allocation of information granularity. A discussion is also made of a set of associated information granularity distribution protocols.

In Fazzolari et al. [39] a fuzzy discretization for the generation of automatic granularities and their associated fuzzy partitions is proposed, to by a multi-objective fuzzy association rule-based classification method which performs a tuning on the rule selection.

In Roh et al. [40] a granular fuzzy classifier based on the concept of information granularity is proposed. This classifier constructs information granules reflective of the geometry of patterns belonging to individual classes; this is done efficiently by improving the class homogeneity from the original information granules and by using a weighting scheme as an aggregation mechanism.

In Lu et al. [41] a method is proposed that partitions an interval information granule-based fuzzy time series for improving its model accuracy. Where predefined interval sizes partition the time series such that information granules are constructed from each interval's data, these constructions takes place from the amplitude-change space belonging to the corresponding trends in the time series. These information granules are continually adjusted to better associate corresponding and adjacent intervals.

In Dick et al. [42] information is translated between two granular worlds via a fuzzy rulebase, where each granular world employs a different granulations, but exist in the same universe of discourse. Translation is done by re-granulating and matching the information into the target world. A first-order interpolation implemented using linguistic arithmetic is used for this purpose.

In Lu et al. [43] a modeling and prediction approach of time series is proposed, based on high-order fuzzy cognitive maps (HFCM) and fuzzy c-means (FCM) algorithms, where the FCM constructs the information granules by transforming the original time series into a granular time series and then generates an HFCM prediction model. A PSO algorithm is used to tune all parameters.

In Sanchez et al. [44] a clustering algorithm based on Newton's law of gravitation is proposed, where the size of each fuzzy information granule is adapted to the context of its numerical evidence. Where this numerical evidence is influenced by gravitational forces during the grouping of information.

In Pedrycz et al. [45] a collaborative fuzzy clustering concept is introduced which has two fundamental optimization features. First, all communication is done via information granules; second, the fuzzy clustering done for individual data must take into account the communication done by the information granules.

In Dong et al. [46] a granular time series concept is introduced which is used for long-term and trend forecasting. A fuzzy clustering technique is used to construct the initial granules afterward a granular modeling scheme is used to construct a better forecasting model, this conforms a time-domain granulation scheme.

In Oh et al. [47] the concept of information granulation and genetic optimization is used a design procedure for a fuzzy system. A c-means clustering is used for initial granulation, where the apex of all granules are found. Afterward a genetic algorithm and a least square method tunes the rest of the fuzzy parameters.

In Pedrycz et al. [48] given a window of granulation, an algorithm is proposed which constructs fuzzy information granules. All fuzzy information granules are legitimized in terms of experimental data while their specificity is also maximized.

In Pedrycz [49] a dynamic data granulation concept done in the presence of incoming data organized in data snapshots is introduced, where each snapshot reveals a structure via fuzzy clustering. An FCM is used as granulation method; information granular splitting is later done by a Conditional FCM to merge two or more neighboring prototypes.

In Dai et al. [50] a fuzzy granular computing method is used to characterize signal dynamics within a time domain of a laser dynamic speckle phenomenon, where sequences of intensity images are obtained in order to evaluate the dynamics of the phenomena. The sequence of characterization is used to identify the underlying activity in each pixel of the intensity images.

In Pedrycz et al. [51] a scheme of risk assessment is introduced which is based on classification results from experimental data from the history of previous threat cases. The granulation process is done via a fuzzy clustering algorithm which reveals an initial structural relationship between information granules. Two criteria are also introduced for the evaluation of information granules, representation capabilities and interpretation aspects.

In Fig. 3, the total number of publications in the category of granular computing research done with T1 FLS is shown, although the amount of research was stable from 1997 to 2005 a temporary increasing trend in research is seen from 2005 to 2011, and again increases in recent years starting from 2012. The total amount of publications done in this area has been 38 research papers for the 1997–2015 time period.

3.2 Higher Type Fuzzy Information Granules Using Interval Type-2 Fuzzy Logic Systems

The following overviews show research done in the area of Granular Computing using Higher Type Fuzzy Information Granules constructed from IT2 FSs and inferred via an FLS.

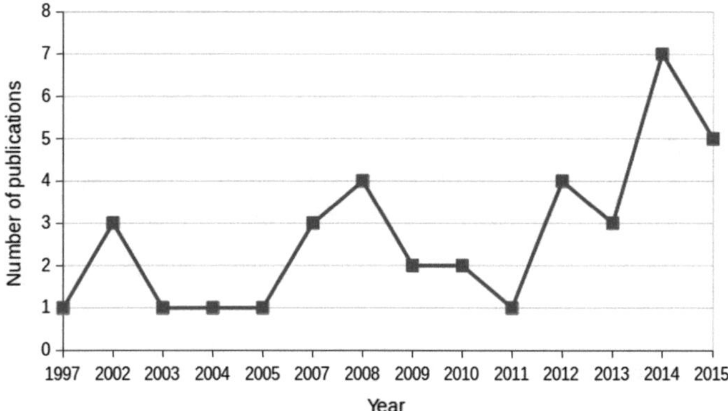

Fig. 3 Total publications per year on granular computing research done with T1 FLS, for the 1997–2015 time period

In Ulu et al. [52] an approach for the formation of Interval Type-2 membership functions is proposed, where the Footprint Of Uncertainty is formed by using rectangular Type-2 fuzzy information granules, providing more design flexibility in Interval Type-2 Fuzzy Logic System design.

In Sanchez et al. [53] a technique for the formation of Higher Type information granules represented by Interval Type-2 membership functions is proposed, which is based on the concept of uncertainty-based information, whereby measuring the difference between information granule samples the Footprint Of Uncertainty is extracted.

In Pedrycz et al. [54] the concept of granular fuzzy decision built on a basis formed by individual decision models is introduced. The principle of justifiable granularity is used to improve the holistic nature of a granular fuzzy decision, alongside an established feedback loop in which the holistic view is adjusted to the individual decision. Various optimization schemes are also discussed along with multiple examples of forming Type-2 and Type-3 Fuzzy Sets.

In Pedrycz [55] the capabilities of Interval Type-2 Fuzzy Sets as Higher Type information granules is discussed. Seen as aggregation of Fuzzy Sets, a framework is proposed which leads to the construction of Interval Type-2 Fuzzy Sets. The principle of justifiable granularity is used as a means to quantify the diversity of membership degrees.

In Fig. 4, the total number of publications in the category of granular computing research done with IT2 FLS is shown, in comparison to the previous research category of using T1 FLS, research done with IT2 FLS is still very lacking, especially since only four research papers can be found for the 2009–2015 time period.

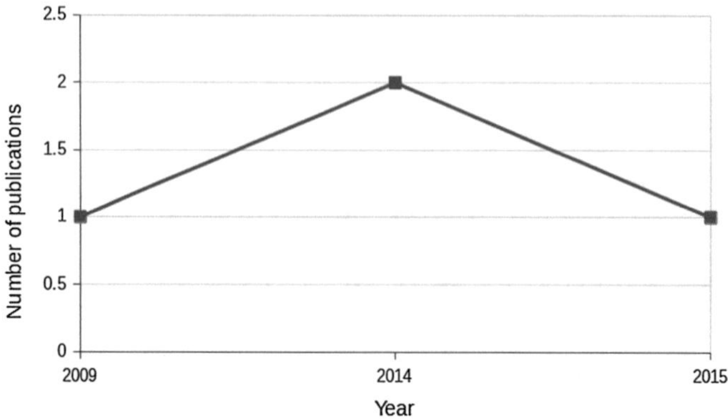

Fig. 4 Total publications per year on granular computing research done with IT2 FLS, for the 2009–2015 time period

3.3 Other Fuzzy Concepts Used in Granular Computing

The following overviews show research done in the area of Granular Computing using various concepts taken from Fuzzy Logic in general, fuzzy hybrid schemes, etc.

In Rubio-Solis et al. [56] a modeling framework for an Interval Type-2 radial basis function neural network is proposed. The functional equivalence of Interval Type-2 Fuzzy Logic Systems to radial basis neural networks is used as a modeling structure. The linguistic uncertainty is used as an advantage in the proposed Interval Type-2 radial basis function neural network. A Granular Computing approach is used as a means to find the Footprint of Uncertainty.

In Nandedkar et al. [57] a granular neural network which can learn to classify granular data is proposed, where granular data is represented by a Hyperbox Fuzzy Set. The proposed granular neural network is capable of learning online using granular or point data.

In Qian et al. [58] a viewpoint is studied which looks to answer the question: "what is the essence of measuring the fuzzy information granularity of a fuzzy granular structure." A binary granular computing model is used to study and demonstrate such viewpoint, altogether five aspects are studied: fuzzy binary granular structure operators, partial order relations, measures for fuzzy information granularity, an axiomatic approach to fuzzy information granularity, and fuzzy information entropies.

In Kundu et al. [59] a fuzzy granular framework for representing social networks is developed. Based on granular computing theory and fuzzy neighborhood systems which provide the framework where fuzzy sets provide the representation of Information granules. New measures are also proposed alongside the provided framework: granular degree, granular betweenness, and granular clustering coefficient of a node, granular embeddedness of a pair of nodes, granular clustering coefficient and granular modularity if a fuzzy granular social network.

In Dick et al. [60] a granular neural network which uses granular values and operations on each neuron is proposed. It is based on a multilayer perceptron architecture, uses fuzzy linguistic terms as neuron connection weights, and has a backpropagation learning algorithm.

In Mencar et al. [61] a survey is provided where they present multiple interpretability constraints for the generation of fuzzy information granules. Their intention is to provide a homogeneous description of all interpretability constants, as well as their respective overviews, and to identify potential different interpretation meanings.

In Pedrycz et al. [62] granular neural networks are introduced. By applying data granulation technique, various aspects of training neural networks with large data sets is addressed. Multiple granular neural network architectures are provided as well as their training scenarios. Several techniques to the design of information granules are also discussed.

In Leite et al. [63] a granular neural network framework is used for evolving fuzzy systems from fuzzy data streams is introduced. This evolving granular neural network obtains local models that are interpretable using fuzzy granular models, fuzzy aggregation neurons, and an online incremental learning algorithm.

In Pal et al. [64] a rough-fuzzy classifier based on granular computing is presented, where class-dependent information granules exist in a fuzzy environment. Feature representation belonging to different classes is done via fuzzy membership functions. The local, and contextual, information from neighbored granules is explored through a selection of rough sets of a subset of granulated features.

In Liu et al. [65] a fuzzy lattice granular computing classification algorithm is proposed, where a fuzzy inclusion relation is computed between information granules by means of a inclusion measure function based on a nonlinear positive valuation function and an isomorphic mapping between lattices.

In Pedrycz [66] a discussion on various research trends regarding Granular Computing is shown, focusing mainly in fuzzy regressions and its future developments, granular fuzzy models, higher order and highertype granular constructs.

In Ganivada et al. [67] a fuzzy rough granular neural network is proposed, based on the multilayer perceptron using backpropagation training for tuning of the network. With an input vector defined by fuzzy information granules and a target vector defined by fuzzy class membership values. The granularity factor is incorporated into the input level and is also used for determining the weights of the proposed network.

In Kaburlasos et al. [68] a fuzzy lattice reasoning classifier which works in granular data domain of fuzzy interval numbers, fuzzy numbers, intervals, and cumulative distribution functions is introduced. The classifier integrates techniques for dealing with imprecision in practice.

In Oh et al. [69] a granular oriented self-oriented hybrid fuzzy polynomial neural network is introduced, based on multilayer perceptrons with polynomial neurons or context-based polynomial neurons. Where a context-based FCM algorithm is used for granulating numeric data for the context-based polynomials neurons, executed in input space.

In Bodjanova [70] axiomatic definitions of granular nonspecificity are proposed. Various approaches to measures of granular nonspecificity are suggested. Other measures are discussed, such as Pawlak's measure, Banerjee and Pal's measure, and Huynh and Nakamori's measure. Furthermore, relationships between roughness, granular nonspecificity, and nonspecificity of a fuzzy set are also discussed.

In Kundu et al. [71] the identification of fuzzy rough communities done via a community detection algorithm is introduced, it proposes that a node with different memberships of their association can be part of many groups and that the granule's node decreases as its distances itself from the granule's center.

In Pedrycz et al. [72] an architecture of functional radial basis function neural network is proposed. One of its core functions is its use of experimental data through a process of information granulation, these granules constructed by a context-driven FCM algorithm. The model values of the fuzzy sets of contexts and the prototypes of the fuzzy clusters from input space are a direct functional character of the proposed network.

In Pal [73] a general overview of data mining and knowledge discovery is shown from a point of view of soft computing techniques, such as computational theory of perceptions, fuzzy granulation in machine and human intelligence, and modeling via rough-fuzzy integration.

In Pedrycz et al. [74] a feature selection based on structure retention concept is proposed, which is quantified by a reconstruction criterion. A combinatorial selection reduced the feature space, where it is used for optimizing the retention of the original structure. An FCM algorithm is used as a granulation technique, and a genetic algorithm optimizes the medium.

In Li et al. [75] a measure for finding the similarity between two linguistic fuzzy rulebases is introduced, based on the concept of granular computing based on linguistic gradients. The linguistic gradient operator is used for the comparison of both linguistic structures.

In Ghaffari et al. [76] an overall flowchart based on information granulation used in the design of rock engineering flowcharts is proposed. Granulation is done by self-organizing neuro-fuzzy inference systems and by self-organizing rough set theory, knowledge rules are then extracted from both sets of granules, crisp granules, and fuzzy rough granules, respectively.

In Gacek [77] a general study is shown of the principles, algorithms and practices of Computational Intelligence and how they are used in signal processing and time series. Although focus is not solely on the scheme of Granular Computing, it is shown in the form of fuzzy granules, where in this regard, fuzzy information granules are found to be excellent in facilitating interpretation while also maintaining a solid modeling base.

In Pedrycz [78] the conceptual role of information granules in system modeling and knowledge management when dealing with collections of models is shown. Various algorithmic schemes are shown for this purpose, among which is the use of fuzzy sets expressed as fuzzy information granules.

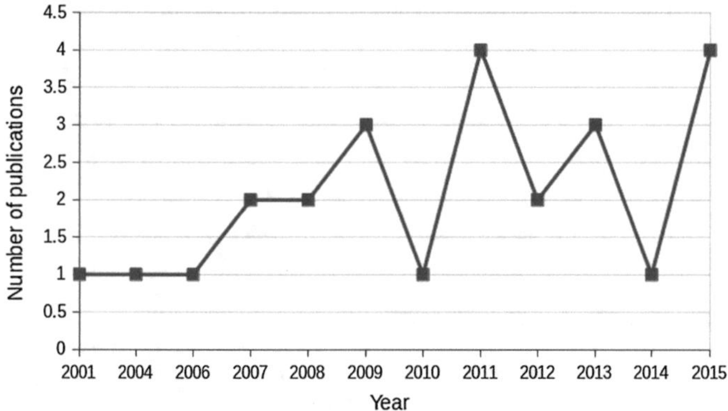

Fig. 5 Total publications per year on granular computing research done with other fuzzy concepts, for the 2001–2015 time period

In Pedrycz et al. [79] fuzzy proximity matrices induced by corresponding partition matrices are introduced, used as an optimization scheme for the variability of granular sources of knowledge. This knowledge networks offers a conceptual framework focused on knowledge aggregation, consensus building, and association building. An FCM algorithm is used as a means of constructing information granules.

In Degang et al. [80] a theory of granular computing based on fuzzy relations is proposed. The concept of granular fuzzy sets based on fuzzy similarity relations is used to describe granular structures of lower and upper approximations of a fuzzy set. Fuzzy rough sets can also be characterized via the proposed fuzzy approximations.

In Fig. 5, the total number of publications in the category of granular computing research done with other Fuzzy concepts is shown; from 2006 to 2015 research has varied in trend. The total amount of publications done in this area has been 25 research papers for the 2001–2015 time period.

4 Trends in Research

The general area of Granular Computing used with fuzzy systems has been steadily increasing in recent years. In Fig. 6, this trend is shown where from 1997 to 2006 the amount of research was fairly steady and minimal, yet after 2006 a steady increase in research exists up to 2015. The total amount of publications done in this area of Fuzzy Granular Computing has been 67 research papers for the 1997–2015 time period.

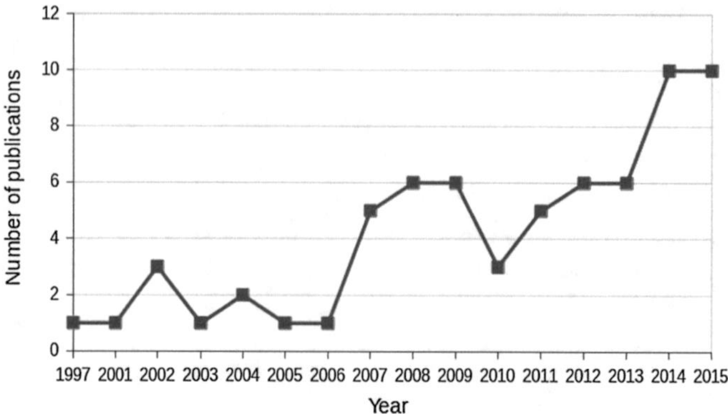

Fig. 6 Total publications per year on granular computing research done with fuzzy logic systems in general, for the 1997–2015 time period

Author-wise statistics show that a total of 97 different authors have participated in publishing papers in the area of Fuzzy Granular Computing, and as such, Table 1 shows a partial list of 21 authors whose names appear more often (2 or more papers), with the most prominent author being Dr. Witold Pedrycz as he surpasses appearance in papers by a considerable amount, 40 research papers from the 67 which were examined. Another trend worth mentioning is that only these 21 authors are recurring researchers in this area. The other 76 unlisted authors have appeared only once on all overviewed research papers.

Table 1 Author appearance in research papers related to granular computing using fuzzy systems in general

Author's name	Number of publications author has appeared in
Witold Pedrycz	40
Sankar K. Pal	5
Sung-Kwun Oh	5
Abdullah Saeed Balamash	4
Ali Morfeq	4
Rami Al-Hmouz	4
Xiaodong Liu	4
George Panoutsos	3
Jianhua Yang	3
Scott Dick	3
Wei Lu	3
Andrzej Bargiela	3
Adam Gacek	2
George Vukovich	2

(continued)

Table 1 (continued)

Author's name	Number of publications author has appeared in
Juan R. Castro	2
Kaoru Hirota	2
Keon-Jun Park	2
Mauricio A. Sanchez	2
Oscar Castillo	2
Soumitra Dutta	2
Suman Kundu	2

As an interesting note, the most widely used mechanism for granulating information is the Fuzzy C-Means and its variations.

5 Conclusion

This paper shows that an increasing trend in research of Granular Computing using Fuzzy Systems exists, be it using Type-1 Fuzzy Systems, Interval type-2 Fuzzy Systems, or any other Fuzzy concepts.

With the oldest research publication done in 1997 and considering that so far only 67 journal research papers exist, and that so far only 97 authors have works in this area, this area is still very alluring as many ideas and concepts can still be explored.

Acknowledgments We thank the MyDCI program of the Division of Graduate Studies and Research, UABC, and Tijuana Institute of Technology the financial support provided by our sponsor CONACYT contract grant number: 314258.

References

1. Pedrycz, W.: Granular Computing - The Emerging Paradigm. J. Uncertain Syst. 1, 38–61 (2007).
2. Pawlak, Z.: Rough sets. Int. J. Comput. Inf. Sci. 11, 341–356 (1982).
3. Zadeh, L.A.: Fuzzy Sets. Inf. Control. 8, 338–353 (1965).
4. Dubois, D., Prade, H.: Rough sets and fuzzy rough sets*. Int. J. Gen. Syst. 17, 191–209 (1990).
5. Zhang, L.Z.L., Zhang, B.Z.B.: Quotient space based multi-granular computing. 2005 IEEE International Conference on Granular Computing. p. 98 (2005).
6. Atanassov, K.T.: Intuitionistic fuzzy sets. Fuzzy Sets Syst. 20, 87–96 (1986).
7. Smarandache, F.: Neutrosophic set - a generalization of the intuitionistic fuzzy set. 2006 IEEE International Conference on Granular Computing. pp. 38–42. IEEE (2006).

8. Smita Sushil Sikchi, S.S.M.S.A.: Design of fuzzy expert system for diagnosis of cardiac diseases. Int. J. Med. Sci. Public Heal. 2, 56–61 (2013).
9. Neshat, M., Adeli, A.: Designing a fuzzy expert system to predict the concrete mix design. 2011 IEEE International Conference on Computational Intelligence for Measurement Systems and Applications (CIMSA) Proceedings. pp. 1–6. IEEE (2011).
10. Goztepe, K.: Designing Fuzzy Rule Based Expert System for Cyber Security, http://www.ijiss.org/ijiss/index.php/ijiss/article/view/3, (2012).
11. Jelleli, T.M., Alimi, A.M.: Automatic design of a least complicated hierarchical fuzzy system. International Conference on Fuzzy Systems. pp. 1–7. IEEE (2010).
12. Joonmin Gil, Chong-Sun Hwang: An automatic design of fuzzy systems based on L-systems. FUZZ-IEEE'99. 1999 IEEE International Fuzzy Systems. Conference Proceedings (Cat. No.99CH36315). pp. 418–423 vol.1. IEEE (1999).
13. Heider, H., Drabe, T.: A cascaded genetic algorithm for improving fuzzy-system design. Int. J. Approx. Reason. 17, 351–368 (1997).
14. Pedrycz, W., Al-Hmouz, R., Balamash, A.S., Morfeq, A.: Designing granular fuzzy models: A hierarchical approach to fuzzy modeling. Knowledge-Based Syst. 76, 42–52 (2015).
15. Pedrycz, W., Homenda, W.: Building the fundamentals of granular computing: A principle of justifiable granularity. Appl. Soft Comput. 13, 4209–4218 (2013).
16. Pedrycz, W.: Relational and directional aspects in the construction of information granules. IEEE Trans. Syst. Man, Cybern. - Part A Syst. Humans. 32, 605–614 (2002).
17. Pedrycz, W., Izakian, H.: Cluster-Centric Fuzzy Modeling. IEEE Trans. Fuzzy Syst. 22, 1585–1597 (2014).
18. Pedrycz, W., Loia, V., Senatore, S.: Fuzzy Clustering with Viewpoints. IEEE Trans. Fuzzy Syst. 18, 274–284 (2010).
19. Bargiela, A., Pedrycz, W.: Granular Mappings. IEEE Trans. Syst. Man, Cybern. - Part A Syst. Humans. 35, 292–297 (2005).
20. Balamash, A., Pedrycz, W., Al-Hmouz, R., Morfeq, A.: An expansion of fuzzy information granules through successive refinements of their information content and their use to system modeling. Expert Syst. Appl. 42, 2985–2997 (2015).
21. Wang, W., Pedrycz, W., Liu, X.: Time series long-term forecasting model based on information granules and fuzzy clustering. Eng. Appl. Artif. Intell. 41, 17–24 (2015).
22. Lu, W., Pedrycz, W., Liu, X., Yang, J., Li, P.: The modeling of time series based on fuzzy information granules. Expert Syst. Appl. 41, 3799–3808 (2014).
23. Pedrycz, W., Song, M.: Granular fuzzy models: a study in knowledge management in fuzzy modeling. Int. J. Approx. Reason. 53, 1061–1079 (2012).
24. Pedrycz, W., Succi, G., Sillitti, A., Iljazi, J.: Data description: A general framework of information granules. Knowledge-Based Syst. 80, 98–108 (2015).
25. Solis, A.R., Panoutsos, G.: Granular computing neural-fuzzy modelling: A neutrosophic approach. Appl. Soft Comput. 13, 4010–4021 (2013).
26. Yu, F., Pedrycz, W.: The design of fuzzy information granules: Tradeoffs between specificity and experimental evidence. Appl. Soft Comput. 9, 264–273 (2009).
27. Pedrycz, A., Hirota, K., Pedrycz, W., Dong, F.: Granular representation and granular computing with fuzzy sets. Fuzzy Sets Syst. 203, 17–32 (2012).
28. Panoutsos, G., Mahfouf, M.: A neural-fuzzy modelling framework based on granular computing: Concepts and applications. Fuzzy Sets Syst. 161, 2808–2830 (2010).
29. Zadeh, L.A.: Toward a theory of fuzzy information granulation and its centrality in human reasoning and fuzzy logic. Fuzzy Sets Syst. 90, 111–127 (1997).
30. Pedrycz, W., Vukovich, G.: Feature analysis through information granulation and fuzzy sets. Pattern Recognit. 35, 825–834 (2002).
31. Kang, X., Li, D., Wang, S., Qu, K.: Formal concept analysis based on fuzzy granularity base for different granulations. Fuzzy Sets Syst. 203, 33–48 (2012).
32. Park, K.-J., Pedrycz, W., Oh, S.-K.: A genetic approach to modeling fuzzy systems based on information granulation and successive generation-based evolution method. Simul. Model. Pract. Theory. 15, 1128–1145 (2007).

33. Choi, J.-N., Oh, S.-K., Pedrycz, W.: Structural and parametric design of fuzzy inference systems using hierarchical fair competition-based parallel genetic algorithms and information granulation. Int. J. Approx. Reason. 49, 631–648 (2008).
34. Novák, V.: Intensional theory of granular computing. Soft Comput. - A Fusion Found. Methodol. Appl. 8, 281–290 (2004).
35. Bargiela, A., Pedrycz, W.: A model of granular data: a design problem with the Tchebyschev FCM. Soft Comput. 9, 155–163 (2003).
36. Pedrycz, W., Al-Hmouz, R., Morfeq, A., Balamash, A.: The design of free structure granular mappings: the use of the principle of justifiable granularity. IEEE Trans. Cybern. 43, 2105–13 (2013).
37. Pedrycz, W., Bargiela, A.: An optimization of allocation of information granularity in the interpretation of data structures: toward granular fuzzy clustering. IEEE Trans. Syst. Man. Cybern. B. Cybern. 42, 582–90 (2012).
38. Pedrycz, W.: Allocation of information granularity in optimization and decision-making models: Towards building the foundations of Granular Computing. Eur. J. Oper. Res. 232, 137–145 (2014).
39. Fazzolari, M., Alcalá, R., Herrera, F.: A multi-objective evolutionary method for learning granularities based on fuzzy discretization to improve the accuracy-complexity trade-off of fuzzy rule-based classification systems: D-MOFARC algorithm. Appl. Soft Comput. 24, 470–481 (2014).
40. Roh, S.-B., Pedrycz, W., Ahn, T.-C.: A design of granular fuzzy classifier. Expert Syst. Appl. 41, 6786–6795 (2014).
41. Lu, W., Chen, X., Pedrycz, W., Liu, X., Yang, J.: Using interval information granules to improve forecasting in fuzzy time series. Int. J. Approx. Reason. 57, 1–18 (2015).
42. Dick, S., Schenker, A., Pedrycz, W., Kandel, A.: Regranulation: A granular algorithm enabling communication between granular worlds. Inf. Sci. (Ny). 177, 408–435 (2007).
43. Lu, W., Yang, J., Liu, X., Pedrycz, W.: The modeling and prediction of time series based on synergy of high-order fuzzy cognitive map and fuzzy c-means clustering. Knowledge-Based Syst. 70, 242–255 (2014).
44. Sanchez, M.A., Castillo, O., Castro, J.R., Melin, P.: Fuzzy granular gravitational clustering algorithm for multivariate data. Inf. Sci. (Ny). 279, 498–511 (2014).
45. Pedrycz, W., Rai, P.: Collaborative clustering with the use of Fuzzy C-Means and its quantification. Fuzzy Sets Syst. 159, 2399–2427 (2008).
46. Dong, R., Pedrycz, W.: A granular time series approach to long-term forecasting and trend forecasting. Phys. A Stat. Mech. its Appl. 387, 3253–3270 (2008).
47. Oh, S.-K., Pedrycz, W., Park, K.-J.: Structural developments of fuzzy systems with the aid of information granulation. Simul. Model. Pract. Theory. 15, 1292–1309 (2007).
48. Pedrycz, W., Gacek, A.: Temporal granulation and its application to signal analysis. Inf. Sci. (Ny). 143, 47–71 (2002).
49. Pedrycz, W.: A dynamic data granulation through adjustable fuzzy clustering. Pattern Recognit. Lett. 29, 2059–2066 (2008).
50. Dai Pra, A.L., Passoni, L.I., Rabal, H.: Evaluation of laser dynamic speckle signals applying granular computing. Signal Processing. 89, 266–274 (2009).
51. Pedrycz, W., Chen, S.C., Rubin, S.H., Lee, G.: Risk evaluation through decision-support architectures in threat assessment and countering terrorism. Appl. Soft Comput. 11, 621–631 (2011).
52. Ulu, C., Güzelkaya, M., Eksin, I.: Granular type-2 membership functions: A new approach to formation of footprint of uncertainty in type-2 fuzzy sets. Appl. Soft Comput. 13, 3713–3728 (2013).
53. Sanchez, M.A., Castillo, O., Castro, J.R.: Information granule formation via the concept of uncertainty-based information with Interval Type-2 Fuzzy Sets representation and Takagi–Sugeno–Kang consequents optimized with Cuckoo search. Appl. Soft Comput. 27, 602–609 (2015).

54. Pedrycz, W., Al-Hmouz, R., Morfeq, A., Balamash, A.S.: Building granular fuzzy decision support systems. Knowledge-Based Syst. 58, 3–10 (2014).
55. Pedrycz, W.: Human centricity in computing with fuzzy sets: an interpretability quest for higher order granular constructs. J. Ambient Intell. Humaniz. Comput. 1, 65–74 (2009).
56. Rubio-Solis, A., Panoutsos, G.: Interval Type-2 Radial Basis Function Neural Network: A Modeling Framework. IEEE Trans. Fuzzy Syst. 23, 457–473 (2015).
57. Nandedkar, A. V, Biswas, P.K.: A granular reflex fuzzy min-max neural network for classification. IEEE Trans. Neural Netw. 20, 1117–34 (2009).
58. Qian, Y., Liang, J., Wu, W.Z., Dang, C.: Information Granularity in Fuzzy Binary GrC Model. IEEE Trans. Fuzzy Syst. 19, 253–264 (2011).
59. Kundu, S., Pal, S.K.: FGSN: Fuzzy Granular Social Networks – Model and applications. Inf. Sci. (Ny). 314, 100–117 (2015).
60. Dick, S., Tappenden, A., Badke, C., Olarewaju, O.: A granular neural network: Performance analysis and application to re-granulation. Int. J. Approx. Reason. 54, 1149–1167 (2013).
61. Mencar, C., Fanelli, A.M.: Interpretability constraints for fuzzy information granulation. Inf. Sci. (Ny). 178, 4585–4618 (2008).
62. Pedrycz, W., Vukovich, G.: Granular neural networks. Neurocomputing. 36, 205–224 (2001).
63. Leite, D., Costa, P., Gomide, F.: Evolving granular neural networks from fuzzy data streams. Neural Netw. 38, 1–16 (2013).
64. Pal, S.K., Meher, S.K., Dutta, S.: Class-dependent rough-fuzzy granular space, dispersion index and classification. Pattern Recognit. 45, 2690–2707 (2012).
65. Liu, H., Xiong, S., Fang, Z.: FL-GrCCA: A granular computing classification algorithm based on fuzzy lattices. Comput. Math. with Appl. 61, 138–147 (2011).
66. Pedrycz, W.: From fuzzy data analysis and fuzzy regression to granular fuzzy data analysis. Fuzzy Sets Syst. (2014).
67. Ganivada, A., Dutta, S., Pal, S.K.: Fuzzy rough granular neural networks, fuzzy granules, and classification. Theor. Comput. Sci. 412, 5834–5853 (2011).
68. Kaburlasos, V.G., Papadakis, S.E.: A granular extension of the fuzzy-ARTMAP (FAM) neural classifier based on fuzzy lattice reasoning (FLR). Neurocomputing. 72, 2067–2078 (2009).
69. Oh, S.-K., Kim, W.-D., Park, B.-J., Pedrycz, W.: A design of granular-oriented self-organizing hybrid fuzzy polynomial neural networks. Neurocomputing. 119, 292–307 (2013).
70. Bodjanova, S.: Granulation of a fuzzy set: Nonspecificity. Inf. Sci. (Ny). 177, 4430–4444 (2007).
71. Kundu, S., Pal, S.K.: Fuzzy-Rough Community in Social Networks. Pattern Recognit. Lett. (2015).
72. Pedrycz, W., Park, H.S., Oh, S.K.: A granular-oriented development of functional radial basis function neural networks. Neurocomputing. 72, 420–435 (2008).
73. Pal, S.K.: Soft data mining, computational theory of perceptions, and rough-fuzzy approach. Inf. Sci. (Ny). 163, 5–12 (2004).
74. Pedrycz, W., Syed Ahmad, S.S.: Evolutionary feature selection via structure retention. Expert Syst. Appl. 39, 11801–11807 (2012).
75. LI, H., DICK, S.: A similarity measure for fuzzy rulebases based on linguistic gradients. Inf. Sci. (Ny). 176, 2960–2987 (2006).
76. Ghaffari, H.O., Sharifzadeh, M., Shahriar, K., Pedrycz, W.: Application of soft granulation theory to permeability analysis. Int. J. Rock Mech. Min. Sci. 46, 577–589 (2009).
77. Gacek, A.: Signal processing and time series description: A Perspective of Computational Intelligence and Granular Computing. Appl. Soft Comput. 27, 590–601 (2015).
78. Pedrycz, W.: Information granules and their use in schemes of knowledge management. Sci. Iran. 18, 602–610 (2011).
79. Pedrycz, W., Hirota, K.: Forming consensus in the networks of knowledge. Eng. Appl. Artif. Intell. 20, 657–666 (2007).
80. Degang, C., Yongping, Y., Hui, W.: Granular computing based on fuzzy similarity relations. Soft Comput. 15, 1161–1172 (2010).

Optimization of Type-2 and Type-1 Fuzzy Integrator to Ensemble Neural Network with Fuzzy Weights Adjustment

Fernando Gaxiola, Patricia Melin, Fevrier Valdez and Juan R. Castro

Abstract In this paper, two bio-inspired methods are applied to optimize the type-2 and type-1 fuzzy integrator used in the neural network with fuzzy weights. The genetic algorithm and particle swarm optimization are used to optimize the type-2 and type-1 fuzzy integrator that work in the integration of the output for the ensemble neural network with three networks. One neural network uses type-2 fuzzy inference systems with Gaussian membership functions to obtain the fuzzy weights; the second neural network uses type-2 fuzzy inference systems with triangular membership functions; and the third neural network uses type-2 fuzzy inference systems with triangular membership functions with uncertainty in the standard deviation. In this work, an optimized type-2 and type-1 fuzzy integrator to manage the output of the ensemble neural network and the results for the two bio-inspired methods are presented. The proposed approach is applied to a case of time series prediction, specifically in Mackey-Glass time series.

Keywords Type-1 Fuzzy Integrator · Type-2 Fuzzy Integrator · Ensemble Neural Network · Fuzzy Weights

F. Gaxiola (✉)
Tijuana Institute of Technology, Tijuana, Mexico
e-mail: fergaor_29@hotmail.com

P. Melin · F. Valdez · J.R. Castro
Autonomous University of Baja California, Tijuana, Mexico
e-mail: pmelin@tectijuana.mx

F. Valdez
e-mail: fevrier@tectijuana.edu.mx

J.R. Castro
e-mail: jrcastror@uabc.edu.mx

P. Melin et al. (eds.), *Nature-Inspired Design of Hybrid Intelligent Systems*,
Studies in Computational Intelligence 667, DOI 10.1007/978-3-319-47054-2_3

1 Introduction

The optimization of type-1 and type-2 fuzzy inference systems (FIST1–FIST2) with bio-inspired methods allows to increase the performance of these FIST2, because this allows to find the optimal structure of the FIST2, meaning the membership functions, rules and others.

In this paper, we optimize the membership functions of a FIST1 and FIST2 integrator, the optimization is performed for triangular or Gaussian membership functions.

We are presenting an ensemble with three neural networks for the experiments. The result for the ensemble was obtained with type-1 and type-2 fuzzy integration. The time series prediction area is the study case for this paper, and particularly the Mackey-Glass time series is used to test the proposed approach.

The use of the neural networks has spread to many areas of research, being particularly a very important method for time series prediction and pattern recognition applications.

This research uses the managing of the weights of a neural networks using type-2 fuzzy inference systems and because these affect the performance of the learning process of the neural network, the used of type-2 fuzzy weights are an important part in the training phase for managing uncertainty.

The weights of a neural network are an important part in the training phase, because these affect the performance of the learning process of the neural network.

This conclusion is based on the practice of neural networks of this type, where some research works have shown that the training of neural networks for the same problem initialized with different weights or its adjustment in a different way but at the end is possible to reach a similar result.

The idea to optimize the type-1 and type-2 fuzzy integrator is based on the fact that obtaining the optimal membership functions empirically or on base at experts is complicated, that is the reason that fundament the utilization of bio-inspired methods to find the optimal membership functions and the footprints of uncertainty of these ones.

The optimization of the type-1 and type-2 fuzzy integrator used for obtaining the output of the ensemble neural network was performed with two bio-inspired methods: genetic algorithm and particle swarm optimization.

The next section presents basic concepts of genetic algorithms, particle swarm optimization, type-2 fuzzy logic and neural networks, and explains background of research about modifications of the backpropagation algorithm, different management strategies of weights in neural networks and optimization with genetics algorithm and particle swarm optimization. Section 3 explains the proposed method and the problem description. Section 4 presents the scheme of optimization of type-1 and type-2 fuzzy integrator for the ensemble neural network with genetic algorithm (GA) and particle swarm optimization (PSO). Section 5 describes the simulation results for the type-1 and type-2 fuzzy integrator optimized with GA and PSO proposed in this paper. Finally, in Sect. 6, some conclusions are presented.

2 Background and Basic Concepts

In this section, a brief review of basic concepts is presented.

2.1 Basic Concepts

A. ***Genetic Algorithms***

The genetic algorithm is a searching procedure using methods that simulate the natural genetic theory, like selection, mutation, and recombination. The individuals used in this algorithm can be binary, integer, or real to represent the possible solutions at the problem of study. The mutation and recombination are used to modify the individuals in searching to obtain better individuals. The selection consists in using objective functions that determine a fitness for all individuals and choose an individual in base at fitness in different ways, like comparison, random or elitist [13].

B. ***Particle Swarm Optimization***

Particle swarm optimization is an algorithm, which simulates the behavior used for the birds flocking or fish schooling looking for food. Each particle represents a solution to the optimization problem and each one has a position and a velocity that performed the search direction in the search space. The particle adjusts the velocity and position according to the best experiences which are called the p_{best} found by itself and g_{best} found by all its neighbors [32, 45, 46].

C. ***Type-2 Fuzzy Logic***

The concept of a type-2 fuzzy set was introduced by Zadeh [50] as an extension of the concept of an ordinary fuzzy set (henceforth called a "type-1 fuzzy set"). A type-2 fuzzy set is characterized by a fuzzy membership function, i.e., the membership grade for each element of this set is a fuzzy set in [0,1], unlike a type-1 set where the membership grade is a crisp number in [0,1] [4, 35]. Such sets can be used in situations where there is uncertainty about the membership grades themselves, e.g., uncertainty in the shape of the membership function or in some of its parameters [39]. Consider the transition from ordinary sets to fuzzy sets. When we cannot determine the membership of an element in a set as 0 or 1, we use fuzzy sets of type-1 [5, 29, 49]. Similarly, when the situation is so fuzzy that we have trouble determining the membership grade even as a crisp number in [0,1], we use fuzzy sets of type-2 [6, 9, 24, 42, 44].

D. ***Neural Network***

An artificial neural network (ANN) is a distributed computing scheme based on the structure of the nervous system of humans. The architecture of a neural network is formed by connecting multiple elementary processors, this being an adaptive system

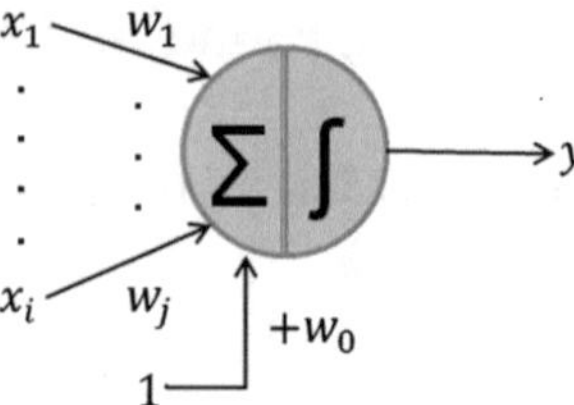

Fig. 1 Schema of an artificial neural network

that has an algorithm to adjust their weights (free parameters) to achieve the performance requirements of the problem based on representative samples [10, 33, 36].

The most important property of artificial neural networks is their ability to learn from a training set of patterns, i.e., they are able to find a model that fits the data [12, 41].

The artificial neuron consists of several parts (see Fig. 1). On the one side are the inputs, weights, the summation, and finally the adapter function. The input values are multiplied by the weights and added: $\sum x_i w_{ij}$. This function is completed with the addition of a threshold amount i. This threshold has the same effect as an entry with value −1. It serves so that the sum can be shifted left or right of the origin. After addition, we have the function f applied to the sum, resulting the final value of the output, also called y_i [40], obtaining the following equation:

$$y_i = f\left(\sum_{i=1}^{n} x_i w_{ij}\right) \tag{1}$$

where f may be a nonlinear function with binary output + −1, a linear function $f(z) = z$, or as sigmoidal logistic function:

$$f(z) = \frac{1}{1 + \mathrm{e}^{-z}}. \tag{2}$$

2.2 Overview of Related Works

In the area of optimization for type-2 fuzzy inference systems with genetic algorithms exists research in different areas of investigation, like that of Hidalgo et al. [25], in which type-2 fuzzy inference systems using gas are designed; Cervantes and Castillo [11], in which GA is used for optimization of membership functions; and other researchers with similar works [23, 34, 38].

In the area of optimization for type-2 fuzzy inference systems with particle swarm optimization exists research in different areas of investigation, like that of Melin et al. [37], in which PSO is used for designing fuzzy classification systems; Valdez et al. [48], in which a hybridization of GA and PSO algorithm is performed and the concept of fuzzy logic is integrated in it; and other researchers with similar works [2, 29].

The backpropagation algorithm applied in the neural network learning is the area of concern for this research.

In many works, the backpropagation algorithm has been modified for obtaining better results, like the momentum method [21, 40] and the adaptive learning rate [14]; Kamarthi and Pittner [30] focused in obtaining a weight prediction of the network at a future epoch using extrapolation; Ishibuchi et al. [26] proposed a fuzzy network where the weights are given as trapezoidal fuzzy numbers; Ishibuchi et al. [27] proposed a fuzzy network where the weights are given as triangular fuzzy numbers. Castro et al. [8] uses interval type-2 fuzzy neurons for the antecedents and interval of type-1 fuzzy neurons for the consequents of the rules.

Gaxiola et al. [15–20] proposed at neural network with type-2 fuzzy weights using triangular membership functions for Mackey-Glass time series.

In addition, recent works on type-2 fuzzy logic have been developed in time series prediction, like that of Castro et al. [7], and other researchers with similar works [1, 31].

3 Proposed Method and Problem Description

The objective of this work is to optimize with genetic algorithm (GA) and particle swarm optimization (PSO) the type-1 and interval type-2 fuzzy integration used for obtaining a result for the ensemble neural network composed of three neural networks with type-2 fuzzy weights. The Mackey-Glass time series (for $\tau = 17$) is utilized for testing the proposed approach.

The three neural networks work with type-2 fuzzy weights [16], one network works with two-sided Gaussian interval type-2 membership functions with uncertain mean and standard deviation in the two type-2 fuzzy inference systems (FIST2) used to obtain the weights (one in the connections between the input and hidden layer and the other between the hidden and output layer); the other two networks work with triangular interval type-2 membership function with uncertain and triangular interval type-2 membership function with uncertain standard deviation, respectively (see Fig. 2).

We work with three neural networks' architecture, and each network consists of 30 neurons in the hidden layer and 1 neuron in the output layer. These neural networks handle type-2 fuzzy weights in the hidden layer and output layer. We are working with a type-2 fuzzy inference system in the hidden layer and output layer of the networks obtaining new weights for the next epoch of the networks [3, 22, 28, 43].

We used two type-2 fuzzy inference systems with the same structure to obtain the type-2 fuzzy weights in the hidden and output layers of the neural network.

The weight managing in the three neural networks will be done with interval type-2 fuzzy weights, performing changes in the way we work internally in the

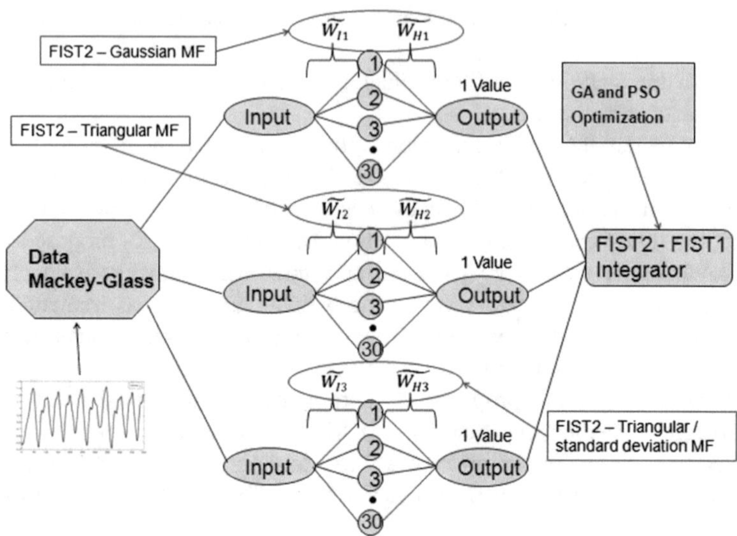

Fig. 2 Proposed optimization with GA and PSO for ensemble neural network architecture with interval type-2 fuzzy weights using type-2 or type-1 fuzzy integration

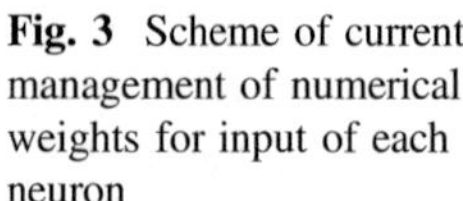

Fig. 3 Scheme of current management of numerical weights for input of each neuron

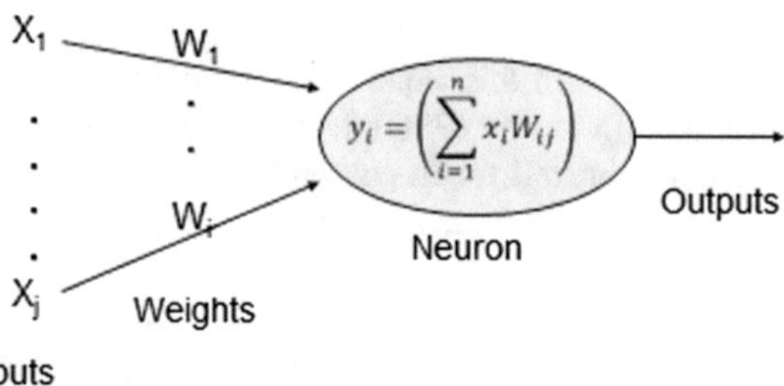

neuron (see Fig. 4), these are differently to the traditional management of weights performed with the backpropagation algorithm (see Fig. 3) [21].

In this research, for the three neural networks, in the neurons of the hidden layer we used the sigmoid function for activation function $f(-)$, and the linear function in the neurons of the output.

The three neural networks used two type-2 fuzzy inference systems with the same structure (see Fig. 5), which have two inputs (the current weight in the actual epoch and the change of the weight for the next epoch) and one output (the new weight for the next epoch).

In the first neural network, the inputs and the output for the type-2 fuzzy inference systems used between the input and hidden layers are delimited with two Gaussian membership functions with their corresponding range (see Fig. 6); and the inputs and output for the type-2 fuzzy inference systems used between the hidden

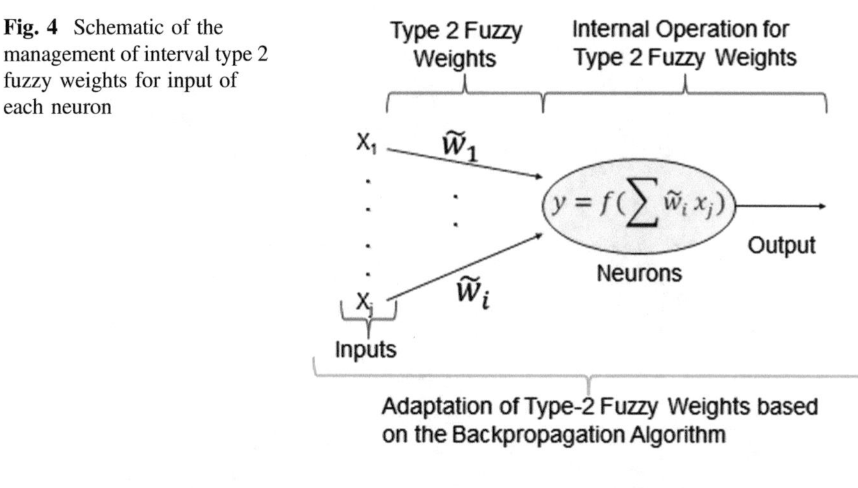

Fig. 4 Schematic of the management of interval type 2 fuzzy weights for input of each neuron

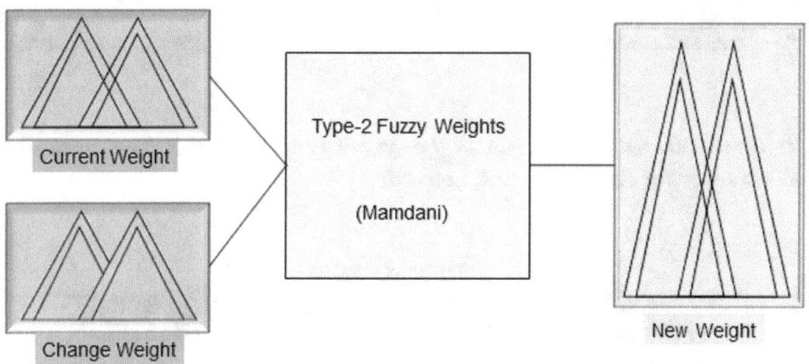

Fig. 5 Structure of the type-2 fuzzy inference systems used in the neural networks

and output layer are delimited with two Gaussian membership functions with their corresponding range (see Fig. 7).

In the second neural network, the inputs and the output for the type-2 fuzzy inference systems used between the input and hidden layers are delimited with two triangular membership functions with their corresponding ranges (see Fig. 8); and the inputs and output for the type-2 fuzzy inference systems used between the hidden and output layers are delimited with two triangular membership functions with their corresponding ranges (see Fig. 9).

In the third neural network, the inputs and the output for the type-2 fuzzy inference systems used between the input and hidden layers are delimited with two triangular membership functions with standard deviation with their corresponding range (see Fig. 10); and the inputs and output for the type-2 fuzzy inference systems used between the hidden and output layers are delimited with two triangular with standard deviation membership functions with their corresponding ranges (see Fig. 11).

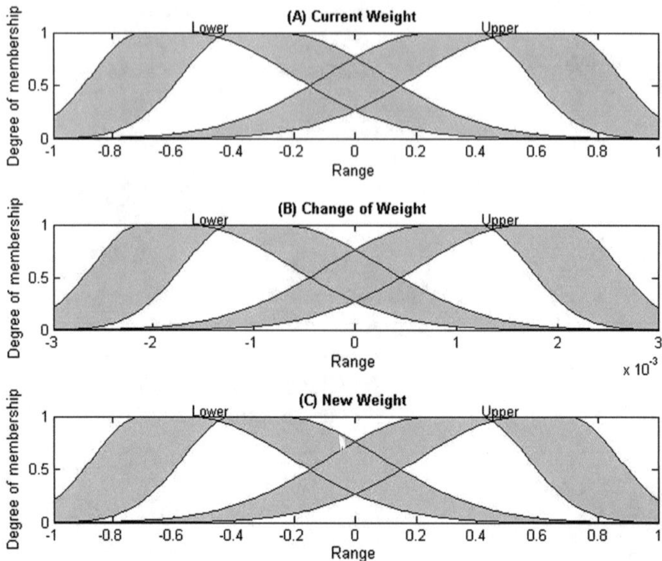

Fig. 6 Inputs (*A* and *B*) and output (*C*) of the type-2 fuzzy inference system used between the input and hidden layer for the first neural network

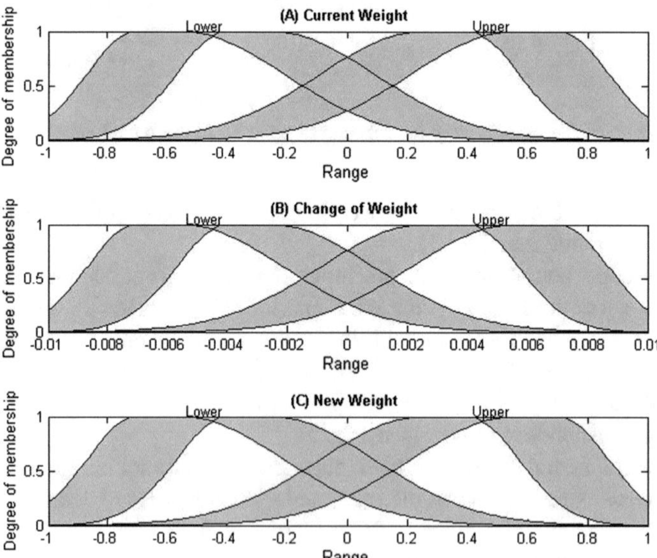

Fig. 7 Inputs (*A* and *B*) and output (*C*) of the type-2 fuzzy inference system used between the hidden and output layer for the first neural network

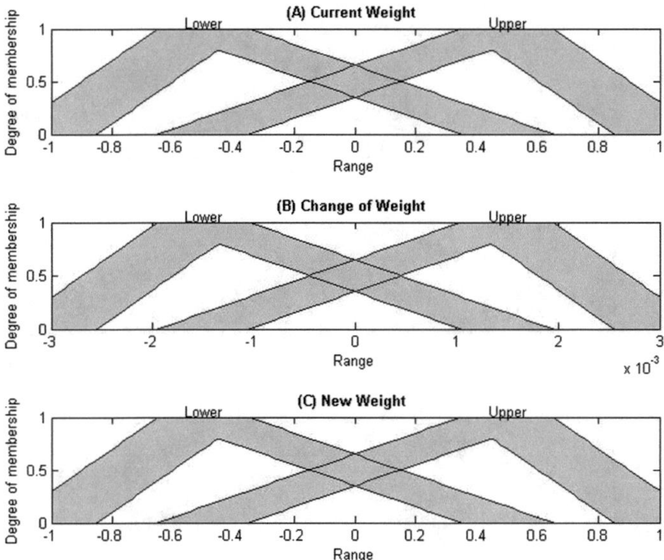

Fig. 8 Inputs (*A* and *B*) and output (*C*) of the type-2 fuzzy inference system used between the input and hidden layer for the second neural network

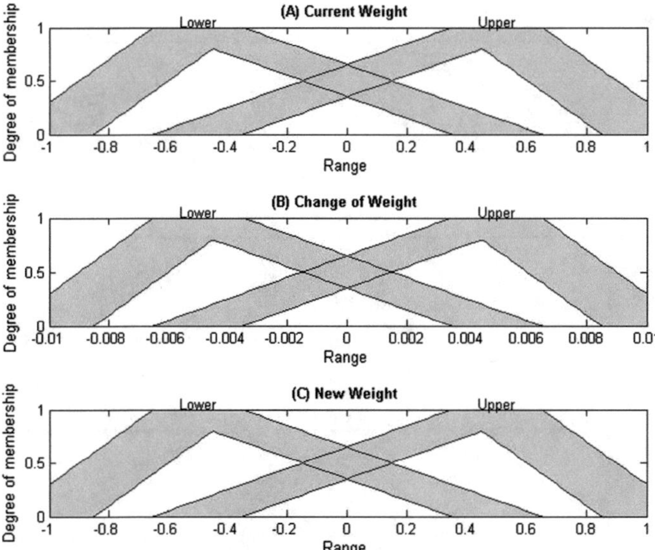

Fig. 9 Inputs (*A* and *B*) and output (*C*) of the type-2 fuzzy inference system used between the input and hidden layer for the second neural network

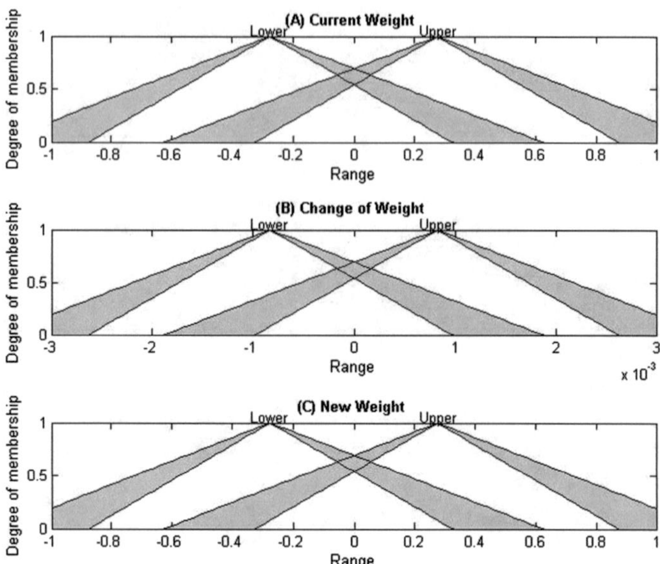

Fig. 10 Inputs (*A* and *B*) and output (*C*) of the type-2 fuzzy inference system used between the input and hidden layer for the third neural network

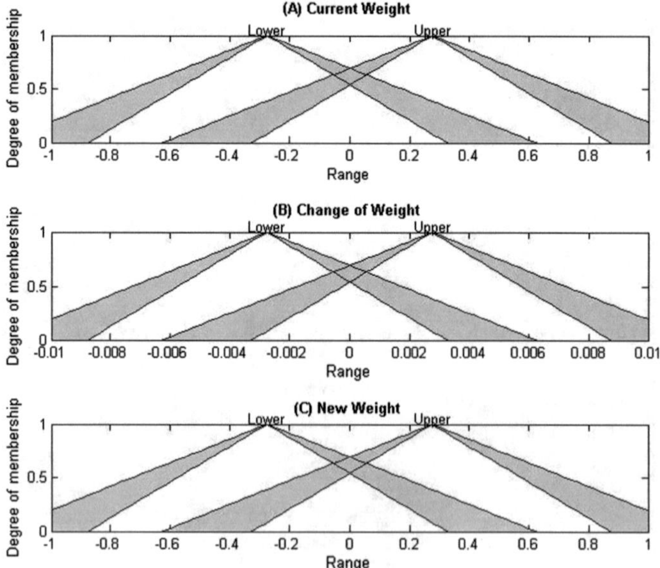

Fig. 11 Inputs (*A* and *B*) and output (*C*) of the type-2 fuzzy inference system used between the input and hidden layer for the third neural network

1. (Current_Weight is lower) and (Change_Weight is lower) then (New_Weight is lower)
2. (Current_Weight is lower) and (Change_Weight is upper) then (New_Weight is lower)
3. (Current_Weight is upper) and (Change_Weight is lower) then (New_Weight is upper)
4. (Current_Weight is upper) and (Change_Weight is upper) then (New_Weight is upper)
5. (Current_Weight is lower) then (New_Weight is lower)
6. (Current_Weight is upper) then (New_Weight is upper)

Fig. 12 Rules of the type-2 fuzzy inference system used in the six FIST2 for the neural networks with type-2 fuzzy weights

The rules for the six type-2 fuzzy inference systems are the same, we used six rules for the type-2 fuzzy inference systems, corresponding to the four combinations of two membership functions and we added two rules for the case when the change of weight is null (see Fig. 12).

We perform a comparison of prediction for the proposed ensemble neural networks with the type-1 and type-2 fuzzy integration optimized with GA and PSO against the monolithic neural network, neural network with type-2 fuzzy weights.

The structures of the type-1 and type-2 fuzzy integration are the same and consist of three inputs: the prediction for the neural network with type-2 fuzzy weights using Gaussian membership functions (MF), triangular MF and triangular with standard deviation MF; and one output: the final prediction of the integration (see Fig. 13).

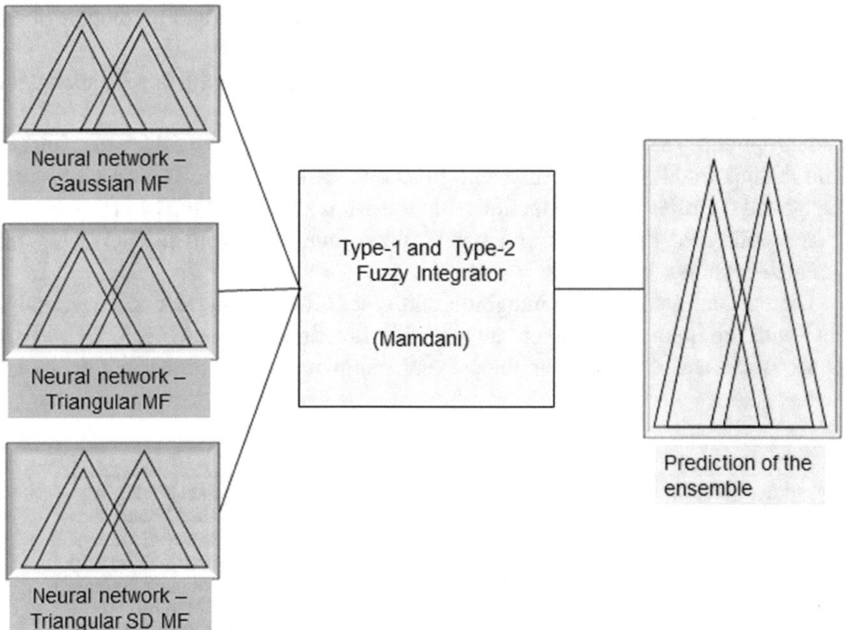

Fig. 13 Structure of the type-1 and type-2 fuzzy integration

In this research, we optimized the type-1 and type-2 fuzzy integrators using GA and PSO for the numbers of membership functions (between two or three), the type of membership functions (between triangular or Gaussian), and the rules established in base at the number of mf's.

4 Optimization of the Type-1 and Type-2 Fuzzy Integration with GA and PSO

The type-1 and type-2 fuzzy integrator (T1FI–T2FI) are optimized with the bio-inspired algorithms: Genetic algorithm (GA) and particle swarm optimization (PSO).

In the type-1 and type-2 fuzzy integrator, the optimization is applied in the number of membership functions for the three inputs and the output of the T1FI and T2FI, optimizing between two or three membership functions (MFs), and the type of membership functions between Gaussians or triangular MFs; besides, in the T2FI, we optimized the footprint of the membership functions (Gaussian or triangular).

In the creation of the type-1 and type-2 fuzzy integrator through chromosomes or particles, the rules are set depending on the number of membership functions (two or three) used. In Fig. 14 we show the rules for two MFs and Fig. 15 show the rules for three MFs (neural network Gaussian MF: NNGMF, neural network triangular MF: NNTMF, neural network triangular SD MF: NNTsdMF and prediction of the ensemble: PE).

We optimized the type-1 and type-2 fuzzy integrator with a GA using the parameters shown in Table 1.

The optimal type-1 fuzzy integrator and type-2 fuzzy integrator achieved using GA with the input and output membership functions shown in Figs. 16 and 17, respectively; and the rules for three membership functions shown in Fig. 12 are used.

We optimized the type-1 and type-2 fuzzy integrator with a PSO using the parameters shown in Table 2.

The optimal type-1 fuzzy integrator and type-2 fuzzy integrator achieved using PSO with the inputs and outputs membership functions shown in Figs. 18 and 19, respectively; and the rules for three membership functions shown in Fig. 14 are used.

1. (NNGMF is Low) and (NNTMF is Low) and (NNTsdMF is Low) then (PE is Low)
2. (NNGMF is Low) and (NNTMF is Low) and (NNTsdMF is High) then (PE is Low)
3. (NNGMF is Low) and (NNTMF is High) and (NNTsdMF is Low) then (PE is Low)
4. (NNGMF is Low) and (NNTMF is High) and (NNTsdMF is High) then (PE is High)
5. (NNGMF is High) and (NNTMF is Low) and (NNTsdMF is Low) then (PE is Low)
6. (NNGMF is High) and (NNTMF is Low) and (NNTsdMF is High) then (PE is High)
7. (NNGMF is High) and (NNTMF is High) and (NNTsdMF is Low) then (PE is High)
8. (NNGMF is High) and (NNTMF is High) and (NNTsdMF is High) then (PE is High)

Fig. 14 Rules for the type-1 and type-2 fuzzy integrator with two membership functions

1. (NNGMF is Low) and (NNTMF is Low) and (NNTsdMF is Low) then (PE is Low)
2. (NNGMF is Low) and (NNTMF is Low) and (NNTsdMF is Middle) then (PE is Low)
3. (NNGMF is Low) and (NNTMF is Low) and (NNTsdMF is High) then (PE is Low)
4. (NNGMF is Low) and (NNTMF is Middle) and (NNTsdMF is Low) then (PE is Low)
5. (NNGMF is Low) and (NNTMF is Middle) and (NNTsdMF is Middle) then (PE is Middle)
6. (NNGMF is Low) and (NNTMF is Middle) and (NNTsdMF is High) then (PE is Middle)
7. (NNGMF is Low) and (NNTMF is High) and (NNTsdMF is Low) then (PE is Low)
8. (NNGMF is Low) and (NNTMF is High) and (NNTsdMF is Middle) then (PE is Middle)
9. (NNGMF is Low) and (NNTMF is High) and (NNTsdMF is High) then (PE is Middle)
10. (NNGMF is Middle) and (NNTMF is Low) and (NNTsdMF is Low) then (PE is Middle)
11. (NNGMF is Middle) and (NNTMF is Low) and (NNTsdMF is Middle) then (PE is Middle)
12. (NNGMF is Middle) and (NNTMF is Low) and (NNTsdMF is High) then (PE is Middle)
13. (NNGMF is Middle) and (NNTMF is Middle) and (NNTsdMF is Low) then (PE is Middle)
14. (NNGMF is Middle) and (NNTMF is Middle) and (NNTsdMF is Middle) then (PE is Middle)
15. (NNGMF is Middle) and (NNTMF is Middle) and (NNTsdMF is High) then (PE is Middle)
16. (NNGMF is Middle) and (NNTMF is High) and (NNTsdMF is Low) then (PE is Middle)
17. (NNGMF is Middle) and (NNTMF is High) and (NNTsdMF is Middle) then (PE is Middle)
18. (NNGMF is Middle) and (NNTMF is High) and (NNTsdMF is High) then (PE is High)
19. (NNGMF is High) and (NNTMF is Low) and (NNTsdMF is Low) then (PE is High)
20. (NNGMF is High) and (NNTMF is Low) and (NNTsdMF is Middle) then (PE is High)
21. (NNGMF is High) and (NNTMF is Low) and (NNTsdMF is High) then (PE is High)
22. (NNGMF is High) and (NNTMF is Middle) and (NNTsdMF is Low) then (PE is High)
23. (NNGMF is High) and (NNTMF is Middle) and (NNTsdMF is Middle) then (PE is Middle)
24. (NNGMF is High) and (NNTMF is Middle) and (NNTsdMF is High) then (PE is Middle)
25. (NNGMF is High) and (NNTMF is High) and (NNTsdMF is Low) then (PE is High)
26. (NNGMF is High) and (NNTMF is High) and (NNTsdMF is Middle) then (PE is High)
27. (NNGMF is High) and (NNTMF is High) and (NNTsdMF is High) then (PE is High)
28. (NNGMF is Low) or (NNTMF is Low) or (NNTsdMF is Low) then (PE is Low)
29. (NNGMF is Middle) or (NNTMF is Middle) or (NNTsdMF is Middle) then (PE is Middle)
30. (NNGMF is High) or (NNTMF is High) or (NNTsdMF is High) then (PE is High)

Fig. 15 Rules for the type-1 and type-2 fuzzy integrator with three membership functions

Table 1 Parameters used in the GA for optimization the type-1 and type-2 fuzzy integrator

Individuals	200
Genes	122 (T2)–62 (T1) (real)
Generations	400
Assign fitness	Ranking
Selection	Stochastic universal sampling
Crossover	Single-point (0.8)
Mutation	1/genes

5 Simulation Results

We accomplished experiments in time series prediction, specifically for the Mackey-Glass time series.

The results for the experiments for the ensemble neural network with type-1 fuzzy integrator optimized with a genetic algorithm (ENNT1FIGA) are shown in Table 3 and Fig. 20. The best prediction error is 0.0212, and the average error is 0.0243.

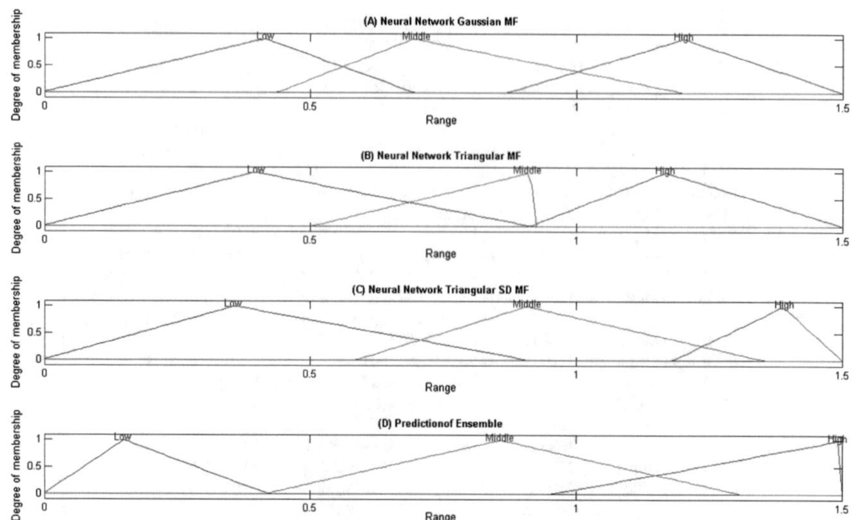

Fig. 16 Inputs (*A*, *B* and *C*) and output (*D*) of the optimal type-1 fuzzy integrator used for the integration of the ensemble neural network optimized using GA

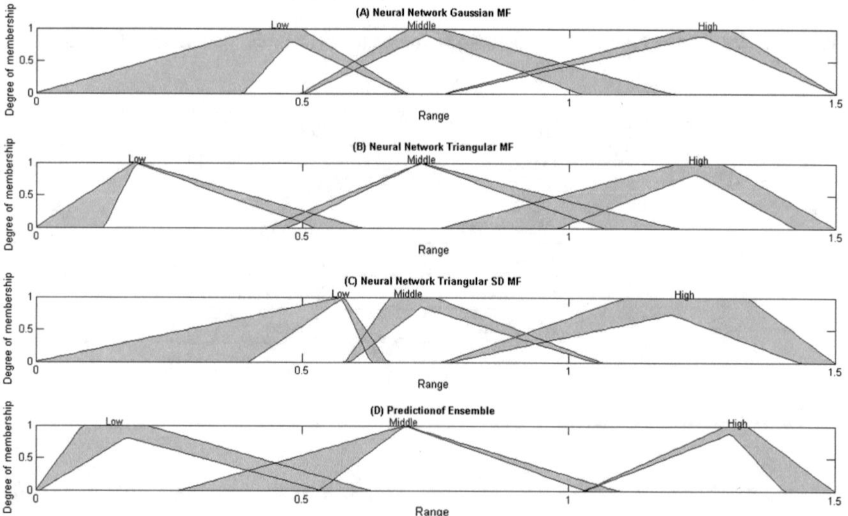

Fig. 17 Inputs (*A*, *B* and *C*) and output (*D*) of the optimal type-2 fuzzy integrator used for the integration of the ensemble neural network optimized using GA

Table 2 Parameters used in the PSO for optimization the type-1 and type-2 fuzzy integrator

Particles	200
Dimensions	122 (T2)–62 (T1) (Real)
Iterations	400
Inertia weight	0.1
Constriction coefficient (*C*)	1
*R*1, *R*2	Random in the range [0,1]
*C*1	Lineal decrement (2–0.5)
*C*2	Lineal increment (0.5–2)

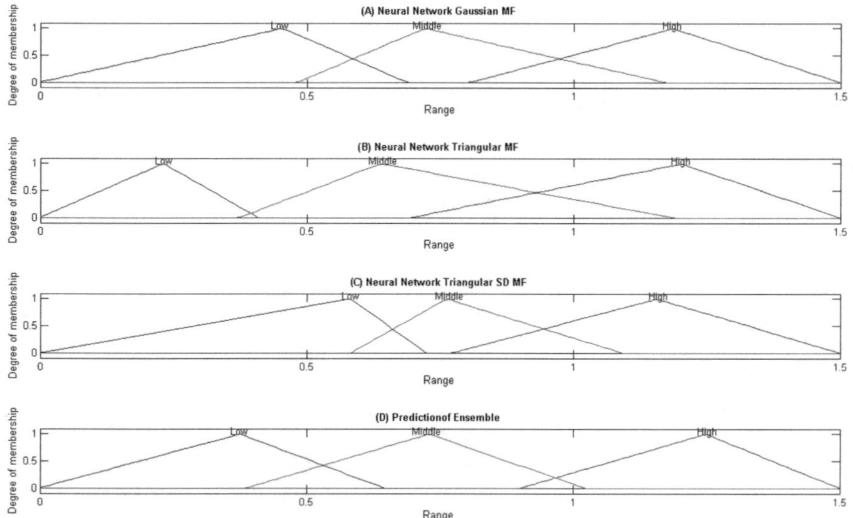

Fig. 18 Inputs (*A*, *B* and *C*) and output (*D*) of the optimal type-1 fuzzy integrator used for the integration of the ensemble neural network optimized using PSO

The results for the experiments for the ensemble neural network with type-2 fuzzy integrator optimized with a genetic algorithm (ENNT2FIGA) are shown in Table 4 and Fig. 21. The best prediction error is 0.0183, and the average error is 0.0198.

The results for the experiments for the ensemble neural network with type-1 fuzzy integrator optimized with a particle swarm optimization (ENNT1FIPSO) are shown in Table 5 and Fig. 22. The best prediction error is 0.0259, and the average error is 0.00333.

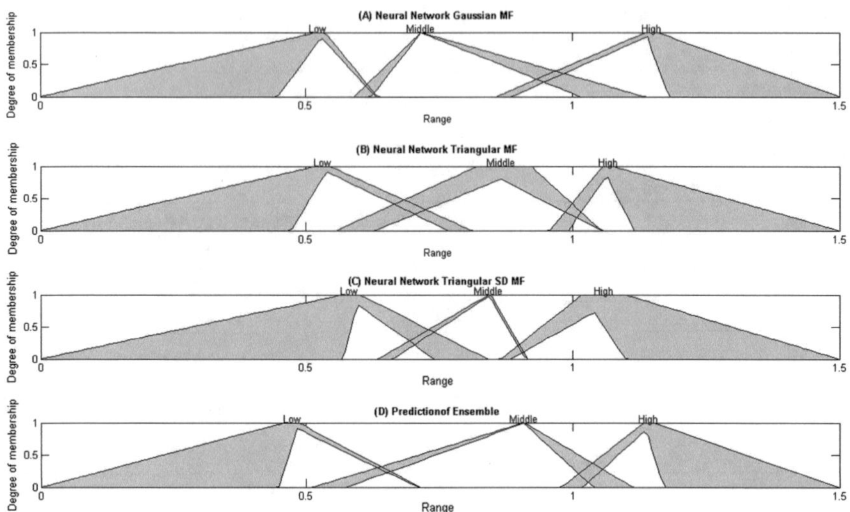

Fig. 19 Inputs (*A*, *B* and *C*) and output (*D*) of the optimal type-2 fuzzy integrator used for the integration of the ensemble neural network optimized using PSO

Table 3 Results for the ensemble neural network with type-1 fuzzy integrator optimized with GA in Mackey-Glass time series

No.	Generations	Individuals	Prediction error
E1	400	200	0.0248
E2	400	200	0.0282
E3	400	200	0.0227
E4	400	200	0.0223
E5	400	200	0.0244
E6	400	200	0.0298
E7	**400**	**200**	**0.0212**
E8	400	200	0.0231
E9	400	200	0.0229
E10	400	200	0.0243
Average prediction error			0.0243

The results for the experiments for the ensemble neural network with type-2 fuzzy integrator optimized with a particle warm optimization (ENNT2FIPSO) are shown in Table 6 and Fig. 23. The best prediction error is 0.0240, and the average error is 0.0274.

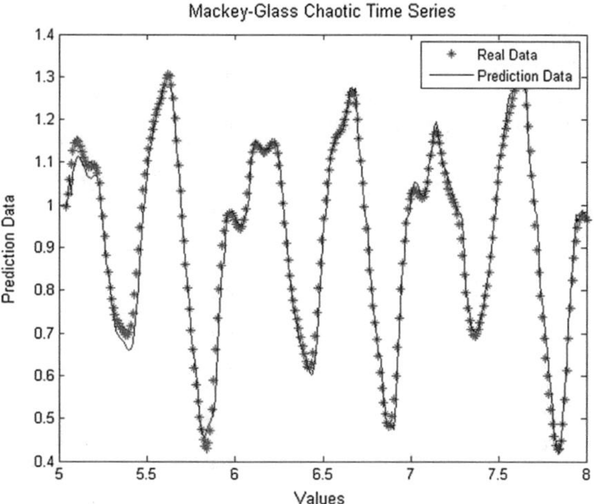

Fig. 20 Graphic of the real prediction data of Mackey-Glass time series against the prediction data achieved with ENNT1FIGA

Table 4 Results for the ensemble neural network with type-2 fuzzy integrator optimized with GA in Mackey-Glass time series

No.	Generations	Individuals	Prediction error
E1	400	200	0.0235
E2	400	200	0.0189
E3	400	200	0.0208
E4	**400**	**200**	**0.0183**
E5	400	200	0.0215
E6	400	200	0.0197
E7	400	200	0.0184
E8	400	200	0.0190
E9	400	200	0.0222
E10	400	200	0.0195
Average prediction error			0.0198

The average prediction errors for all the results are obtained for 30 experiments.

In Table 7, a comparison for the prediction for the Mackey-Glass time series between the results for the monolithic neural network (MNN), the neural network with type-2 fuzzy weights (NNT2FW), the ensemble neural network with

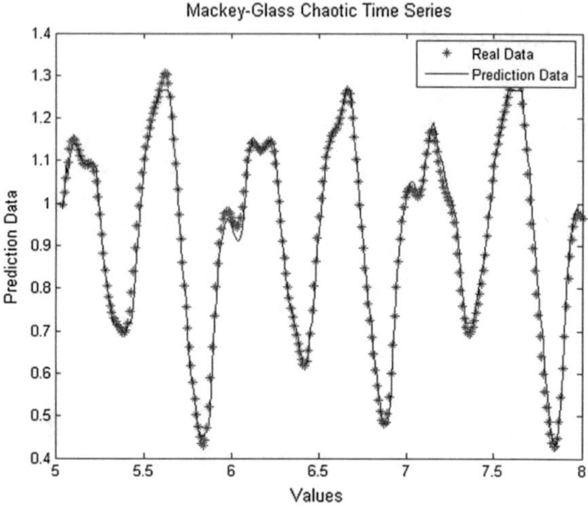

Fig. 21 Graphic of the real prediction data of Mackey-Glass time series against the prediction data achieved with ENNT2FIGA

Table 5 Results for the ensemble neural network with type-1 fuzzy integrator optimized with PSO in Mackey-Glass time series

No.	Iterations	Particles	Prediction error
E1	400	200	0.0344
E2	400	200	0.0288
E3	400	200	0.0419
E4	400	200	0.0267
E5	**400**	**200**	**0.0259**
E6	400	200	0.0326
E7	400	200	0.0287
E8	400	200	0.0307
E9	400	200	0.0472
E10	400	200	0.0361
Average prediction error			0.0333

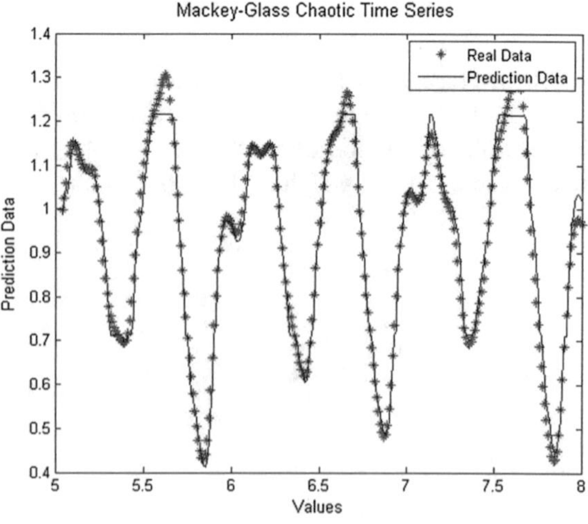

Fig. 22 Graphic of the real prediction data of Mackey-Glass time series against the prediction data achieved with ENNT1FIPSO

Table 6 Results for the ensemble neural network with type-2 fuzzy integrator optimized with PSO in Mackey-Glass time series

No.	Iterations	Particles	Prediction error
E1	400	200	0.0269
E2	400	200	0.0247
E3	400	200	0.0291
E4	400	200	0.0241
E5	400	200	0.0321
E6	400	200	0.0341
E7	400	200	0.0255
E8	400	200	0.0282
E9	**400**	**200**	**0.0240**
E10	400	200	0.0310
Average prediction error			0.0274

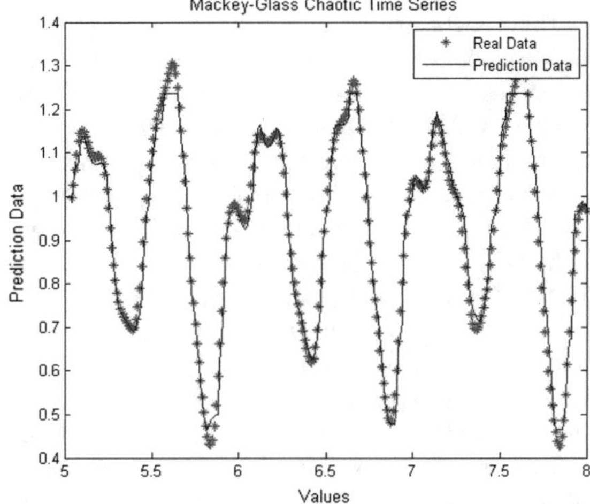

Fig. 23 Graphic of the real prediction data of Mackey-Glass time series against the prediction data achieved with ENNT2FIPSO

Table 7 Comparison of results for the ensemble neural network with type-1 and type-2 fuzzy integrator optimized with GA and PSO in Mackey-Glass time series

Method	Prediction error
MNN [16]	0.0530
NNT2FW [16]	0.0390
ENNT1FIGA	0.0212
ENNT1FIPSO	0.0259
ENNT2FIGA	0.0183
ENNT2FIPSO	0.0240

type-1fuzzy integrator optimized with GA (ENNT1FIGA) and the ensemble neural network with type-1 fuzzy integrator optimized with PSO (ENNT1FIPSO), the ensemble neural network with type-2 fuzzy integrator optimized with GA (ENNT2FIGA), and the ensemble neural network with type-2 fuzzy integrator optimized with PSO (ENNT2FIPSO) are presented.

6 Conclusions

In this work, in based on the experiments, we observe that by optimizing the type-1 and type-2 fuzzy integrator for the ensemble neural network using genetic algorithms and particle swarm optimization, we can achieve better results than the monolithic neural network and the neural network with type-2 fuzzy weights for the Mackey-Glass time series.

This conclusion is based on the fact that for the prediction error of 0.0183, for the ensemble neural network with type-2 fuzzy integrator optimized with GA is better than predictions error of the others methods: 0.0530 for the monolithic neural network, 0.0390 for neural network with type-2 fuzzy weights, 0.0212 for the ensemble neural network with type-1 fuzzy integrator optimized with GA, 0.0259 for the ensemble neural network with type-2 fuzzy integrator optimized with PSO and 0.0240 for the ensemble neural network with type-2 fuzzy integrator optimized with PSO; besides, the average errors (30 experiments) of 0.0198 is better than 0.0770, 0.0610, 0.0243, 0.0333, and 0.0274.

The optimization in the type-2 fuzzy integrator with genetic algorithms presents better results in almost all the experiments than the optimization of the type-1 fuzzy integrator with genetic algorithm or particle swarm optimization, or type-2 fuzzy integrator optimized with particle swarm optimization.

These results show that using bio-inspired methods for optimized type-2 fuzzy inference systems, such as genetic algorithm and particle swarm optimization, we can obtain better results.

References

1. Abiyev R., "A Type-2 Fuzzy Wavelet Neural Network for Time Series Prediction", Lecture Notes in Computer Science, vol. 6098, pp. 518-527, 2010.
2. Castillo O., Martínez-Marroquín R., Melin P., Soria J., Valdez F., "Comparative Study of Bio-Inspired Algorithms Applied to the Optimization of Type-1 and Type-2 Fuzzy Controllers for an Autonomous Mobile Robot", Information Sciences Vol. 192 (1), pp. 19-38, 2012.
3. Castillo O.,Melin P., "A review on the design and optimization of interval type-2 fuzzy controllers". Applied Soft Computing, vol. 12 (4), pp. 1267-1278, 2012.
4. Castillo O., Melin P., "Soft Computing for Control of Non-Linear Dynamical Systems", Springer-Verlag, Heidelberg, Germany, 2001.

5. Castillo O., Melin P., "Type-2 Fuzzy Logic Theory and Applications," Springer Verlag, Berlin. pp. 29-43, 2008.
6. Castro J., Castillo O., MelinP.,"An Interval Type-2 Fuzzy Logic Toolbox for Control Applications". FUZZ-IEEE, pp. 1-6, 2007.
7. Castro J., Castillo O., Melin P., Mendoza O., Rodríguez-Díaz A., "An Interval Type-2 Fuzzy Neural Network for Chaotic Time Series Prediction with Cross-Validation and Akaike Test". Soft Computing for Intelligent Control and Mobile Robotics, pp. 269-285, 2011.
8. Castro J., Castillo O., Melin P., Rodríguez-Díaz A., "A Hybrid Learning Algorithm for a Class of Interval Type-2 Fuzzy Neural Networks", Information Sciences, vol. 179 (13), pp. 2175-2193, 2009.
9. Castro J., Castillo O., Melin P., Rodriguez-Diaz A."Building Fuzzy Inference Systems with a New Interval Type-2 Fuzzy Logic Toolbox", Transactions on Computational Science 1, pp. 104-114, 2008.
10. Cazorla M. and Escolano F., "Two Bayesian Methods for Junction Detection", IEEE transaction on Image Processing, vol. 12, Issue 3, pp. 317-327, 2003.
11. Cervantes L., Castillo O., "Genetic Optimization of Membership Functions in Modular Fuzzy Controllers for Complex Problems". Recent Advances on Hybrid Intelligent Systems, pp. 51-62, 2013.
12. De Wilde O., "The Magnitude of the Diagonal Elements in Neural Networks", Neural Net-works, vol. 10 (3), pp. 499-504, 1997.
13. Eiben A. E., and Smith J. E., "Introduction to Evolutionary", Ed. Springer, pp. 304, 2007.
14. Feuring T., "Learning in Fuzzy Neural Networks", Neural Networks, IEEE International Conference on, vol. 2, pp. 1061-1066, 1996.
15. Gaxiola F., Melin P., Valdez F., "Backpropagation Method with Type-2 Fuzzy Weight Adjustment for Neural Network Learning", Fuzzy Information Processing Society (NAFIPS), Annual Meeting of the North American, pp. 1-6, 2012.
16. Gaxiola F., Melin P., Valdez F., Castillo O., "Interval type-2 fuzzy weight adjustment for backpropagation neural networks with application in time series prediction", Information Sciences, vol. 260, pp. 1-14, 2014.
17. Gaxiola F., Melin P., Valdez F., Castillo O., "Neural Network with Type-2 Fuzzy Weights Adjustment for Pattern Recognition of the Human Iris Biometrics", MICAI'12 Proceedings of the 11th Mexican international conference on Advances in Computational Intelligence, vol. 2, pp. 259-270, 2012.
18. Gaxiola F., Melin P., Valdez F., Castillo O., "Optimization of type-2 Fuzzy Weight for Neural Network using Genetic Algorithm and Particle Swarm Optimization", 5th World Congress on Nature and Biologically Inspired Computing, pp. 22-28, 2013.
19. Gaxiola F., Melin P., Valdez F., "Genetic Optimization of Type-2 Fuzzy Weight Adjust-ment for Backpropagation Ensemble Neural Network", Recent Advances on Hybrid Intelli-gent Systems, pp. 159-171, 2013.
20. Gaxiola F., Melin P., Valdez F., "Neural Network with Lower and Upper Type-2 Fuzzy Weights using the Backpropagation Learning Method", IFSA World Congress and NAFIPS Annual Meeting (IFSA/NAFIPS), pp. 637-642, 2013.
21. Hagan M.T., Demuth H.B. and Beale M.H., "Neural Network Design", PWS Publishing, Boston, pp. 736, 1996.
22. Hagras H., "Type-2 Fuzzy Logic Controllers: A Way Forward for Fuzzy Systems in Real World Environments", IEEE World Congress on Computational Intelligence, pp. 181-200, 2008.
23. Hidalgo D., Castillo O., Melin P., "Type-1 and Type-2 Fuzzy Inference Systems as Integration Methods in Modular Neural Networks for Multimodal Biometry and its Optimization with Genetic Algorithms", Information Sciences, vol. 179 (13), pp. 2123-2145, 2009.
24. Hidalgo D., Castillo O., Melin P., "Type-1 and Type-2 Fuzzy Inference Systems as Integration Methods in Modular Neural Networks for Multimodal Biometry and Its

Optimization with Genetic Algorithms", Soft Computing for Hybrid Intelligent Systems, pp. 89-114, 2008.
25. Hidalgo D., Melin P., Castillo O., "An Optimization Method for Designing Type-2 Fuzzy Inference Systems Based on the Footprint of Uncertainty Using Genetic Algorithms", Expert Systems with Application, vol. 39, pp 4590-4598, 2012.
26. Ishibuchi H., Morioka K. and Tanaka H., "A Fuzzy Neural Network with Trapezoid Fuzzy Weights, Fuzzy Systems", IEEE World Congress on Computational Intelligence, vol. 1, pp. 228-233, 1994.
27. Ishibuchi H., Tanaka H. and Okada H., "Fuzzy Neural Networks with Fuzzy Weights and Fuzzy Biases", Neural Networks, IEEE International Conference on, vol. 3, pp. 1650-165, 1993.
28. Islam M.M. and Murase K., "A New Algorithm to Design Compact Two-Hidden-Layer Artificial Neural Networks", Neural Networks, vol. 14 (9), pp. 1265-1278, 2001.
29. Jang J.S.R., Sun C.T., Mizutani E., "Neuro-Fuzzy and Soft Computing: a Computational Approach to Learning and Machine Intelligence", Ed. Prentice Hall, 1997.
30. Kamarthi S. and Pittner S., "Accelerating Neural Network Training using Weight Extrapolations", Neural Networks, vol. 12 (9), pp. 1285-1299, 1999.
31. Karnik N. and Mendel J., "Applications of Type-2 Fuzzy Logic Systems to Forecasting of Time-Series", Information Sciences, vol. 120 (1-4), pp. 89-111, 1999.
32. Kennedy J., and Eberhart R., "Particle swarm optimization", In Proceedings of IEEE international conference on neural networks pp. 1942–1948, 1995.
33. Martinez G., Melin P., Bravo D., Gonzalez F. and Gonzalez M., "Modular Neural Networks and Fuzzy Sugeno Integral for Face and Fingerprint Recognition", Advances in Soft computing, vol..34, pp. 603-618, 2006.
34. Martínez-Soto R., Castillo O., Aguilar L., Melin P., "Fuzzy Logic Controllers Optimization Using Genetic Algorithms and Particle Swarm Optimization". Advances in Soft Computing - 9th Mexican International Conference on Artificial Intelligence, pp. 475-486, 2010.
35. Melin P., Castillo O., "Hybrid Intelligent Systems for Pattern Recognition Using Soft Computing", Springer-Verlag, Heidelberg, pp 2-3, 2005.
36. Melin P., "Modular Neural Networks and Type-2 Fuzzy Systems for Pattern Recognition", Springer, pp. 1-204, 2012.
37. Melin P., Olivas F., Castillo O., Valdez F., Soria J., García-Valdez J.M., "Optimal design of fuzzy classification systems using PSO with dynamic parameter adaptation through fuzzy logic". Expert Systems with Application, vol. 40(8), pp. 3196-3206, 2013.
38. Melin P., Sánchez D., Castillo O., "Genetic Optimization of Modular Neural Networks with Fuzzy Response Integration for Human Recognition", Information Sciences, vol. 197, pp. 1-19, 2012.
39. Okamura M., Kikuchi H., Yager R., Nakanishi S., "Character diagnosis of fuzzy systems by genetic algorithm and fuzzy inference", Proceedings of the Vietnam-Japan Bilateral Symposium on Fuzzy Systems and Applications, Halong Bay, Vietnam, pp. 468-473, 1998.
40. Phansalkar V.V. and Sastry P.S., "Analysis of the Back-Propagation Algorithm with Momentum", IEEE Transactions on Neural Networks, vol. 5 (3), pp. 505-506, 1994.
41. Salazar P.A., Melin P. and Castillo O., "A New Biometric Recognition Technique Based on Hand Geometry and Voice Using Neural Networks and Fuzzy Logic", Soft Computing for Hybrid Intelligent Systems, pp. 171-186, 2008.
42. Sanchez D., Melin P., "Optimization of modular neural networks and type-2 fuzzy integrators using hierarchical genetic algorithms for human recognition", IFSA World Congress, Surabaya-Bali, Indonesia, OS-414, 2011.
43. Sepúlveda R., Castillo O., Melin P., Montiel O., "An Efficient Computational Method to Implement Type-2 Fuzzy Logic in Control Applications", Analysis and Design of Intelligent Systems using Soft Computing Techniques, pp. 45-52, 2007.
44. Sepúlveda R., Castillo O., Melin P., Rodriguez A., Montiel O., "Experimental study of intelligent controllers under uncertainty using type-1 and type-2 fuzzy logic", Information Sciences, Vol. 177, No. 11, pp. 2023-2048, 2007.

45. Shi Y., and Eberhart R., "A modified particle swarm optimizer", In: Proceedings of the IEEE congress on evolutionary computation, pp. 69–73, 1998.
46. Shi Y., and Eberhart R., "Empirical study of particle swarm optimization", In: Proceedings of the IEEE congress on evolutionary computation, pp. 1945–1950, 1999.J. Clerk Maxwell, A Treatise on Electricity and Magnetism, 3rd ed., vol. 2. Oxford: Clarendon, 1892, pp.68–73.
47. Valdez F., Melin P., Castillo O., "An Improved Evolutionary Method with Fuzzy Logic for Combining Particle Swarm Optimization and Genetic Algorithms", Applied Soft Computing, vol. 11 (2), pp. 2625-2632, 2011.
48. Valdez F., Melin P., Castillo O., "Evolutionary method combining particle swarm optimization and genetic algorithms using fuzzy logic for decision making", Proceedings of the IEEE International Conference on Fuzzy Systems, pp. 2114–2119, 2009.
49. Wang W., Bridges S., "Genetic Algorithm Optimization of Membership Functions for Mining Fuzzy Association Rules", Department of Computer Science Mississippi State University, 2000.
50. Zadeh L. A., "Fuzzy Sets", Journal of Information and Control, Vol. 8, pp. 338–353, 1965.

Interval Type-2 Fuzzy Possibilistic C-Means Optimization Using Particle Swarm Optimization

Elid Rubio and Oscar Castillo

Abstract In this paper, we present optimization of the Interval Type-2 Fuzzy Possibilistic C-Means (IT2FPCM) algorithm using Particle Swarm Optimization (PSO), with the goal of automatically finding the optimal number of clusters and the optimal lower and upper limit of Fuzzy and Possibility exponents of weight of the of the IT2FPCM algorithm, and also the centroids of clusters of each dataset tested with the IT2FPCM algorithm optimized using PSO.

Keywords Fuzzy clustering · Type-2 Fuzzy Logic Techniques · Interval Type-2 Fuzzy Possibilistic C-Means · Particle Swarm Optimization

1 Introduction

Patterns or groupings of data with similar characteristics make possible the proposal of clustering algorithms. Among the most used algorithms are fuzzy clustering algorithms, this is because this kind of clustering algorithms allows the data set partitioning into a homogeneous fuzzy clusters.

The popularity of the fuzzy clustering algorithms [2–7] is due to the fact that allow a datum belong to different data clusters into a given data set. Fuzzy Clustering algorithms has been widely used satisfactory in different research areas like pattern recognition [1], data mining [8], classification [9], image segmentation [10, 11], data analysis and modeling [12], and also has been the base to develop other clustering algorithm.

Although this kind of algorithm has been used in different areas of research with good results, this algorithm has three notables inconvenient that are mentioned below:

E. Rubio (✉) · O. Castillo
Tijuana Institute of Technology, Tijuana, Mexico
e-mail: elid.rubio@hotmail.com

O. Castillo
e-mail: ocastillo@tectijuana.mx

P. Melin et al. (eds.), *Nature-Inspired Design of Hybrid Intelligent Systems*,
Studies in Computational Intelligence 667, DOI 10.1007/978-3-319-47054-2_4

1. Finding the optimal number of clusters within a data set is difficult, that because the number of clusters has to be set arbitrarily, i.e., the number of clusters to be created by the FCM must be set manually on each algorithm execution, this is done again and again until finding the optimal number of clusters.
2. Fuzzification exponent influences the performance of FCM algorithm, this factor is denoted by the parameter m and normally is fixed $m = 2$, and work to find the optimal number of clusters in some datasets but in others not, this mean to each data set the fuzzification exponent different.
3. FCM algorithm is very susceptible to noise and is incapable of manage large amounts of uncertainty in the data set.

However, the fuzzy clustering methods are not able to handle uncertainty in a given data set during the process of data clustering; due to this problem, some clustering algorithms like FCM, PCM, FPCM have been extended using Type-2 Fuzzy Logic Techniques [16], with the intention to improves the performance of FCM algorithm and making this algorithms less susceptible to noise and able to handle uncertainty, the improvement of these algorithms was called Interval Type-2 Fuzzy C-Means (IT2FCM) [13, 14], Interval Type-2 Possibilistic C-Means (IT2PCM) [3, 5], Interval Type-2 Fuzzy Possibilistic C-Means (IT2PCM). These extensions have been applied to the creation of membership functions [15, 16], classification [17], image processing [18, 19], and Designing of Fuzzy Inference Systems. However, the extensions before mentioned only makes it less prone to noise and allows you to manage uncertainty to FCM handle uncertainty.

With the intention to overcome the inconveniences above mentioned, Rubio and Castillo in [20], presented the implementation of Particle Swarm Optimization (PSO) for the optimization of Interval Type-2 Fuzzy C-Means algorithm, this in order to automatically find the optimal number of clusters c and the interval of fuzzification or weight exponent [$m1$, $m2$], using as objective function for optimization algorithm an index cluster validation. In this work, we pretend to automatically finding the optimal number c of clusters, the optimal interval of the fuzzification exponent [$m1$, $m2$], and also the optimal position for the lower and upper limits of the interval to the centers of clusters using the PSO. In Sect. 2, we show a brief background of the Interval Type-2 Fuzzy C-Means (IT2 FCM) and thePSO algorithms is presented in Sect. 3.

2 Interval Type-2 Fuzzy Possibilistic C-Means Algorithm

The Interval Type-2 Fuzzy-Possibilistic C-Means algorithm is a proposed extension of the FPCM algorithm using Type-2 Fuzzy Logic Techniques. This algorithm produces in the same way that FPCM algorithm both membership and possibility using weights as exponents m and η for the fuzziness and possibility, respectively, which may be represented by a range rather than a precise value, i.e., $m = [m_1, m_2]$, where m_1 and m_2 represent the lower and upper limit of weighting exponent for fuzziness and $\eta = [\eta_1, \eta_2]$, where m_1 and m_2 represent the lower and upper limit of weighting exponent for possibility.

Because the m value is represented by an interval, the fuzzy partition matrix $\mu_i(x_k)$ must be calculated for the interval $[m_1, m_2]$, for this reason $\mu_i(x_k)$ would be given by belonging interval $[\underline{\mu}_i(x_k), \overline{\mu}_i(x_k)]$ where $\underline{\mu}_i(x_k)$ and $\overline{\mu}_i(x_k)$ represent the lower and upper limit of the belonging interval of datum x_j to a clustering v_i, updating the lower and upper limits of the range of the fuzzy membership matrix can be expressed as:

$$\underline{\mu}_i(x_k) = \min\left\{\left(\sum_{j=1}^{c}\left(\frac{d_{ik}}{d_{jk}}\right)^{\frac{2}{m_1-1}}\right)^{-1}, \left(\sum_{j=1}^{c}\left(\frac{d_{ik}}{d_{jk}}\right)^{\frac{2}{m_2-1}}\right)^{-1}\right\} \tag{1}$$

$$\overline{\mu}_i(x_k) = \max\left\{\left(\sum_{j=1}^{c}\left(\frac{d_{ik}}{d_{jk}}\right)^{\frac{2}{m_1-1}}\right)^{-1}, \left(\sum_{j=1}^{c}\left(\frac{d_{ik}}{d_{jk}}\right)^{\frac{2}{m_2-1}}\right)^{-1}\right\} \tag{2}$$

Because the η value is represented by an interval, fuzzy partition matrix $\tau_i(x_k)$ must be calculated for the interval $[\eta_1, \eta_2]$, for this reason $\tau_i(x_k)$ would be given by the belonging interval $[\underline{\tau}_i(x_k), \overline{\tau}_i(x_k)]$ where $\underline{\tau}_i(x_k)$ and $\overline{\tau}_i(x_k)$ represent the lower and upper limit of the belonging interval of datum x_j to a clustering v_i, updating the lower and upper limits of the range of the fuzzy membership matrix can be expressed as:

$$\underline{\tau}_i(x_k) = \min\left\{\left(\sum_{j=1}^{c}\left(\frac{d_{ik}}{d_{jk}}\right)^{\frac{2}{\eta_1-1}}\right)^{-1}, \left(\sum_{j=1}^{c}\left(\frac{d_{ik}}{d_{jk}}\right)^{\frac{2}{\eta_2-1}}\right)^{-1}\right\} \tag{3}$$

$$\overline{\tau}_i(x_k) = \max\left\{\left(\sum_{j=1}^{c}\left(\frac{d_{ik}}{d_{jk}}\right)^{\frac{2}{\eta_1-1}}\right)^{-1}, \left(\sum_{j=1}^{c}\left(\frac{d_{ik}}{d_{jk}}\right)^{\frac{2}{\eta_2-1}}\right)^{-1}\right\} \tag{4}$$

Updating the positions of the centroids of clusters should take into account the degree of belonging interval of the fuzzy and possibilistic matrices, resulting in a range of coordinates of the positions of the centroids of the clusters. The procedure for updating cluster prototypes in IT2FPCM requires calculating the centroids for the lower and upper of the limit of the interval using the fuzzy and possibilistic membership matrices, these centroids will be given by the following equations:

$$\underline{v}_i = \frac{\sum_{j=1}^{n}\left(\underline{\mu}_i(x_j) + \underline{\tau}_i(x_j)\right)^{m_1} x_j}{\sum_{j=1}^{n}\left(\underline{\mu}_i(x_j) + \underline{\tau}_i(x_j)\right)^{m_1}} \tag{5}$$

$$\overline{v}_i = \frac{\sum_{j=1}^{n} \left(\overline{\mu}_i(x_j) + \overline{\tau}_i(x_k)\right)^{m_1} x_j}{\sum_{j=1}^{n} \left(\overline{\mu}_i(x_j) + \overline{\tau}_i(x_k)\right)^{m_1}} \tag{6}$$

The centroid calculation for the lower and upper limit of the interval results in an interval of coordinates of positions of the centroids of the clusters. Type-reduction and defuzzification use the type-2 fuzzy operations. The centroids matrix and the fuzzy partition matrix are obtained by the type-reduction as shown in the following equations:

$$v_j = \frac{\underline{v}_j + \overline{v}_j}{2} \tag{7}$$

$$\mu_i(x_j) = \frac{\left(\underline{\mu}_i(x_j) + \underline{\tau}_i(x_j)\right) + \left(\overline{\mu}_i(x_j) + \overline{\tau}_i(x_j)\right)}{2} \tag{8}$$

Based on all this, the IT2 FPCM algorithm consists of the following steps:

1. Establish c, m_1, m_2, η_1, η_2.
2. Initialize randomly centroids for the lower and upper bounds of the interval.
3. Calculate the update of the fuzzy partition matrices for lower and upper bound of the interval using Eqs. (1) and (2), respectively.
4. Calculate the update of the possibilistic partition matrices for lower and upper bound of the interval using Eqs. (3) and (4), respectively.
5. Calculate the centroids for the lower and upper fuzzy partition matrix using Eqs. (5) and (6), respectively.
6. Type reduction of the fuzzy partition matrix and centroid, if the problem requires using Eqs. (7) and (8), respectively.
7. Repeat Steps 3–5 until $|\tilde{J}_{m,\eta}(t) - \tilde{J}_{m,\eta}(t-1)| < \varepsilon$.

This extension on the FPCM algorithm is intended to realize that this algorithm is capable of handling uncertainty and is less susceptible to noise.

3 Particle Swarm Optimization Algorithm

Particle Swarm Optimization (PSO) proposed by Eberhart and Kennedy [20, 21, 22] in 1995, is a bio-inspired optimization method based in the behavior of biological communities that exhibits individual and social behavior [23–26], examples of these communities are groups of birds, schools of fish, and swarms of bees.

PSO algorithm is based fundamentally in social science and computer science, this algorithm uses the swarm intelligence concept, which is a property of a system whereby the collective behavior of unsophisticated agents who are interacting locally with their environment creates consistent global functional patterns [23].

The cornerstones of PSO are social concepts, principles of swarm intelligence, and computational characteristics.

In PSO, particles (potential solutions), fly over the search space of the problem, which iteratively less optimal particles flying toward optimal particles, until all particles converge toward the same point (solution). To make a cloud particles or particle swarm converge toward the same point, the PSO algorithm needs to apply two kinds of knowledge: cognitive knowledge and social knowledge. Cognitive knowledge is the experience that each particle acquires along the optimization of this process is called cognitive component. Social knowledge is the experience that all the particle cloud acquired by exchanging information between particles during the optimization process is called social component, i.e., the particles fly through a multidimensional search space, where the position of each particle is adjusted according to its own experience and that of your neighborhood [25].

The position of the particle is changed by adding a velocity, $v_i(t)$ for the current position [20, 21] [23–26]. As shown in the following equation:

$$x_i(t+1) = x(t) + v_i(t+1) \tag{9}$$

With $x_i(0) \sim U(x_{\min}, x_{\max})$

Velocity vector which carries the optimization process and reflects both the experimental knowledge of the particle and exchange social information of the particles in the neighborhood. The calculation of the velocity vector is given by the following equation:

$$v_i(t+1) = v(t) + c_1 r_{1j}(t)[y_{ij} - x_{ij}(t)] + c_2 r_{2j}(t)[\hat{y}_{ij} - x_{ij}(t)] \tag{10}$$

where the variables of the equation represent the following:

- $v_{ij}(t)$ is the velocity of the particle i in dimension $j = 1, \ldots, n$ in the time step t.
- $x_{ij}(t)$ is the position of the particle i in dimension j at time step t.
- $c1$ and $c2$ are positive acceleration constants used to scale the contribution of cognitive and social components, respectively.
- $r_{1j}(t)$ and $r_{2j}(t) \sim U(0,1)$ are random values in the range [0,1], samples from a uniform distribution. These random values to introduce a stochastic element algorithm.

Best personal position y_i, associated with the particle i, is best position of the particle that has visited since the first time step. Considering minimization problems, personal best position at the next time step $t + 1$, is calculated as

$$y_i(t+1) = \begin{cases} y_i(t) & \text{if} f(x_i(t+1)) \geq f(y_i(t)) \\ x_i(t+1) & \text{if} f(x_i(t+1)) \geq f(y_i(t)) \end{cases} \tag{11}$$

where $f{:}\mathbb{R}^n \rightarrow \mathbb{R}$ is the fitness function, which measures how close the solution corresponding to the optimal, i.e., fitness function quantifies the performance or quality of a particle (solution). The best global position $\hat{y}_i(t)$ at the time step t, is defined as

$$\hat{y}_i(t) \in \{y_0(t),\ldots,y_{n_s}(t)\}|f(\hat{y}_i(t)) = \min\{f(y_0(t)),\ldots,f(y_{n_s}(t))\} \qquad (12)$$

where n_s is the total number of particles in the swarm. Note that the definition in Eq. (12) says that $\hat{y}$ is the best position found by any particle so far—which is calculated as the best personal position. The best global position can also be selected from the current swarm, in which case

$$\hat{y}_i(t) = \min\{f(x_0(t)),\ldots,f(x_{n_s}(t))\} \qquad (13)$$

4 Optimization of Interval Type-2 Fuzzy C-Means Using Particle Swarm Optimization

In this section, we show how to implement the Particle Swarm Optimization (PSO) algorithm to automatically find the optimal number of clusters, the optimal fuzzy, and possibility weight and also the optimal position of the centers of clusters. The IT2 FPCM algorithm is an extension of FPCM algorithm proposed by Rubio et al. [20], which allow handle uncertainty and less prone to noise, IT2FPCM algorithm is a improvesation of FPCM algorithm, which allows handle uncertainty and less prone to noise, but even have disadvantage that number of clusters and the interval of fuzzification and possibilities are to be placed manually, i.e., the optimal number of groups for a given data set is a trial and error repeatedly executing the algorithm with different number of clusters for the given data set at each execution.

Change the number of clusters and the interval of fuzzification and possibilities in each run of the algorithm IT2FPCM becomes tedious; this is a common problem on fuzzy clustering algorithms, which is why optimization is done. The optimization of IT2 FPCM clustering algorithm was performed by PSO algorithm which has been successfully implemented in different research [27], but we can use any optimization algorithm.

The operation of the optimization using PSO algorithm is given by a sequence of steps, which are:

1. Generate the initial swarm.
2. Evaluate the particles of the swarm.
3. Updatie the particle velocity.
4. Calculate new positions of the particles.

From Step 2 to Step 4, begins an iterative process until a stopping criterion is satisfied. In Fig. 1, we can see the outline of the algorithm of particle swarm optimization. To optimize the IT2 FPCM algorithm following the outline of the

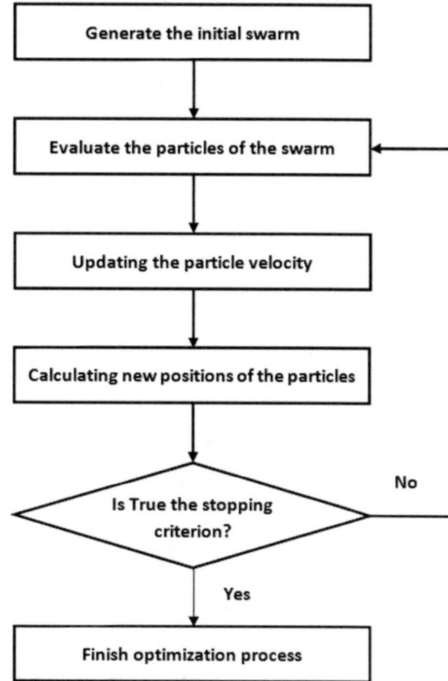

Fig. 1 Block diagram of the PSO algorithm

algorithm of PSO, the first thing is to create a particle swarm to create the cloud that has to choose the representation of candidate solutions (particles), for which we choose a real (continue values) representation.

Once we have the representation of candidate solutions (particles), set the number of variables to indicate the position of the particle or possible solution to our problem, number of clusters, the weight exponents of fuzzification and possibilities, and the centers positions optimal for a given dataset.

The optimization of the algorithm IT2 FPCM the representation of the particle is composed of the number of clusters, lower and upper limits of the range of the weight exponents of fuzzification and possibilities, as shown in Fig. 2. Set the number of variables that will be composed by the particle, the initial population is generated randomly within a set range for each variable according to upper and lower limits of the search space for each variable. The creation of the initial swarm of particles is given according to

$$\mathrm{Swarm}_{ij} = \mathrm{LI}_j + (\mathrm{LS}_j - \mathrm{LI}_j) * r_{ij}, \quad i = 1, 2, \ldots \mathrm{NP}, \tag{14}$$

where LI and LS are the lower limit and upper limit of the space search, respectively, r is a random number between [0,1], i is the number of individual, j is the number of dimensions of the search space, and NP is the number of particles within the swarm. Particles swarm generated proceeds to evaluate each particle cloud created, the evaluation of the particles is performed by the IT2 FPCM algorithm,

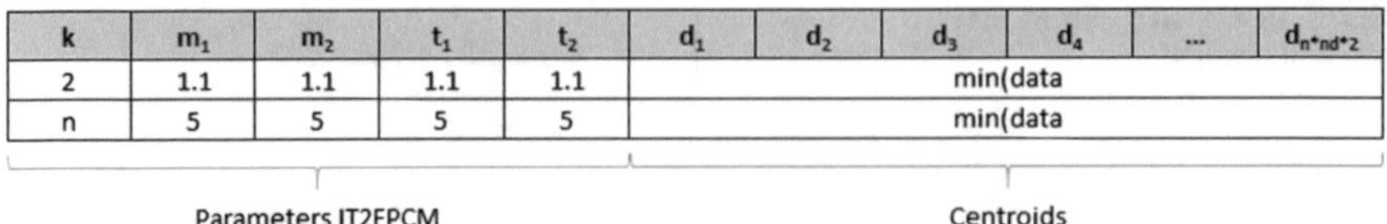

k
m_1
m_2
t_1
t_2
d_1
d_2
d_3
d_4
...
d_{n*nd*2}
2
1.1
1.1
1.1
1.1
min(data
n
5
5
5
5
min(data
Parameters IT2FPCM
Centroids

Fig. 2 Particle representation

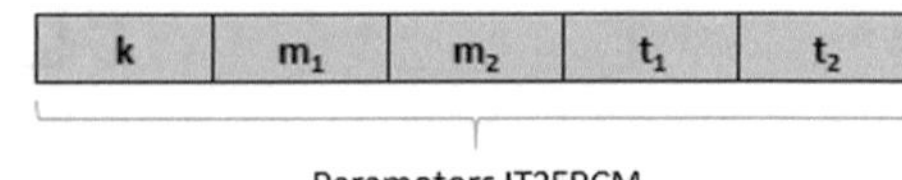

Fig. 3 Representation of the parameters of the IT2FPCM into a particle

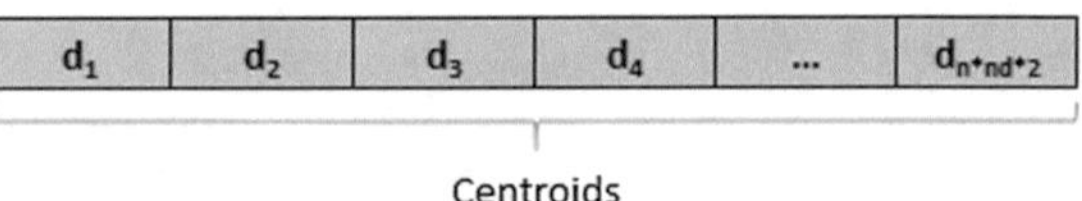

Fig. 4 Representation of the lower and upper limit of the interval of the centroids into a particle

which creates upper and lower fuzzy partition matrices according to the parameters contained in the particle of the swarm.

Each individual is a vector represented by two parts the parameters to compute the membership partitions and the positions of the centroids data clusters c, as shown in Fig. 2.

To evaluate the centroids of the particle, we need to transform this part of the vector into a matrix, but we can take into account that this part of vector depends of the value that represent n and the number of features nd, i.e., if $n = 8$ and $nd = 4$ the length of this part is equal to $n * 2 * nd = 64$ positions into a vector to centroids of cluster to represent the lower and upper limits of the interval, then reshape this vector into a matrix $n * 2 \times nd$. In Fig. 4, we can observe a graphical representation of this step.

Also we need to calculate fuzzy partition matrices and possibilistic partition matrices, but only take into account the number of clusters k of the particle, i.e., assuming that the maximal number of clusters n is 4 and the data set features nd is 4 according to Fig. 5; we obtain a matrix of $n * 2 \times nd$ as shown in Fig. 6 where we can observe that the lower and upper matrices of centroids take all the possible centroids, but we only extract the number of vectors that be indicated by k of the particle for the lower and upper matrices of centroids to create the fuzzy and possibilistic partitions matrices for the centers positions indicated by the particle evaluated, as shown in Fig. 7.

After obtaining the vectors k indicated for the particle to evaluate, we compute fuzzy possibilistic partition matrices for the lower and upper limit of the interval for the number of clusters indicated by the particle using the parameters that shown in Fig. 3 to perform this task.

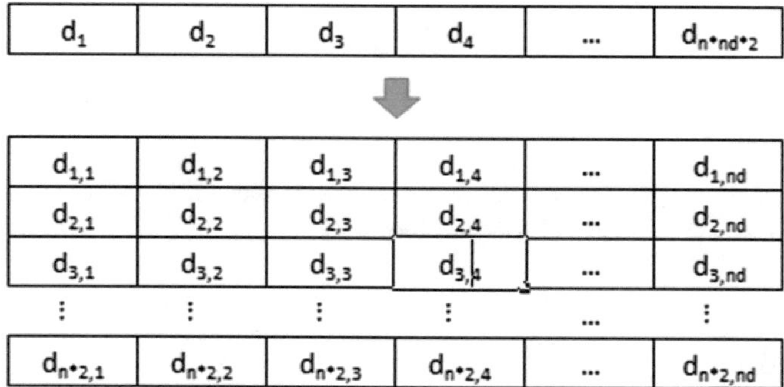

Fig. 5 Representation of the reshape vector to a matrix

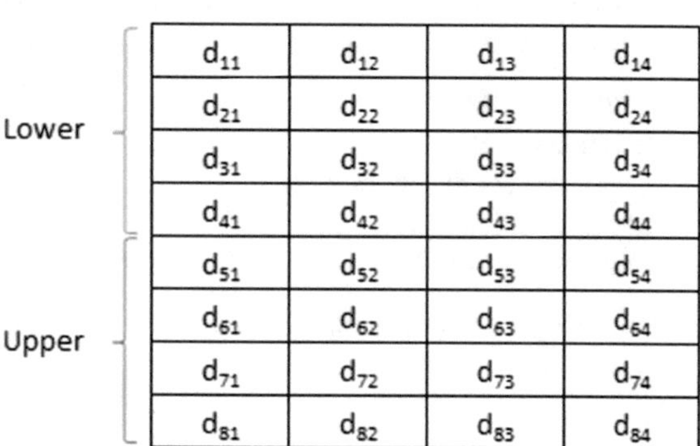

Fig. 6 Representation of the maximal number of cluster for the lower and upper limit of the interval of the centroid matrices

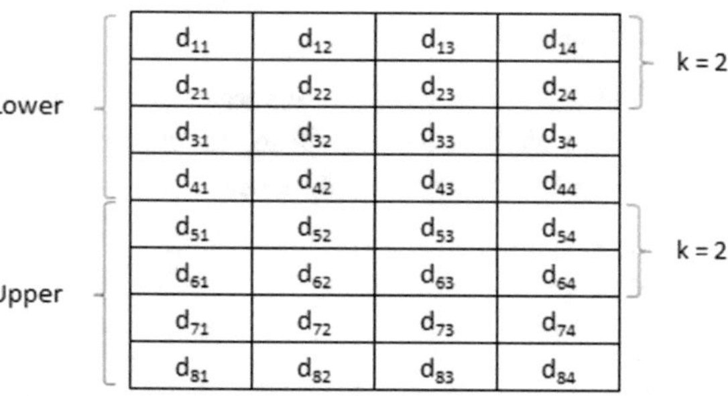

Fig. 7 Centroid to be evaluated is indicated by *k*

However, it is necessary to evaluate accuracy of the algorithm to find the parameters optimal of the clustering algorithm, and exist multiple index validation [17–19] [28–30] to perform this part of the process for clustering algorithms that take into account the matrices of partition, the distance between centers and other values obtained by the clustering algorithm, in this case we use only the mean of the sum of distance between centers of the clusters to measure the accuracy of the optimization proposal with the objective to know if the number of clusters indicated by the particle is the optimal, to perform this evaluation we use the following equation:

$$\text{fitness} = \frac{\sum_{\substack{i,j=1 \\ i \neq j}}^{k} \left\| c_i - c_j \right\|^2}{k} \tag{15}$$

In the next section, we present some results of the optimization of the Interval Type-2 Fuzzy Possibilistic C-Means algorithm.

5 Results

In this section we present the results obtained by the experiments realized to the optimization of the Interval Type-2 Fuzzy Possibilistic C-Means, we perform 30 independent experiments using the following parameters to the PSO algorithm:

- Particles = 100
- Iterations = 50
- Cognitive coefficient = 2
- Social coefficient = 2
- Constriction = 0.01

For each of the following datasets:

- WDBC
- Wine
- Iris
- Abalone
- Ionosphere

In Fig. 8, we show the convergence of the PSO algorithm during the optimization of IT2FPCM algorithm, to clustering process of the WDBC data set contain the following characteristics:

- 30 dimensions
- 2 clusters
- 569 samples

In Tables 1 and 2, we shown the parameter found by the PSO for the IT2FPCM algorithm and some positions of the centers found by the algorithm PSO for the centers of clusters Wine dataset.

Table 1 shows the optimal parameters found by the PSO algorithm for the IT2FPCM algorithm, where we can observe the number of cluster found in average is 2.9613, which is incorrect for the WDBC dataset according to the characteristics of the data set. The average indicates that most of 30 experiments performed found the incorrect number of clusters. Table 2 shows the centers found by the PSO to some dimensions of the WDBC dataset according to the best solution found by the optimization algorithm for this dataset.

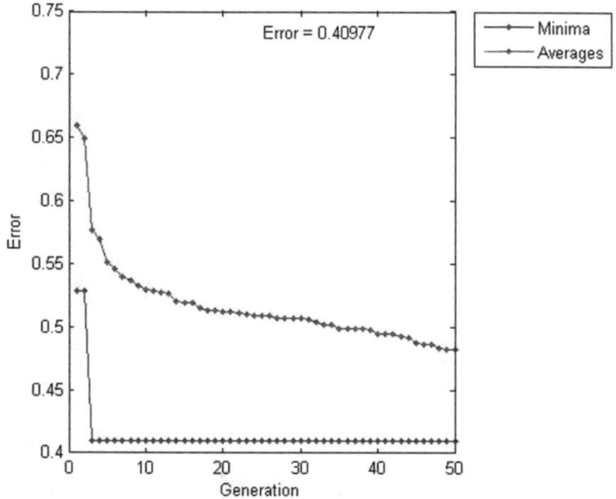

Fig. 8 Convergence of the PSO algorithm during the optimization of IT2FPCM to WDBC dataset

Table 1 Parameters found by the PSO for the IT2FPCM algorithm to WDBC dataset

Parameters				
Centroids	m_1	t_1	m_2	t_2
2.9613	1.4978	1.9238	2.2429	2.5871

In Fig. 9, we show the convergence of the PSO algorithm during the optimization of the IT2FPCM algorithm, for the clustering process of the Wine dataset containing the following characteristics:

- 13 dimensions
- 3 clusters
- 178 samples

In Tables 3 and 4, we show the parameter found by the PSO for the IT2FPCM algorithm and the some positions of the centers found by the algorithm PSO for the centers of clusters to Wine dataset.

Table 3 shows the optimal parameters found by the PSO algorithm for the IT2FPCM algorithm, where we can observe the number of cluster found in average is 3, which is correct number of clusters for the Wine dataset according to the characteristics of the data set. Table 4 shows the centers found by the PSO for some dimensions of the Wine dataset according to the best solution the best solution found by the optimization algorithm for this dataset. In Fig. 10, we show the convergence of the PSO algorithm during the optimization of the IT2FPCM algorithm, for the clustering process of the Iris Flower dataset.

In Tables 5 and 6, we show the parameters found by the PSO for the IT2FPCM algorithm and the some positions of the centers found by the algorithm PSO for the centers of clusters to Iris Flower dataset.

Table 5 shows the optimal parameters found by the PSO algorithm for the IT2FPCM algorithm, where we can observe the number of cluster found in average is 2.9560, which is incorrect number of clusters for the Iris flower dataset, but the

Table 2 Centers found by the PSO for the IT2FPCM algorithm to WDBC dataset

Inferior	0.14944089	0.04010856	0.18769583	0.04238179	0.20093396	0.14940005	0.04224307
	0.47556032	0.01498528	0.0956788	0.01679527	0.10345179	0.56533107	0.01489458
	0.60975479	0.0171366	0.1984427	0.66525477	0.01409628	0.55281627	0.01641679
Superior	17.3640891	12.5284487	18.0674062	0.15481733	12.0648743	18.0926744	11.8249198
	25.3552341	152.510457	22.5420271	0.49059792	159.571688	25.0143643	157.781377
	108.50101	23.0683117	108.472241	23.4340769	0.18717535	106.782934	23.1647945

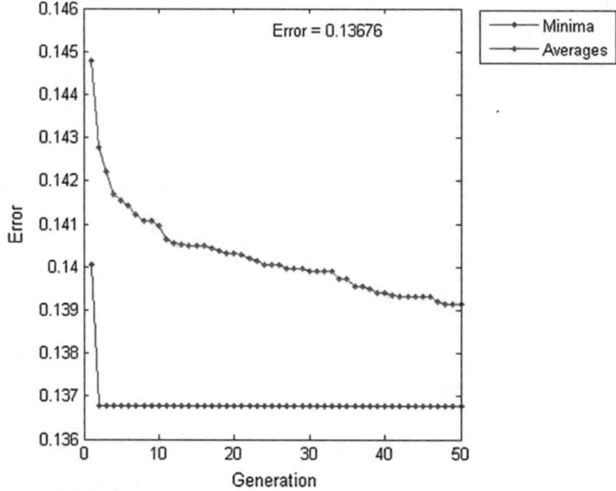

Fig. 9 Convergence of the PSO algorithm during the optimization of IT2FPCM to Wine dataset

Table 3 Parameters found by the PSO for the IT2FPCM algorithm to Wine dataset

Parameters				
Centroids	m_1	t_1	m_2	t_2
3	1.4278	1.8809	2.2752	2.5844

Table 4 Centers found by the PSO for the IT2FPCM algorithm to Wine dataset

Inferior	12.2919	12.3043	12.3310	12.3060	12.2396	12.4566	12.2552
	3.1360	3.3287	3.0457	2.9906	3.4308	3.5056	3.1438
	2.2673	2.4304	2.3969	2.3185	2.4753	2.3792	2.3352
Superior	12.6271	12.3053	12.3739	12.4847	12.3403	12.5740	12.5045
	3.1513	3.4827	4.0061	3.4076	3.5771	3.7050	3.1877
	2.4294	2.4612	2.4006	2.4023	2.5101	2.4201	2.5336

Fig. 10 Convergence of the PSO algorithm during the optimization of IT2FPCM to Iris Flower dataset

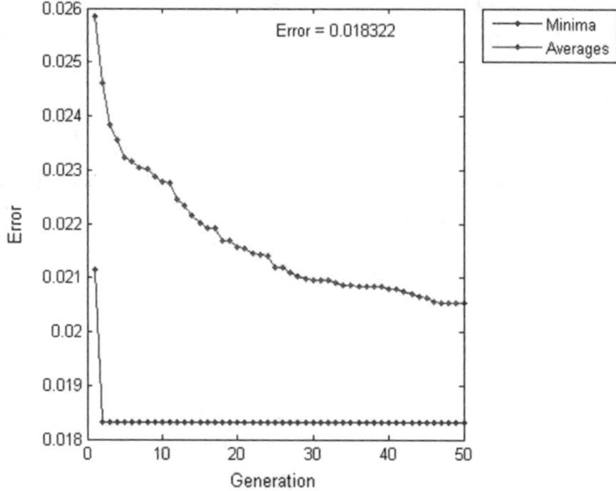

Table 5 Parameters found by the PSO for the IT2FPCM algorithm to Iris Flower dataset

Parameters				
Centroids	m_1	t_1	m_2	t_2
2.9560	1.5071	1.8782	2.1956	2.5740

Table 6 Centers found by the PSO for the IT2FPCM algorithm to Iris Flower dataset

Inferior	5.3798	5.2673	5.3786	5.4392
	3.3876	3.5167	3.4317	3.5864
	4.0777	4.1201	4.2579	4.3158
Superior	5.4443	5.3943	5.4499	5.5366
	3.5173	3.5450	3.5803	3.5971
	4.2747	4.3467	4.3448	4.3718

average indicates that most of 30 experiments performed found the correct number of clusters. Table 6 shows the centers found by the PSO for some dimensions of the Iris flower dataset according to the best solution the best solution found by the optimization algorithm for this dataset.

6 Conclusion

Optimization of the IT2FPCM algorithm using PSO was performed with the objective to find automatically the parameters of the IT2FPCM algorithm as well as the centers of cluster of each of the datasets used in this work.

During the optimization process the parameters and clusters centers was find automatically, but the parameters found by the PSO are not optimal in some experiment performed and, therefore, the number of clusters was incorrect, but we think that behavior during the optimization process is due not exist a cluster validation index that can validate the structure of each dataset existent and determine the optimal number of clusters.

All indices validation considering one or more parameters (compaction, separation, similarity, entropy, etc.) to validate the cluster found by the algorithm but this measures are combined in one equation obtaining one numerical value, we think that the cluster validation problem is a multi-objective problem, this because in a well data clustering we have well separated and compact groups, the separation and compaction are different constraint and good data clustering depends of this constrains evaluates separately.

References

1. J. Bezdek, "Pattern Recognition with Fuzzy Objective Function Algorithms", Plenum, 1981.
2. D. E. Gustafson and W. C. Kessel, "Fuzzy clustering with a fuzzy covariance matrix," in Proc. IEEE Conf. Decision Contr., San Diego, CA, pp. 761–766, 1979.

3. R. Krishnapuram and J. Keller, "A possibilistic approach to clustering," IEEE Trans. Fuzzy Sys., vol. 1, no. 2, pp. 98-110, May 1993.
4. R. Krishnapuram and J. Keller, "The possibilistic c-Means algorithm: Insights and recommendations," IEEE Trans. Fuzzy Sys., vol. 4, no. 3, pp. 385-393, August 1996.
5. N. R. Pal, K. Pal, J. M. Keller and J. C. Bezdek, "A Possibilistic Fuzzy c-Means Clustering Algorithm," IEEE Trans. Fuzzy Sys., vol. 13, no. 4, pp. 517-530, August 2005.
6. J. Yen; R. Langari; "Fuzzy Logic: Intelligence, Control, and Information," Upper Saddle River, New Jersey; Prentice Hall, 1999.
7. R. Kruse, C. Döring, M. J. Lesot; "Fundamentals of Fuzzy Clustering," In: Advances in Fuzzy Clustering and its Applications; John Wiley & Sons Ltd, The Atrium, Southern Gate, Chichester, West Sussex PO19 8SQ, England, 2007, Pages 3-30
8. K. Hirota, W. Pedrycz, "Fuzzy Computing for data mining," Proceeding of the IEEE, Vol 87 (9), 1999, pp 1575-1600.
9. N. S. Iyer, A. Kendel, and M. Schneider, "Feature-based fuzzy classification for interpretation of mamograms," Fuzzy Sets and Systems, Vol. 114, 2000, pp. 271-280.
10. W.E. Philips, R.P. Velthuinzen, S. Phuphanich, L.O. Hall, L.P Clark, and M. L Sibiger, "Aplication of fuzzy c-means segmentation technique for tissue deifferentation in MR images of hemorrhagic gliobastomamultifrome," Magnetic Resonance Imaging, Vol 13(2), 1995, pp. 277-290.
11. Miin-Shen Yang, Yu-Jen Hu, Karen Chia-Ren Lin, and Charles Chia-Lee Lin, "Segmentation techniques for tissue differentiation in MRI of Ophthalmology using fuzzy clustering algorithms," Magnetic Resonance Imaging, Vol. 20, 2002, pp. 173-179.
12. X. Chang, Wei Li, and J. Farrell, "A C-means clustering based fuzzy modeling method," Fuzzy Systems, 2000. FUZZ IEEE 2000. The Ninth IEEE International Conference on, vol.2, 2000, pp. 937-940.
13. C. Hwang, F. Rhee, "Uncertain fuzzy clustering: interval type-2 fuzzy approach to C-means", IEEE Transactions on Fuzzy Systems 15 (1) (2007) 107–120.
14. B. Choi, F. Rhee, "Interval type-2 fuzzy membership function generation methods for pattern recognition," Information Sciences, Volume 179, Issue 13, 13 June 2009, Pages 2102-2122,
15. L. A. Zadeh, "The concept of a linguistic variable and its application to approximate reasoning-I," Inform. Sci., vol. 8, no. 3, pp. 199-249, 1975.
16. J. Mendel, "Uncertain Rule-Based Fuzzy Logic Systems: Introduction and New Directions," Prentice Hall, 2001.
17. K. L. Wu, M. S. Yang; A cluster validity index for fuzzy clustering, Pattern Recognition Letters, Volume 26, Issue 9, 1 July 2005, Pages 1275-1291.
18. M. K. Pakhira, S. Bandyopadhyay, U. Maulik, A study of some fuzzy cluster validity indices, genetic clustering and application to pixel classification, Fuzzy Sets and Systems, Volume 155, Issue 2, 16 October 2005, Pages 191-214.
19. W. Wang, Y. Zhang; On fuzzy cluster validity indices, Fuzzy Sets and Systems, Volume 158, Issue 19, Theme: Data Analysis, 1 October 2007, Pages 2095-2117, ISSN 0165-0114.
20. E. Rubio and O. Castillo; "Optimization of the Interval Type-2 Fuzzy C-Means using Particle Swarm Optimization". Nabic 2013, pages 10-15.
21. R. Eberhart, J. Kennedy, "A new optimizer using particle swarm theory", in proc. 6th Int. Symp. Micro Machine and Human Science (MHS), Oct. 1995, pages: 39-43.
22. R. Eberhart, Y. Shi, "Particle swarm optimization: Developments, applications and resources", in Procceding of the IEEE Congress on Evolutionary Computation, May 2001, vol. 1, pages: 81–86.
23. J. Kennedy, R. Eberhart, "Particle Swam Optimization", in Proc. IEEE Int. Conf. Neural Network (ICNN), Nov. 1995, vol. 4, pages: 1942-1948.
24. Y. del Valle, G.K. Venayagamoorthy, S. Mohagheghi a J.-C. Hernandez and Harley R.G., "Particle Swarm Optimization: Basic Concepts, Variants and Applications in Power Systems", Evolutionary Computation, IEEE Transactions on, Apr 2008, pages: 171-195.
25. R. Eberhart, Y. Shi and J. Kennedy, "Swam Intelligence", San Mateo, California. Morgan Kaufmann, 2001.

26. A. P. Engelbrecht, “Fundamentals of Computational Swarm Intelligence”, John Wiley & Sons, 2006.
27. Escalante H. J., Montes M., Sucar L. E., “Particle Swarm Model Selection”, Journal of Machine Learning Research 10, 2009, pages: 405-440.
28. K. L. Wu, M. S. Yang, “A cluster validity index for fuzzy clustering”, Pattern Recognition Letters, Volume 26, Issue 9, 1 July 2005, Pages 1275-1291.
29. Y. Zhang, W. Wang, X. Zhang, Y. Li, “A cluster validity index for fuzzy clustering”, Information Sciences, Volume 178, Issue 4, 15 February 2008, Pages 1205-1218.
30. E. Rubio, O. Castillo, and P. Melin; “A new validation index for fuzzy clustering and its comparisons with other methods”. SMC 2011, pages 301-306.

Choquet Integral and Interval Type-2 Fuzzy Choquet Integral for Edge Detection

Gabriela E. Martínez, D. Olivia Mendoza, Juan R. Castro, Patricia Melin and Oscar Castillo

Abstract In this paper, a method for edge detection in digital images based on morphological gradient technique in combination with Choquet integral, and the interval type-2 Choquet integral is proposed. The aggregation operator is used as a method to integrate the four gradients of the edge detector. Simulation results with real images and synthetic images are presented and the results show that the interval type-2 Choquet integral is able to improve the detected edge.

Keywords Choquet integral · Fuzzy measures · Fuzzy logic · Edge detector · Morphological gradient · Interval type-2 Choquet integral

1 Introduction

One of the main steps which are performed in many applications of digital images processing is the extraction of patterns or significant elements. Among the main features that are usually obtained are the edges or outlines of objects of the image, which provide information of great importance for later stages.

Edge detection is a process in digital image analysis that detects changes in light intensity, and it is an essential part of many computer vision systems. The edge detection process is useful for simplifying the analysis of images by dramatically reducing the amount of data to be processed [1]. An edge may be the result of changes in light absorption (shade/color/texture, etc.) and can delineate the boundary between two different regions in an image.

The resultant images of edge detectors preserve more details of the original images, which is a desirable feature for a pattern recognition system. In the literature, there are various edge detectors, amongst them, the best known operators

G.E. Martínez · D. Olivia Mendoza · J.R. Castro
University of Baja California, Tijuana, Mexico

P. Melin (✉) · O. Castillo
Tijuana Institute of Technology, Tijuana, Mexico
e-mail: pmelin@tectijuana.mx

P. Melin et al. (eds.), *Nature-Inspired Design of Hybrid Intelligent Systems*,
Studies in Computational Intelligence 667, DOI 10.1007/978-3-319-47054-2_5

Table 1 Edge detection methods

Traditional methods	Computational intelligence techniques
Morphological gradient	Morphological gradient with fuzzy system
Sobel	Morphological gradient with interval type-2 fuzzy system
Prewitt	
Roberts	Sobel with fuzzy system
Laplacian	
Canny	Sobel with interval type-2 fuzzy system
Kirsch	
LoG	Sugeno integral
Zero crossing	Interval type-2 Sugeno Integral

are morphological gradient, Sobel [16], Prewitt [14], Robert [15], Canny [1], and Kirsch [4]. There are also edge detection methods that combine traditional detection methods with different techniques or even intelligent methods which have been successful such as type-1 fuzzy systems [3], interval type-2 fuzzy systems combined with the Sobel operator [9], interval type-2 fuzzy systems and morphological gradient [7, 10], Sugeno integral and interval type-2 Sugeno integral [11], among others. In Table 1, a summary of edge detection existing methods is presented.

The Choquet integral is an aggregation operator, which offers a more flexible and realistic approach to model uncertainty, this operator has been successfully used in a variety of applications as [5, 12, 18].

The main goal of this paper is to use the Choquet integral and interval type-2 Choquet integral as an aggregation method of the morphological gradient detector edges. In addition to the visual comparison of the results using real images, we also applied in synthetic images the Pratt's Figure of Merit performance index to determine the quality of the resulting image.

This paper is organized as follows, in the next section, we show the technique of edge detection based on the morphological gradient, Sect. 3 presents the concepts of fuzzy Measures, Choquet integral, and the calculations of your intervals. Section 4 describes the simulation results, in Sect. 5 are shown the Results and discussion, and finally in Sect. 6 the Conclusions are presented.

2 Edge Detection

The edges of a digital image can be defined as transitions between two regions of significantly different gray levels, and they provide valuable information about the boundaries of objects and can be used to recognize patterns and objects, segmenting an image, etc. When edge detection is applied over an image we can reduce significantly the amount of data to be processed, getting preserve the properties of the objects in the image. If the application of edge detector was satisfactory, it can greatly reduce the process of image interpretation, however, not always from the actual images ideal edges are obtained. The edge detection process is one of the key elements in image analysis, image processing and computer vision techniques.

2.1 Morphological Gradient

Gradient operators are based on the idea of using the first or second derivative of the gray level. Based on an image $f(x, y)$, the gradient of point (x, y) is defined as a gradient vector (∇f) and is calculated as follows:

$$\nabla f = \begin{bmatrix} Gx \\ Gy \end{bmatrix}$$
$$\nabla f = \begin{bmatrix} \frac{\partial f}{\partial x} \\ \frac{\partial f}{\partial y} \end{bmatrix} \tag{1}$$

where, the gradient magnitude vector (mag(∇f)) is calculated with

$$\text{mag}(\nabla f) = \left[\left(\frac{\partial f}{\partial x} \right)^2 + \left(\frac{\partial f}{\partial y} \right)^2 \right]^{1/2}$$
$$\text{mag}(\nabla f) = \left[G_x^2 + G_y^2 \right]^{1/2} \tag{2}$$

In this case, we are going to use G_i instead of $\text{mag}(\nabla f)$, if we apply (2) for a matrix of 3 × 3 as it is shown in Fig. 1. The values for z_i are obtained using (3), and the possible direction of edge G_i with (4–7). The G gradient can be calculated using (8).

$$\begin{array}{ll} z_1 = (x-1, y-1) & z_6 = (x, y+1) \\ z_2 = (x-1, y) & z_7 = (x+1, y-1) \\ z_3 = (x-1, y) & z_8 = (x+1, y) \\ z_4 = (x, y-1) & z_9 = (x+1, y+1) \\ z_5 = (x, y) & \end{array} \tag{3}$$

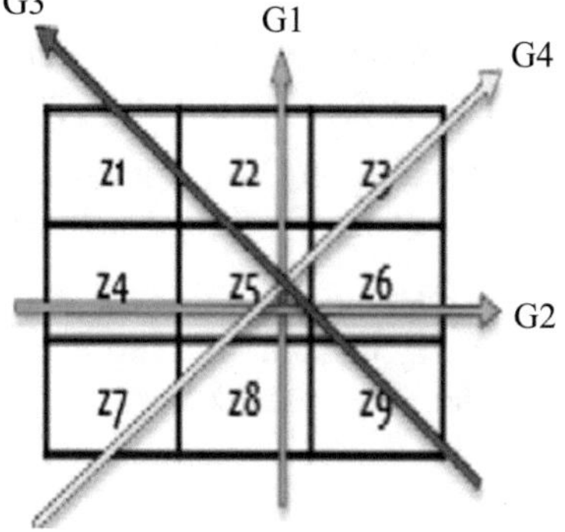

Fig. 1 Matrix of 3 × 3 of the index Z_i, that indicating the calculation of the gradient in the four directions

$$G1 = \sqrt{(z_5 - z_2)^2 + (z_5 - z_8)^2} \tag{4}$$

$$G2 = \sqrt{(z_5 - z_4)^2 + (z_5 - z_6)^2} \tag{5}$$

$$G3 = \sqrt{(z_5 - z_1)^2 + (z_5 - z_9)^2} \tag{6}$$

$$G4 = \sqrt{(z_5 - z_7)^2 + (z_5 - z_3)^2} \tag{7}$$

$$G = G1 + G2 + G3 + G4 \tag{8}$$

3 Fuzzy Logic, Fuzzy Measures, and Choquet Integral

Since the introduction of fuzzy logic in 1965 [19], a considerable improvement to traditional logic has been performed, as having only two truth values {0, 1}, can now be considered any value between [0, 1]. First came the type-1 fuzzy systems which are capable of representing imprecision or vagueness, however, then came the type-2 fuzzy systems by intervals which are capable of representing a degree of uncertainty (which is the central element in this paper proposed method).

In 1974, the concepts of "fuzzy measure and fuzzy integral" were defined by Michio Sugeno [17] in order to define sets that have not well-defined boundaries. A fuzzy measure is a nonnegative monotone function of values defined in "classical sets." Currently, when referring to this topic, the term "fuzzy measures" has been replaced by the term "monotonic measures," "non-additive measures "or" generalized measures" [4, 6, 20]. When fuzzy measures are defined on fuzzy sets, we speak of fuzzified monotonic measures [6].

3.1 *Fuzzy Measures and the Choquet Integral*

A fuzzy measure can be defined as a fuzzy measure μ with respect to the dataset X, and must satisfy the following conditions:

(1) $\mu(X) = 1$; $\mu(\varnothing) = 0$ Boundary conditions
(2) *Si* $A \subset B$, then $\mu(A) \leq \mu(B)$ Monotonicity

In condition two, A and B are subsets of X.

A fuzzy measure is a Sugeno measure or λ-fuzzy, if it satisfies the condition (9) of addition for some $\lambda > -1$.

$$\mu(\mathrm{A} \cup \mathrm{B}) = \mu(\mathrm{A}) + \mu(\mathrm{B}) + (*\mu(\mathrm{A}) * \mu(\mathrm{B})) \tag{9}$$

Where λ can be calculated with (10)

$$f(\lambda) = \left\{ \prod_{i=1}^{n} (1 + M_i(x_i)\lambda) \right\} - (1 + \lambda) \tag{10}$$

The value of the parameter λ is determined by the conditions of Theorem 1.

Theorem 1 *Let* $\mu(\{x\}) < 1$ *for each* $x \in X$ *and let* $\mu(\{x\}) > 0$ *for at least two elements of* X, *then* (10) *determines a unique parameter* λ *in the following way*:

$$\textit{If} \sum_{x \in X} \mu(\{x\}) < 1, \text{ then } \lambda \text{ is in the interval } (0, \infty).$$

$$\textit{If} \sum_{x \in X} \mu(\{x\}) = 1, \text{ then } \lambda = 0; \text{That is the unique root of the equation.}$$

$$\textit{If} \sum_{x \in X} \mu(\{x\}) > 1, \text{ then } \lambda \text{ se is in the interval } (-1, 0).$$

The fuzzy measure represents the importance or relevance of the information sources when computing the aggregation. The method to calculate the Sugeno measures carries out the calculation in a recursive way, using (11) and (12).

$$\mu(A_1) = \mu(M_i) \tag{11}$$

$$\mu(A_i) = \mu(A_{(i-1)}) + \mu(M_i) + \left(\lambda * \mu(M_i) * \mu(A_{(i-1)})\right) \tag{12}$$

where A_i represents the fuzzy measure and M_i represents the fuzzy density determined by an expert in the domain of application, where M_i should be permuted with respect to the descending order of their respective $\mu(A_i)$.

In the literature, there are two types of integrals that used these measures, the integral of Sugeno and the Choquet integral.

3.2 Choquet Integral

The Choquet integral was proposed by Choquet in 1953 [2], and is a very popular data aggregation approach with respect to a fuzzy measure. The generalized Choquet integral with respect to a signed fuzzy measure can act as an aggregation tool, which is especially useful in many applications [8, 9, 12].

The Choquet integral can be calculated as follows:

$$\text{Choquet} = \sum_{i=1}^{n} \left\{ \left[A_i - A_{(i-1)} \right] * D_i \right\} \tag{13}$$

with $A_0 = 0$, where i indicates the indices that must be permuted as $0 \le D_{(1)} \le D_{(2)} \le \cdots \le D_{(n)} \le 1$, in this case A_i represents the fuzzy measurement associated with a data D_i.

3.2.1 Pseudocode of Choquet Integral

The series of steps needed to calculate the Choquet Integral for each information source are given as follows:

INPUT: Number of information sources n; information sources $x_1, x_2, \ldots, x_n$; fuzzy densities of information sources $M_1, M_2, \ldots, M_n$ that are in the interval (0,1).
OUTPUT: Choquet integral.
STEP 1: Calculate finding the root of the function using (10).
STEP 2: Fuzzify the variable x_i to obtain D_i.
STEP 3: Reorder M_i with respect to $D(x_i)$ in descending order
STEP 4: Calculate fuzzy measures for each data with (11) and (12).
STEP 5: Calculate Choquet integral with (13).
STEP 6: OUTPUT Choquet.

STOP

3.3 *Interval Type-2 for Fuzzy Densities*

For the estimation of the fuzzy densities of each information source, we take the maximum value of each X_i, and where an interval of uncertainty is added. We need to add an uncertainty footprint or FOU which will create an interval based on the fuzzy density. Equation (14) can be used to approximate the center of the interval for each fuzzy density, and (15) and (16) are used to estimate left and right values of the interval for each fuzzy density. Note that the domain for $\mu_L(x_i)$ and $\mu_U(x_i)$ is given in Theorem 1 [19].

The calculation of the fuzzy densities is performed with

$$\mu_c(x_i) = \max(X_i) \tag{14}$$

$$\mu_L(x_i) = \begin{cases} \mu_c(x_i) - \text{FOU}_\mu/2; & \text{if } \mu_c(x_i) > \text{FOU}_\mu/2 \\ 0.0001 & \text{otherwise} \end{cases} \tag{15}$$

$$\mu_U(x_i) = \begin{cases} \mu_c(x_i) + \text{FOU}_\mu/2; & \text{if} \mu_c(x_i) < (1 - \text{FOU}_\mu/2) \\ 0.9999 & \text{otherwise} \end{cases} \tag{16}$$

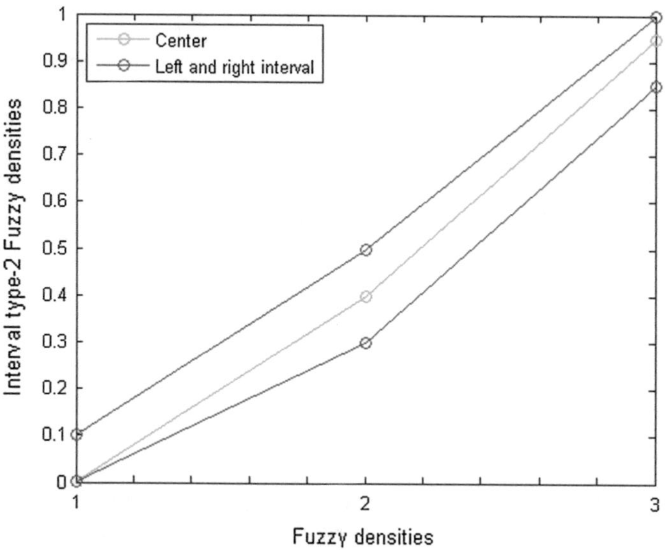

Fig. 2 Interval type-2 fuzzy densities for specific values $\boldsymbol{\mu}_{c(x_1)} = 0.02$, $\boldsymbol{\mu}_{c(x_2)} = 0.4$, $\boldsymbol{\mu}_{c(x_3)} = 0.95$ and FOU $\boldsymbol{\mu} = 0.2$

For a better understanding, Fig. 2 shows an interval generated with specific values of the fuzzy densities $\mu_{c(x_1)} = 0.02$, $\mu_{c(x_2)} = 0.4$, $\mu_{c(x_3)} = 0.95$ and FOU $\mu = 0.2$.

After this is necessary to perform the calculation of the λ_L and λ_U parameters for either side of the interval with

$$\lambda_L + 1 = \prod_{i=1}^{n} (1 + \lambda_L \mu_L(\{x_i\})) \tag{17}$$

$$\lambda_U + 1 = \prod_{i=1}^{n} (1 + \lambda_U \mu_U(\{x_i\})) \tag{18}$$

Once the λ_U, λ_L are obtained left and right fuzzy measures can be calculated by extending the recursive formulas (11) and (12) obtaining

$$\mu_L(A_1) = \mu_L(x_1) \tag{19}$$

$$\mu_L(A_i) = \mu_L(x_i) + \mu_L(A_{i-1}) + \lambda_L \mu_L(x_i) \mu_L(A_{i-1}) \tag{20}$$

$$\mu_U(A_1) = \mu_U(x_1) \tag{21}$$

$$\mu_U(A_i) = \mu_U(x_i) + \mu_U(A_{i-1}) + \lambda_U \mu_U(x_i) \mu_U(A_{i-1}) \tag{22}$$

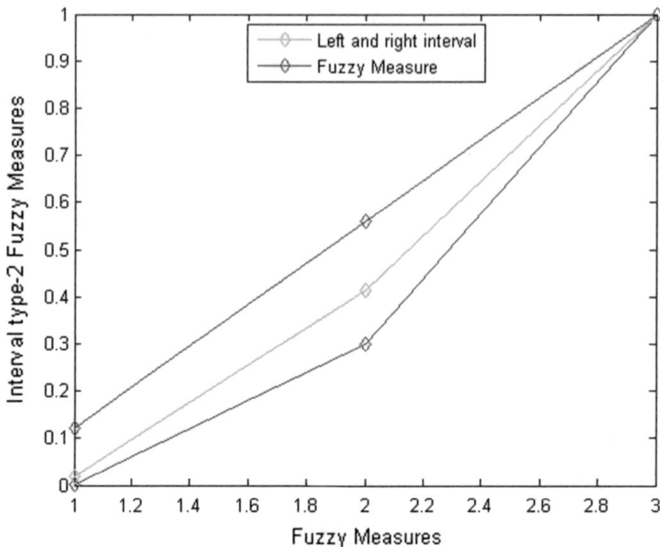

Fig. 3 Interval type-2 fuzzy measures for the specific values $\mu(A_1) = 0.02$, $\mu(A_2) = 0.4$, $\mu(A_3) = 0.95$ and FOU $\boldsymbol{\mu} = 0.2$

In Fig. 3 we can find an interval generated with specific values for the fuzzy measures.

3.4 *Interval Type-2 Fuzzy Logic for Choquet Integral*

The Choquet integral can be calculated using (13), and if we extend the equation using intervals we obtain the following expression:

$$\text{Choquet}_U = \sum_{i=1}^{n} \left\{ \left[A_{U(i)} - A_{U(i-1)} \right] * D_{U(i)} \right\} \tag{23}$$

with $A_{U(0)} = 0$ and

$$\text{Choquet}_L = \sum_{i=1}^{n} \left\{ \left[A_{L(i)} - A_{L(i-1)} \right] * D_{L(i)} \right\} \tag{24}$$

with $A_{L(0)} = 0$, where Choquet_U and Choquet_L represent the upper and lower interval of the Choquet integral, the next step is to calculate the average of both values to obtain the Choquet integral using

$$\text{Choquet} = (\text{Choquet}_U + \text{Choquet}_L)/2 \tag{25}$$

4 Simulation Results

In this section, the simulation results of the proposed method applied to synthetic images and real images are presented. To carry out the aggregation process we use G_i as the information sources which must be aggregate, in this case instead of using the traditional method of MG (8), we use Choquet integral (CHMG) with Eq. (13) and the Interval type-2 Choquet integral (IT2CHMG) using the formulas (23)–(25) to detect the edges in the images.

The aggregation process of the IT2CHMG is shown in Fig. 4.

Once edge detection with a particular technique is performed, it is necessary to use some evaluation method to determine whether the result is good or better than other edge detection methods existing in digital images.

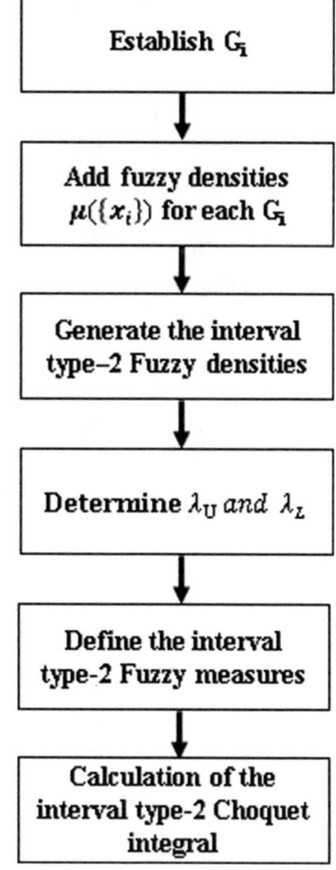

Fig. 4 Diagram that represent the integration of gradients

In the literature, we can find different metrics to evaluate the detected edges, to perform these evaluations, a measure of differences that is frequently used is the quadratic average distance proposed by Pratt [20], and it is also called Figure of Merit (FOM). This measure represents the variation that exists from a real point (calculated) from the ideal edge.

$$\text{FOM} = \frac{1}{\max(I_I, I_A)} \sum_{i=1}^{I_A} \frac{1}{1 + \propto d_i^2} \tag{26}$$

where I_A represents the number of detected edge points, I_I is the number of points on the ideal edge, $d(i)$ is the distance between the current pixel and its correct position in the reference image and α is a scaling parameter (normally 1/9). To apply (26) we require the synthetic image and its reference. A FOM = 1 corresponds to a perfect match between the ideal edge and the detected edge points.

In Tables 2 and 3, we can observe the parameters used to implement the CHMG method in the synthetic images of the donut and sphere respectively. Each of the four gradients is considered by the aggregation operator as a source of information to which it should be assigned a fuzzy density, which represents the degree of membership or level of importance of that data. In this case, to each of the gradients it was assigned the same fuzzy density.

Various tests in which the FOU parameter was varied from 0.1 to 0.9 were made. After this, for each test the calculation of the lambda parameter was performed to carry out the aggregation of the gradients by detecting edges. Finally, was proceeded to the evaluation of the results using FOM (26).

In Table 2 we can find that for the image of the donut, using the CHMG a FOM of 0.8787 was obtained, while for the image of the sphere, a FOM of 0.8236 was achieved; this can be found in Table 3. In both cases, the best result was achieved by assigning a fuzzy density of 0.59 to each gradient.

Table 2 FOM using CHMG in the synthetic image of the donut

Fuzzy density CHMG	λ	FOM
0.1	6.608	0.8661
0.2	0.7546	0.8712
0.3	−0.4017	0.8736
0.4	−0.7715	0.8732
0.5	−0.9126	0.8771
0.6	−0.9694	0.8785
0.7	−0.9912	0.4824
0.8	−0.9984	0.4611
0.9	−0.9999	0.4387
0.59	−0.9657	0.8787
0.55	−0.9474	0.8777

Table 3 FOM using CHMG in the synthetic image of the Sphere

Fuzzy density CHMG	λ	FOM
0.1	6.608	0.8132
0.2	0.7546	0.8183
0.3	−0.4017	0.8207
0.4	−0.7715	0.8194
0.5	−0.9126	0.8227
0.6	−0.9694	0.8228
0.7	−0.9912	0.4667
0.8	−0.9984	0.4453
0.9	−0.9999	0.4183
0.59	−0.9657	0.8236
0.55	−0.9474	0.8233

Figures 5 and 6 show the original synthetic images of the donut and the sphere, the reference image is needed to perform the evaluation of the detected edge using the PRATT metric, the image obtained after applying the traditional method of MG using (8) and finally presents the detected edges making variations in the fuzzy densities of each of the information sources from 0.1 to 0.9.

Visually it is difficult to appreciate the difference between the detected edges and for that reason, we use the metric (26) that is described in Sect. 4 is made.

Figure 7 shows the results obtained after applying the morphological gradient edge detector and IT2CHMG as the aggregation operator for real images.

In the first column, we can find the actual images used, the second column represents the images obtained after applying the MG methodology. Of the third to the seventh column, the results obtained after using IT2CHMG are shown. For each of these tests it was assigned a FOU of 0.1 to 0.9. In the images we can find that the edges detected with proposed operator tend to be more defined than those detected with the traditional method.

Next, the results of tests on the synthetic image of the donut are presented. Table 4 shows the evaluation performed with the PRAT metric (FOM) using integration edge detector CHMG through. For each of the gradients calculated by the method of MG in the four directions, a fuzzy density ranging from 0.1 to 0.9 is assigned. In the case of the IT2CHMG edge detector, also we made a variation of the FOU of 0.1 0.4.

Table 5 shows the results obtained for the same image, however in this case the FOU was varying from 0.5 to 0.9.

As we can observe, the assigned value to the fuzzy density as the FOU play an important role at the time of the evaluation of the detected edge in the digital image, which we can see reflected in the FOM obtained. When using the CHMG edge detector, the best FOM was of 0.8821 using a fuzzy density of 0.6. However, when using the IT2CHMG edge detector, the FOM highest value was of 0.9078 using a fuzzy density of 0.6 and FOU 0.1.

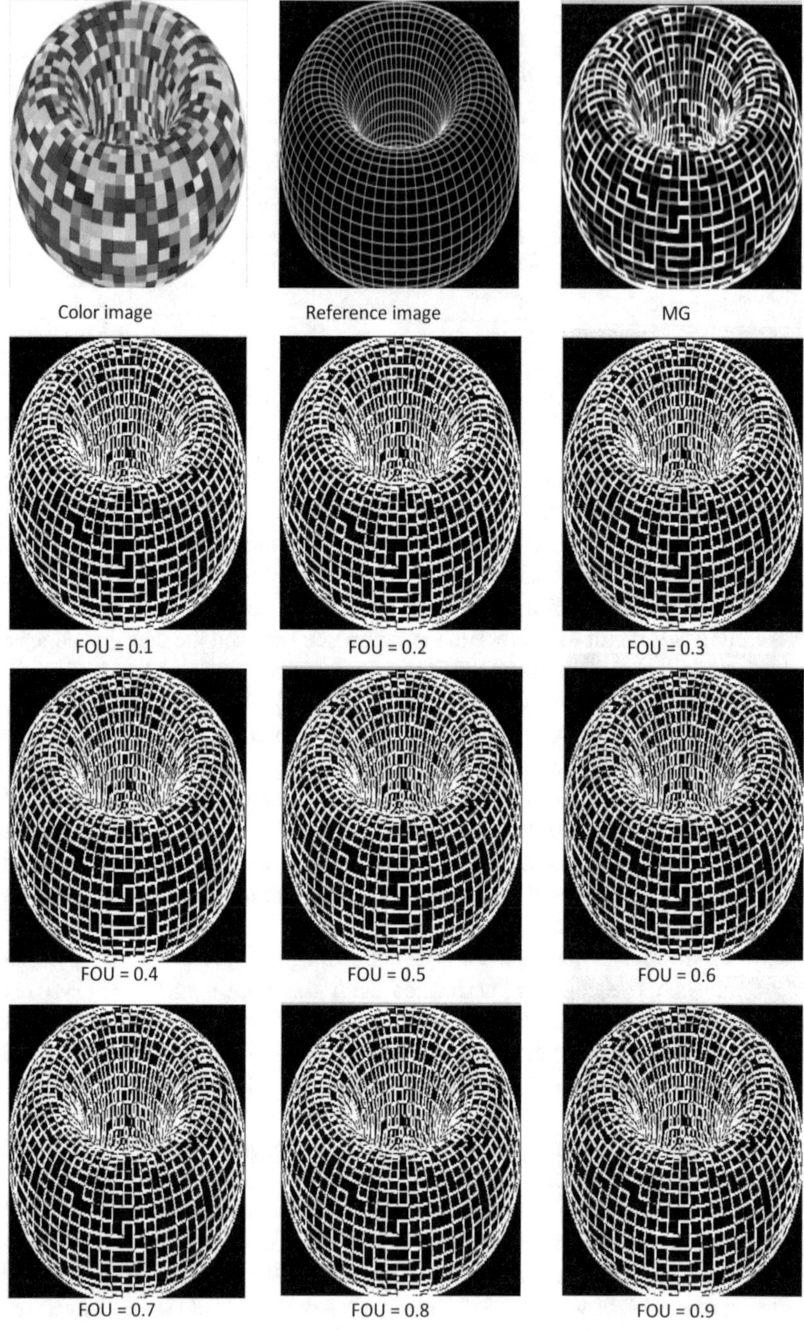

Fig. 5 Simulation results applying MG and CHMG in the donut image

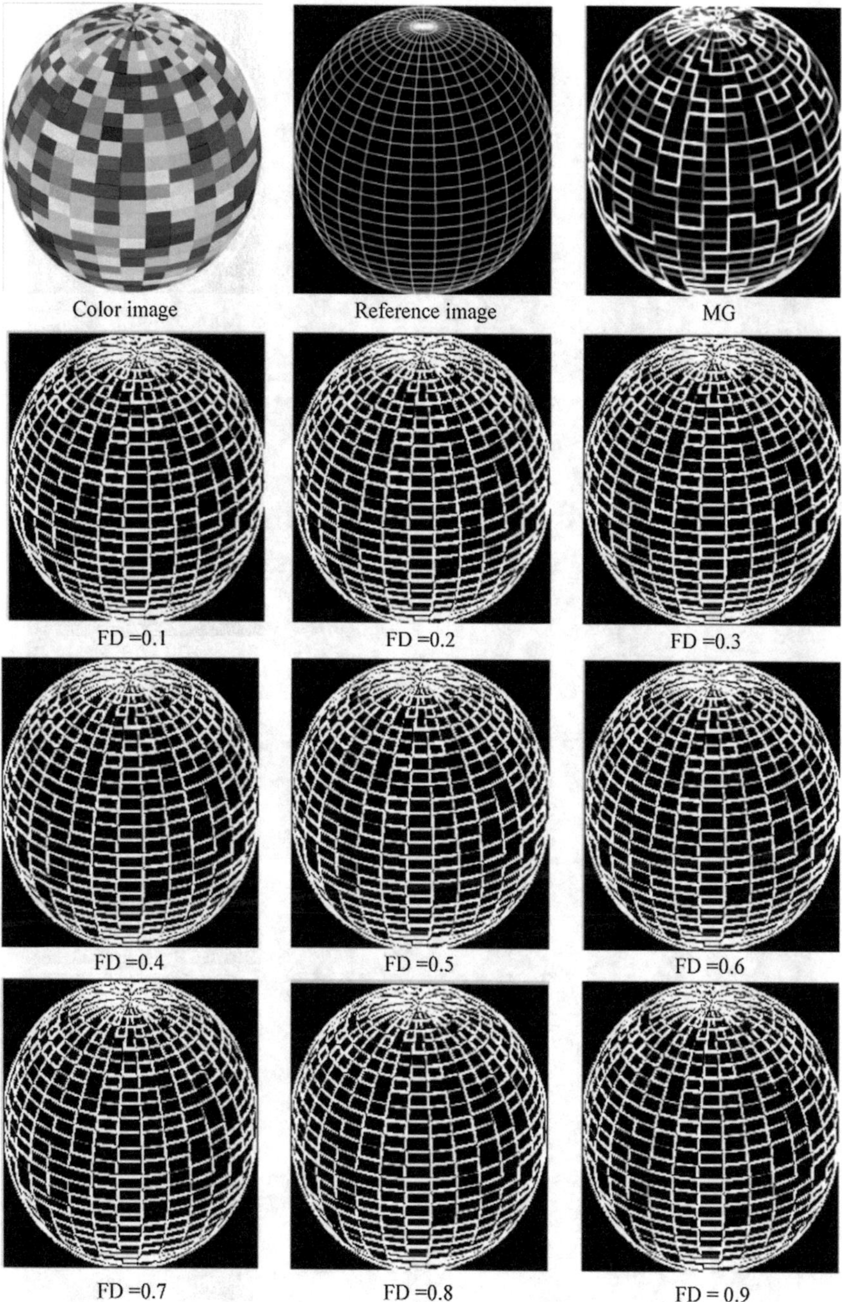

Fig. 6 Simulation results applying MG and CHMG in the sphere image

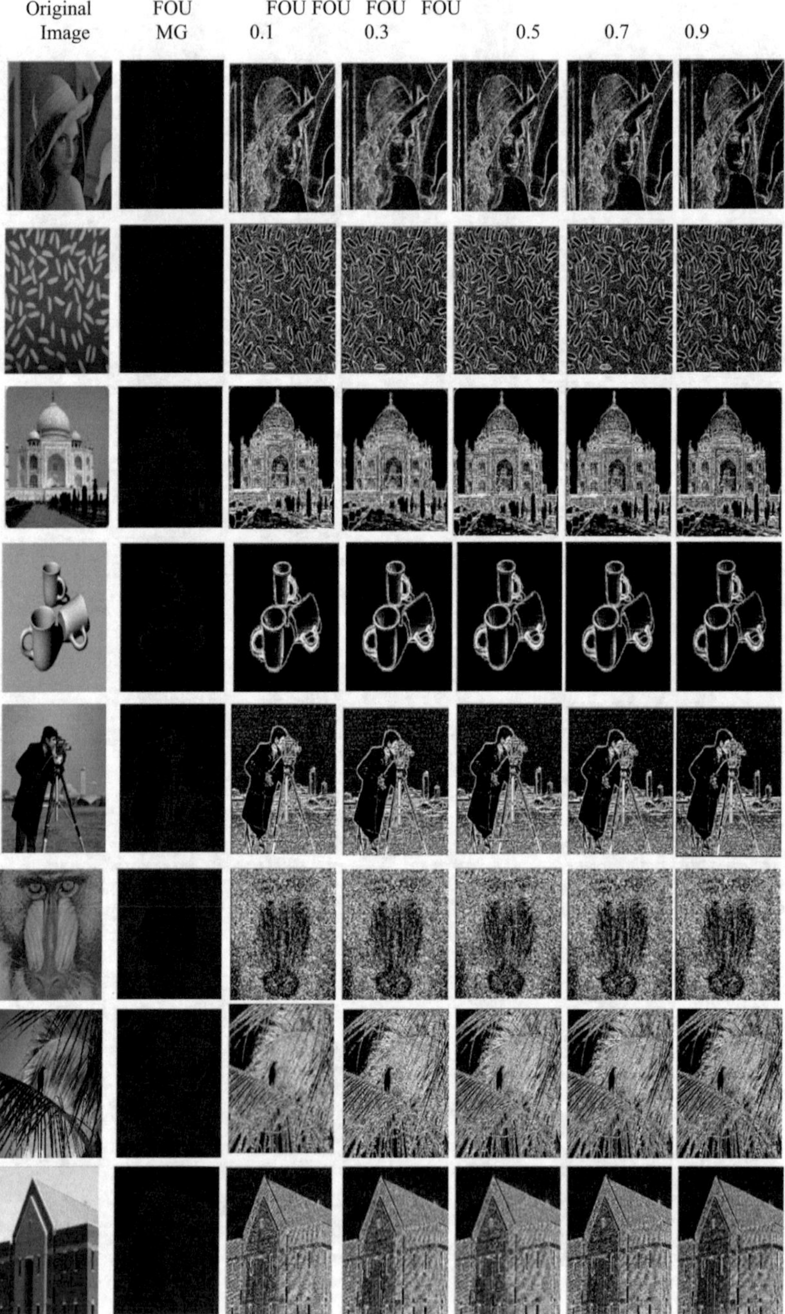

Fig. 7 Simulation results applying MG and CHMG in real images

Table 4 FOM using IT2CHMG in the Donut image with FOU of 0.1 at 0.4

Fuzzy densities	IT2-Choquet integral				
	Choquet	FOU 0.1	FOU 0.2	FOU 0.3	FOU 0.4
[0.1]	0.8697	0.8634	0.8634	0.8634	0.8629
[0.2]	0.8748	0.8727	0.8805	0.8814	0.8812
[0.3]	0.8772	0.8812	0.8823	0.8814	0.8812
[0.4]	0.8768	0.888	0.8823	0.8814	0.8812
[0.5]	0.8807	0.8988	0.8823	0.8814	0.8812
[0.6]	0.8821	0.9078	0.8823	0.8814	0.8812
[0.7]	0.4844	0.8789	0.8823	0.8814	0.8812
[0.8]	0.463	0.8805	0.8823	0.8814	0.8812
[0.9]	0.4406	0.8823	0.8823	0.8814	0.8812

Table 5 FOM using IT2CHMG in the Donut image with FOU of 0.5 at 0.9

Fuzzy densities	IT2-Choquet integral				
	FOU 0.5	FOU 0.6	FOU 0.7	FOU 0.8	FOU 0.9
[0.1]	0.862	0.8614	0.8595	0.8581	0.8561
[0.2]	0.8802	0.8778	0.878	0.8767	0.8753
[0.3]	0.8802	0.879	0.878	0.8767	0.8753
[0.4]	0.8802	0.879	0.878	0.8767	0.8753
[0.5]	0.8802	0.879	0.878	0.8767	0.8753
[0.6]	0.8802	0.8595	0.878	0.8767	0.8753
[0.7]	0.8802	0.8595	0.878	0.8767	0.8753
[0.8]	0.8802	0.8595	0.878	0.8767	0.8753
[0.9]	0.8802	0.8595	0.878	0.8767	0.8753

Table 6 FOM using IT2CHMG in the Sphere image with FOU of 0.1 at 0.4

Fuzzy densities	IT2-Choquet integral				
	Choquet	FOU 0.1	FOU 0.2	FOU 0.3	FOU 0.4
[0.1]	0.817	0.8109	0.8109	0.8108	0.8098
[0.2]	0.8222	0.8197	0.8197	0.8196	0.8329
[0.3]	0.8245	0.827	0.827	0.8268	0.8274
[0.4]	0.8232	0.8336	0.8336	0.8512	0.8274
[0.5]	0.8266	0.8431	0.8431	0.828	0.8274
[0.6]	0.8267	0.8515	0.8515	0.828	0.8274
[0.7]	0.4689	0.8257	0.8257	0.828	0.8274
[0.8]	0.4474	0.8277	0.8277	0.828	0.8274
[0.9]	0.4202	0.8286	0.8286	0.828	0.8274

Tables 6 and 7 present the results to evaluate the image of the sphere with the detectors CHMG and IT2CHMG. In the case of CHMG, the optimal parameters

Table 7 FOM using IT2CHMG in the Sphere image with FOU of 0.5 at 0.9

Fuzzy densities	IT2-Choquet integral				
	FOU 0.5	FOU 0.6	FOU 0.7	FOU 0.8	FOU 0.9
[0.1]	0.8093	0.8091	0.8071	0.8057	0.8042
[0.2]	0.842	0.8241	0.8484	0.8162	0.8205
[0.3]	0.8266	0.8261	0.8251	0.8237	0.8224
[0.4]	0.8266	0.8261	0.8251	0.8237	0.8224
[0.5]	0.8266	0.8261	0.8251	0.8237	0.8224
[0.6]	0.8266	0.8261	0.8251	0.8237	0.8224
[0.7]	0.8266	0.8261	0.8251	0.8237	0.8224
[0.8]	0.8266	0.8261	0.8251	0.8237	0.8224
[0.9]	0.8266	0.8261	0.8251	0.8237	0.8224

assigned to the fuzzy density was of 0.6, while for IT2CHMG, the best result was obtained by assigning a FOU both of 0.1 and 0.2, and a fuzzy density of 0.6.

In Fig. 8 we can observe the images obtained after performing the edge detection with the three aggregation operators. In the first column, the real images are shown, the second column presented the edge detection using the MG traditional method, in the third column the result of the aggregation of the MG method is observed through Choquet integral and finally, the last column shows the results of applying the interval type-2 Choquet integral.

5 Results and Discussion

In Table 8 it is possible to observe the FOM obtained with each edge detector (MG, CHMG, and IT2CHMG) used on synthetic images of the donut and sphere. In the case of IT2CHMG the best results were obtained when we assigned a FOU of 0.6 and a fuzzy density of 0.6 for each of the four gradients that were calculated. To each gradient was assigned the same diffuse density, so it is necessary to test different values for the information sources to analyze the behavior of the method.

For the synthetic image of the sphere, the FOM obtained with the classical MG edge detector was 0.6914, with the CHMG was 0.8267 while for the IT2CHMG was of 0.8515; for the synthetic image of the donut the behavior of the evaluation was similar, because the best result was obtained with IT2CHMG with a FOM of 0.9078.

Figure 9 shows the FOM for both synthetic images to help appreciate more clearly this comparison using the three aggregation methods. In this image we can see that according to the metric of FOM, the best results are obtained using IT2CHMG.

Finally, we also made some tests in which we made variations of the α parameter of Eq. (26) to determine whether the metric produce better results, however it can be seen in Table 9, that the larger is the α parameter, the FOM results tend to be lower.

Fig. 8 Simulation results applying MG, CHMG and IT2CHMG in real images

Table 8 FOM rates with MG, CHMG and IT2CHMG

Aggregation method	Images	
	Sphere	Donut
FOM-MG	0.8247	0.8781
FOM-CHMG	0.8267	0.8821
FOM-IT2CHMG	0.8515	0.9078

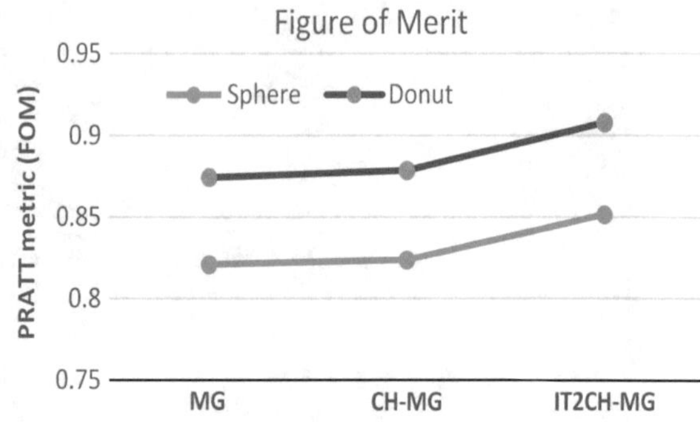

Fig. 9 Figure of merit of Pratt (FOM) of Table 8

Table 9 Variation of ∝ parameter

A	1/9	1/8	1/7	1/5
FOM-GM	0.8247	0.8215	0.8174	0.8054
FOM-CH	0.8266	0.8231	0.8187	0.8057

6 Conclusions

The main contribution of this paper is to perform the aggregation process of the morphological gradient edge detection method using the Choquet integral and interval type-2 Choquet integral.

As can be noted in Fig. 9 and in the results of Table 8, when the proposed IT2CHMG method is applied, better results are obtained, and the reason is that the interval type-2 Choquet integral can improve the performance of the morphological gradient edge detector, based on gradient measures.

The aggregation process of the interval type-2 Choquet integral, improves the performance when there is uncertainty in the original gradients, so that the proposed method can represent this uncertainty.

In consequence, the use of interval type-2 Choquet integral is viable to improve image processing methods based on gradient measures. However, it is necessary to use a method that optimizes the value of the Sugeno measures assigned to the calculation of each gradient, because at this time these values are assigned arbitrarily. As future work is considered an optimization of the proposed method using some evolutionary method as genetic algorithms, ant colony optimization or the particle swarm optimization, for find another way to generate an interval of uncertainty in the data, fuzzy measures, value of lambda and fuzzy densities. Also we consider use the traditional method of the Choquet integral and the interval type-2 Choquet integral with the Sobel edge detector to perform the aggregation process of the gradients.

Acknowledgment We thank the MyDCI program of the Division of Graduate Studies and Research, UABC, Tijuana Institute of Technology, and the financial support provided by our sponsor CONACYT contract grant number: 189350.

References

1. Canny J., A computational approach to edge detection, IEEE Transactions on Pattern Analysis and Machine Intelligence, vol. 8, no. 2, pp. 679–698, 1986.
2. Choquet, G. Theory of capacities. Ann. Inst. Fourier, Grenoble 5: 131–295. 1953.
3. Hua L., Cheng H, and Zhanga M., A high performance edge detector base d on fuzzy inference rules, Journal of Information Sciences, Elsevier Science, Vol. 177, No. 21, pp. 4768-4784, 2007.
4. Kirsch R., Computer determination of the constituent structure of biological images, Computers and Biomedical Research, vol. 4, pp.315–328, 1971.
5. Kirsch R., Computer determination of the constituent structure of biological images, Computers and Biomedical Research, vol. 4, pp. 315–328, 1971.
6. Melin P., Gonzalez C., Bravo D., Gonzalez F., and Martínez G., Modular Neural Networks and Fuzzy Sugeno Integral for Pattern Recognition: The case of human face and fingerprint in Hybrid Intelligent Systems
7. Melin P., Mendoza O., Castillo O., An improved method for edge detection based on interval type-2 fuzzy logic, Expert Systems with Applications, Vol. 37, pp. 8527-8535, 2010.
8. Melin P., Mendoza O., Castillo O., Face Recognition with an Improved In-terval Type-2 Fuzzy Logic Sugeno Integral and Modular Neural Networks, IEEE Transactions on systems, man, and cybernetics-Part A: systems and humans, vol. 41, no. 5, pp. 1001-1012, 2011.
9. Mendoza O., Melin P., Licea G., A new method for edge detection in image processing using interval type-2 fuzzy logic, IEEE international conference on granular computing (GRC 2007). CA, USA: Silicon Valley, 2007.
10. Mendoza O., Melin P., Licea G., Interval type-2 fuzzy logic for edge detec-tion in digital images. International Journal of Intelligent System 24 (11), pp.1115-1134, 2009.
11. Mendoza O., Melin P., Sugeno Integral for Edge Detection in Digital Im-ages, International Journal of Intelligent Systems, vol. 24, pp. 1115–1133, 2009.
12. Mikhail Timonin. Robust optimization of the Choquet Integral. Fuzzy Sets and Systems, vol. 213, pp. 27–46, 2013.
13. Pratt, W.K., Digital Image Processing. John Wiley and Sons, New York, 1978.
14. Prewitt J. M. S., Object enhancement and extraction, B.S. Lipkin, A. Rosen-feld (Eds.), Picture Analysis and Psychopictorics, Academic Press, New York, NY, pp. 75–149, 1970.
15. Roberts, L, Machine Perception of 3-D Solids, Optical and Electro-optical Information Processing, MIT Press, 1965.
16. Sobel I, Camera Models and Perception, Ph.D. thesis, Stanford University, Stanford, CA, 1970.
17. Sugeno M, Theory of fuzzy integrals and its applications. Thesis Doctoral, Tokyo Institute of Technology, Tokyo, Japan, 1974.
18. Yanga W., Chena Z., New aggregation operators based on the Choquet Integral and 2-tuple linguistic information. Expert Systems with Applica-tions, vol. 39, num. 3, pp. 2662–2668, 2012.
19. Zadeh, L.A., Fuzzy Sets. Inf. Control 8, 338-353, 1965.
20. Zhou L. G., Chen H. Y., Merigó J. M., Gil-Lafuente A. M. Uncertain genera-lized aggregation operators. Expert Systems with Applications, vol. 39, pp. 1105–1117, 2012.

Bidding Strategies Based on Type-1 and Interval Type-2 Fuzzy Systems for Google AdWords Advertising Campaigns

Quetzali Madera, Oscar Castillo, Mario Garcia and Alejandra Mancilla

Abstract Google AdWords has a bidding price optimization method for its campaigns, where the user establishes the maximum bidding price, and AdWords adapts the final bidding price according to the performance of a campaign. This chapter proposes a bidding price controller based on a fuzzy inference system. Specifically, two approaches are considered: a type-1 fuzzy inference system, and an interval type-2 fuzzy inference system. The results show that the proposed methods are superior to the AdWords optimization method, and that there is not enough statistical evidence to support the superiority of the interval type-2 fuzzy inference system against the type-1 fuzzy inference system, although type-2 is slightly better.

Keywords Fuzzy inference system · Type-1 fuzzy logic · Interval type-2 fuzzy logic

1 Introduction

The Google AdWords tool is one of the most popular advertising methods in the Internet, and in the overall marketing industry. The tool relies on the positioning of small advertisements on the Google Search result web pages, as well as in websites that participate in the Google AdSense program.

In order to determine what advertisement unit is going to be shown to a user who is performing a web search using Google's search engine, or in certain website's web page, a bidding process takes places beforehand. This process—explained in a very basic manner—involves the auctioning of a position within the web pages that the user is visiting. Those advertisement units who have a higher bidding price will have a higher chance of appearing to the user (it must be noted that the actual

Q. Madera · O. Castillo (✉) · M. Garcia · A. Mancilla
Tijuana Institute of Technology, Tijuana, BC, Mexico
e-mail: ocastillo@tectijuana.mx

P. Melin et al. (eds.), *Nature-Inspired Design of Hybrid Intelligent Systems*,
Studies in Computational Intelligence 667, DOI 10.1007/978-3-319-47054-2_6

process is, apparently, far more complex, as Google takes into consideration the performance of the unit, as well as other factors).

There are various parameters that a user can establish in the Google AdWords interface that directly impacts the performance and behavior of an advertisement campaign. In this work, focus is given to the maximum bidding price a campaign is allowed to use.

A user can let Google use an automated method for determining the bidding price, given a maximum established by the user. For example, a maximum bidding price of $1 USD can be used, and Google will automatically establish what is an appropriate bidding price for a campaign, based on its performance (i.e., Google can take $0.70 USD as the bidding price, resulting in costs per click that oscillate between $0 and $0.70 USD). As an alternative, the user can manually enter a bidding price. For example, a bidding price of $1 USD can be used for a campaign, and as a result, the maximum bidding price will always be $1 USD (i.e., the cost for each click will oscillate between $0 and $1 USD).

A well-performing advertisement campaign can have several meanings, as this depends on what factors are being taken into consideration to measure such performance. For example, a campaign can be described in terms of how many clicks it is obtaining, on its Cost Per Click (CPC), on its Click Through Rate (CTR), etc. It must be noted that establishing what factors are going to be considered in order to explain the performance is very important. For example, if one only considers the number of clicks, it would not matter if this high volume of clicks was achieved using a very high bidding price.

The proposed method considers the performance to be expressed in terms of CPC, i.e., if a campaign has a lower CPC than another, it is said to be performing better. Thus, the controllers proposed have the objective of obtaining campaigns with lower CPCs than those optimized by using Google AdWords' method.

It is also important to note that this work proposes an automated process, which works in real time, i.e., the process does not require any human evaluation, as this knowledge can be completely represented by fuzzy linguistic variables and fuzzy rules, and it has the capability of proposing the maximum bidding price at any moment (it does not require a fixed time period to finish in order to calculate a new price).

The present chapter is organized as follows: in Sect. 2, a brief description about type-2 fuzzy logic is given, in order to understand the differences between the interval type-2 and type-1 fuzzy inference system controllers; in Sect. 3, extensive related work is given to understand the direction and fundamentals of the proposed method; in Sect. 4, the proposed method is described; in Sect. 5, the design of the experiments is explained, and Sect. 6 shows the results, along with several hypothesis tests to sustain the claim that a fuzzy inference system should lower the cost-per-click of an advertisement campaign; finally, in Sects. 7 and 8, the conclusions and future work are presented.

2 Type-2 Fuzzy Logic

Type-2 fuzzy sets were first described by Zadeh as an extension to the type-1 fuzzy sets. The difference between a type-1 fuzzy set and a type-2 fuzzy set is that, in the latter, the grades of membership are described by fuzzy sets instead of real numbers. This fuzzy set (secondary membership function), can be equal to any subset in the primary membership function, and every primary grade of membership is now described by a secondary grade of membership. A type-2 fuzzy set is an extension of a type-1 fuzzy set, as any type-1 fuzzy set can be described with grades of membership equal to a fuzzy set with only one element.

There are two common implementations of type-2 fuzzy sets: interval and generalized. An interval type-2 fuzzy set is one where all the secondary grades of membership are equal to 1, whereas a generalized type-2 fuzzy set can hold varying secondary grades of membership in its primary membership function. Generalized type-2 fuzzy sets are an extension of interval type-2 fuzzy sets.

A type-1 fuzzy set can be described by (1)

$$A = \{(x, \mu_A(x)) \mid \forall x \in X\} \tag{1}$$

where A denotes a fuzzy set consisting of tuples formed by a real number x, and its grade of membership $\mu_A(x)$. Following what was explained about a type-2 fuzzy set, and extending (1), we can define a type-2 fuzzy set as in (2)

$$\widetilde{A} = \{((x, u), \mu_{\widetilde{A}}(x, u)) \mid \forall x \in X, \forall u \in J_x \subseteq [0, 1]\} \tag{2}$$

where $\tilde{A}$ denotes a type-2 fuzzy set consisting of tuples formed by pair of a real number x and its primary membership u, and a grade of membership $\mu_{\widetilde{A}}(x, u)$.

By having the concept of a type-2 fuzzy set, one can also extend the concept of a type-1 fuzzy inference system (FIS). As in a type-1 FIS, a type-2 FIS implements a fuzzifier method, a rule base, a fuzzy inference engine, and a defuzzifier. As an additional step, the output of a type-2 FIS needs to be reduced to a type-1 fuzzy set first, and then defuzzified by the defuzzifier. The rules structure present in a type-1 FIS can be used for a type-1 FIS. The only difference between the design of a type-1 FIS and a type-2 FIS is that the membership functions are now type-2 fuzzy sets.

Type-2 sets and FISs have been used in decision-making, solving fuzzy relation equations, survey processing, time-series forecasting, function approximation, time-varying channel equalization, control of mobile robots, and preprocessing of data.

In a type-1 FIS, the outputs of each of the fired rules are represented as type-1 fuzzy sets, which are then combined using a defuzzifier method, and are converted to a crisp number using a specified operation in the design of the FIS, such as the centroid of the combined fuzzy sets. In a type-2 FIS, the output fuzzy sets are also combined using type-2 fuzzy operations, but before defuzzifying the output, a type-reduction must take place. After performing a type-reduction operation over the output fuzzy sets, the type-2 fuzzy set become a type-1 fuzzy set, and the traditional crispifying methods can operate over these sets.

The type-reduction procedure in a FIS is usually very computational expensive. In a generalized type-2 FIS, this process can be so time consuming, that, to date, most of the research involving type-2 FIS avoids it, and the interval type-2 FIS are dominant in the literature [1].

3 Related Work

There have been several works in the past where bidding price controllers are proposed. In the end, these controllers form part of what is known as a bidding strategy. In this work, focus is given to bidding price controllers in Internet ad auctions. For a comprehensive guide on Internet ad auctions, please refer to the work of Muthukrishnan [2].

A general antecedent for the proposed method is those tools designed for the creation and optimization of online advertising campaigns. As a first example, one can read the thesis by Thomaidou [3], where he proposes a system to automatically create, monitor, and optimize advertising campaigns to achieve cost efficiency with budget constraints. As additional tools, he proposes a mechanism that extracts keywords from a web page to automatically generate the keywords for the advertising campaign, and also proposes a method for the ad-text generation (a field that has been previously tackled by the authors in works such as [4, 5]).

Two additional examples for such general tools that manage advertising campaigns are the works of Liu et al. [6], and Thomaidou et al. [7]. In the former, the authors examine the problem of advertising campaign optimization from the advertiser's point of view, by selecting better ad-texts which are taken from a pool of ad-text lines generated by an also proposed method in the same work. In the latter, the optimization involves the selection of the better-performing ad units from different advertising campaigns in order to create better campaigns for the subsequent testing periods.

Similar in spirit to this work, but different on its point of view comes the work by Zainal-Abidin and Wang [8]. The main concern in this work is to maximize the clicks obtained by an advertising campaign, while having the constraints of maximum average CPC, and overall budget. This work differs in that the main focus is to lower the CPC, and the maximization of the clicks occurs as a product of the minimization of this feature.

In the work by Özlük and Cholette [9], another approach is taken in order to optimize an advertising campaign. Rather than manipulating the ad-text, the bidding price, or obtaining the better-performing ad units from different campaigns, in order to maximize the clicks or lower the CPC, these authors decided to perform an optimization processes involving a selection of keywords (the ad units will be shown to the user when the user performs a search in a search engine involving these keywords). After deciding which keywords will be used to increase the performance, another process takes place that determines how much money is going to be allocated for each keyword.

As this work involves the direct manipulation of the maximum bidding price to decrease the CPC with an overall budget constraint, the work by Even-Dar et al. [10] (who are Google researchers) proved to be useful in order to understand how the bidding price can affect the performance of an advertising campaign. In their work, they explore how a minimum price for the position auctions affects the quality of the ads that appear on the web pages. Although their objective is not directly related to the objective of the proposed method, their work served as valuable insights.

Two works that propose autonomous bidding strategies for advertising campaign position auctions are the ones by Berg et al. [11], and Zhou et al. [12]. The relationship of these two works with the proposed method is in its autonomous nature. The method of this work can be left running indefinitely, and its fuzzy-based controller will automatically recalculate a maximum bidding price that should increase the number of clicks while obtaining a low CPC. This also means that the proposed method works in real time. Other papers, similar in nature to this work, that propose optimization algorithms that work in real time are the ones by Lee et al. [13]; and Zhang et al. [14]. In the first paper, the authors explore a solution to the problem of the clicks delay (the statistics shown in an advertising platform are usually delayed, as presenting information in real time would be very expensive in computational resources) by presenting an algorithm which adjusts the bid price based on the prior performance distribution in an adaptive manner by distributing the budget across time. In the second paper, real-time bidding is analyzed and a suggestion is derived which states that optimal bidding strategies should try to bid more impressions rather than focus on a small set of high valued impressions, as this lowers the cost of an advertising campaign.

Finally, the reader can find other related works, similar in the objectives of advertising campaign optimization in the following references: for overall budget optimization, the works by Muthukrishnan et al. [15]; Feldman et al. [16]; and Yang et al. [17]; for bid price optimization, the works by Even Dar et al. [18]; Borgs et al. [19]; and Abhishek and Hosanagar [20].

4 Proposed Method

This chapter proposes the use of Fuzzy Inference Systems (FIS) as a campaign's bidding price controller. The decision to use FIS as a means to control the bidding price results from the versatility of a FIS to be tuned according to the information gathered from experts, which is usually expressed in fuzzy terms, such as "very low," "medium," or "very high."

The current architecture of the FIS uses two-input variables (the current bidding price a campaign is using, and the current CTR) and one-output variable (the new bidding price the campaign must use).

Two types of FIS are used and are compared against the Google optimization method: a type-1 FIS (T1-FIS) and an interval type-2 FIS (IT2-FIS). The IT2-FIS uses the architecture of the T1-FIS as its base, and it adds uncertainty by including a Footprint of Uncertainty (FOU). This FOU enables the system to interpret more uncertainty in its input and output variables (e.g., is this CTR value truly high?). Both of these FIS are Mamdani systems, and have the following rule base:

1. If CTR is LOW and PRICE is LOW then NPRICE is HIGH.
2. If CTR is LOW and PRICE is MEDIUM then NPRICE is MEDIUM.
3. If CTR is LOW and PRICE is HIGH then NPRICE is LOW.
4. If CTR is MEDIUM and PRICE is LOW then NPRICE is MEDIUM.
5. If CTR is MEDIUM and PRICE is MEDIUM then NPRICE is HIGH.
6. If CTR is MEDIUM and PRICE is HIGH then NPRICE is MEDIUM.
7. If CTR is HIGH and PRICE is LOW then NPRICE is LOW.
8. If CTR is HIGH and PRICE is MEDIUM then NPRICE is MEDIUM.
9. If CTR is HIGH and PRICE is HIGH then NPRICE is MEDIUM.

If the CTR for an advertisement unit is low, we want to increase the price if it is low to try to increase the CTR by displaying the unit more; if it has a medium price, we want to keep it there; if it has a high price, we want to get it to low, because not even a high price was able to increase the CTR of the unit. If the CTR is medium, then we want to bring the price to medium, unless it is already in medium, where we want to bring it to higher prices to see if it increases the CTR. Finally, if the CTR is high, and the price is low, we want to keep it at low as it means it is performing very well with a low budget, and we want to keep it at medium otherwise.

5 Experiments

Two experiments were designed in order to prove the efficacy of a type-1 Fuzzy Controller and an Interval type-2 Fuzzy Controller against the default optimization algorithm provided by Google AdWords. In both experiments, each of the tools being tested consisted on 30 campaigns that were being run at the same time as the other campaigns were being optimized by the other tools.

In the first experiment, the Google optimization algorithm is compared against a Type-1 Fuzzy Controller, and in the second, it is compared against an Interval Type-2 Fuzzy Controller.

All the campaigns had a cost limit of five Mexican Pesos. Due to the nature of Google AdWords, this price rarely acted as a "hard limit," as most of the campaigns exceeded it. As Google needs to be checking on millions of campaigns, stopping a campaign automatically before hitting the established limit is improbable. Nevertheless, this did not affect the experiments, as this configuration only worked as a means to reduce the overall cost of the experiments.

The maximum allowed bidding price for a click is 0.50 Mexican Pesos. This means that a controller can establish a bidding price from 0 to 0.50 Pesos. The objective of each method is to establish the maximum bidding price for an ad, in order to obtain a high performance. In this chapter, performance is defined as a campaign having a low average CPC. For example, if an ad is performing very well, and the bidding price is very low, the controller could decide to increase the bidding price in order to increase the performance even further.

5.1 Type-1 Fuzzy Controller

The Type-1 Fuzzy Controller was constructed using a very basic design. It consists on a two-input and one-output Fuzzy Inference System, where each of the inputs and outputs has three linguistic variables: low, medium, and high. These variables are represented by Gaussian membership functions. The first input (Fig. 1) represents the current bidding price, and the second input (Fig. 2) represents the current CTR for the campaign. As the output, a new bidding price is calculated.

The membership functions for the CTR input variable have the following parameters: low, mean = 0, SD = 2; medium, mean = 5, SD = 2; high, mean = 10, SD = 2. For the current bidding price: low, mean = 0, SD = 0.13; medium, mean = 0.25, SD = 0.13; high, mean = 0.5, SD = 0.13. The new price variable is represented by the same membership functions as the current price variable.

The membership functions for the CTR input variable are shown in Fig. 1. The membership functions for the current bidding price are shown in Fig. 2. The

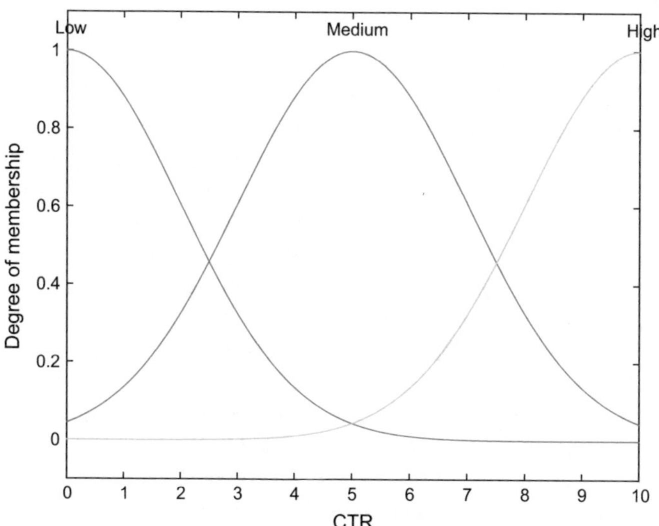

Fig. 1 The membership functions for the CTR input variable

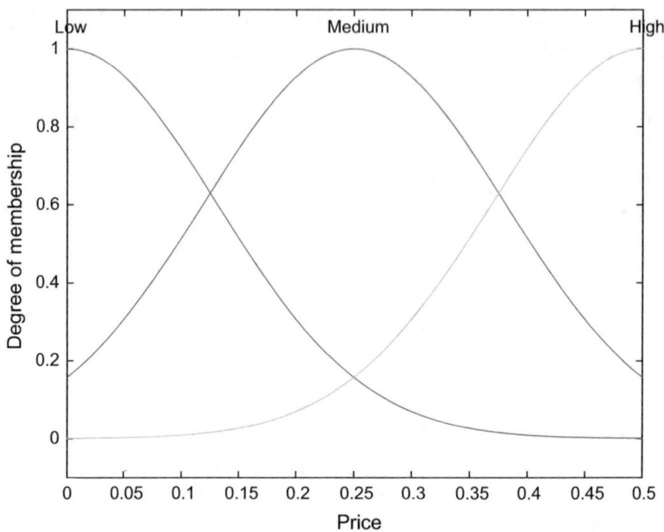

Fig. 2 The membership functions for the price input variable

membership function for the new bidding price (the output variable) is not showed, as it currently uses exactly the same configuration as the membership functions in the current bidding price variable.

5.2 *Type-2 Fuzzy Controller*

The second experiment consisted on the use of an Interval Type-2 Fuzzy Inference System (IT2-FIS). This IT2-FIS used the same architecture as the T1-FIS in the first experiment, but with an increase in its uncertainty through the use of a Footprint of Uncertainty (FOU). A Gaussian membership function in a IT2-FIS can have uncertainty in its standard deviation or in its mean. For this experiment, the Gaussian membership functions have uncertainty in their means.

The membership functions for the CTR input variable have the following parameters: low, mean1 = −1, mean2 = 1, SD = 2; medium, mean1 = 4, mean2 = 6, SD = 2; high, mean1 = 9, mean2 = 11, SD = 2. For the current bidding price: low, mean1 = −0.05, mean2 = 0.05, SD = 0.13; medium, mean1 = 0.2, mean2 = 0.3, SD = 0.13; high, mean1 = 0.45, mean2 = 0.55, SD = 0.13. As with the T1-FIS in the first controller, the new price variable is represented by the same membership functions as the current price variable.

The membership functions for the CTR input variable are shown in Fig. 3. The membership functions for the current bidding price are shown in Fig. 4. The membership function for the new bidding price (the output variable) is not showed, as it currently uses exactly the same configuration as the membership functions in the current bidding price variable.

For this experiment, 90 campaigns were run: 30 campaigns that were being optimized by the Google AdWords method, 30 campaigns being optimized by the T1-FIS controller, and 30 campaigns being optimized by the IT2-FIS. This had to be done, as the first experiment was performed weeks before this last experiment was conducted. The environment where the first experiment was conducted was different than the one for this experiment, and a comparison of the IT2-FIS against the T1-FIS was not going to be reliable.

6 Simulation Results

6.1 First Experiment (Google Versus T1-FIS)

The results for the first experiment are shown in Table 1. One can see the means and standard deviations for each of the features that were being compared, for each of the controllers. These statistic values are the result of performing 30 campaigns for the Google controller, and 30 campaigns for the T1-FIS controller.

In Table 2, the results of three hypothesis tests are shown. The alternative hypothesis in the CTR case is defined as T1-FIS > G, meaning that we want to prove that the T1-FIS controller achieved a higher CTR than the Google controller.

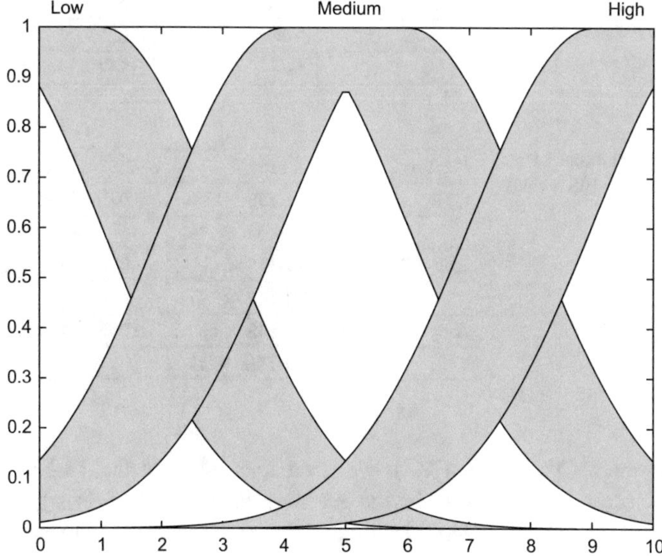

Fig. 3 The membership functions for the CTR input variable

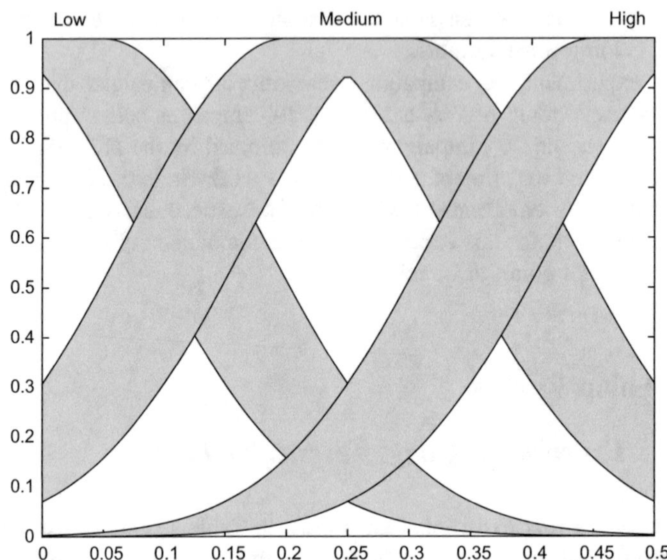

Fig. 4 The membership functions for the price input variable

Table 1 First experiment; Google versus T1-FIS

Controller	Feature	Mean	St. Dev.
Google	CTR	1.835	0.423
	Avg. CPC	0.25	0.028
	Avg. Pos.	1.526	0.089
T1-FIS	CTR	1.929	0.371
	Avg. CPC	0.22	0.029
	Avg. Pos.	1.532	0.093

Table 2 Second experiment; Google versus T1-FIS versus IT2-FIS

Feature	Hypothesis	*t*-value	Result
CTR	Ha: T1-FIS > G	0.915	H0 not rejected
	H0: T1-FIS $\leq$ G		
Avg. CPC	Ha: T1-FIS < G	−4.076	H0 rejected
	H0: T1-FIS $\geq$ G		
Avg. Pos.	Ha: T1-FIS < G	0.255	H0 not rejected
	H0: T1-FIS $\geq$ G		

For the Average CPC (Avg. CPC), we want to prove that the T1-FIS controller achieved a lower cost. Finally, for the Average Position (Avg. Pos.), we want to prove that the T1-FIS achieved a lower position (the lowest is 1, meaning that the ad unit is shown at the first place in a web page).

As a conclusion to this experiment, one can note that the T1-FIS outperforms the Google controller at the Average CPC, but performed badly regarding the two other features.

6.2 *Second Experiment (Google Versus T1-FIS Versus IT2-FIS)*

For this second experiment, a comparison among three controllers was performed: Google, T1-FIS, and IT2-FIS controllers. The results from the first experiment could not be used, as it was proven (see Sect. 6.3) that the controllers achieve different results depending on the time the campaigns were running.

Similarly to the first experiment, the alternative hypotheses are the same (see Sect. 6.1), but the comparisons differ in that the IT2-FIS controller is compared against the T1-FIS controller, and both FIS controllers are compared against the Google controller.

In Table 3, one can find the means and standard deviations for each of the features, for each controller. Using this information, one can perform hypothesis tests between two different controllers, and arrive at the conclusions shown in Tables 4, 5 and 6.

Table 3 Means and standard deviations

Controller	Feature	Mean	St. Dev.
Google	CTR	3.32	0.805
	Avg. CPC	0.231	0.027
	Avg. Pos.	1.603	0.063
T1-FIS	CTR	2.667	0.711
	Avg. CPC	0.132	0.019
	Avg. Pos.	1.982	0.283
IT2-FIS	CTR	2.076	0.274
	Avg. CPC	0.127	0.012
	Avg. Pos.	1.921	0.134

Table 4 First experiment; TI-FIS versus Google conclusions

Feature	Hypothesis	*t*-value	Result
CTR	Ha: T1-FIS > G	−3.33	H0 not rejected
	H0: T1-FIS $\leq$ G		
Avg. CPC	Ha: T1-FIS < G	−16.424	H0 rejected
	H0: T1-FIS $\geq$ G		
Avg. Pos.	Ha: T1-FIS < G	7.16	H0 not rejected
	H0: T1-FIS $\geq$ G		

Table 5 Second experiment; IT2-FIS versus Google conclusions

Feature	Hypothesis	*t*-value	Result
CTR	Ha: IT2-FIS > G	−8.012	H0 not rejected
	H0: IT2-FIS $\leq$ G		
Avg. CPC	Ha: IT2-FIS < G	−19.279	H0 rejected
	H0: IT2-FIS $\geq$ G		
Avg. Pos.	Ha: IT2-FIS < G	11.763	H0 not rejected
	H0: IT2-FIS $\geq$ G		

Table 6 Second experiment; TI-FIS versus TI-FIS conclusions

Feature	Hypothesis	*t*-value	Result
CTR	Ha: IT2 > T1	−4.248	H0 not rejected
	H0: IT2 $\leq$ T1		
Avg. CPC	Ha: IT2 < T1	−1.218	H0 not rejected
	H0: IT2 $\geq$ T1		
Avg. Pos.	Ha: IT2 < T1	−1.067	H0 not rejected
	H0: IT2 $\geq$ T1		

As a general conclusion to this experiment, one can see that both the T1-FIS and IT2-FIS controllers outperform the Google controller in terms of Average CPC, but perform badly regarding the two other features, and there was not enough statistical evidence to conclude that the IT2-FIS performs better than the T1-FIS in either of the features.

As a surprise, one can actually use the data from Table 6 to conclude that the T1-FIS controller performed better than the IT2-FIS, regarding the CTR feature.

6.3 Comparison of the First Experiment Against Second Experiment

Tables 7 and 8 show the hypothesis tests used to prove that one cannot use the data gathered from one experiment performed in certain period of time against the data gathered from another experiment performed in another period of time. The conclusion of this is that in order to prove other configurations of the FIS controllers against each other or against the Google controller, one has to run each of the competing controllers at the same time.

A possible explanation for this is that the crowd's behavior (the users clicking on the ads) changes significantly depending on several time factors, i.e., were the experiments run on a holiday? What time of the day? What was the season?

Table 7 Testing for consistency of the results over time for the Google controller

Feature	Hypothesis	*t*-value	Result
CTR	Ha: CTR1 ≠ CTR2	−8.944	H0 rejected
	H0: CTR1 = CTR2		
Avg. CPC	Ha: CPC1 ≠ CPC2	2.675	H0 rejected
	H0: CPC2 = CPC2		
Avg. Pos.	Ha: POS1 ≠ POS2	−3.867	H0 rejected
	H0: POS1 = POS2		

Table 8 Testing for consistency of the results over time for the it2-fis controller

Feature	Hypothesis	*t*-value	Result
CTR	Ha: CTR1 ≠ CTR2	−5.04	H0 rejected
	H0: CTR1 = CTR2		
Avg. CPC	Ha: CPC1 ≠ CPC2	13.90	H0 rejected
	H0: CPC2 = CPC2		
Avg. Pos.	Ha: POS1 ≠ POS2	−8.274	H0 rejected
	H0: POS1 = POS2		

7 Conclusions

As the results of the experiments shown, the proposed method can successfully decrease the CPC of a campaign. Both of the fuzzy controllers show enough evidence to prove their efficacy against the Google AdWords optimization method.

There was not enough evidence to conclude that the Interval Type-2 Fuzzy Inference System (IT2-FIS) performed better than the Type-1 Fuzzy Inference System (T1-FIS), but the authors of this chapter believe that minor changes to the parameters of the IT2-FIS would be enough to statistically surpass the T1-FIS. One can have a superficial look at the results obtained by the IT2-FIS, and could arrive to the conclusion that it was superior to the T1-FIS. This leads the authors to the idea that maybe the configuration is adequate, but more experiments had to be conducted in order to provide more evidence to arrive to a satisfactory statistical conclusion.

8 Future Work

As was mentioned in Sect. 7, more experiments need to be conducted, or the configuration of the IT2-FIS needs to be changed in order to achieve better results with this controller that can handle more uncertainty. Moreover, the authors think that a more correct way of comparing these two controllers would be by optimizing their parameters and comparing their optimized versions against each other. This solution would be really costly, but sub-optimized versions could be used in this study.

As another future work, and one that the authors are already working on, is the implementation of a Generalized Type-2 Fuzzy Inference System as a bidding price controller. This new controller should be able to handle even more uncertainty than the IT2-FIS, and has the potential to outperform the proposed controllers of this work.

References

1. Q. Liang and J. M. Mendel, "Interval type-2 fuzzy logic systems: theory and design," *Fuzzy Systems, IEEE Transactions on* 8.5, pp. 535-550, 2000.
2. S. Muthukrishnan, "Internet ad auctions: Insights and directions," *Automata, Languages and Programming*. Springer Berlin Heidelberg, pp. 14-23, 2008.
3. S. Thomaidou, "Automated Creation and Optimization of Online Advertising Campaigns," Ph.D. dissertation. Department of Informatics, Athens University of Economics and Business, 2014.
4. Q. Madera *et al.*, "Fuzzy Logic for Improving Interactive Evolutionary Computation Techniques for Ad Text Optimization," *Novel Developments in Uncertainty Representation and Processing*. Springer International Publishing, pp. 291-300, 2016.
5. Q. Madera *et al.*, "A Method Based on Interactive Evolutionary Computation for Increasing the Effectiveness of Advertisement Texts," *Proceedings of the Companion Publication of the 2015 on Genetic and Evolutionary Computation Conference*, ACM, 2015.
6. W. Liu, *et al.*, "Online advertisement campaign optimisation," *International Journal of Services Operations and Informatics* 4.1, pp. 3-15, 2008.
7. S. Thomaidou *et al.*, "Toward an integrated framework for automated development and optimization of online advertising campaigns," *arXiv preprint* arXiv:1208.1187, 2012.
8. A. Zainal-Abidin and J. Wang, "Maximizing clicks of sponsored search by integer programming," *Proceedings of the ADKDD,* 2010.
9. Ö. Özlük and S. Cholette, "Allocating expenditures across keywords in search advertising," *Journal of Revenue & Pricing Management* 6.4, pp. 347-356, 2007.
10. E. Even-Dar *et al.*, "Position auctions with bidder-specific minimum prices," *Internet and Network Economics*. Springer Berlin Heidelberg, pp. 577-584, 2008.
11. J. Berg, *et al.*, "A first approach to autonomous bidding in ad auctions," *Workshop on Trading Agent Design and Analysis at the 11th ACM Conference on Electronics Commerce*, 2010.
12. Y. Zhou *et al.*, "Budget constrained bidding in keyword auctions and online knapsack problems," *Internet and Network Economics*. Springer Berlin Heidelberg, pp. 566-576, 2008.
13. K. C. Lee *et al.*, "Real time bid optimization with smooth budget delivery in online advertising," *Proceedings of the Seventh International Workshop on Data Mining for Online Advertising*, ACM, 2013.
14. W. Zhang *et al.*, "Optimal real-time bidding for display advertising," *Proceedings of the 20th ACM SIGKDD international conference on Knowledge discovery and data mining*, ACM, 2014.
15. S. Muthukrishnan *et al.*, "Stochastic models for budget optimization in search-based advertising," *Internet and Network Economics*. Springer Berlin Heidelberg, pp. 131-142, 2007.
16. J. Feldman *et al.*, "Budget optimization in search-based advertising auctions," *Proceedings of the 8th ACM conference on Electronic commerce*, ACM, 2007.
17. Y. Yang *et al.*, "A budget optimization framework for search advertisements across markets," *Systems, Man and Cybernetics, Part A: Systems and Humans, IEEE Transactions on* 42.5, pp. 1141-1151, 2012.

18. E. Even Dar *et al.*, "Bid optimization for broad match ad auctions," *Proceedings of the 18th international conference on World wide web*, ACM, 2009.
19. C. Borgs *et al.*, "Dynamics of bid optimization in online advertisement auctions," *Proceedings of the 16th international conference on World Wide Web*, ACM, 2007.
20. V. Abhishek and K. Hosanagar, "Optimal bidding in multi-item multislot sponsored search auctions," *Operations Research* 61.4, pp. 855-873, 2013.

On the Graphical Representation of Intuitionistic Membership Functions for Its Use in Intuitionistic Fuzzy Inference Systems

Amaury Hernandez-Aguila, Mario Garcia-Valdez and Oscar Castillo

Abstract This work proposes an approach for graphically representing intuitionistic fuzzy sets for their use in Mamdani fuzzy inference systems. The proposed approach is used and plots for several membership and non-membership functions are presented, including: triangular, Gaussian, trapezoidal, generalized bell, sigmoidal, and left-right functions. Plots of some operators used in fuzzy logic are also presented, i.e., union, intersection, implication, and alpha-cut operators. The proposed approach should produce plots that are clear to understand in the design of an intuitionistic fuzzy inference system, as the membership and non-membership functions are clearly separated and can be plotted in the same figure and still be recognized with ease.

Keywords Fuzzy inference systems · Intuitionistic fuzzy logic · Membership functions

1 Introduction

Fuzzy sets have been used as the building blocks of many other areas and applications since they were conceived by L.A. Zadeh in 1965 [23]. One of the most prominent areas is fuzzy logic and its wide area of application of control. As a consequence of its usefulness in the industry, many works have been dedicated to improving the architectures and theory used in the construction of control systems. A remarkable example of this is the extension of fuzzy sets to the concept of type-2 fuzzy sets by L.A. Zadeh in 1975 [24] and the extension from traditional fuzzy sets

---

A. Hernandez-Aguila (✉) · M. Garcia-Valdez · O. Castillo
Tijuana Institute of Technology, Tijuana, BC, Mexico
e-mail: amherag@tectijuana.mx

M. Garcia-Valdez
e-mail: mario@tectijuana.mx

O. Castillo
e-mail: ocastillo@tectijuana.mx

P. Melin et al. (eds.), *Nature-Inspired Design of Hybrid Intelligent Systems*,
Studies in Computational Intelligence 667, DOI 10.1007/978-3-319-47054-2_7

to intuitionistic fuzzy sets (IFS) by K.T. Atanassov in 1986 [3]. These extensions focus on increasing the uncertainty a fuzzy set can model, and as a consequence, these fuzzy sets can help in the creation of better controls where the input data has high levels of noise. Specifically, type-2 fuzzy sets enable the membership of an element in a fuzzy set to be described with another fuzzy set, and IFSs enable the description of an element in terms of both membership and non-membership.

A common application of fuzzy sets is in inference systems, and these systems are called Fuzzy Inference Systems (FIS). FISs use fuzzy sets to associate different degrees of membership to the inputs of the system and work as the antecedents of the inference system. Other fuzzy sets are used to model the consequents of the inference system (as in a Mamdani type FIS). Traditional fuzzy sets (also called type-1 fuzzy sets), are commonly defuzzified in order to obtain a scalar value as the output of the system, instead of a fuzzy set. When a FIS uses type-2 fuzzy sets as its antecedents and consequents, the output of the system needs to be reduced to a type-1 fuzzy set first, and then this type-1 fuzzy set is defuzzified to a scalar value.

One of the drawbacks of working with type-2 fuzzy sets in a FIS is that the type reduction procedure is very time consuming, due to the high quantity of steps involved in the process, and the use of a type-2 FIS (T2-FIS) is slower than the use of a type-1 FIS (T1-FIS). The type reduction and defuzzification processes are described by N.N. Karnik and J.M. Mendel in [14]. This drawback is a possible explanation of why many controllers still rely on the use of T1-FIS instead of T2-FIS. Furthermore, most of the T2-FIS use a special case of type-2 fuzzy sets called interval type-2 fuzzy sets, as these fuzzy sets require less steps in their type reduction procedure, and thus, interval type-2 FIS (IT2-FIS) are faster than a general type-2 FIS (GT2-FIS).

IFSs can also be used to construct a Mamdani type FIS, as in [13]. In contrast to a T2-FIS, an intuitionistic FIS (IFIS) does not require a type reduction procedure, and an implementation of an IFIS should work nearly as fast as a T1-FIS. The advantage of an IFIS over a T2-FIS is that the inference system can handle more uncertainty without a high penalty in time.

At the time of writing this paper, there are not many works involving Mamdani type intuitionistic inference systems yet, and the authors of this work are only aware of the work by Castillo et al. [7], and Hernandez–Aguila and Garcia–Valdez [13]. As a consequence of this lack of works involving Mamdani FISs, there is not a common way to graphically represent IFS for its use as membership and non-membership functions for an IFIS. Additionally, there is not a common way to graphically represent the architecture of an IFIS, as the one presented in Matlab's Fuzzy Logic Toolbox. Having a standardized way of representing these components of an IFIS should ease the description of such systems in future research.

This work proposes an approach to construct graphical representations of IFSs for their use in IFISs, which is described in detail in Sect. 4. Some preliminaries can be found in Sect. 2, which are needed in order to understand the proposed approach. In Sect. 3, one can find a number of related works, which describe other ways of representing IFSs. Finally, in Sects. 5 and 6 one can find the conclusions and the future work.

2 Preliminaries

An IFS, as defined by K.T. Atanassov in [4], is represented by a capital letter with superscript star. An example of this notation is A^*. The definition of an IFS is also defined in [4], and is described in Eq. (1). In this equation, x is an element in set A^*, $\mu_{A^*}(x)$ is the membership of x, and $\upsilon_{A^*}(x)$ is the non-membership of x. For every triplet in A^*, (2) must be satisfied.

$$A^* = \{\langle x, \mu_A(x), \upsilon_A(x)\rangle | x \in E\} \quad (1)$$

$$0 \leq \mu_A(x) + \upsilon_A(x) \leq 1 \quad (2)$$

An IFS is a generalization of a traditional fuzzy set, meaning that a traditional fuzzy set can be expressed using the terminology of an IFS. An example of this is found in (3).

$$A^* = \{\langle x, \mu_A(x), 1 - \mu_A(x)\rangle | x \in E\} \quad (3)$$

If $0 \leq \mu_A(x) + \upsilon_A(x) < 1$ is true for an IFS, it is said that indeterminacy exists in the set. This concept of indeterminacy can also be found in the literature as hesitancy or nondeterminacy, and it is described in Eq. (4).

$$\pi_A(x) = 1 - \mu_A(x) - \upsilon_A(x) \quad (4)$$

3 Related Work

The most common approach to graphically represent an IFS is by lattices. Examples of this type of representation can be found in the works by Despi et al. [12], and Deschrijver et al. [11]. This is a popular approach to graphically represent an IFS as it enables more compact and concise mathematical expressions. Another representation that is suitable for mathematical processes is that of a matrix, and is discussed in detail in the works by Parvathi et al. [16], Çuvalcioglu et al. [9], and Yilmaz et al. [22].

IFSs have been graphically represented like membership functions are usually represented in Mamdani FISs, and some example works are the ones by Angelov [2], Atanassov [4], and Davarzani and Khorheh [10]. This notation can be suitable for representing an architecture of an IFIS, but if the plot is in black and white, or in grayscale, the reader can get confused by the membership and non-membership plots. This problem can be alleviated by plotting the membership and non-membership functions in separate plots is in the works by Castillo et al. [7], and Akram et al. [1].

There are several other graphical representations of IFSs, such as by radar charts, as in the work by Atanassova [5], and by geometrical representations, orthogonal projections, and three-dimensional representations as can be found in the work by Szmidt and Kacprzyk [19].

Some applications of IFSs in the area of medical sciences can be found in the works by Szmidt and Kacprzyk [20], Own [15], and Chakarska and Antonov [8]. In the area of group decision making, we have an example in the work by Xu [21]. IFSs have also been used in word recognition, in the area of artificial vision, as in the example work of Baccour et al. [6].

This work proposes that IFSs, in a Mamdani IFIS, should follow an approach similar to that found in the work by Atanassov [4], where the membership is plotted as is commonly done in a traditional FIS, but the non-membership function should be plotted as $1 - \upsilon_A$. The reason behind this decision is that the non-membership function should be easily differentiated from the membership function, while seeing both functions in the same plot. An implementation of an IFIS that uses this approach for representing IFSs for a Mamdani IFIS can be found in the work by Hernandez–Aguila and Garcia–Valdez [13].

4 Proposed Approach

What follows is a series of graphical representations of several commonly used membership functions in FISs, as well as graphical representations of common operators used in the construction of these systems, such as the union, intersection, and implication between two IFSs.

In Fig. 1, one can see how a traditional fuzzy set can be constructed using the proposed approach. A Gaussian membership function with mean of 50 and a standard deviation of 15 is depicted.

Figure 2 is the first case of an IFS that cannot be considered a traditional fuzzy set. The red line represents the membership function, while the blue line represents the non-membership function. As can be seen, the Gaussian membership function does not have a kernel, meaning that its highest valued member does not equal to 1. In this case, its highest valued member equals to 0.7, and for the non-membership function, its highest valued member equals to 0.3. The Gaussian membership function is constructed with a mean of 50 and a standard deviation of 15. For the non-membership function, it is constructed with a mean of 30 and standard deviation of 30.

The triangular membership function is depicted in Fig. 3. The membership function is constructed with the following points: 30, 50, and 80, meaning that, from left to right, the last 0 valued member is at 30, the first and only 1 valued member is at 50, and the first 0 valued member after the previous series of nonzero valued members is at 80. In the same fashion, the non-membership function is constructed with the following points: 40, 60, and 80. The highest valued member for the membership function is equal to 0.8, and for the non-membership function, it equals to 0.2.

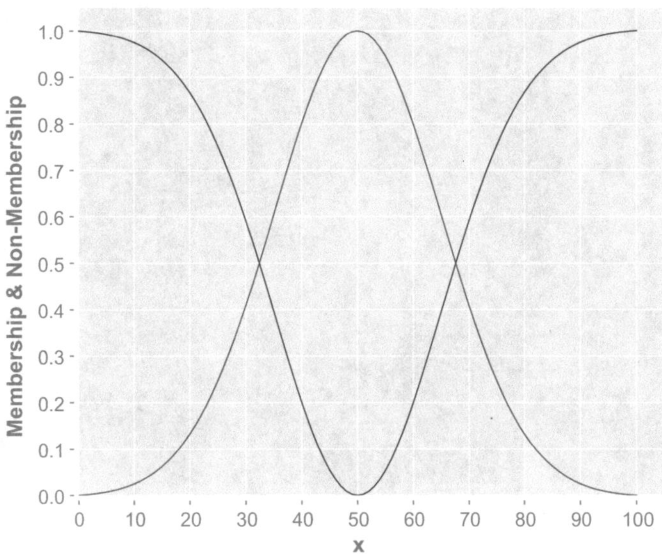

Fig. 1 A traditional fuzzy set represented as an intuitionistic fuzzy set

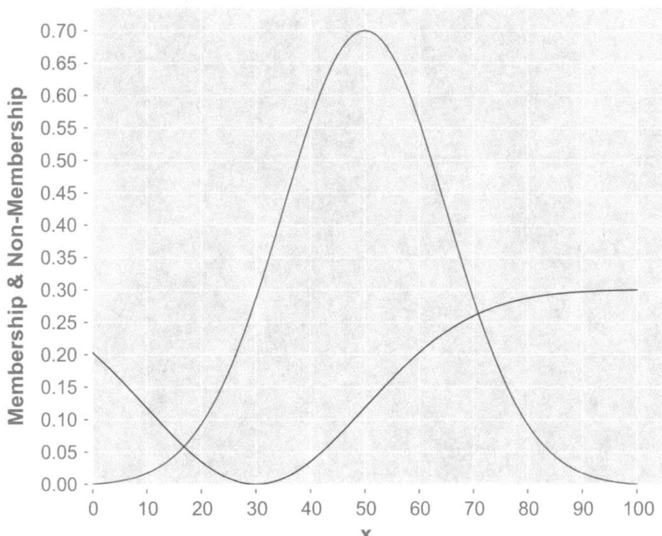

Fig. 2 Example of an intuitionistic fuzzy set

The trapezoidal membership function is shown in Fig. 4. The membership function is constructed using the following points: 20, 40, 60, and 80. These points mean that, from left to right, the first nonzero valued member will be at 20, and a

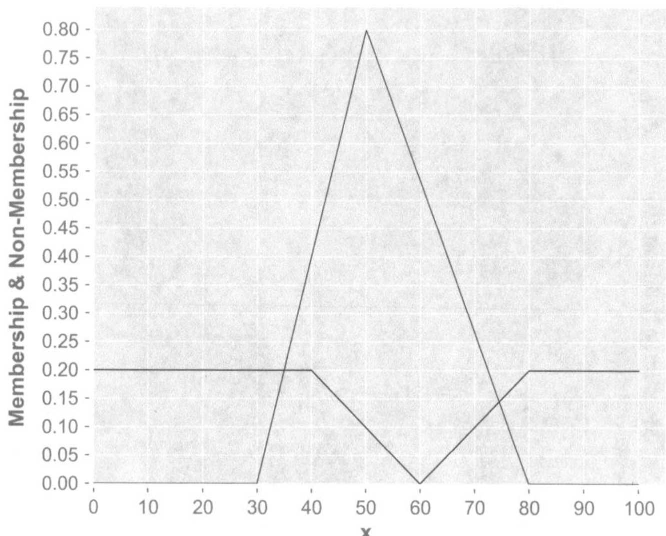

Fig. 3 Example of triangular membership and non-membership functions

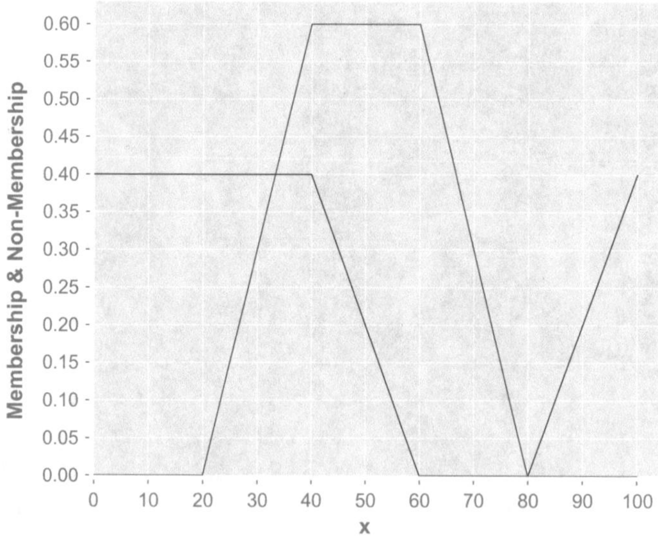

Fig. 4 Example of trapezoidal membership and non-membership functions

line will be drawn from 20 to 40. All members from 40 to 60 will have a value of 1, and then a line will be drawn to 80. In the same fashion, the non-membership function is drawn using the following points: 40, 60, 80, and 100. The highest

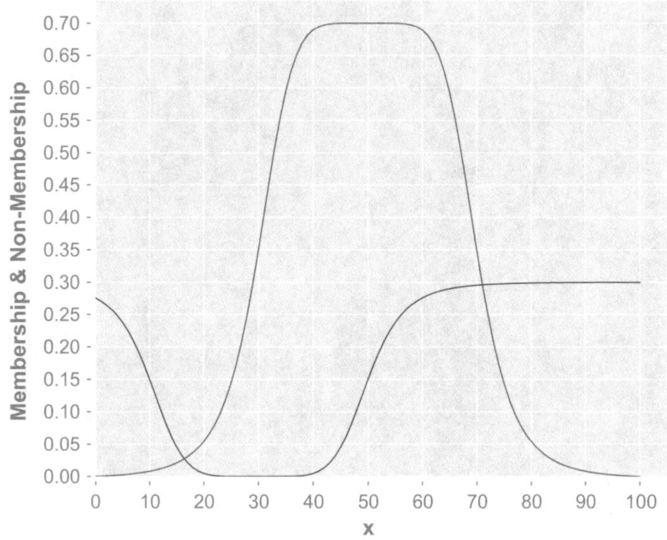

Fig. 5 Example of generalized bell membership and non-membership functions

valued members in the membership function equal to 0.6, while in the non-membership function these members equal to 0.4.

In Fig. 5 a generalized bell membership function is plotted. The membership function is constructed with a center of 50, a width of 20, and the parameter that determines the roundness of the corners of the bell is set at 3. Considering (5), which is the equation to generate a generalized bell, the membership function would be constructed with $a = 20$, $b = 3$, $c = 50$. In the case of the non-membership function, it would be constructed with $a = 20$, $b = 3$, $c = 30$. The highest valued member in the membership function is equal to 0.7, while the highest valued member in the non-membership function is equal to 0.3.

$$f(x; a, b, c) = \frac{1}{1 + \left|\frac{x-c}{a}\right|^{2b}} \tag{5}$$

Figure 6 shows an example of a left-right membership function and a non-membership function. Considering (6), the membership function is constructed with the following parameters: $c = 60$, $\alpha = 10$, $\beta = 65$, and the non-membership function is constructed with the same parameters. Both the membership and the non-membership functions have highest valued members that equal to 1, and in this case we are depicting a traditional fuzzy set represented as an IFS.

$$LF(x; c, \alpha, \beta) = \begin{cases} F_L\left(\frac{c-x}{\alpha}\right), & x \leq c \\ F_R\left(\frac{x-c}{\beta}\right), & x \geq c \end{cases} \tag{6}$$

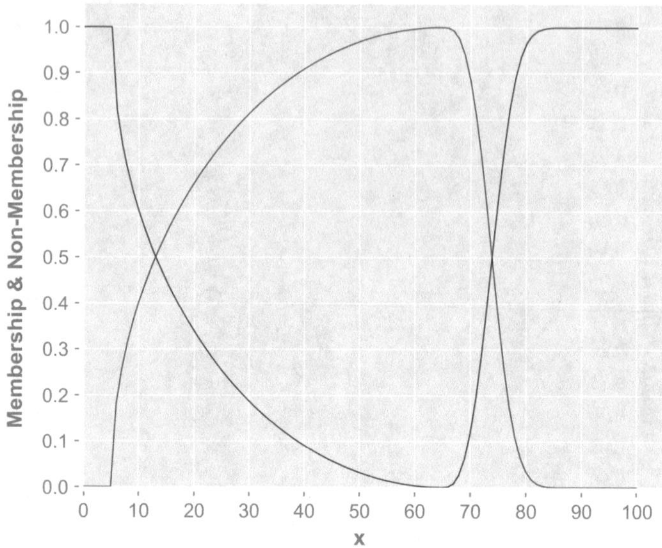

Fig. 6 Example of left-right membership and non-membership functions

The last membership function presented in this Section is the sigmoidal membership function. Figure 7 presents a sigmoidal membership function that is constructed with the equation presented in (7). The membership and non-membership functions are constructed by using the same parameters, which are $a = 0.3$, $b = 50$. As in the example of the left-right membership and non-membership functions, we are depicting a traditional fuzzy set represented as an intuitionistic fuzzy set.

$$\mathrm{sig}\,(x; a, c) = \frac{1}{1 + \exp[-a\,(x - c)]} \tag{7}$$

Figure 8 shows the result of applying the union operator between two IFSs. Figure 9 shows the result of the intersection operator, and Fig. 10 the result of the implication operator, both applied to two IFSs.

Lastly, Fig. 11 shows an example of an alpha-cut performed over Gaussian membership and non-membership functions. Some other works, as in the ones by Sharma [17, 18], describe procedures for performing an alpha-cut to an IFS. Nevertheless, the authors of the present work could not find works that show a graphical representation of an alpha-cut IFS.

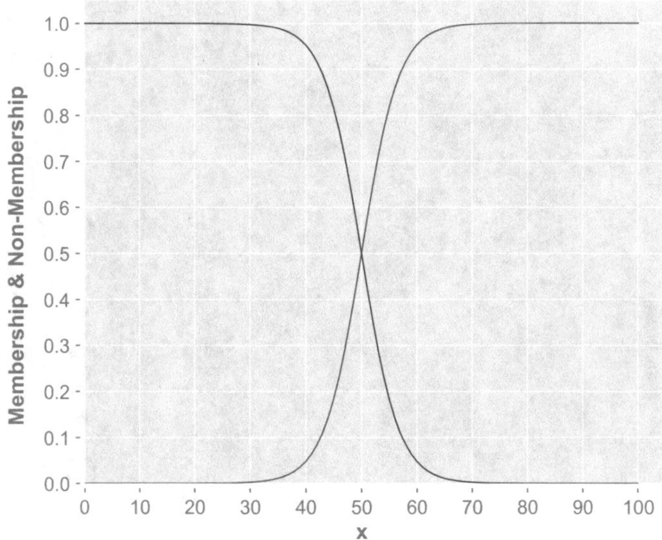

Fig. 7 Example of sigmoidal membership and non-membership functions

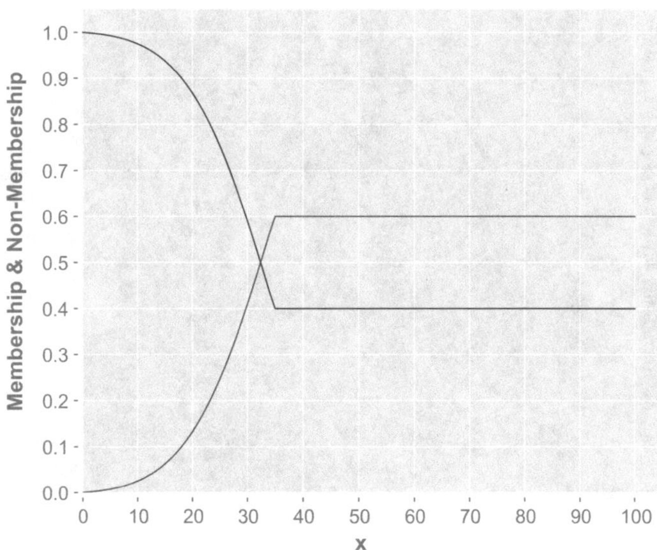

Fig. 8 Example of the union of two intuitionistic fuzzy sets

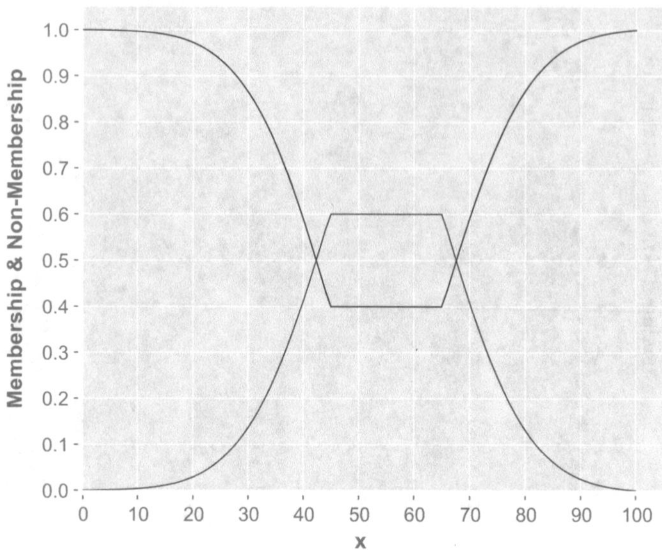

Fig. 9 Example of the intersection of two intuitionistic fuzzy sets

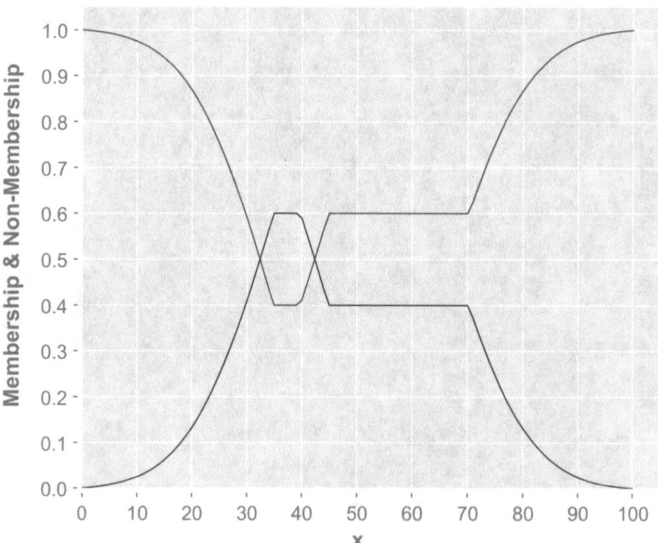

Fig. 10 Example of implication of two intuitionistic fuzzy sets

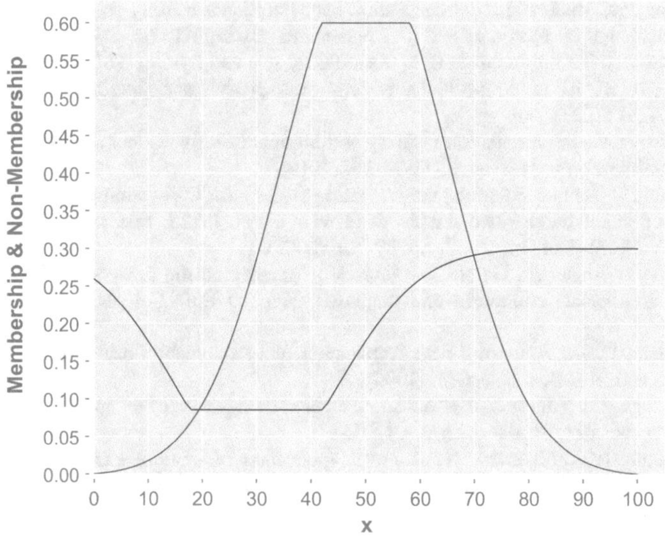

Fig. 11 Example of an alpha-cut of an intuitionistic fuzzy set

5 Conclusions

This work proposes an approach to graphically represent IFSs. The approach is focused on providing plots where the membership and the non-membership functions are easily recognized, and they can conveniently be used in IFISs. The work presents plots of the most common membership functions, along with their non-membership functions, and this way the reader can then decide if the use of this approach is convenient to represent the antecedents and consequents in the construction of an IFIS.

6 Future Work

The authors of this paper are currently working on an implementation of a graphical user interface (GUI) to create IFISs. This GUI will be similar in spirit to the one included in Matlab for its Fuzzy Logic Toolbox.

References

1. Akram M., Habib S., Javed I.: Intuitionistic Fuzzy Logic Control for Washing Machines. Indian Journal of Science and Technology, 7(5), 654-661. (2014).

2. Angelov P.:Crispification: defuzzification over intuitionistic fuzzy sets. Bulletin for Studies and Exchanges on Fuzziness and its Applications, BUSEFAL, 64, 51-55. (1995).
3. Atanassov K. T.: Intuitionistic fuzzy sets. Fuzzy sets and Systems,20(1), 87-96. (1986).
4. Atanassov K. T.: Intuitionistic fuzzy sets: past, present and future. In EUSFLAT Conf. (pp. 12-19) (2003).
5. Atanassova V.: Representation of fuzzy and intuitionistic fuzzy data by Radar charts. Notes on Intuitionistic Fuzzy Sets, 16(1), 21-26. (2010).
6. Baccour L., Kanoun S., Maergner V., Alimi A. M.: An application of intuitionistic fuzzy information for handwritten Arabic word recognition. In12th International Conference on IFSs (NIFS08) (Vol. 14, No. 2, pp. 67-72). (2008).
7. Castillo O., Alanis A., Garcia M., Arias H.: An intuitionistic fuzzy system for time series analysis in plant monitoring and diagnosis. Applied Soft Computing, 7(4), 1227-1233. (2007).
8. Chakarska D. D., Antonov L. S.: Application of Intuitionistic Fuzzy Sets in Plant Tissue Culture and in Invitro Selection. (1995).
9. Cuvalcioglu G., Yilmaz S., Bal A.: Some algebraic properties of the matrix representation of intuitionistic fuzzy modal operators. (2015).
10. Davarzani H., Khorheh M. A.: A novel application of intuitionistic fuzzy sets theory in medical science: Bacillus colonies recognition. Artificial Intelligence Research, 2(2), p 1. (2013).
11. Deschrijver G., Cornelis C., Kerre E. E.: On the representation of intuitionistic fuzzy t-norms and t-conorms. Fuzzy Systems, IEEE Transactions on, 12(1), 45-61. (2004).
12. Despi I., Opris D., Yalcin E.: Generalised Atanassov Intuitionistic Fuzzy Sets. In Proceeding of the Fifth International Conference on Information, Process, and Knowledge Management (2013).
13. Hernandez-Aguila A., Garcia-Valdez M.: A Proposal for an Intuitionistic Fuzzy Inference System. IEEE World Congress on Computational Intelligence (To be published) (2016).
14. Karnik N. N., Mendel J. M.: Centroid of a type-2 fuzzy set. Information Sciences, 132(1), 195-220. (2001).
15. Own C. M.: Switching between type-2 fuzzy sets and intuitionistic fuzzy sets: an application in medical diagnosis. Applied Intelligence, 31, 283-291. (2009).
16. Parvathi R., Thilagavathi S., Thamizhendhi G., Karunambigai M. G.: Index matrix representation of intuitionistic fuzzy graphs. Notes on Intuitionistic Fuzzy Sets, 20(2), 100-108. (2014).
17. Sharma P. K.: Cut of intuitionistic fuzzy groups. In International mathematical forum (Vol. 6, No. 53, pp. 2605-2614). (2011).
18. Sharma P. K.: Cut of Intuitionistic fuzzy modules. International Journal of Mathematical Sciences and Applications, 1(3), 1489-1492. (2011).
19. Szmidt E., Kacprzyk J.: Distances between intuitionistic fuzzy sets.Fuzzy sets and systems, 114(3), 505-518. (2000).
20. Szmidt E., Kacprzyk J.: Intuitionistic fuzzy sets in some medical applications. In Computational Intelligence. Theory and Applications (pp. 148-151). Springer Berlin Heidelberg. (2001).
21. Xu Z.: Intuitionistic preference relations and their application in group decision making. Information sciences, 177(11), 2363-2379. (2007).
22. Yilmaz S., Citil M., Cuvalcioglu G.: Some properties of the matrix representation of the intuitionistic fuzzy modal operators. Notes on Intuitionistic Fuzzy Sets, Volume 21, 2015, Number 2, pages 19—24. (2015).
23. Zadeh L. A.: Fuzzy sets. Information and control, 8(3), 338-353. (1965).
24. Zadeh L. A.: The concept of a linguistic variable and its application to approximate reasoning —I. Information sciences, 8(3), 199-249. (1975)

A Gravitational Search Algorithm Using Type-2 Fuzzy Logic for Parameter Adaptation

Beatriz González, Fevrier Valdez and Patricia Melin

Abstract In this paper, we are presenting a modification of the Gravitational Search Algorithm (GSA) using type-2 fuzzy logic to dynamically change the alpha parameter and provide a different gravitation and acceleration to each agent in order to improve its performance. We test this approach with benchmark mathematical functions. Simulation results show the advantage of the proposed approach.

Keywords Type-2 fuzzy logic · Gravitational search algorithm, · GSA · Parameter

1 Introduction

In recent years, it has been shown by many researchers that these algorithms are well suited to solve complex problems. For example, genetic algorithms (GAs) are inspired on Darwinian evolutionary theory [7, 15], ant colony optimization (ACO) mimics the behavior of ants foraging for food [9], and particle swarm optimization (PSO) simulates the behavior of flocks of birds [5, 6]. On the other hand, the bat algorithm (BA) is inspired by the echolocation behavior of microbats [13], etc. In this paper, we consider the gravitational search algorithm (GSA), which is a metaheuristic optimization method based on the laws of gravity and mass interactions [4].

There exists different works concerning the gravitational search algorithm, but only the most important and relevant for this paper will be reviewed [3, 10–12]. There also exists a previous proposal of a fuzzy gravitational search algorithm (FGSA), and its application to the optimal design of multimachine power system

B. González · F. Valdez · P. Melin (✉)
Tijuana Institute of Technology, Calzada Tecnologico s/n, Tijuana, Mexico
e-mail: pmelin@tectijuana.mx

B. González
e-mail: betygm8@hotmail.com

F. Valdez
e-mail: fevrier@tectijuana.mx

P. Melin et al. (eds.), *Nature-Inspired Design of Hybrid Intelligent Systems*,
Studies in Computational Intelligence 667, DOI 10.1007/978-3-319-47054-2_8

stabilizers (PSSs). The FGSA based-PSS design is validated for two multimachine systems: a 3-machine 9-bus system and a 10-machine 39-bus. In this case, fuzzy logic is used to speed up the final stages of the process and find the accurate results in a shorter time, even for very large problems. In this proposed GSA, the fuzzy-based mechanism and fitness sharing are applied to aid the decision-maker to choose the best compromise solution from the Pareto front [2]. However, we proposed in this work in the, type-2 fuzzy logic in the gravitational search algorithm (FGSA) to dynamically change the alpha parameter values and provide a different gravitation and acceleration values to each agent in order to improve its performance, which was originally presented with type-1 fuzzy logic in [1].

The meaning of optimization is finding the parameter values in a function that makes a better solution. All appropriate values are possible solutions and the best value is the optimal solution [8].

The rest of the paper describe this approach in detail and is organized as follows. In Sect. 2, we show some of the previous works and basic concepts. In Sect. 3 experimental results are presented and in Sect. 4 the conclusions are offered.

2 Background and Basic Concepts

2.1 Previous Work

In the work of Rashedi et al., Hossein Nezamabadi-pour, Saeid Saryazdi they proposed a GSA: A Gravitational Search Algorithm as a new optimization algorithm based on the law of gravity and mass interactions [4]. In the proposed algorithm, the searcher agents are a collection of masses which interact with each other based on the Newtonian gravity and the laws of motion.

In their work, Sombra et al., M. Fevrier Valdez, Patricia Melin and Oscar Castillo proposed a new gravitational search algorithm using fuzzy logic to parameter adaptation. In their work they proposed a new Gravitational Search Algorithm (GSA) using fuzzy logic to change alpha parameter and give a different gravitation and acceleration to each agent in order to improve its performance We use this new approach for mathematical functions and present a comparison with original approach [1].

2.2 The Law of Gravity and Second Motion Law

Isaac Newton proposed the law of gravity stating that "The gravitational force between two particles is directly proportional to the product of their masses and inversely proportional to the square of the distance between them" [4]. The gravity force is present in each object in the universe and its behavior is called "action at a distance," this means gravity acts between separated particles without any intermediary and without any delay. The gravity law is represented by the following equation:

$$F = G\frac{M_1 M_2}{R^2} \tag{1}$$

where:

F	is the magnitude of the gravitational force,
G	is gravitational constant,
M_1 and M_2	are the mass of the first and second particles, respectively and,
R	is the distance between the two particles

The Gravitational search algorithm furthermore to be based on Newtonian gravity law is also based on Newton's second motion law, which says "The acceleration of an object is directly proportional to the net force acting on it and inversely proportional to its mass" [4]. The second motion law is represented by the following equation:

$$\alpha = \frac{F}{M} \tag{2}$$

where:

α	is the magnitude of acceleration,
F	is the magnitude of the gravitational force and,
M	is the mass of the object.

2.3 Gravitational Search Algorithm

The approach was proposed by E. Rashedi et al., where they introduce a new algorithm for finding the best solution in problem search spaces using physical rules. Based on populations, at the same time it takes as fundamental principles the law of gravity and second motion law, its principal features are that agents are considered as objects, and their performance is measured by their masses; all these objects are attracted to each other by the gravity force, and this force causes a global movement of all objects; the masses cooperate using a direct form of communication; through gravitational force, an agent with heavy mass corresponds to a good solution therefore it moves more slowly than lighter ones. Finally its gravitational and inertial masses are determined using a fitness function [4].

We can notice that in Eq. (1) the gravitational constant G appears, this is a physic constant which determines the intensity of the gravitational force between the objects and it is defined as a very small value. The equation by which G is defined is given as follows:

$$G(t) = G(t_0)\, x \left(\frac{t_0}{t}\right)^{\beta},\ \beta < 1 \tag{3}$$

where:

$G\ (t)$ is the value of the gravitational constant at time t and,
$G_O\ (t)$ is the value of the gravitational constant at the first cosmic quantum-interval of time t_O

The way in which the position of a number N of agents is represented by

$$X_i = \left(X_i^1, \ldots, X_i^d, \ldots, X_i^n\right) \text{ for } i = 1, 2, \ldots, N, \tag{4}$$

where X_i^d presents the position of the ith agent in the dth dimension.

Now, Eq. (1) with new concepts of masses is defined as following: the force acting on mass i from mass j in a specific time t, is

$$F_{ij}^d(t) = G(t)\frac{M_{pi}(t)\, x\, M_{aj}}{R_{ij}(t) + \varepsilon}\left(x_j^d(t) - d_j^d(t)\right) \tag{5}$$

where M_{aj} is the active gravitational mass related to agent j, M_{pi} is the passive gravitational mass related to agent I, $G(t)$ is gravitational constant at time t, ε is a small constant, and $R_{ij}(t)$ is the Euclidian distance between two agents i and j

$$R_{ij}(t) = \left\|X_i(t), X_j(t)\right\| \tag{6}$$

The stochastic characteristic of this algorithm is based on the idea of the total force that acts on agent i in a dimension d be a randomly weighted sum of dth components of the forces exerted from other agents,

$$F_i^d(t) = \sum_{j=i, j \neq i}^{N} \text{rand}_j\, F_{ij}^d(t) \tag{7}$$

where rand_j is a random number in the interval [0, 1]. The acceleration now is expressed as

$$a_i^d(t) = \frac{F_i^d(t)}{M_{ii}(t)} \tag{8}$$

where M_{ii} is the inertial mass of ith agent. To determine the velocity of an agent we consider that is as a fraction of its current velocity added to its acceleration.

$$v_i^d(t+1) = \text{rand}_i \, x \, v_i^d(t) + a_i^d(t) \tag{9}$$

The position of agents could be calculated as the position in a specific time t added to its velocity in a time t + 1 as follows:

$$x_i^d(t+1) = x_i^d(t) + v_i^d(t+1) \tag{10}$$

In this case, the gravitational constant G is initialized at the beginning and will be reduced with time to control the search accuracy. Its equation is

$$G(t) = G(G_0, t) \tag{11}$$

This is because G is a function of the initial value G_0 and time t. As mentioned previously, gravitational and inertia masses are simply calculated by the fitness evaluation and a heavier mass means a more efficient agent. The update of the gravitational and inertial masses is performed with the following equations:

$$M_{ai} = M_{pi} = M_{ii} = M_i, \; i = 1, 2, \ldots, N, \tag{12}$$

$$m_i(t) = \frac{\text{fit}_i(t) - \text{worst}(t)}{\text{best}_t(t) - \text{worst}(t)} \tag{13}$$

$$M_i(t) = \frac{m_i(t)}{\sum_{j=1}^{N} m_i(t)} \tag{14}$$

the fitness value of the agent i at time t is defined by fit$_i$(t), and best(t) and worst (t) are represented as

$$\text{best}(t) = \min_{j \in \{1,\ldots,N\}} \text{fit}_j(t) \tag{15}$$

$$\text{worst}(t) = \max_{j \in \{1,\ldots,N\}} \text{fit}_j(t) \tag{16}$$

If we want to use GSA for a maximization problem, we only have to change Eqs. (15) and (16) as following:

$$\text{best}(t) = \max_{j \in \{1,\ldots,N\}} \text{fit}_j(t) \tag{17}$$

$$\text{worst}(t) = \min_{j \in \{1,\ldots,N\}} \text{fit}_j(t) \tag{18}$$

The gravitational search algorithm has a kind of elitism in order that only a set of agents with the bigger mass apply their force to the other. This is with an objective to have a balance between exploration and exploitation with lapse of time that is achieved by only the *Kbest* agents that will attract the others; *Kbest* is a function of time, with the initial value K_0 at the beginning and decreasing with time. In such a way, at the beginning, all agents apply the force, and as time passes; *Kbest* is decreased linearly and at the end there will be just one agent applying force to the others. For this reason Eq. (7), can be modified as follows:

$$F_i^d(t) = \sum_{j \in Kbest,\, j \neq 1} \text{rand}_i\, F_{ij}^d(t) \tag{19}$$

where *Kbest* is the set of first *K* agents with the best fitness value and largest mass. A better representation of the GSA process is shown in Fig. 1, and is the principle of this algorithm.

First, an initial population is generated, next the fitness of each agent is evaluated, thereafter we update the gravitational constant *G*, *best* and *worst* of the population; the next step is calculating the mass and acceleration of each agent, until meeting end of iterations. In this case, the maximum of iterations then returns the best solution, else executes the same steps starting from the fitness evaluation. It is in the third step, where we apply the modification in this algorithm, we propose changing alpha parameter to update *G* and help to GSA a better performance.

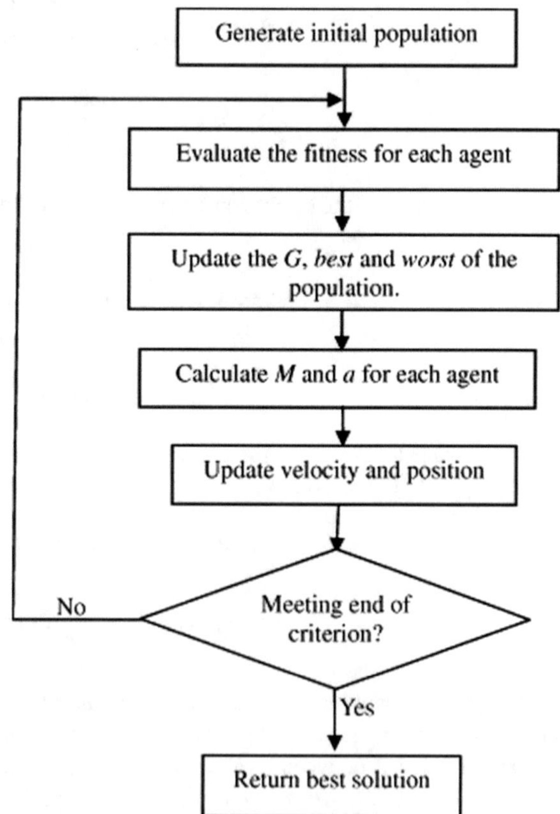

Fig. 1 General principle of GSA. Taken of [4]

Fig. 2 Proposal for change alpha using type-2 fuzzy logic

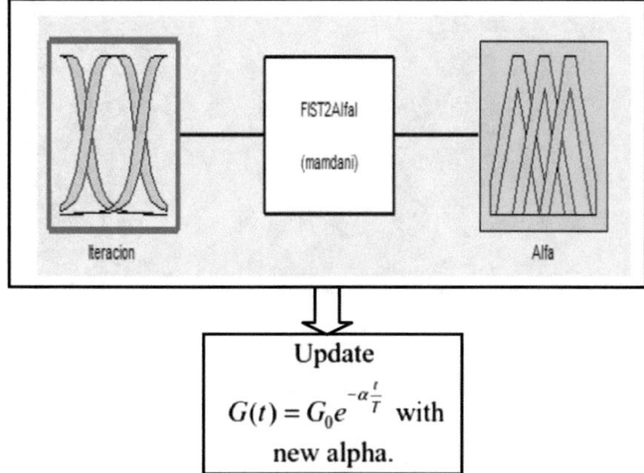

Now we have focused the third block, particularly the update of *G*, Fig. 2 is a representation of our idea, where we use a fuzzy system to obtain the values of alpha and thus updating the value of *G*,

If we change the alpha value along the iterations, we can make the algorithm apply a different *G* value and at the same time influence the gravitational force to each of the agents and finally change its acceleration providing an opportunity to agents to explore others good solutions in the search space and improve the final result.

But otherwise, the variables that determine the gravitational constant *G* (alpha and G_0) affect significantly the performance of the algorithm because they work jointly, since once *G* was established we can use it to determine the magnitude of the gravity force and at the same time determines the agent acceleration. It has chosen the alpha parameter to be modified throughout the execution of the algorithm because Eq. (11) in this case was defined as follows

$$G(t) = G_0 \, \mathrm{e}^{-\alpha t / T} \tag{20}$$

and alpha as a very small negative exponent impacts greatly the result of *G* for being just an exponent. Meanwhile G_0 is considered an initial value of *G*, so that, effectively has impact in the final result, but it is not as drastic change as it is, if we make a change in alpha value.

We have as input variable the elapsed number of iterations and it was granulated as follows: (Fig. 3).

Otherwise the output is the alpha value, and it was granulated as follows (Fig. 4).

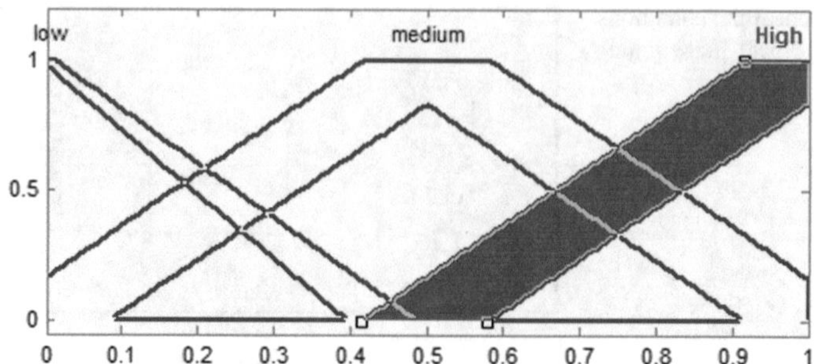

Fig. 3 Fuzzy system input for increase alpha value

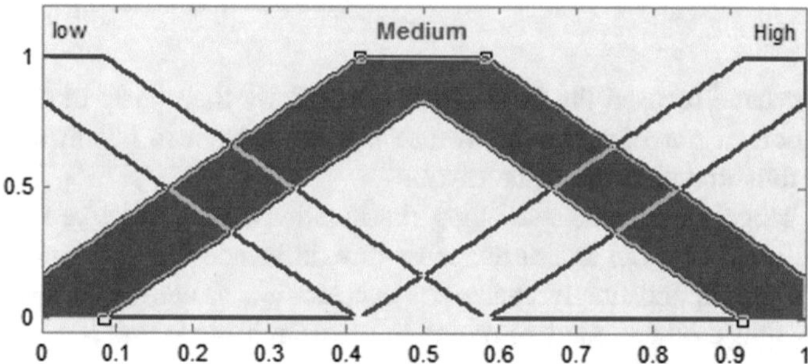

Fig. 4 Fuzzy system output for increase alpha value

The rules are

1. If (Iteration is Low) then (Alpha is Low)
2. If (Iteration is Medium) then (Alpha is Medium)
3. If (Iteration is High) then (Alpha is High).

With this new approach, the rules were designed based on, "if iterations low alpha value should be low, because we need more gravity and acceleration for better search space exploration." Moreover, in high iterations we need a large alpha value to reduce the gravitation and acceleration of the agents, in order to exploit the search space.

2.4 Benchmark Functions

In the field of evolutionary computation, it is common to compare different algorithms using a large test set, especially when the test set involves function optimization. We have made a previous study of the functions to be optimized for constructing a test set with benchmark function selection [14] (Table 1).

Table 1 Benchmark functions

Expression	s
$F_1(x) = \sum_{i=1}^{n} x_i^2$	$[-100, 100]^n$
$F_2(x) = \sum_{1=1}^{n} \lvert x_i \rvert + \prod_{i=1}^{n} \lvert x_i \rvert$	$[-10, 10]^n$
$F_3(x) = \sum_{i=1}^{n} \left(\sum_{j=1}^{i} x_j \right)^2$	$[-100, 100]^n$
$F_4(x) = \max\{\lvert x_i \rvert, \ 1 \le i \le n\}$	$[-100, 100]^n$
$F_5(x) = \sum_{i=1}^{n-1} [100\,(x_{x+1} - x_i^2)^2 + (x_i - 1)^2]$	$[-30, 30]^n$
$F_6(X) = \sum_{i=1}^{n} ([x_i + 0.5])^2$	$[-100, 100]^n$
$F_7(x) = \sum_{i=1}^{n} ix_i^4 + \text{random}\,[0,\ 1]$	$[-1.28, 1.28]^n$
$F_8(x) = \sum_{i=1}^{n} -x_i \sin\left(\sqrt{\lvert x_i \rvert}\right)$	$[-500, 500]^n$
$F_9(x) = \sum_{i=1}^{n} [x_i^2 - 10\cos(2\pi x_i) + 10]$	$[-5.12, 5.12]^n$
$F_{10}(x) = -20\exp\left(-0.2\sqrt{\frac{1}{n}\sum_{i=1}^{n} x_i^2}\right) - \exp\left(\frac{1}{n}\sum_{i=1}^{n}\cos(2\pi x_i)\right) + 20 + e$	$[32, 32]^n$

3 Experimental Results

We made 30 experiments for each function on the same conditions as the original proposal, with number of agents = 50, maximum of iterations = 1000, dimensions = 30, G_0 = 100 and a = 20. We show a comparative Table 2 with the average of the best-so-far solution for original GSA, FGSA with type-1 fuzzy logic, FGSA with type-2 fuzzy logic and PSO.

Table 2 Comparative table with the average of the best-so-far solution for original GSA, FGSA with type-1 fuzzy logic, FGSA with type-2 fuzzy logic. and PSO

Function	GSA [26]	FGSA_T1_I [29]	FGSA_T2_I	PSO [26]
F1	7.3×10^{-11}	8.85×10^{-34}	2.4×10^{-17}	1.8×10^{-3}
F2	4.03×10^{-5}	1.15×10^{-10}	2.41×10^{-08}	2
F3	$0.16 \times 10^{+3}$	$4.68 \times 10^{+2}$	$2.008 \times 10^{+02}$	$4.1 \times 10^{+3}$
F4	3.7×10^{-6}	0.0912	3.3×10^{-09}	8.1
F5	25.16	61.2473	26.66	$3.6 \times 10^{+4}$
F6	8.3×10^{-11}	0.1	0	1.0×10^{-3}
F7	0.018	0.0262	0.017	0.04
F8	$-2.8 \times 10^{+3}$	-2.6×10^{3}	$-2.7 \times 10^{+03}$	$-9.8 \times 10^{+3}$
F9	15.32	17.1796	14.59	55.1
F10	6.9×10^{-6}	6.33×10^{-15}	3.7×10^{-09}	9.0×10^{-3}

We show a comparative Table 3 with the average of the best-so-far solution FGSA with type-2 fuzzy logic with FM triangular (increase), FGSA with type-2 fuzzy logic with FM triangular (Decrement), and FGSA with type-2 fuzzy logic with FM Gauss (increase).

We show a comparative Table 4 with the average of the best-so-far solution FGSA with type-2 fuzzy logic with FM triangular (increase), FGSA with type-2 fuzzy logic with FM triangular (Decrement), and FGSA with type-2 fuzzy logic with FM Gauss (increase).

We show a comparative Table 5 with the average of the best-so-far solution FGSA with type-2 fuzzy logic with FM triangular (increase), FGSA with type-2 fuzzy logic with FM triangular (Decrement), and FGSA with type-2 fuzzy logic with FM Gauss (increase).

We show a comparative Table 6 with the average of the best-so-far solution FGSA with type-2 fuzzy logic with FM triangular (increase), FGSA with type-2 fuzzy logic with FM triangular (Decrement), and FGSA with type-2 fuzzy logic with FM Gauss (increase).

Table 3 Rastrigin function

Dimensions	FGSA_Inc_Triang	FGSA_Dec_Triang	FGSA__Inc_FM_Gauss
30	14.5	**12.93**	14.59
100	**71.31**	72	75.5

Table 4 Rosenbrock function

Dimensions	FGSA_Inc_Triang	FGSA_Dec_Triang	FGSA__Inc_FM_Gauss
30	26.6	**26.09**	26.10
100	$7.84 \times 10^{+02}$	**26.10**	$1.758 \times 10^{+3}$

Table 5 Ackley function

Dimensions	FGSA_Inc_Triang	FGSA_Dec_Triang	FGSA__Inc_FM_Gauss
30	3.7×10^{-09}	3.6×10^{-09}	$\mathbf{3.52 \times 10^{-09}}$
100	$\mathbf{5.9 \times 10^{-01}}$	1.2	9.76×10^{-01}

Table 6 Sum squared function

Dimensions	FGSA_Inc_Triang	FGSA_Dec_Triang	FGSA_Inc_FM_Gauss
30	2.41×10^{-08}	2.46×10^{-08}	$\mathbf{2.38 \times 10^{-08}}$
100	$\mathbf{4.74 \times 10^{-01}}$	1.20	1.08

4 Conclusion

As can be seen in our experiments with different gravitational search algorithm modifications, we can improve the results in mathematical functions; therefore we check that a change in alpha value along algorithm iterations helps in a better convergence. We realize that the idea of changing the alpha value and specifically with a fuzzy system to increase its value is a good concept because in some cases the FGSA with type-2 fuzzy provides a better solution than the traditional algorithm. This is because when you make a change in the alpha value, recalculates the value of the gravitational constant G, which in turn changes the force of gravity between agents and thereafter changes its acceleration. Once we do the above, we achieved to give more opportunity for agents to accelerate them in another direction and could be able to be attracted by another agent with a heavy mass and reach a better solution. A comparison with the average of the best-so-far solution for original GSA, FGSA with logic fuzzy type1, FGSA with type-2 fuzzy logic and PSO applied to Benchmark mathematical functions using 30 dimensions. The numerical results indicated that FGSA with type-2 fuzzy logic method offers much higher performance to update the existing methods on optimization problems. A comparison with the average of the best-so-far solution FGSA with type-2 fuzzy logic with FM triangular (increase), FGSA with type-2 fuzzy logic with FM triangular (Decrement), and FGSA with type-2 fuzzy logic with FM Gauss (increase) applied to Benchmark mathematical functions using 30 and 100 dimensions.

Acknowledgments We would like to express our gratitude to CONACYT, Tijuana Institute of Technology for the facilities and resources granted for the development of this research.

References

1. A. Sombra, F. Valdez, P. Melin, A new gravitational search algorithm using fuzzy logic to parameter adaptation, in: IEEE Congress on Evolutionary Computation, Cancun, México, 2013, pp. 1068–1074.
2. A. Ghasemi, H. Shayeghi, H. Alkhatib, Robust design of multimachine power system stabilizers using fuzzy gravitational search algorithm, Int. J. Electr. Power Energy Syst. 51 (2013)190–200.
3. Dowlatshahi, M., & Nezamabadi-pour, H. (2014). GGSA: A grouping gravitational search algorithm for data clustering. Engineering Applications of Artificial Intelligence, 36, 114–121.
4. E. Rashedi, H. Nezamabadi-pour, S. Saryazdi, GSA: a gravitational search algorithm, Inf. Sci.179(13)(2009)2232–2248.
5. F.V.D. Bergh, A.P. Engelbrecht, A study of particle swarm optimization particle trajectories, Inf. Sci. 176(2006), 937–971.
6. J. Kennedy, R.C. Eberhart, Particle swarm optimization, Proc. IEEE Int. Conf. Neural Netw. 4 (1995) 1942–1948.
7. K.S. Tang, K.F. Man, S. Kwong, Q. He, Genetic algorithms and their applications, IEEE Signal Process. Mag. 13(6)(1996)22–37.

8. Liu, Y., Passino, K.M.: Swarm intelligence: a survey. In: International Conference of Swarm Intelligence (2005)
9. M. Dorigo, V. Maniezzo, A. Colorni, The ant system: optimization by a colony of cooperating agents, IEEE Trans. Syst., Man, Cybern. B 26 (1) (1996) 29–41.
10. M. Dowlatshahi, H. Nezamabadi, M. Mashinchi, A discrete gravitational search algorithm for solving combinatorial optimization problems, Inf. Sci. 258 (2014) 94–107.
11. S. Mirjalili, S. Mohd, H. Moradian, Training feedforward neural networks using hybrid particle swarm optimization and gravitational search algorithm, Appl. Math. Comput. 218(22) (2012)11125–11137.
12. S. Yazdani, H. Nezamabadi, S. Kamyab, A gravitational search algorithm for multimodal optimization, Swarm Evol. Comput. 14 (2014) 1–14.
13. X. Yang, Bat algorithm: a novel approach for global engineering optimization, Eng. Comput.: Int. J. Comput. Aided Eng. Softw. 29 (5) (2012) 464–483.
14. X. Yao, Y. Liu, G. Lin, Evolutionary programming made faster, IEEE Transactions on Evolutionary Computation 3 (1999) 82–102.
15. Tang K. S., Man K. F., Kwong S. and He Q., Genetic algorithms and their applications, IEEE Signal Processing Magazine 13 (6) (1996) 22–37.

Part II
Neural Networks Theory and Applications

Particle Swarm Optimization of the Fuzzy Integrators for Time Series Prediction Using Ensemble of IT2FNN Architectures

Jesus Soto, Patricia Melin and Oscar Castillo

Abstract This paper describes the construction of intelligent hybrid architectures and the optimization of the fuzzy integrators for time series prediction; interval type-2 fuzzy neural networks (IT2FNN). IT2FNN used hybrid learning algorithm techniques (gradient descent backpropagation and gradient descent with adaptive learning rate backpropagation). The IT2FNN is represented by Takagi–Sugeno–Kang reasoning. Therefore this TSK IT2FNN is represented as an adaptive neural network with hybrid learning in order to automatically generate an interval type-2 fuzzy logic system (TSK IT2FLS). We use interval type-2 and type-1 fuzzy systems to integrate the output (forecast) of each Ensemble of ANFIS models. Particle Swarm Optimization (PSO) was used for the optimization of membership functions (MFs) parameters of the fuzzy integrators. The Mackey-Glass time series is used to test of performance of the proposed architecture. Simulation results show the effectiveness of the proposed approach.

Keywords Time series · IT2FNN · Particle swarm optimization · Fuzzy integrators

1 Introduction

The analysis of the time series consists of a (usually mathematical) description of the movements that compose it, then building models using movements to explain the structure and predict the evolution of a variable over time [3, 4]. The fundamental procedure for the analysis of a time series is described below

J. Soto · P. Melin (✉) · O. Castillo
Tijuana Institute of Technology, Tijuana, Mexico
e-mail: pmelin@tectijuana.mx

J. Soto
e-mail: jesvega83@gmail.com

O. Castillo
e-mail: ocastillo@tectijuana.mx

P. Melin et al. (eds.), *Nature-Inspired Design of Hybrid Intelligent Systems*,
Studies in Computational Intelligence 667, DOI 10.1007/978-3-319-47054-2_9

1. Collecting data of the time series, trying to ensure that these data are reliable.
2. Representing the time series qualitatively noting the presence of long-term trends, cyclical variations, and seasonal variations.
3. Plot a graph or trend line length and obtain the appropriate trend values using the method of least squares.
4. When seasonal variations are present, obtain these and adjust the data rate to these seasonal variations (i.e., data seasonally).
5. Adjust the seasonally adjusted trend.
6. Represent the cyclical variations obtained in step 5.
7. Combining the results of steps 1–6 and any other useful information to make a prediction (if desired) and if possible discuss the sources of error and their magnitude.

Therefore the above ideas can assist in the important problem of prediction in the time series. Along with common sense, experience, skill and judgment of the researcher, such mathematical analysis can, however, be of value for predicting the short, medium, and long term.

As related work we can mention: Type-1 Fuzzy Neural Network (T1FNN) [15, 18, 19, 29] and Interval Type-2 Fuzzy Neural Network (IT2FNN) [13, 23–25, 44]; type-1 [1, 8, 16, 33, 45] and type-2 [11, 35, 41, 47] fuzzy evolutionary systems are typical hybrid systems in soft computing. These systems combine T1FLS generalized reasoning methods [18, 28, 34, 42, 43, 48, 51] and IT2FLS [21, 30, 46] with neural networks learning capabilities [12, 14, 18, 37] and evolutionary algorithms [2, 5, 9–11, 29, 35–37, 41] respectively.

This paper reports the results of the simulations of three main architectures of IT2FNN (IT2FNN-1, IT2FNN-2 and IT2FNN-3) for integrating a first-order TSK IT2FIS, with real consequents (A2C0) and interval consequents (A2C1), are used. Integration strategies to process elements of TSK IT2FIS are analyzed for each architecture (fuzzification, knowledge base, type reduction, and defuzzification). Ensemble architectures have three choices IT2FNN-1, IT2FNN-2, and IT2FNN-3. Therefore the output of the Ensemble architectures are integrated with a fuzzy system and the MFs of the fuzzy systems are optimized with PSO. The Mackey-Glass time series is used to test the performance of the proposed architecture. Prediction errors are evaluated by the following metrics: root mean square error (RMSE), mean square error (MSE), and mean absolute error (MAE).

In the next section, we describe the background and basic concepts of the Mackey-Glass time series, Interval type-2 fuzzy systems, Interval Type-2 Fuzzy Neural-Networks, and Particle Swarm Optimization. Section 3 presents the general proposed architecture. Section 4 presents the simulations and the results. Section 5 offers the conclusions.

2 Background and Basic Concepts

This section presents the basic concepts that describe the background in time series prediction and basic concepts of the Mackey-Glass time series, Interval type-2 fuzzy systems, Interval Type-2 Fuzzy Neural-Networks, and Particle Swarm Optimization.

2.1 Mackey-Glass Time Series

The problem of predicting future values of a time series has been a point of reference for many researchers. The aim is to use the values of the time series known at a point $x = t$ to predict the value of the series at some future point $x = t + P$. The standard method for this type of prediction is to create a mapping from D points of a Δ spaced time series, is $(x\,(t - (D - 1)\,\Delta) \ldots x\,(t - \Delta), x\,(t))$, to a predicted future value $x\,(t + P)$. To allow a comparison with previous results in this work [11, 19, 29, 41] the values $D = 4$ and $\Delta = P = 6$ were used.

Chaotic time series data used is defined by the Mackey-Glass [26, 27] time series, whose differential equation is given by Eq. (1)

$$x(t) = \frac{0.2x(t - \tau)}{1 - x^{10}(t - \tau)} - 0.1x(t - \tau) \quad (1)$$

For obtaining the values of the time series at each point, we can apply the Runge–Kutta method [17] for the solution of Eq. (1). The integration step was set at 0.1, with initial condition $x(0) = 1.2$, $\tau = 17$, $x(t)$ is then obtained for $0 \leq t \leq 1200$, (Fig. 1) (we assume $x(t) = 0$ for $t < 0$ in the integration).

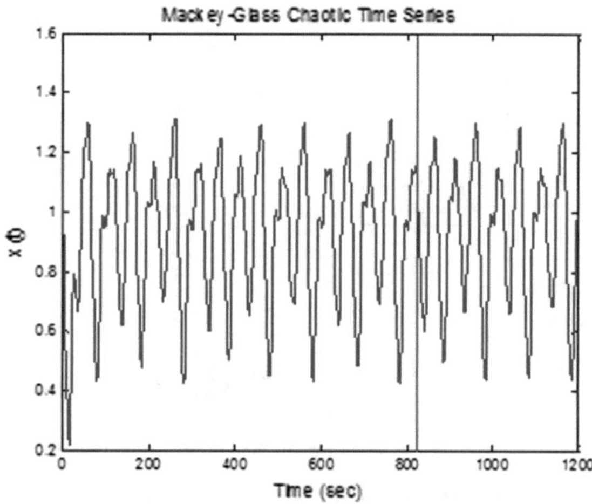

Fig. 1 The Mackey-Glass time series

2.2 Interval Type-2 Fuzzy Systems

Type-2 fuzzy sets are used to model uncertainty and imprecision; originally they were proposed by Zadeh [49, 50] and they are essentially "fuzzy–fuzzy" sets in which the membership degrees are type-1 fuzzy sets (Fig. 2).

The basic structure of a type-2 fuzzy system implements a nonlinear mapping of input to output space. This mapping is achieved through a set of type-2 if-then fuzzy rules, each of which describes the local behavior of the mapping.

The uncertainty is represented by a region called footprint of uncertainty (FOU). When $\mu_{\tilde{A}}(x, u) = 1, \forall u \in l_x \subseteq [0, 1]$; we have an interval type-2 membership function [5, 7, 20, 31] (Fig. 3).

The uniform shading for the FOU represents the entire interval type-2 fuzzy set and it can be described in terms of an upper membership function $\bar{\mu}_{\tilde{A}}(x)$ and a lower membership function $\underline{\mu}_{\tilde{A}}(x)$.

A fuzzy logic systems (FLS) described using at least one type-2 fuzzy set is called a type-2 FLS. Type-1 FLSs are unable to directly handle rule uncertainties, because they use type-1 fuzzy sets that are certain [6, 7, 46]. On the other hand, type-2 FLSs are very useful in circumstances where it is difficult to determine an exact certainty value, and there are measurement uncertainties.

2.3 Interval Type-2 Fuzzy Neural Networks (IT2FNN)

One way to build interval type-2 fuzzy neural networks (IT2FNN) is by fuzzifying a conventional neural network. Each part of a neural network (the activation function, the weights, and the inputs and outputs) can be fuzzified. A fuzzy neuron is

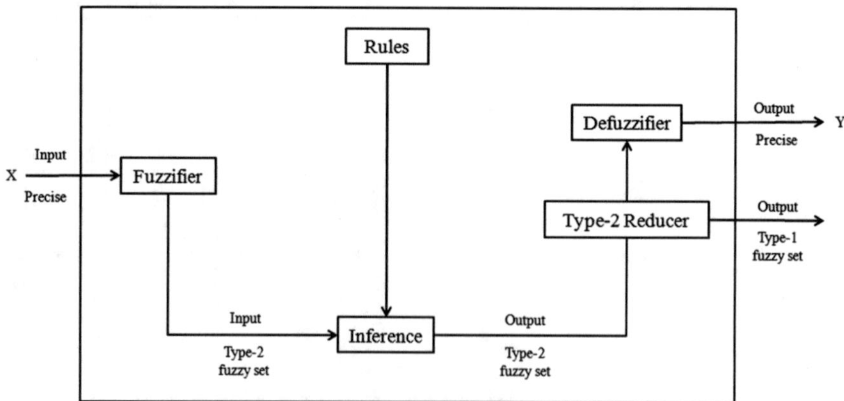

Fig. 2 Basic structure of the interval type-2 fuzzy logic system

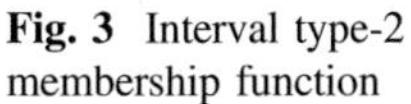

Fig. 3 Interval type-2 membership function

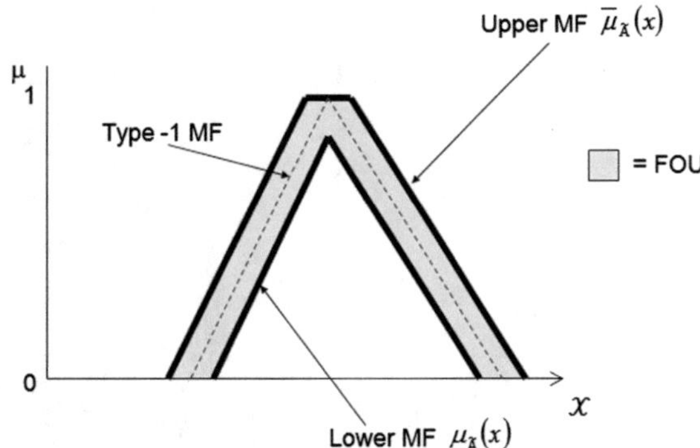

basically similar to an artificial neuron, except that it has the ability to process fuzzy information.

The interval type-2 fuzzy neural network (IT2FNN) system is one kind of interval Takagi–Sugeno–Kang fuzzy inference system (IT2-TSK-FIS) inside neural network structure. An IT2FNN is proposed by Castro [6], with TSK reasoning and processing elements called interval type-2 fuzzy neurons (IT2FN) for defining antecedents, and interval type-1 fuzzy neurons (IT1FN) for defining the consequents of rules.

An IT2FN is composed by two adaptive nodes represented by squares, and two non-adaptive nodes represented by circles. Adaptive nodes have outputs that depend on their inputs, modifiable parameters, and transference function while non-adaptive, on the contrary, depend solely on their inputs, and their outputs represent lower $\underline{\mu}_{\tilde{A}}(x)$ and upper $\bar{\mu}_{\tilde{A}}(x)$ membership functions (Fig. 4).

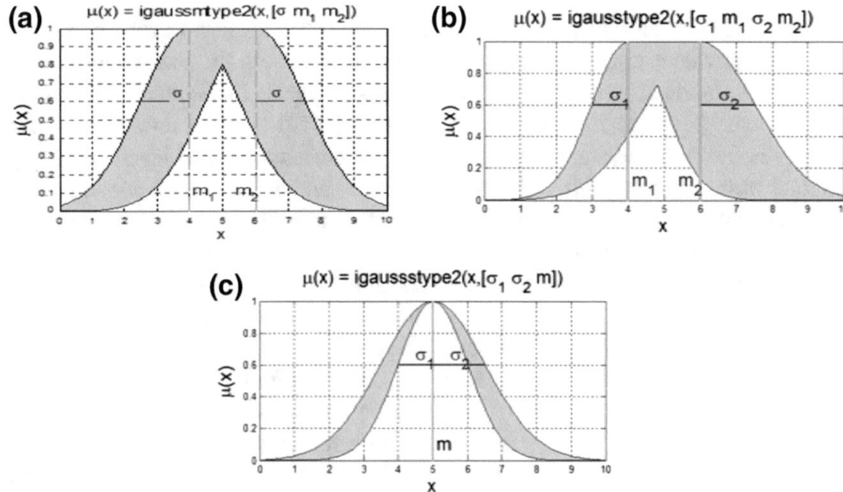

Fig. 4 The MFs used for training the IT2FNN architecture

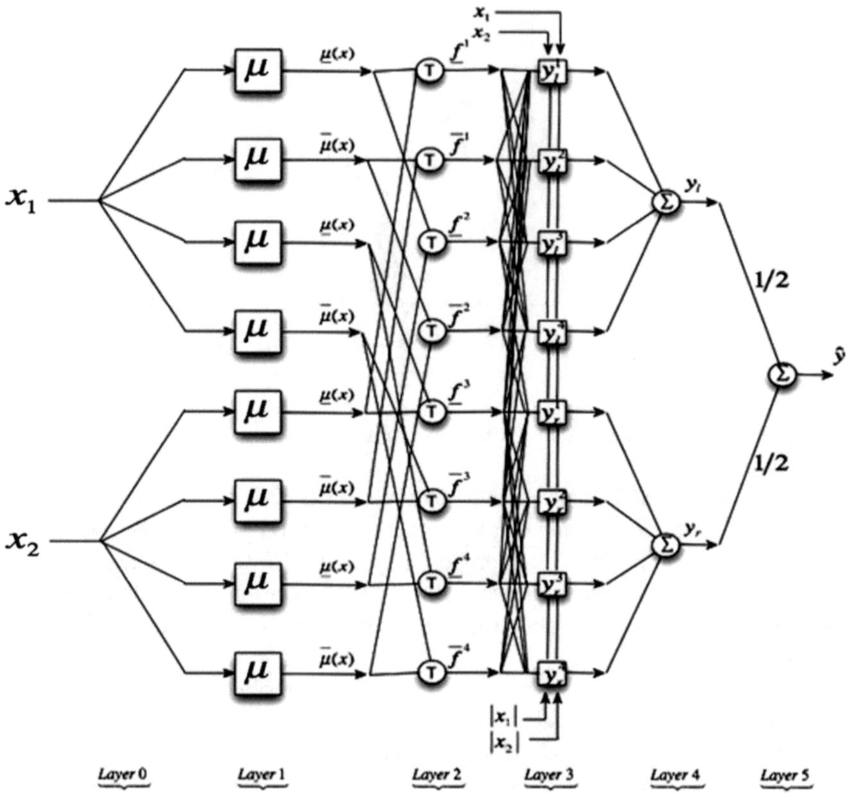

Fig. 5 IT2FNN1 architecture

The IT2FNN-1 architecture has five layers (Fig. 5), consists of adaptive nodes with an equivalent function to lower-upper membership in fuzzification layer (layer 1). Non-adaptive nodes in the rules layer (layer 2) interconnect with fuzzification layer (layer 1) in order to generate TSK IT2FIS rules antecedents. The adaptive nodes in consequent layer (layer 3) are connected to input layer (layer 0) to generate rules consequents. The non-adaptive nodes in type-reduction layer (layer 4) evaluate left-right values with KM algorithm [19–21]. The non-adaptive node in defuzzification layer (layer 5) average left-right values.

The IT2FNN-2 architecture has six layers (Fig. 6 and uses IT2FN for fuzzifying inputs (layers 1–2). The non-adaptive nodes in the rules layer (layer 3) interconnect with lower-upper linguistic values layer (layer 2) to generate TSK IT2FIS rules antecedents. The non-adaptive nodes in the consequents layer (layer 4) are connected with the input layer (layer 0) to generate rule consequents. The non-adaptive nodes in type-reduction layer (layer 5) evaluate left-right values with KM algorithm. The non-adaptive node in defuzzification layer (layer 6) averages left-right values.

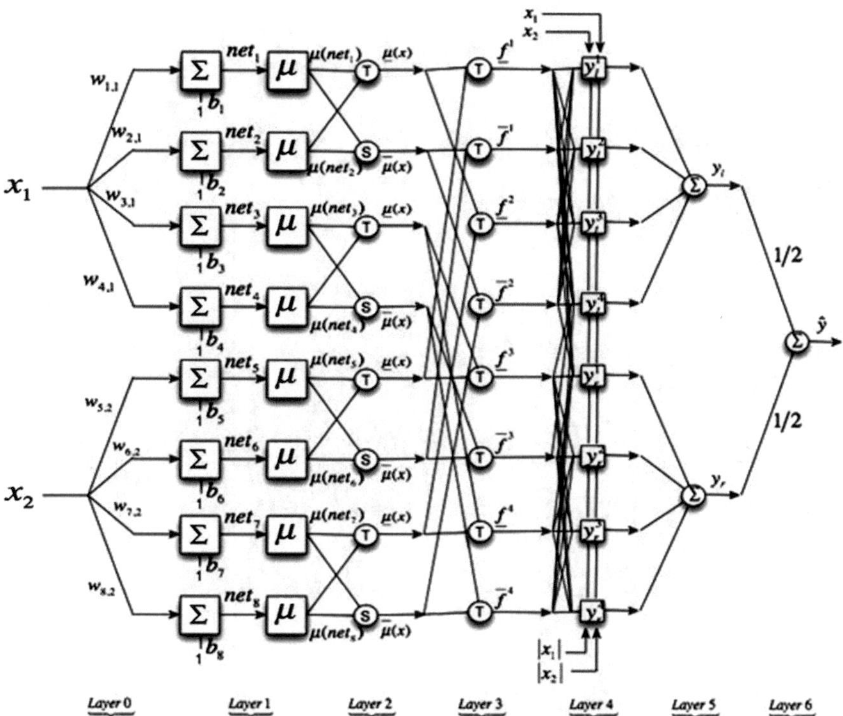

Fig. 6 ITFNN2 architecture

IT2FNN-3 architecture has seven layers (Fig. 7). Layer 1 has adaptive nodes for fuzzifying inputs; layer 2 has non-adaptive nodes with the interval fuzzy values. Layer 3 (rules) has non-adaptive nodes for generating firing strength of TSK IT2FIS rules. Layer 4, lower and upper values the rules firing strength are normalized. The adaptive nodes in layer 5 (consequent) are connected to layer 0 for generating the rules consequents. The non-adaptive nodes in layer 6 evaluate values from left-right interval. The non-adaptive node in layer 7 (defuzzification) evaluates average of interval left-right values.

2.4 Particle Swarm Optimization

Particle Swarm Optimization (PSO) is a metaheuristic search technique based on a population of particles (Fig. 8). The main idea of PSO comes from the social behavior of schools of fish and flocks of birds [22, 32]. In PSO, each particle moves

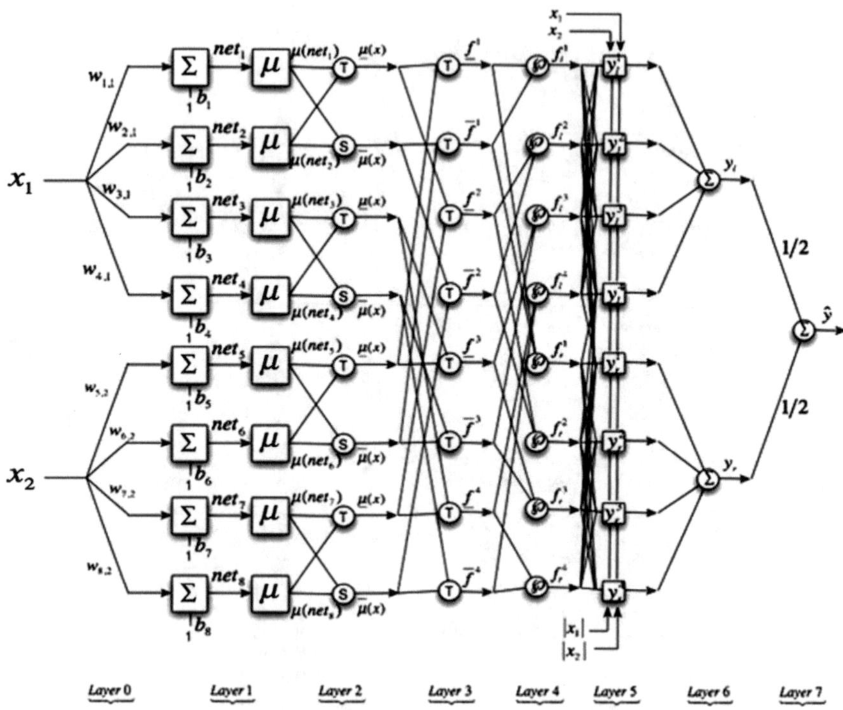

Fig. 7 IT2FNN3 architecture

in a D-dimensional space based on its own past experience and those of other particles. Each particle has a position and a velocity represented by the vectors $x_i = (x_{i1}, x_{i2}, \ldots, x_{iD})$ and $V_i = (v_{i1}, v_{i2}, \ldots, v_{iD})$ for the i-th particle. At each iteration, particles are compared with each other to find the best particle [32, 38]. Each particle records its best position as $P_i = (p_{i1}, p_{i2}, \ldots, p_{iD})$. The best position of all particles in the swarm is called the global best, and is represented as $G = (G_1, G_2, \ldots, G_D)$. The velocity of each particle is given by Eq. (2).

$$V_{id} = w v_{id} + C_1 \cdot \text{rand}_1() \cdot (p\text{best}_{id} - x_{id}) + C_2 \cdot \text{rand}_2() \ldots \cdot (g\text{best} - x_{id}) \quad (2)$$

In this equation, i = 1, 2, …, M, d = 1, 2, …,D, C_1 and C_2 are positive constants (known as acceleration constants), $\text{rand}_1()$ and $\text{rand}_2()$ are random numbers in [0,1], and w, introduced by Shi and Eberhart [39] is the inertia weight. The new position of the particle is determined by Eq. (3)

$$x_{id} = x_{id} + v_{id} \quad (3)$$

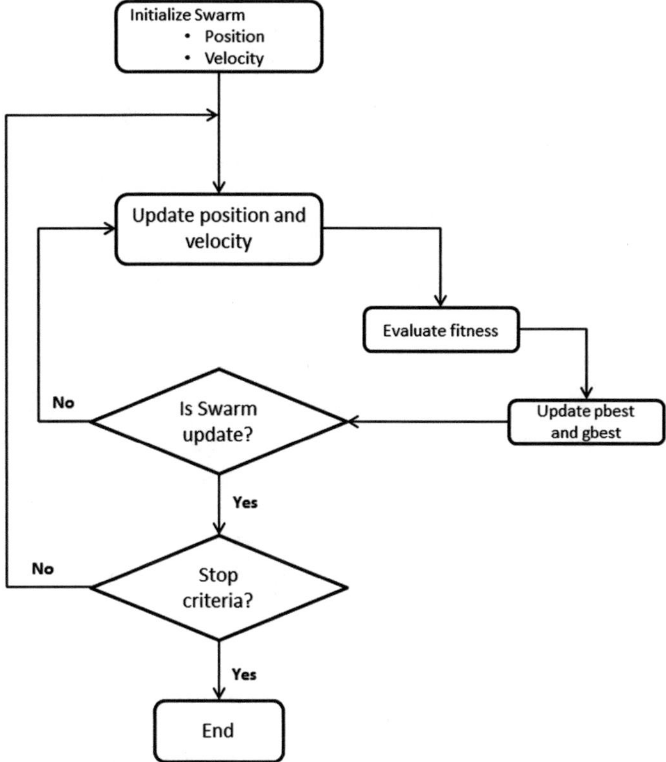

Fig. 8 Flowchart of the PSO algorithm

3 Problem Statement and Proposed Architecture

The general proposed architecture combines the ensemble of IT2FNN models and the use of fuzzy systems as response integrators using PSO for time series prediction (Fig. 9).

This architecture is divided into four sections, where the first phase represents the database to simulate in the Ensemble [40] of IT2FNN, which in this case is the historical data of the Mackey-Glass [26, 27] time series. From the Mackey-Glass time series we used 800 pairs of data points (Fig. 1), similar to [35, 36].

We predict $x(t)$ from three past (delays) values of the time series, that is, $x(t-18)$, $x(t-12)$, and $x(t-6)$. Therefore the format of the training and checking data is

$$\lfloor x(t-18), x(t-12), x(t-6); x(t) \rfloor \tag{4}$$

where $t = 19$–818 and $x(t)$ is the desired prediction of the time series.

In the second phase, training (the first 400 pairs of data are used to train the IT2FNN architecture) and validation (the second 400 pairs of data are used to

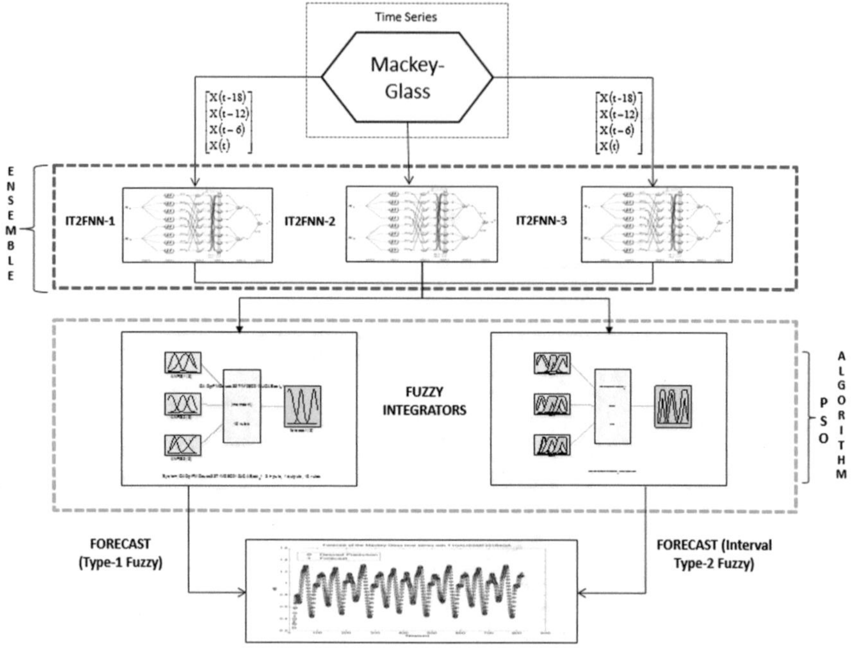

Fig. 9 The general proposed architecture

validate the ITFNN architecture) is performed sequentially in each IT2FNN, in this case we are dealing with a set of 3 IT2FNN (IT2FNN-1, IT2FNN-2, and IT2FNN-3) in the Ensemble. Therefore each IT2FNN architecture has three input variables $(x(t-18), x(t-12), x(t-6))$ and one output variable $(x(t))$ is the desired prediction.

In the fourth phase, we integrate the overall results of each Ensemble of IT2FNN which are (IT2FNN-1, IT2FNN-2, and IT2FNN-3) architecture, and such integration will be done by the fuzzy inference system (type-1 and interval type-2 fuzzy system) of Mamdani type; but each fuzzy integrators will be optimized with PSO of the MFs parameters. Finally the forecast output determined by the proposed architecture is obtained and it is compared with desired prediction.

3.1 Design of the Fuzzy Integrators

The design of the type-1 and interval type-2 fuzzy inference systems integrators are of Mamdani type and have three inputs (IT2FNN1, IT2FNN2, and IT2FNN3) and one output (Forecast), so each input is assigned two MFs with linguistic labels "Small and Large" and the output will be assigned three MFs with linguistic labels "OutIT2FNN1, Out IT2FNN2 and Out IT2FNN3" (Fig. 10) and have eight if-then

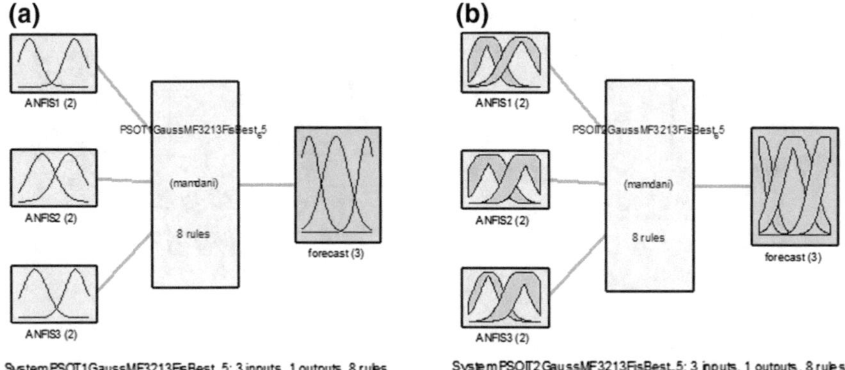

Fig. 10 Structure of the type-1 FIS (**a**) and interval type-2 FIS (**b**) integrators

rules. The design of the if-then rules for the fuzzy inference system depends on the number of membership functions used in each input variable using the system [e.g., our fuzzy inference system uses three input variables which each entry contains two membership functions, therefore the total number of possible combinations for the fuzzy rules is 8 (e.g., 2*2*2 = 8)], therefore we used eight fuzzy rules for the experiments (Fig. 11) because the performance is better and minimized the prediction error of the Mackey-Glass time series.

In the type-1 FIS integrators, we used different MFs (Gaussian, Generalized Bell, and Triangular) Fig. 12a and for the interval type-2 FIS integrators we used different MFs (igaussmtype2, igbelltype2, and itritype2) Fig. 12b [7] to observe the behavior of each of them and determine which one provides better forecast of the time series.

3.2 *Design of the Representation for the Particle Swarm Optimization*

The PSO is used to optimize the parameters values of the MFs in each of the type-1 and interval type-2 fuzzy integrators. The representation in PSO is of Real-Values

1. If (ANFIS1 is small) and (ANFIS2 is small) and (ANFIS3 is small) then (forecast is OutANFIS1) (1)
2. If (ANFIS1 is small) and (ANFIS2 is small) and (ANFIS3 is large) then (forecast is OutANFIS1) (1)
3. If (ANFIS1 is small) and (ANFIS2 is large) and (ANFIS3 is small) then (forecast is OutANFIS2) (1)
4. If (ANFIS1 is small) and (ANFIS2 is large) and (ANFIS3 is large) then (forecast is OutANFIS2) (1)
5. If (ANFIS1 is large) and (ANFIS2 is small) and (ANFIS3 is small) then (forecast is OutANFIS2) (1)
6. If (ANFIS1 is large) and (ANFIS2 is small) and (ANFIS3 is large) then (forecast is OutANFIS2) (1)
7. If (ANFIS1 is large) and (ANFIS2 is large) and (ANFIS3 is small) then (forecast is OutANFIS3) (1)
8. If (ANFIS1 is large) and (ANFIS2 is large) and (ANFIS3 is large) then (forecast is OutANFIS3) (1)

Fig. 11 If-then rules for the fuzzy integrators

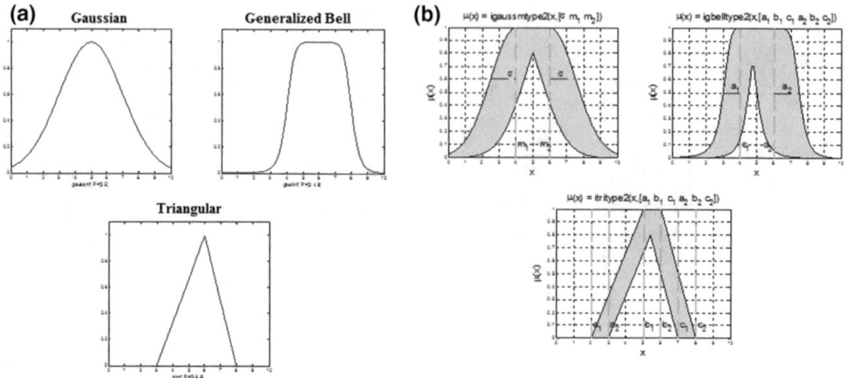

Fig. 12 Type-1 MFs (**a**) and interval type-2 MFs (**b**) for the fuzzy integrators

and the particle size will depend on the number of MFs that are used in each design of the fuzzy integrators.

The objective function is defined to minimize the prediction error as follows in Eq. (5)

$$f(t) = \sqrt{\frac{\sum_{t=1}^{n} (a_t - p_t)^2}{n}} \tag{5}$$

where a, corresponds to the real data of the time series, p corresponds to the output of each fuzzy integrators, t is de sequence time series, and n is the number of data points of time series.

The general representation of the particles represents the utilized membership functions. The number of parameters varies according to the kind of membership function of the type-1 fuzzy system (e.g., two parameter are needed to represent a Gaussian MF's are "sigma and mean") Fig. 13a and interval type-2 fuzzy system (e.g., three parameter are needed to represent "igaussmtype2" MF's are "sigma, mean1 and mean2") Fig. 13b. Therefore the number of parameters that each fuzzy inference system integrator has depends of the MFs type assigned to each input and output variables.

The parameters of particle swarm optimization used for optimizing the type-1 and interval type-2 fuzzy inference systems integrators are shown on Table 1.

We performed experiments in time series prediction, specifically for the Mackey-Glass time series in ensembles of IT2FNN architectures using fuzzy integrators optimized with PSO.

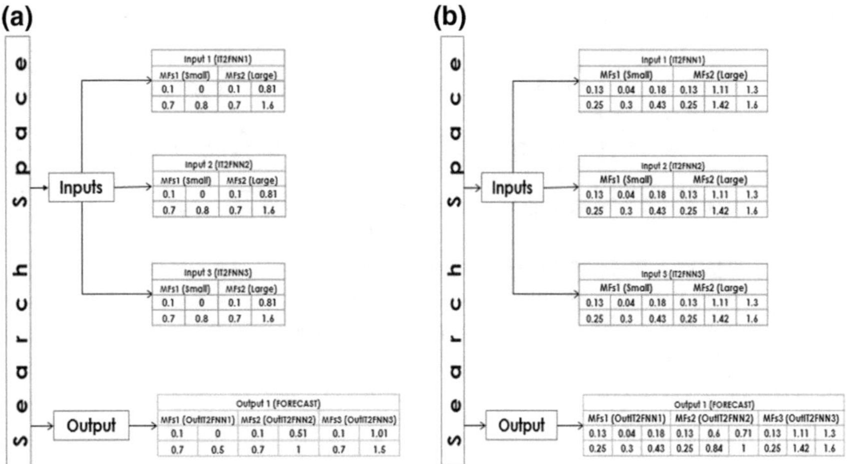

Fig. 13 Representation of the particles structure of the type-1 (**a**) and interval type (**b**) fuzzy integrators

Table 1 Parameters of PSO

Parameters	Value
Particles	100
Iterations	65
Inertia Weight "ω"	Linear decrement [0.88−0]
Constriction "C"	Linear increment [0.01–0.9]
$r1$, $r2$	Random
$c1$	Linear decrement [2–0.5]
$c2$	Linear increment [0.5–2]

4 Simulations Results

This section presents the results obtained through experiments on the architecture for the optimization of the fuzzy integrators in ensembles of IT2FNN architectures for time series prediction, which show the performance that was obtained from each experiment to simulate the Mackey-Glass time series.

The best errors were produced by the type-1 fuzzy integrator (using Generalized Bell MFs) with PSO are shown on Table 2. The RMSE is 0.035228102 and the average RMSE is 0.047356657, the MSE is 0.005989357 and the MAE is 0.056713089, respectively. The MFs optimized with PSO are presented in Fig. 14a, the forecasts in Fig. 14b, and the evolution errors in Fig. 14c are obtained for the proposed architecture.

The best errors were produced by the interval type-2 fuzzy integrator (using igbelltype2 MFs) with PSO are shown on Table 2. The RMSE is 0.023648414 and the average RMSE is 0.024988012, the MSE is 0.00163873 and the MAE is

Table 2 PSO results for the optimization of the fuzzy integrators

Metrics	Type-1 MFs			Interval type-2 MFs		
	Gaussian	Generalized Bell	Triangular	Igaussmtype2	Igbelltype2	Itritype2
RMSE (best)	0.035946912	**0.035228102**	0.0797536	0.024183221	**0.023648414**	0.0251151
RMSE (average)	0.044289015	**0.047356657**	0.0928408	0.026416896	**0.024988012**	0.0286363
MSE	0.008587152	**0.005989357**	0.0147635	0.003288946	**0.00163873**	0.0022031
MAE	0.065247859	**0.056713089**	0.0968261	0.039278738	**0.028366955**	0.0326414
Time (HH: MM:SS)	00:19:06	00:18:10	00:17:33	02:23:06	02:12:10	02:59:21

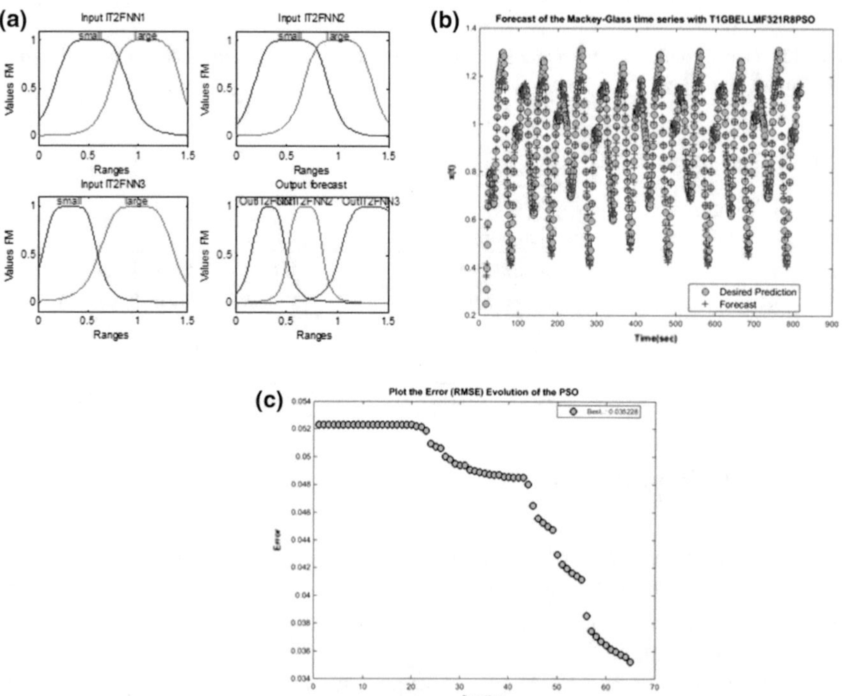

Fig. 14 Variables inputs/output MFs (**a**), forecast (**b**), and evolution errors of "65 iterations" (**c**), are generated for the optimized of the type-1 FIS integrator with PSO

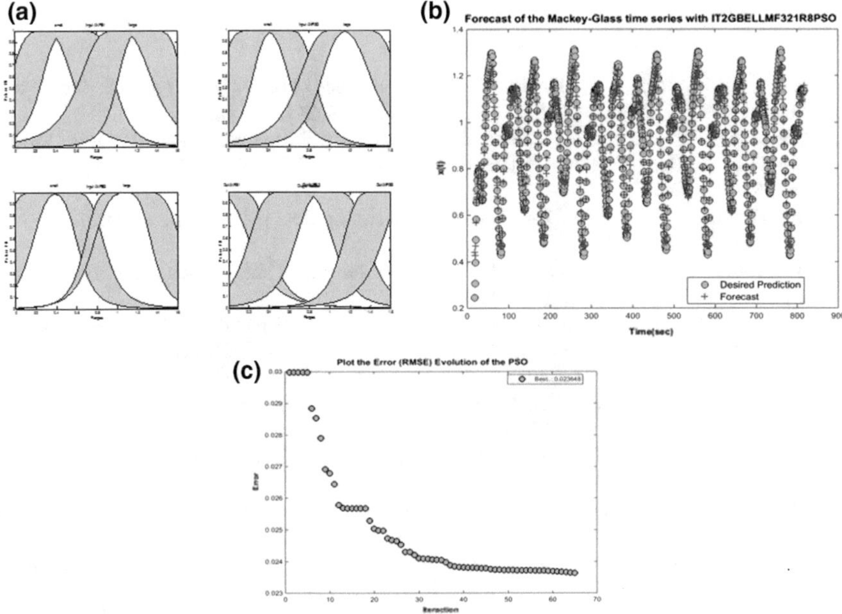

Fig. 15 Variables inputs/output MFs (**a**), forecast (**b**), and evolution errors of "65 iterations" (**c**), are generated for the optimized of the interval type-2 FIS integrator with PSO

0.028366955 respectively. The MFs optimized with PSO are presented in Fig. 15a, the forecasts in Fig. 15b, and the evolution errors in Fig. 15c are obtained for the proposed architecture.

5 Conclusion

Particle swarm optimization of the fuzzy integrators for time series prediction using ensembles of IT2FNN architecture was proposed in this paper.

The best result generated for the optimization the interval type-2 FIS (using igbelltype2 MFs) integrator is with a prediction error of 0.023648414 (98 %).

The best result generated for the optimization of type-1 FIS (using Generalized Bell MFs) integrator with a prediction error of 0.035228102 (97 %).

These results showed efficient results in the prediction error of the time series Mackey-Glass generated by proposed architecture.

References

1. Ascia, G., Catania, V., Panno, D.: An Integrated Fuzzy-GA Approach for Buffer Management. IEEE Trans. Fuzzy Syst. 14(4), pp. 528–541. (2006).
2. Bonissone, P.P., Subbu, R., Eklund, N., Kiehl, T.R.: Evolutionary Algorithms + Domain Knowledge = Real-World Evolutionary Computation. IEEE Trans. Evol Comput. 10(3), pp. 256–280. (2006).
3. Brocklebank J. C., Dickey, D.A.: SAS for Forecasting Series. SAS Institute Inc. Cary, NC, USA, pp. 6-140. (2003).
4. Brockwell, P. D., Richard, A.D.: Introduction to Time Series and Forecasting. Springer-Verlag New York, pp 1-219. (2002).
5. Castillo, O., Melin, P.: Optimization of type-2 fuzzy systems based on bio-inspired methods: A concise review, Information Sciences, Volume 205, pp. 1-19. (2012).
6. Castro J.R., Castillo O., Melin P., Rodriguez A.: A Hybrid Learning Algorithm for Interval Type-2 Fuzzy Neural Networks: The Case of Time Series Prediction. Springer-Verlag Berlin Heidelberg, Vol. 15a, pp. 363-386. (2008).
7. Castro, J.R., Castillo, O., Martínez, L.G.: Interval type-2 fuzzy logic toolbox. Engineering Letters, 15(1), pp. 89–98. (2007).
8. Chiou, Y.-C., Lan, L.W.: Genetic fuzzy logic controller: an iterative evolution algorithm with new encoding method. Fuzzy Sets Syst. 152(3), pp. 617–635. (2005).
9. Deb, K.: A population-based algorithm-generator for real-parameter optimization. Springer, Heidelberg. (2005).
10. Engelbrecht, A.P.: Fundamentals of computational swarm intelligence. John Wiley & Sons, Ltd., Chichester. (2005).
11. Gaxiola, F., Melin, P., Valdez, F., Castillo, O.: Optimization of type-2 fuzzy weight for neural network using genetic algorithm and particle swarm optimization. Nature and Biologically Inspired Computing (NaBIC). World Congress on, vol., no., pp. 22-28. (2013).
12. Hagan, M.T., Demuth, H.B., Beale, M.H.: Neural Network Design. PWS Publishing, Boston. (1996).
13. Hagras, H.: Comments on Dynamical Optimal Training for Interval Type-2 Fuzzy Neural Network (T2FNN). IEEE Transactions on Systems Man And Cybernetics Part B 36(5), pp. 1206–1209. (2006).
14. Haykin, S.: Adaptive Filter Theory. Prentice Hall, Englewood Cliffs. (2002) ISBN 0-13-048434-2.
15. Horikowa, S., Furuhashi, T., Uchikawa, Y.: On fuzzy modeling using fuzzy neural networks with the backpropagation algorithm. IEEE Transactions on Neural Networks 3, (1992).
16. Ishibuchi, H., Nozaki, K., Yamamoto, N., Tanaka, H.: Selecting fuzzy if-then rules for classification problems using genetic algorithms. IEEE Trans. Fuzzy Syst. 3, pp. 260–270. (1995).
17. Jang J.S.R.: Fuzzy modeling using generalized neural networks and Kalman fliter algorithm. Proc. of the Ninth National Conference on Artificial Intelligence. (AAAI-91), pp. 762-767. (1991).
18. Jang, J.S.R., Sun, C.T., Mizutani, E.: Neuro-fuzzy and Soft Computing. Prentice-Hall, New York. (1997).
19. Jang, J.S.R.: ANFIS: Adaptive-network-based fuzzy inference systems. IEEE Trans. on Systems, Man and Cybernetics. Vol. 23, pp. 665-685 (1992).
20. Karnik, N.N., Mendel, J.M., Qilian L.: Type-2 fuzzy logic systems. Fuzzy Systems, IEEE Transactions on. vol.7, no.6, pp. 643,658. (1999).
21. Karnik, N.N., Mendel, J.M.: Applications of type-2 fuzzy logic systems to forecasting of time-series. Inform. Sci. 120, pp. 89–111. (1999).
22. Kennedy, J., Eberhart, R.: Particle swarm optimization. Neural Networks. Proceedings., IEEE International Conference on. vol. 4. pp. 1942-1948. (1995).

23. Lee, C.H., Hong, J.L., Lin, Y.C., Lai, W.Y.: Type-2 Fuzzy Neural Network Systems and Learning. International Journal of Computational Cognition 1(4), pp. 79–90. (2003).
24. Lee, C.-H., Lin, Y.-C.: Type-2 Fuzzy Neuro System Via Input-to-State-Stability Approach. In: Liu, D., Fei, S., Hou, Z., Zhang, H., Sun, C. (eds.) ISNN 2007. LNCS, vol. 4492, pp. 317–327. Springer, Heidelberg (2007).
25. Lin, Y.-C., Lee, C.-H.: System Identification and Adaptive Filter Using a Novel Fuzzy Neuro System. International Journal of Computational Cognition 5(1) (2007).
26. Mackey, M.C., Glass, L.: Oscillation and chaos in physiological control systems. Science, Vol. 197, pp. 287-289. (1997).
27. Mackey, M.C.: Mackey-Glass. McGill University, Canada, http://www.sholarpedia.org/-article/Mackey-Glass_equation, September 5th, (2009).
28. Mamdani, E.H., Assilian, S.: An experiment in linguistic synthesis with a fuzzy logic controller. Int. J. Man-Mach. Stud. 7, pp. 1–13. (1975).
29. Melin, P., Soto, J., Castillo, O., Soria, J.: A New Approach for Time Series Prediction Using Ensembles of ANFIS Models. Experts Systems with Applications. Elsevier, Vol. 39, Issue 3, pp 3494-3506. (2012).
30. Mendel, J.M.: Uncertain rule-based fuzzy logic systems: Introduction and new directions. Ed. USA: Prentice Hall, pp 25-200. (2000).
31. Mendel, J.M.: Why we need type-2 fuzzy logic systems. Article is provided courtesy of Prentice Hall, By Jerry Mendel. (2001).
32. Parsopoulos, K.E., Vrahatis, M.N.: Particle Swarm Optimization Intelligence: Advances and Applications. Information Science Reference. USA. pp. 18-40. (2010).
33. Pedrycz, W.: Fuzzy Evolutionary Computation. Kluwer Academic Publishers, Dordrecht. (1997).
34. Pedrycz, W.: Fuzzy Modelling: Paradigms and Practice. Kluwer Academic Press, Dordrecht. (1996).
35. Pulido M., Melin P., Castillo O.: Particle swarm optimization of ensemble neural networks with fuzzy aggregation for time series prediction of the Mexican Stock Exchange. Information Sciences, Volume 280,, pp. 188-204. (2014).
36. Pulido, M., Mancilla, A., Melin, P.: An Ensemble Neural Network Architecture with Fuzzy Response Integration for Complex Time Series Prediction. Evolutionary Design of Intelligent Systems in Modeling, Simulation and Control, pp. 85-110. (2009).
37. Russell, S., Norvig, P.: Artificial Intelligence: A Modern Approach. Prentice-Hall, NJ. (2003).
38. Shi, Y., Eberhart, R.: A modified particle swarm optimizer. In: Proceedings of the IEEE congress on evolutionary computation, pp. 69-73. (1998).
39. Shi, Y., Eberhart, R.: Empirical study of particle swarm optimization. In: Proceedings of the IEEE congress on evolutionary computation, pp. 1945-1950. (1999).
40. Sollich, P., Krogh, A.: Learning with ensembles: how over-fitting can be useful. in: D.S. Touretzky M.C. Mozer, M.E. Hasselmo (Eds.). Advances in Neural Information Processing Systems 8, Denver, CO, MIT Press, Cambridge, MA, pp. 190-196. (1996).
41. Soto, J., Melin, P., Castillo, O.: Time series prediction using ensembles of ANFIS models with genetic optimization of interval type-2 and type-1 fuzzy integrators. International Journal Hybrid Intelligent Systems Vol. 11(3): pp. 211-226. (2014).
42. Takagi T., Sugeno M.: Derivation of fuzzy control rules from human operation control actions.Proc. of the IFAC Symp. on Fuzzy Information, Knowledge Representation and Decision Analysis, pp. 55-60. (1983).
43. Takagi, T., Sugeno, M.: Fuzzy identification of systems and its applications to modeling and control. IEEE Trans. Syst., Man, Cybern. 15, pp. 116–132. (1985).
44. Wang, C.H., Cheng, C.S., Lee, T.-T.: Dynamical optimal training for interval type-2 fuzzy neural network (T2FNN). IEEE Trans. on Systems, Man, and Cybernetics Part B: Cybernetics 34(3), pp. 1462–1477. (2004).
45. Wang, C.H., Liu, H.L., Lin, C.T.: Dynamic optimal Learning rate of A Certain Class of Fuzzy Neural Networks and Its Applications with Genetic Algorithm. IEEE Trans. Syst. Man, Cybern. 31(3), pp. 467–475. (2001).

46. Wu, D., Mendel, J.M.: A Vector Similarity Measure for Interval Type-2 Fuzzy Sets and Type-1 Fuzzy Sets. Information Sciences 178, pp. 381–402. (2008).
47. Wu, D., Wan Tan, W.: Genetic learning and performance evaluation of interval type-2 fuzzy logic controllers. Engineering Applications of Artificial Intelligence 19(8), pp. 829–841. (2006).
48. Xiaoyu L., Bing W., Simon Y.: Time Series Prediction Based on Fuzzy Principles. Department of Electrical & Computer Engineering FAMU-FSU College of Engineering, Florida State University Tallahassee, FL 32310, (2002).
49. Zadeh L. A.: Fuzzy Logic = Computing with Words. IEEE Transactions on Fuzzy Systems, 4 (2), 103, (1996).
50. Zadeh L. A.: Fuzzy Logic. Computer, Vol. 1, No. 4, pp. 83-93. (1988).
51. Zadeh, L.A.: Fuzzy Logic, Neural Networks and Soft Computing. Communications of the ACM 37(3), pp. 77–84. (1994).

Long-Term Prediction of a Sine Function Using a LSTM Neural Network

Magdiel Jiménez-Guarneros, Pilar Gómez-Gil, Rigoberto Fonseca-Delgado, Manuel Ramírez-Cortés and Vicente Alarcón-Aquino

Abstract In the past years, efforts have been made to improve the efficiency of long-term time series forecasting. However, when the involved series is highly oscillatory and nonlinear, this is still an open problem. Given the fact that signals may be approximated as linear combinations of sine functions, the study of the behavior of an adaptive dynamical model able to reproduce a sine function may be relevant for long-term prediction. In this chapter, we present an analysis of the modeling and prediction abilities of the "Long Short-Term Memory" (LSTM) recurrent neural network, when the input signal has a discrete sine function shape. Previous works have shown that LSTM is able to learn relevant events among long-term lags, however, its oscillatory abilities have not been analyzed enough. In our experiments, we found that some configurations of LSTM were able to model the signal, accurately predicting up to 400 steps forward. However, we also found that similar architectures did not perform properly when experiments were repeated, probably due to the fact that the LSTM architectures got over trained and the learning algorithm got trapped in a local minimum.

M. Jiménez-Guarneros (✉) · P. Gómez-Gil · R. Fonseca-Delgado
Department of Computer Science, Instituto Nacional de Astrofísica, Óptica y Electrónica, Luis Enrique Erro No. 1, C.P. 72840 Santa María Tonantzintla, Puebla, Mexico
e-mail: magdiel.jg@inaoep.mx

P. Gómez-Gil
e-mail: pgomez@inaoep.mx

R. Fonseca-Delgado
e-mail: rfonseca@inaoep.mx

M. Ramírez-Cortés
Department of Electronics, Instituto Nacional de Astrofísica, Óptica y Electrónica, Luis Enrique Erro No. 1, C.P. 72840 Santa María Tonantzintla, Puebla, Mexico
e-mail: jmram@inaoep.mx

V. Alarcón-Aquino
Department of Electronics, Universidad de las Américas, Cholula, Puebla, Mexico
e-mail: vicente.alarcon@udlap.mx

P. Melin et al. (eds.), *Nature-Inspired Design of Hybrid Intelligent Systems*, Studies in Computational Intelligence 667, DOI 10.1007/978-3-319-47054-2_10

Keywords Long short-term memory · Long-term prediction · Recurrent neural networks · Time series

1 Introduction

Time series forecasting is a problem with a great interest in domains such as medicine [1], finances [2], ecology [3], and economy [4], among others. Forecasting is referred as estimating one or several unknown future values of a time series, based on past values, [5, 6], which is done using data uniformly sampled in a continuous signal. However, it is well known that predicting multiple future values (long-term prediction) in highly nonlinear time series is still an open problem [6].

A way to address long-term prediction is using recursive prediction [7–9]. This consists on estimating one element at a time, so that the estimated value is recurrently fed back to the predictor input, in order to estimate the next value. This implies that an accurate model for one-step prediction must be built; otherwise the prediction accuracy will strongly diminish as the horizon of prediction increases [10].

Several methods have been proposed for addressing long-term prediction of highly nonlinear time series. Among them, neural networks have shown promising results. For example, Park and colleagues [11] proposed a Multiscale Bi-Linear Recurrent Neural Network (M-BLRNN) based on wavelet analysis, which was applied for the prediction of network traffic. Junior and Barreto [12] used a Nonlinear Autoregressive Model with Exogenous inputs (NARX) network to predict several steps ahead of univariate time series, such as the benchmark laser time series and a variable bit rate (VBR) video traffic. Alarcón-Aquino and Barria [13] proposed a multi-resolution Finite-impulse-response (FIR) neural network based on maximal overlap discrete wavelet transform (MODWT); such a model is also applied to network traffic prediction. In [10], Gómez-Gil and colleagues presented the "Hybrid-Connected Complex Neural Network," composed of small recurrent neural networks, called "harmonic generators," which are part of the hidden layer of a structure with feedforward connections. Once trained, harmonic generators are able to accurately reproduce sine waves. Such model was used for the long-time prediction of the Mackey–Glass time series and an electrocardiogram (EGC) signal. For a complete analysis and assessment of existing strategies for multi-step ahead forecasting, we refer the reader to [6].

In this chapter, we present an analysis of the behavior of two extensions of the neural network known as Long Short-Term Memory (LSTM) [14], when used for predicting large horizons. LSTM is a recurrent neural network that allows learning relevant events among long time lags, overcoming the problem known as "Vanish gradient". Such problem refers to the rapid decay of the backpropagated error during the learning phase, for long time lags. LSTM-based networks have shown a good performance in several fields, such as sequential labeling [15], handwriting recognition [16], speech recognition [17], and even short-term prediction in time series [18]. Here, we focused on the long-term prediction of a discrete representation of a sine function. The underlying idea is that many signals can be fairly approximated by linear combinations of sine waves. In this way, if a model may accurately predict discrete sine waves for long horizons, this could be an aid to predict a signal. For our experiments, we evaluated two extensions of the classical LSTM architecture proposed in [18, 19], trained with the algorithm known as RPROP minus [20]. Our experiments were tested using Monte Carlo Cross-Validation (MCCV) [21]. Predictions were performed using a recursive fashion, which consists on modeling a one-step predictor and using each single estimation as inputs for predicting the next one [8]. Prediction horizons were as long as two times the size of the training set signals.

This chapter is organized as follows: Sect. 2 describes LSTM architectures and the training algorithm used in our experiments. Section 3 shows the results obtained by evaluating the LSTM networks using MCCV. Finally, Sect. 4 presents some conclusions and future work.

2 Long Short-Term Memory

This section describes the LSTM architectures implemented for the experiments reported here, starting with the original model. Afterwards, two extensions of such model are detailed and their differences are highlighted. Last, we explain the algorithm used to train the LSTM networks.

2.1 Architectures

LSTM architecture was originally proposed by Hochreiter and Schmidhuber [14] and it is based on memory cells as basic units. A memory cell has a linear function unit with a recurrent self-connection, called the "Constant Error Carousel" (CEC) and a pair of gates, which control the information flow in the cell (Fig. 1). When the gates are closed, the error signals flow through the cell without changing its state, so the backpropagated error remains constant for long time lags. Input gate protects the CEC unit from perturbations generated by irrelevant inputs; output gate protects other memory cells from perturbations stored in the current cell. Both gates are activated using two multiplicative operators [shown as black circles in (Fig. 1)] and their values are normalized using two sigmoid functions: g and h.

For our analysis, we used two extensions of the architecture shown in Fig. 1. The first one, proposed by Gers et al. [19], contains a forget gate to restart occasionally the cell state, at each start of new input streams. In this way, continuous input streams are segmented in subsequences, which may allow avoiding saturating the function h. This behavior eventually reduces a LSTM cell to a simple Recurrent Neural Network unit. The second extension to this model, also proposed in [19],

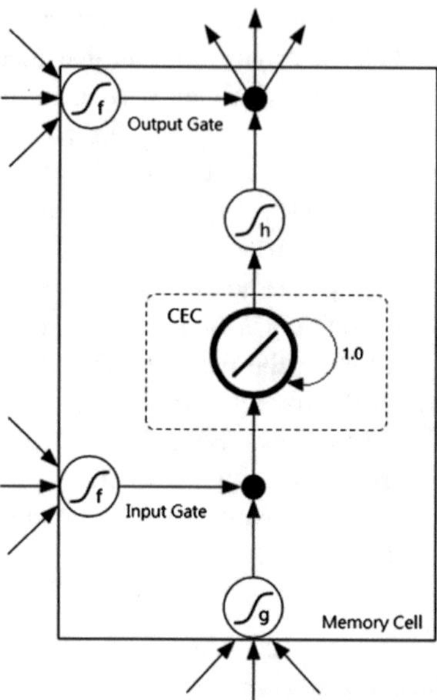

Fig. 1 A LSTM memory cell architecture [14]. It is composed of one unit with a linear function and a self-recurrent connection, with a weight value of 1.0

adds weighted "peephole" connections from the CEC unit to the gates in the memory cell. These connections aim to identify the duration intervals for specific tasks, which allow generating pattern sequences at exact intervals. Figure 2 shows both extensions.

Figure 3 shows an example of a LSTM architecture using peephole connections, which is composed of input, hidden, and output layers. In this example, input and output layers have four and five neurons, respectively, while hidden layer contains two LSTM memory cells (not all connections are shown in the figure).

The algorithm known as "Resilient back-propagation minus" (RPROP minus) was used to train the LSTM neural networks [22, 23]. The main characteristic of this algorithm is that the size of each weight change is determined only by its update value and by the sign of the gradient.

It is well known that gradient descent algorithms are most popular for supervised learning. Their basic idea is to compute the influence of each weight in the network, with respect to an error function E [23, 24]. For each iteration, weights are modified according to

$$w_{ij}(t+1) = w_{ij}(t) + \Delta w_{ij}(t),$$

where w_{ij} denotes the weight from neuron i to j; $\Delta w_{ij}(t)$ is calculated as follows:

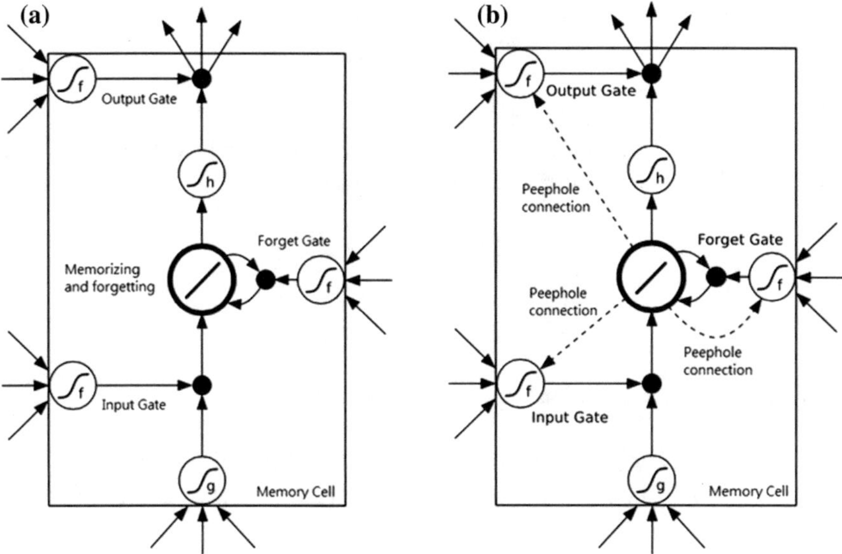

Fig. 2 LSTM architectures proposed in [18, 19]: **a** Memory cell with a forget gate and **b** Memory cell with peephole connections

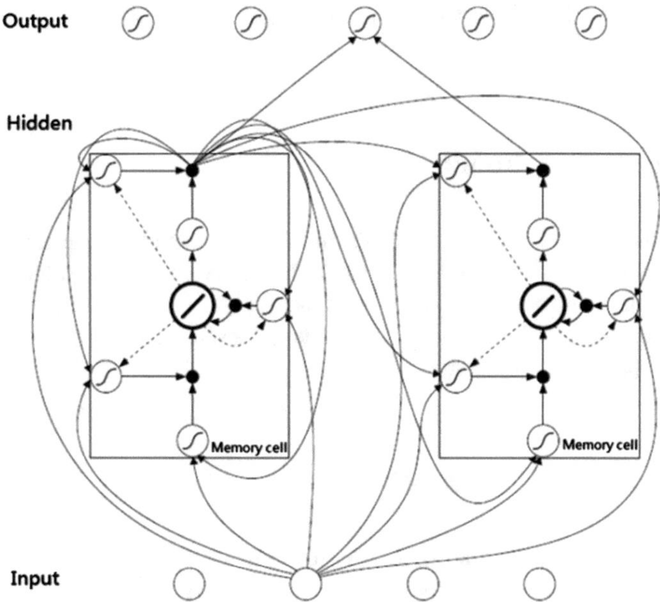

Fig. 3 An example of LSTM neural network using peephole connections [19] with three layers and two memory cells in the hidden layer (not all connections are shown)

$$\Delta w_{ij}(t) = -\alpha \frac{\partial E}{\partial w_{ij}}(t)$$

being α the value of learning rate.

On the other hand, RPROP minus only considers the sign of the partial derivative $\frac{\partial E}{\partial w_{ij}}$ to indicate the direction of the weight update [23], which is defined as

$$\Delta w_{ij}(t) = \begin{cases} +\Delta_{ij}(t), & \text{if } \frac{\partial E}{\partial w_{ij}}(t) > 0 \\ -\Delta_{ij}(t), & \text{if } \frac{\partial E}{\partial w_{ij}}(t) < 0 \\ 0 & \text{otherwise,} \end{cases}$$

where each new update values $\Delta_{ij}(t)$ are computed as

$$\Delta_{ij}(t) = \begin{cases} \min\left(\eta^{+} \cdot \Delta_{ij}(t-1), \Delta_{\max}\right) & \text{if } \frac{\partial E}{\partial w_{ij}}(t-1) \cdot \frac{\partial W}{\partial w_{ij}}(t) > 0 \\ \min\left(\eta^{-} \cdot \Delta_{ij}(t-1), \Delta_{\min}\right) & \text{if } \frac{\partial E}{\partial w_{ij}}(t-1) \cdot \frac{\partial E}{\partial w_{ij}}(t) < 0 \\ \Delta_{ij}(t-1) & \text{otherwise,} \end{cases}$$

being $0 < \eta^{-} < 1 < \eta^{+}$. When the partial derivative for the weight w_{ij} changes its sign, it means that the last update was too big and the algorithm has jumped over a local minimum. Then, the update value $\Delta_{ij}(t)$ is decreased by η^{-}. Conversely, if the derivative keeps the same sign, the update value is increased to accelerate convergence. Notice that the update values are bounded by $\Delta_{\min}$ and $\Delta_{\max}$. In addition, all Δ_{ij} are initialized to a constant value Δ_0.

3 Experimental Results

This section presents several experimental results obtained for the long-term prediction of a discrete sine function, using the LSTM architectures with a forget gate and peephole connections, which were described in Sect. 2. First, we detail the experimental setup; afterward, we show the achieved results and present a brief discussion about these outcomes.

3.1 Experimental Description

In order to analyze the ability of LSTM for remembering and reproducing cyclic behaviors in a time series, we trained two types of LSTM architectures. Performances were evaluated using Monte Carlo Cross-Validation (MCCV) [21], which works as follows: first, an initial position in the time series is randomly selected, which defines the initial position of a training series with a specific size.

After training, a portion of h values following the training series are used as test data; h corresponds to the prediction horizon. This process is repeated multiple times, generating randomly new training and testing partitions on each repetition. For the experiments shown here, we used training series of 200 values, a prediction horizon of 400 values and a MCCV with 1000 repetitions. The initial positions of each training set were randomly selected in the first 100 points of the series. The original time series was created by sampling 700 values of a sine function, using a sampling frequency of 7; the amplitude of the signal was normalized to values from 0.2 to 0.8. All experiments were implemented using the neural network library known as PyBrain [25], written in Python.

Several LSTM networks were analyzed, containing from one to six memory cells; weights were randomly initialized. The training algorithm RPROP Minus was applied using a learning rate of 1E-4, looking for reaching a mean square error of at least 1E-5; the maximum number of epochs was set to 5000. Default values were kept for $\Delta_{\min}$, $\Delta_{\max}$, and Δ_0.

To predict an element in the time series, we used three past values. Once that the network is trained, the first prediction is done using three known values. After that, each predicted element was repeatedly fed back to the input network, to predict a new element until the prediction horizon is covered (Recursive prediction). Accuracy was measured using Symmetric Mean Absolute Percentage Error (SMAPE) [26, 27]:

$$\text{SMAPE} = \frac{1}{n}\sum_{t=1}^{n} \frac{|F_t - A_t|}{(|A_t| + |F_t|)/2} * 100$$

Notice that SMAPE scores are in [0,200] where 0 corresponds to the best possible value.

3.2 Results and Discussion

Table 1 shows the average SMAPE obtained for several LSTM networks using forget gates with several numbers of cells in the hidden layer; Table 2 shows the same, but using LSTM's with peepholes connections. As we explained before, this performance was calculated using 1000 repetitions in a Monte Carlo Cross-Validation. Both tables also show the average number of epochs that each architecture required to reaching the minimum training error. Notice that both LSTM architectures are trained faster as the number of cells is increased, that is, the average number of epochs decreased as the number of cells increased. Third column in both tables contains the number of training executions that reached the maximum number of epochs before reaching the minimum training error. This number also decreased as the number of cells increased. Last column in both tables shows the standard deviation of average SMAPE. For both cases, standard deviations increase

Table 1 Monte Carlo cross-validation for LSTM architecture with a forget gate, using from one to six memory cells

Cells	Average epochs	Executions with a maximum epochs	Average SMAPE	SMAPE standard deviation
1	1584	129	55.31	19.97
2	982	51	54.41	32.28
3	673	51	53.44	37.02
4	401	20	54.49	40.09
5	357	24	56.74	42.16
6	236	11	58.19	43.15

Table 2 Monte Carlo cross-validation for LSTM architecture with peephole connections, using from one to six memory cells

Cells	Average epochs	Executions with a maximum epochs	Average SMAPE	SMAPE standard deviation
1	2643	372	57.35	28.56
2	1633	173	57.95	36.12
3	958	85	59.27	41.73
4	762	79	58.82	42.13
5	584	64	58.76	44.01
6	217	5	56.53	42.34

as the number of cells increases. In our opinion, this behavior suggests that, even though a more complicated LSTM architecture obtains an average better results, unfortunately as more cells are used, more dispersed performances are obtained. That is, more scores with a good prediction are achieved, but also more bad predictions results are achieved. This is also noticed in Figs. 4 and 5, where the frequency distributions of SMAPE values are shown.

Figure 6 shows box plots of SMAPEs obtained for both LSTM architectures with 1–6 cells. In both architectures, we observe that SMAPEs do not follow a normal distribution, but a positive asymmetric distribution. In other words, values concentrate in low scores of SMAPE (below the median) while the highest scores are more scattered (above the median); such asymmetric distribution is more noticeable with large numbers of cells.

Table 3 compare the best and the worst SMAPE's obtained for all networks, showing the Mean Square Error (MSE) achieved during training and the number of epochs for such executions. All best SMAPE scores were obtained when minimum error for training was achieved. Two observations highlight our attention: first, for some executions, the minimum training error was not achieved using the maximum number of epochs, that is, the network was trapped in a local minimum. Second, for several executions, the training error reached the minimum, but the prediction results were poor. Such conditions indicate that, for these experiments, the LSTM networks were over trained, which results in that they were not able to generalize.

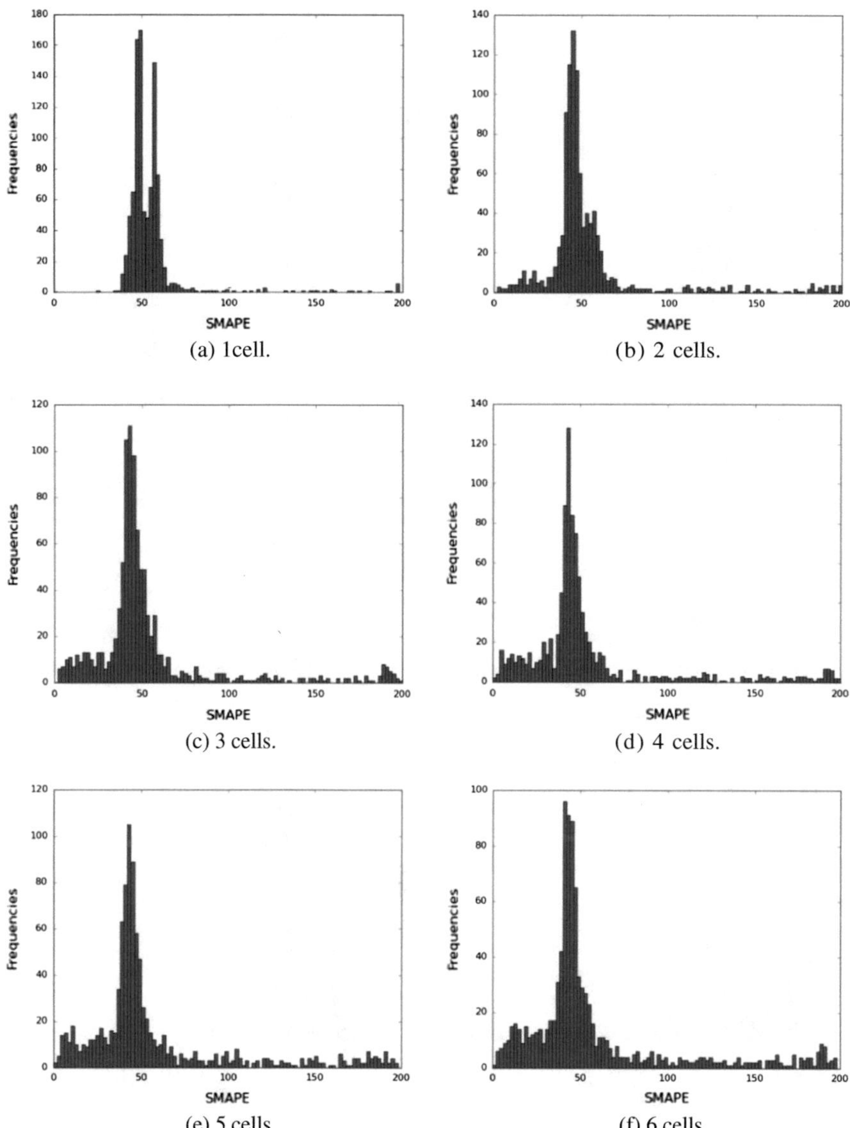

Fig. 4 Frequency histograms of SMAPE measures built with 1000 executions. Predictors were built using **a** 1, **b** 2, **c** 3, **d** 4, **e** 5, and **f** 6 memory cells with a forget gate

Finally, it should be noticed that LSTM architectures with a forget gate and peephole connections showed similar SMAPE results. This may indicate that, in order to obtain better performances, the conditions of overtraining and getting trapped in local minimum should be solved.

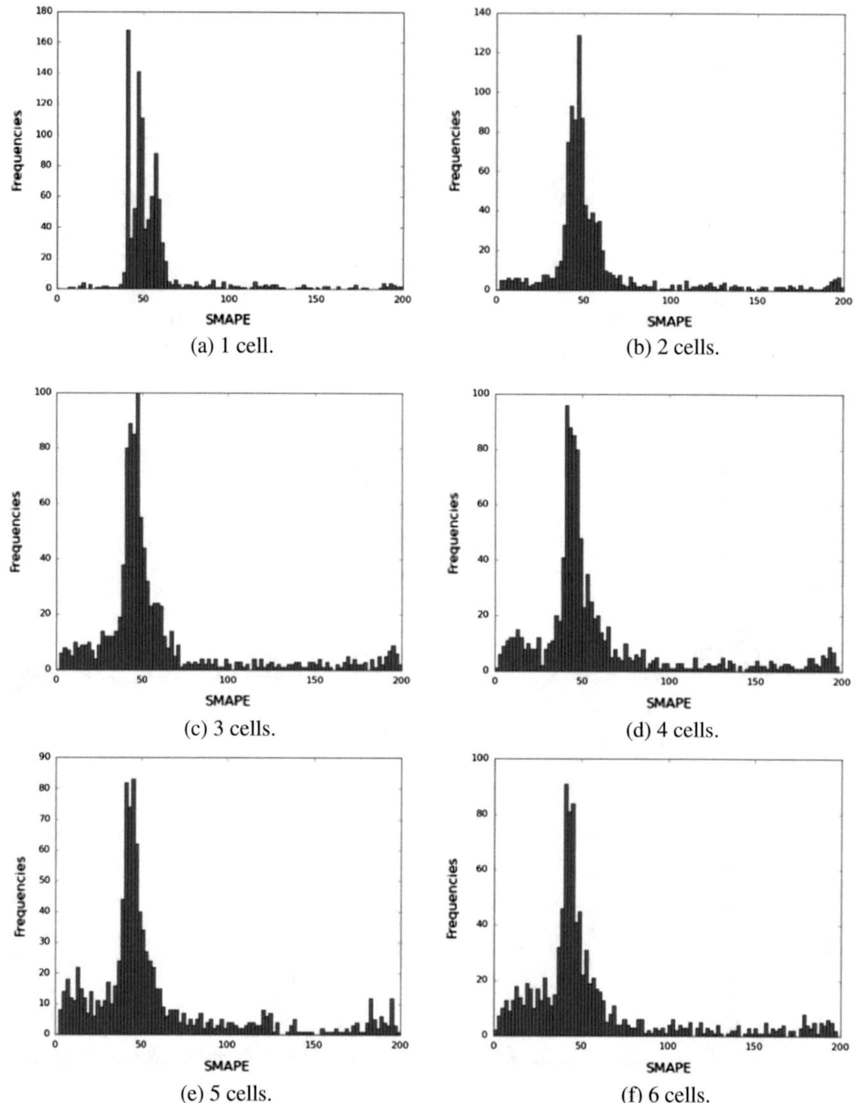

Fig. 5 Frequency histograms of SMAPE measures built with 1000 executions. Predictors were built using **a** 1, **b** 2, **c** 3, **d** 4, **e** 5, and **f** 6 LSTM memory cells with peephole connections

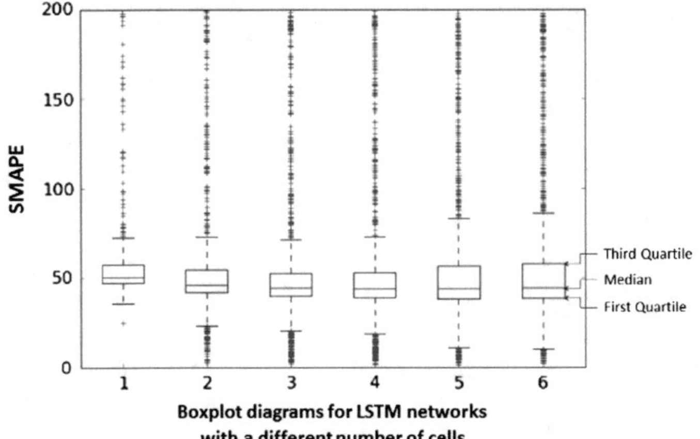

(a) Boxplot diagrams for LSTM architectures with a forget gate.

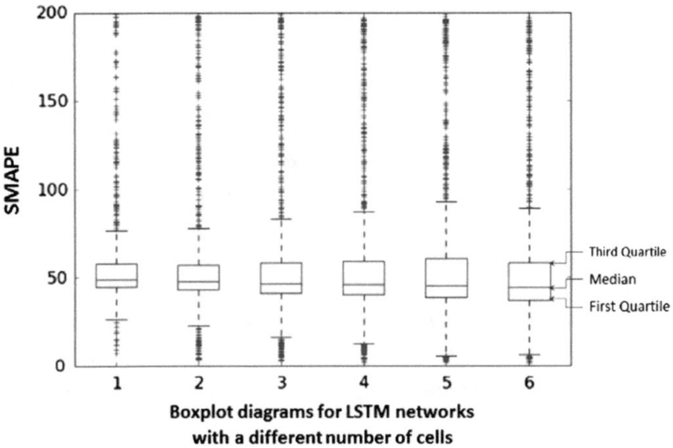

(b) Boxplot diagrams for LSTM architectures with peephole connections.

Fig. 6 Boxplot diagrams for LSTM architectures with **a** a forget gate and **b** peephole connections using from one to six memory cells

Table 3 Best and worst results for LSTM architecture with a forget gate

Cells	Best results			Worst results		
	Training error	Epochs	SMAPE	Training error	Epochs	SMAPE
1	9.99e−6	1294	24.83	9.99e–5	5000	197.94
2	9.98e−6	793	2.27	0.2243	5000	200.00
3	9.94e−6	289	2.00	9.77e–6	130	198.43
4	9.95e−6	163	1.68	0.0376	5000	199.10
5	9.96e−6	322	1.03	9.56e–6	133	197.96
6	9.98e−6	128	1.73	9.60e–6	191	197.80

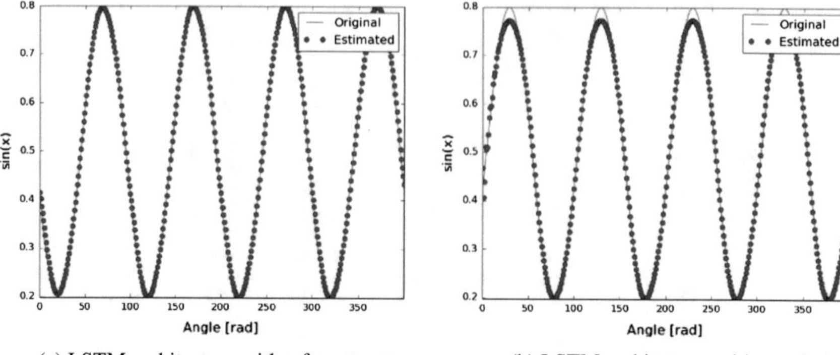

(a) LSTM architecture with a forget gate.

(b) LSTM architecture with peephole connections.

Fig. 7 The best prediction achieved by LSTM architectures with **a** a forget gate and **b** peephole connections, using six memory cells

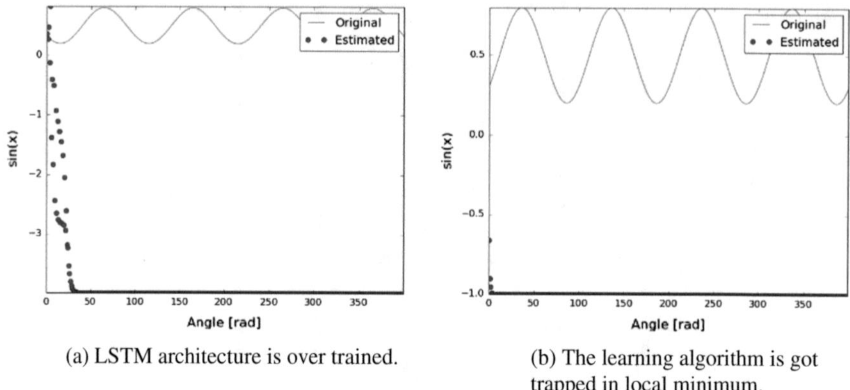

(a) LSTM architecture is over trained.

(b) The learning algorithm is got trapped in local minimum.

Fig. 8 Worst predictions obtained by LSTMs using a forget gate. **a** A LSTM with 3 memory cells is over trained. **b** A LSTM with 3 memory cells got trapped in local minimum

Figure 7 plots the best prediction results achieved by both LSTM architectures with six memory cells. Such executions show a very good forecast of up to two times the size of the training set signals. However, Figs. 8 and 9 show the worst results obtained for the two LSTM extensions, when they were over trained and the learning algorithm got trapped in local minimum.

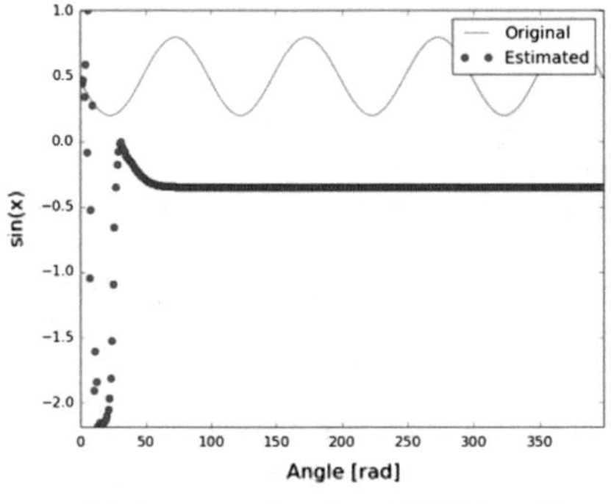

(a) An example where LSTM architecture is over trained.

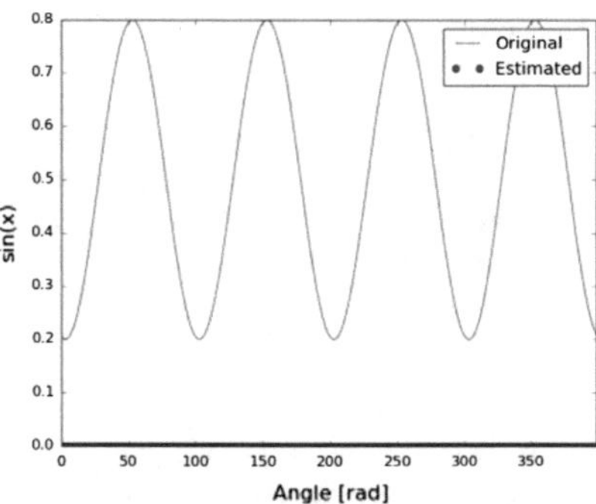

(b) An example where the learning algorithm is got trapped in local minimun.

Fig. 9 Worst prediction obtained by LSTMs using peephole connections. **a** ALSTM with 4 cells is overtrained. **b** ALSTM with one cell got trapped in local minimum

4 Conclusions

In this chapter, we present an analysis of two types of LSTM networks, when they were trained for the long-term prediction of time series, built with a discrete sine function. Given the fact that oscillation or quasi-oscillation is a characteristic frequently found in nonlinear time series, the ability of such behavior in adaptive predictors is an important issue to be explored. LSTM networks have shown good abilities to memorize long-term events, which make them suitable for long-term predictions of time series. Two extensions of the original LSTM architecture were studied: forget gate and peephole connections. The analysis was carried out using from 1 to 6 memory cells in the hidden layer of each network. The best average SMAPE was 53.44 ± 37.02, which has achieved by a LSTM with 3 memory cells and forget gates. Our results also found that, in most cases, LSTM networks improve the prediction as the number of cells is increased. However, the number of experiments with low predictions also increased as the number of cells increased, which may be due to over training of the networks.

Currently we are exploring the use of weight decay [28] and other techniques as strategies for tackling over training. In addition, we propose as future work to evaluate the performance of LSTM networks for long-term prediction of chaotic time series. The behavior of other LSTM extensions as LSTM Bidirectional [29] and Gated Recurrent Unit (GRU) [30] are also a possible branch to be explored.

Acknowledgments This research has been supported by *Consejo Nacional de Ciencia y Tecnología* (CONACYT) México, grant No. CB-2010-155250.

References

1. Strauss, D.G., Poole, J.E., Wagner, G.S., Selvester, R.H., Miller, J.M., Anderson, J., Johnson, G., McNulty, S.E., Mark, D.B., Lee, K.L., et al.: An ECG index of myocardial scar enhances prediction of defibrillator shocks: an analysis of the sudden cardiac death in heart failure trial. Heart Rhythm 8(1) (2011) 38–45.
2. Pavlidis, N., Tasoulis, D., Vrahatis, M.N.: Financial forecasting through unsupervised clustering and evolutionary trained neural networks. In: Evolutionary Computation, 2003. CEC'03. The 2003 Congress on. Volume 4, IEEE (2003) 2314–2321.
3. Cao, Q., Ewing, B.T., Thompson, M.A.: Forecasting wind speed with recurrent neural networks. European Journal of Operational Research 221(1) (2012) 148–154.
4. Pilinkiene, V.: Selection of market demand forecast methods: Criteria and application. Engineering Economics 58(3) (2015).
5. De Gooijer, J.G., Hyndman, R.J.: 25 years of time series forecasting. International journal of forecasting 22(3) (2006) 443–473.
6. Taieb, S.B., Bontempi, G., Atiya, A.F., Sorjamaa, A.: A review and comparison of strategies for multi-step ahead time series forecasting based on the NN5 forecasting competition. Expert systems with applications 39(8) (2012) 7067–7083.
7. Judd, K., Small, M.: Towards long-term prediction. Physica D: Nonlinear Phenomena 136(1) (2000) 31–44.
8. Crone, S.F., Hibon, M., Nikolopoulos, K.: Advances in forecasting with neural networks Empirical evidence from the NN3 competition on time series prediction. International Journal of Forecasting 27(3) (2011) 635–660.
9. Cheng, H., Tan, P.N., Gao, J., Scripps, J.: Multistep-ahead time series prediction. In: Advances in knowledge discovery and data mining. Springer (2006) 765–774.
10. Gómez-Gil, P., Ramírez-Cortes, J.M., Hernández, S.E.P., Alarcón-Aquino, V.: A neural network scheme for long-term forecasting of chaotic time series. Neural Processing Letters 33 (3) (2011) 215–233.
11. Park, D.C., Tran, C.N., Lee, Y.: Multiscale bilinear recurrent neural networks and their application to the long-term prediction of network traffic. In: Advances in Neural Networks-ISNN 2006. Springer (2006) 196–201.
12. Menezes, J.M.P., Barreto, G.A.: Long-term time series prediction with the narx network: an empirical evaluation. Neurocomputing 71(16) (2008) 3335–3343.
13. Alarcon-Aquino, A., Barria, J.A.: Multiresolution FIR neural-network-based learning algorithm applied to network traffic prediction. IEEE Transactions on Systems, Man, and Cybernetics, Part C: Applications and Reviews 36(2) (2006) 208–220.
14. Hochreiter, S., Schmidhuber, J.: Long short-term memory. Neural Comput. 9(8) (November 1997) 1735–1780.
15. Graves, A., Rahman Mohamed, A., Hinton, G.: Speech recognition with deep re- current neural networks (2013).
16. Graves, A., Liwicki, M., Fernández, S., Bertolami, R., Bunke, H., Schmidhuber, J.: A novel connectionist system for unconstrained handwriting recognition. IEEE Trans. Pattern Anal. Mach. Intell. 31(5) (May 2009) 855–868.
17. Sak, H., Senior, A.W., Rao, K., Irsoy, O., Graves, A., Beaufays, F., Schalkwyk, J.: Learning acoustic frame labeling for speech recognition with recurrent neural networks. In: 2015 IEEE International Conference on Acoustics, Speech and Signal Processing, ICASSP 2015, South Brisbane, Queensland, Australia, April 19-24, 2015. (2015) 4280–4284.
18. Gers, F.: Long short-term memory in recurrent neural networks. Thesis No. 2366. Ecole Polytechnique Federale de Lausanne. Doctoral Thesis. Lausane, EPFL.(2001).
19. Gers, F.A., Schraudolph, N.N., Schmidhuber, J.: Learning precise timing with LSTM recurrent networks. The Journal of Machine Learning Research 3 (2002) 115–143.

20. Riedmiller, M., Braun, H.: A direct adaptive method for faster backpropagation learning: the RPROP algorithm. In: Neural Networks, 1993, IEEE International Conference on. (1993) 586–591 vol. 1.
21. Picard, R.R., Cook, R.D.: Cross-validation of regression models. Journal of the American Statistical Association 79(387) (1984) 575–583.
22. Igel, C., Husken, M.: Improving the RPROP learning algorithm. In: Proceedings of the second international ICSC symposium on neural computation (NC 2000). Volume 2000, Citeseer (2000) 115–121.
23. Igel, C., Husken, M.: Empirical evaluation of the improved RPROP learning algorithms. Neurocomputing 50 (2003) 105 – 123.
24. Riedmiller, M., Braun, H.: A direct adaptive method for faster backpropagation learning: The rprop algorithm. In: Neural Networks, 1993, IEEE International Conference on, IEEE (1993) 586–591.
25. Schaul, T., Bayer, J., Wierstra, D., Sun, Y., Felder, M., Sehnke, F., Ruckstie, T., Schmidhuber, J.: PyBrain. Journal of Machine Learning Research 11 (2010) 743–746.
26. Tessier, T.H.: Long range forecasting: From crystal ball to computer. Journal of Accountancy (pre-1986) 146(000005) (1978) 87.
27. Andrawis, R.R., Atiya, A.F., El-Shishiny, H.: Forecast combinations of computational intelligence and linear models for the nn5 time series forecasting competition. International Journal of Forecasting 27(3) (2011) 672–688.
28. Rognvaldsson, T.S.: A simple trick for estimating the weight decay parameter. In: Neural networks: Tricks of the trade. Springer (1998) 71–92.
29. Graves, A., Fernández, S., Gómez, F., Schmidhuber, J.: Connectionist temporal classification: labelling unsegmented sequence data with recurrent neural networks. In: Proceedings of the 23rd international conference on Machine learning, ACM (2006) 369–376.
30. Cho, K., Van Merriënboer, B., Gülçehre, C., Bahdanau, D., Bougares, F., Schwenk, H., Bengio, Y.: Learning phrase representations using RNN encoder–decoder for statistical machine translation. In: Proceedings of the 2014 Conference on Empirical Methods in Natural Language Processing (EMNLP), Doha, Qatar, Association for Computational Linguistics (October 2014) 1724–1734.

UAV Image Segmentation Using a Pulse-Coupled Neural Network for Land Analysis

Mario I. Chacon-Murguia, Luis E. Guerra-Fernandez and Hector Erives

Abstract This chapter presents a pulse-coupled neural network architecture, PCNN, to segment imagery acquired with UAV images. The images correspond to normalized difference vegetation index values. The chapter describes the image analysis system design, the image acquisition elements, the original PCNN architect, the simplified PCNN, the automatic parameter setting methodology, and qualitative and quantitative results of the proposed method using real aerial images.

Keywords UAV · Image segmentation · Pulse coupled neural network

1 Introduction

Monitoring of agricultural fields for nutrient levels, water deficiency, fertilization, infestation, and plant health inspection are all activities that have been done traditionally by qualified individuals, who know the origin and solution to these problems. Furthermore, these activities are usually carried out regularly on the ground and afoot, and sometimes from the air. Despite all these efforts to inspect agricultural fields in a timely manner, the results from the inspection process are often times not on time, inaccurate, and costly. Precision agriculture is a viable solution to these problems. A branch of the precision agriculture is remote sensing, which consists of the acquisition and analysis of images of the land via airborne or space-borne platforms, which may be used to estimate plan health based on the

M.I. Chacon-Murguia (✉) · L.E. Guerra-Fernandez
Visual Perception Applications on Robotic Lab, Chihuahua Institute of Technology, Chihuahua, Mexico
e-mail: mchacon@itchihuahua.edu.mx

L.E. Guerra-Fernandez
e-mail: luis.e.g.@ieee.org

H. Erives
New Mexico Tech, NM, USA
e-mail: erives@ee.nmt.edu

P. Melin et al. (eds.), *Nature-Inspired Design of Hybrid Intelligent Systems*, Studies in Computational Intelligence 667, DOI 10.1007/978-3-319-47054-2_11

visible spectrum reflected and absorbed by the plant. Airborne platforms, like Unmanned Air Vehicles or UAVs, have been proven to be valuable tools for the analysis of land [1–3]. These platforms are usually equipped with a variety of sensors, including thermal and visible multispectral and hyperspectral cameras, which are able to acquire and record ground temperatures and upwelling radiation emitted by the soil and plants.

Although it is difficult to detect stressed vegetation using only multispectral images, through the years a number of vegetation indexes have been developed to automate the classification of vegetation in remotely sensed imagery. However, remote-sensed imagery undergoes a calibration and registration process (and possibly geo-registration) before it is ready for classification. Image registration step is particularly important when the final product (vegetation index) requires a ratio of multispectral bands; an error may be introduced into the imagery when miss-alignments exist [4–6]. A detail account of the process can be found in [7]. A well-known vegetation index used in remote sensing systems is the Normalized Difference Vegetation Index (NDVI) [8–10]. This index is popular because it can be computed quickly and efficiently, and has proven to be a good indicator of the status of vegetation, including the detection of stressed vegetation when it is subjected to adverse conditions.

The final step in the process consists of the segmentation of the imagery, which will yield the final product. In this chapter a pulse coupled neural network architecture, PCNN, is proposed to perform segmentation of imagery corresponding to NDVI information obtained with a UAV system.

The organization of the chapter is as follows. The UAV system is described in Sect. 2. The original PCNN and the simplified PCNN models are presented in Sects. 3 and 4 respectively. Section 5 reports the experimental results, and Sect. 6 exposes the conclusions.

2 UAV System

2.1 System Design

The design of the land UAV image analysis system implicated several stages that are illustrated in Fig. 1. The video acquisitions system is in charge of acquiring the land images. The components of the acquisition system like cameras and filters had to be characterized in order to validate and generate the correct NDVI information. The multispectral stage involved the generation of multiband images in order to be able to compute the NDVI index. The next stage is related to video analysis which involves image preprocessing, computation of the vegetative index, and land tile generation. The last stage corresponds to the analysis of the NDVI tile information which is achieved by a modified pulse neural network.

2.2 Image Acquisition System

The UAV system employed in this work involves a DJI Phantom quadcopter, wireless color cameras CMOS 640 × 480, and a NGB filter. This UAV system was used to acquire videos from different crop fields. The images generated by the UAV system correspond to NDVI images. NDVI comes from Normalized Difference Vegetation Index (NDVI), defined by

$$\mathrm{NDVI} = \frac{(\mathrm{NIR} - R)}{(\mathrm{NIR} + R)} \tag{1}$$

where NIR stands for near infrared, and R for red. The technique used to generate the NDVI images involves a NGB (NIR, Green, BLUE) filter specifically designed for this application. Since the red color channel is a high-pass filter, when the red information is suppressed this is substituted by the acquired NIR information. In this form, the NDVI is computed by the difference between the blue or green channels and the red channel (NIR).

3 PCNN Model

Once the UAV has provided the land NDVI information it is necessary to analyze it in the land tile to determine possible maintenance zones. This is achieved through image segmentation. The proposed image segmentation technique is based on a pulse neural network architecture, PCNN. It is know that the PCNN architecture has been frequently used in image processing [11–14] task. A PCNN is a model derived from a neural mammal model [15–18]. Current works with PCNN document how the PCNN can be used to perform important image processing task; edge detection, segmentation, feature extraction, and image filtering [16–21]. Because of this kind of performance the PCNN is considered a good preprocessing element. The basic model of a neuron element of a PCNN has three main modules: the dendrite tree, the linking, and the pulse generator [15]. The dendrite tree includes two special regions of the neuron element, the linking and the feeding. Neighborhood information is incorporated through the linking. The input signal information is obtained through the feeding. The pulse generator module compares the internal activity, linking plus feeding activity, with a dynamic threshold to decide if the neuron element fires or not. Figure 1 illustrates the basic model of the PCNN. A PCNN mathematical definition is given by (1)–(5). Equation (1) corresponds to the feeding region of the neural element, where S is the input image, α_{Ft} is the time constant of the leakage filter of the feeding region, $Y(n)$ is the neuron output at time t, and W is the feeding kernel. The outputs $Y(t)$ of the PCNN can be observed as output images called pulsed images of the PCNN. Equation (2) describes the linking activity. Hereα_L is the time constant of the leakage filter of the linking region and M is the

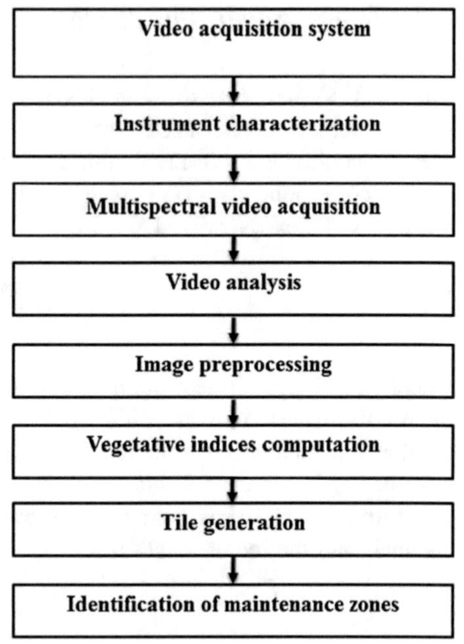

Fig. 1 Land UAV image analysis system stages

linking kernel. Equation (3) corresponds to the internal activity of the neuron element. The internal activity depends on the linking and feeding activity. In (3) β is the linking coefficient. β defines the amount of modulation of the feeding due to the linking activity. The dynamic threshold is implemented by (4), where α_T is the time constant of the leakage filter of the threshold and V_{T} is the threshold gain. Finally the output of the neuron is defined by (5). In the case of an image processing task, each pixel is related to a neural element. For more information on how a PCNN works consult [15].

$$F_{i,j}[n] = \mathrm{e}^{-\alpha_{\mathrm{F}}} F_{i,j}[n-1] + V_{\mathrm{F}} \sum_{k,l} W_{i,j,k,l} Y_{i,j}[n-1] + S_{i,j} \tag{1}$$

$$L_{i,j}[n] = \mathrm{e}^{-\alpha_{\mathrm{L}}} L_{i,j}[n-1] + V_{\mathrm{L}} \sum_{k,l} m_{i,j,k,l} Y_{i,j}[n-1] \tag{2}$$

$$U_{i,j}[n] = F_{i,j}[n]\left(1 + \beta L_{i,j}[n]\right) \tag{3}$$

$$T_{i,j}[n] = \mathrm{e}^{-\alpha_{\mathrm{T}}} T_{i,j}[n-1] + V_T Y_{i,j}[n] \tag{4}$$

$$Y_{i,j}[n] = \begin{cases} 1 & U_{i,j}[n] > T_{i,j}[n] \\ 0 & \text{otra forma} \end{cases} \tag{5}$$

4 SPCNN Model

A special issue regarding the original PCNN is parameter settings. The results of the PCNN are highly dependent on its parameters values, and at the moment there is not a general training algorithm to determine those parameters. However, there exist some works like [22, 23] where a scheme to automatically determine the PCNN parameters for specific task are described. In this work we follow the method proposed by Chen and et al. However, the present work includes a modification to improve the results. In 2011, Chen. et al. developed a method to determine the PCNN parameters of a Simplified PCNN, SPCNN, for segmentation purposes. The method is based on; derivation of the dynamic threshold, internal activity, according to the dynamic properties of neurons and, image intensity standard deviation and optima histogram threshold. The new architecture corresponds to a simplified model of PCNN, SPCNN, based on a Spiking Cortical Model (SCM) given by

$$U_{ij}[n] = e^{-\alpha_F} U_{ij}[n-1] + S_{ij}\left(1 + \beta V_L \sum_{kl} W_{ijkl} Y_{kl}[n-1]\right) \tag{6}$$

$$Y_{ij}[n] = \begin{cases} 1, & si \quad U_{ij}[n] > T_{ij}[n-1] \\ 0, & \text{otherwise} \end{cases} \tag{7}$$

$$T_{ij}[n] = e^{-\alpha_T} T_{ij}[n-1] + V_T Y_{ij}[n] \tag{8}$$

This SPCNN model has five parameters: α_F, β, V_L, V_T, and α_T.

During the iteration process, the neurons with higher intensity stimuli always fire prior to the ones with lower intensity stimuli in PCNN-based models, and the minimum intensity neuron has the longest fire period T_{lowest}, which is referred as "a pulsing cycle." The pulses produced by higher intensity neurons and those by lower ones would mix together. This phenomenon indicates that any single pixel with relatively high intensity may be grouped into two or more segments, which goes against the purpose of image segmentation, therefore this must be corrected. This correction consists of the following, once a neuron fires twice, it would be prevented from firing again by setting its internal activity U_{ij} to zero.

The process considers a hierarchical segmentation process. The image is first divided into two regions. The region with the highest intensities is considered the first segment, meanwhile the region with the lowest intensities is recursively segmented into smaller regions on each pulsation.

4.1 Automatic Parameter Update

This section describes the automatic parameter process to update the five parameters of the SPCNN proposed by Chen et al. [22].

The parameter α_F affects the range of the internal activity $U[n]$, $U_{S\max}[n] - U_{\min}[n]$. As α_F decreases, the wider the range of the internal activity distribution. From experimentation it was found that there is an inverse relation between α_F and the standard deviation of the image σ given by

$$\alpha_F = \log\left(\frac{1}{\sigma(I)}\right) \tag{9}$$

The normalized σ is always less than 1, then $\alpha_F > 0$, which assures that $U_{ij}[n]$ does not increases exponentially with n.

The parameter β represents the neuron interaction. Given a V_L value, a neuron will be more affected by its neighbor neurons as β increases, yielding a drastic fluctuation in $Uij[n]$. The range of β for a possible perfect segmentation of an image was determined by Kuntimad and Ranganath [24]. In this work it was decided to use $\beta_{\max}$, since it requires less computation of a priori values,

$$\beta = \beta_{\max} = \frac{\left(\frac{S_{R\max}}{S_{B\max}}\right) - 1}{L_{B_{\max}}} \tag{10}$$

where $S_{R\max}$ and $S_{B\max}$ denote the maximum intensities of the region with the highest values and lower values respectively. $L_{B\max}$ is the maximum value of the linking inputs in the region of lower intensity values.

As commented before, at the beginning, the image is segmented into two regions. The key to obtain $\beta_{\max}$ is to find the highest intensity of the region with lower values $S_{B\max}$. Since the value $S_{B\max}$ change for each image, $\beta_{\max}$ may be determined by the optimal threshold S', which is obtained by the Otsu method. $S_{R\max}$ can take the maximum intensity value of the image $S_{\max}$, and $L_{B\max}$ can be the maximum value of the linking input $L = 6V_L$, giving

$$\beta = \frac{\left(\frac{S_{\max}}{S'}\right) - 1}{6V_L} \tag{11}$$

The value of V_L is the amplitude of the linking input. In fact V_L can be arbitrary and since the term βV_L acts as one factor in the SPCNN, the specific value of V_L does not affect the final result of the segmentation, thus V_L is fixed to 1.

α_T is the decaying exponential coefficient of the dynamic threshold. As α_T increases, the segmentation result is less precise. V_T is the scalar of the dynamic threshold, T. α_T and V_T can be deduced from the expression of the first intensity range S considering that the first segment of the image is obtained in the second decaying step. The first intensity range can be expressed as

$$\frac{T[1]}{M[1]} \geq S \geq \frac{T[2]}{M[3]} \quad M[2] = \mathrm{e}^{-\alpha_{\mathrm{F}}} + 1 + 6\beta V_{\mathrm{L}} \quad M[3] = \frac{1 - \mathrm{e}^{-3\alpha_{\mathrm{F}}}}{1 - \mathrm{e}^{-\alpha_{\mathrm{F}}}} + 6\beta V_{\mathrm{L}} \mathrm{e}^{-\alpha_{\mathrm{F}}} \tag{12}$$

The region with the highest intensities corresponds to the first segment, thus the first intensity range will be the maximum intensity, 1, up to the highest intensity of the region with lower intensity values $1 \geq S > S_{B\max}$ and $S_{B\max} = S'$, then (12) can be rewritten as

$$\frac{T[1]}{M[2]} \geq 1 \geq S > S' \geq \frac{T[2]}{M[3]} \tag{13}$$

Finally substituting $T[1] = V_{\mathrm{T}}$ and $T[2] = V_{\mathrm{T}}\,\mathrm{e}^{-\alpha T}$ we have

$$\begin{aligned} &\frac{T[1]}{M[2]} \geq 1 \rightarrow \frac{V_T}{M[2]} \geq 1 \rightarrow V_{\mathrm{T}} \geq M[2] \\ &\frac{T[2]}{M[3]} \leq S' \rightarrow \frac{V_{\mathrm{T}} \mathrm{e}^{-\alpha_{\mathrm{T}}}}{M[3]} \leq S' \rightarrow \alpha_{\mathrm{T}} \geq \ln\left(\frac{V_{\mathrm{T}}}{S' M[3]}\right) \end{aligned} \tag{14}$$

where

$$M[2] = \mathrm{e}^{-\alpha_{\mathrm{F}}} + 1 + 6\beta V_{\mathrm{L}} \quad M[3] = \frac{1 - \mathrm{e}^{-3\alpha_{\mathrm{F}}}}{1 - \mathrm{e}^{-\alpha_{\mathrm{F}}}} + 6\beta V_{\mathrm{L}} \mathrm{e}^{-\alpha_{\mathrm{F}}}$$

The unique values for V_T and α_T correspond to the lower limits

$$V_{\mathrm{T}} = M[2] \quad \alpha_{\mathrm{T}} = \ln \frac{V_{\mathrm{T}}}{S' M[3]} \tag{15}$$

The proposed method presents the issue that the region with the highest intensity is not segmented again, even more, the method does not warranty that the region with the highest intensity would be the smaller. Therefore, the following modification was included. The complement of the image is computed

$$\overline{S_{ij}} = 1 - S_{ij} \tag{16}$$

and an SPCNN is applied to each image, original and complement.

In PCNNs is important to know when to stop the pulsing process. In this case the process is kept if $R > \varepsilon$

Table 1 SPCNN quantitative performance

Video	Sensitivity	Specificity
1	0.6472	0.8594
2	0.7070	0.6141
3	0.7606	0.8071
4	0.9453	0.4887
5	0.1789	0.9432
6	0.9360	0.5554
7	0.9688	0.6591
8	0.4385	0.9919
Average	0.8018	0.7399

where

$$R = \frac{\mathrm{Num}_{\mathrm{fired}}}{\mathrm{Num}_{\mathrm{total}}} \tag{17}$$

and R is computed in each pulsation. The previous condition indicates that the current segmented region can be still divided into smaller regions.

5 Results

The system developed was tested with 8 different videos. Figure 2 illustrates some examples of NDVI tiles and the SPCNN segmentation.

Quantitative results using the sensitivity and specificity metrics were obtained by

$$\text{Sensitivity} = \frac{TP}{P} = \frac{TP}{(TP + FN)} \tag{18}$$

$$\text{Specifisity} = \frac{TN}{N} = \frac{TN}{(TN + FP)} \tag{19}$$

The ground truths were manually generated for each tile. Table 1 shows the results of the two metrics for each of the eight tiles.

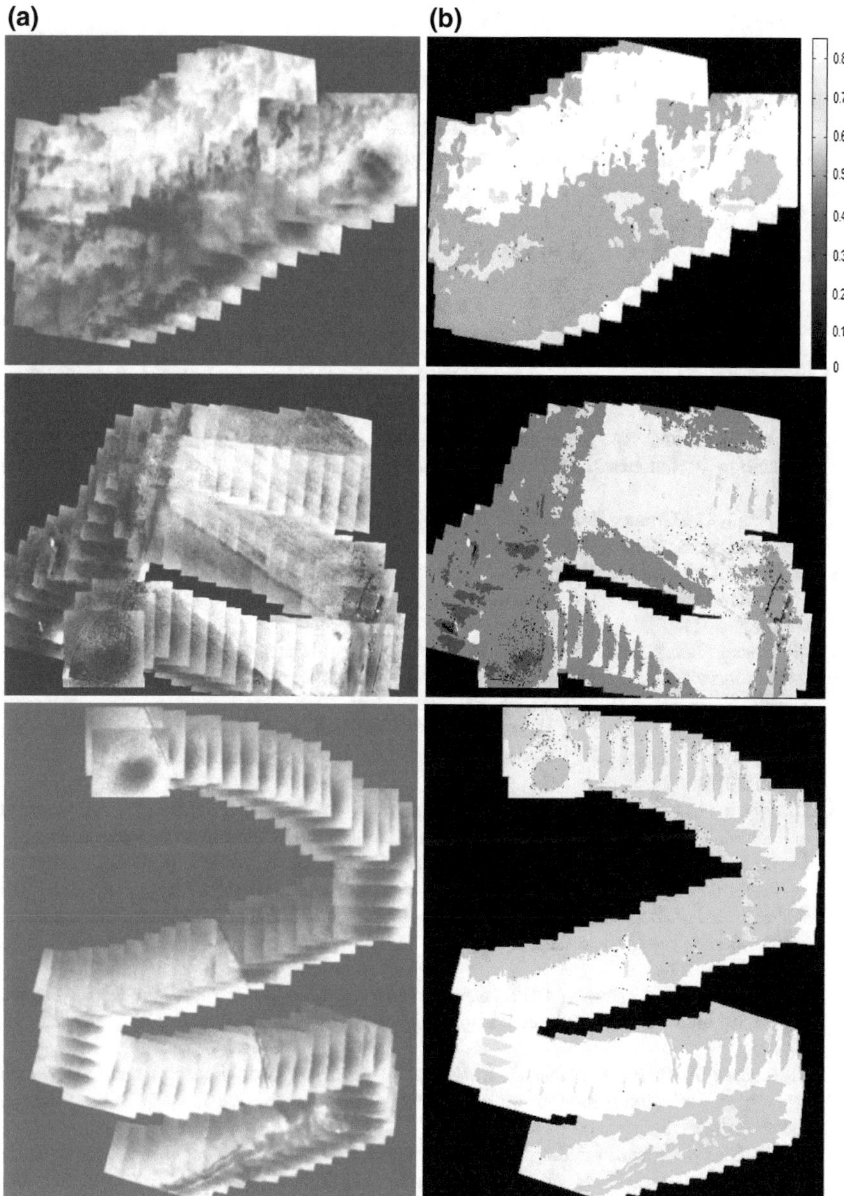

Fig. 2 SPCNN segmentation. **a** NDVI image. **b** Segmentation

6 Conclusions

Regarding the experimental results, we can say that the SPCNN achieved acceptable results to segment crop land from land only. The SPCNN was also able to segment subregions inside of the main regions, which correspond to small zones of poor condition vegetation, dry zones, houses, and other type of objects. The low sensitivity and specificity cannot be adjudged to the SPCNN, but also to the information quality yielded in the tile generation process, e.g., filter distortion.

References

1. C. A. Rokhmana.: The potential of UAV-based remote sensing for supporting precision agriculture in Indonesia. Procedia Environmental Sciences. 24. DOI:10.1016, December 2015.
2. E. Salami, C. Barrado, E. Pastor.: UAV flight experiments applied to the remotes sensing of vegetated areas, Remote Sensing. **6**(11), pp 11051-11081, November 2014.
3. H. Xiang, L. Tian.: Development of a low cost agricultural remote sensing system based on autonomous unmanned aerial vehicle (UAV), Biosystems Engineering. **108**(2), pp 174-190, February 2011.
4. M. Hasan, J. Xiuping, A. Robles-Kelly, J. Zhou.: Multi-spectral remote sensing image registration via spatial relationship analysis on shift key points, IEEE Geoscience and Remote Sensing Symposium (IGARSS). pp 1011-1014, July 2010.
5. Q. Zhang, Z. Cao, Z. Hu, Y. Jia.: Joint image registration and fusion for panchromatic and multispectral images. IEEE Geoscience and Remote Sensing Letters, **12**(3), pp 467-471, March 2016.
6. H. Erives G. J. Fitzgerald.: Automated registration of hyperspectral images for precision agriculture. Computers and Electronics in Agriculture. **47**(2), pp 103-119, April 2005.
7. L. E. Guerra-Fernandez.: Analysis and acquisition of multispectral aerial images of agricultural fields by a UAV, M.S. Thesis, Chihuahua Institute of Technology, August 2015.
8. A. Vina, A. A. Gitelson, A. L. Nguy-Robertson, Y. Peng.: Comparison of different vegetation indices for the remote assessment of green leaf are index of crops. **115**(12), pp. 3468-3478, September 2011.
9. A. K. Bhandari, a. Kumar, G. K. Singh.: Feature extraction using normalized difference vegetation index: a case study at Jabalpur City. Procedia Technology. **6**(2012), pp. 612-621, (2012).
10. V. Poenaru, A. Badea, S. Mihai, A. Irimescu.: Multi-temporal multi-spectral and radar remote sensing for agricultural monitoring in the Braila Plain. Agriculture and Agricultural Science Procedia. **6**(2015). pp. 506-516, (2015).
11. L. Yi, T. Qinye and F. Yingle.:Texture image segmentation using pulse couple neural networks. In: Proceedings of Industrial Electronics and Applications, 2007. pp. 355-359, September 2007.
12. H. Rughooputh and S. Rughooputh.: Spectral recognition using a modified Eckhorn neural network model. Image and Vision Computing. 18(14), pp. 1101-1103, (2000).
13. M.I. Chacon-Murguia and A. Zimmeman.: PCNNP: A pulse-coupled neural network processor. In: Proceedings of the IEEE IJCNN. pp. 1581-1585, (2002).
14. M.I. Chacon-Murguia and A. Zimmeman.: Image processing using the PCNN time matrix as a selective filter. In: Proceedings of the IEEE ICIP. pp. 1195-1199, (2003).

15. T. Lindbland and J. Kinser.: Image processing using pulse-coupled neural networks. Springer, (1998).
16. J.L. Johnson and M.L. Padgett.: PCNN models and applications. IEEE Transactions on Neural Networks **10**(3). pp. 480-498, (1999).
17. J. Johnson.: Pulse-Coupled Neural Nets: Translation, rotation, scale, distortion, and intensity signal invariance for images. Applied Optics. 30(26), pp. 6239-6253, (1994).
18. D.G. Kuntimad and H. Ranganath: Perfect image segmentation using pulse coupled neural networks. IEEE Transactions on Neural Networks. **10**(3), pp. 591-598, (1999).
19. E. Keller and J. Johnson.: Pulse-coupled neural network for medical analysis. In: Proceedings of SPIE Applications and science of Computation Intelligence II. pp. 444- 451, (1999).
20. F. Ranganath and G. Kuntimad.: Object detection using pulse coupled neural networks. IEEE Transactions on Neural Networks. **10**(3), pp. 615-620, (1999).
21. H. Ranganath and G. Kuntimad.: Image segmentation using pulse coupled neural networks. In: Proceedings of IEEE World Congress on Computational Intelligence. pp. 1285-1290, (1994).
22. Y. Chen, S.K. Park, Y. Ma and R. Ala.: A new automatic parameter setting method of a simplified PCNN for image segmentation. IEEE transactions on neural networks. **22**(6), pp. 880-892, (2011).
23. X. Deng, and Y. Ma.: PCNN Automatic parameters determination in image segmentation based on the analysis of neuron firing time. Foundations of Intelligent Systems Series Advances in Intelligent and Soft Computing.. **122**(2012), pp 85-91, (2012).
24. H. S. Ranganath, G. Kuntimad and J. L. Johnson.: Pulse coupled neural networks for image processing. In: Proceedings of IEEE Southeast con. pp. 37-43, (1995).

Classification of Arrhythmias Using Modular Architecture of LVQ Neural Network and Type 2 Fuzzy Logic

Jonathan Amezcua and Patricia Melin

Abstract In this paper, a new model for arrhythmia classification using a modular LVQ neural network architecture and a type-2 fuzzy system is presented. This work focuses on the implementation of a type-2 fuzzy system to determine the shortest distance in a LVQ neural network competitive layer. In this work, the MIT-BIH arrhythmia database with 15 classes was used. Results show that using five modules architecture could be a good approach for classification of arrhythmias.

Keywords Classification · Fuzzy systems · LVQ · Neural networks

1 Introduction

In this work a new model based on a modular architecture of LVQ neural network and a type-2 fuzzy system is presented. LVQ [11, 13] is an adaptive-learning classification method based on neural networks and used to solve many problems; this algorithm uses supervised training, but also applies unsupervised data-clustering techniques, to preprocess the dataset and obtain the clusters centers.

Furthermore, fuzzy systems have been used in many areas such as robotics, automatic control, expert systems, classification, etc. Fuzzy systems are based on the theory of fuzzy sets, if-then fuzzy rules and a fuzzy reasoning mechanism. The fuzzy sets express a degree to which an element belongs to a set; consequently the characteristic function of a fuzzy set is allowed to have values between 0 and 1, which denotes the degree of membership of a given element in a given set. In the next sections concepts on LVQ neural networks and fuzzy systems are thoroughly described.

J. Amezcua · P. Melin (✉)
Tijuana Institute of Technology, Tijuana, Mexico
e-mail: pmelin@tectijuana.edu.mx

J. Amezcua
e-mail: jonathan.aguiluz@yahoo.com

P. Melin et al. (eds.), *Nature-Inspired Design of Hybrid Intelligent Systems*,
Studies in Computational Intelligence 667, DOI 10.1007/978-3-319-47054-2_12

The rest of this paper is organized as follows, in Sect. 2 some basic concepts of LVQ are described, Sect. 3 presents some concepts of fuzzy systems, Sect. 4 shows the proposed model, Sect. 5 presents the simulation results and finally in Sect. 6 some conclusions are presented.

2 LVQ

LVQ (Learning Vector Quantization) is an adaptive method that is used for data classification [2–4]. It consists of two layers, a linear layer with supervised training method, and a competitive layer which employs unsupervised data-clustering techniques to preprocess the dataset and obtain the cluster centers. When there is no available information regarding the desired outputs, unsupervised techniques update weights only on the basis of the input patterns. In these cases competitive learning is the popular scheme to achieve this task.

The LVQ [19, 20] algorithm consists of two steps. In the first step, the unsupervised data-clustering method is used to locate the cluster centers, the number of cluster centers can either be determined a priori or with a cluster technique capable of adding new cluster centers when necessary [8, 10].

In the second step of the LVQ algorithm, the cluster centers are fine-tuned to approximate to the desired position. In the learning process, first a cluster center **w** that is closest to an input vector **x** must be found. If both **w** and **x** belong to the same class, **w** is moved toward **x**, otherwise **w** is moved away from **x**. The LVQ algorithm is as follows:

1. Initialize the cluster centers using a clustering method.
2. Label each cluster.
3. Randomly select a training input vector x and find k such that $\|x - w_k\|$ is a minimum.
4. If x and w_k belong to the same class, update w_k by

$$\Delta w_k = \eta(x - w_k)$$

   else

$$\Delta w_k = -\eta(x - w_k)$$

5. If the maximum number of iterations is reached, stop. Otherwise return to 3.

In the above description, η is the learning rate, which is a small constant that should decrease with each iteration. Figure 1 shows the architecture for a LVQ network.

Research regarding classification with LVQ includes [5, 12, 13, 17]. Besides, this algorithm has been applied on various areas such as galaxy classification based on their shape.

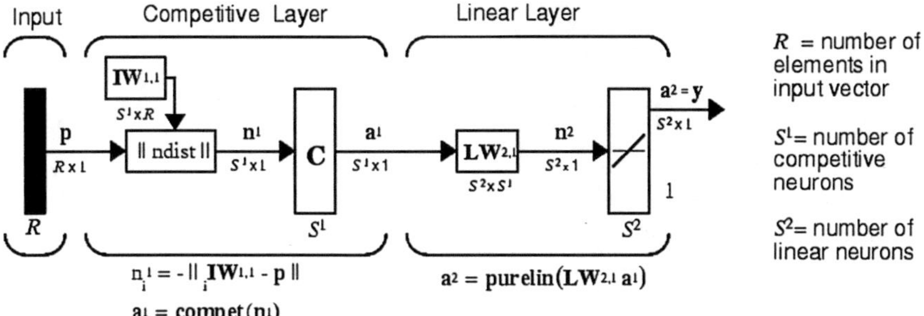

Fig. 1 Architecture of LVQ network

3 Fuzzy Systems

Nowadays there is a variety of fuzzy systems applications, such as cameras, washing machines, medical instrumentation, and industrial process control, among others. Fuzzy systems are based on the theory of fuzzy sets, fuzzy if-then rules and fuzzy reasoning.

Unlike the classical sets, a fuzzy set expresses the degree of membership of an element to a set, therefore the function of a fuzzy set takes values between 0 and 1, which denotes the degree of membership of a given element to a set. Although the classic sets have proven to be an important tool in the area of computing, they do not represent the nature of the concepts used by humans, which tend to be vague.

As an extension of the definition of classical sets, a fuzzy set can be defined as follows: If Z is a collection of objects generically denoted as z, then a fuzzy set X in Z is defined as a set of ordered pairs

$$X = \{(z, \mu_x(z)) \mid z \in Z\}$$

where $\mu_x(z)$ is the membership function for the fuzzy set X. Here the membership function maps each element of Z to a membership grade between 0 and 1 [8]. This definition is just an extension of the definition of a classical set.

There are three models of fuzzy systems, which are Mamdani, Sugeno, and Tsukamoto; the difference between these models lays in the form of the consequents of their fuzzy rules, hence their aggregation and defuzzification processes are also different.

3.1 Type-2 Fuzzy Systems

Type-2 fuzzy systems were proposed by Zadeh as an extension of fuzzy systems. Also known as Interval Type-2 Fuzzy Systems, these have proved to be successful when the data contains much more noise.

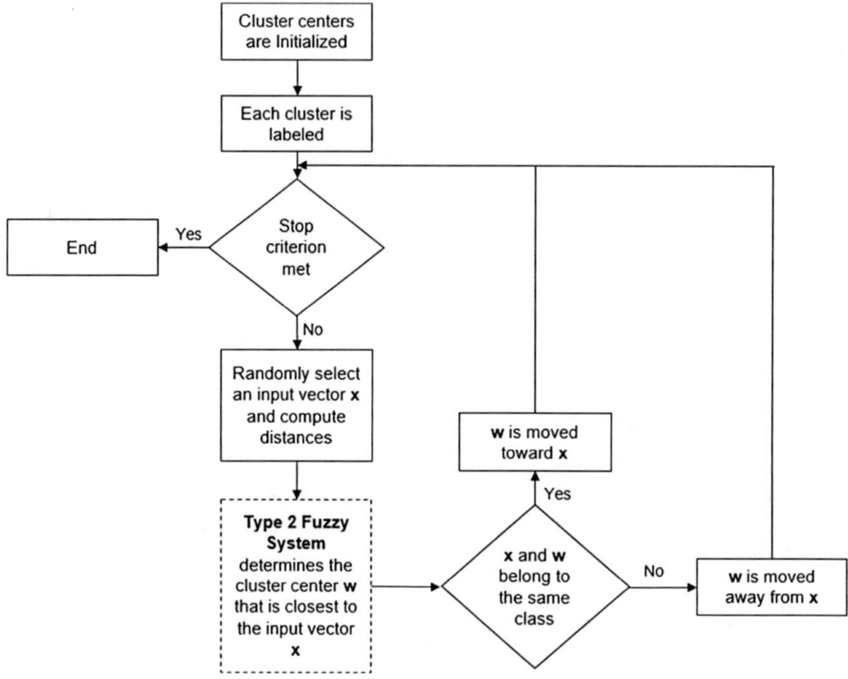

Fig. 2 Flow chart for the proposed model

Frequently, the knowledge used to build the fuzzy rules in a fuzzy inference system is uncertain. The consequents, obtained from experts, are different because experts do not necessarily agree; in fuzzy systems these inconsistencies are interpreted as membership functions with uncertain antecedents and consequents, which can be defined as *fuzzy membership functions*. Hence, the difference between type-1 Fuzzy Systems and Type-2 Fuzzy Systems is related to the membership functions, where Type-2 Fuzzy Systems membership functions have degrees of membership, which define a primary membership function and a secondary membership function. Figure 2 shows an example of this concept.

Due to the good results that are achieved by Type-2 Fuzzy Systems handling noise in data, these have become an interesting research area, some of the works include logic controllers, optimization, neural networks, etc. [1, 14, 15, 18].

4 Proposed Model

In this work, the implementation of a Type-2 Fuzzy System in a modular LVQ network architecture for classification is presented. Our Type-2 Fuzzy System model is focused on determining the smallest distances between an input vector **x** and each of the cluster centers (denoted as **w** in Sect. 2).

In the third step of the LVQ algorithm, the distances are computed, and these data are used to dynamically create the type-2 fuzzy system, and at the same time these distances are the input of the type-2 fuzzy system. The LVQ algorithm is an iterative process; hence in each iteration of the LVQ method, a new type-2 fuzzy system for finding the smallest distance is created. In Fig. 2 the flow chart for this model is presented.

The output of the type-2 fuzzy system is then the cluster center **w**, which is closest to the input vector **x**. An example of the fuzzy rules for this model is as follows:

- If (dist1 is small) and (dist2 is not small) and (dist3 is not small) then (Closest Center is C1).
- If (dist1 is not small) and (dist2 is small) and (dist3 is not small) then (Closest Center is C2).
- If (dist1 is not small) and (dist2 is not small) and (dist3 is small) then (Closest Center is C3).

Notice that since we are interested in the smallest distance between an input vector and the defined cluster centers, it is important to have the negation of the respectively distances in each of the rules. As mentioned before, this process is performed dynamically, depending on the user-defined number of cluster centers. Regarding the membership functions used in the fuzzy system architecture, triangular membership functions are used, and these are also dynamically created. In Fig. 3 an example of one of these functions is presented. Figure 3a shows an example of the membership functions for an input variable, where F1, F2, and F3 represent the granularity for Small, Medium, and Large distance, respectively. Figure 3b shows the membership functions for the output variable, in this example with 4 possible cluster centers, represented by C1, C2, C3, C4, and C5.

This type-2 fuzzy system model was applied to a five-module architecture LVQ neural network for arrhythmia classification. Each of the modules works with three different classes of arrhythmias. The arrhythmia dataset is described in more detail below.

4.1 Arrhythmia Dataset

Arrhythmias are expressed as changes in the normal sequences of electrical hearth impulses, which may happen too fast, to slowly or erratically, and can be measured

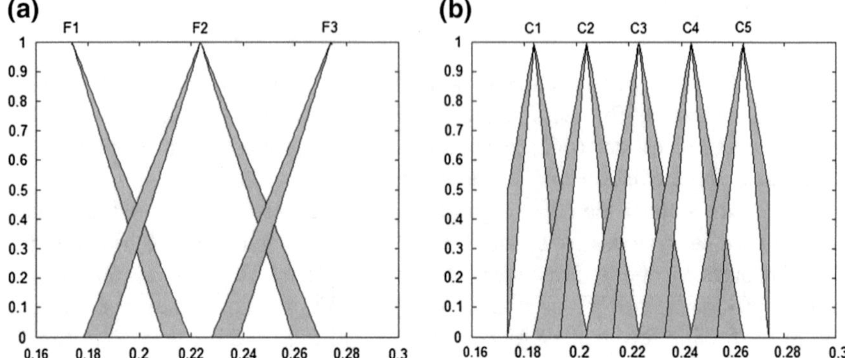

Fig. 3 Example of **a** an input membership function and **b** an output membership function

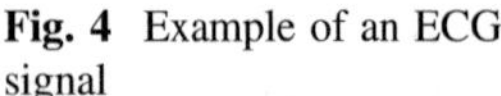

Fig. 4 Example of an ECG signal

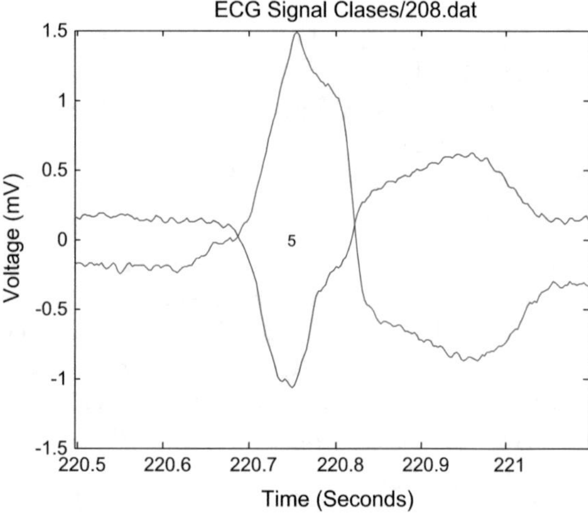

with a Holter device in ECG signals [6]. Figure 4 shows an example of one of these ECG signals [7, 9, 21].

The MIT-BIH [16] arrhythmia dataset was used for this research. This database consists of 15 different types of arrhythmias. This database contains 48 half-hour excerpts of ECG recordings, obtained from 47 subjects. The recordings were digitized at 360 samples per second per channel with 11-bit resolution over a 10 mV range. The recordings were preprocessed as follows:

- Analog outputs were filtered using a passband from 0.1 to 100 Hz. The passband-filtered signals were digitized at 360 Hz per signal.
- Taking the R point [27] as reference, in the obtained signals was located the start and end point for each wave.

Table 1 Simulation results

	Clusters	LR	Accuracy	Time
1	6	0.01	97.88	00:57
2	8	0.01	98.89	01:02
3	10	0.01	98.89	01:15
4	5	0.01	93.01	00:59
5	8	0.01	98.89	01:31
6	7	0.01	97.88	01:09
7	9	0.01	98.89	01:26
8	10	0.01	98.89	01:43
9	6	0.01	97.88	01:06
10	8	0.01	98.89	01:19
11	6	0.01	97.88	01:11
12	9	0.01	98.89	01:26
13	10	0.01	98.89	01:28
14	7	0.01	97.88	01:14
15	10	0.01	98.89	01:39

- The 38 higher voltages and 38 lower voltages of each vector were taken, resulting in vectors of 76 V.

5 Simulation Results

A set of 15 experiments were performed using this new classification method, based on LVQ neural networks and type-2 fuzzy systems, Table 1 shows the obtained results, where ***Time*** is expressed in the HH:MM format. Experiments were performed in a Windows 7 PC × 64, Quad-Core i7 processor and 16 GB of RAM.

6 Conclusions

In this paper, a classification method based on a modular LVQ [30] network architecture for classification of arrhythmias is presented. Although the used approach is very simple for a type-2 fuzzy system, and the obtained results are good enough to state that the combination of type-2 fuzzy logic and LVQ networks to solve classification problems could lead to the development of more robust and reliable classification systems.

As a future work we plan to use new approaches for applying type-2 fuzzy logic to different architectures of LVQ networks for classification.

References

1. Amador-Angulo L., Castillo O. Statistical Analysis of Type-1 and Interval Type-2 Fuzzy Logic in dynamic parameter adaptation of the BCO. IFSA-EUSFLAT, pp. 776-783, 2015.
2. Amezcua J., Melin P. A modular LVQ Neural Network with Fuzzy response integration for arrhythmia Classification. IEEE 2014 Conference on Norbert Wiener in the 21st century. Boston. June 2014.
3. Biswal B., Biswal M., Hasan S., Dash P.K. Nonstationary power signal time series data classification using LVQ classifier. Applied Soft Computing, Elsevier. Vol. 18, May 2014. pp. 158-166.
4. Castillo O., Melin P., Ramirez E., Soria J. Hybrid intelligent system for cardiac arrhythmia classification with Fuzzy K-Nearest Neighbors and neural networks combined with a fuzzy system, Journal of Expert Systems with Applications, vol. 39 (3) 2012 pp. 2947-2955.
5. Grbovic M., Vucetic S. Regression Learning Vector Quantization. 2009 Ninth IEEE International Conference on Data Mining. Miami, December 2009.
6. Hu Y.H., Palreddy S., Tompkins W. A patient adaptable ECG beat classifier using a mixture of experts approach, IEEE Trans. Biomed. Eng. (1997) 891–900.
7. Hu Y.H., Tompkins W., Urrusti J.L., Afonso V.X. Applications of ANN for ECG signal detection and classification, J. Electrocardiol. 28 (1994) 66–73.
8. Jang J., Sun C., Mizutani E. Neuro-Fuzzy and Soft Computing. Prentice Hall, New Jersey, 1997.
9. Kim J., Sik-Shin H., Shin K., Lee M. Robust algorithm for arrhythmia classification in ECG using extreme learning machine, Biomed. Eng. Online, October 2009.
10. Kohonen T. Improved versions of learning vector quantization. International Joint Conference on Neural Networks, pages Vol. 1: 545-550, San Diego, 1990.
11. Learning Vector Quantization Networks. Site: http://www.mathworks.com/help/nnet/ug/bss4b_l-15.html. Last Access: 06/24/2014.
12. Martín-Valdivia M.T., Ureña-López L.A., García-Vega M. The learning vector quantization algorithm applied to automatic text classification tasks, Neural Networks 20 (6) (2007) 748–756.
13. Melin P., Amezcua J., Valdez F., Castillo O. A new neural network model based on the LVQ algorithm for multi-class classification of arrhythmias. Information sciences. 279 (2014) 483–497.
14. Melin P., Castillo O. A review on type-2 fuzzy logic applications in clustering, classification and pattern recognition. Applied Soft Computing. Vol. 21, August 2014, pp. 568-577.
15. Mendoza O., Melin P., Castillo O. Interval type-2 fuzzy logic and modular neural networks for face recognition applications. Applied Soft Computing. Vol. 9, September 2009, pp. 1377-1387.
16. MIT-BIH Arrhythmia Database. PhysioBank, Physiologic Signal Archives for Biomedical Research, site: http://www.physionet.org/physiobank/database/mitdb/. Last access: 06/24/2014.
17. Nasiri J.A., Naghibzadeh M., Yazdi H.S., Naghibzadeh B. ECG arrhythmia classification with support vector machines and genetic algorithm, in: Third UKSim European Symposium on Computer Modeling and Simulation, 2009.
18. Olivas F., Valdez F., Castillo O., Melin P. Dynamic parameter adaptation in particle swarm optimization using interval type-2 fuzzy logic. Soft Computing. 2009, pp. 1057-1070.
19. Pedreira C. Learning Vector Quantization with Training Data Selection. IEEE Transactions on Pattern Analysis and Machine Intelligence, January 2006. Vol 28. pp. 157-162.
20. Torrecilla J.S., Rojo E., Oliet M., Domínguez J.C., Rodríguez F. Self-organizing maps and learning vector quantization networks as tools to identify vegetable oils and detect adulterations of extra virgin olive oil, Comput. Aided Chem. Eng. 28 (2010) 313–318.
21. Tsipouras M.G., Fotiadis D.I., Sideris D. An arrhythmia classification system based on the RR-interval signal, Artif. Intell. Med. (2005) 237–250.

A New Method Based on Modular Neural Network for Arterial Hypertension Diagnosis

Martha Pulido, Patricia Melin and German Prado-Arechiga

Abstract In this paper, a method is proposed to diagnose the blood pressure of a patient (Astolic pressure, diastolic pressure, and pulse). This method consists of a modular neural network and its response with average integration. The proposed approach consists on applying these methods to find the best architecture of the modular neural network and the lowest prediction error. Simulations results show that the modular network produces a good diagnostic of the blood pressure of a patient.

Keywords Modular neural networks · Systolic · Diastolic · Pulse · Time series · Diagnosis

1 Introduction

At the beginning of the twentieth century a cardiovascular disease was a rare cause of death and disability worldwide but by the end of this century such conditions were set up as a major deaths and permanent damage. In 2001 in the adult population was the number one cause of death in five of the six world regions suggested by the World Health Organization. According to data of this organization, 30 % of deaths in the world, that is, 17 million people a year are by this disease. The fast increasing mortality from cardiovascular diseases in this relatively short time is attributed to various factors, such as changes in diet, lack of physical activity and increased life expectancy, which characterize the development of industrialized societies. High prevalence of hypertension in different populations has contributed significantly to the current pandemic diseases, such as the estimated five million of the previous deaths are caused by cerebral vascular events, which is closely related to the presence of hypertension [1–3].

M. Pulido · P. Melin (✉) · G. Prado-Arechiga
Tijuana Institute of Technology, Tijuana, Mexico
e-mail: pmelin@tectijuana.mx

P. Melin et al. (eds.), *Nature-Inspired Design of Hybrid Intelligent Systems*,
Studies in Computational Intelligence 667, DOI 10.1007/978-3-319-47054-2_13

This paper is organized as follows: Sect. 2 describes the concepts of neural networks, in Sect. 3, we describe the concepts of Hypertension, in Sect. 4 the problem statement and proposed method are presented, Sect. 6 shows the simulation results of the proposed method, and Sect. 7 offers conclusions and future work.

2 Neural Networks

Neural networks are composed of many elements (Artificial Neurons), grouped into layers that are highly interconnected (with the synapses). This structure has several inputs and outputs, which are trained to react (or give values) in a way we want to input stimuli (R values). These systems emulate in some way, the human brain. Neural networks are required to learn to behave (Learning) and someone should be responsible for the teaching or training (Training), based on prior knowledge of the problem environment [4].

Artificial neural networks are inspired by the architecture of the biological nervous system, which consists of a large number of relatively simple neurons that work in parallel to facilitate rapid decision-making [5].

A neural network is a system of parallel processors connected as a directed graph. Schematically each processing element (neuron) of the network is represented as a node. These connections establish a hierarchical structure that is trying to emulate the physiology of the brain as it looks for new ways of processing to solve the real-world problems. What is important in developing the techniques of an NN is if it is useful to learn the behavior, recognize, and apply relationships between objects and plots of real-world objects themselves. In this sense, artificial neural networks have been applied to many problems of considerable complexity. Its most important advantage is in solving problems that are too complex for conventional technologies, problems that have no solution or an algorithm of for solution is very difficult to find [6].

3 Hypertension Arterial

Hypertension is a disease present in almost all human groups currently inhabiting the planet. It is a very common disease and because cardiovascular complications that accompany it, is a global health problem. Recently, a study in order to develop policy for the prevention and control of this disease was published. In this study, national surveys on the prevalence of hypertension reported for the different countries were used. In most of them the mercury sphygmomanometer was used to measure blood pressure and on a single occasion the numbers reported overestimates the Measurements made on two separate visits in three studies.

The social consequences of a disease such as hypertension are many and may have different approaches. First, it can be concluded mortality generated, which is a very strong epidemiological criterion, although the discussion can be extended to cover different topics, among which are the business days and years of productive life lost, and finish by doing some considerations on the cost of treatment.

Some of the major risk factors for hypertension include: obesity, Lack of exercise, Smoking, Salt consumption, Alcohol, the stress level, Age, Sex, Ethnicity, and Genetic factors.

It is very important to note that the pressure varies from beat to beat, day and night, facing everyday situations, such as walking, talking on the phone or exercise. Therefore, variation in blood pressure is a normal phenomenon. The tension is not related to the day and night, but with the activity and rest (sleep) [7, 8].

4 Pulse Pressure

The pulse is the number of heart beats per minute, the procedure for how the test is performed and the pulse can be measured in the following regions of the human body: The back of the knees, the groin, the neck, Temple, top, or inner side of the foot.

In these areas, an artery passes close to the skin. To measure the pulse at the wrist, place the index and middle fingers on the front of the opposite wrist, below the thumb. Press with the fingers until a pulse is sensed to measure the pulse on the neck, place the index, and middle fingers to the side of the Adam's apple in the soft, hollow, and press gently until we locate the pulse. Note: sit or lie down before taking the pulse of the neck. Neck arteries in some people are sensitive to pressure and fainting may occur or decreased heartbeat. Also, do not take the pulse on both sides of the neck at the same time. Doing this can reduce the flow of blood to the head and lead to fainting. Once we find the pulse, we count the beats for a full minute or for 30 s and multiply by two, which will give the beats per minute. In exam preparation to determine the resting heart rate, the person must have been resting for at least 10 min.

4.1 Exam Preparation

If we are going to determine the resting heart rate, the person must have been resting for at least 10 min, take heart rate during exercise while training and a persona slight finger pressure is made why the test is performed. Pulse measurement provides important information about the health of a person. Any change from normal heart rate can indicate a medical condition (Bernstein). The rapid pulse can be a sign of an infection or dehydration. In emergency situations, the pulse rate can help determine if the patient's heart is pumping. During or immediately after the pulse provides information about your fitness level and health) [9–12].

4.2 Normal Values

For the resting heart rate the normal values are: Newborns (0–1 month old): 70–190 beats per minute. Babies (1–11 months old): 80–160 beats per minute, Children (1–2 years old): 80–130 beats per minute, Children (3–4 years old): 80–120 beats per minute, Children (5–6 years old): 75–115 beats per minute, Children (7–9 years old): 70–110 beats per minute, Children 10 years and older and adults (including elderly): 60–100 beats per minute, trained athletes: 40–60 beats per minute.

The resting heart rates that are continually high (tachycardia) may indicate a problem and that the person needs to consult a doctor. Also about resting heart rates that are below the normal values (bradycardia). Also, the doctor should check a pulse to observe that is very firm (bounding pulse) and that lasts longer than a few minutes. An irregular pulse can also indicate a problem. A pulse that is hard to find may mean there is blockage in the artery. These blockages are common in people with diabetes or atherosclerosis from high cholesterol. In this case the doctor may order a test known as a Doppler study to assess obstructions.

4.3 Bradycardia

Bradycardia is characterized by slow heart rate usually below 60 beats per minute, while the normal resting rate is 60–100 beats per minute [13, 14].

4.4 Sinus Tachycardia

In Cardiology, sinus tachycardia is a heart rhythm disorder characterized by an increased frequency of cardiac impulses originating from the sinus node is the natural pacemaker of the heart, and defined with a heart rate greater than 100 beats per minute in an average adult. When the normal frequency is 60–100, in adult-although rarely exceeds 200 bpm. Typically, sinus tachycardia begins and ends gradually in contrast with supraventricular tachycardia, which appears gradually and may end abruptly [15, 16].

5 Problem Statement and Proposed Method

This section describes the problem statement and the proposed method for pressure blood diagnostic based on modular neural networks with average response integration. One of the main goals is to implement a Modular Neural Network, where are number of modules is 3, the first module is for the Astolic Pressure, in the

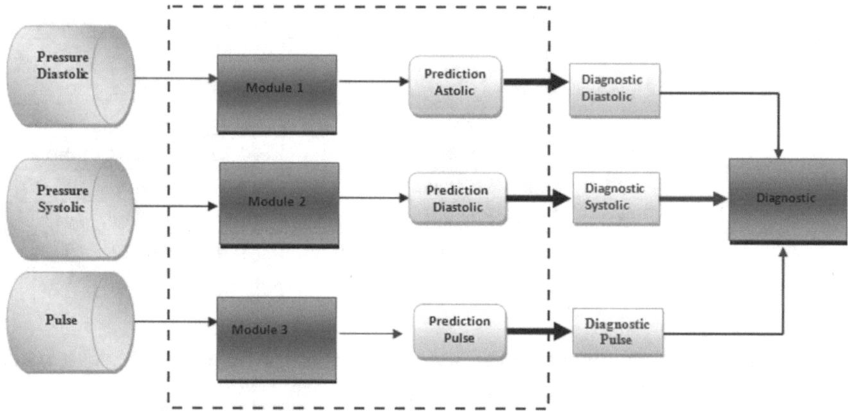

Fig. 1 The proposed modular neural network

second module we have the Diastolic Pressure and third module is for the Pulse in order to diagnose the blood pressure of a person.

Figure 1 illustrates the modular neural network model, where we have first the historical data, for the modular neural network, which in this case has three modules, as well as the number of layers that this module that could be from 1 to 3 and the number of neurons by layer that could be from 1 to 3. Then after the responses of the modular neural network ensemble are obtained, the integration is performed with the Average Integration method.

In Fig. 1 each of the modules is a modular neural network and the corresponding calculations are performed as follows.

The net input $\bar{x}$ of a node is defined by the weighted sum of the incoming signals plus a bias term. For instance, the net input of the output of the node j.

$$\bar{x}_j = \sum\nolimits_i \omega_{ij} + \omega_j,$$

$$x = f(\bar{x}_j) = \frac{1}{1 + \exp(-\bar{x}_j)}, \tag{1}$$

where x_i is the output of node i located in any of the previous layers, ω_{ij} is the weight associated with the link connecting nodes i and j, and ω_j is the bias of node j. Since the weights ω_{ij} are actually internal parameters associated with each node j and changing the weights of the node will alter behavior of the node and in turn alter behavior of the whole backpropagation MLP.

A squared error measure for the pth input-output pair is defined as:

$$E_P = \sum (d_k - x_k)^2, \tag{2}$$

Table 1 Definitions and classification of the blood pressure levels (mmHg)

Category	Systolic		Diastolic
Optimal	<120	And	<80
Normal	120–129	And/or	80–84
High normal	130–139	And/or	85–89
Grade 1 hypertension	140–159	And/or	90–99
Grade 2 hypertension	160–179	And/or	100–109
Grade 3 hypertension	≥ 180	And/or	≥ 110
Isolated systolic hypertension	≥ 140	And/or	<90

where d_k is the desired output for the node x_k is the actual output for the node when the input part of the *p*th data pair is presented.

Historical data of the diastolic pressure historical data were obtained from 16 people with 45 samples, for systolic pressure and pulse were also the data of the same people.

In Table 1 the blood pressure (BP) category is defined by the highest level of BP, whether systolic or diastolic. And should be graded as 1, 2, or 3 according to the systolic or diastolic BP value. Isolated systolic hypertension it is according to the systolic BP value in the ranges indicated.

6 Simulation Results

This section shows results of the optimization of the modular neural network. The main goal is the diagnosis of high blood pressure, where experiments with 16 persons are presented.

Table 2 shows the results of the modular neural network applied to the diagnosis of high blood pressure.

Table 3 shows the average of 30 experiments of the modular network for each of the 16 people, the average time, Diastolic, Systolic, and pulse are represented.

Figure 2 shows results of the best architecture of the Modular Network, where for the first module of Diastolic Pressure, Systolic Pressure, and Pulse, 25 neurons were used in the first layer and 30 neurons for the second layer for each the modules, and the Target Error was of 0.002 and 500 epoch were used.

Figure 3 shows the plot of the diagnostic error of the modular neural network for the Systolic Pressure.

Figure 4 shows the plot of the diagnostic error of the modular neural network for the Diastolic Pressure.

Table 2 Results of modular neural network

No. persons	Ėpoch	Target error	Number of layers	Number of neurons	Time	Systolic	Diastolic	Pulse	Diagnostic
Persona 1	500	0.02	2	15,13	00:05:06	116	74	67	Optimal
Persona 2	500	0.02	2	15,13	00:05:06	106	72	77	Optimal
Persona 3	500	0.02	2	15,13	00:05:06	114	70	80	Optimal
Persona 4	500	0.02	2	15,13	00:05:06	119	71	72	Optimal
Persona 5	500	0.02	2	15,13	00:05:06	145	84	75	Grado 1 hypertension
Persona 6	500	0.02	2	15,13	00:05:06	104	61	90	Optimal
Persona 7	500	0.02	2	15,13	00:05:06	125	87	97	Normal
Persona 8	500	0.02	2	15,13	00:05:06	109	64	97	Optimal
Persona 9	500	0.02	2	15,13	00:05:06	129	74	73	Normal
Persona 10	500	0.02	2	15,13	00:05:06	122	78	57	Normal
Persona 11	500	0.02	2	15,13	00:05:06	136	63	65	High normal
Persona 12	500	0.02	2	15,13	00:05:06	136	81	70	High normal
Persona 13	500	0.02	2	15,13	00:05:06	120	75	78	Normal
Persona 14	500	0.02	2	15,13	00:05:06	110	62	77	Optimal
Persona 15	500	0.02	2	15,13	00:05:06	121	69	70	Normal
Persona 16	500	0.02	2	15,13	00:05:06	132	82	80	High normal

Table 3 Average results of the modular neural network

Person	Time	Systolic	Diastolic	Pulse
Person 1	00:13:09	115	73	67
Person 2	00:13:09	105	72	77
Person 3	00:13:09	114	70	80
Person 4	00:13:09	119	71	72
Person 5	00:13:09	145	84	75
Person 6	00:13:09	104	61	90
Person 7	00:13:09	125	87	96
Person 8	00:13:09	109	64	73
Person 9	00:13:09	129	73	56
Person 10	00:13:09	122	77	65
Person 11	00:13:09	136	63	70
Person 12	00:13:09	136	81	71
Person 13	00:13:09	120	74	78
Person 14	00:13:09	110	62	77
Person 15	00:13:09	119	68	70
Person 16	00:13:09	131	82	80

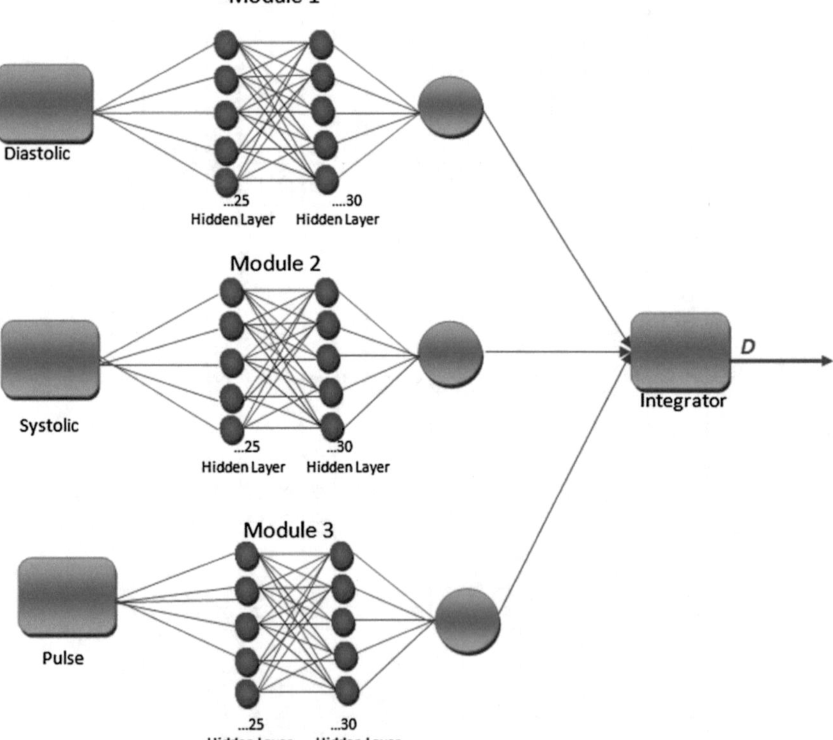

Fig. 2 The best architecture for the modular neural network

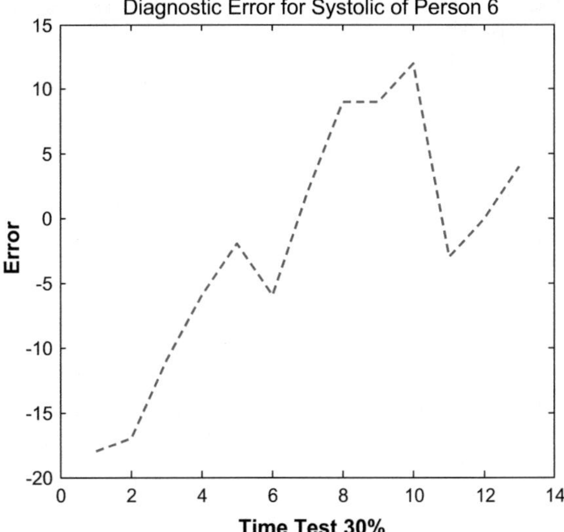

Fig. 3 Diagnostic error of the systolic pressure with the modular neural network

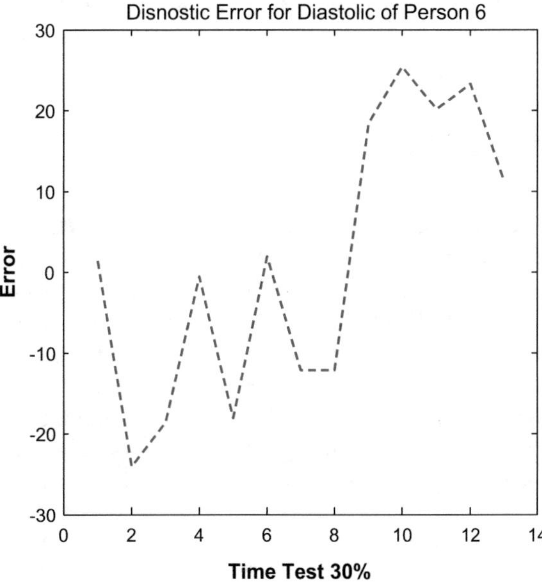

Fig. 4 Diagnostic error of the diastolic pressure with the modular neural network

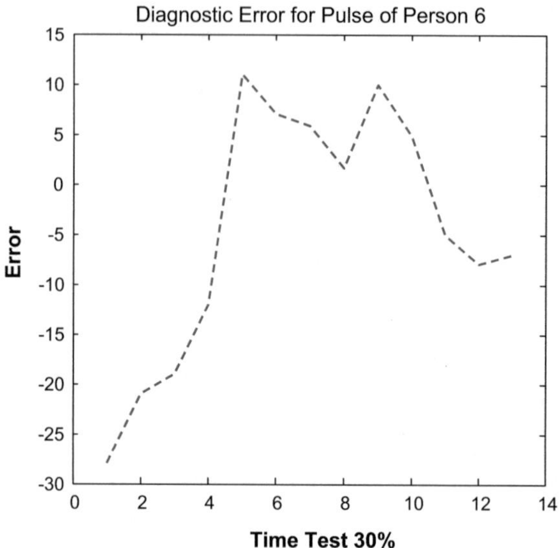

Fig. 5 Diagnostic error of the pulse with the modular neural network

Figure 5 shows the plot of the diagnostic error of the modular neural network for the Pulse.

7 Conclusions

We can conclude that with the proposed method based on modular neural networks good results are obtained in diagnosing the risk of hypertension of person. As we can note in this work, modular neural networks have proven to be a reliable and accurate technique compared to conventional statistical methods for this problem. As is the case in this work, the results obtained in each of the models can be considered good because the margin of error that is obtained is very small and another important part of this paper is the training method that is used which is the Levenberg–Marquardt algorithm (trainlm). This is a good method as is able to solve diagnosis problems faster and easier and as it guarantees a high learning speed.

Acknowledgments We would like to express our gratitude to the CONACYT, Tijuana Institute of Technology for the facilities and resources granted for the development of this research.

References

1. The WHO MONICA Project: Geographical variation in the major risk factors of coronary heart disease in men and women aged 35–64 years. Rapp Trimest Statis Sanit Mond 1988; 41 115-139.
2. Mendis S., Lindholm L. H., Mancia G., Whiwort J., Alderman M et al.: World Health Organization (WHO) and International Society of Hypertension (ISH) risk prediction charts: assessment of cardiovascular risk for prevention and control of cardiovascular disease in low and middle income countries. J Hypertension 2007; 25(8):1578-1582.
3. Sacco RL: The 2006 William Feinberg Lecture: Shifting the paradigm from stroke to global vascular risk estimation. Stroke 2007;38 1980-1987.
4. Brockwell, P.D. and Davis R. A. "Introduction to Time Series and Forecasting", Springer-Verlag New York, 2002, pp. 1-219.
5. Beevers G, Lip GYH y O'Brien E. 2001. Blood pressure measurement Part I—Sphygmomanometry: factors common to all techniques. *British Medical Journal* 322 (7292): 981-985.
6. Cowpertwait P. and Metcalfe A. "Time Series", Introductory Time Series with R., Springer Dordrecht Heidelberg London New York, 2009, pp. 2–5.
7. Kearney P.M., Whelton M., Reynaldos K., Whelton P. K., Worldwide prevalence of hypertension: a systematic review. J Hypertens 2004; 22(1): 21-24.
8. Agustino R.B., Russel M.W., Huse D.M., Ellison C., Silbershatz et al: Primary and subsequent coronary risk appraisal: new results The Framing Study. Am Heart J 2000: 139: 272-281.
9. Samant, R., and Srikantha R. "Evaluation of Artificial Neural Networks in Prediction of Essential Hypertension." International Journal of Computer Applications 2013. vol. 14, pp. 11-21.
10. Samant, R. and Srikantha R. "Performance of alternate structures of artificial neural networks in prediction of essential hypertension." International Journal of Advanced Technology & Engineering Research 2013. vol. 10, pp. 1-11.
11. Shehu, N., S. U. Gulumbe, and H. M. Liman. "Comparative study between conventional statistical methods and neural networks in predicting hypertension status." Advances in Agriculture, Sciences and Engineering Research 2013.
12. Simel DL. Approach to the patient: history and physical examination. In: Goldman L, Schafer AI, eds. *Goldman's Cecil Medicine.*
13. Djam, X. Y., and Y. H. Kimbi. "Fuzzy expert system for the management of hypertension." The Pacific Journal of Science and Technology 2011. vol. 11, pp. 1.
14. Harrison Principios de Medicina Interna 16a edición (2006). «Capítulo 214. Taquiarritmias». *Harrison online en español.* McGraw-Hill.
15. Azian A., Abdullah, Zakaria Z., and Mohammad NurFarahiyah. "Design and development of Fuzzy Expert System for diagnosis of hypertension." International Conference on Intelligent Systems, Modelling and Simulation, Univ. Malaysia Perils, Jejawi, Malaysia, IEEE 2011. vol. 56, no. 5-6, pp.
16. Azian A., Abdullah, Zakaria Z., and Mohammad NurFarahiyah. "Design and development of Fuzzy Expert System for diagnosis of hypertension." International Conference on Intelligent Systems, Modelling and Simulation, IEEE 2011. vol. 10, pp. 131-141.

Spectral Characterization of Content Level Based on Acoustic Resonance: Neural Network and Feedforward Fuzzy Net Approaches

Juan Carlos Sanchez-Diaz, Manuel Ramirez-Cortes, Pilar Gomez-Gil, Jose Rangel-Magdaleno, Israel Cruz-Vega and Hayde Peregrina-Barreto

Abstract Free vibration occurs when a mechanical system is disturbed from equilibrium by an external force and then it is allowed to vibrate freely. In free vibrations, the system oscillates under the influence of inherent forces on the system itself. Free vibrations are associated with natural frequencies that are properties of the oscillating system, quantified in parameters such as mass, shape, and stiffness distribution. A number of these mechanical characteristics can be inferred from vibration patterns or from the generated sound using the adequate sensors. It is well known that liquid level inside a container modifies its natural frequencies. Unfortunately, other container characteristics such as shape, composition, temperature, and pressure modifies the natural frequencies of vibration making the task of level measurement nontrivial. Preliminary experiments aiming to do measurement of liquid content level and container characterization are presented in this work. Spectral analysis in Fourier domain is used to perform feature extraction, with the feature vectors containing information about the frequencies having the greatest amplitude in the respective spectral analysis. Classification has been carried out using two computational intelligence techniques for comparison purposes: neural network classification and a fuzzy logic inference system built using singleton fuzzifier, product inference rule, Gaussian membership functions and center average defuzzifier. Preliminary results showed a better performance when using the neural network-based approach in comparison to the fuzzy logic-based approach, obtaining in average a MSE of 0.02 and 0.09, respectively.

J.C. Sanchez-Diaz · M. Ramirez-Cortes · J. Rangel-Magdaleno · I. Cruz-Vega
Department of Electronics, National Institute of Astrophysics,
Optics and Electronics, Santa María Tonantzintla, Puebla, Mexico

P. Gomez-Gil (✉) · H. Peregrina-Barreto
Department of Computer Science, National Institute of Astrophysics,
Optics and Electronics, Santa María Tonantzintla, Puebla, Mexico
e-mail: pgomez@inaoep.mx

P. Melin et al. (eds.), *Nature-Inspired Design of Hybrid Intelligent Systems*,
Studies in Computational Intelligence 667, DOI 10.1007/978-3-319-47054-2_14

Keywords Acoustic resonance · Vibration · Spectral analysis · Neural network · Fuzzy system

1 Introduction

Measurement of liquid level inside a non-transparent closed container is a required common task in industrial, commercial, and domestic environments. For that purpose, different methods and techniques have been recently developed, such as the use of capacitive sensors, optical sensors, ultrasound sensors, differential pressure sensors, and microwave sensors, among others [1, 2]. Unfortunately, in many situations, these measurements are required to be done in a noninvasive way and in a short period of time, such as in a production line or during quality or security inspections, for instance. This requirement limits the available methods to those where sensors are placed outside the container, and the sensors have minimum contact with container's surface, or even better using contactless procedures. Recent research has focused the attention in developing new methods for liquid level measurement inside opaque containers or to improve the existent ones. Nakagawa et al. [2] used a combination of Doppler sensors and a piezoelectric vibrator to improve the accuracy for liquid level measurement inside opaque containers. Results showed a nonlinear error in measurement of ± 0.5 mm. However, the method is suitable to be used just with plastic or non-metallic containers and water or other similar fluid as contained liquid because the use of millimeter waves. Furthermore, the need of sensor displacement along container surface to take a measurement requires an amount of time that depends of the speed of this process.

The use of acoustic resonators is another promising field of research that has been widely studied. An acoustic resonator consists of an acoustic chamber with a constant diameter and a total free length which depends on the liquid level inside the container. A loudspeaker is located inside the resonator to generate acoustic waveforms that are captured by a microphone inside the container. Donglagic et al. [3] used an acoustic resonator and a neural network-based approach to build a container level detector. Results showed an accuracy ranging from 2 mm in the best case to 1–3 cm in the worst case. An alternative method based on resonators developed for content level measurement by the same authors is presented in [4]. In this article, the authors propose an algorithm to estimate the level of content based on frequency response of the resonator incorporating a sweep in frequency. In that paper several considerations are included, such as impact of the airing phenomenon (presence of holes in a closed resonator), thickness of resonator walls and foaming, obtaining in average a resolution of 0.12 %. Jung et al. [5] performed a similar experiment, with two different loudspeaker-microphone configurations, and two approaches: cepstral analysis, and Fourier spectrum. In their concluding remarks, the authors describe that using the loudspeaker in the lid of the resonator and the microphone in side one (c-configuration) got better accuracy than microphone and loudspeaker in opposite sides (s-configuration). Also, frequency spectrum was more effective than cepstrum, with an accuracy of 0.5 mm for a 601 mm resonator, in the best case. Webster and Davies [6] present another method of content level measurement based on Helmoltz resonance. A Helmholtz resonator consists of a port tied to a volumetric chamber. Port's length and diameter and volume of the chamber determine the resonance frequency for the resonator. Once the volume is determined, liquid level inside the container can be easily calculated. Frequency swept was used as input, and microphones as sensors. The reported accuracy is in the order of ± 0.1 %. In the work of Brunnader and Holler [7], a Helmholtz resonator model is used to estimate the free volume of an arbitrary shape container corresponding to an automobile gas tank. Pressure sensors are used in the experimental setup, and an impedance equivalent model is obtained. Despite accuracy and simplicity, these methods have disadvantages which make them not directly applicable in the problem presented in this work. These methods are within the category of invasive, as insertion of a resonator inside the container is required. As an additional drawback, dynamic response of these methods is slow as they require in the range of 40 s to several minutes in order to obtain a single measurement. Other methods involve the use of flexural waves. Flexural waves appear on plate's surfaces when an excitation force is supplied. Lucklum and Jacobi [8] use planar coils and special plate resonators to make a device in which mechanic resonance can be converted in electrical impedance. Using this device coupled to a fluidic chamber, authors generate vibration modes leading to the radiation of compressional longitudinal waves into the liquid. Longitudinal waves travel through the liquid content and are reflected to the surface, producing interference. The interference modifies the electric impedance according to liquid level in the chamber, and then the level can be measured. An accuracy of 1.5 μm for a total height of 6.38 mm is reported. This is a noninvasive method, however, the sensor must be coupled to the container surface and carefully calibrated to be placed in a perpendicular position with respect to the content surface. Another approach employing flexural waves is described in a patent presented by Dam and Austerlitz [9] using two different transducers. The first one excites a single pulse of sonic energy that produces a flexural or elastic wave in the wall and the second one receives the flexural wave to produce an electrical signal waveform. In order to perform the liquid level measurement, phase delay or time delay between flexural waves at two different times is measured. The method requires information about thickness and material of the container wall, and it is sensitive to wall imperfections.

Another approach which has been recently reported is the use of acoustic resonance, which is a condition arising when an external force interacts with a body producing a vibratory effect caused by body stiffness and inherent forces of the body itself. If the applied force is impulsive these vibrations are called free vibrations. When this force is periodic, the result is called forced vibration. Once the body has been excited vibrations of different frequencies are produced. Some of them will be canceled and others will be amplified due to mutual interference according to different body characteristics. Those that are amplified represent the so-called resonant frequencies. Sinha in [10] presents an introduction to Acoustic Resonance Spectrography (ARS). In this technique, the system is fed with a frequency sweep and the returned power distribution in the spectral domain is measured, showing the resonant frequencies. The article focuses on liquid level measurement, but additional examples of applications in other areas are presented. Because using of frequency sweep, instant energy provided to the system remains low against the impulse method, becoming this technique attractive when unstable contents are measured. However, frequency swept takes much time according to desired resolution.

As an alternative to ARS, free vibration resonance phenomenon has been studied for liquid level measurement purposes. Costley et al. [11] employs free resonance phenomenon to estimate pressure and level of content in sealed storage drums. Information of vibration is extracted and processed using FFT and spectral analysis from signals obtained through microphones and accelerometers mounted on the container lid as sensors, different test are performed, measuring characteristics of drums like pressure, level of fill, wall, and lid thickness against frequency. Behavior of fundamental and axisymmetric modes against pressure is also presented. Conclusion shows the viability to calculate pressure and fill level using free acoustic resonance information. Hsieng-Huang and Zong-Hao [12] propose a model and an experimental procedure based on acoustic resonance to calculate the filling level for Liquefied Propane Gas (LPG) containers. The mathematic model is based on Bernoulli–Euler beam partially loaded with a distributed mass. Test on different LPG containers are performed to obtain a set of magnitude frequency-weight plots. Conclusions describe an agreement of obtained plots with pinned-pinned and clamped-clamped models of Bernoulli–Euler beam. A patent presented by Sides [13] describes a method to determine the fill level in a LPG using similar principles employed in [12] but using the first large peak after 900 Hz. A smartphone is used as processing device. The main disadvantages of the system are the lack of accuracy, high dependency of LGP characteristics, and the required strong calibration steps. It can be seen that the approach based on free acoustic resonance accomplishes with both original requirements; it is not invasive and the measurement can be done in a short period of time. However, this technique has a strong dependence on many characteristics of container, such as shape, material, wall thickness, pressure, temperature, and others. The work described in this paper includes preliminary results of a study aiming to develop a simplified model for a noninvasive level content measurement based on free acoustic resonance, relying on filling level ratio between free height of the container and filling level, and frequency spectral analysis. Acoustic information from several containers is obtained using a semiautomatic process under controlled conditions. Once audio information is available, spectral analysis using FFT is performed on each audio signal. A feature vector is built using the first five predominant frequencies. Feature vectors database is divided two feature vectors sets in order to train and test for comparison purposes a neural network and a fuzzy system, with the output corresponding to the liquid level inside the container.

2 Motion of a Beam in Transverse Vibration

In Ref. [12], Huang et al., present a method for liquid level estimation inside a LPG using acoustic resonance. In this work, the authors describe a model based on Bernoulli–Euler beam theory in support of a description aimed to describe the behavior of dominant frequencies related to liquid content level inside the studied containers. Moreover, the model presented in the article is a modification of the models described by Chan et al. [14] and Jacobs et al. [15], used to estimate the liquid level inside containers that are partially supported or fully supported, respectively. It is evident that Bernoulli–Euler theory for think beams is a relevant framework for understanding the resonance phenomenon on partially filled containers.

Experimental results in the presented work show that some containers behave like ones described in Ref. [14] and others in a similar way that those described in Ref. [15], so in looking for a more generalized model this information is considered. Modeling of vibration in containers is developed around beam in transverse vibration principle [14–16]. A thin beam subjected to a transverse force [16], is shown in Fig. 1.

A free body diagram for the same beam of length dx is shown in Fig. 2.

In Fig. 2, $M(x, t)$ represents the bending moment, $V(x, t)$ is the shear force and $f(x, t)$ denotes the external transverse force. Inertia force acting on the element of the beam is expressed in Eq. (1) as

$$\rho A(x)\mathrm{d}x\frac{\partial^2 w}{\partial t^2}(x,t) \tag{1}$$

Equation (2) expresses the force equation of motion in the z direction

$$(-V+\mathrm{d}V)+f(x,t)+V=\rho A(x)\mathrm{d}x\frac{\partial^2 w}{\partial t^2}(x,t), \tag{2}$$

where ρ is the mass density and $A(x)$ the cross-sectional area of the beam. The moment equilibrium equation on the y axis passing through point P in Fig. 2, leads to

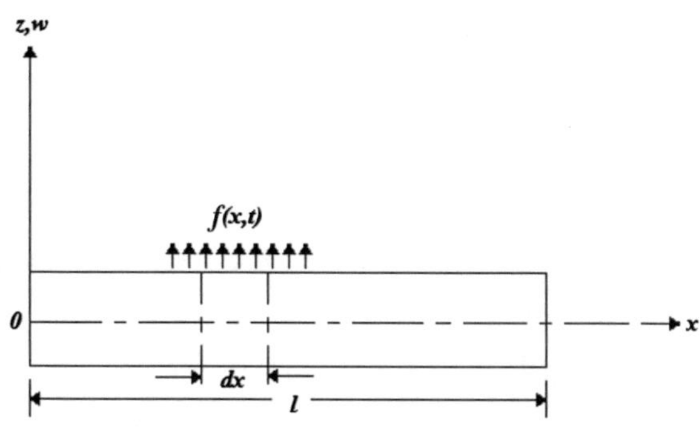

Fig. 1 Thin beam subjected to a transverse force

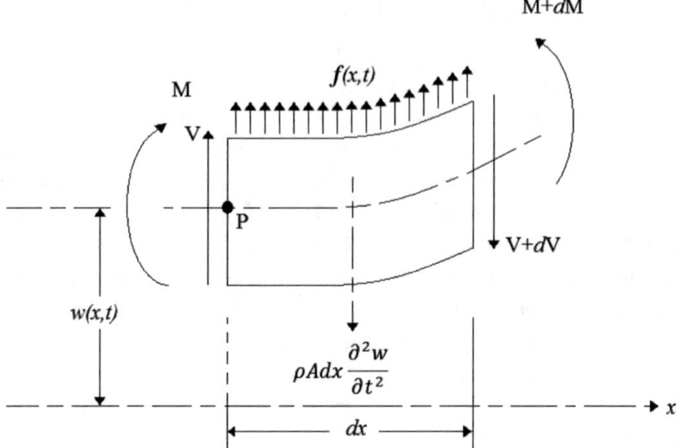

Fig. 2 Free body diagram for a beam of length dx

$$(M + \mathrm{d}M) - (V + \mathrm{d}V)\mathrm{d}x + f(x,t)\mathrm{d}x\frac{\mathrm{d}x}{2} - M = 0 \tag{3}$$

Writing the differential terms as $\mathrm{d}V = \frac{\mathrm{d}V}{\partial x}\mathrm{d}x$ and $\mathrm{d}M = \frac{\mathrm{d}M}{\partial x}\mathrm{d}x$, and disregarding terms involving second powers in dx, the last equations can be written as

$$-\frac{\partial V}{\partial x}(x,t) + f(x,t) = \rho A(x)\frac{\partial^2 w}{\partial t^2}(x,t) \tag{4}$$

$$-\frac{\partial M}{\partial x}(x,t) - V(x,t) = 0 \tag{5}$$

Using the relation $V = \frac{\partial M}{\partial x}$, it can be obtained

$$-\frac{\partial^2 M}{\partial x^2}(x,t) + f(x,t) = \rho A(x)\frac{\partial^2 w}{\partial t^2}(x,t) \tag{6}$$

From Euller–Bernoulli or thin beam theory, the relationship between bending moment and deflection can be expressed as

$$M(x,t) = EI(x)\frac{\partial^2 w}{\partial x^2}(x,t), \tag{7}$$

where E is Young's modulus and $I(x)$ is the moment of inertia of the beam cross-section on y axis.

Combining these relations, the equation of motion for the forced lateral vibration of a non-uniform beam can be obtained as

$$\frac{\partial^2}{\partial x^2}\left[EI(x)\frac{\partial^2 w}{\partial x^2}(x,t)\right] + \rho A(x)\frac{\partial^2 w}{\partial t^2}(x,t) = f(x,t) \tag{8}$$

For a uniform beam, Eq. (9) reduces to

$$EI\frac{\partial^4 w}{\partial x^4}(x,t) + \rho A\frac{\partial^2 w}{\partial t^2}(x,t) = f(x,t) \tag{9}$$

This equation can be solved using the proper initial and boundary conditions. For example, if beam has an initial displacement $w_o(x)$ and an initial velocity $\dot{w}_o(x)$, the initial conditions can be expressed as

$$w(x, t = 0) = w_o(x) \tag{10}$$

$$\frac{dw}{dt}(x, t = 0) = \dot{w}_o(x) \tag{11}$$

If the beam is fixed at $x = 0$ and pinned at $x = l$, the deflection slope will be zero at $x = 0$ and the deflection and the bending moment will be zero at $x = l$. Hence, the boundary conditions are given by

$$w(x = 0, t) = 0 \quad t > 0 \tag{12}$$

$$\frac{dw}{dt}(x = 0, t) = 0 \quad t > 0 \tag{13}$$

$$w(x = l, t) = 0 \quad t > 0 \tag{14}$$

$$\frac{dw^2}{dx^2}(x = l, t) = 0 \quad t > 0 \tag{15}$$

Further application and solution details can be found in [13, 14]. In the work presented by Chan et al. [14], a detailed study of free vibrations in a simply supported beam partially loaded with distributed mass is presented. Plots of frequency ratios against length of added mass to beam ratios and relative amplitude of oscillation versus distance ratio from loaded end for different mass densities and vibration modes are described. Jacobs et al. [15] uses the same theory for calculations of free vibrations in a steel capillary partially filled with mercury firmly supported in both ends. Practical results are obtained using impedance measurements. In the work developed by Hsieng-Huang and Zong-Hao [12], a method used for liquid level estimation inside a LPG using acoustic resonance is presented. In this work, the author builds a model based on Bernoulli–Euler beam theory to describe the behavior of maximum amplitude frequency found against liquid level inside the containers. Moreover, the model presented in the article is a modification of the models presented by Chan et al. [14] and Jacobs et al. [15], used to estimate the liquid level inside a container partially supported or fully supported respectively. In this way, analysis of Bernoulli–Euler theory for think beams can be a key for understanding the resonance phenomenon on partially filled containers.

Experimental results on our work showed that some containers behave like ones described in [14] and others in the same way that those described in [15], so in looking for a more generalized model this information have to be considered.

3 Fuzzy System with Gradient Descent Training

In this work, an inference fuzzy system is used as universal approximator [17], in order to provide an estimation of the content level based on spectral feature information. Optimal membership parameters are computed through the described algorithm. The proposed fuzzy system to be used as universal approximator consists of a Takagi–Sugeno structure with a product inference engine, singleton fuzzifier, and center average defuzzifier with Gaussian membership functions. The system is designed following the form expressed in Eq. (16)

$$f(x) = \frac{\sum_{l=1}^{M} \bar{y}^l \left(\prod_{i=1}^{n} \exp\left(\left(-\frac{x_i - \bar{x}_i^l}{\sigma_i^l} \right)^2 \right) \right)}{\prod_{i=1}^{n} \exp\left(-\left(\frac{x_i - \bar{x}_i^l}{\sigma_i^l} \right)^2 \right)}, \tag{16}$$

where M is a representation of the number of rules in the fuzzy system, n is the number of inputs $\bar{x}_i^l, \sigma_i^l$ are the mean and standard deviation for each input membership and $\bar{y}^l$ is the mean for each output membership function. The fuzzy system can be represented as a feedforward network as shown in Fig. 3, where parameters $\bar{y}^l, \bar{x}_i^l, \sigma_i^l$ are to be determined.

Input **x** passes through a product Gaussian operator to become $z^l = \prod_{i=1}^{n} \exp\left(-\left(\frac{x_i - \bar{x}_i^l}{\sigma_i^l} \right)^2 \right)$; then, the z^l are passed through a summation operation and a weighted summation operator to obtain $a = \sum_{l=1}^{m} \bar{y}^l z^l$ and $b = \sum_{l=1}^{m} z^l$ and finally, the output of the system is computed as $f(x) = a/b$. The task now is to design a system $f(x)$ in the form of Fig. 3 such that matching error $e^p = \frac{1}{2}\left[f\left(x_0^p \right) - y_0 \right]^2$ is minimized. A gradient descent algorithm is used to determine the parameters. Specifically, Eq. (18) is used to obtain $\bar{y}$ through an iterative process.

Fig. 3 Feedforward network fuzzy system

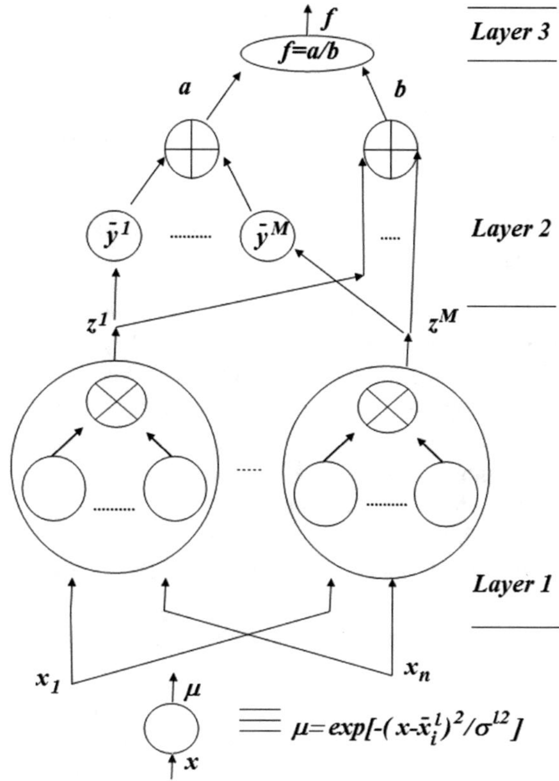

$$\bar{y}^l(q+1) = \bar{y}^l - \alpha \frac{\partial e}{\partial \bar{y}^l}\bigg|_q, \tag{17}$$

where $l = 1, 2\ldots, M$, $q = 0, 1, 2\ldots$ and α is a constant stepsize. Using the chain rule, we have

$$\frac{\partial e}{\partial \bar{y}^l} = (f - y)\frac{\partial f}{\partial a}\frac{\partial a}{\partial \bar{y}^l} = (f - y)\frac{1}{b}z^l \tag{18}$$

Substituting Eq. (19) in Eq. (18)

$$\bar{y}^l(q+1) = \bar{y}^l - \alpha\frac{(f - y)}{b}z^l \tag{19}$$

where $l = 1, 2\ldots, M$ and $q = 0, 1, 2\ldots.$

Applying the same process, $\bar{x}_i^l$ and σ_i^l are determined as:

$$\bar{x}_i^l(q+1) = \bar{x}_i^l(q) - \alpha\frac{(f-y)}{b}(\bar{y}^l - f)z^l\frac{2(x_{0i}^p - \bar{x}_i^l(q))}{\sigma_i^{l2}(q)} \tag{20}$$

$$\sigma_i^l(q+1) = \sigma_i^l(q+1) - \alpha\frac{(f-y)}{b}(\bar{y}^l - f)z^l\frac{2(x_{0i}^p - \bar{x}_i^l(q))^2}{\sigma_i^{l3}(q)} \tag{21}$$

Using these equations, the optimal values for $\bar{y}^l, \bar{x}_i^l, \sigma_i^l$ can be obtained. The method is summarized in the next steps:

Step 1. Structure determination and initial parameter setting. Chose the fuzzy system in the described form as universal approximator, and determine M (number of rules). Larger M results in more parameters and more computation but it provides a better approximation accuracy. Specify the initial parameters $\bar{y}^l(0)$, $\bar{x}_i^l(0)$ and $\sigma_i^l(0)$. These initial parameters may be determined according to the linguistic rules by a human expert, or it can be chosen in such a way that the corresponding membership functions uniformly cover the input and output spaces.
Step 2. Enter data input and calculate the output of the fuzzy system. For a given input–output pair $(\bar{x}_0^p, \bar{y}_0^p)$, $p = 0, 1, 2\ldots$ and the q'th stage of training, $q = 0, 1, 2\ldots$, enter $\bar{x}_0^p$ to the input layer of the fuzzy system and compute the outputs of the layers, as expressed in Eqs. (22)–(25).

$$z^l = \prod_{i=1}^{n}\exp\left(-\left(\frac{\bar{x}_0^p - \bar{x}_i^l(q)}{\sigma_i^l(q)}\right)^2\right) \tag{22}$$

$$b = \sum_{l=1}^{m} z^l \tag{23}$$

$$a = \sum_{l=1}^{m} \bar{y}^l(q)z^l \tag{24}$$

$$f(x) = a/b \tag{25}$$

Step 3. Update parameters. Use the training algorithm equations to compute $\bar{y}^l(q+1)$, $\bar{x}_i^l(q+1)$ and $\sigma_i^l(q+1)$ where $y = y_0^p$
Step 4. Go to step 2 with $q = q + 1$ until the error $|f - y_0^p|$ is less than a predefined number ϵ or until q equals a prespecified number.
Step 5. Go to step 2 with $p = p + 1$; that implies a parameters update using the input–output pair $(\bar{x}_0^{p+1}, \bar{y}_0^{p+1})$.

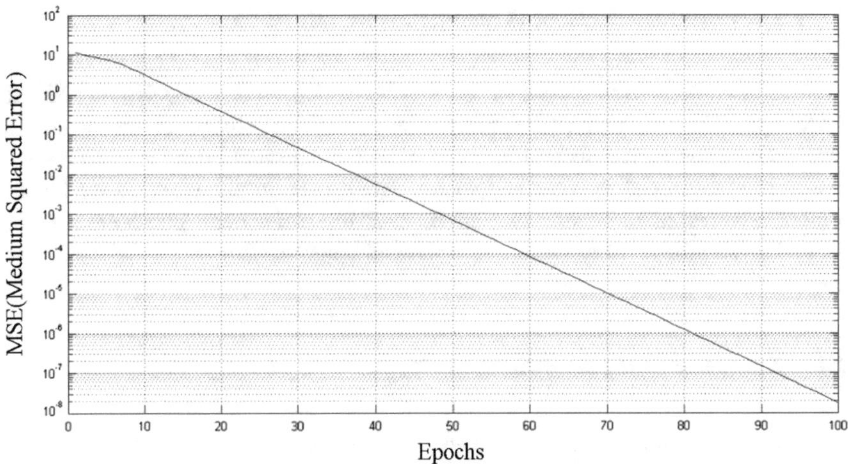

Fig. 4 MSE after 100 epochs for spherical function

Figure 4 shows an example of MSE descent during a training phase after 100 epochs.

3.1 Neural Network Overview

The neural network model used in this work is a feedforward multilayer perceptron (MLP) with three layers of neurons: five neurons in the input layer, one hidden layer with 25 neurons and one output layer with a single neuron [18]. A sigmoid is used as activation function in the hidden layer, while in the output layer a linear activation function was employed. When a multilayer perceptron network has n input nodes, one hidden layer of m neurons, and two output neurons, the output of the network is given by Eq. (26).

$$y_i = f_i\left(\sum_{k=1}^{m} w_{ki} f_k\left(\sum_{j=1}^{n} w_{kj}\, x_j\right)\right) \tag{26}$$

f_k, f_i Activation functions of the hidden layer and the output layer neurons
w_{ki}, w_{kj} Weights connected to the output layer and to the hidden layer
x_j Input feature vector.

Training was performed using Levenberg-Marquardt backpropagation algorithm. The number of epochs in the training phase differs from one exercise to another; however, the Levenberg-Marquardt backpropagation algorithm provided consistently a fast convergence.

4 Experimental Setup; Methods and Materials

There are two parts in this work to be described: acoustic data acquisition and data processing. The first part corresponds to a general description of the developed experimental setup. The second part describes the data processing approach based on spectral features of the obtained sound signals used in this work, as well as the content level estimators using two approaches for comparison purposes: a neural network system and fuzzy system as universal approximator.

4.1 Acoustic Data Acquisition Stage

Acoustic information was obtained using 10 different glass containers. Total height of containers was uniformly splitted in ten levels of filling. For each level, 100 audio captures were taken giving a total of 1000 audio recordings per container. The containers used for testing in the described experiment are shown in Fig. 5. Container characteristics are summarized in Table 1.

Sampling rate was set to 44,000 samples per second and a total of 88,000 samples were recorded in each capture (2 s). To ensure repeatability of audio captures in controlled conditions, an experimental setup with a semiautomatic device was designed. This device consists of a solenoid aimed to work as a hammer

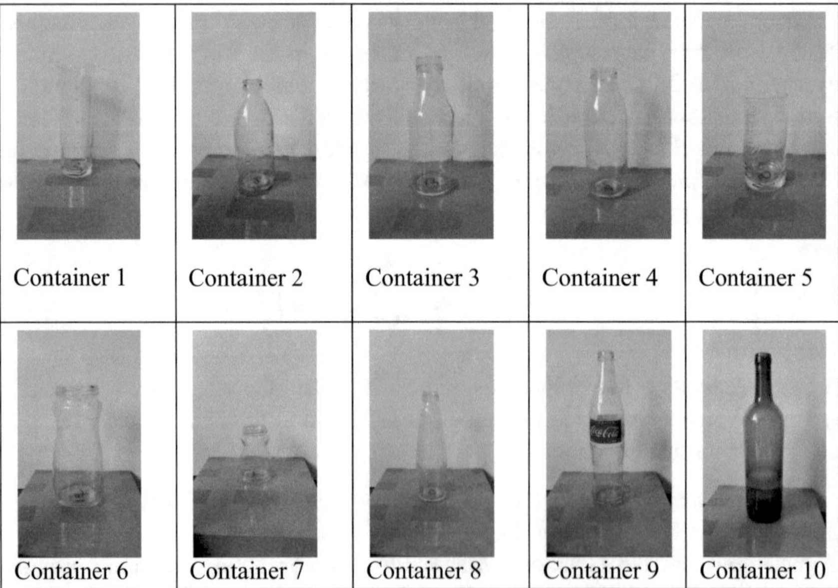

Fig. 5 Images of test containers

Table 1 Physical characteristics of studied containers

Container no.	Volume (ml)	Total height (mm)	Mean wall thickness (mm)
1	316	156	1.5
2	269	147	2.5
3	474	186	1.5
4	439	174	1.7
5	346	121	1.1
6	929	195	3
7	209	100	2
8	322	187	1.4
9	515	256	3
10	789	285	2.5

Fig. 6 Experimental setup for data acquisition of audio samples

controlled through an interface circuit connected to a PC. An application program developed in MATLAB allows the system to acquire audio data in a fast and controlled way. The experimental setup is shown in Fig. 6.

4.2 Data Processing Stage

Data processing stage of this work can be summarized in the block diagram of Fig. 7. The first block corresponds to the audio data acquisition of the acoustic resonance phenomenon previously described. In the second block, the spectral analysis through FFT is performed in order to obtain the associated power spectrum of the acoustic resonance data. The third block represents the feature extraction process in which the five predominant spectral components are detected. This information is used to construct a feature vector to be used in further stages.

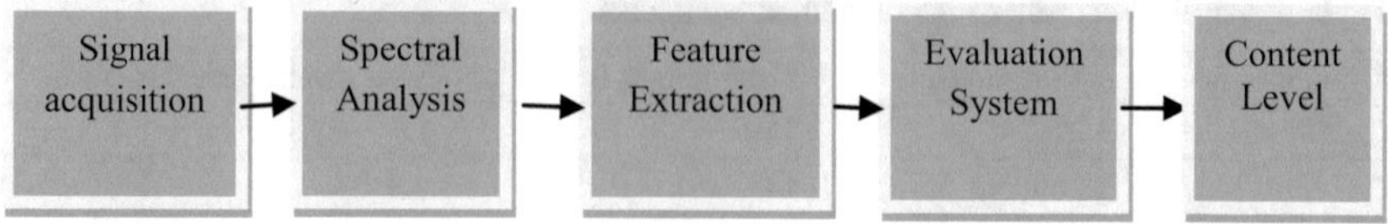

Fig. 7 Data processing diagram block

The last block represents the data processing approach aiming to evaluate the condition of each container in terms of the level content. In this work, two computational intelligence approaches are proposed for comparison purposes; a neural network and a fuzzy system in a configuration of universal approximator trained with a gradient descent technique. The neural network structure consists of a 5 node input layer, a 25 nodes hidden layer, and one output layer with one node. Data set is divided in three groups to be used during training, validation and testing stages. Training is performed using 70 % of available input data, validation uses 15 % of available data and test uses the last 15 %. The second system employs a fuzzy system with a configuration of universal approximator using gradient descent training with an M factor of 130, according to the detailed description included in the previous section. Training is performed using 60 % of available data while testing is performed using 40 %.

5 Experimental Results

Figure 8 shows in detail the power spectra of both containers in empty conditions. This observation shows that, in general, using just the highest amplitude frequency in spectra in order to characterize liquid level content inside a container can lead us to wrong results. Figure 9 shows a plot of the maximum amplitude frequency

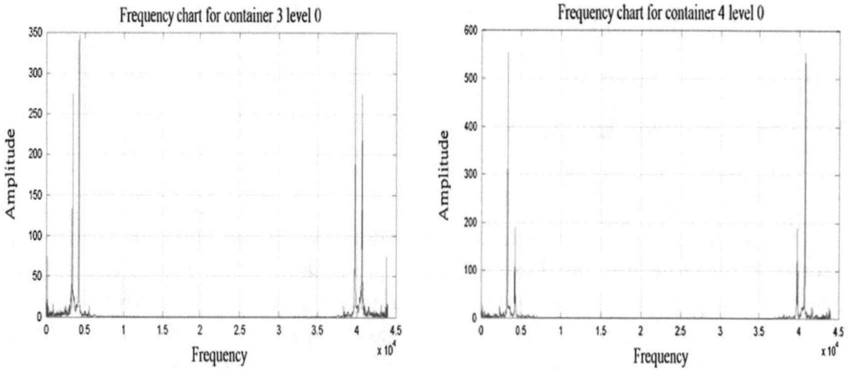

Fig. 8 Frequency spectra of containers 3 and 4 at level 0 of filling

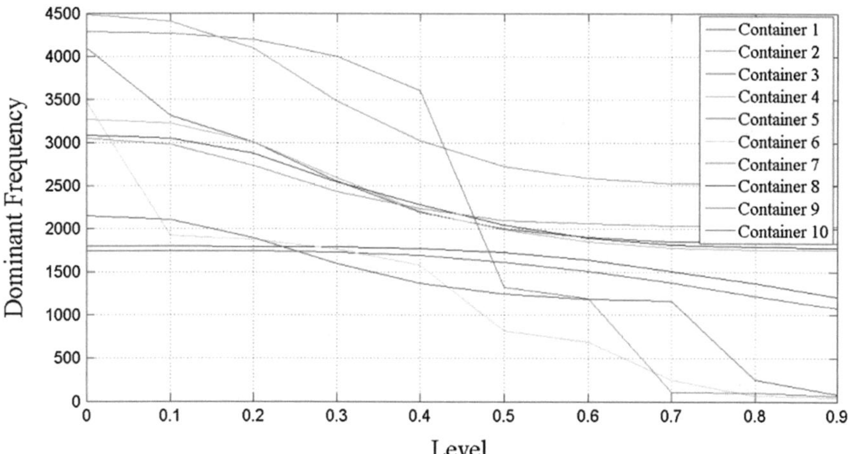

Fig. 9 Plot of maximum amplitude frequency versus liquid level for all tested containers

versus liquid content level for all containers used in the experiment. Several characteristics can be inferred from the plot, for instance, although containers 3 and 4 are quite similar, their dominant frequency when containers are empty occurs at different value, although the evolution as the container is filled behaves accordingly. Table 2 shows the estimated MSE according to liquid level and container obtained from the neural network.

It can be seen that in average, the neural network presented an overall MSE of 0.0177. From Table 2, it is possible to verify than liquid levels in the extremes corresponding to levels 0–0.9 had the worst behavior. An analysis of their power spectrum indicates that their highest amplitude frequency were located very close to the adjacent components causing errors. In the other hand, containers 7 and 10 presented good approximations but more variations, which could be the result of irregularities on those containers with a spectral behavior very different to the rest.

Table 3 shows the estimated MSE according to liquid content level and container for the fuzzy system configured as an approximator with gradient descent training. The overall MSE corresponding to the fuzzy system was 0.0918 in average. In this case, highest levels of fill and containers 7 and 5 had the worst performance.

A set of experiments was carried out by increasing the number of rules used in the described fuzzy system. Table 4 shows the overall MSE and computing time in each case. These experiments were implemented using a PC with an A8 quad-core processor at 1.7 GHz and 8 Gb of RAM. As it can be seen, although the computing time increased with the number of rules as expected, the obtained MSE did not show a consistent improvement.

Table 2 Liquid level and container MSE for the NN approach

		Content level									
		0	0.1	0.2	0.3	0.4	0.5	0.6	0.7	0.8	0.9
Container	1	0.010	0.020	0.013	0.006	0.005	0.006	0.007	0.008	0.006	0.018
	2	0.007	0.008	0.012	0.016	0.012	0.013	0.033	0.016	0.006	0.024
	3	0.039	0.024	0.012	0.024	0.008	0.004	0.003	0.003	0.006	0.002
	4	0.029	0.019	0.002	0.016	0.009	0.013	0.005	0.001	0.034	0.015
	5	0.020	0.005	0.021	0.015	0.006	0.002	0.003	0.008	0.002	0.003
	6	0.049	0.027	0.008	0.004	0.026	0.016	0.002	0.017	0.021	0.053
	7	0.021	0.010	0.012	0.013	0.007	0.035	0.013	0.009	0.031	0.173
	8	0.014	0.013	0.010	0.009	0.005	0.005	0.011	0.010	0.037	0.074
	9	0.036	0.025	0.012	0.042	0.009	0.035	0.008	0.007	0.014	0.026
	10	0.117	0.016	0.006	0.013	0.018	0.006	0.011	0.027	0.001	0.004

Table 3 Liquid level and container MSE for the fuzzy system approach

		Content level									
		0	0.1	0.2	0.3	0.4	0.5	0.6	0.7	0.8	0.9
Container	1	0.02	0.01	0.02	0.01	0.03	0.02	0.03	0.17	0.14	0.24
	2	0.01	0.01	0.02	0.05	0.02	0.02	0.04	0.09	0.19	0.23
	3	0.10	0.03	0.07	0.04	0.03	0.00	0.00	0.00	0.05	0.11
	4	0.03	0.01	0.03	0.04	0.03	0.02	0.08	0.20	0.36	0.15
	5	0.12	0.15	0.15	0.03	0.03	0.02	0.01	0.03	0.15	0.65
	6	0.07	0.03	0.02	0.00	0.01	0.02	0.06	0.10	0.19	0.30
	7	0.05	0.02	0.02	0.02	0.07	0.09	0.22	0.24	0.31	0.53
	8	0.09	0.05	0.01	0.05	0.07	0.12	0.03	0.17	0.07	0.28
	9	0.04	0.04	0.02	0.03	0.02	0.05	0.06	0.13	0.19	0.35
	10	0.10	0.06	0.04	0.01	0.04	0.00	0.05	0.27	0.12	0.16

Table 4 Overall performance and computing time for fuzzy system using different number of rules

Rules	10	20	30	40	50	60	70
MSE	0.09	0.19	0.17	0.19	0.08	0.08	0.20
C. time	192.26	200.73	213.62	227.13	235.25	246.76	256.41
Rules	80	90	100	110	120	130	140
MSE	0.11	0.07	0.07	0.11	0.09	0.23	0.28
C. time	268.52	278.61	303.93	310.31	313.82	322.23	333.40
Rules	150	160	170	180	190	200	210
MSE	0.08	0.18	0.08	0.08	0.12	0.13	0.10
C. time	354.76	355.66	367.29	377.14	388.61	399.62	410.62
Rules	220	230	240	250	260	270	280
MSE	0.10	0.09	0.16	0.16	0.08	0.14	0.12
C. time	422.67	433.60	441.53	453.76	465.98	476.10	485.37
Rules	290	300	310	320	330	340	350
MSE	0.09	0.10	0.14	0.06	0.22	0.08	0.09
C. time	497.44	507.73	520.33	530.79	540.57	550.59	564.06

6 Conclusions

In this work, a series of experiments aiming to do measurement of liquid content level and container characterization from spectral analysis have been presented. Two evaluation techniques were implemented for comparison purposes: neural network classification and a fuzzy logic inference system as approximator, obtaining in average a MSE of 0.02 and 0.09 respectively. Preliminary results indicate a good approximation in measuring content level, however additional work should be developed in order to improve these results. Exploring of alternatives in spectral decomposition techniques for finding common patterns in spectral content of acoustic resonance captures, such as wavelets, DCT, and EMD, are currently in progress.

Acknowledgments The first author acknowledges the financial support from the Mexican National Council for Science and Technology (CONACYT), scholarship No. 235187.

References

1. Creus, S.: Medición de nivel: Instrumentación Industrial, Alfaomega - Marcombo, (1998).
2. Nakagawa, T., Kogo, K., and Kurata, H.: Contactless liquid level measurement with frequency-modulated millimeter wave through opaque container. Sensors Journal, **13**(3), 926-933 (2013).
3. Donlagic, D., Zabrsznik, M., and Donlagic, D.: Low frequency resonance level detector with neural network classification. Sensors and Actuators **55**(2), 99-106 (1996).
4. Donlagic, D., Kumperskac, V., and Zabrsnik, M.: Low frequency acoustic resonance level gauge. Sensors and Actuators **57**(3), 209-215 (1996).
5. Jung, S., Cho, S., and Kim, Y.T.: Level gauge by using the acoustic resonance frequency. Journal of the Korean Physical Society 43(5), 727 – 731 (2003).
6. Webster, E., and Davies, C.: The use of helmholtz resonance for measuring the volume of liquid and solids. Sensors, **10**(12), 10663 – 10672 (2010).
7. Brunnader, R., and Holler, G.: Electroacustic model based pneumatic fill - level measurement for fluids and bulk solids. Advancement in Sensing Technology, **1**, 165-180 (2013).
8. Lucklum, F., and Jakoby, B.: Principle of a non-contact liquid level sensor using electromagnetic-acoustic resonators. Elektrotechnik und Informations technik **126**(1), 3-7 (2009).
9. Dam N., Austerlitz, H.P. System and method of non-invasive, discreet, continuous and multi-point level liquid sensing using flexural waves, U.S. patent 6 631 639, (2003).
10. Shina, D.: Acoustic Resonance Spectroscopy. IEEE Potentials **11** (1992).
11. Costley, R., Henderson, M., Patel, H., Jones, E.W., Boudreaux, G.M., Plodinec, M.J.: The measurement of pressure and level of fill in sealed storage drums. NDT&E International, **40** (4), 300-308(2007).
12. Hsieng-Huang, P., Zong-Hao, Y.: Liquid level detector for sealed gas tank based on spectral analysis. In: Proceedings of the 19th International Conference on Digital Signal Processing, Hong Kong, China, (2014).
13. Sides, P.: Aparatus and method for determining the liquid level in an un-modified tank. U.S. patent 13/859087, (2013).
14. Chan, K. Leung, T., Wong, W.O.: Free vibration on simply supported beam partially loaded with distributed mass. Journal of Sound and Vibration, **191**(4), 590-597(1996).

15. Jacobs, M., Breeuwer, R., Kemmere, M.F., Keurentjes, J.T.F.: Contactless Liquid Detection in a partly filled tube by resonance. Journal of Sound and Vibration, **285**(4), 1039-1048 (2005).
16. Rao, S., Equation of Motion of a Beam in Transverse Vibration. Vibrations of Continuous Systems, John Willey & Sons, (2007).
17. Wang, L., Design of Fuzzy Systems Using Gradient Descendent Training, A course in Fuzzy Systems and Control, Prentice-Hall, (1996).
18. Haykin, S., Neural Networks and Learning Machines, Third Edition, A Comprehensive Foundation, Prentice-Hall, (2009).

Comparison of Optimization Techniques for Modular Neural Networks Applied to Human Recognition

Daniela Sánchez, Patricia Melin, Juan Carpio and Hector Puga

Abstract In this paper a comparison of optimization techniques for a Modular Neural Network (MNN) with a granular approach is presented. A Hierarchical Genetic Algorithm, a Firefly Algorithm (FA), and a Grey Wolf Optimizer are developed to perform a comparison of results. These algorithms design optimal MNN architectures, where their main task is the optimization of some parameters of MNN such as, number of sub modules, percentage of information for the training phase and number of hidden layers (with their respective number of neurons) for each sub module and learning algorithm. The MNNs are applied to human recognition based on iris biometrics, where a benchmark database is used to perform the comparison, having as objective function in each optimization algorithm the minimization of the error of recognition.

1 Introduction

Hybrid intelligent systems are combination of different intelligent techniques, and these kinds of systems have emerged to overcome the limitations of each intelligent technique has individually, creating powerful intelligent systems [23, 34]. Nowadays, there are many works performing these systems, where good results have been shown, such as in [1, 2, 9, 11, 20, 26]. Some of these techniques are fuzzy logic, neural networks (NNs), and genetic algorithms, among these techniques, there are a

D. Sánchez (✉) · P. Melin
Tijuana Institute of Technology, Tijuana, Mexico
e-mail: danielasanchez.itt@hotmail.com

P. Melin
e-mail: pmelin@tectijuana.mx

J. Carpio · H. Puga
León Institute of Technology, León, Mexico
e-mail: juanmartin.carpio@itleon.edu.mx

H. Puga
e-mail: pugahector@yahoo.com

P. Melin et al. (eds.), *Nature-Inspired Design of Hybrid Intelligent Systems*,
Studies in Computational Intelligence 667, DOI 10.1007/978-3-319-47054-2_15

lot of soft computing techniques dedicated to performing the optimization problems such as: Genetic Algorithm [12], Particle Swarm Optimization [13], Ant Colony Optimization [7], Cuckoo Optimization Algorithm [22] and Bee Colony Optimization [14], among others. In this paper different intelligent techniques are combined such as NNs, genetic algorithms, firefly algorithm (FA) and grey wolf optimizer (GWO). The proposed method was applied to human recognition based on iris biometrics. Optimizing some parameters of Modular Neural Network (MNN) such as; number of sub modules, percentage of information for the training phase and number of hidden layers (with their respective number of neurons) for each sub module and learning algorithm, having as objective function the minimization of the error of recognition. This paper is organized as follows: The basic concepts used in this work are presented in Sect. 2. On the other hand, Sect. 3 contains the general architecture of the proposed method. Section 4 presents experimental results and in Sect. 5, the conclusions of this work are presented.

2 Basic Concepts

In this section the basic concepts used in this research work are presented.

2.1 *Modular Neural Networks*

A NN is integrated by many artificial neurons and its objective is to convert the inputs into significant outputs. A NN has a large number of features similar to the brain due to its constitution and its foundations, as it can be to learn from the experience [8]. The learning mode can be supervised or unsupervised [24]. The concept of modularity is an extension of the principle of divide and conquers. If the computation performed by the network can be decomposed into two or more modules (subsystems) is said to be modular, where each NN works independently in its own domain and each of those NNs is build and trained for a specific subtask. The simpler subtasks are then accomplished by a number of the specialized local computational systems or models which are integrated together via an integrating unit [3, 16].

2.2 *Hierarchical Genetic Algorithms*

A Genetic algorithm (GA) is an optimization and search technique based on the principles of genetics and natural selection. This is a Darwinian approach to the idea of the survival of the fittest. In computational terms this has to be measured in terms of an evaluation function called fitness function [10, 18]. Genetic algorithms belong to the larger class of evolutionary algorithms (EA), which generate solutions to optimization problems using techniques inspired by natural evolution, such as mutation, selection, and crossover [5]. Hierarchical genetic algorithm (HGA) was introduced in [25] and it is a kind of genetic algorithm. HGA is a type of genetic algorithm. Its structure is more flexible than the conventional GA. The main difference between HGA and GAs is the structure of the chromosome. The basic idea under HGA is that for some complex systems which cannot be easily represented, this type of GA can be a better choice [27, 28].

2.3 Granular Computing

Granular computing (GrC) originally proposed by Zadeh [33] plays a fundamental role in human reasoning and problem solving. Its three basic issues are information granulation, organization, and causation. A granule may be interpreted as one of the numerous small particles forming a larger unit. The philosophy of thinking in terms of levels of granularity, and its implementation in more concrete models, would result in disciplined procedures that help to avoid errors and to save time for solving a wide range of complex problems [32]. The information granulation involves decomposition of whole into parts; the organization involves integration of parts into whole; and the causation involves association of causes with effects. They have been applied in relevant fields such as bioinformatics, e-Business, security, machine learning, data mining, interval analysis, cluster analysis, databases and knowledge discovery [21] in terms of efficiency, effectiveness, robustness, and uncertainty [4].

2.4 Firefly Algorithm

FA was introduced in [29, 30]. This optimization algorithm is based on fireflies (their flashing patterns and behavior). FA is based on three rules: (1) Fireflies are unisex. A firefly can attract to other fireflies independently of their sex. (2) The attractiveness is proportional to the brightness, and they both decrease as their distance increases, where for any two flashing fireflies; the firefly with less brightness will move toward the brighter one. If there is no brighter one than a particular firefly, it will move randomly. (3) The brightness of a firefly is determined by objective function [29–31]. It is important to say, in this work the attractiveness of a firefly is represented by the objective function, and it is also associated with its brightness.

2.5 Grey Wolf Optimizer

GWO was introduced in [17]. Grey wolves mostly prefer to live in a pack; the group size is approximately between 5 and 12 grey wolves. Each grey wolf has an important role. They are divided into four categories: alpha, beta, delta, and omega. The alpha wolf is the dominant wolf and his/her orders should be followed by the pack. The second level is beta [15]. The betas are subordinate wolves that help the alpha in decision-making or other pack activities. Delta wolves have to submit to alphas and betas, but they dominate the omega. Scouts, sentinels, elders, hunters, and caretakers belong to this category. The omega plays the role of scapegoat. Omega wolves always have to submit to all the other dominant wolves. They are the last wolves that are allowed to eat. In addition to the social hierarchy of wolves, group hunting is another interesting social behavior of grey wolves. The main phases of grey wolf hunting are as follows: Tracking, chasing, and approaching the prey, pursuing, encircling, and harassing the prey until it stops moving and attack toward the prey [19].

3 Proposed Method

The proposed method consists in designing optimal architectures of modular granular neural networks (MGNNs) using different optimization techniques to perform a comparison among them. The main idea is to find an optimal granulation of the information (data) and optimal MGNNs architectures minimizing the error of recognition. The granulation of the information is implemented and represented in two parts. First, the data per module is represented by the number of persons in each module (there will be different number of persons in each one). Second, the percentage of data for training is represented using the number of images used for the training phase. The optimization algorithms determine the optimal number of sub modules or granules, having as search space up to "*m*" sub modules or granules (the size of each granule can be different among themselves), percentage of information for the training phase, number of hidden layers (with their respective number of neurons) for each sub module and learning algorithm. Figure 1 shows the architecture of proposed method for MGNNs optimization.

The minimum and maximum values used for the MGNNs optimization are shown in Table 1. Those parameters are used to establish the search space of each optimization technique, and these parameters can be decreased or increased without any problem and the biometric measure can be also changed, for this work, the database used is describe later. The optimized algorithms developed to perform the optimization describe above are: HGA, FA, and GWO, and their parameters are established in Table 2.

For the learning algorithm, three backpropagation algorithms are used to perform the MNNs simulations: Gradient descent with scaled conjugate gradient, Gradient

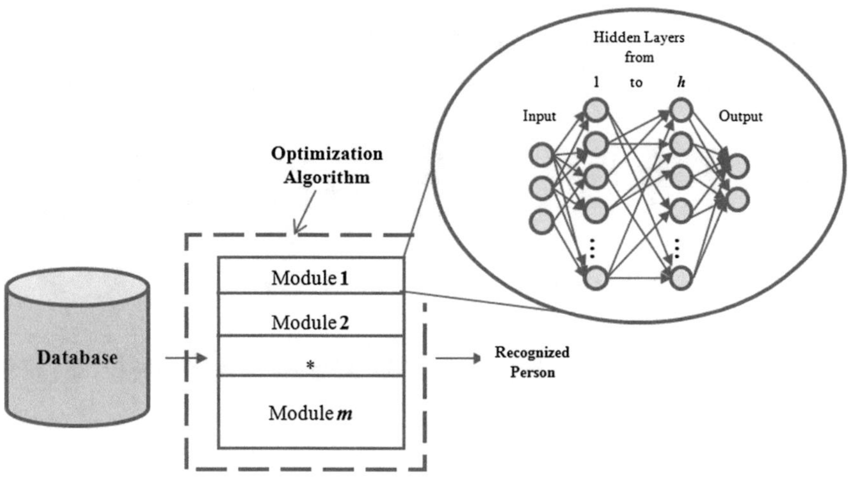

Fig. 1 Architecture of proposed method for MGNN optimization

Table 1 Values for MGNNs

Parameters of MNNs	Minimum	Maximum
Modules (*m*)	1	10
Percentage of data for training	20	80
Error goal	0.000001	0.001
Learning algorithm	1	3
Hidden layers (*h*)	1	5
Neurons for each hidden layers	20	400

Table 2 Values for optimization algorithms

Parameters of	Number
Individuals (HGA)	10
Fireflies (FA)	
Agents (GWO)	
Generations (HGA)	30
Iterations (FA and GWO)	

descent with adaptive learning and momentum (GDX), and Gradient descent with adaptive learning (GDA).

3.1 Iris Database

The database of human iris from the Institute of Automation of the Chinese Academy of Sciences was used [6]. It contains of 14 images (7 for each eye) per person. For this work, only 77 persons were used. The image dimensions are

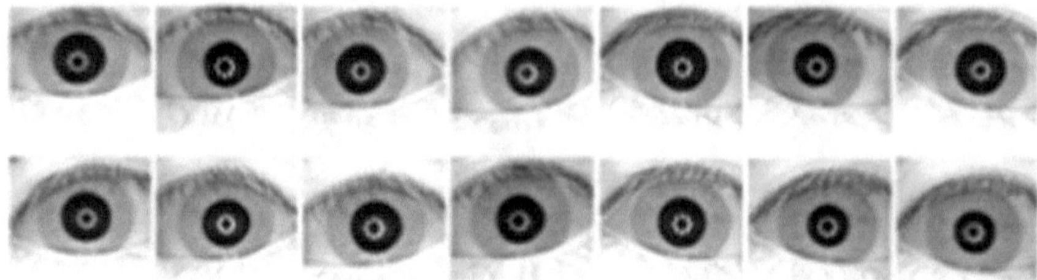

Fig. 2 Examples of the human iris images from CASIA database

320 × 280, JPEG format. Figure 2 shows examples of the human iris images from CASIA database.

4 Experimental Results

The optimized results using the different optimization techniques are shown and compared in this section. Twenty evolutions for each technique were performed, and only the five best results of each one are shown.

4.1 Hierarchical Genetic Algorithm

The five best results obtained by the HGA are shown in Table 3. In evolution #13, the best results are obtained, where by using 7 % of data for the training phase a 99.68 of recognition rate is obtained. In Fig. 3, the convergence of the best evolution is shown.

4.2 Firefly Algorithm

The five best results obtained by this optimization technique are shown in Table 4. In evolution #9, the best results is obtained, where by using 59 % of data for the training phase a 99.13 of recognition rate is obtained. In Fig. 4, the convergence of the best evolution is shown.

Table 3 The first five results for iris (HGA)

Evolution	Images for training	Number of neurons per hidden layer	Persons per module	Rec. rate/error
HGA4	69 % (1, 2, 3, 4, 5, 6, 8, 12, 13 and 14)	287, 36, 155	Module #1 (1–3)	99.35 % (0.0065)
		297, 184, 251, 26	Module #2 (4–15)	
		225, 31, 23	Module #3 (16–29)	
		162	Module #4 (30–42)	
		93, 118, 34	Module #5 (43–47)	
		157, 181	Module #6 (48–49)	
		163, 286, 145, 85	Module #7 (50–56)	
		87, 50, 167	Module #8 (57–61)	
		60	Module #9 (62–77)	
HGA11	79 % (1, 2, 3, 4, 5, 6, 7, 8, 10, 11 and 14)	245	Module #1 (1–6)	99.57 % (0.0043)
		130	Module #2 (7–20)	
		272	Module #3 (21–24)	
		171	Module #4 (25–38)	
		211	Module #5 (39–58)	
		280	Module #6 (59–77)	
HGA13	71 % (1, 2, 3, 5, 6, 8, 9, 11, 13 and 14)	168	Module #1 (1–14)	99.68 % (0.0032)
		120, 164	Module #2 (15–20)	
		286, 232	Module #3 (21–22)	
		78, 134	Module #4 (23–30)	
		296, 293	Module #5 (31–36)	
		152, 33, 216, 160	Module #6 (37–40)	

(continued)

Table 3 (continued)

Evolution	Images for training	Number of neurons per hidden layer	Persons per module	Rec. rate/error
		184, 55, 64, 48	Module #7 (41–46)	
		114, 240	Module #8 (47–62)	
		230, 27	Module #9 (63–77)	
HGA16	79 % (1, 2, 3, 5, 6, 7, 8, 11, 12, 13 and 14)	209	Module #1 (1–5)	99.57 % (0.0043)
		133	Module #2 (6–10)	
		238	Module #3 (11–22)	
		203	Module #4 (23–32)	
		118	Module #5 (33–36)	
		117	Module #6 (37–46)	
		147	Module #7 (47–55)	
		46	Module #8 (56–71)	
		67	Module #9 (72–77)	
HGA19	75 % (1, 3, 4, 5, 6, 8, 9, 10, 11, 13 and 14)	75	Module #1 (1–11)	99.57 % (0.0043)
		300	Module #2 (12–24)	
		211	Module #3 (25–37)	
		257	Module #4 (38–46)	
		234	Module #5 (47–60)	
		139	Module #6 (61–77)	

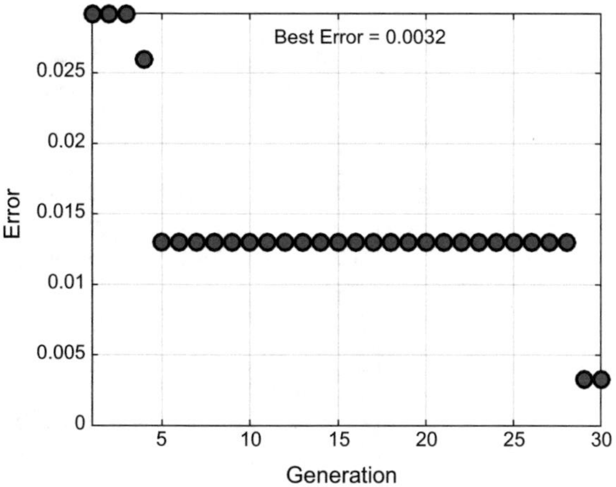

Fig. 3 Convergence of evolution #13 (HGA)

Table 4 The first five results for iris (FA)

Evolution	Images for training	Number of neurons per hidden layer	Persons per module	Rec. rate/error
FA9	59 % (2, 3, 4, 5, 6, 7, 8 y 11)	101, 47, 67, 139, 42	Module #1 (1 a 6)	99.13 % (0.0087)
		93, 164, 147, 44, 25	Module #2 (7 a 21)	
		143, 139, 67, 156, 90	Module #3 (22 a 41)	
		110, 52, 77	Module #4 (42 a 44)	
		74, 21, 92, 96	Module #5 (45 a 48)	
		85, 89, 63, 158, 179	Module #6 (49 a 58)	
		143, 168, 58, 143, 146	Module #7 (59 a 65)	
		117, 123, 181	Module #8 (66 a 68)	
		119, 62, 86, 107	Module #9 (69 a 77)	

(continued)

Table 4 (continued)

Evolution	Images for training	Number of neurons per hidden layer	Persons per module	Rec. rate/error
FA12	61 % (1, 3, 4, 5, 6, 8, 9, 12 y 14)	64, 164, 89, 154, 176	Module #1 (1 a 5)	98.70 % (0.0130)
		111, 148, 142, 141, 85	Module #2 (6 a 17)	
		107, 98, 108, 89, 65	Module #3 (18 a 34)	
		104, 70, 115, 147	Module #4 (35 a 46)	
		110, 98, 126, 40, 128	Module #5 (47 a 58)	
		85, 85, 129, 188	Module #6 (59 a 64)	
		141, 76, 84	Module #7 (65 a 68)	
		89, 136, 110, 129	Module #8 (69 a 74)	
		69, 90, 95	Module #9 (75 a 77)	
FA16	69 % (1, 2, 3, 4, 5, 6, 8, 9, 11 y 13)	85, 66, 87, 25	Module #1 (1 a 11)	99.03 % (0.0097)
		102, 59, 107, 97, 82	Module #2 (12 a 23)	
		117, 98, 109, 83	Module #3 (24 a 40)	
		105, 110, 68	Module #4 (41 a 42)	
		113, 126, 109	Module #5 (43 a 45)	
		102, 116, 113, 116	Module #6 (46 a 61)	
		149, 98, 95, 177	Module #7 (62 a 77)	
FA18	63 % (2, 3, 4, 5, 6, 8, 10, 11 y 13)	151, 137, 79, 138, 135	Module #1 (1 a 5)	98.70 % (0.0130)
		125, 38, 122	Module #2 (6 a 26)	
		109, 140, 101, 81, 66	Module #3 (27 a 32)	
		124, 88, 193, 130, 173	Module #4 (33 a 48)	
		99, 181, 82, 82, 22	Module #5 (49 a 59)	
		153, 68, 112, 81, 136	Module #6 (60 a 77)	

(continued)

Table 4 (continued)

Evolution	Images for training	Number of neurons per hidden layer	Persons per module	Rec. rate/error
FA20	68 % (1, 3, 4, 5, 6, 7, 8, 9, 11 y 14)	116, 100, 174, 101, 144	Module #1 (1 a 4)	99.03 % (0.0097)
		71, 86, 129, 111, 171	Module #2 (5 a 17)	
		73, 86, 103, 102, 151	Module #3 (18 a 32)	
		124, 96, 88, 190	Module #4 (33 a 45)	
		91, 105, 140, 112, 56	Module #5 (46 a 54)	
		90, 109, 49, 114, 31	Module #6 (55 a 69)	
		120, 103, 113	Module #7 (70 a 77)	

Fig. 4 Convergence of evolution #9 (FA)

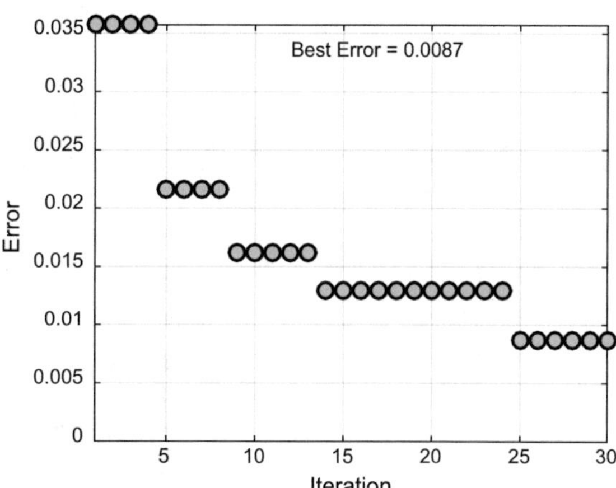

4.3 Grey Wolf Optimizer

The five best results obtained by this optimization technique are shown in Table 5. In evolution #17, the best results are obtained, where by using 79 % of data for the training phase a 100 of recognition rate is obtained, in only 22 iterations. In Fig. 5, the convergence of the best evolution is shown.

Table 5 The first five results for iris (GWO)

Evolution	Images for training	Number of neurons per hidden layer	Persons per module	Rec. rate/error
GWO 7	78 % (1, 2, 3, 4, 5, 6, 8, 9, 10, 13 and 14)	123, 65, 73, 172, 180	Module #1 (1–6)	99.57 % (0.0043)
		105, 102, 94, 116, 50	Module #2 (7–14)	
		73, 109, 132, 118, 80	Module #3 (15–24)	
		60, 76, 66, 86, 139	Module #4 (25–26)	
		94, 105, 177, 136, 99	Module #5 (27–38)	
		114, 141, 137, 122, 122	Module #6 (39–42)	
		103, 123, 52, 32, 113	Module #7 (43–51)	
		164, 94, 58, 124, 115	Module #8 (52–54)	
		68, 95, 34, 64, 135	Module #9 (55–67)	
		61, 122, 151, 101, 148	Module #10 (68–77)	
GWO 8	75 % (1, 2, 3, 4, 5, 6, 7, 8, 11, 13 and 14)	140, 48, 78, 88, 63	Module #1 (1–4)	99.57 % (0.0043)
		107, 48, 99, 93, 68	Module #2 (5–18)	
		112, 71, 142, 31, 92	Module #3 (19–36)	
		69, 134, 124, 65, 130	Module #4 (37–47)	
		48, 109, 163, 116, 87	Module #5 (48–49)	
		114, 93, 42, 104, 59	Module #6 (50–61)	
		95, 128, 69, 52, 164	Module #7 (62–63)	
		80, 57, 79, 154, 89	Module #8 (64–68)	
		77, 94, 186, 99, 59	Module #9 (69–70)	
		54, 71, 38, 113, 154	Module #10 (71–77)	

(continued)

Table 5 (continued)

Evolution	Images for training	Number of neurons per hidden layer	Persons per module	Rec. rate/error
GWO 10	78 % (1, 2, 3, 4, 5, 6, 9, 10, 11, 13 and 14)	185, 153, 103, 150, 58	Module #1 (1–3)	99.57 % (0.0043)
		71, 59, 56, 80, 89	Module #2 (4–18)	
		184, 109, 146, 92, 88	Module #3 (19–27)	
		77, 84, 28, 87, 82	Module #4 (28–31)	
		76, 123, 76, 48, 178	Module #5 (32–43)	
		109, 145, 38, 53, 118	Module #6 (44–45)	
		66, 95, 150, 74, 149	Module #7 (46–59)	
		95, 127, 164, 119, 73	Module #8 (60–61)	
		55, 174, 131, 50, 95	Module #9 (62–63)	
		65, 88, 129	Module #10 (64–77)	
GWO 13	78 % (1, 2, 3, 4, 5, 6, 8, 9, 11, 13 and 14)	85, 104, 54, 135, 173	Module #1 (1–10)	99.57 % (0.0043)
		102, 88, 123, 110, 172	Module #2 (11–22)	
		60, 105, 106, 104, 121	Module #3 (23–28)	
		77, 134, 93, 154	Module #4 (29–45)	
		126, 37, 59, 169, 107	Module #5 (46–55)	
		104, 44, 38, 48, 69	Module #6 (56–64)	
		92, 83, 66, 95, 58	Module #7 (65–72)	
		158, 65, 79, 128, 78	Module #8 (73–74)	
		74, 134, 80, 96, 68	Module #9 (75–77)	

(continued)

Table 5 (continued)

Evolution	Images for training	Number of neurons per hidden layer	Persons per module	Rec. rate/error
GWO 17	79 % (1, 2, 3, 5, 6, 8, 9, 10, 11, 13 and 14)	200, 100, 73, 142, 41	Module #1 (1–7)	100 % (0)
		58, 179, 74, 129, 46	Module #2 (8–20)	
		185, 96, 137, 91, 121	Module #3 (21–23)	
		59, 103, 51, 38, 116	Module #4 (24–35)	
		74, 160, 98, 140, 83	Module #5 (36–42)	
		148, 121, 62, 100, 153	Module #6 (43–55)	
		87, 100, 141, 81, 169	Module #7 (56–62)	
		127, 100, 105, 148, 129	Module #8 (63–64)	
		84, 43, 73, 66, 127	Module #9 (65–77)	

4.4 Comparison of Results

In Table 6, the best, the average and the worst result obtained for each optimization technique is shown. It can be observed that, the GWO achieves better results than the other techniques compared in this research work.

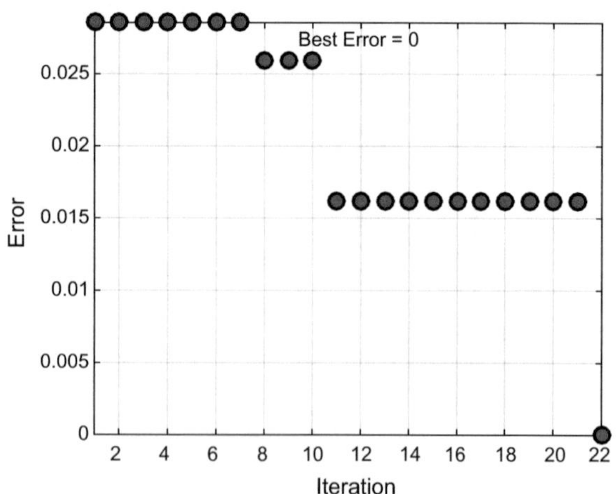

Fig. 5 Convergence of evolution #17 (GWO)

Table 6 Comparison of results

Method	Best	Average	Worst
HGA	99.68	98.68	97.40
	(0.0032)	(0.0132)	(0.0260)
FA	99.13	98.22	96.59
	(0.0087)	(0.0178)	(0.0341)
GWO	100.00	99.04	97.84
	(0)	(0.0096)	(0.0216)

5 Conclusions

In this paper, a comparison among different optimization techniques was performed, where each technique optimized some parameters of MNNs applied to human recognition based on iris biometrics. The chosen techniques to achieve this comparison were: HGA, FA, and GWO, where each technique optimized MNNs architectures and some parameters were number of submodules, percentage of information for the training phase, number of hidden layers (with their respective number of neurons) for each sub module and learning algorithm. Twenty evolutions for each optimization techniques were performed, only the five best results were shown of each one of them.

As the obtained results show that better results are achieved when the GWO is used, but also it is important to say that an analysis about the number of data for the training phase must be performed, because for example when the FA is used a 99.13 of recognition rate is obtained using only 8 images, compared with the best results (with GWO), where 11 images are used for this phase (with a 100 of recognition rate). So, the question would be—which is more important, less information for the training phase or simply more recognition rate? To answer that question, as future work, multiobjective approaches of these techniques will be used to optimize also the number of data for the training phase.

References

1. Abdallah H., Emara H. M., Dorrah H. T., Bahgat A., "Using Ant Colony Optimization algorithm for solving project management problems". Expert Syst. Appl. 36(6), 2009, pp. 10004-10015.
2. Amezcua J., Melin P., "Optimization of the LVQ Network Architecture with a Modular Approach for Arrhythmia Classification Using PSO". Design of Intelligent Systems Based on Fuzzy Logic, Neural Networks and Nature-Inspired Optimization 2015, pp. 119 - 126.
3. Azamm F., "Biologically Inspired Modular Neural Networks", PhD thesis, Virginia Polytechnic Institute and State University, Blacksburg, Virginia. May, 2000.
4. Bargiela A., Pedrycz W., "The roots of granular computing", IEEE International Conference on granular computing (GrC), 2006, pp. 806-809.

5. Chowdhury S., Das S. K., Das A., "Application of Genetic Algorithm in Communication Network Security", International Journal of Innovative Research in Computer and Communication Engineering, Vol. 3, Issue 1, 2015, pp. 274 - 280.
6. Database of Human Iris. Institute of Automation of Chinese Academy of Sciences (CASIA). Found on the Web page: http://www.cbsr.ia.ac.cn/english/IrisDatabase.asp (Accessed 12 November 2015).
7. Dorigo M., Gambardella L. M., "Ant colony system: a cooperative learning approach to the traveling salesman problem", IEEE Trans. On Evolutionary Computation, Volume 1(1), 1997, pp. 53-66.
8. Funes E., Allouche Y., Beltrán G., Jiménez A., "A Review: Artificial Neural Networks as Tool for Control Food Industry Process", Journal of Sensor Technology, 2015, 5, pp. 28-43.
9. Gaxiola F., Melin P., Valdez F., Castro J. R., Castillo O., "Optimization of type-2 fuzzy weights in backpropagation learning for neural networks using GAs and PSO". Appl. Soft Comput. 38, 2016, pp. 860 - 871.
10. Haupt R., Haupt S., "Practical Genetic Algorithms", Wiley-Interscience, 2 edition, 2004, pp. 42-43.
11. Hidalgo D., Melin P., Castillo O., "An optimization method for designing type-2 fuzzy inference systems based on the footprint of uncertainty using genetic algorithms". Expert Syst. Appl. 39(4), 2012, pp. 4590 – 4598.
12. Holland J.H., "Adaptation in Natural and Artificial Systems", MI: University of Michigan Press, 1975.
13. Kennedy J., Eberhart R. C., "Particle Swarm Optimization", In Proceedings of the IEEE international Joint Conference on Neuronal Networks, IEEE Press, 1995, pp. 1942 - 1948.
14. Lucic P., Teodorovic D., "Bee system: Modeling Combinatorial Optimization Transportation Engineering Problems by Swarm Intelligence", Preprints of the TRISTAN IV Triennial Symposium on Transportation Analysis, 2001, pp. 441 - 445.
15. Mech L. D., "Alpha status, dominance, and division of labor in wolf packs," Canadian Journal of Zoology, vol. 77, 1999, pp. 1196-1203.
16. Melin P., Castillo O., "Hybrid Intelligent Systems for Pattern Recognition Using Soft Computing: An Evolutionary Approach for Neural Networks and Fuzzy Systems", Springer, 1st edition, 2005, pp. 119-122.
17. Mirjalili S., Mirjalili S. M., Lewis A., "Grey Wolf Optimizer", Advances in Engineering Software , vol. 69, 2014, pp. 46 - 61.
18. Mitchell M., "An Introduction to Genetic Algorithms". A Bradford Book, 3rd edition, 1998.
19. Muro C., Escobedo R., Spector L., Coppinger R., "Wolf-pack (Canis lupus) hunting strategies emerge from simple rules in computational simulations," Behavioral processes, vol. 88, 2011, pp. 192-197.
20. Nawi N. M., Khan A., Rehman M. Z., "A New Back-Propagation Neural Network Optimized with Cuckoo Search Algorithm", ICCSA 2013, Part I, LNCS 7971, 2013, pp. 413 - 426.
21. Qian Y., Zhang H., Li F., Hu Q., Liang J., "Set-based granular computing: A lattice model", International Journal of Approximate Reasoning 55, 2014, pp. 834 - 852.
22. Rajabioun R., "Cuckoo Optimization Algorithm", Applied Soft Computing journal, Volume 11, 2011, pp. 5508 - 5518.
23. Sánchez D., Melin P., "Hierarchical Genetic Algorithms for Type-2 Fuzzy System Optimization Applied to Pattern Recognition and Fuzzy Control", Recent Advances on Hybrid Approaches for Designing Intelligent Systems 2014, pp. 19 – 35.
24. Saravanan K., Sasithra S., "Review On Classification Based On Artificial Neural Networks", International Journal of Ambient Systems and Applications (IJASA) Volume 2, No.4, 2014, pp. 11-18.
25. Tang K. S., Man K. F., Kwong S., Liu Z. F., "Minimal Fuzzy Memberships and Rule Using Hierarchical Genetic Algorithms", IEEE Trans. Ind. Electron., Vol. 45, No. 1, 1998, pp. 162–169.

26. Vázquez J. C., Valdez F., Melin P., "Comparative Study of Particle Swarm Optimization Variants in Complex Mathematics Functions". Fuzzy Logic Augmentation of Nature-Inspired Optimization Metaheuristics 2015, pp. 163 – 178.
27. Wang C., Soh Y. C., Wang H., Wang H., " A Hierarchical Genetic Algorithm for Path Planning in a Static Environment with Obstacles", Electrical and Computer Engineering, 2002. IEEE CCECE 2002. Canadian Conference on, vol.3, 2002, pp. 1652 - 1657.
28. Worapradya K., Pratishthananda S., "Fuzzy supervisory PI controller using hierarchical genetic algorithms", Control Conference, 2004. 5th Asian, Vol.3, 2004, pp. 1523 - 1528.
29. Yang X. S., "Firefly algorithms for multimodal optimization", Proc. 5th Symposium on Stochastic Algorithms, Foundations and Applications, (Eds. O. Watanabe and T. Zeugmann), Lecture Notes in Computer Science, 5792, 2009, pp. 169 - 178.
30. Yang X. S., "Nature-Inspired Metaheuristic Algorithms", Luniver Press, UK, 2008.
31. Yang X.S., He X., "Firefly Algorithm: Recent Advances and Applications", Int. J. of Swarm Intelligence, Volume 1, No.1, 2013 pp. 36 - 50.
32. Yao Y.Y., "On Modeling Data Mining with Granular Computing", 25th International Computer Software and Applications Conference, (COMPSAC), 2001, pp. 638-649.
33. Zadeh L. A., "Toward a theory of fuzzy information granulation and its centrality in human reasoning and fuzzy logic", Fuzzy Sets and Systems, Volume 90, Issue 2, 1 September 1997, pp. 111 – 127.
34. Zhang Z., Zhang C., "An Agent-Based Hybrid Intelligent System for Financial Investment Planning", 7th Pacific Rim International Conference on Artificial Intelligence (PRICAI),2002, pp. 355 - 364.

A Competitive Modular Neural Network for Long-Term Time Series Forecasting

Eduardo Méndez, Omar Lugo and Patricia Melin

Abstract In this paper, a modular neural network (MNN) architecture based on competitive clustering and a winner-takes-all strategy is proposed. In this case, the modules are obtained from clustering the training data with a competitive layer. And each module consists of a single hidden layer nonlinear autoregressive neural network. This MNN architecture can be used for short-term and long-term time series forecasting.

Keywords Modular neural network · Competitive learning · Forecasting · Time series

1 Introduction

Some problems are so complex that are difficult to model or solve using a single approach. For this reason, divide and conquer strategies are often used to deal with these type of problems.

A Modular Neural Network (MNN) decomposes a complex task into subtasks, and each subtask is handled by a simple, fast, and efficient module (monolithic ANN). Typically, these modules are decoupled, which means there is no communication between modules in the training phase. Once a MNN is trained, in the simulation phase each module obtains a sub-solution, then a final solution is obtained from these sub-solutions using a decision-making strategy [1]. Common approaches for computing the final solution are either by integration, election, or competition.

In this paper, a Competitive Modular Neural Network (CMNN) architecture is proposed for long-term forecasting of chaotic time series. This architecture is based on competitive learning [15], which is an unsupervised learning paradigm used for clustering. A competitive learning layer is utilized in the training phase to designate

E. Méndez · O. Lugo · P. Melin (✉)
Tijuana Institute of Technology, Tijuana, Mexico
e-mail: pmelin@tectijuana.mx

P. Melin et al. (eds.), *Nature-Inspired Design of Hybrid Intelligent Systems*,
Studies in Computational Intelligence 667, DOI 10.1007/978-3-319-47054-2_16

each module a subset of the training data, and also in the simulation phase, as a gating (competition) mechanism to obtain the final solution.

Forecasting refers to a process by which the future behavior of a dynamical system is estimated based on our understanding and characterization of the system [14].

Long-term forecasting also known as recursive prediction consists of calculating future values of a series based on past values of the series and values computed by the predictor itself.

A *time series* is a collection of observations made sequentially through time:

$$\mathbf{y} = y_1, y_2, \ldots, y_t. \tag{1}$$

The task of forecasting future values consists of estimating y_{t+k}, which its forecast from the time t to k steps into the future. Forecasting future values of a time series is a problem of interest in many areas such as meteorology, planning, sales, inventory control, etc.

Classical forecasting techniques are linear autoregressive models, like the Box–Jenkins method using ARMA or ARIMA models [3, 4]. These types of models are based on the linear autocorrelation of the time series. Most of the natural phenomenon's and industrial processes are nonlinear, so a nonlinear modeling is necessary. The competitive neural network presented in this paper is based on nonlinear autoregressive modeling, since each module is a Nonlinear Autoregressive Neural Network (NAR-NN).

This paper is organized as follows. In Sect. 2 the nonlinear autoregressive model is described. Section 3 is about the CMNN architecture. In Sect. 4 the training process is described and exemplified. A comparative study using the Mackey-Glass time series is presented in Sect. 5 and, finally in Sect. 6 some conclusions and future work are given.

2 Nonlinear Autoregressive Model

Nonlinear regression is the modeling of the static dependence of a response variable called the regressand $y \in Y \subset R$ on the regression vector $\mathbf{x} = [x_1, x_2, \ldots, x_p]$ over some domain $X \subset R^p$. The elements of the regression vector are called the *regressors* and the domain X the *regressor space* [2]. Given a regression vector $\mathbf{x}$, a regressand y can be estimated by:

$$y \approx f(\mathbf{X}) \tag{2}$$

The aim of regression is to construct a function that can serve as a reasonable approximation of $f(\cdot)$ for the entire domain X. Typically, $f(\cdot)$ is constructed using a basis function expansion, for example, a linear combination of affine functions, radial basis functions, Fourier series, or a wavelets expansion. In fact, any universal function approximation can be used, and in this paper, we propose a MNN architecture to address the problem of nonlinear regression applied to time series forecasting.

A time series forecasting problem can be modeled by a Nonlinear Autoregressive (NAR) model, which establish a relation between past values of a time series and a future value at the instant $k + 1$ as:

$$\hat{y}(k+1) = f(y(k), y(k-2), \ldots, y(k-n_y+1), \theta), \tag{3}$$

where k denotes discrete time samples, n_y is an integer related to the system order (number of past values), $f(\cdot)$ denotes a nonlinear parametric function, and the vector, $\boldsymbol{\theta} = [\theta_1, \theta_2, \ldots, \theta_m]$, contains the parameters or weights which characterize $f(\cdot)$.

In order to use a NAR model, it is necessary to transform a given time series, $\mathbf{y} = [y(0), \ldots, y(t)]$, into regression vectors as follows:

$$\mathbf{x}(k) = \left[y(k), y(k-1), \ldots, y(k-n_y+1)\right]^{\mathrm{T}}, \tag{4}$$

where $k > n_y$. Depending on the problem, a time delay d_y can be added to the regression vectors as shown below:

$$\mathbf{x}(k) = \left[y(k-d_y+1), y(k-d_y), \ldots, y(k-n_y-d_y+2)\right]^{\mathrm{T}}. \tag{5}$$

Once we have a set of regression vectors, the time series forecasting model can be described as:

$$\hat{y}(k+1) = f(\mathbf{x}, \theta), \tag{6}$$

where $f(\cdot)$ is a Non-Linear Autoregressive Neural Network (NAR-NN) with weights θ, which takes a regression vector $\mathbf{x}$ as input. Using Eq. (6) with a proper NAR-NN architecture, the problem consists of learning the weights of the neural network.

3 Competitive Modular Neural Network

The proposed CMNN architecture utilizes a competitive layer to designate which module belongs to one given regression vector $\mathbf{x}$. In other words, given an input only one module of the CMNN is activated. The above statement has some implications: first, the training set is clustered into c-subsets, which implies that each module is trained with a different subset. And also, it means that final output is computed only from the activated module. In the proposed CMNN architecture, only one module is activated which implies a *winner takes all* strategy, but if more than one module are to be activated a *k-winner takes all* approach could be implemented using a weighted integrator, like those used in [13, 14, 17].

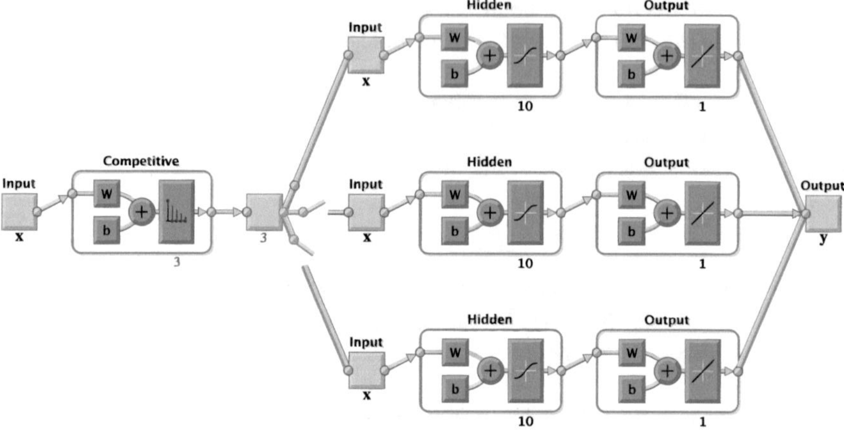

Fig. 1 A competitive modular neural network (CMNN) architecture with 3 modules

The input layer consists of n_y values, which is the size of a regression vector $\mathbf{x}$, and is connected to the competitive layer and also to each module. The modules of the proposed CMNN consist of a single hidden layer NAR-NN, but a more complex architecture could also be used. In Fig. 1, a diagram of the CMNN architecture with 3 modules is shown.

4 Training

The steps for the training of a CMNN and the forecast simulation are detailed below, using the Mackey-Glass time series as an example.

First, a time series $\mathbf{y}$ is transformed into a matrix of regression vectors $\mathbf{X}$ using Eq. (5). Figure 2 shows the time series on the left and its regression vectors on the right.

Next, the time series $\mathbf{y}$ and its regression vectors $\mathbf{X}$ are split into training and test sets, as shown in Figs. 3 and 4, respectively. For the training of the CMNN, the series $\mathbf{y}_{\text{train}}$ is used as the targets and $\mathbf{X}_{\text{train}}$ as the inputs. For forecasting the series $\mathbf{y}_{\text{test}}$ is only used for comparisons with the forecasted values and $\mathbf{X}_{\text{test}}$ is used as the inputs for single-step prediction and only the first d_y vectors of $\mathbf{X}_{\text{test}}$ are used for long-term forecasting, since the other ones will be generated recursively using the forecasted values.

Next for the training of the CMNN, the regression vectors $\mathbf{X}$ are clustered using the competitive layer as shown in the Fig. 5. After clustering the data, each module is trained in a decoupled fashion with its corresponding training subset, so each module learns a different behavior of the time series, as shown in Fig. 6.

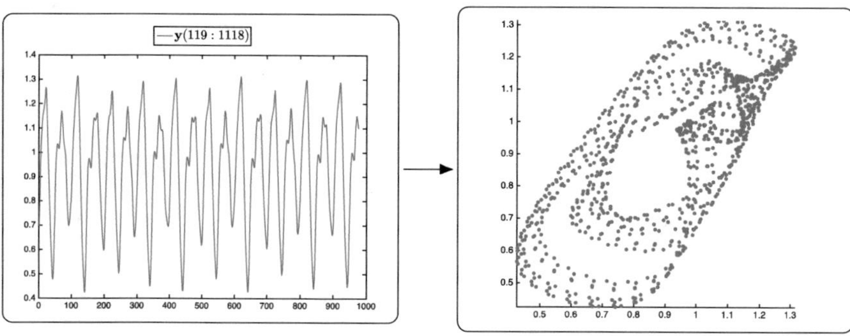

Fig. 2 Mackey-Glass time series and its regression vectors **X**

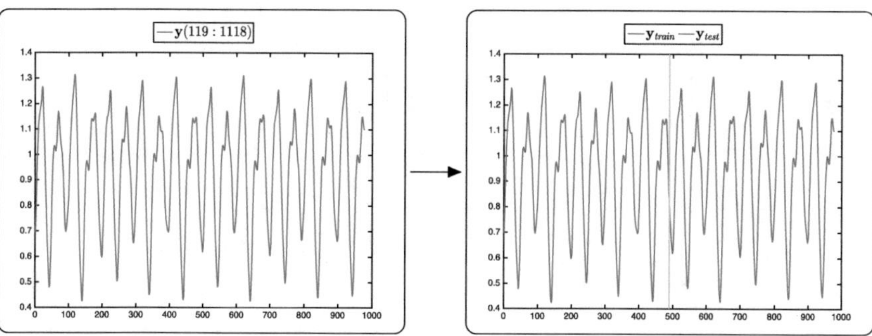

Fig. 3 Mackey-Glass time series **y** split into $\mathbf{y}_{\text{train}}$ and $\mathbf{y}_{\text{test}}$

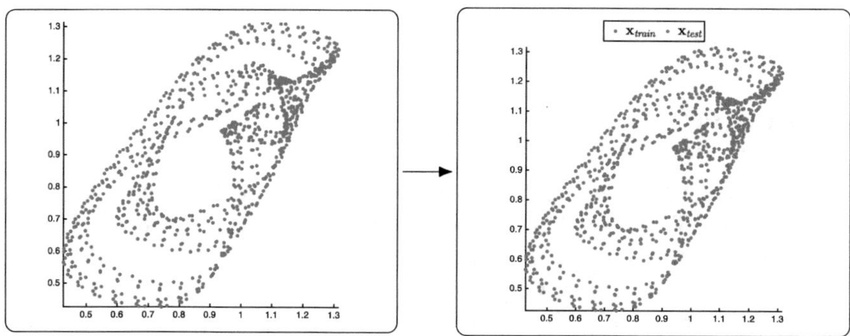

Fig. 4 Mackey-Glass regression vectors **X** split into $\mathbf{X}_{\text{train}}$ and $\mathbf{X}_{\text{test}}$

In Fig. 7 a forecast of $\mathbf{y}_{\text{train}}$ is shown, in which we can observe that every module learns to do something different and is activated only in specific areas of the time series.

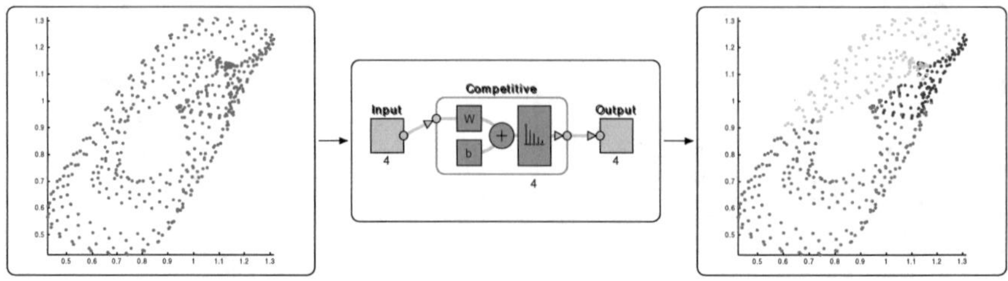

Fig. 5 Clustering the training set $\mathbf{X}_{\text{train}}$ into 4 clusters

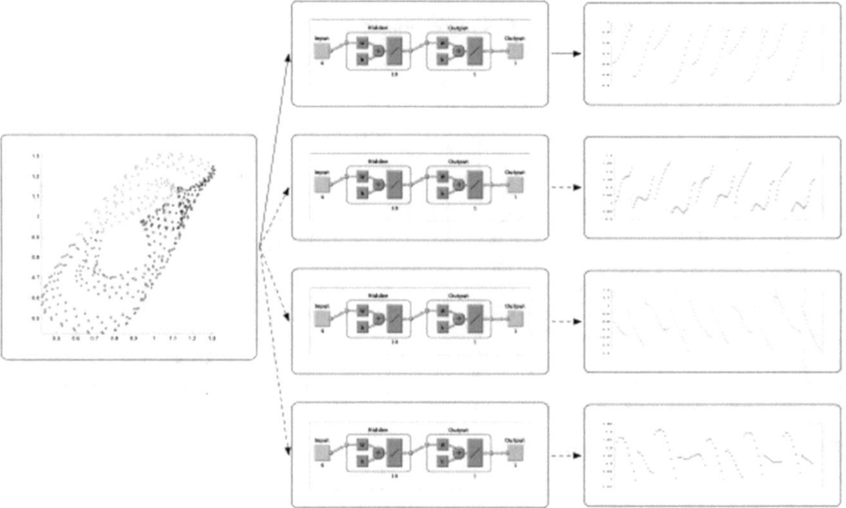

Fig. 6 Each module learns a different behavior of the time series

5 Experiments

Short-term and long-term simulations were performed on the Mackey-Glass chaotic time series [11]. For all experiments, the first 50 % of the data was used for training and the remaining 50 % was used for testing. Regression vectors of size $n_y = 4$ were used for all the experiments. The root-mean-squared-error (RMSE) and the mean-squared-errors (MSE) were taken as measures of performance. Finally, each experiment consists of 100 samples.

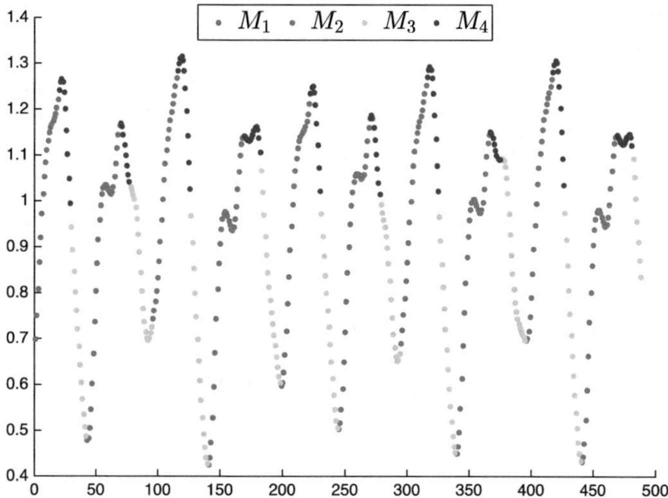

Fig. 7 Forecast of the training series $\mathbf{y}_{\text{train}}$

The short-term forecast simulations are of a single-step and the long-term forecast were simulated using a time horizon equivalent to the length of the corresponding test series $\mathbf{y}_{\text{test}}$.

Two training methods were compared Levenberg–Marquardt (LM) [8, 9, 12] and Bayesian Regularization (BR) [6, 10], and architectures of 1–5 modules were also compared.

A Mackey-Glass time series with $x(0) = 1.2$ and $\tau = 17$ was used, and the first 118 values were ignored (see Fig. 3).

The statics utilized for the comparison of the training methods and architectures are for the RMSE and MSE measures, the sampling mean $(\bar{\mu})$, the standard deviation $(\bar{\sigma})$, the minimum (Best), and the maximum (Worst). And also the sampling mean of the Epochs Per Module (EPM).

Single-step forecasting results are shown in Tables 1 and 2. The results for long-term forecast of 488 steps are shown in Tables 3 and 4. From the results, we can conclude that the CMNN trained with Bayesian regularization outperformed the ones trained with Levenberg–Marquardt. Also, although it seems that the more modules the better, using more modules would mean that the training will take much longer and also that depending on the number of epochs used for the training of the competitive layer, some modules would never activate if the epochs are not enough.

Table 1 Single-step forecasting results for the Mackey-Glass series ordered by $\bar{\mu}_{\text{MSE}}$

Training	Modules	$\bar{\mu}_{\text{MSE}}$	$\bar{\sigma}_{\text{MSE}}$	BEST_{MSE}	$\text{Worst}_{\text{MSE}}$	$\bar{\mu}_{\text{EPM}}$
BR	5	5.02×10^{-7}	2.17×10^{-7}	1.92×10^{-7}	1.22×10^{-6}	543
BR	4	5.52×10^{-7}	3.01×10^{-7}	2.84×10^{-7}	2.39×10^{-6}	552
BR	3	6.79×10^{-7}	1.49×10^{-7}	4.36×10^{-7}	1.24×10^{-6}	537
LM	3	1.29×10^{-6}	2.36×10^{-6}	5.10×10^{-7}	2.36×10^{-5}	843
BR	2	1.35×10^{-6}	2.88×10^{-7}	8.54×10^{-7}	2.00×10^{-6}	558
LM	4	1.52×10^{-6}	2.27×10^{-6}	3.19×10^{-7}	1.39×10^{-5}	777
LM	2	1.73×10^{-6}	5.73×10^{-7}	8.65×10^{-7}	4.27×10^{-6}	908
LM	5	2.48×10^{-6}	5.89×10^{-6}	3.46×10^{-7}	4.47×10^{-5}	640
BR	1	5.40×10^{-6}	4.33×10^{-7}	4.57×10^{-6}	6.44×10^{-6}	533
LM	1	5.74×10^{-6}	1.12×10^{-6}	4.09×10^{-6}	1.41×10^{-5}	925

Table 2 Single-step forecasting results for the Mackey-Glass series ordered by $\bar{\mu}_{\text{RMSE}}$

Training	Modules	$\bar{\mu}_{\text{RMSE}}$	$\bar{\sigma}_{\text{RMSE}}$	$\text{Best}_{\text{RMSE}}$	$\text{Worst}_{\text{RMSE}}$	$\bar{\mu}_{\text{EPM}}$
BR	5	6.95×10^{-4}	1.41×10^{-4}	4.38×10^{-4}	1.10×10^{-3}	543
BR	4	7.26×10^{-4}	1.60×10^{-4}	5.33×10^{-4}	1.55×10^{-3}	552
BR	3	8.19×10^{-4}	8.60×10^{-5}	6.61×10^{-4}	1.11×10^{-3}	537
LM	3	1.04×10^{-3}	4.56×10^{-4}	7.14×10^{-4}	4.86×10^{-3}	843
LM	4	1.09×10^{-3}	5.74×10^{-4}	5.64×10^{-4}	3.73×10^{-3}	777
BR	2	1.16×10^{-3}	1.25×10^{-4}	9.24×10^{-4}	1.41×10^{-3}	558
LM	5	1.28×10^{-3}	9.20×10^{-4}	5.88×10^{-4}	6.68×10^{-3}	640
LM	2	1.30×10^{-3}	2.09×10^{-4}	9.30×10^{-4}	2.07×10^{-3}	908
BR	1	2.32×10^{-3}	9.25×10^{-5}	2.14×10^{-3}	2.54×10^{-3}	533
LM	1	2.39×10^{-3}	2.06×10^{-4}	2.02×10^{-3}	3.76×10^{-3}	925

Table 3 Long-term forecasting results for the Mackey-Glass series ordered by μ_{MSE}

Training	Modules	$\bar{\mu}_{\text{MSE}}$	$\bar{\sigma}_{\text{MSE}}$	Best_{MSE}	$\text{Worst}_{\text{MSE}}$	$\bar{\mu}_{\text{EPM}}$
BR	5	3.93×10^{-4}	2.89×10^{-4}	2.63×10^{-5}	1.46×10^{-3}	543
BR	4	6.91×10^{-4}	6.57×10^{-4}	7.90×10^{-5}	3.75×10^{-3}	552
BR	3	8.40×10^{-4}	6.79×10^{-4}	1.03×10^{-4}	3.35×10^{-3}	537
LM	4	1.03×10^{-3}	1.16×10^{-3}	1.29×10^{-4}	7.36×10^{-3}	777
LM	3	1.30×10^{-3}	1.94×10^{-3}	8.75×10^{-5}	1.75×10^{-2}	843
BR	2	1.72×10^{-3}	1.10×10^{-3}	1.72×10^{-4}	7.48×10^{-3}	558
LM	2	1.76×10^{-3}	1.24×10^{-3}	1.18×10^{-4}	6.49×10^{-3}	908
LM	1	3.32×10^{-3}	1.33×10^{-3}	1.19×10^{-3}	7.59×10^{-3}	533
BR	1	3.90×10^{-3}	2.53×10^{-3}	6.16×10^{-4}	1.12×10^{-2}	925
LM	5	1.77×10^{-2}	1.47×10^{-1}	4.60×10^{-5}	1.46	640

Table 4 Long-term forecasting results for the Mackey-Glass series ordered by μ_{RMSE}

Training	Modules	$\bar{\mu}_{\text{RMSE}}$	$\bar{\sigma}_{\text{RMSE}}$	$\text{Best}_{\text{RMSE}}$	$\text{Worst}_{\text{RMSE}}$	$\bar{\mu}_{\text{EPM}}$
BR	5	1.86×10^{-2}	6.74×10^{-3}	5.12×10^{-3}	3.82×10^{-2}	543
BR	4	2.40×10^{-2}	1.07×10^{-2}	8.89×10^{-3}	6.12×10^{-2}	552
BR	3	2.72×10^{-2}	1.01×10^{-2}	1.01×10^{-2}	5.79×10^{-2}	537
LM	4	2.88×10^{-2}	1.42×10^{-2}	1.14×10^{-2}	8.58×10^{-2}	777
LM	3	3.24×10^{-2}	1.60×10^{-2}	9.36×10^{-3}	1.32×10^{-1}	843
BR	2	3.97×10^{-2}	1.22×10^{-2}	1.31×10^{-2}	8.65×10^{-2}	558
LM	2	3.97×10^{-2}	1.37×10^{-2}	1.08×10^{-2}	8.06×10^{-2}	908
LM	5	4.62×10^{-2}	1.25×10^{-1}	6.78×10^{-3}	1.21	640
BR	1	5.64×10^{-2}	1.18×10^{-2}	3.45×10^{-2}	8.71×10^{-2}	533
LM	1	5.94×10^{-2}	1.95×10^{-2}	2.48×10^{-2}	1.06×10^{-1}	925

Table 5 Single-step forecast comparison between the CMNN and other methods

Method	Length ($\mathbf{y}$)	Length ($\mathbf{y}_{\text{test}}$)	Measure	Error	CMNN error
DECS	1000	500	RMSE	2.89×10^{-2}	1.33×10^{-3}
E-ANFIS	800	400	RMSE	1.73×10^{-2}	2.03×10^{-3}
MNN-T1FIS	800	300	MAE	5.83×10^{-2}	6.96×10^{-4}
NN-IT2FW	800	300	MAE	4.13×10^{-2}	6.96×10^{-4}
NN-GT2FW	800	300	MAE	5.48×10^{-2}	6.96×10^{-4}

Next, a single-step forecast comparison with other methods using different set-ups and measures of performance is presented. The compared methods are the following:

- Dynamic evolving computation system (DECS) [5].
- Ensemble of adaptive neuro fuzzy inference systems (E-ANFIS) [17].
- MNN with type-1 fuzzy inference system (MNN-T1FIS) [16].
- Neural network with interval type-2 fuzzy weights (NN-IT2FW) [7].
- Neural network with generalized type-2 fuzzy weights (NN-GT2FW) [7].

Table 5 shows the results of this comparison, and we can observe that the CMNN performed better than the other methods. (Figs. 8 and 9).

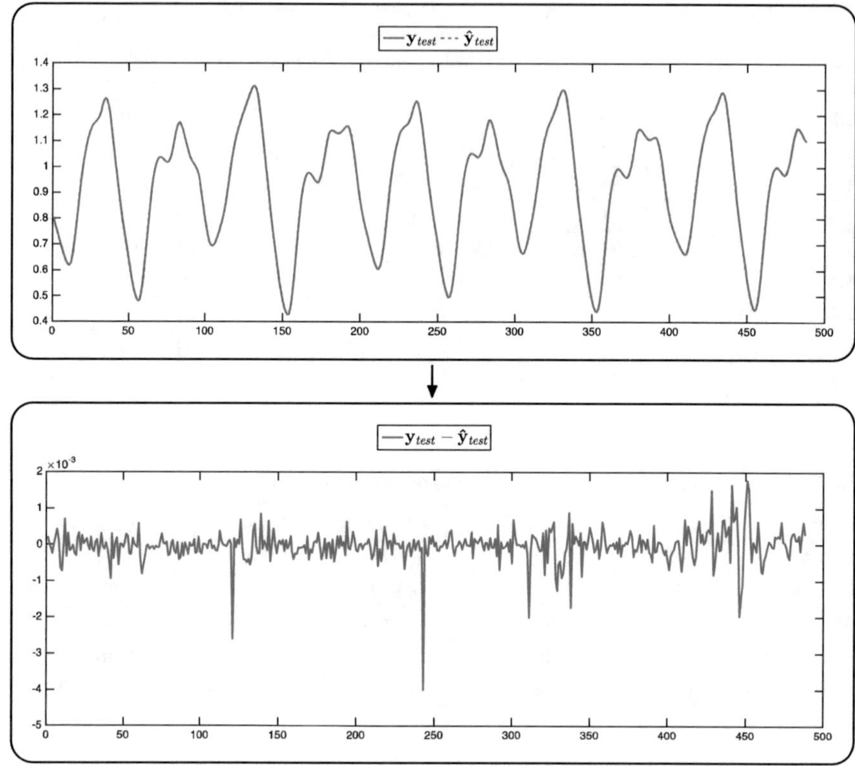

Fig. 8 Single-step forecast of the Mackey-Glass test series $\mathbf{y}_{\text{test}}$

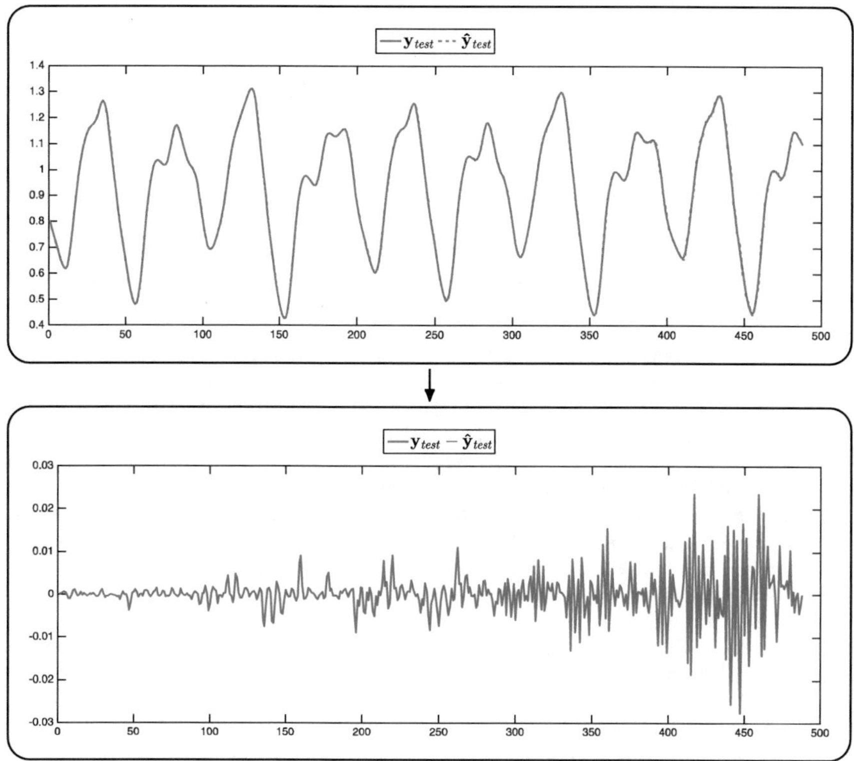

Fig. 9 Long-term forecast of the Mackey-Glass test series $\mathbf{y}_{\text{test}}$.

6 Conclusions

From the work done we reached the following conclusions. By using modular networks, it is possible to solve complex problems such as the forecast of chaotic time series. Also, the use of the competitive clustering is an excellent choice to identify the modules of a MNN, especially when it is not very clear how to split the dataset into modules. The CMNN presented in this paper, outperformed similar architectures based on MNNs, or ensemble neural networks. Although the obtained forecast results are good, they could be improved using a *k-winners takes all* strategy based on a fuzzy integrator.

References

1. Auda, G., Kamel, M.: Modular Neural Network Classifiers: A Comparative Study. J. Intell. Robot. Syst. 21, 2, 117–129 (1998).
2. Babuška, R.: Fuzzy Modeling for Control (International Series in Intelligent Technologies). Kluwer Academic Publishers, USA (1998).
3. Box, G.E.P., Jenkins, G.: Time Series Analysis, Forecasting and Control. Holden-Day, Incorporated (1990).
4. Chatfield, C.: The analysis of time series: an introduction. Chapman & Hall/CRC (2004).
5. Chen, Y.M., Lin, C.-T.: Dynamic Parameter Optimization of Evolutionary Computation for On-line Prediction of Time Series with Changing Dynamics. Appl. Soft Comput. 7, 4, 1170–1176 (2007).
6. Foresee, F.D., Hagan, M.T.: Gauss-Newton approximation to Bayesian learning. In: Neural Networks,1997., International Conference on. pp. 1930–1935 vol.3 (1997).
7. Gaxiola, F. et al.: Generalized type-2 fuzzy weight adjustment for backpropagation neural networks in time series prediction. Inf. Sci. 325, 159–174 (2015).
8. Hagan, M.T., Menhaj, M.B.: Training feedforward networks with the Marquardt algorithm. IEEE Trans. Neural Networks. 5, 6, 989–993 (1994).
9. Levenberg, K.: A method for the solution of certain non-linear problems in least squares. Q. J. Appl. Mathmatics. II, 2, 164–168 (1944).
10. MacKay, D.J.C.: Bayesian Interpolation. Neural Comput. 4, 3, 415–447 (1992).
11. Mackey, M.C., Glass, L.: Oscillation and chaos in physiological control systems. Science197 (4300), 287–289 (1977)
12. Marquardt, D.W.: An algorithm for least-squares estimation of nonlinear parameters. SIAM J. Appl. Math. 11, 2, 431–441 (1963).
13. Melin, P. et al.: A Hybrid Modular Neural Network Architecture with Fuzzy Sugeno Integration for Time Series Forecasting. Appl. Soft Comput. 7, 4, 1217–1226 (2007).
14. Melin, P. et al.: Forecasting Economic Time Series Using Modular Neural Networks and the Fuzzy Sugeno Integral as Response Integration Method. In: IJCNN. pp. 4363–4368 IEEE (2006).
15. Rumelhart, D.E., Zipser, D.: Feature Discovery by Competitive Learning*. Cogn. Sci. 9, 1, 75–112 (1985).
16. Sánchez, D., Melin, P.: Modular Neural Networks for Time Series Prediction Using Type-1 Fuzzy Logic Integration. In: Design of Intelligent Systems Based on Fuzzy Logic, Neural Networks and Nature-Inspired Optimization. pp. 141–154 (2015).
17. Soto, J. et al.: A new approach for time series prediction using ensembles of ANFIS models with interval type-2 and type-1 fuzzy integrators. In: 2013 IEEE Conference on Computational Intelligence for Financial Engineering Economics (CIFEr). pp. 68–73 (2013).

Part III
Fuzzy Metaheuristics

Differential Evolution Using Fuzzy Logic and a Comparative Study with Other Metaheuristics

Patricia Ochoa, Oscar Castillo and José Soria

Abstract This paper proposes an improvement to the algorithm differential evolution (DE) using fuzzy logic. The main contribution of this work is to dynamically adapt the parameter of mutation (F) using a fuzzy system, with the aim that the fuzzy system calculates the optimal parameters of the DE algorithm for obtaining better solutions, in this way arriving to the proposed new fuzzy differential evolution (FDE) algorithm. In this paper, experiments are performed with a set of mathematical functions using the proposed method to show the advantages of the FDE algorithm.

Keywords Differential evolution · Fuzzy logic · Dynamic parameters · Mutation F

1 Introduction

The use of fuzzy logic in evolutionary computing is becoming a common approach to improve the performance of the algorithms [19, 24, 25]. In most of the cases the parameters involved in the algorithms are determined by trial and error. In this aspect we propose in this paper the application of fuzzy logic, which can then be responsible of performing the dynamic adjustment of the mutation and crossover parameters in the DE algorithm. This has the goal of providing a better performance to DE with a fuzzy logic augmentation of this algorithm.

Fuzzy logic or multi-valued logic is based on the fuzzy set theory proposed by Zadeh in 1965, which can helps us with modeling knowledge, through the use of if-then fuzzy rules. Fuzzy set theory provides a systematic calculus to deal with linguistic information, and improves the numerical computation by using linguistic labels stipulated by membership functions [11]. DE is one of the latest evolutionary algorithms that have been proposed. It was created in 1994 by Price and Storn in an attempt to solve the Chebychev polynomial problem. The following years these two

P. Ochoa · O. Castillo (✉) · J. Soria
Tijuana Institute of Technology, Tijuana, Mexico
e-mail: ocastillo@tectijuana.mx

P. Melin et al. (eds.), *Nature-Inspired Design of Hybrid Intelligent Systems*, Studies in Computational Intelligence 667, DOI 10.1007/978-3-319-47054-2_17

authors proposed the DE for optimization of nonlinear and non-differentiable functions on continuous spaces.

The DE algorithm is a stochastic method of direct search, which has proven to be effective, efficient, and robust in a wide variety of applications, such as the learning of a neural network, a filter design of IIR, aerodynamically optimization. The DE has a number of important features, which makes it attractive for solving global optimization problems, among them are the following: it has the ability to handle non-differentiable, nonlinear, and multimodal objective functions, usually converges to the optimal with few control parameters, etc.

The DE belongs to the class of evolutionary algorithms that is based on populations. It uses two evolutionary mechanisms for the generation of descendants: mutation and crossover; finally a replacement mechanism, which is applied between the father vector and son vector determining who survives into the next generation. There exist works, where they are currently using fuzzy logic to optimize the performance of metaheuristic algorithms. For example just to name a few, papers such as: optimization of membership functions for type-1 and type 2 fuzzy controllers of an autonomous mobile robot using PSO [1], optimization of a fuzzy tracking controller for an autonomous mobile robot under perturbed torques by means of a chemical optimization paradigm [2], design of fuzzy control systems with different PSO variants [4], a method to solve the traveling salesman problem using ant colony optimization variants with ant set partitioning [6], evolutionary optimization of the fuzzy integrator in a navigation system for a mobile robot [7], fuzzy differential evolution (FDE) algorithm [29], optimal design of fuzzy classification systems using PSO with dynamic parameter adaptation through fuzzy logic [8], dynamic fuzzy logic parameter tuning for ACO and its application in TSP problems [9], bio-inspired optimization methods on graphic processing unit for minimization of complex mathematical functions [27], a new gravitational search algorithm using fuzzy logic to parameter adaptation [21], evolutionary method combining particle swarm optimization, and genetic algorithms using fuzzy logic for decision-making [26], an improved evolutionary method with fuzzy logic for combining particle swarm optimization and genetic algorithms [28].

Similarly, there are papers on differential evolution (DE) applications that use this algorithm to solve real-world problems. For example, just to mention a few: there is a work on fuzzy logic control using a DE algorithm aimed at modeling the financial market dynamics [5], another one is the design of optimized cascade fuzzy controllers based on DE [10], also eliciting transparent fuzzy model using DE [3], a harmony search algorithm comparison with genetic algorithms [12], DE [15], opposition-based DE [17], cooperative co-evolutionary DE for function optimization [19], an adaptive DE algorithm with novel mutation and crossover strategies for global numerical optimization [20], on the usage of DE for function optimization [22], journal of global optimization [23], DE using a neighborhood-based mutation operator [24], a new evolutionary algorithm for global optimization [25] real-time deterministic chaos control by means of selected evolutionary techniques [30] and finally assessment of human operator functional state using a novel DE optimization-based adaptive fuzzy model [18, 20].

The main contribution of this paper is the proposed FDE approach that is based on using fuzzy systems to dynamically adapt the parameters of the DE algorithm to improve the exploration and exploitation abilities of the method. The proposed FDE approach is different from existing works in the literature and for this reason is the main contribution of this paper.

This paper is organized as follows: Sect. 2 shows the concept of the DE algorithm. Section 3 describes the proposed method. Section 4 presents the fuzzy system to dynamically modify F, Sect. 5 shows the Benchmark functions, Sect. 6 outlines the Wilcoxon test statistics and Sect. 7 offers the conclusions.

2 Differential Evolution

DE is an optimization method belonging to the category of evolutionary computation that can be applied in solving complex optimization problems.

The DE is composed of four basic steps:

- Initialization.
- Mutation.
- Crossover.
- Selection.

This is a nondeterministic technique based on the evolution of a vector population (individuals) of real values representing the solutions in the search space. The generation of new individuals is carried out by the differential crossover and mutation operators [15].

The operation of the algorithm is explained in more detail below:

2.1 Population Structure

The DE algorithm maintains a pair of vector populations, both of which contain Np D-dimensional vectors of real-valued parameters [16].

$$P_{x,g} = \left(\mathbf{x}_{i,g}\right),\ i = 0, 1, \ldots, \mathrm{Np},\ g = 0, 1, \ldots, g_{\max} \tag{1}$$

$$\mathbf{x}_{i,g} = \left(x_{j,i,g}\right),\ j = 0, 1, \ldots, D - 1 \tag{2}$$

where:

P_x current population.

$g_{\max}$ maximum number of iterations.

i index population.
j parameters within the vector.

Once the vectors are initialized, three individuals are selected randomly to produce an intermediate population, $P_{v,g}$, of Np mutant vectors, $\boldsymbol{v}_{i,g}$.

$$P_{v,g} = \left(\boldsymbol{v}_{i,g}\right), i = 0, 1, \ldots, \mathrm{Np} - 1, g = 0, 1, \ldots, g_{\max} \tag{3}$$

$$\boldsymbol{v}_{i,g} = \left(v_{j,I,g}\right), j = 0, 1, \ldots, D - 1 \tag{4}$$

Each vector in the current population is recombined with a mutant vector to produce a trial population, P_u, the NP, mutant vector $\mathbf{u}_{i,g}$:

$$P_{v,g} = \left(\boldsymbol{u}_{i,g}\right), i = 0, 1, \ldots, \mathrm{Np} - 1, g = 0, 1, \ldots, g_{\max} \tag{5}$$

$$\boldsymbol{u}_{i,g} = \left(u_{j,I,g}\right), j = 0, 1, \ldots, D - 1 \tag{6}$$

2.2 Initialization

Before initializing the population, the upper and lower limits for each parameter must be specified. These 2D values can be collected by two D-dimensional, initialized vectors, b_L and b_U, for which the subscripts L and U indicate the lower and upper limits, respectively. Once the initialization limits have been specified a number generator randomly assigns each parameter in every vector a value within the set range. For example, the initial value ($g = 0$) of the jth vector parameter is:

$$x_{j,i,0} = \mathrm{rand}_j(0, 1) \cdot \left(b_{j,U} - b_{j,L}\right) + b_{j,L} \tag{7}$$

2.3 Mutation

In particular, the differential mutation uses a random sample equation showing how to combine three different vectors chosen randomly to create a mutant vector.

$$\boldsymbol{v}_{i,g} = \mathbf{x}_{r0,g} + F \cdot \left(\mathbf{x}_{r1,g} - \mathbf{x}_{r2,g}\right) \tag{8}$$

The scale factor, $F \in (0, 1)$ is a positive real number that controls the rate at which the population evolves. While there is no upper limit on F, the values are rarely greater than 1.0.

2.4 Crossover

To complement the differential mutation search strategy, DE also uses uniform crossover. This is sometimes known as a discrete recombination (dual). In particular, DE crosses each vector with a mutant vector:

$$U_{i,g} = \left(u_{j,i,g}\right) = \begin{cases} v_{j,i,g} & \text{if}\left(\text{rand}_j(0,1) \leq \text{Cr or } j = j_{\text{rand}}\right) \\ x_{j,i,g} & \text{otherwise.} \end{cases} \tag{9}$$

2.5 Selection

If the test vector, $\boldsymbol{U}_{i,g}$ has a value of the objective function equal to or less than, its target vector, $\boldsymbol{X}_{\text{i,g}}$, replaces the target vector in the next generation; otherwise, the target retains its place in population for at least another generation [15]:

$$X_{i,g+1} = \begin{cases} U_{i,g} & \text{if} f\left(U_{i,g}\right) \leq f\left(X_{i,g}\right) \\ X_{i,g} & \text{otherwise.} \end{cases} \tag{10}$$

The process of mutation, recombination, and selection are repeated until the optimum is found, or terminating precriteria specified is satisfied. DE is a simple, but powerful search engine that simulates natural evolution, combined with a mechanism to generate multiple search directions based on the distribution of solutions in the current population. Each vector I in the population at generation G, $\mathbf{x}_{i,G}$, called at this moment of reproduction as the target vector will be able to generate one offspring, called trial vector ($\mathbf{u}_{i,G}$). This trial vector is generated as follows: First of all, a search direction is defined by calculating the difference between a pair of vectors $r1$ and $r2$, called "*differential vectors,*" both of them chosen at random from the population. This difference vector is also scaled by using a user-defined parameter called "$F \geq 0$." This scaled difference vector is then added to a third vector $r3$, called "*base vector.*" As a result, a new vector is obtained, known as the mutation vector. After that, this mutation vector is recombined with the target vector (also called parent vector) by using discrete recombination (usually binomial crossover) controlled by a crossover parameter $0 \leq \text{CR} \leq 1$ whose value determines how similar the trial vector will be with respect to the target vector. There are several DE variants. However, the most known and used is DE/rand/1/bin, where the base vector is chosen at random, there is only a pair of differential vectors and a binomial crossover is used.

3 Proposed Method

The DE Algorithm is a powerful search technique used for solving optimization problems. In this paper, a new algorithm called FDE with dynamic adjustment of parameters is proposed. The main goal is that the fuzzy system can dynamically provide with the optimal parameters during execution for the best performance of the DE algorithm.

We propose exploring algorithm by first modifying the mutation, in two different forms will make two fuzzy system for each of the variables, this means that to dynamically modify the variable *F* (mutation) we will have a fuzzy system that modify *F* in decrease, expecting that this fuzzy system us best results and so to compare us with other algorithms that have a fuzzy system as well as the algorithm that we propose.

In this case the parameter that the fuzzy system optimizes is the mutation, as shown in Fig. 1.

4 Fuzzy System to Dynamically Modify the *F* Parameter

Then the fuzzy system, in which F is dynamically decreased is described as follows:

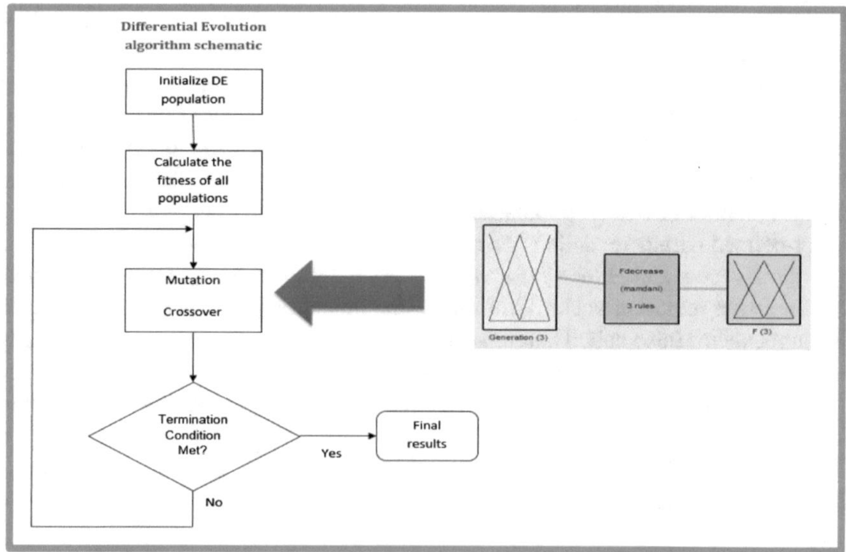

Fig. 1 The proposed differential evolution (DE) algorithm by integrating a fuzzy system to dynamically adapt parameter

- Contains one input and one output.
- Is of Mamdani type.
- All functions are triangular.
- The input of the fuzzy system is the number of generations and divided into three membership functions and they are: MF1 = 'Low' [−0.5 0 0.5], MF2 = 'Medium' [0 0.5 1], MF3 = 'High' [0.5 1 1.5].
- The output of the fuzzy system is the *F* parameter and is divided into three membership functions which are: MF1 = 'Low', [−0.5 0 0.5], MF2 = 'Medium', [0 0.5 1] MF3 = 'High', [0.5 1 1.5].
- The fuzzy system uses three rules and what it does is to decrease the value of the *F* parameter in a range of (0.1).

The fuzzy rules are presented in Fig. 2.

5 Benchmark Function

We performed experiments with the proposed FDE algorithm, where *F* decrease because it is the way in which better results obtained with the set of functions previously used Benchmark. Table 1 shows the new set of functions used.

In Fig. 3 we can find the set of Benchmark functions listed in Table 1.

6 Wilcoxon Test Statistics

We decided to check our proposed method algorithm of FDE with two fuzzy algorithms, for this we use fuzzy harmony search algorithm (FHS) [13] and fuzzy bat algorithm (FBA) [14] since these two algorithms use fuzzy logic for parameters adaptation as well as our proposed algorithm.

Fig. 2 Rules of the fuzzy system

1. - If (Generations is Low) then (F is High) (1)
2. - If (Generations is Medium) then (F is Medium) (1)
3. - If (Generations is High) then (F is Low) (1)

Table 1 Benchmark functions

Function	Search domain	*f* min
Sphere	$-5.12 \leq x_i \leq 5.12$	0
Rosenbrock	$-5 \leq x_i \leq 10$	0
Ackley	$-15 \leq x_i \leq 30$	0
Rastrigin	$-5.12 \leq x_i \leq 5.12$	0
Zakharov	$-5 \leq x_i \leq 10$	0
Sum Squared	$-10 \leq x_i \leq 10$	0

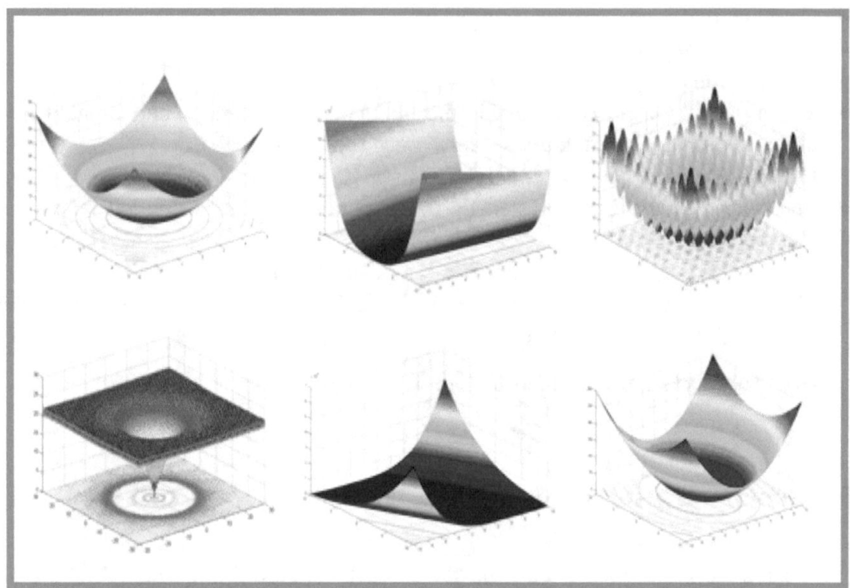

Fig. 3 Benchmark mathematical functions

Table 2 Parameters of the experiments

Parameters
NP = 10, 20, 30, 40, 45 and 50
D = 10
CR = 0.1
GEN = 100

Table 2 shows the parameters used for experiments, where F changes dynamically in decrease; the used search space is that of each function listed in Table 1.

Experiments were carried out per number of population, 30 experiments to population NP = 10, 30 for number of population NP = 20, up to the number of population of 50, later we obtained average and we can observe these in Table 3.

Taking into account the previous experiments we performed statistical Wilcoxon testing. The first test we did is with FBA, the statistical test used for comparison is the Wilcoxon matched pairs test was used to analyses the data, whose parameters are given in Table 4.

The alternative hypothesis states that the average of the results of the FDE algorithm is different than the average performance of the FBA, and therefore the null hypothesis tells us that the average of the results of the FDE algorithm is equal to the average of the FBA.

To test the hypothesis, first, the absolute values $|Z_i|\ldots|Z_n|$ are sorted and assigned its range Rank, Sign column indicates that all values obtained are positive, the column signed rank indicates the order of these values from lowest to highest.

Table 3 Average results for each function

Average by function						
Function	No. of population					
	10	20	30	40	45	50
Sphere	1.8111E−11	6.935E−43	1.971E−43	9.608E−22	9.60E−22	2.99E−43
Rosenbrock	0.40765257	0	0	0	0	0
Ackley	0.00079817	4.440E−15	4.204E−15	4.20E−15	4.20E−15	4.32E−15
Rastrigin	1.39463747	0.0331653	0	0	0	0
Zakharov	2.44E−12	4.41E−58	1.63E−59	3.86E−61	3.31E−61	1.39E−61
Sum Square	1.028E−06	4.339E−22	2.189E−21	1.171E−21	9.60E−22	1.29E−21

Table 4 Parameters for the statistical test

Function	No.	F1	F2	Difference	Abs (Difference)	Rank	Sign	Signed rank
		FBA	FDE					
Spherical	1	0.039735	0.00	0.0397352	0.0397352	1	1	1
Rosenbrock	2	0.685333	0.067942094	0.6173912	0.6173912	6	1	6
Rastrigin	3	0.3676021	0.237967	0.1296350	0.1296350	2	1	2
Ackley	4	0.3655000	0.000133	0.3653669	0.3653669	4	1	4
Zakharov	5	0.3318333	0.00	0.3318333	0.3318333	3	1	3
Sum Square	6	0.4398333	0.00	0.4398331	0.4398331	5	1	5

The formula of the statistical test is given as:

$$W^{+} = \sum_{\approx i > 0} R_i \tag{11}$$

That is, the sum of the ranges R_i corresponding to positive values Z_i.

The value of W^{+} is the sum of the positive ranks, the value W^{-} is the sum of the negative ranks, W represents the differences between two data samples, $W0$ indicates the value of the table for a two-tailed test using 30 samples.

The test to evaluate is as follows:

If $W \leq W0$,

Then reject Ho.

Table 5 shows the statistical test applied to the two fuzzy methods. With a confidence level of 95 % and a value of $W = 0$ and $W0 = 1$. So the statistical test results are that: for the fuzzy harmony search, there is a significant evidence to

Table 5 Values of parameters for the statistical test

W^{-}	W^{+}	W	Level significance	m = degrees of FREEDOM	$W0 = W\alpha{,}m$ =
0	21	0	0.05	6	1

Table 6 Parameters for the statistical test

Function	No.	F1	F2	Difference	Abs (Difference)	Rank	Sign	Signed rank
		FHS	FDE					
Spherical	1	0.00001379	0.00	0.000013	0.0000137	3	1	3
Rosenbrock	2	0.00000953	0.067942	0.067932	0.0679325	5	0	−5
Rastrigin	3	0.00000000	0.237967	0.237967	0.2379671	6	0	−6
Ackley	4	0.00004729	0.000133	0.000085	0.0000857	4	0	−4
Zakharov	5	0.00000001	0.00	0.000000	0.0000000	1	1	1
Sum square	6	0.00000253	0.00	0.000002	0.0000023	2	1	2

Table 7 Values of parameters for the statistical test

W^-	W^+	W	Level of Significance	m = degrees of freedom	$W0 = W\alpha$,m =
15	6	6	0.05	6	1

reject the null hypothesis and the alternative hypothesis is accepted mentioning that the average FDE is different than the average performance of the FBA.

The following comparison is with the FHS, and Table 6 shows the parameters used in this case.

The alternative hypothesis states that the average of the results of the FDE algorithm is different than the average performance of the FHS, and therefore the null hypothesis tells us that the average of the results of the FDE algorithm is equal to the average of the FHS.

The value of W^+ is the sum of the positive ranks, the value W^- is the sum of the negative ranks, W is the differences between two data samples, $W0$ indicates the value of the table for a two-tailed test using 30 samples.

The test to evaluate is as follows:

If $W \leq W0$, then fails to reject Ho.

Table 7 shows a statistical test applied to the two fuzzy methods. With a confidence level of 95 % and a value of $W = 0$ and $W0 = 1$.

So the statistical test results are that:

There is not enough evidence to reject the null hypothesis therefore we cannot accept the alternative hypothesis and this means that the FDE algorithm and the FHS are identical.

7 Conclusions

We can conclude that dynamically setting the parameters of a method of evolutionary optimization (in this case the DE algorithm), can improve the quality of the results. In our work using fuzzy logic to dynamically change the F parameters of the algorithm.

However using the FDE algorithm changing F in decrease we can observe that when performing the statistical test of Wilcoxon with respect to other two fuzzy algorithms the proposed algorithm is competitive, although we are still working on a way in which our proposed algorithm can be even better.

We can conclude that only with the modification of F by dynamically changing in the algorithm provides good results, in a matter of generations the proposed algorithm produces better results in fewer generations than the original DE algorithm, with at the runtime of the algorithm proposed by us is better. In general, we can state that our proposed method was what we expected, we have achieved good results.

References

1. Aguas-Marmolejo S. J., Castillo O.: Optimization of Membership Functions for Type-1 and Type 2 Fuzzy Controllers of an Autonomous Mobile Robot Using PSO. Recent Advances on Hybrid Intelligent Systems 2013: 97-104.
2. Astudillo L., Melin P., Castillo O.: Optimization of a Fuzzy Tracking Controller for an Autonomous Mobile Robot under Perturbed Torques by Means of a Chemical Optimization Paradigm. Recent Advances on Hybrid Intelligent Systems 2013: 3-20.
3. Eftekhari M., Katebi S.D., Karimi M., A.H. Jahanmir: Eliciting transparent fuzzy model using differential evolution, School of Engineering, Shiraz University, Shiraz, Iran, Applied Soft Computing 8 (2008) 466–476.
4. Fierro R., Castillo O., Design of Fuzzy Control Systems with Different PSO Variants. Recent Advances on Hybrid Intelligent Systems 2013: 81-88.
5. Hachicha N., Jarboui B., Siarry P.: A fuzzy logic control using a differential evolution algorithm aimed at modelling the financial market dynamics, Institut Supérieur de Commerce et de Comptabilité de Bizerte, Zarzouna 7021, Bizerte, Tunisia, Information Sciences 181 (2011) 79–91.
6. Lizárraga E., Castillo O., Soria J.: A Method to Solve the Traveling Salesman Problem Using Ant Colony Optimization Variants with Ant Set Partitioning. Recent Advances on Hybrid Intelligent Systems 2013: 237-2461.
7. Melendez A., Castillo O.: Evolutionary Optimization of the Fuzzy Integrator in a Navigation System for a Mobile Robot. Recent Advances on Hybrid Intelligent Systems 2013: 21-31.
8. Melin P., Olivas F., Castillo O., Valdez F., Soria J., GarcíaJ.: Optimal design of fuzzy classification systems using PSO with dynamic parameter adaptation through fuzzy logic. Expert Syst. Appl. 40(8): 3196-3206 (2013).
9. Neyoy H., Castillo O., José Soria: Dynamic Fuzzy Logic Parameter Tuning for ACO and Its Application in TSP Problems. Recent Advances on Hybrid Intelligent Systems 2013: 259-271.
10. Oh S.-K., Kim W.-D., Pedrycz W., Design of optimized cascade fuzzy controller based on differential evolution: Simulation studies and practical insights, Department of Electrical Engineering, The University of Suwon, Engineering Applications of Artificial Intelligence 25 (2012) 520–532.
11. Olivas F., Castillo O.: Particle Swarm Optimization with Dynamic Parameter Adaptation Using Fuzzy Logic for Benchmark Mathematical Functions. Recent Advances on Hybrid Intelligent Systems 2013: 247-258.
12. Peraza C., Valdez F. and Castillo O., A Harmony Search Algorithm Comparison with Genetic Algorithms,, Fuzzy Logic Augmentation of Nature-Inspired Optimization Metaheuristics, Studies in Computational Intelligence 574, Springer International Publishing Switzerland 2015.

13. Peraza Cinthia, Valdez Fevrier, Castillo Oscar, An Improved Harmony Search Algorithm Using Fuzzy Logic for the Optimization of Mathematical Functions, Design of Intelligent Systems Based on Fuzzy Logic, Neural Networks and Nature-Inspired Optimization,605-615, Springer International Publishing(2015).
14. Perez Jonathan J., Valdez F., Castillo O.: A new bat algorithm with fuzzy logic for dynamical parameter adaptation and its applicability to fuzzy control design, Fuzzy logic augmentation of nature-inspired optimization meta heuristics, pp. 65-80, Springer (2015).
15. Price, Storn R., Lampinen J. A., Differential Evolution, Kenneth V., Springer 2005.
16. Price K., Storn R. and Lampinen J. A. Differential Evolution: A Practical Approach to Global Optimization (Natural Computing Series), 2005: Springer.
17. Rahnamayan S., Tizhoosh, H. R., Salama M. M. A. Opposition-Based Differential Evolution, Evolutionary Computation, IEEE Transaction on (Volume: 12, Issue: 1), 2008, pp. 64-79..
18. Raofen W., Zhang J., Zhang Y., Wang X.: Assessment of human operator functional state using a novel differential evolution optimization based adaptive fuzzy model, Lab for Brain-Computer Interfaces and Control, East China University of Science and Technology, Shanghai 200237, PR China, Biomedical Signal Processing and Control 7 (2012) 490– 498.
19. Shi Y.-J., Teng H.-F., and Li Z.-Q., "Cooperative Co-evolutionary differential evolution for function optimization", Proc. 1st Int. Conf. Advances in Natural Comput, pp. 1080 -1088, 2005.
20. Sk. I., Swagatam D., Saurav G., Subhrajit R., Ponnuthurai N. S., An Adaptive Differential Evolution Algorithm With Novel Mutation and Crossover Strategies for Global Numerical Optimization, IEEE Transactions on systems, man, and Cybernetics—Part b: Cybernetics, vol. 42, no. 2, April 2012.
21. Sombra A., Valdez F., Melin P., Castillo O.: A new gravitational search algorithm using fuzzy logic to parameter adaptation. IEEE Congress on Evolutionary Computation 2013: 1068-1074.
22. Storn R. "On the usage of differential evolution for function optimization", Proc. Biennial Conf. North Amer. Fuzzy Inf. Process. Soc., pp. 519 -523 1996.
23. Storn R. and Price K. Journal of Global Optimization 11, pp. 341 -359 1997.
24. Swagatam D., Ajith A., Uday K. C., Amit K., Differential Evolution Using a Neighborhood-based Mutation Operator, Department of Electronics and Telecommunication Engineering, Jadavpur University, Kolkata 700032, India.
25. Sun J., Zhang Q. and Tsang E., "DE/EDA: A new evolutionary algorithm for global optimization", *Inf. Sci.*, vol. 169, pp. 249 -262 2004.
26. Valdez F., Melin P., Castillo O., Evolutionary method combining particle swarm optimization and genetic algorithms using fuzzy logic for decision making, in: Proceedings of the IEEE International Conference on Fuzzy Systems, 2009, pp. 2114–2119.
27. Valdez F., Melin P., Castillo O.: Bio-inspired Optimization Methods on Graphic Processing Unit for Minimization of Complex Mathematical Functions. Recent Advances on Hybrid Intelligent Systems 2013: 313-322.
28. Valdez F., P. Melin, O. Castillo: An improved evolutionary method with fuzzy logic for combining Particle Swarm Optimization and Genetic Algorithms. Applied Soft Computing 11 (2011) 2625–2632.
29. Vucetic D., Fuzzy differential evolution algorithm, The University of Western Ontario, London, Ontario, Canada, 2012.
30. Zelinka I. Real-time deterministic chaos control by means of selected evolutionary techniques, Engineering Applications of Artificial Intelligence 22 (2009) 283–297.

An Adaptive Fuzzy Control Based on Harmony Search and Its Application to Optimization

Cinthia Peraza, Fevrier Valdez and Oscar Castillo

Abstract This paper develops a new fuzzy harmony search algorithm (FHS) for solving optimization problems. FHS employs a novel method using fuzzy logic for adaptation of parameter the pitch adjustment (*PArate*) that enhances accuracy and convergence of harmony search (HS) algorithm. In this paper the impact of constant parameters on harmony search algorithm are discussed and a strategy for tuning these parameters is presented. The FHS algorithm has been successfully applied to various benchmarking optimization problems. Numerical results reveal that the proposed algorithm can find better solutions when compared to HS and other heuristic methods and is a powerful search algorithm for various benchmarking optimization problems.

Keywords Harmony search · Fuzzy logic · Dynamic parameter adaptation

1 Introduction

A meta-heuristic algorithm, mimicking the improvisation process of music players, has been recently developed and named harmony search (HS). harmony search algorithm had been very successful in a wide variety of optimization problems, presenting several advantages with respect to traditional optimization techniques such as the following [3]: (a) HS algorithm imposes fewer mathematical requirements and does not require initial value settings of the decision variables. (b) As the HS algorithm uses stochastic random searches, derivative information is also unnecessary (c).

C. Peraza (✉) · F. Valdez · O. Castillo
Tijuana Institute of Technology, Tijuana, BC, Mexico
e-mail: cinthia_sita@hotmail.com

F. Valdez
e-mail: fevrier@tectijuana.mx

O. Castillo
e-mail: ocastillo@tectijuana.mx

P. Melin et al. (eds.), *Nature-Inspired Design of Hybrid Intelligent Systems*,
Studies in Computational Intelligence 667, DOI 10.1007/978-3-319-47054-2_18

The HS algorithm generates a new vector, after considering all of the existing vectors, whereas the genetic algorithm (GA) only considers the two parent vectors. These features increase the flexibility of the HS algorithm and produce better solutions. HS is good at identifying the high performance regions of the solution space at a reasonable time, but gets into trouble in performing local search for numerical applications. In order to improve the fine-tuning characteristic of HS algorithm, FHS employs a new method using the fuzzy logic, which is responsible in performing the dynamic adjustment the parameter of pitch adjustment in the Harmony Search algorithm (HS) [16]. The FHS algorithm has the power of the HS algorithm with the fine-tuning feature of mathematical techniques and can outperform either one individually. To show the great power of this method, FHS algorithm is applied to various standard benchmarking optimization problems. Numerical results reveal that the proposed algorithm is a powerful search algorithm for various optimization problems.

In addition, we carefully conducted a set of experiments to reveal the impact of control parameters applied to benchmark mathematical functions and comparison with other method: fuzzy bat algorithm (FBA) [13].

Similarly, there are papers on Harmony Search algorithm (HS) applications that use this algorithm to solve real problems. To mention a few: a parameter setting free harmony search algorithm in [4], a Tabu Harmony Search Based Approach to Fuzzy Linear Regression in [7], a Novel Global Harmony Search Algorithm presented in [1], Global best harmony search in [8], an improved harmony search algorithm for solving optimization problems presented in [9], a new meta-heuristic algorithm for continuous engineering optimization harmony search theory and practice in [3], harmony search algorithms for structural design optimization in [5], and Benchmarking of heuristic optimization methods presented in [15].

There are many algorithms for solving problems. Some of them are special, some are more general. Many problems cannot be solved by deterministic algorithm, the algorithm used heuristically.

In the set of meta heuristic algorithms can find:

Fuzzy PSO and GA [16].
Gravitational search algorithm [14].
Self-Adaptive HS [17].
Differential evolution algorithm [11]
A new optimizer using particle swarm theory [2].

In order to get a clear overview of the classification and analysis presented in the following sections: Sect. 2 shows the concept of the harmony search algorithm as applied to the technique for parameter optimization. Section 3 describes the proposed method. Section 4 shows the methodology for parameter adaptation. Section 5 offers the simulation results comparison HS and FHS applied to benchmarking optimization problems. Section 6 describes simulation results comparison FHS with self-adaptive harmony search algorithm. Section 7 offers a Statistical comparison. Section 8 describes the conclusions.

2 Harmony Search Algorithm

This section describes the simple harmony search algorithm (HS). Harmony search is a relatively new metaheuristic optimization algorithm inspired music and was first developed by ZW Geem et al. in 2001 [6].

This algorithm can be explained in more in detail with the process of improvisation that a musician uses, which consists of three options:

Play any song you have in your memory.
Play a similar composition to an existing.
Play a new song or randomly.

If we formalize these three options for optimization, we have three corresponding components: harmony memory accepting, pitch adjustment and randomization [10].

2.1 *Operators Process of Improvisation the Harmony Memory*

2.1.1 Memory in the Harmony Search Algorithm (HMR)

The use of harmony memory is important because it is similar to choosing the best individuals in GAs. This will ensure the best harmonies will be transferred to the new memory harmony. In order to use this memory more effectively, we can assign a parameter r_{accept} € [0, 1] called acceptance rate memory [18].

$$r_{\text{accept}} \in [0, 1] \tag{1}$$

2.1.2 Pitch Adjustment (PArate)

To adjust the pitch slightly in the second component, we have to use such a method can adjust the frequency efficiently. In theory, the pitch adjustment can be adjusted linearly or nonlinearly, but in practice the linear approach is used. If the current solution is X_{old} (or pitch), then the new solution (tone) is generated X_{new}.

$$X_{\text{new}} = X_{\text{old}} + b_{\text{p}}(2\text{rand} - 1) \tag{2}$$

where "rand"is a random number drawn from a uniform distribution [0, 1]. Here is it bandwidth, which controls the local range of tone adjustment in fact, we can see that the pitch adjustment (2) is a random step.

The Pitch adjustment setting is similar to the mutation operator in GAs [18].

2.1.3 Randomization

The third component is a randomization component (3) that is used to increase the diversity of the solutions. The use of randomization can further push the system to explore various regions with high diversity solution in order to find the global optimum [18]. So we have:

$$P_{\mathrm{a}} = P_{\text{lower limit}} + P_{\text{range}} * \text{rand} \tag{3}$$

where "rand" is a generator of random numbers in the range of 0 and 1. (Search space).

2.2 *Pseudo Code for Harmony Search*

The basic steps of the harmony algorithm, can be summarized as the pseudo code for HS is presented below:

```
Objective function f (x), x = (x1, ...., xn) T
Initial generate harmonics (matrices of real numbers)
Define pitch adjustment rate (rpa) and limits of tone
Define acceptance rate of the harmony memory (r accept)
        while (t <Maximum number of iterations)
              Generate a new harmony and accept the best harmonies
              Setting the tone for new harmonies (solutions)
     if (rand> raccept)
              Choose an existing harmony randomly
     else if (rand> rpa)
     Setting the tone at random within a bandwidth      (2)
     else
              Generate a new harmony through a randomization (3)
     End if
              Accepting new harmonies (solutions) best
     End while
To find the best solutions.
```

3 Proposed Method

This section describes the proposed fuzzy harmony search (FHS) algorithm. A brief overview of the modification procedures of the proposed FHS algorithm is also presented.

The *PArate* parameter introduced in the fuzzy system help the algorithm find globally improved solutions, respectively.

PArate and *HMR* in HS algorithm are very important parameters in fine-tuning of optimized solution vectors, and can be potentially useful in adjusting convergence rate of algorithm to optimal solution. So fine adjustment of these parameters are of great interest. The traditional HS algorithm uses fixed value for both *HMR* and *PArate*. In the HS method *HMR* and *PArate* values adjusted in initialization step and cannot be changed during new generations. The main drawback of this method appears in the number of iterations the algorithm needs to find an optimal solution. Small *HMR* values are selected only the best harmonies and converged very slowly. Furthermore large *HMR* values almost all the harmonies are used in memory of harmony, then other harmonies are not well explored, leading to potentially erroneous solutions. Small *PArate* values with large values can cause to poor performance of the algorithm and considerable increase in iterations needed to find optimum solution. Although small values in final generations increase the fine-tuning of solution vectors, but in early generations must take a bigger value to enforce the algorithm to increase the diversity of solution vectors. Furthermore large *PArate* values with small values usually cause the improvement of best solutions in final generations which algorithm converged to optimal solution vector.

The main difference between FHS and traditional HS method is in the way of adjusting *PArate*. To improve the performance of the HS algorithm and eliminate the drawbacks lies with fixed values of *HMR*, FHS algorithm use variable *PArate* in improvisation step. FHS using a fuzzy system for it to be responsible to change dynamically the parameter *PArat*e in the range 0.7 to 1 in each iteration number as shown in Fig. 1 and expressed as follow:

4 Methodology for Parameter Adaptation

In Sects. 2 and 3 we show the most important parameters of the algorithm, based on the literature, we decided to use the parameter *Parate* as an outlet for our fuzzy system and must be in the range of 0.7 and 1, plus it is also suggested that changing the parameter *PArate* dynamically during the execution of this algorithm can produce better results.

In addition it is also found that the algorithm can have Performance measures, such as: the iterations, needs to be considered to run the algorithm, among others. In our work all the above are taken in consideration for the fuzzy systems to modify

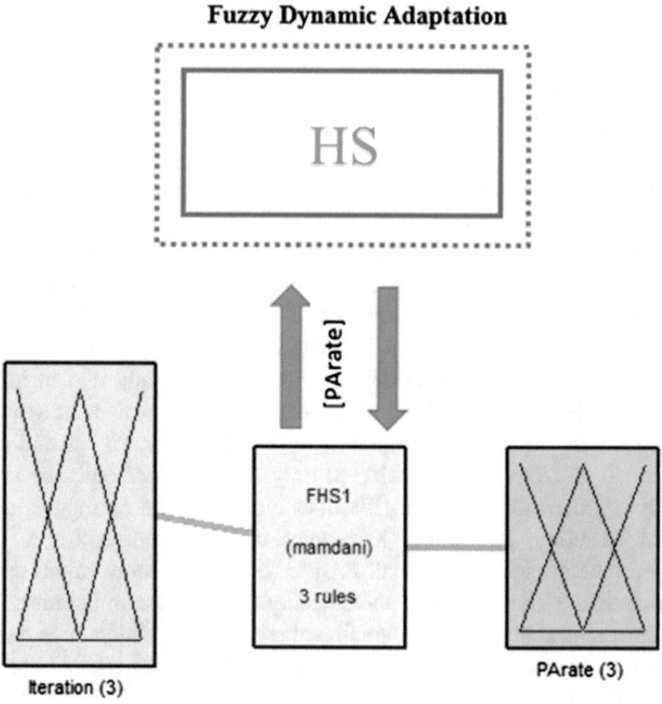

Fig. 1 Variation of PArate using the fuzzy system

the parameter pitch adjustment dynamically changing these parameters in each iteration of the algorithm.

For measuring the iterations of the algorithm, it was decided to use a percentage of iterations, i.e. when starting the algorithm the iterations will be considered "low", and when the iterations are completed it will be considered "high" or close to 100 %, the design of the fuzzy system can be seen in Fig. 2 it has a one input and one output. To represent this idea we use [12]:

$$\text{Iteration} = \frac{\text{Current iteration}}{\text{Maximum of iterations}} \quad (4)$$

The design of the input variable can be appreciated in Fig. 3, which shows the input iteration, that input is granulated into three triangular membership functions. For the output variables, as mentioned above, the recommended values for *PArate* are between 0.7 and 1, so that the output variables were designed using this range of values. The output is granulated in three triangular membership functions. The design of the output variable can be seen in Fig. 4.

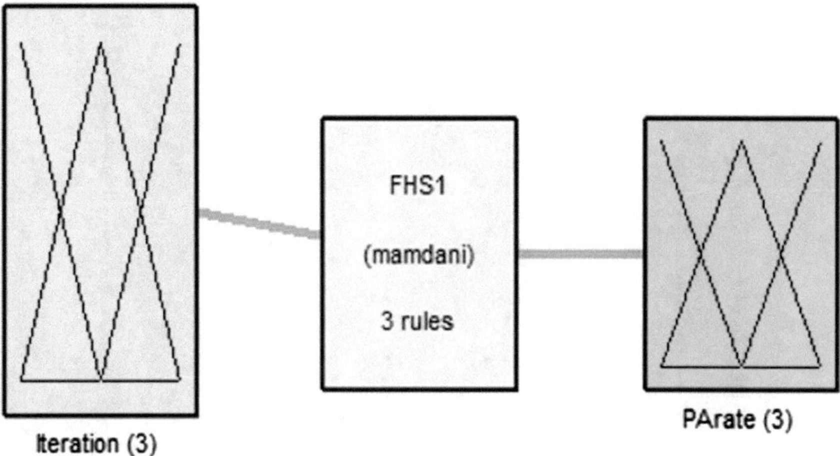

Fig. 2 Fuzzy system for parameter adaptation

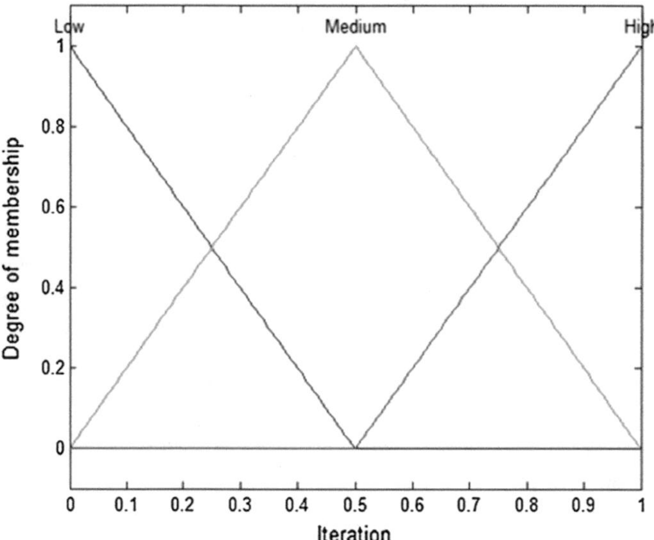

Fig. 3 Input 1: iteration

The surface of fuzzy system shows the parameter *PArate* changes dynamically with the increasing iterations shown in Fig. 5. Based on the behavior of the parameter *PArate* we decide to use rules to increase first exploration of the search space and eventually exploitation, this in order to find better solutions the rules for the fuzzy system are shown in Fig. 6.

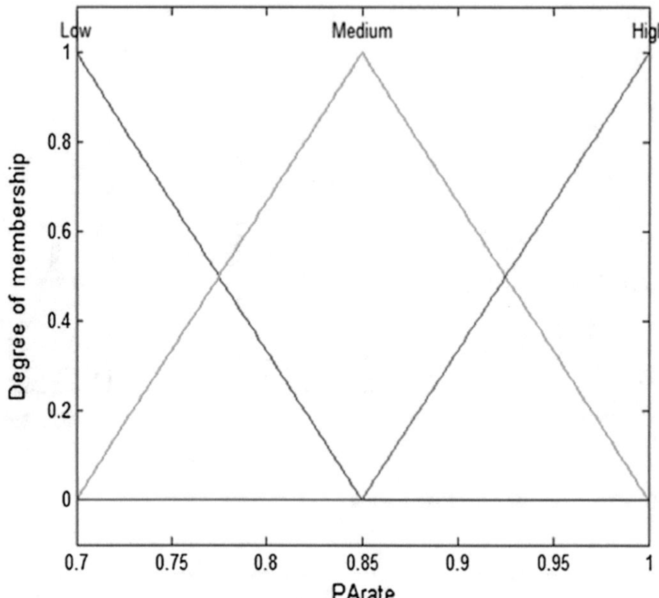

Fig. 4 Output 1: PArate

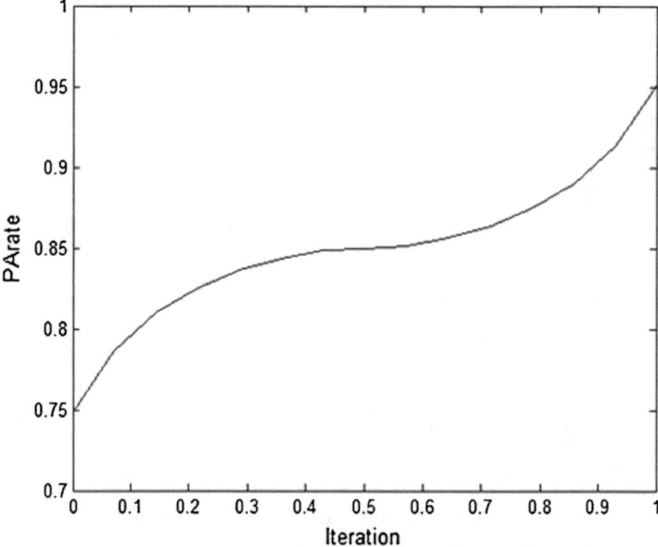

Fig. 5 Surface fuzzy system

Fig. 6 Rules for fuzzy system FHS

1. If (Iteration is Low) then (PArate is Low)
2. If (Iteration is Medium) then (PArate is Medium)
3. If (Iteration is High) then (PArate is High

5 Simulation Results Comparison Harmony Search Algorithm with Fuzzy Harmony Search Algorithm

In this section the comparison of the FHS algorithm is made against the simple harmony search algorithm for benchmarking optimization problems. In each of the algorithms 7 Benchmark mathematical functions were considered separately and all functions the global minimum is 0 can be appreciated in Table 1, a dimension of 8, 16, 32, 64, 128 variables was used with 50 runs for each function varying the parameters of the algorithms.

From Table 1 shows the mathematical functions, search domain and local minimum of each function that was used.

The parameters used in each method can be appreciated in Table 2:

From Table 2 shows the parameters used in each method, Harmony memory (HMS), pitch adjustment (*PArate*), Harmony memory accepting (*HMR*) and dimensions.

We conducted experiments for HS method and our method, where HMS = 4, 5, 10, 20, 30, 40 (6 levels) and *PArate* = 0.75 for the Hs method and Dynamic adaptation using a fuzzy system for the FHS method and the parameter *HMR* uses fixed value for both methods. We conducted sets of experiments on each test functions in 8, 16, 32, 64, 128 dimensions, and the results are presented in Table 3 shows the simulation results for the experiments in each method.

Table 3 HS and FHS algorithm is applied to various standard benchmarking optimization problems. We can see the overall averages obtained in each

Table 1 Benchmark functions

Function	Search domain	$f_{\min}$
Sphere	$-5.12 \leq x_i \leq 5.12$	0
Rosenbrock	$-5 \leq x_i \leq 10$	0
Ackley	$-15 \leq x_i \leq 30$	0
Rastrigin	$-5.12 \leq x_i \leq 5.12$	0
Zakharov	$-5 \leq x_i \leq 10$	0
Griewank	$-600 \leq x_i \leq 600$	0
Dixon & Price	$-10 \leq x_i \leq 10$	0

Table 2 Parameters used for test problems

Methods	HMS	PArate	HMR	Dimensions
Simple HS	4–40	0.75	0.95	8, 16, 32, 64, 128
Fuzzy HS	4–40	Dynamic	0.95	8, 16, 32, 64, 128

Table 3 Benchmark optimization results for each method

Functions	HS	FHS
Sphere	7.63E+00	3.37E−10
Rosenbrock	2.37E−01	5.82E−04
Ackley	1.20E−04	5.30E−09
Rastrigin	2.93E−08	1.97E−08
Zakharov	1.04E+02	2.19E−03
Griewank	9.04E−02	1.17E−03
Dixon & Price	1.28E+01	3.06E−01

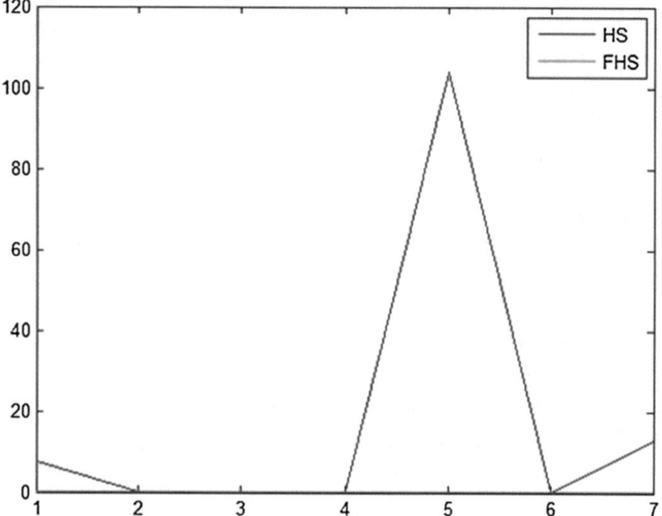

Fig. 7 Graphic of the experiments of the simple Hs (*Blue* line) and FHS (*Green* line). All results have been averageover 50 runs

mathematical function for each method, and the best results are with our proposed method.

The graphic of the experiments for each method simple harmony search algorithm (HS) and FHS algorithm for all functions as shown in Fig. 7.

6 Simulation Results Comparison Harmony Search Algorithm with Fuzzy Bat Algorithm

In this section the comparison of the FHS algorithm is made against the FBA for optimization [13]. In each of the algorithms 6 Benchmark mathematical functions were considered separately and all functions the global minimum is 0 can be appreciated in Table 4, a dimension of 10 variables was used with 30 runs for each function varying the parameter of the algorithm [2, 15].

From Table 4 shows the mathematical functions it was applied for each method, search domain and local minimum of each functions that was used.

The parameters used in each method can be appreciated in Table 5:

From Table 5 shows the parameters used in each method. We conducted sets of experiments on each test functions in 2, 5, 10, 20, 30 and 40 harmony's and Bats, 10 dimensions it was applied for each mathematical function in each method.

In this section we show the experimental results obtained by the FHS algorithm and FBA. Table 6 shows the simulation results for the 10 dimensions in each function.

From Table 6 it can be appreciated the overall averages in each mathematical function for each method, and the best results are with our proposed method FHS. The results were averaged over 30 runs. The Graphic of the all experiments used 10 dimensions for each function in each method as shown in Fig. 8.

Table 4 Benchmark functions

Function	Search domain	$f_{\min}$
Sphere	$-5.12 \le x_i \le 5.12$	0
Ackley	$-15 \le x_i \le 30$	0
Rosenbrock	$-2.048 \le x_i \le 2.048$	0
Zakharov	$-5 \le x_i \le 10$	0
Rastrigin	$-5.12 \le x_i \le 5.12$	0
Sum Squared	$-10 \le x_i \le 10$	0

Table 5 Default parameters for all methods

Methods	Harmonies and Bats	Dimensions
Fuzzy bat Algorithm [13]	2, 5, 10, 20, 30, 40	10
Fuzzy HS	2, 5, 10, 20, 30, 40	10

Table 6 Benchmark function optimization results in 10 dimensions

10 Dimensions			
FBA [13]		FHS	
Sphere	3.97E−02	Sphere	1.38084E−05
Ackley	6.86E−01	Ackley	9.54688E−06
Rosenbrock	3.68E−01	Rosenbrock	1.76111E−11
Zakharov	3.66E−01	Zakharov	4.72388E−05
Rastrigin	3.32E−01	Rastrigin	1.07782E−08
Sum Squared	4.40E−01	Sum Squared	2.52934E−06

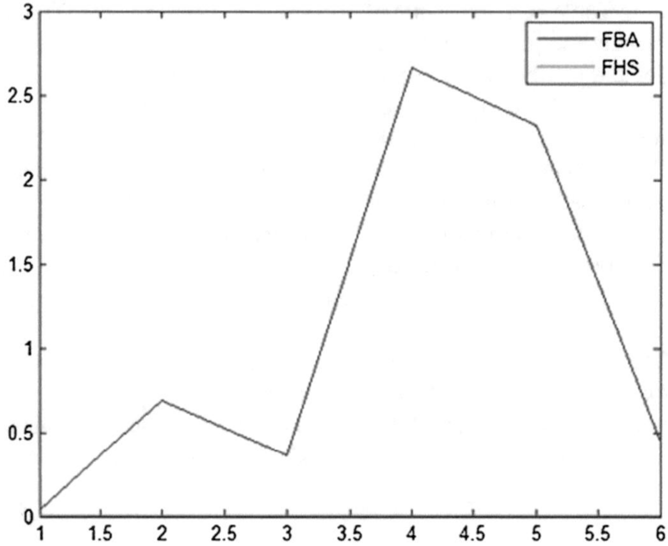

Fig. 8 Comparisons among two methods on six test functions in 10 dimensions. The *green* lines indicate the results of the proposed method (FHS). The *blue* lines indicate the results of the Fuzzy Bat algorithm (FBA). They are to be distinct. Respectively indicate the best result, the average result. All results have been averaged over 30 runs

7 Statistical Comparison

To perform the statistical comparison, we have:

2 methods to compare against the Fuzzy HSPArate and FBA, 6 benchmark mathematical functions, 30 experiments were performed for each method by each function, and so it has a total of 180 experiments for each method. Of this total, we took a random sample of 30 experiments for each method for statistical comparison.

The statistical test used for comparison is the *z*-test, whose parameters are defined in Table 7. With the results contained in Table 6, we applied the statistical *z*-test, obtaining the results contained in Table 8.

Table 7 Parameters for statistical *z*-test

Parameter	Value
Level of significance	95 %
Alpha	0.05 %
H_a	$\mu_1 < \mu_2$
H_0	$\mu_1 \geq \mu_2$
Critical value	−1.645

Table 8 Results of applying statistical *z*-test

Algorithm	Number of samples	Mean	S.D.	Alpha
FHS	30	1.41E−05	3.08E−05	
FBA	30	3.58E−01	2.96E−01	0.05

The alternative hypothesis states that the average of the results of the FHS algorithm is less than the average of the results of FBA, and therefore the null hypothesis tells us that the average algorithm fuzzy search for harmony is greater than or equal to the average of the FBA with rejection region for all values lower than −1.645.

The equation of the statistical test applied was:

$$Z = \frac{(\overline{X_1} - \overline{X_2}) - (\mu_1 - \mu_2)}{\sigma_{\overline{X_1} - \overline{X_2}}} \tag{5}$$

In applying the statistic *z*-test, with significance level of 95 % and with a value of $Z = -6.6262$, and the alternative hypothesis says that the average of the proposed method FHS is lower than the FBA, and of course the null hypothesis tells us that the average of the proposed method is greater than or equal to the average of the FBA, with a rejection region for all values fall below −1.645. So the statistical test results are that: for the FHS with 10 dimensions in all functions, there is significant evidence to reject the null hypothesis.

The Graphic of the statistical test is shown in Fig. 9.

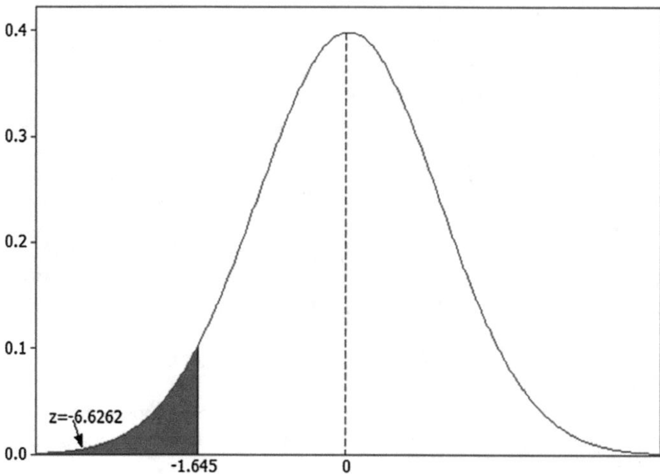

Fig. 9 Graphic of the statistical test where we can see that the value of *z* falls in the rejection of the null hypothesis

8 Conclusions

In this paper, we proposed a new FHS with dynamic adjustment of parameters using a fuzzy system. In this case we applied a fuzzy system for it to be responsible to dynamically adjust the parameter of pitch adjustment and it's applied to benchmark mathematical functions.

A comparison with two methods was performed the Fuzzy HS and FBA, applied to 6 Benchmark mathematical functions using 10 dimensions.

Based on these observations, the proposed method was compared with the FBA. The numerical results indicated that our method offers much higher performance to update the existing methods on three optimization problems. Besides, it is interesting to note that in original methods can get better performance obtained with the parameters from the experiments. Furthermore, our method in contrast to others has a better diversity and always converges to a better solution eventually.

Acknowledgment We would like to express our gratitude to CONACYT and Tijuana Institute of Technology for the facilities and resources granted for the development of this research.

References

1. Dexuan Z., Yanfeng G., Liqun G., Peifeng W.: A Novel Global Harmony Search Algorithm for Chemical Equation Balancing, In Computer Design and Applications (ICCDA), 2010 International Conference on, Vol. 2, IEEE (2010).
2. Eberhart R., Kennedy J.: A new optimizer using particle swarm theory, In Proceedings of the sixth international symposium on micro machine and human science Vol. 1, pp. 39-43., pp. 39–43, IEEE (1995).
3. Geem Z., Lee K.: A new meta-heuristic algorithm for continuous engineering optimization harmony search theory and practice, Computer methods in applied mechanics and engineering, pp. 3902-3933. Elsevier, Maryland, USA (2004).
4. Geem Z., Sim K., Parameter setting free harmony search algorithm, Applied Mathematics and Computation, pp. 3881-3889. Elsevier, Chung Ang, China (2010).
5. Geem Z.: Harmony search algorithms for structural design optimization, Studies in computational intelligence, Vol. 239, pp. 8–121. Springer, Heidelberg, Germany (2009).
6. Geem Z.: Music inspired harmony Search Algorithm theory and applications, Studies in computational intelligence pp. 8–121, Springer, Heidelberg, Germany (2009).
7. Hadi M., Mehmet A., Mashinchi M., Pedrycz W.:A Tabu Harmony Search Based Approach to Fuzzy Linear Regression, Fuzzy Systems, IEEE Transactions on, pp. 432-448. IEEE, New Jersey, USA (2011).
8. Mahamed G., Mahdavi M.: Global best harmony search, Applied Mathematics and Computation, pp. 1–14. Elsevier, Amsterdam, Holland (2008).
9. Mahdavi M., Fesanghary M., Damangir E.: An improved harmony search algorithm for solving optimization problems, applied Mathematics and Computation, pp. 1567–1579. Elsevier, Amsterdam, Holland (2007).
10. Manjarres D., Torres L., Lopez S., DelSer J, Bilbao M., Salcedo S., Geem Z.: A survey on applications of the harmony search algorithm, Engineering Applications of Artificial Intelligence, pp. 3–14, Elsevier, Amsterdam, Holland (2013).

11. Ochoa P., Castillo O., Soria J., Differential evolution with dynamic adaptation of parameters for the optimization of fuzzy controllers, Recent Advances on Hybrid Approaches for designing intelligent systems, pp. 275–288. Springer, Heidelberg, Germany (2013).
12. Olivas F., Melin P., Castillo O., et. al, Optimal design of fuzzy classification systems using PSO with dynamic parameter adaptation through fuzzy logic, pp, 2-11, Elsevier (2013).
13. Perez J., Valdez F., Castillo O.: A new Bat Algorithm with fuzzy logic for Dynamical Parameter adaptation and its applicability to fuzzy control design, Fuzzy Logic augmentation of nature inspired optimization metaheuristics, Volume 574, pp. 65-79, Springer (2015).
14. Sombra A., Valdez F., Melin P., Castillo O.: A new gravitational search algorithm using fuzzy logic to parameter adaptation. IEEE Congress on Evolutionary Computation, pp. 1068–1074 (2013).
15. Štefek A.: Benchmarking of heuristic optimization methods, Mechatronika 14th International Symposium, pp 68-71, IEEE (2011).
16. Valdez F.., Melin P., Castillo O.: Fuzzy Control of Parameters to Dynamically Adapt the PSO and GA Algorithms, Fuzzy Systems International Conference, pp. 1-8, IEEE, Barcelona, Spain (2010).
17. Wang C., Huang Y..: Self-adaptive harmony search algorithm for optimization, Expert Systems with Applications Volume 37, pp. 2826-2837, Elsevier (2010).
18. Yang X.: Nature Inspired Metaheuristic Algorithms, Second Edition, University of Cambridge, United Kingdom, pp 73-76, Luniver Press (2010).

A Review of Dynamic Parameter Adaptation Methods for the Firefly Algorithm

Carlos Soto, Fevrier Valdez and Oscar Castillo

Abstract The firefly algorithm is a bioinspired metaheuristic-based on the firefly's behavior. This paper shows previous works on parameters analysis and dynamical parameter adjustment, using different approaches and fuzzy logic.

Keywords Firefly algorithm · Parameter adaptation · Optimization problems · Mathematical functions · Fuzzy adaptation

1 Introduction

The metaheuristic algorithms for search and optimization usually divide their work into two tasks. The first consist in making an exploration or localization of promising areas, where the best solutions may be, and the second task, the exploitation, which consists on concentrating in the areas where the best solutions were found to continue with the search.

The firefly algorithm (FA) has been proved to be very efficient in solving multimodal, nonlinear, global optimization problems, classification problems, and image processing. The fireflies are beetles from the Lampyridae family; which the main characteristic are their wings. There exist over 2000 species of fireflies; their brightness comes from the special luminal organs under the abdomen. The flashing light of the fireflies shines in a specific way for each species. Each shining way is an optical sign that helps the fireflies to find their couple. The light can also work as a defense mechanism. This characteristic inspired Xin-She Yang in 2008 to design the FA [1].

This paper is organized as follows. Section 2 describes the FA. Section 3 is about parameter tuning and dynamic adjustment. Section 4 presents a summary of

C. Soto · F. Valdez · O. Castillo (✉)
Tijuana Institute of Technology, Tijuana, BC, Mexico
e-mail: ocastillo@tectijuana.mx

P. Melin et al. (eds.), *Nature-Inspired Design of Hybrid Intelligent Systems*,
Studies in Computational Intelligence 667, DOI 10.1007/978-3-319-47054-2_19

previous works with FA and where a dynamical adjustment of the FA parameters is carried out. Section 5 presents some final comments.

2 Firefly Algorithm

The FA uses three idealized rules

1. Fireflies are unisex so that one firefly will be attracted to other fireflies regardless of their sex.
2. The attractiveness is proportional to the brightness, and decreases when the distance increases between two fireflies. Thus for any two flashing fireflies, the less brighter one will move toward the brighter one. If there is no brighter one than a particular firefly, then it will move randomly.
3. The brightness of a firefly is determined by the landscape of the objective function.

As shown in Fig. 1, the algorithm begins by defining the number of fireflies to use and the number of iterations. The parameters α, β, and γ, controlling the exploration and exploitation, must be initialized, and the optimization function must be defined. The next step in the algorithm is the main cycle. The FA is based on the attraction between the fireflies depending on its light intensity, in the algorithm one firefly is selected and then its light intensity is compared with all the other fireflies. If the light intensity of a firefly A is less brighter than the firefly B, then the firefly A moves towards the firefly B. If not, the firefly has only a random movement. This process continues until all the iterations are over or a stop criterion is reached.

2.1 Firefly's Movement

The firefly moves according to the following equation.

$$x_{i+1} = x_i + \beta_0 \, \mathrm{e}^{-\gamma r_{ij}^2}(x_i - x_j) + \alpha\varepsilon_1 \tag{1}$$

The FA movement consists of three terms to determine the next position of the firefly. The actual position of the firefly is represented by the vector x_i, the second term manages the exploitation where β_0 is the initial attraction of the firefly, γ is the constriction factor, r is the Euclidean distance between the position of the firefly x_i and the firefly x_j. The last term manages the exploration where α is the parameter that controls how much randomness is allowed the firefly to have in its movement and ε_1 is a vector that contains random numbers drawn from a Gaussian distribution or uniform distribution at time t. If $\beta_0 = 0$, it becomes a simple random walk, or β_0

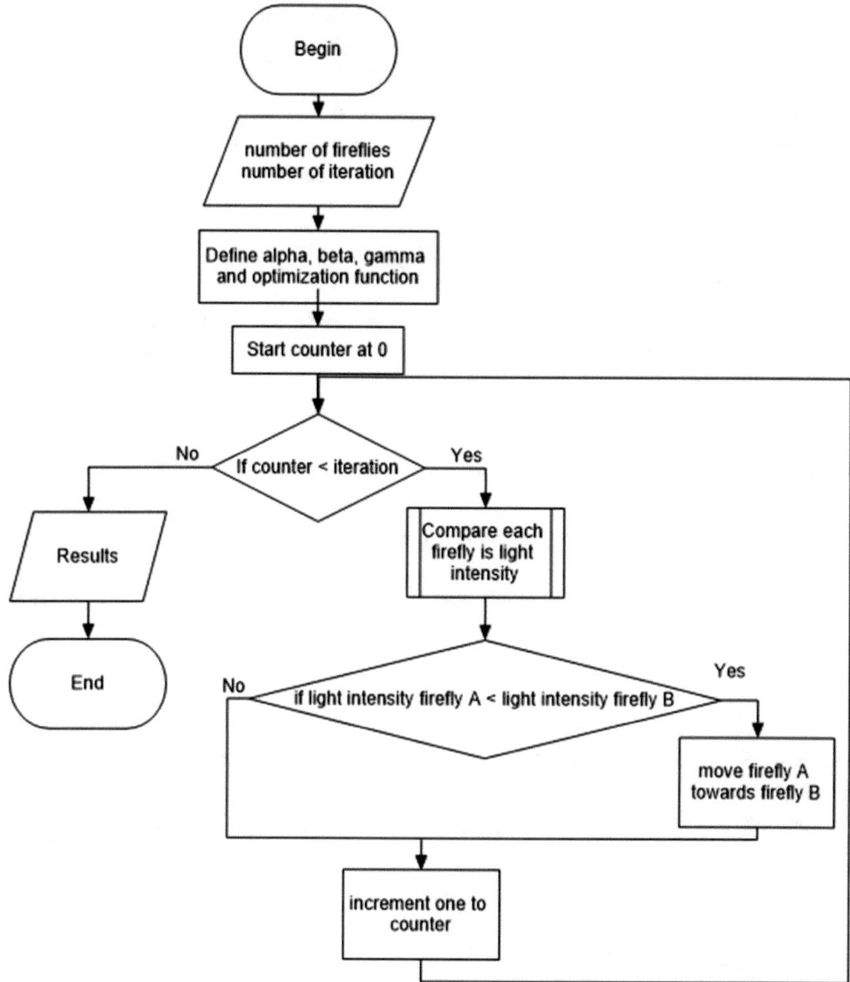

Fig. 1 FA flowchart

is the attractiveness at the distance $r = 0$. On the other hand, if $\gamma = 0$, FA reduces to a variant of particle swarm optimization.

2.2 *Light Intensity and Attractiveness*

The variation of light intensity and formulation of the attractiveness are very important in the FA. The attractiveness of a firefly is determined by its brightness, which can be associated with the encoded fitness function.

As a firefly's attractiveness is proportional to the light intensity that is seen by adjacent fireflies, we can now define the variation of attractiveness β with the distance r_{ij} between firefly i and firefly j. The light intensity decreases with the distance from its source, and light is also absorbed in the media, this is why the attractiveness varies with the degree of absorption γ.

The light intensity is presented in the equation below

$$I(r) = \frac{I_s}{r^2}, \tag{2}$$

where I_s is the intensity at the source. For a given medium with a fixed light absorption coefficient γ, the light intensity I varies with the distance r.

$$I = I_0\, \mathrm{e}^{-\gamma r}, \tag{3}$$

where I_0 is the original light intensity and combining Eqs. (1) and (2), we have

$$I(r) = I_0\, \mathrm{e}^{-\gamma r^2}, \tag{4}$$

As a firefly's attractiveness is proportional to the light intensity that is seen by adjacent fireflies, we can now define the attractiveness β of a firefly by

$$\beta = \beta_0\, \mathrm{e}^{-\gamma r^2}, \tag{5}$$

where β_0 is the attractiveness at $r = 0$. Some studies suggest $\beta_0 = 1$ can be used for most applications.

The attractiveness function $\beta(r)$ can be any monotonically decreasing functions [1].

2.3 Restriction Coefficient

The γ parameter is the absorption coefficient for the light and controls the variation of the attractiveness (and light intensity), and its values determine the speed of the convergence of the FA. In theory, $\gamma \in [0, \infty]$. But if γ is very large, then the light intensity decreases too quickly and would result in stagnation, thus the second term (4) becomes negligible. On the other hand if γ is very small then the exponential factor:$\mathrm{e}^{-\gamma r_{ij}^2} \to 1$ and would suffer from premature convergence [1]. But this rule is not always true as was report in [2] were the experiments show that with $\gamma = 0.008$ the highest success rate was achieved.

In [3] is mentioned that for most applications its value varies from 0.01 to 100, but they used $\gamma = 1$ for their simulations. Other experiments set the light absorption coefficient as $0.00006 \leq \gamma \leq 0.4$ and they recommend the fixed value of 0.02 of their method [4].

However, we can set $\gamma = \sqrt{L}$, where L is the scale. Or we can set $\gamma = O(1)$ if the scaling variations are not significant, then.

For the population size n we can use $15 \leq n \leq 100$, or $25 \leq n \leq 40$ [5]. But there are some cases when a small number of fireflies is more efficient as in [4] although increasing the population helps to improve the search because more fireflies are distributed throughout the search space.

2.4 Distance

The variable r in the second term on (1) is the distance between any two fireflies i and j at vector x_i and vector x_j, is the Cartesian distance,

$$r_{ij} = \sqrt{\sum_{k=1}^{d} (x_{i,k} - x_{j,k})^2}, \tag{6}$$

where $x_{i,k}$ is the kth component of the spatial coordinate x_i of the ith firefly. Although the distance r is not limited to the Euclidean distance. If necessary another type of distance can be use in the n-dimensional hyperspace depending on the type of problem of our interest. Any measure that can effectively characterize the quantities of interest in the optimization problem can be used as the distance r [1]. For example [6] uses distance r as the difference between the scores of two fireflies (solutions). The Hamming distance is used in [2] to represent the structural differences between candidate (firefly or solution) genotypes.

2.5 Randomization

The component of uncertainty in (1) is the third term in the equation. This component adds randomization to the algorithm that helps to explore various regions of the search space and has diversity of solutions [5]. With the adequate control of exploration, the algorithm can jump out of any local optimum and the global optimum can be reached.

Some experiments suggest $\alpha \rightarrow 1$ at the start of the iterations and finishing with a $\alpha \rightarrow 0$. The formula presented below is representing this idea.

$$\alpha_t = \alpha_0 \delta^t, 0 < \delta < 1 \tag{7}$$

where α_0 is the initial randomness scaling factor, and δ is a cooling factor, and it can be use as $\delta = 0.95$–0.97. If α_0 is associated with the scalings of design variables FA will be more efficient. Let L be the average scale of the problem of interest, we can set $\alpha_0 = 0.01L$ initially. The factor 0.01 comes from the fact that random walks requires a number of steps to reach the target while balancing the local exploitation without jumping too far in a few steps, for most problems 0.001–0.01 can be used [5].

Another formula for the control of randomness was presented in [1], and the implementation is as follows

$$\begin{aligned} \Delta &= 1 - \left(\frac{10^{-4}}{0.9}\right)^{\frac{1}{\mathrm{MAX\_GEN}}}, \\ \alpha^{(t+1)} &= (1 - \Delta) * \alpha^{t}, \end{aligned} \tag{8}$$

where Δ determines the step size of the random walk, MAX_GEN is the maximum number of iterations and t is the generation counter. The parameter $\alpha^{(t+1)}$ descends with the increasing of the generation counter.

In [3] they replaced the α by $\alpha * S_j$, where the scaling parameters S_j in the d dimensions are determined by the actual scales of the problem of interest, and is calculated by

$$S_j = u_j l_j \tag{9}$$

where $j = 1, 2, \ldots, d$, u_j and l_j are the lower and upper bound. Also in this paper it is reported that value of the parameter α less than 0.01 do not affect the optimization results.

With the goal of giving a better exploration behavior to the FA the following formula is proposed

$$\propto = \propto_{\infty} + (\propto_0 - \propto_{\infty})\mathrm{e}^{-t}, \tag{10}$$

where $t \in [0, \text{Max\_}textIteration]$ is the time for simulations and Max_Iteration is the maximum number of generation. $\propto_0$ is the initial randomization parameter while $\propto_{\infty}$ is the final value [7].

Referring to the vector of random numbers ε_1 it can be drawn from a Gaussian distribution or Uniform distribution [1]. However, the appropriate distribution depends on the problem to be solved, more precisely, on a fitness landscape that maps each position in the search space into fitness value. When the fitness landscape is flat, uniform distribution is more preferable for the stochastic search process, whilst in rougher fitness landscapes Gaussian distribution should be more appropriate. Various probability distributions were used to study their impact on the algorithm; they used Uniform, Gaussian, Lévi flights, chaotic maps, and the Random sampling in a turbulent fractal cloud. Here is concluded that in some cases the selection of the more appropriate randomized method can even significantly improve the results of the original FA (Gaussian). The best results were observed by the Random Sampling in Turbulent Fractal Cloud and Kent chaotic map [8].

The original FA does not consider the current global best "*gbest*", and adding an extra term can make an improvement [1].

$$\lambda \in_2 (\text{gbest} - x_i) \tag{11}$$

In [5] an analysis on the number of iterations needed to achieve a certain level of accuracy is showed, here we see that the number of iterations it is not affected much by dimensionality and for higher dimensional problems the number of iterations does not increase significantly. The analysis above mentioned was on the algorithm worst-case scenario.

3 Parameter Control

Parameter control is an open question problem, a quest for the right technique that can show mathematically how the performance of the algorithm is affected by the parameters and use this information to make the right adjustment for improvement. The works that have been done until this moment propose very simple cases, strict conditions, and sometimes unrealistic assumptions, but there are no theoretical results, so the problem remains unsolved [5].

3.1 Parameter Tuning

The convergence rate of the algorithms is related to the eigenvalues that control the parameters and the randomness when we represent as a vector equation once establishing an algorithm as a set of interacting Markov chains. The difficulty of finding this eigenvalue makes the tuning of parameters a very hard problem. The aim of parameter tuning is to find the best parameter setting so that an algorithm can perform well for a wider range of problems, at the end, this optimization problem requires a higher level of optimization methods [5]. One approach to develop a successful framework for self-tuning algorithms was presented in [9] having good results. Here the FA algorithm is used to tune itself.

3.2 Parameter Control

After tuning the parameters of an algorithm very often, they remain fixed during iterations, on the other hand for dynamic parameter control they vary during the iterations, searching for the global optimal. This is also an optimization problem unresolved. In the next lines, we are going to point out the considered factors to construct a fuzzy controller in other studies.

The use of fuzzy logic has been widely use for controlling the parameters of metaheuristics to get an improvement in the performance. Knowing the difficult task of choosing the correct values of the parameters to have a good balance between exploration and exploitation, [10] uses ACO for doing this work focusing on the alpha parameter who has a big effect on the diversity and can control the convergence, for the fuzzy control inputs they use the error and change of error with respect to an average branching factor reference level. And an improvement was observed but when they try to attack optimization problems with benchmark functions their proposed strategy fails due to the lack of heuristic information.

Another proposal that improves the performance of the algorithm in this case PSO is presented in [11] where three approaches using fuzzy control were presented, here two parameters, the cognitive factor, and social factor, are changed dynamically during execution via a control using fuzzy logic; the inputs considered were the iteration, diversity and error. The results show two of the fuzzy controllers help to improve the algorithm. In another study, only using three tests functions, the inputs considered are the current best performance evaluation and the current inertia weight; the output variable is the inertia weight [12].

The importance of knowing how much influence a parameter has in the algorithm is crucial for implementing an efficient fuzzy control system [13]. In [14] only one input (generations) is use for the fuzzy control system and one parameter its values is dynamically decreasing.

4 FA Applications

FA is potentially more powerful than other existing algorithms such as PSO, and it is being used in many areas, for example: benchmark reliability-redundancy optimization problems [15], benchmark engineering problems [16], path planning [17], dynamic environment optimization problems [18–20], neural network training [21], image and data clustering [22, 23], and generate optimized fuzzy models [24]. Some study leaves the door open for more research where varying an added parameter can make an improvement on the convergence of the algorithm [25].

4.1 Firefly Algorithm Parameter Adjustment

The strategy for setting the optimal values of the adjustable parameters on [6] is trial and error, the results show a Hybridizing Firefly Algorithm improvement and the fitness of the optimal solution was obtained with a significant reduction of the execution time.

Using the combination of different techniques we can obtain good results: Learning Automata for adapting a parameter, Genetic Algorithm for sharing information between the populations [26].

For a job scheduling application the FA's parameters were set doing experiments and statistical analysis, in this case (the values may be specific to this problem) the best values were: 100 fireflies, 25 generations, $\alpha = 0.5$, $\beta = 1$ and $\gamma = 0.1$. They found the most influencing factor was α, followed by β, a number of fireflies, generations and finally γ [27].

The virtue of FA is a natural attraction and how it works for a firefly and his close neighbors.

4.2 *Fuzzy Control for Parameter Adjustment*

An important factor to consider in the design of a fuzzy parameter controller is the iteration, diversity, and error which are used in [28]. But also choosing the right combination of parameters is important, in this case, the results obtained do not improve the original algorithm, this may be because they only focus on controlling the parameters that are responsible for the convergence but forgot to reduce the parameter responsible for the randomness.

Different analysis leaves to different implementations, [29] considers the parameters α (control the exploration) and γ (coefficient of light restriction) to be the ones to control, although in Sect. 2.3 of this paper is mentioned that its influence is not significant. As inputs to the fuzzy control system they use the variable count for referring to the generations and Delta that is defined in Eq. (12)

$$\begin{aligned} \text{Delta (count)} = {} & \text{Fitness of the best solution (count)} \\ & - \text{fitness of best solution (till count)} \end{aligned} \tag{12}$$

With the proposed method they reported an increased performance in FA for solving traveling salesman problems.

5 Conclusions

There is no correct technique for tuning or dynamically controlling the parameters of an algorithm, so following a framework is very helpful. It is worth pointing out the need for investigating how to improve the frameworks that already exist or create new ones.

For parameter tuning, experiments and statistical studies is often used and proven to improve the performance so it should be the first technique to try, but the disadvantage is setting will only work for this specific problem. In the case of parameter control a combination of fuzzy logic or another metaheuristic or computational intelligent technique, such as neural network or learning automata improves the performance.

References

1. Yang, X.-S.: Nature-Inspired Metaheuristic Algorithms. (2008).
2. Husselmann, A. V, Hawick, K.A.: Cuckoo Search and Firefly Algorithm: Theory and Applications. Presented at the (2014).
3. Brajevic, I., Tuba, M.: Cuckoo Search and Firefly Algorithm: Theory and Applications. Presented at the (2014).
4. Yousif, A., Nor, S.M., Abdullah, A.H., Bashir, M.B.: Cuckoo Search and Firefly Algorithm: Theory and Applications. Presented at the (2014).
5. Yang, X.-S. ed: Cuckoo Search and Firefly Algorithm. Springer International Publishing, Cham (2014).
6. Salomie, I., Chifu, V.R., Pop, C.B.: Cuckoo Search and Firefly Algorithm: Theory and Applications. Presented at the (2014).
7. Yang, X.-S.: Engineering Optimization: An Introduction with Metaheuristic Applications. (2010).
8. Fister, I., Yang, X.-S., Brest, J.: Cuckoo Search and Firefly Algorithm: Theory and Applications. Presented at the (2014).
9. Yang, X.-S., Deb, S., Loomes, M., Karamanoglu, M.: A framework for self-tuning optimization algorithm. Neural Comput. Appl. 23, 2051–2057 (2013).
10. Neyoy, H., Castillo, O., Soria, J.: Fuzzy Logic Augmentation of Nature-Inspired Optimization Metaheuristics: Theory and Applications. Presented at the (2015).
11. Olivas, F., Valdez, F., Castillo, O.: Fuzzy Logic Augmentation of Nature-Inspired Optimization Metaheuristics. Springer International Publishing, Cham (2015).
12. Eberhart, R.C.: Fuzzy adaptive particle swarm optimization. In: Proceedings of the 2001 Congress on Evolutionary Computation (IEEE Cat. No.01TH8546). pp. 101–106. IEEE (2001).
13. Pérez, J., Valdez, F., Castillo, O.: Fuzzy Logic Augmentation of Nature-Inspired Optimization Metaheuristics: Theory and Applications. Presented at the (2015).
14. Castillo, O., Melin, P., Pedrycz, W., Kacprzyk, J. eds: Recent Advances on Hybrid Approaches for Designing Intelligent Systems. Springer International Publishing, Cham (2014).
15. dos Santos Coelho, L., de Andrade Bernert, D.L., Mariani, V.C.: A chaotic firefly algorithm applied to reliability-redundancy optimization. In: 2011 IEEE Congress of Evolutionary Computation (CEC). pp. 517–521. IEEE (2011).
16. Yang, X.S.: Firefly algorithm, stochastic test functions and design optimisation. Int. J. Bio-Inspired Comput. 2, 78 (2010).
17. Wang, G., Guo, L., Duan, H., Liu, L., Wang, H.: A Modified Firefly Algorithm for UCAV Path Planning. Int. J. Hybrid Inf. Technol. 5, 123–144.
18. Abshouri, A.A., Meybodi, M.R., Bakhtiary, A.: New Firefly Algorithm based On Multi swarm & Learning Automata in Dynamic Environments.
19. Farahani, S.M., Nasiri, B., Meybodi, M.R.: A multiswarm based firefly algorithm in dynamic environments. 3, 68–72 (2011).
20. Nasiri, B., Meybodi, M.R.: Speciation based firefly algorithm for optimization in dynamic environments, http://www.ceser.in/ceserp/index.php/ijai/article/view/2359, (2012).
21. Nandy, S., Sarkar, P.P., Das, A.: Analysis of a Nature Inspired Firefly Algorithm based Back-propagation Neural Network Training. CoRR. abs/1206.5, (2012).
22. Image Clustering using Fuzzy-based Firefly Algorithm| Parisut Jitpakdee - Academia.edu, https://www.academia.edu/5870258/Image_Clustering_using_Fuzzy-based_Firefly_Algorithm.
23. Jitpakdee, P., Aimmanee, P., Uyyanonvara, B., Ritthipakdee, A.: Fuzzy-Based Firefly Algotithm for Data Clustering. (2013).
24. Kumar, S., Kaur, P., Singh, A.: Fuzzy Model Identification: A Firefly Optimization Approach. Int. J. Comput. Appl. 58, 1–8.
25. Yang, X.-S.: Firefly Algorithm, Levy Flights and Global Optimization. 10 (2010).

26. Farahani, S.M., Abshouri, A.A., Nasiri, B., Meybodi, M.: Some hybrid models to improve firefly algorithm performance. 8, 97–117 (2012).
27. Khadwilard, A., Chansombat, S., Thepphakorn, T., Thapatsuwan, P., Chainate, W., Pongcharoen, P.: Application of Firefly Algorithm and Its Parameter Setting for Job Shop Scheduling. J. Ind. Technol. 8, (2012).
28. Solano-Aragón, C., Castillo, O.: Fuzzy Logic Augmentation of Nature-Inspired Optimization Metaheuristics: Theory and Applications. Presented at the (2015).
29. Bidar, M., Rashidy Kanan, H.: Modified firefly algorithm using fuzzy tuned parameters. In: 2013 13th Iranian Conference on Fuzzy Systems (IFSC). pp. 1–4. IEEE (2013).

Fuzzy Dynamic Adaptation of Parameters in the Water Cycle Algorithm

Eduardo Méndez, Oscar Castillo, José Soria and Ali Sadollah

Abstract This paper describes the enhancement of the water cycle algorithm (WCA) using a fuzzy inference system to dynamically adapt its parameters. The original WCA is compared in terms of performance with the proposed method called WCA with dynamic parameter adaptation (WCA-DPA). Simulation results on a set of well-known test functions show that the WCA is improved with a fuzzy dynamic adaptation of the parameters.

Keywords WCA · Fuzzy logic · Optimization

1 Introduction

Dynamic parameter adaptation can be performed in many ways, the most common approaches being linearly increasing or decreasing a parameter, and other approaches include nonlinear or stochastic functions. In this paper, a different approach is taken, which is using a fuzzy inference system (FIS) to replace a function or to change its behavior, with the final purpose of improving the performance of the water cycle algorithm (WCA). The WCA is a population-based and nature-inspired metaheuristic, which is inspired on a simplified form of the water cycle process [2, 12].

Using a FIS to enhance global-optimization algorithms is an active area of research; some works of enhancing particle swarm optimization are PSO-DPA [9], APSO [17] and FAPSO [13]. Since the WCA has some similarities with PSO, a FIS similar to the one in [9] was developed.

E. Méndez · O. Castillo (✉) · J. Soria
Tijuana Institute of Technology, Tijuana, Mexico
e-mail: ocastillo@tectijuana.mx

A. Sadollah
Nanyang Technological University, Singapore, Singapore
e-mail: ali_sadollah@yahoo.com

P. Melin et al. (eds.), *Nature-Inspired Design of Hybrid Intelligent Systems*,
Studies in Computational Intelligence 667, DOI 10.1007/978-3-319-47054-2_20

A comparative study was conducted which highlights the similarities and differences with other hierarchy-based metaheuristics. In addition, a performance study between the proposed water cycle algorithm with Dynamic Parameter Adaptation (WCA-DPA) and the original WCA was also conducted, using ten well-known test functions frequently used in tin the literature.

This paper is organized as follows. In Sect. 2 the WCA is described. In Sect. 3, some similarities with other metaheuristics are highlighted. Section 4 is about how to improve the WCA with fuzzy parameter adaptation. A comparative study is also presented in Sect. 5 and, finally in Sect. 6 some conclusions and future work are given.

2 Nonlinear the Water Cycle Algorithm

The WCA is a population-based and nature-inspired metaheuristic, where a population of streams is formed from rainwater drops. This population of streams follows a behavior inspired on the hydrological cycle. In which streams flows downhill, then they form rivers, which also flow downhill towards the sea. This process in which streams flows toward rivers and rivers towards the sea is a simplified form of the *runoff* process of the hydrologic cycle. Some of those streams are evaporated and some new streams are formed from rain as part of the hydro-logic cycle.

2.1 The Landscape

There are a number of landforms involved in the hydrologic cycle, for example: streams, rivers, lakes, valleys, mountains, and glaciers. But in the WCA only three of them are considered, which are streams, rivers, and seas, and in fact there is only one sea. In this subsection the structure, preprocessing and initialization of the algorithm are described.

In the WCA an individual (a.k.a. stream), is an object which consist of n variables grouped as a n-dimensional column vector

$$\mathbf{x}_k = [x_{k1}, \ldots, x_{kn}]^{\mathrm{T}} \in \mathbb{R}^n. \quad (1)$$

And the whole population of N streams is denoted by

$$\mathbf{X} = \{\mathbf{x}_k | k = 1, 2, \ldots, N\}, \quad (2)$$

which is often represented as a $N \times n$ matrix:

$$\mathbf{X} = \begin{bmatrix} \mathbf{x}_1^{\mathrm{T}} \\ \vdots \\ \mathbf{x}_N^{\mathrm{T}} \end{bmatrix} = \begin{bmatrix} x_{11} & \cdots & x_{1n} \\ \vdots & \ddots & \vdots \\ x_{N1} & \cdots & x_{Nn} \end{bmatrix}. \tag{3}$$

In short each row is an individual stream and their columns are their variables. From the whole population some of those streams will become rivers and another one will become the sea. The number of streams and rivers are defined by the following equations:

$$N_{\mathrm{sr}} = \text{Number of Rivers} + \underbrace{1}_{\text{sea}}, \tag{4}$$

$$N_{\mathrm{streams}} = N - N_{\mathrm{sr}}, \tag{5}$$

where N_{sr} is a value established as a parameter of the algorithm and N_{streams} is the number of remaining streams. Which individuals become rivers or sea will depend on the fitness of each stream. To obtain the fitness, first we need an initial population matrix X, and this matrix is initialized with random values as follows:

$$\mathbf{x}_i = \mathbf{b}_{\mathrm{lower}} + \mathbf{r} \cdot (\mathbf{b}_{\mathrm{lower}} - \mathbf{b}_{\mathrm{upper}}), \quad \text{for} \quad i = 1, 2, \ldots, N, \tag{6}$$

where $\mathbf{b}_{\mathrm{lower}}$, $\mathbf{b}_{\mathrm{upper}}$ are vectors in $\mathbb{R}^n$ with the lower and upper bounds for each dimension which establish the search-space, and $\mathbf{r}$ is an n-dimensional vector of independent and identically distributed (i.i.d) values, that follows a uniform distribution:

$$\mathbf{r} \sim U(0,1)^n. \tag{7}$$

Once an initial population is created the fitness of each stream $\mathbf{x}_i$ is obtained by:

$$\mathbf{f}_i = f(\mathbf{x}_i) = f(x_{i1}, x_{i2}, \ldots x_{in}), \quad \text{for} \quad i = 1, 2, 3, \ldots, N, \tag{8}$$

where $f(\cdot)$ is a problem dependent function to estimate the fitness of a given stream. This fitness function it is what the algorithm tries to optimize.

Sorting the individuals by fitness and in ascending order using:

$$[\mathbf{X}, \mathbf{f}] \leftarrow \text{sort}](\mathbf{X}, \mathbf{f}), \tag{9}$$

the first individual becomes the sea, the next $N_{\mathrm{sr}-1}$ the rivers, and the following N_{streams} individuals turn into the streams who flow toward the rivers or sea, as show in:

$$\mathbf{X} = \begin{array}{l} sea\{ \\ rivers\left\{ \begin{array}{l} \\ \\ \\ \end{array} \right. \\ streams\left\{ \begin{array}{l} \\ \\ \\ \end{array} \right. \end{array} \begin{bmatrix} \mathbf{x}_1^{\mathsf{T}} \\ \mathbf{x}_2^{\mathsf{T}} \\ \vdots \\ \mathbf{x}_{N_{sr}}^{\mathsf{T}} \\ \mathbf{x}_{N_{sr}+1}^{\mathsf{T}} \\ \vdots \\ \mathbf{x}_{N_{sr}+N_{streams}}^{\mathsf{T}} \end{bmatrix} = \begin{bmatrix} x_{11} & \cdots & x_{1n} \\ \vdots & \ddots & \vdots \\ x_{N1} & \cdots & x_{Nn} \end{bmatrix}. \tag{10}$$

Each of the N_{streams} is assigned to a river or sea, this assignment can be done randomly. But the *stream order* [14, 15], which is the number of streams assigned to each river/sea is calculated by:

$$\mathbf{so}_i \left\lfloor \left| \frac{\mathbf{f}_i}{\sum_{j=1}^{N_{\text{sr}}} \mathbf{f}_j + \varepsilon} \right| \cdot N_{\text{streams}} \right\rfloor, \quad n = 1, 2, \ldots, N_{\text{sr}}, \tag{11}$$

$$\mathbf{so}_1 \leftarrow \mathbf{so}_1 + \left(N_{\text{streams}} - \sum_{i=1}^{N_{\text{sr}}} \mathbf{so}_i \right), \tag{12}$$

where $\varepsilon \approx 0$. The idea behind the Eq. (11) is that the amount of water (streams) entering a river or sea varies so when a river is more abundant (has a better fitness) than another, it means that more streams flow into the river, hence the discharge (stream-flow) is higher. This means, the streamflow magnitude of rivers is inversely proportional to its fitness in the case of minimization problems.

The Eq. (11) has been changed from the original proposed in [2], the round function was replaced by a floor function, a value of ε was added to the divisor, and Eq. (12) was also added to handle the remaining streams. These changes are for the implementation purposes and an alternative to the method proposed in [12].

After obtaining the stream order of each river and sea, the streams are randomly distributed between them.

2.2 The Run-off Process

The run-off process is one of the three processes considered in the WCA, which handles the way water flows in form of streams and rivers towards the sea. The following equations describe how the flow of streams and rivers are simulated at a given instant (iteration):

$$\mathbf{x}_{\text{stream}}^{i+1} = \mathbf{x}_{\text{stream}}^{i} + \mathbf{r} \cdot C \cdot (\mathbf{x}_{\text{sea}}^{i} - \mathbf{x}_{\text{stream}}^{i}), \tag{13}$$

$$\mathbf{x}_{\text{stream}}^{i+1} = \mathbf{x}_{\text{stream}}^{i} + \mathbf{r} \cdot C \cdot (\mathbf{x}_{\text{river}}^{i} - \mathbf{x}_{\text{stream}}^{i}), \tag{14}$$

$$\mathbf{x}_{\text{river}}^{i+1} = \mathbf{x}_{\text{river}}^{i} + \mathbf{r} \cdot C \cdot (\mathbf{x}_{sea}^{i} - \mathbf{x}_{\text{river}}^{i}), \tag{15}$$

for $i = 1, 2, \ldots, N_{it}$, where N_{it} and C are parameters of the algorithm, and $\mathbf{r}$ is a vector with i.i.d values defined by Eq. (7), although any other distribution could be used.

The Eq. (13) defines the movement of streams who flow directly to the sea, Eq. (14) is for streams who flow toward the rivers, and Eq. (15) is for the rivers flow toward the sea. A value of $C > 1$ enables streams to flow in different directions toward the rivers or sea. Typically, the value of C is chosen from the range (1,2] being 2 the most common.

2.3 *Evaporation and Precipitation Processes*

The runoff process of the WCA basically consists of moving indirectly toward the global best (sea). Algorithms focused on following the global best although they are really fast, tend to premature convergence or stagnation. The way in which WCA deals with exploration and convergence is with the evaporation and precipitation processes. So when streams and rivers are close enough to the sea, some of those streams are evaporated (discarded) and then new streams are created as part of the precipitation process. This type of reinitialization is similar to the cooling down and heating up reinitialization process of the simulated annealing algorithm [3].

The evaporation criterion is: if a river is close enough to the sea, then the streams assigned to that river are evaporated (discarded) and new streams are created by raining around the search space. To evaporate the streams of a given river the following condition must be satisfied:

$$\underbrace{|\mathbf{x}_{\text{sea}} - \mathbf{x}_{\text{river}}| < d_{\max}}_{\text{evaporation criterion}}, \tag{16}$$

where $d_{\max} \approx 0$ is a parameter of the algorithm. This condition must be applied to every river, and if its satisfied each stream who flow toward this river must be replaced as:

$$\underbrace{\mathbf{x}_{\text{stream}} = \mathbf{b}_{\text{lower}} + \mathbf{r} \cdot (\mathbf{b}_{\text{lower}} - \mathbf{b}_{\text{upper}})}_{\text{raining around the search space}}, \tag{17}$$

a high value of $d_{\max}$ will favor the exploration and a low one will favor the exploitation.

To increase the exploration and exploitation around the sea an especial evaporation criterion is used for the streams, which flow directly to sea:

$$\underbrace{|\mathbf{x}_{\text{sea}} - \mathbf{x}_{\text{seastream}}| < d_{\max}}_{\text{evaporation criterion}}, \tag{18}$$

where $\mathbf{x}_{\text{seastream}}$ is a stream which flows directly to the sea. If this criterion defined by the inequality (18) is satisfied then, the stream is evaporated and a new one is created using:

$$\underbrace{\mathbf{x}_{\text{seastream}} = \mathbf{x}_{\text{sea}} + \mathbf{g}}_{\text{raining around the sea}}, \tag{19}$$

where $\mathbf{g}$ is an n-dimensional vector of independent and identically distributed (i.i.d) values, that follow a normal distribution:

$$\mathbf{g} \sim \mathcal{N}(\mu = 0, \sigma^2 = 0.01)^n. \tag{20}$$

2.4 Steps of WCA

The steps of WCA are summarized as an algorithm in Fig. 1.

Algorithm: Water Cycle Algorithm

input : f, $N_{sr}, d_{\max}, N, \mathbf{b}_{lower}, \mathbf{b}_{upper}, it_{\max}$

1: Generate the initial population of streams (raindrops) using eq. (3).
2: Calculate the fitness of each stream by Eq. (8).
3: Sort the population of streams by fitness to determine streams, rivers and sea, as in (10).
4: Calculate the intensity of flow for rivers and sea using (11) and designate which streams flows towards each river or sea.
5: Streams flow towards the sea and rivers by Eqs. (13) and (14) respectively, and rivers flow to the sea by Eq. (15).
6: Update the fitness of each stream by Eq. (8). After each update check:
If a stream new fitness is better than his assigned river/sea, exchange positions.
If a river new fitness is better than the sea, exchange positions.
7: Check the evaporation conditions for both rivers and streams, using Eqs. (16) and (18) respectively, and start the evaporation and raining processes using Eqs. (17) and (19).
8: Check the convergence criterion. If satisfied stop the algorithm, otherwise return to the step 5.

output: The individual with the best fitness also known as the sea.

Fig. 1 The water cycle algorithm (WCA)

3 Similarities and Differences with Other Metaheuristics

The WCA has some similarities with other metaheuristics, but yet is different from those. Some of the similarities and differences have already been studied, for example: In [2], differences and similarities with particle swarm optimization (PSO) [7] and genetic algorithms (GA) [4] are explained. In [11], WCA is compared with the Imperialist Competitive Algorithm (ICA) [1] and PSO [7]. In [12], the WCA is compared with two nature-inspired metaheuristics: the intelligent water drops [5] and water wave optimization [18]. So far similarities and differences with population-based and nature-inspired metaheuristics have been studied, in this subsection, WCA is compared with two metaheuristics who use a hierarchy.

3.1 Hierarchical Particle Swarm Optimization

In hierarchical particle swarm optimization (H-PSO), particles are arranged in a regular tree hierarchy that defines the neighborhoods structure [6]. This hierarchy is defined by a height h and a branching degree d, this is similar to the landscape (hierarchy) of the WCA, in fact the WCA would be like a tree of height $h = 3$ (sea, rivers, streams), but with varying branching degrees, since the level-2 consist of N_{sr} branches (rivers) and the level-3 depends of the stream orders (**so**), so WCA hierarchy it is not a nearly regular tree like in the H-PSO.

Another difference is that H-PSO uses velocities to update the positions of the particles just like in standard PSO. But a similarity is that instead of moving towards the global best like in PSO they move toward their parent node, just like streams flow to rivers and rivers flow to the sea. As in WCA, in H-PSO particles move up and down the hierarchy, and if a particle at a child node has found a solution that is better than the best so far solution of the particle at the parent node, the two particles are exchanged. This is similar yet different to the runoff process of the WCA, the difference being that WCA uses only the social component to update the positions, and H-PSO uses both the social and cognitive components, and also the velocity with inertia weight. The cognitive component and the inertia weight are the ways in which H-PSO deals with exploration and exploitation. The WCA uses the evaporation and precipitation processes for those phases.

3.2 Grey Wolf Optimizer

The Grey Wolf Optimizer (GWO) algorithm mimics the leadership hierarchy and hunting mechanism of grey wolves in nature. Four types of grey wolves are simulated: alpha, beta, delta, and omega, which are employed for simulating the leadership hierarchy [10]. This social hierarchy is similar to the WCA hierarchy with a $N_{sr} = 3$, where the alpha could be seen as the sea, the beta and delta as the rivers and the omegas as the streams. Although the hierarchy is similar, the way in which the GWO algorithm updates the positions of the individuals is different. GWO position update depends of the hunting phases: searching for prey, encircling prey, and attacking prey. Those hunting phases are the way in which the GWO deals with exploration and exploitation. As mentioned before, the WCA uses the evaporation and precipitation process, which are very different to the hunting phases.

4 Fuzzy Parameter Adaptation

The objective of dynamic parameter adaptation is to improve the performance of an algorithm by adapting its parameters. Dynamic parameter adaptation can be done in many ways, the most common being linearly increasing or decreasing a parameter, usually the acceleration coefficients. Other approaches include using nonlinear or stochastic functions. In this paper, a different approach is taken, which is using a fuzzy system to replace a function or to change its behavior.

In the WCA there are two parameters which can be adapted dynamically, that is while the algorithm is running. One is the parameter *C*, which is used in Eqs. (13)–(15) for updating the positions of the streams. The other one is the parameter $d_{\max}$ used in Eq. (16) as a threshold for the evaporation criterion. In [12] the parameter $d_{\max}$ is linearly decreased and a stochastic evaporation rate for every river is introduced, together both changes improve the performance of the WCA.

Since there are already improvements with the parameter $d_{\max}$, the subject of study in this paper is the *C* parameter.

4.1 Mamdani's Fuzzy Inference System

A single-input and multiple-output (SIMO) Mamdani's FIS [8] was developed. The system consists of the *Iteration* input and the outputs *C* and C_{rivers}. In Fig. 2 the layout of the FIS is shown. The idea is to use different values of the parameter *C*, one for Eqs. (13) and (14) which are for the flow of streams and another one (C_{rivers}) for Eq. (15) which is for the flow of rivers.

Before going into the FIS, the Iteration input is scaled to the interval [0, 1] by the following equation:

$$\text{Iteration} = \frac{i}{N_{it}}, \quad \text{for} \quad i = 1, 2, \ldots, N_{it}. \tag{21}$$

The membership functions for the iteration input are shown in Fig. 3. The range of the first output *C* had been chosen as the interval (1.8, 3.7), and for the output

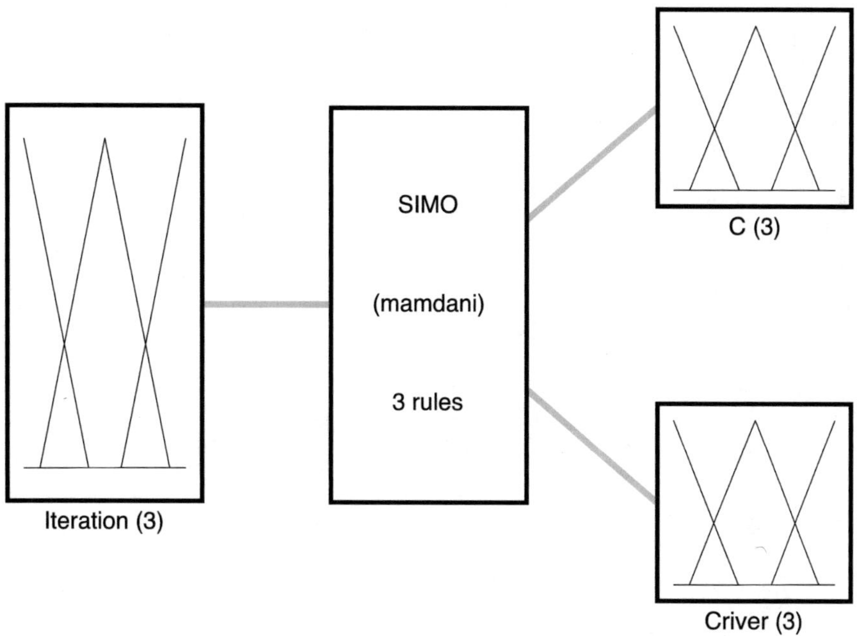

Fig. 2 Single-input and multiple-output Mamdani's fuzzy inference system

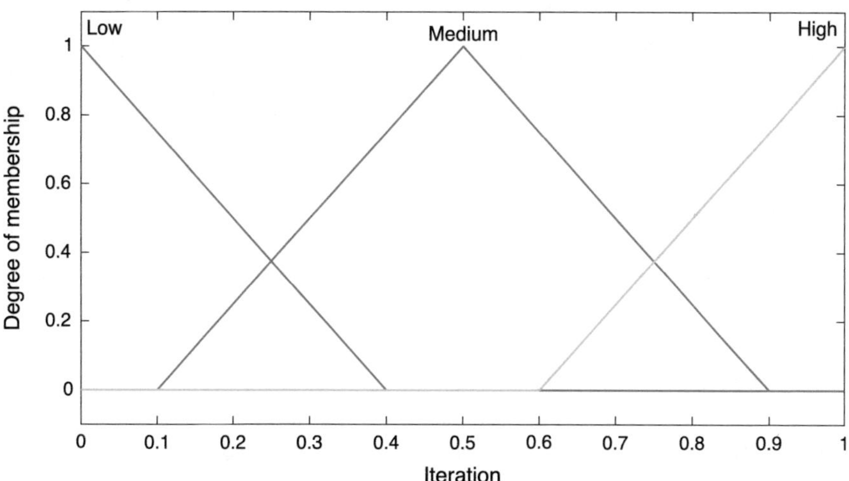

Fig. 3 Input: iteration

C_{rivers} the interval (2, 10), the details of the membership functions are shown in Figs. 4 and 5. The idea of using higher values for these parameters is to favor the exploration in the runoff process at an early stage. Since having a greater value than

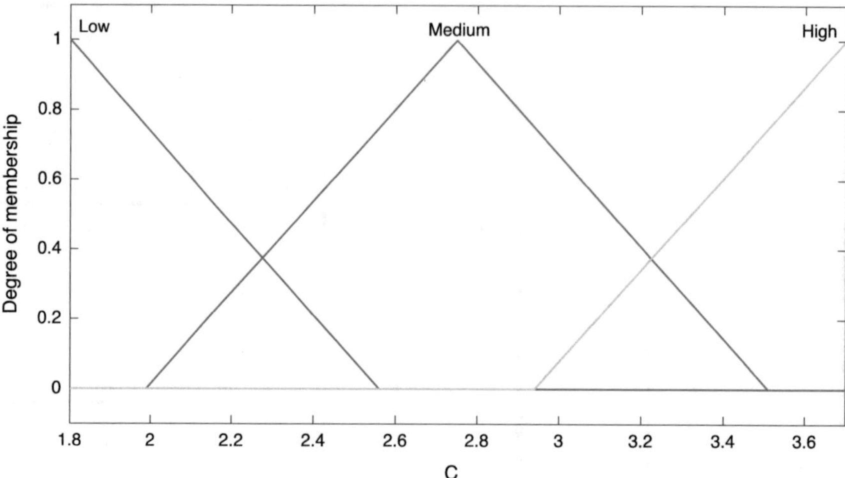

Fig. 4 Output: C

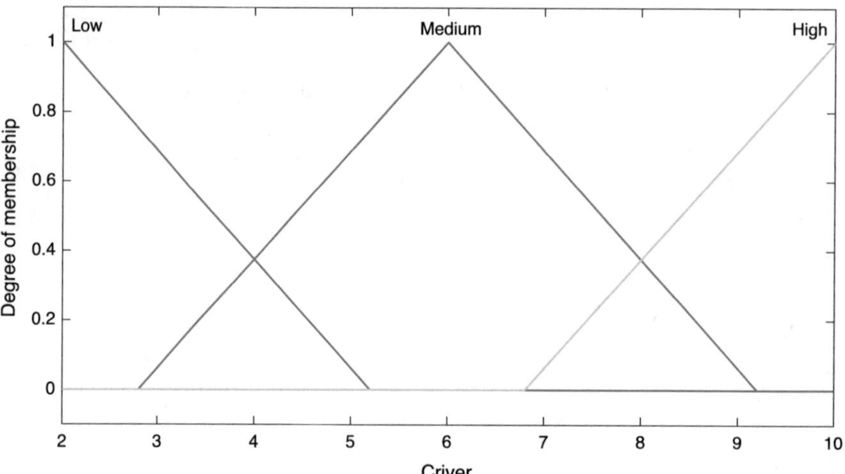

Fig. 5 Output: C_{rivers}

2 means that there is a higher probability of moving beyond the rivers or sea. For the special case of C_{rivers} it also helps to prevent the evaporation and precipitation processes. Figure 6 shows a flowchart of the WCA with the SIMO-FIS integrated.

5 Experiments and Comparisons

For the comparison between the WCA and the WCA-DPA, a subset of test functions was used. Although in the literature there is no agreed set of test functions for measuring the performance of algorithms, a diverse subset of 10 test functions who had been used before for some bio-inspired algorithms was chosen. In Table 1 the specifications of those functions are summarized and in Fig. 7 the plots for 2-dimensions viewed from a 3D perspective and from above are shown. In [16] there is a more detailed description of those functions.

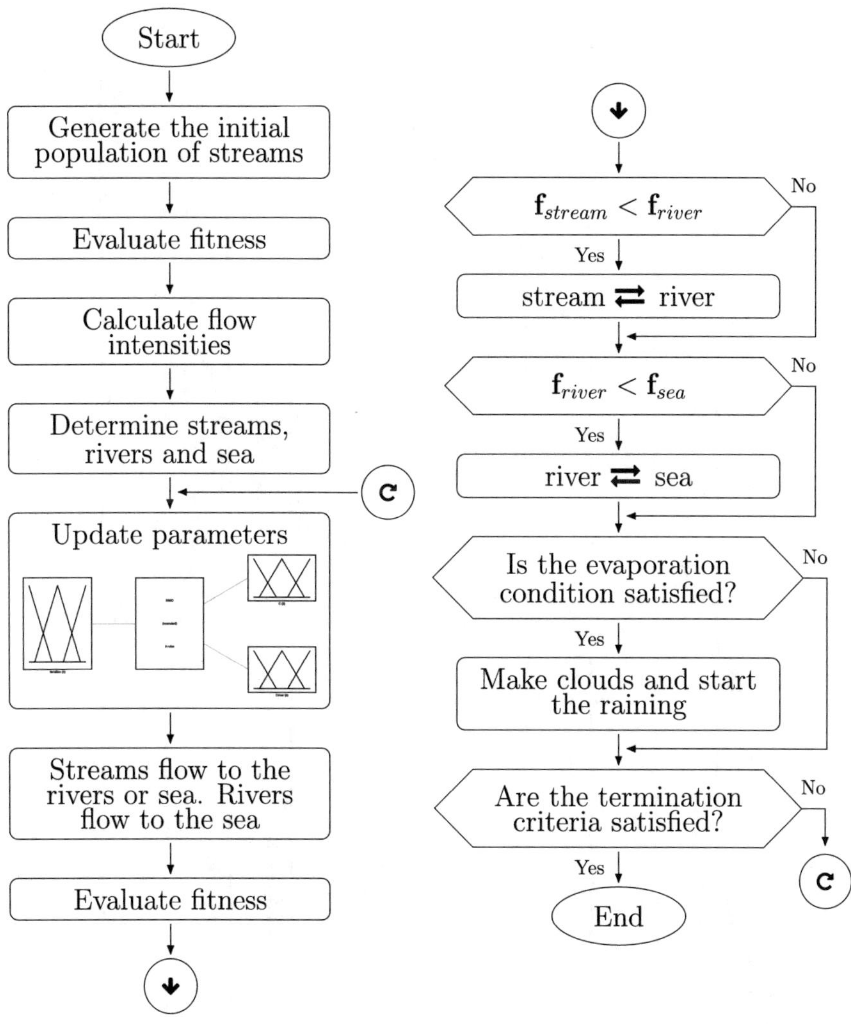

Fig. 6 Flowchart of the water cycle algorithm with fuzzy dynamic adaptation of parameters

Table 1 Test functions used for comparisons

No.	Function name	Equation	Domain
1	Michalewicz	$f(x) = -\sum_{i=1}^{n} \sin(x_i) \cdot \left[\sin\left(\frac{ix_i^2}{\pi}\right)\right]^{2m}, \quad m = 10$	$0 \le x_i \le \pi$
2	Rosenbrock	$f(x) = \sum_{i=1}^{n-1} [(x_i - 1)^2 + 100(x_{i+1} - x_i^2)^2]$	$-5 \le x_i \le 5$
3	De Jong	$f(x) = \sum_{i=1}^{n-1} x_i^2$	$-5.12 \le x_i \le 5.12$
4	Schwefel	$f(x) = -\sum_{i=1}^{n} x_i \sin\left(\sqrt{\lvert x_i \rvert}\right)$	$-500 \le x_i \le 500$
5	Ackley	$f(x) = -20 \exp\left(-\frac{1}{5}\sqrt{\frac{1}{n}\sum_{i=1}^{n} x_i^2}\right) - \exp\left(-\frac{1}{n}\sum_{i=1}^{n} \cos(2\pi x_i)\right) + 20 + \exp(1)$	$-32.768 \le x_i \le 32.768$
6	Rastrigin	$f(x) = 10n + \sum_{i=1}^{n} [x_i^2 - 10\cos(2\pi x_i)]$	$-5.12 \le x_i \le 5.12$
7	Easom	$f(x) = -\cos(x)\cos(y)\exp[-(x-\pi)^2 + (y-\pi)^2]$	$-100 \le x, y \le 100$
8	Griewank	$f(x) = \frac{1}{4000}\sum_{i=1}^{n} x_i^2 - \prod_{i=1}^{n} \cos\left(\frac{x_i}{\sqrt{i}}\right) + 1$	$-600 \le x_i \le 600$
9	Shubert	$f(x) = \left[\sum_{i=1}^{5} i\cos(i + (i+1)x)\right] \cdot \left[\sum_{i=1}^{5} i\cos(i + (i+1)y)\right]$	$-10 \le x, y \le 10$
10	Yang	$f(x) = \left(\sum_{i=1}^{n} x_i\right) \exp\left[-\sum_{i=1}^{n} \sin(x_i^2)\right]$	$-2\pi \le x_i \le 2\pi$

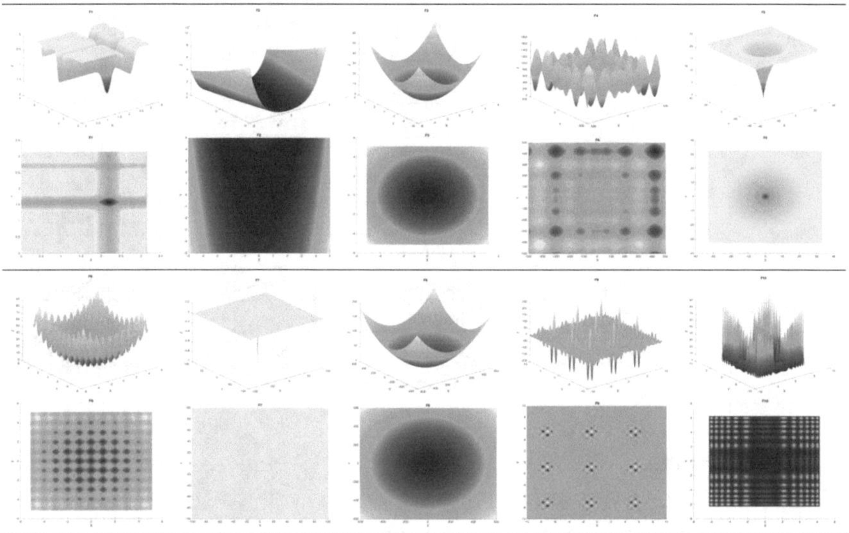

Fig. 7 Test functions plots for 2-dimensions, shown from a 3D perspective and from above

Table 2 MAE of 100 experiments for each test function in 2-dimensions

No.	Function	WCA	WCA-DPA
1	Michalewicz	6.28×10^{-14}	2.66×10^{-15}
2	Rosenbrock	1.36×10^{-9}	2.92×10^{-11}
3	De Jong	1.72×10^{-35}	3.18×10^{-119}
4	Schwefel	6.16×10^{-13}	4.90×10^{-13}
5	Ackley	3.43×10^{-8}	3.77×10^{-15}
6	Rastrigin	7.81×10^{-14}	1.42×10^{-16}
7	Easom	1.26×10^{-14}	5.55×10^{-18}
8	Griewank	8.88×10^{-4}	8.14×10^{-4}
9	Shubert	1.78×10^{-8}	1.92×10^{-44}
10	Yang	8.83×10^{-6}	8.83×10^{-6}

$N_{pop} = 50$, $N_{sr} = 7$, $d_{max} = 1 \times 10^{-6}$, $N_{it} = 4000$

Experiments in 2 and 30 dimensions were performed. The experiments consisted of 100 samples, taking the mean absolute error (MAE) as the measure of performance. Optimization results for the experiments in 2 and 30 dimensions are listed in Tables 2 and 3, respectively. Also, the parameters used are listed at the end of each table. From the results obtained in 2 dimensions (Table 2), we can observe improvements in most of the test functions. However, for 30 dimensions (Table 3), although it performed better for most of the test functions the results are only slightly better.

Table 3 MAE of 100 experiments for each test function in 30-dimensions

No.	Function	WCA	WCA-DPA
1	Michalewicz	1.13×10^1	7.38
2	Rosenbrock	4.53	7.88
3	De Jong	1.68×10^{-9}	3.24×10^{-8}
4	Schwefel	4.25×10^3	3.94×10^3
5	Ackley	4.13	2.12
6	Rastrigin	1.07×10^2	1.02×10^2
7	Easom	–	–
8	Griewank	2.75×10^{-2}	1.32×10^{-2}
9	Shubert	–	–
10	Yang	7.82×10^{-13}	1.14×10^{-13}

$N_{pop} = 50$, $N_{sr} = 7$, $d_{max} = 1 \times 10^{-6}$, $N_{it} = 4000$

6 Conclusions

From the experiments, it can be concluded that dynamically adapting the parameter C, can help to improve the performance of the WCA. But establishing an interval for the outputs, to work with any number of dimensions for the type of FIS developed could be a difficult or impossible task. It would be more appropriate not to bind the outputs to a given interval. Instead, could be better to have as outputs an increment or decrement of the parameter. Similar to the improved versions of particle swarm optimization: APSO or FAPSO developed in [13, 17], respectively. Another alternative could be to add a velocity component to the equations that update the positions of the streams, and then adapt an inertia weight instead of actual value C.

References

1. Atashpaz-Gargari, E., Lucas, C.: Imperialist competitive algorithm: An algorithm for optimization inspired by imperialistic competition. In: IEEE Congress on Evolutionary Computation. pp. 4661–4667 IEEE (2007).
2. Eskandar, H. et al.: Water Cycle Algorithm - A Novel Metaheuristic Optimization Method for Solving Constrained Engineering Optimization Problems. Comput. Struct. 110-111, 151–166 (2012).
3. Gelatt, C.D., Vecchi, M.P.: Optimization by simulated annealing. Science (80-.). 220, 4598, 671–680 (1983).
4. Goldberg, D.E.: Genetic Algorithms in Search, Optimization and Machine Learning. Addison-Wesley Longman Publishing Co., Inc., Boston, MA, USA (1989).
5. Hosseini, H.S.: Problem solving by intelligent water drops. In: Evolutionary Computation, 2007. CEC 2007. IEEE Congress on. pp. 3226–3231 (2007).
6. Janson, S., Middendorf, M.: A hierarchical particle swarm optimizer and its adaptive variant. IEEE Trans. Syst. Man, Cybern. Part B. 35, 6, 1272–1282 (2005).
7. Kennedy, J., Eberhart, R.C.: Particle swarm optimization. In: Proceedings of the IEEE International Conference on Neural Networks. pp. 1942–1948 (1995).

8. Mamdani, E.H., Assilian, S.: An experiment in linguistic synthesis with a fuzzy logic controller. Int. J. Mach. Stud. 7 (1), 1–13 (1975).
9. Melin, P. et al.: Optimal Design of Fuzzy Classification Systems Using PSO with Dynamic Parameter Adaptation Through Fuzzy Logic. Expert Syst. Appl. 40, 8, 3196–3206 (2013).
10. Mirjalili, S. et al.: Grey Wolf Optimizer. Adv. Eng. Softw. 69, 46–61 (2014).
11. Sadollah, A. et al.: Water cycle algorithm for solving multi-objective optimization problems. Soft Comput. 19, 9, 2587–2603 (2015).
12. Sadollah, A. et al.: Water cycle algorithm with evaporation rate for solving constrained and unconstrained optimization problems. Appl. Soft Comput. 30, 58–71 (2015).
13. Shi, Y., Eberhart, R.C.: Fuzzy adaptive particle swarm optimization. In: Evolutionary Computation, 2001. Proceedings of the 2001 Congress on. pp. 101–106 vol. 1 (2001).
14. Shreve, R.L.: Infinite Topologically Random Channel Networks. J. Geol. 75, 2, 178–186 (1967).
15. Shreve, R.L.: Statistical Law of Stream Numbers. J. Geol. 74, 1, 17–37 (1966).
16. Yang, X.-S.: Flower Pollination Algorithm for Global Optimization, (2013).
17. Zhan, Z.H. et al.: Adaptive Particle Swarm Optimization. IEEE Trans. Syst. Man, Cybern. Part B. 39, 6, 1362–1381 (2009).
18. Zheng, Y.-J.: Water wave optimization: {A} new nature-inspired metaheuristic. Comput. {&} {OR}. 55, 1–11 (2015).

Fireworks Algorithm (FWA) with Adaptation of Parameters Using Fuzzy Logic

Juan Barraza, Patricia Melin, Fevrier Valdez and Claudia González

Abstract The main goal of this paper is to improve the performance of the fireworks algorithm (FWA). This improvement is based on fuzzy logic, which means we implemented different fuzzy inference systems into the FWA with the intent to convert parameters that were usually constant in dynamic parameters. After having studied the performance of the FWA, we concluded that two parameters are key of the performance the algorithm (FWA), the parameters that we comment are: the number of sparks and explosion amplitude of each firework, these parameters were adjusted using fuzzy logic, and this adjustment we called Fuzzy Fireworks Algorithm and we denoted as FzFWA. We can justify this adjustment of parameters with simulation results obtained in evaluating six mathematical benchmark functions.

Keywords Fuzzy logic · Exploration · Exploitation · Optimizer · FWA · FzWA

1 Introduction

Nowadays computer science is solving problems through their different areas, depending on the complexity of the problem and the results expected from the idea to optimize (minimize or maximize) the solution to a problem.

In recent years, swarm intelligence (SI), has been very popular among researchers working in optimization problems around the world [1].

Some examples in SI are the following algorithms:

- PSO (Particle Swarm Optimization).
- ACO (Ant Colony Optimization).
- ABC (Artificial Bee Colony).
- GA (Genetic Algorithm).

J. Barraza · P. Melin (✉) · F. Valdez · C. González
Tijuana Institute of Technology, Tijuana, Mexico
e-mail: pmelin@tectijuana.mx

P. Melin et al. (eds.), *Nature-Inspired Design of Hybrid Intelligent Systems*,
Studies in Computational Intelligence 667, DOI 10.1007/978-3-319-47054-2_21

We have mentioned these types of algorithms (SI), and the fireworks algorithm (FWA) can be viewed as belonging to the same set of optimization methods [2–4].

There are some methodologies in computer science, which allow talking about the existence of artificial intelligence such as: neural networks, fuzzy logic, evolutionary computation, to name a few [5]. We based on fuzzy logic to improvement the FWA, this methodology we can describe as a methodology that provides a simple way to obtain a conclusion from ambiguous, imprecise, or linguistic information inputs.

Overall, fuzzy logic mimics a person, a control system, etc., and make decisions, based on information (input) with the features mentioned in the previous paragraph [6].

We focused on controlling the amplitude coefficient, which is a constant parameter on the calculi of the explosion amplitude (Eq. 4) for each firework based on functions evaluations and number of sparks, and with this control we will have an enhanced (FWA), which we denoted FzFWA.

This paper is organized as follows: Sect. 2 shows the FWA, the next Section (Sect. 3) shows the proposed method (FzFWA). Section 4 describes the Benchmark Functions, Sect. 4 shows the experiments and results, and Sect. 5 shows the Conclusions.

2 Fireworks Algorithm (FWA)

FWA is a SI algorithm, which was developed by Ying Tan and Yuanchum Zhu in 2009, based on the simulation of the explosion of fireworks [7, 8].

For each firework, we begin a process of explosion and a shower of sparks fill the local space around the firework. The recently generated sparks represent solutions in the search space [9, 10].

The FWA is composed of four general steps

- Initialization of location.
- Number of Sparks.
- Explosion Amplitude.
- Selection.

2.1 FWA

Within this algorithm we can modify or change the following parameters: numbers of sparks, explosion amplitude, and generation of sparks, to mention the most important.

2.2 Number of Sparks

At first we selected the search space where the sparks (possible solutions) will work and are determined in the following way:

$$\text{Minimize} f(x) \in R,\ x\_\text{min} \leq x \leq x\_\text{max} \tag{1}$$

where $\boldsymbol{x} = \boldsymbol{x}_1, \boldsymbol{x}_2, \boldsymbol{x}_3, \ldots, \boldsymbol{x}_d$ indicates a location in the search space, $x_{\min} \leq x \leq x_{\max}$ represents the bounds of the same space and $f(x)$ refers to objective function [11].

After, using the following equation the number of sparks for each firework is generated [12].

$$S_i = m \cdot \frac{y_{\max} - f(x_i) + \in}{\sum_{i=1}^{n} (y_{\max}) - f(x_i)) + \in} \tag{2}$$

where m represent a constant parameter, which controls the numbers of sparks of the n fireworks, $y_{\max} = \max(\mathrm{f}(x_i))$ $(i = 1, 2, 3, \ldots, n)$, is the maximum value, that is, the worst value of the objective function in the n fireworks and $\in$ indicates a constant with the smallest number in the computer, and it is utilized with the goal that a mistake with the division of zero cannot be done [13].

To avoid not having balance of the Fireworks, the bounds are defined for S_i.

$$\widehat{S}_i = \begin{cases} \text{round } (a \cdot m) & \text{if } S_i < am \\ \text{round } (b \cdot m) & \textit{if } S_i > bm,\ a < b < 1, \\ \text{round } (S_i) & \text{otherwise} \end{cases} \tag{3}$$

where a and b are constant parameters [14].

2.3 Explosion Amplitude

On the contrary to calculate the number of sparks, an explosion is better if the amplitude is small. The explosion amplitude for each firework is calculated as follows:

$$A_i = \widehat{A} \cdot \frac{f(x_i) - y_{min} + \in}{\sum_{i=1}^{n} (f(x_i) - y_{min}) + \in} \tag{4}$$

where $\widehat{A}$ is a constant parameter that controls the maximum amplitude of each firework, $y_{\min} = \min(f(x_i))$ $(i = 1, 2, 3, \ldots, n)$, indicate the minimum value (best) of the objective function among n fireworks and $\in$ indicates the smaller constant in the computer, and it is utilized with the goal that there cannot exist an error of division by zero [15].

2.4 Selection of Locations

Starting in each iteration explosion, n locations should be selected for the next explosion of fireworks. In the FWA, the current best location x^* in which the objective function $f(x^*)$ is optimal between the current locations always is kept of next iteration explosion. After that, $n - 1$ locations are selected in function their distance to other places with the purpose of keep the diversity of sparks. The general distance between a location of a x_i and the other places is defined as follows [16]:

$$R(x_i) = \sum_{j \in K} d(x_i, x_j) = \sum_{j \in K} ||x_i - x_j|| \tag{5}$$

where K is set of all current locations form both fireworks. Then the probability for selection of location at x_i is defined as:

$$p(x_i) = \frac{R(x_i)}{\sum_{j \in K} R(x_j)} \tag{6}$$

When calculating the distance, any distance measure can be utilized including the Manhattan distance, Euclidean distance, Angle-based distance.

The next algorithm (Algorithm 3) shows the pseudo code of the conventional FWA [16].

Algorithm 3 Pseudocode of the FWA

Algorithm 3. Pseudo code of the FWA

Randomly select:n locations for Fireworks;
While stop criteria = false **do**
Set off$x$ fireworks at the n locations:
for each fireworkx_i**do**
Calculate the number of sparks$\widehat{S_i}$ to the fireworks, Eq. 3
Obtain locations of $\widehat{S_i}$ sparks using Algorithm 1;
end for
fork = 1:$\widehat{m}$**do**
Randomly select a firework x_j;
Generate sparks (Gauss) using Algorithm 2;
end for
Select the best location and keep it for next explosion;
Randomly select $n - 1$ locations from the two types of sparks and de current fireworks according to the probability given in Eq. 7
end while

The performance of the algorithm 1 and 2 we can see in paper [13] or [17], both algorithms are used to generate the sparks for each firework, the unique difference is that in the algorithm 1 will generate normal sparks and the algorithm 2 will generate Gaussian sparks.

3 Proposed Method

Our proposed method is to keep the equilibrium between exploration and exploitation [18, 19], in the previous paper "Fuzzy FWA with dynamic adaptation of parameters" [17], we discussed this method.

Figure 1 shows the enhancement that we made of the FWA.

We can see in the Fig. 1 the yellow block (Fuzzy System for control of explosion amplitude), those block added into the flow chart of FWA with the objective the improvement the performance of the algorithm FWA. This improvement we achieved with a fuzzy system, the characteristics of this fuzzy system are explained as follows:

- Type: Mamdani.
- Input variables: evaluations [0, 1] and sparks [2, 40].
- Output variables: amplitude coefficient [0, 45].
- Membership functions: trapezoidal (3).

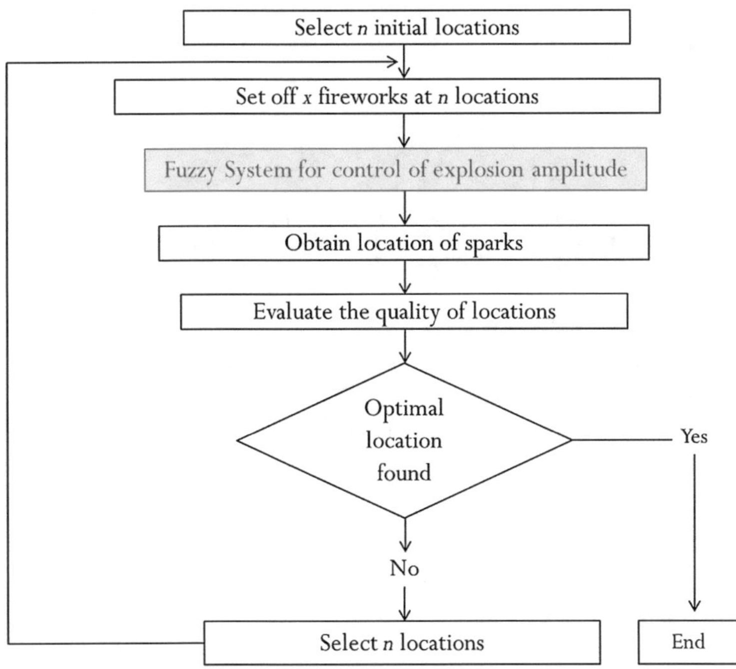

Fig. 1 Flow chart of FzFWA

1. If (**Evaluation** is **Low**) and (**sparks** is **Low**) then (**AmpliudeCoefficient** is **Big**)
2. If (**Evaluation** is **Low**) and (**sparks** is **Medium**) then (**AmpliudeCoefficient** is **Big**)
3. If (**Evaluation** is **Low**) and (**sparks** is **High**) then (**AmpliudeCoefficient** is **Big**)
4. If (**Evaluation** is **Medium**) and (**sparks** is **Low**) then (**AmpliudeCoefficient** is **Medium**)
5. If (**Evaluation** is **Medium**) and (**sparks** is **Medium**) then (**AmpliudeCoefficient** is **Medium**)
6. If (**Evaluation** is **Medium**) and (**sparks** is **High**) then (**AmpliudeCoefficient** is **Medium**)
7. If (**Evaluation** is **High**) and (**sparks** is **Low**) then (**AmpliudeCoefficient** is **Small**)
8. If (**Evaluation** is **High**) and (**sparks** is **Medium**) then (**AmpliudeCoefficient** is **Small**)
9. If (**Evaluation** is **High**) and (**sparks** is **High**) then (**AmpliudeCoefficient** is **Small**)

Fig. 2 Fuzzy rules

The rules used for controlling the explosion amplitude are presented in Fig. 2

4 Benchmark Functions

In this paper we consider 6 Benchmark functions, which are briefly explained below [20].

4.1 Ackley Function

Number of variables: n variables.

Definition

$$f(x) = -20\exp\left(-0.2\sqrt{\frac{1}{n}\sum_{i=1}^{n}x_i^2}\right) - \exp\left(\frac{1}{n}\sum_{i=1}^{n}\cos(2\pi x_i)\right) + 20 + e \quad (7)$$

Search domain: $-32 \le x_i \le 32$, $i = 1, 2, \ldots, n$.

Number of local minima: no local minimum except the global one.

The global minima: $x^* = (0, \ldots, 0)$, $f(x^*) = 0$.

Function graph: for $n = 2$ is presented in Fig. 3.

4.2 Griewank Function

Number of variables: n variables.

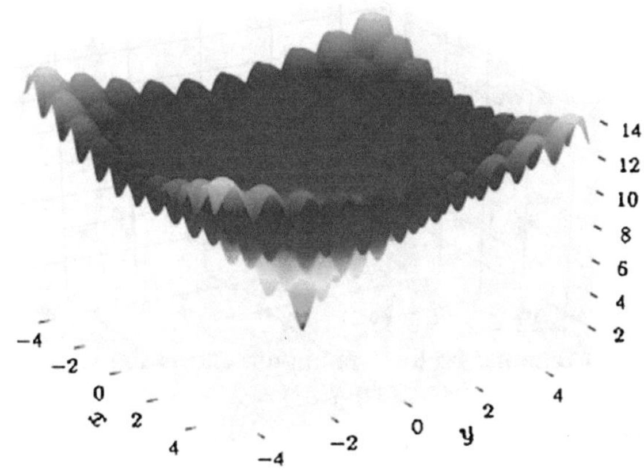

Fig. 3 Ackley function

Definition

$$f(x) = \frac{1}{4000}\sum_{i=1}^{n} x_1^2 - \prod_{i=1}^{n} \cos\left(\frac{x_i}{\sqrt{i}}\right) + 1 \tag{8}$$

Search domain: $-600 \le x_i \le 600$, $i = 1, 2, \ldots, n$.
Number of local minima: no local minimum except the global one.
The global minima: $x^* = (0, \ldots, 0)$, $f(x^*) = 0$.
Function graph: for $n = 2$ is presented in Fig. 4.

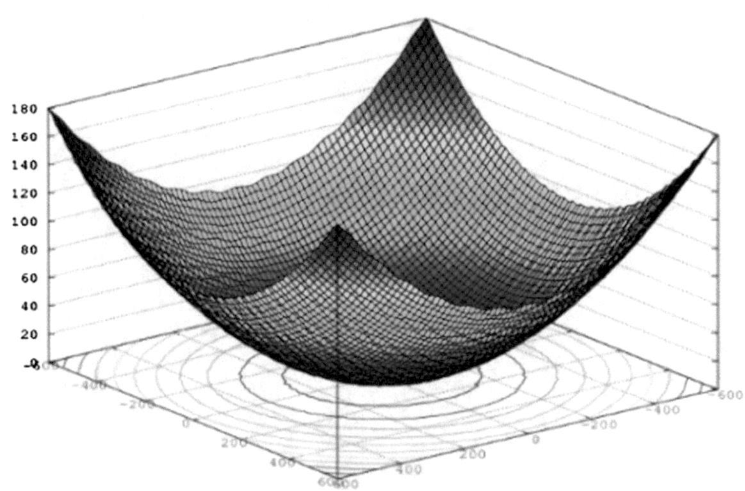

Fig. 4 Griewank function

4.3 Rastringin Function

Number of variables: n variables.
Definition

$$f(x) = \sum_{i=1}^{n} \left[x_i^2 - 10\cos(2\pi x_i) + 10\right] \quad (9)$$

Search domain: $-5.12 \leq x_i \leq 5.12$, $i = 1, 2, \ldots, n$.
Number of local minima: no local minimum except the global one.
The global minima: $x^* = (0, \ldots, 0)$, $f(x^*) = 0$.
Function graph: for $n = 2$ is presented in Fig. 5.

4.4 Rosenbrock Function

Number of variables: n variables.
Definition

$$f(x) = \sum_{i=1}^{n-1} \left[100(x_{i+1} - x_i^2)^2 + (x_1 - 1)^2\right] \quad (10)$$

Search domain: $-30 \leq x_i \leq 30$, $i = 1, 2, \ldots, n$.

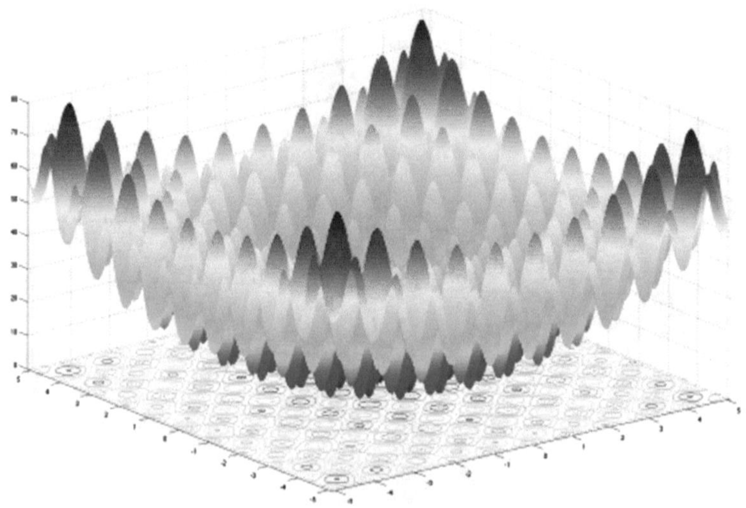

Fig. 5 Rastrigin function

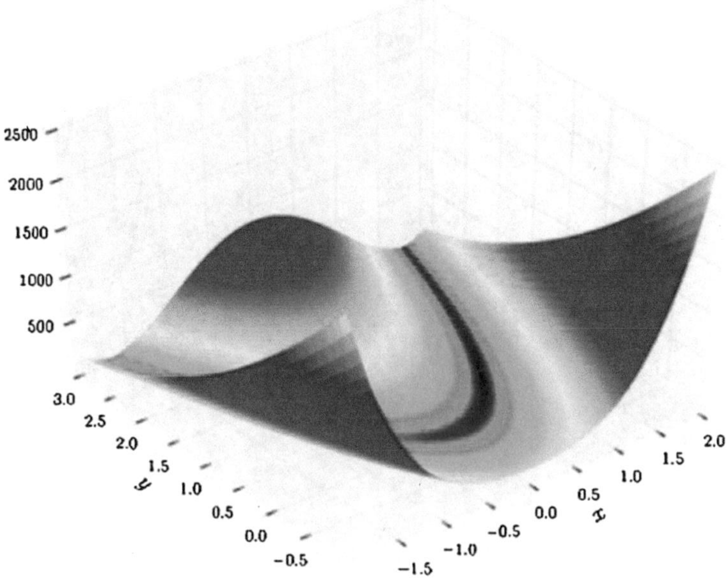

Fig. 6 Rosenbrock function

Number of local minima: no local minimum except the global one.
The global minima: $x^* = (0, \ldots, 0)$, $f(x^*) = 0$.
Function graph: for $n = 2$ is presented in Fig. 6.

4.5 Schwefel Function

Number of variables: n variables.
Definition

$$f(x) = \sum_{i=1}^{n} -x_i \sin\left(\sqrt{|x_i|}\right) \tag{11}$$

Search domain: $-500 \le x_i \le 500$, $i = 1, 2, \ldots, n$.
Number of local minima: no local minimum except the global one.
The global minima: $x^* = (0, \ldots, 0)$, $f(x^*) = 0$.
Function graph: for $n = 2$ is presented in Fig. 7.

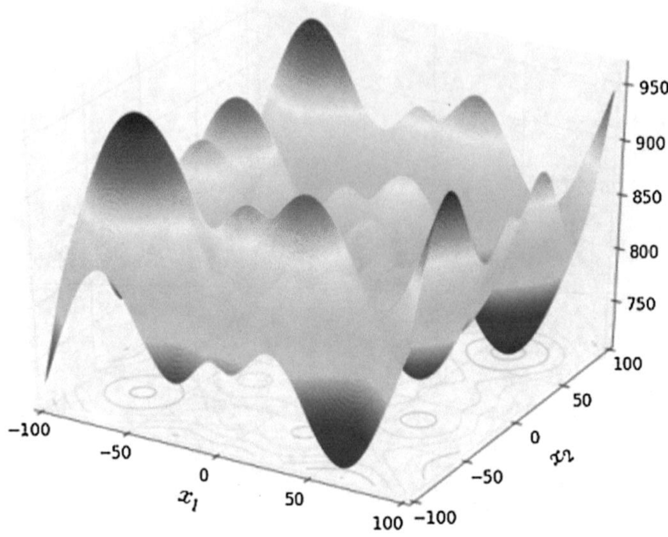

Fig. 7 Schwefel function

4.6 Sphere Function

Number of variables: n variables.
Definition

$$f(x) = \sum_{i=1}^{n} x_i^2 \tag{12}$$

Search domain: $-100 \leq x_i \leq 100$, $i = 1, 2, \ldots, n$.
Number of local minima: no local minimum except the global one.
The global minima: $x^* = (0, \ldots, 0)$, $f(x^*) = 0$.
Function graph: for $n = 2$ is presented in Fig. 8.

5 Experiments and Results

To show the performance of the FzFWA, we test with six benchmark functions, all with a range of $[-100, 100]^D$. We used those functions, firstly for test the performance of algorithm and after, the most important to comparing with original algorithm.

In Table 1 we enlist the name of every function utilized, as well as their ranges of initialization and number of dimensions.

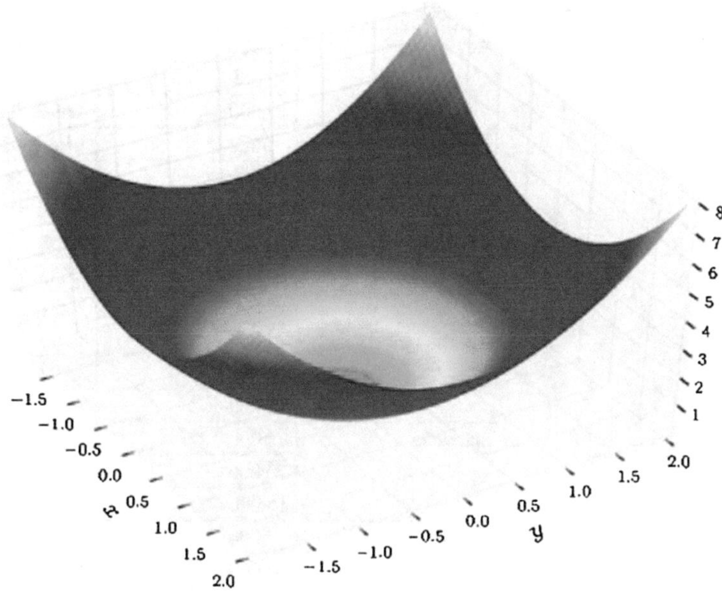

Fig. 8 Sphere function

Table 1 Benchmark functions to simulate results

Function	Range of initialization	Dimensions
Ackley	$[80, 100]^D$	30
Griewank	$[30, 50]^D$	30
Rastrigin	$[80, 100]^D$	30
Rosenbrock	$[30, 50]^D$	30
Schwefel	$[15, 30]^D$	30
Sphere	$[80, 100]^D$	30

In Tables 2 and 3, we show the results obtained with 5000, 10,000 and 15,000 functions evaluations, respectively with the 6 benchmark functions mentioned above.

The parameters utilized to prove the FWA with adaptation of parameters using fuzzy logic (FzFWA) were the same that were used by the author of FWA, only that we modified the coefficient of amplitude, that is to say, as we mentioned above, a static parameter was changed to dynamic (in a range). The parameters will be follows:$n = 5$, $m = 50$, $a = 0.04$, $b = 0.8$, $\widehat{A} = 40$, $\widehat{\mathrm{FA}} = [0, 45]$ and $D = 30$. These parameters were utilized to prove and show the results.

Tables 2 and 3 are shown in the following pages.

The above results (Table 2 and 3) were obtained of the means of 30 independent runs, we take the best results in each iteration, and with these results we calculated the mean of every iteration and finally we calculate the mean of 30 means obtained.

Table 2 Simulation results of FWA

Function	Evaluations	FWA
Ackley	5000	6.64E−02
	10,000	5.00E−02
	15,000	4.69E−02
Griewank	5000	1.80E−03
	10,000	8.00E−04
	15,000	6.00E−04
Rastrigin	5000	5.168
	10,000	2.635
	15,000	2.251
Rosenbrock	5000	30.937
	10,000	30.937
	15,000	28.077
Schwefel	5000	2.20E−02
	10,000	8.40E−03
	15,000	7.00E−04
Sphere	5000	4.75E−02
	10,000	3.08E−02
	15,000	2.50E−02

Table 3 Simulation results of FzWA

Function	Evaluations	FzFWA
Ackley	5000	1.63E−01
	10,000	1.17E−01
	15,000	9.38E−02
Griewank	5000	8.30E−03
	10,000	8.00E−04
	15,000	6.00E−04
Rastrigin	5000	9.122
	10,000	4.224
	15,000	3.031
Rosenbrock	5000	28.025
	10,000	25.242
	15,000	17.855
Schwefel	5000	2.22E−02
	10,000	9.60E−03
	15,000	5.60E−03
Sphere	5000	2.87E−02
	10,000	3.08E−02
	15,000	2.50E−02

Table 4 Comparing of results between FWA and FzFWA

Function	Evaluations	FWA	FzFWA
Ackley	5000	6.64E−02	1.63E−01
	10,000	5.00E−02	1.17E−01
	15,000	4.69E−02	9.38E−02
Griewank	5000	**1.80E−03**	**8.30E−03**
	10,000	**8.00E−04**	**8.00E−04**
	15,000	**6.00E−04**	**6.00E−04**
Rastrigin	5000	5.168	9.122
	10,000	2.635	4.224
	15,000	2.251	3.031
Rosenbrock	5000	30.858	**28.025**
	10,000	30.937	**25.242**
	15,000	28.077	**17.855**
Schwefel	5000	**2.20E-02**	**2.22E−02**
	10,000	8.40E−03	9.60E−03
	15,000	7.00E−04	5.60E−03
Sphere	5000	4.75E−02	**2.87E−02**
	10,000	3.08E−02	**1.70E−02**
	15,000	2.50E−02	**1.31E−02**

Bold shows the best results we obtained to comparing FWA and FzFWA

Table 5 Hypothesis test of FWA vs FzFWA

Function	Ackley	Griewank	Rastrigin	Rosenbrock	Schwefel	Sphere
Evaluations	10,000	10,000	10,000	10,000	10,000	10,000
FWA mean	5.00E−02	8.00E−04	2.635	30.937	8.40E−03	3.08E−02
STD	1.19E−02	7.26E−04	1.126	7.279	8.77E−03	3.38E−02
z-value	−1.645	−1.645	−1.645	−1.645	−1.645	−1.645
FzWA mean	6.29E−02	8.00E−04	**1.923**	**26.688**	**2.60E−03**	**2.65E−02**
STD	1.95E−02	5.76E−04	2.096	**6.447**	**7.24E−03**	**1.31E−02**
z-value	3.092	**0**	−1.639	**−2.393**	**−2.793**	−0.6497

Bold shows the best results we obtained to comparing FWA and FzFWA

Table 4 shows the comparison between two algorithms, in specific, we compared the means of the best solutions of every benchmark function, and the calculations were based on 30 independent runs.

Table 5 shows a hypothesis test of the conventional FWA and the variant we made (FzFwa) with 10,000 function evaluations and 30 dimensions, with of z-value equal to −1.645, that is, 95 % degree of confidence or 5 % significance level.

6 Conclusions

We conclude that in a study of the FWA algorithm we can find the parameters which can be optimized using a dynamic method, that's why we decided to make the dynamic setting called coefficient amplitude ($\widehat{A}$), which is responsible for controlling the explosion amplitude of each firework, and the adjustment was made based on two parameters (evaluations and number of sparks) as antecedents to obtain a consequent (amplitude coefficient) within a fuzzy system.

After observing the tables of results we can say that we improved the performance of the FWA, the above, we can justify with Table 5 which shows a hypothesis test between conventional algorithm fireworks (FWA) and the proposed method (FzFWA). As future work, will need to perform more experiments with variations in membership functions of fuzzy systems and also test other mathematical functions benchmark with optimal value different to 0.

References

1. S. Das, A. Abraham and A. Konar "Swarm intelligence algorithms in bioinformatics". Studies in Computational Intelligence 94 (2008), 113–147.
2. K. Ding, S. Zheng and Y. Tan. "A GPU-based Parallel Fireworks Algorithm for Optimization" GECCO'13, Amsterdam, the Netherlands, July 6-10, 2013.
3. J. Kennedy and R.C. Eberhart. "Particle swarm optimization". In: Proceedings of IEEE International Conference on Neural Networks (1995), vol. 4, pp. 1942–1948.
4. M. Dorigo, V. Maniezzo and A. Colorni. "Ant system: optimization by a colony of cooperating agents". IEEE Transactions on Systems, Man, and Cybernetics (1996), Part B: Cybernetics 26(1), 29–41.
5. L.A. Zadeh "Knowledge Representation in Fuzzy Logic".IEEE transactions on knowledge and data engineering, vol. I, no. I, march 1989,pp. 89-0084.
6. M. Simoes, K. Bose and J. Spiegel: "Fuzzy Logic Based Intelligent Control of a Variable Speed Cage Machine Wind Generation System". IEEE transactions on power electronics, vol. 12, no. 1, January 1997,pp. 87–95.
7. Y. Zheng, X. Xu, H. Ling and Sheng-Yong Chen. "A hybrid fireworks optimization method with differential evolution operators", Neurocomputing 148 (2015) 75–82.
8. Y. Pei, S. Zheng, Y. Tan, and T. Hideyuki, "An empirical study on influence of approximation approaches on enhancing fireworks algorithm," in Proceedings of the 2012 IEEE Congress on System, Man and Cybernetics. IEEE, 2012, pp. 1322–1327.
9. A. Mohamed and M. Kowsalya. "A new power system reconfiguration scheme for power loss minimization and voltage profile enhancement using Fireworks Algorithm", Electrical Power and Energy Systems 62 (2014) 312–322.
10. Y. Zheng, Qin Song, S.-Y Chen. "Multiobjective fireworks optimization for variable-rate fertilization in oil crop production", Applied Soft Computing 13 (2013) 4253–4263.
11. J.Li and S.Z. "Adaptive Fireworks Algorithm". IEEE Congress on Evolutionary Computation 2014 (CEC),, 3214-3221.
12. Y. Tan, "Fireworks Algorithm", Springer-Verlag Berlin Heidelberg 2015, 355–364.
13. Y. Tan and Y. Zhu, "Fireworks Algorithm for Optimization," Springer-Verlag Berlin Heidelberg 2010, pp. 355–364.

14. Y. Tan and S. Z. "Enhanced Fireworks Algorithm". IEEE Congress on Evolutionary Computation (2013), 2069-2077.
15. Y.Tan and S. Z. "Dynamic Search in Fireworks Algorithm. Evolutionary Computation" (CEC 2014).
16. N. H. Abdulmajeed and M. Ayob, "A Firework Algorithm for Solving Capacitated Vehicle Routing Problem", International Journal of Advancements in Computing Technology, January 2014, (IJACT), Volume 6, Number 1, 79-86.
17. J. Barraza, P. Melin, F. Valdez "Fuzzy FWA with dynamic adaptation of parameters", IEEE CEC 2016,"accepted for publication".
18. M., Liu, S.H., and Mernik. "Exploration and exploitation in evolutionary algorithms": Asurvey. ACM Comput. Surv. 2013, 45, 3, 35:32.
19. J. Liu, S. Zheng, and Y. Tan, "The improvement on controlling exploration and exploitation of firework algorithm," in Advances in Swarm Intelligence. Springer, 2013, pp. 11–23.
20. L. Rodriguez, O. Castillo, J. Soria "Grey Wolf Optimizer (GWO) with dynamic adaptation of parameters using fuzzy logic", IEEE CEC 2016, "accepted for publication".

Imperialist Competitive Algorithm with Dynamic Parameter Adaptation Applied to the Optimization of Mathematical Functions

Emer Bernal, Oscar Castillo and José Soria

Abstract In this paper, we describe an imperialist competitive algorithm with dynamic adjustment of parameters using fuzzy logic to adjust the Beta and Xi parameters. We are considering different fuzzy systems to measure the performance of the algorithm with six benchmark mathematical functions with different number of decades and performing 30 experiments for each case. The results demonstrate the efficiency of the fuzzy ICA algorithm in optimization problems and give us the guidelines for future work.

Keywords Imperialist competitive algorithm · Fuzzy logic · Optimization

1 Introduction

Swarm intelligence techniques have become increasingly popular over the past two decades time due to their ability to find a relatively optimal solutions for complex combinatorial optimization problems. They have been applied in the fields of engineering, economics, management science, industry, etc. The problems that benefit from the application of swarm intelligence techniques are generally very difficult to solve optimally in the sense that there is no exact algorithm to solve them in polynomial time. These optimization problems are also known as NP—difficult problems [12].

Swarm intelligence techniques are approximate algorithms that incorporate a wide range of intelligent algorithms largely inspired by natural processes, such as the Particle Swarm Optimization (PSO), Ant Colony Optimization (ACO), Genetic Algorithm (GA), and Artificial Honey Bee (AHB) method [12].

An algorithm that is well recognized in the field of metaheuristics is the imperialist competitive algorithm (ICA), which was presented by Atashpaz-Gargari and Lucas in 2007. ICA was inspired by the concept of imperialism [21]; where

E. Bernal · O. Castillo (✉) · J. Soria
Tijuana Institute of Technology, Tijuana, B.C., Mexico
e-mail: ocastillo@tectijuana.mx

P. Melin et al. (eds.), *Nature-Inspired Design of Hybrid Intelligent Systems*,
Studies in Computational Intelligence 667, DOI 10.1007/978-3-319-47054-2_22

powerful countries try to make a colony of other countries. These algorithms have recently been used in various engineering applications [3].

The study of the algorithm is performed in order to show the effectiveness of the imperialist competitive algorithm (ICA) when applied to optimization problems, and taking the original ICA as a basis for a modification to the algorithm for dynamically adjusting some of its parameters. This has already been proven in other algorithms which use adjustment along the iterations and helps to improve the results, with respect to when fixed parameters are used.

To make this modification we use fuzzy logic, we will proceed with the implementation of the algorithm for mathematical benchmark functions. The idea is to observe the results obtained since the ICA algorithm is relatively new and it is expected to achieve good or better results than with other metaheuristic algorithms.

We describe the ICA with dynamic adjustment of parameters using fuzzy logic. We propose different fuzzy systems for the Beta and Xi parameters to measure the performance of fuzzy ICA against the original algorithm. This algorithm was tested with mathematical benchmark functions.

The paper is organized as follows: in Sect. 2 a description about the ICA is presented; in Sect. 3, a description of the mathematical functions is presented; in Sect. 4, our proposed method is described; in Sect. 5, the implemented fuzzy systems in the ICA algorithm are described; in Sect. 6, the experiments results are presented and we can appreciate the ICA algorithm behavior to implement the fuzzy systems in Beta and Xi; in Sect. 7, the conclusions obtained after the study of the ICA versus mathematical functions are presented.

Currently in the literature there are papers dealing with the imperialist competitive algorithm, like the paper of the Imperialist competitive algorithm for minimum bit error rate beamforming [2]. In this paper the imperialist competitive algorithm (ICA) is used to design an optimal antenna array which minimizes the error probability. In the paper of a modified imperialist competitive algorithm based on attraction and repulsion concepts for reliability-redundancy optimization [5], an improvement in the ICA by implementing an attraction and repulsion concept during the search for better solutions is proposed. In the paper of an imperialist competitive algorithm optimized artificial neural networks for UCAV global path planning [7], a novel hybrid method for the globally optimal path planning of UCAV is proposed, based on an artificial neural network (ANN) trained by imperialist competitive algorithm (ICA). In the paper of a template matching using chaotic imperialist competitive algorithm [8], a novel template matching method based on chaotic ICA is presented. Based on the introduction of the principle of ICA, the correlation function used in this approach is proposed. The chaos can improve the global convergence of ICA. In the paper of a hybrid imperialist competitive algorithm for minimizing make span in a multi-processor open shop [9]. A hybrid imperialist competitive algorithm (ICA) with genetic algorithm (GA) is presented to solve this problem, in the paper Imperialist competitive algorithm with PROCLUS classifier for service time optimization in cloud computing service composition [14]. An improved imperialist competitive algorithm is employed to select more suitable service providers for the required unique services, in the paper the application of an imperialist competitive algorithm to the design of a linear induction motor [15]. In this paper, a novel optimization algorithm based on imperialist competitive algorithm (ICA) is used for the design of a low-speed single-sided linear induction motor (LIM). In the paper of an Imperialist competitive algorithm combined with refined high-order weighted fuzzy time series (RHWFTS–ICA) for short-term load forecasting [18]. This method is proposed to perform efficiently under short-term load forecasting (STLF). First, autocorrelation analysis was used to recognize the order of the fuzzy logical relationships. Next, the optimal coefficients and optimal intervals of adaption were obtained by means of an imperialist competitive algorithm in the training dataset, in the paper D-FICCA: A density-based fuzzy imperialist competitive clustering algorithm for intrusion detection in wireless sensor networks [19]. In this paper, a hybrid clustering method is introduced, the imperialist competitive algorithm (ICA) is modified with a density-based algorithm and fuzzy logic for optimum clustering in WSNs, in the paper Colonial competitive algorithm: A novel approach for PID controller design in MIMO distillation column process [1]. This paper aims to describe colonial competitive algorithm (CCA), how it is used to solve real-world engineering problems by applying it to the problem of designing a multivariable proportional-integral- derivative (PID) controller.

In other works shown below on the use of others metaheuristics as GA and ACO among other, like the paper of an ant colony optimization for the traveling purchaser problem [4]. In this work, we address the solution of the TPP with an ant colony optimization procedure. We combine it with a local-search scheme exploring a new neighborhood structure, in the paper ant colony optimization theory: A Survey [6]. This paper provides a survey on theoretical results on ant colony optimization. First, some convergence results are reviewed. Then relations between ant colony optimization algorithms and other approximate methods for optimization are discussed in the paper Convergence Criteria for Genetic Algorithms [10]. In this paper, convergence properties for genetic algorithms are discussed. By looking at the effect of mutation on convergence, we show that by running the genetic algorithm for a sufficiently long time we can guarantee convergence to a global optimum with any specified level of confidence.

2 Imperialist Competitive Algorithm

In the field of evolutionary computation, the novel ICA algorithm is based on human social and political advancements [12], unlike other evolutionary algorithms, which are based on the natural behaviors of animals or physical events.

ICA starts with an initial randomly generated population, in which the individuals are known as countries. Some of the best countries are considered imperialists, whereas the other countries represent the imperialist colonies [12].

All the colonies of the initial population are divided among the mentioned imperialists based on their power. The power of an empire, which is the counterpart of the fitness value in GA and is inversely proportional to its cost [17].

2.1 Moving the Colonies of an Empire Toward the Imperialist (Assimilation)

As shown in Fig. 1, the colony moves a distance x along with the d direction toward its imperialist. The process of moving a distance x is a random number generated by random distribution within the interval (0, βd) [12].

$$x \sim U(0, \beta d), \tag{1}$$

Where β is a number greater than 1 and d is the distance between the colony and the imperialist.

As shown in Fig. 2, to search for different locations around the imperialist we add a random amount of deviation to the direction of motion, which is given by [3]:

$$\theta \sim U(-\gamma, \gamma), \tag{2}$$

where θ is a random number with uniform distribution and γ is a parameter that adjusts the deviation from the original direction.

2.2 Total Power of an Empire

The power of an empire is calculated based on the power of its imperialist and a fraction of the power of its colonies. This fact has been modeled by defining the total cost as given by [12]:

$$\mathrm{TC}_n = \mathrm{Cost(Imp)} + \xi\,\mathrm{mean}\{\mathrm{Cost(Col)}\} \tag{3}$$

where TC_n is the total cost of nth Empire and ξ is a positive number between 0 and 1.

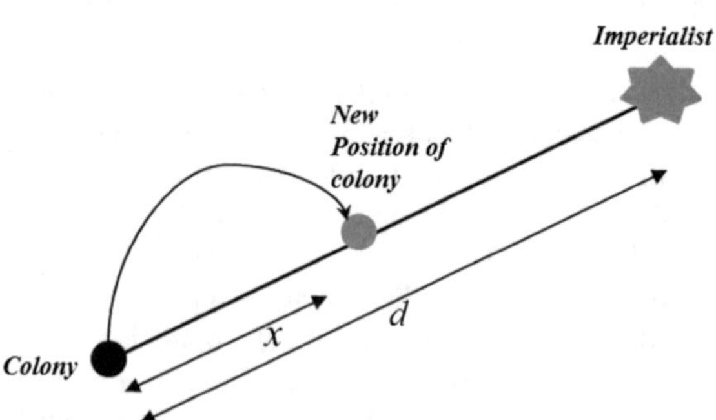

Fig. 1 Movement of the colonies toward the imperialist

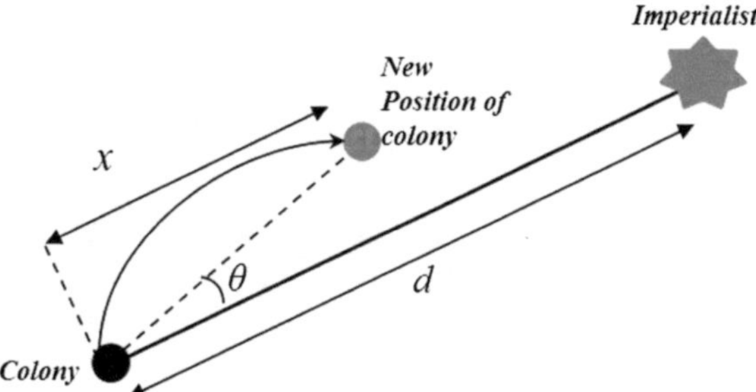

Fig. 2 Movement of the colonies toward their relevant imperialist in a randomly direction deviation

2.3 Pseudocode of ICA

The pseudocode of ICA is defined as follows:

1. *Select some random points on the function and initialize the empires.*
2. *Move the colonies toward their relevant imperialist (assimilation).*
3. *If there is a colony in an empire which has lower cost than that of imperialist, exchange the positions of that colony and the imperialist.*
4. *Calculate the total cost of all empires (related to the power of both imperialist and its colonies).*
5. *Pick the weakest colony (colonies) from the weakest empire and give it (them) to the empire that has the most likelihood to possess it (imperialistic competition).*
6. *Eliminate the powerless empires.*
7. *If there is just one empire, stop, if not go to Step 2.*

3 Benchmark Mathematical Functions

In this section, the benchmark functions that were used in the tests are listed for evaluating the performance of the ICA algorithm by dynamically adjusting its parameters and analyzing the results obtained.

In the area of metaheuristics for optimization, it is common to use mathematical functions, and these are used in this work, that is, a modification to an optimization algorithm based on imperialism known as ICA, where the dynamic adjustment of the Beta and Xi parameters is made and therefore the variation of the results was analyzed. The mathematical functions are listed below [11, 20]:

- **Sphere**

$$f(x) = \sum_{j=1}^{n_x} x_j^2 \tag{4}$$

Search Domain $x_j \in [-5.12, 5.12]$ and $f^*(x) = 0.0$

- **Quartic**

$$f(x) = \sum_{i=1}^{n} ix_i^4 \tag{5}$$

Search Domain $x_i \in [-1.28, 1.28]$ and $f^*(x) = 0.0$

- **Rosenbrock**

$$f(x) = \sum_{j=1}^{n_z/2} [100(x_{2j} - x_{2j-1}^2)^2 + (1 - x_{2j-1})^2] \tag{6}$$

Search Domain $x_j \in [-2.048, 2.048]$ and $f^*(x) = 0.0$

- **Rastrigin**

$$f(x) = \sum_{j=1}^{n_x} (x_j^2 - 10\cos(2\pi x_j) + 10) \tag{7}$$

Search Domain $x_j \in [-5.12, 5.12]$ and $f^*(x) = 0.0$

- **Griewank**

$$f(x) = 1 + \frac{1}{400}\sum_{i=1}^{n} x_i^2 - \prod_{i=1}^{n} \cos\left(\frac{xi}{\sqrt{i}}\right) \tag{8}$$

Search Domain $x_i \in [-600, 600]$ and $f^*(x) = 0.0$

- **Ackley**

$$f(x) = -20e^{-0.2\sqrt{\frac{1}{n_x}\sum_{j=1}^{n_x} x_j^2 - \frac{1}{e^{n_x}}\sum_{j=1}^{n_x}\cos(2\pi x_j)}} + 20 + e \tag{9}$$

Search Domain $x_j \in [-30, 30]$ and $f^*(x) = 0.0$

4 Proposed Method

The ICA is a search technique that was inspired by the concept of imperialism, where powerful countries try to make a colony of other countries and has recently been used to solve complex combinatorial optimization problems. A new algorithm called fuzzy imperialist competitive algorithm (FICA) with dynamic adjustment of parameters applied to the optimization of mathematical functions is proposed in this paper.

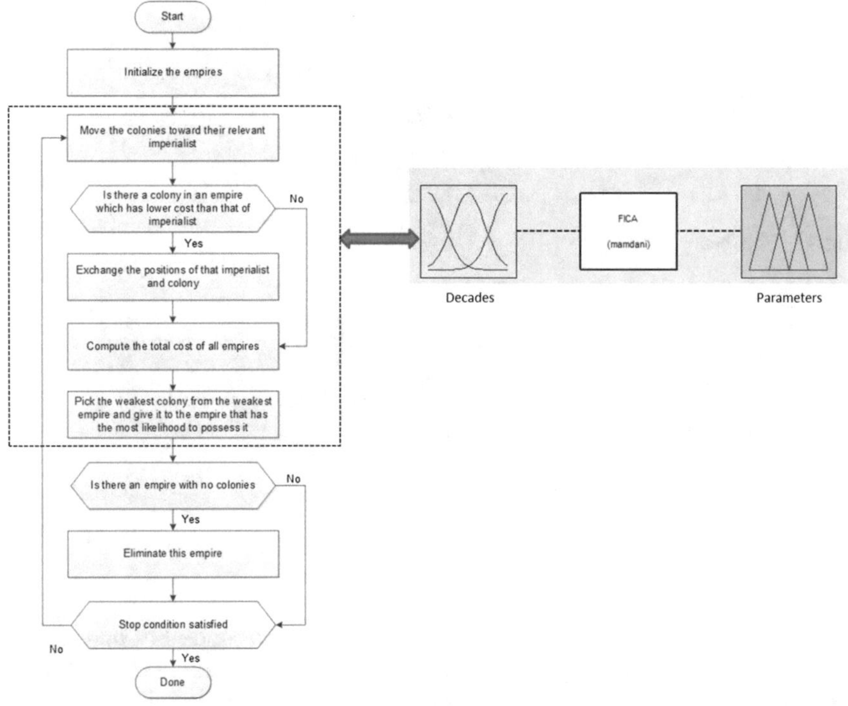

Fig. 3 The proposed FICA

The main objective is that the fuzzy systems give us the optimal values of the parameters for the best performance the ICA algorithm. The parameters that the fuzzy systems are dynamically adjusting are Beta and Xi, as shown Fig. 3.

To measure the decades of the algorithm, we decided to use a percentage of decades, that is, when the algorithm begins the decades will be considered "low", as the decades have been completed they will be considered "high" or close to 100 %. This idea is represented as follows [16]:

$$\text{Decades} = \frac{\text{Current Decade}}{\text{Total Number Of Decades}} \tag{10}$$

5 Fuzzy System

This paper mentions two fuzzy systems with which the experiments were conducted. We have a fuzzy system for the Beta parameter and another for the Xi parameter.

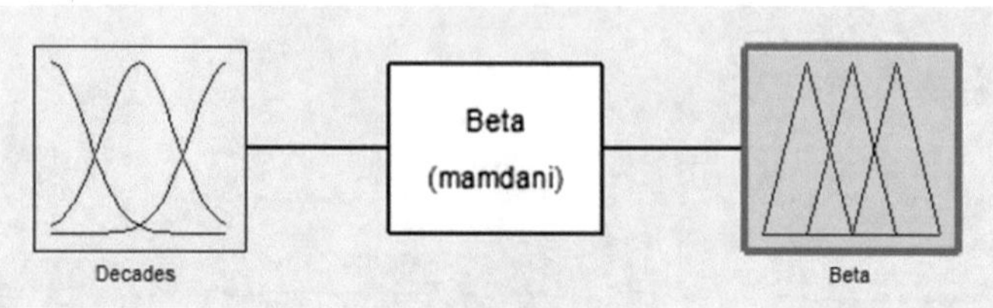

Fig. 4 FICA beta

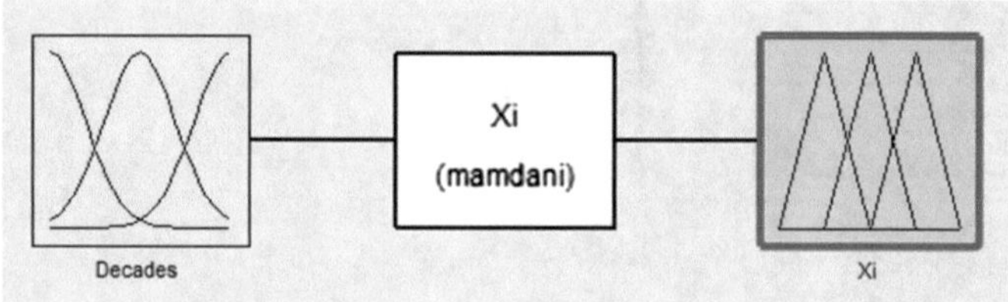

Fig. 5 FICA Xi

The ICA is a search technique that was inspired by the concept of imperialism, where powerful countries try to make a colony of other countries and has recently been used to solve complex combinatorial optimization problems. In this paper, a new algorithm called FICA with dynamic adjustment of parameters applied to the optimization of mathematical functions is proposed.

The fuzzy systems [13], shown in Figs. 4 and 5 are of the Mamdani type with the input as the decades and one output, respectively; the first is the parameter Beta and the second with the parameter Xi.

The design of the input variable which represents the decades; the input variable is granulated into three triangular membership functions, Low, Medium and High, as shown in Fig. 6.

The output variables of fuzzy systems are shown in Figs. 7 and 8 are Beta and Xi, are equally distributed in three triangular membership functions called Low, Medium and High.

To design the rules of each fuzzy system, it was decided that in the first decades of the ICA algorithm we must explore and then eventually exploit. The rules of the fuzzy systems were designed with the idea that they were on an increase behavior as shown below:

Rules of the fuzzy system for Beta:

1. If (Decades is Low) then (Beta is Low).
2. If (Decades is Middle) then (Beta is Medium).
3. If (Decades is Alto) then (Beta is Alto).

Rules of the fuzzy system for Xi:

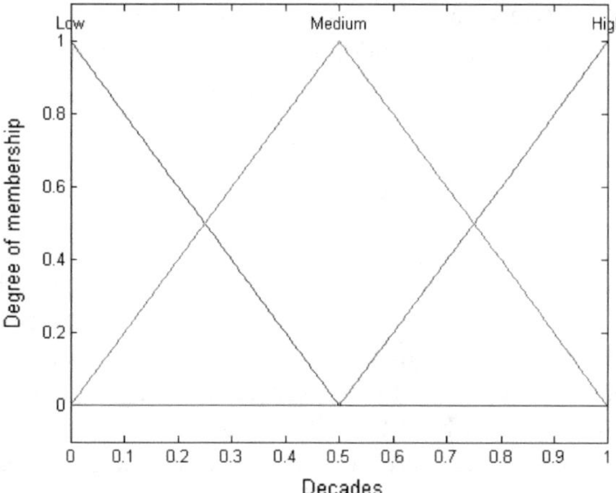

Fig. 6 Input as the decades

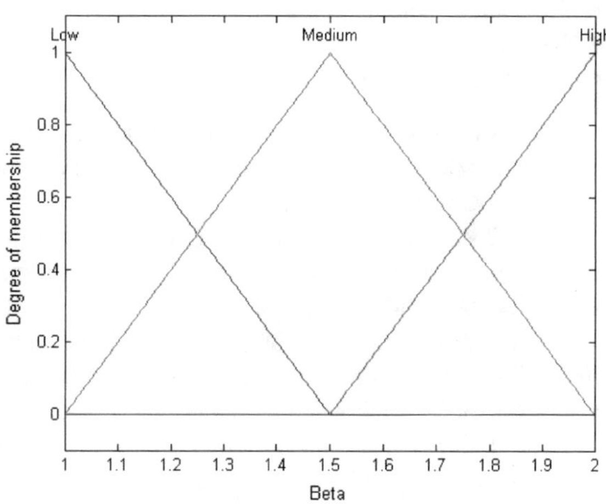

Fig. 7 Output beta

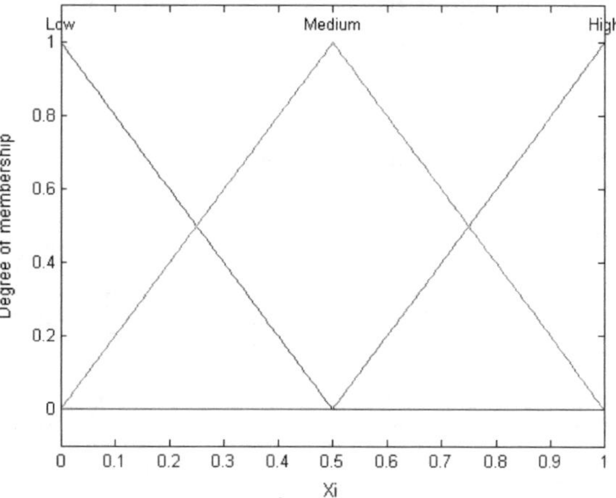

Fig. 8 Output Xi

1. If (Decades is Low) then (Xi is Low).
2. If (Decades is Middle) then (Xi is Medium).
3. If (Decades is Alto) then (Xi is Alto).

6 Simulation Results

In this section, the ICA is implemented with six benchmark mathematical functions with 30 variables for each of the Beta and Xi parameters, and the results obtained by the ICA algorithm and our proposed method are shown in separate tables by each function. Each table contains the average, obtained after 30 runs for algorithm ranging from 1000 to 5000 decades.

The parameters used in the ICA and the FICAs are as follows:

- Number of variables: 30
- Number of countries: 200
- Number of imperialist: 10
- Revolution rate: 0.2

Table 1 shows that after the execution of the proposed algorithm by the ICA 30 times, dynamically adjusting the Beta and Xi parameters, we can find the average results for the sphere function.

Table 2 shows that after the execution of the proposed algorithm by the ICA 30 times, dynamically adjusting the Beta and Xi parameters, we can find the average results for the quartic function.

Table 3 shows that after the execution of the proposed algorithm by the ICA 30 times, dynamically adjusting the Beta and Xi parameters, we can find the average results for the Rosenbrock function.

Table 4 shows that after the execution of the proposed algorithm by the ICA 30 times, dynamically adjusting the Beta and Xi parameters, we can find the average results for the Rastrigin function.

Table 5 shows that after the execution of the proposed algorithm by the ICA 30 times, dynamically adjusting the Beta and Xi parameters, we can find the average results for the Griewank function.

Table 1 Results for the Sphere function

Sphere function				
Decades		ICA	Increase beta	Increase Xi
			1–2	0–1
1000	Average	2.51E−21	2.28E−25	2.50E−19
2000	Average	1.76E−49	5.08E−52	1.21E−28
3000	Average	4.43E−54	1.53E−77	2.52E−26
4000	Average	5.56E−73	9.51E−99	3.84E−49
5000	Average	1.50E−122	1.42E−21	3.30E−17

Table 2 Results for the Quartic function

Quartic function				
Decades		ICA	Increase beta	Increase Xi
			1–2	0–1
1000	Average	9.76E−41	2.97E−39	2.39E−42
2000	Average	4.39E−89	1.85E−86	1.26E−90
3000	Average	2.14E−148	4.98E−130	1.75E−149
4000	Average	4.52E−197	1.00E−162	2.05E−197
5000	Average	5.04E−242	6.52E−197	8.34E−247

Table 3 Results for the Rosenbrock function

Rosenbrock function				
Decides		ICA	Increase beta	Increase Xi
			1–2	0–1
1000	Average	18.32843006	17.30207724	18.7086048
2000	Average	10.70282924	9.026534957	10.5753621
3000	Average	5.009639063	3.526645451	5.11616193
4000	Average	2.518922008	0.81218439	2.50972633
5000	Average	2.032004713	1.078276786	1.57959886

Table 4 Results for the Rastrigin function

Rastrigin function				
Decides		ICA	Increase beta	Increase Xi
			1–2	0–1
1000	Average	131.0164965	95.8100497	125.297209
2000	Average	116.4624694	55.95472621	115.942776
3000	Average	100.2171148	103.0026754	97.9436946
4000	Average	101.1418009	90.6758531	99.6304181
5000	Average	101.0967699	93.80047281	109.224593

Table 5 Results for the Griewank function

Griewank function				
Decides		ICA	Increase beta	Increase Xi
			1–2	0–1
1000	Average	0.35910253	0.50333873	0.41852252
2000	Average	0.34491885	0.4949299	0.36873076
3000	Average	0.25614264	0.50039122	0.4444043
4000	Average	0.46998072	0.46775839	0.33053626
5000	Average	0.53988701	0.66356327	0.29744525

Table 6 Results for the Ackley function

Ackley function				
Decides		ICA	Increase beta	Increase Xi
			1–2	0–1
1000	Average	5.00997147	4.6910324	5.53751731
2000	Average	5.03586405	2.91434453	5.28942225
3000	Average	5.62259386	2.91120991	4.86681416
4000	Average	4.69447289	2.37875375	5.18774309
5000	Average	5.25589516	3.2539677	5.137672

Table 6 shows that after the execution of the proposed algorithm by the ICA 30 times, dynamically adjusting the Beta and Xi parameters, we can find the average results for the Ackley function.

7 Conclusions

After analyzing the results obtained with the ICA and implemented the fuzzy systems for the Beta and Xi parameters, it could be noted that the fuzzy system in beta exceeds the fuzzy system in Xi and the original algorithm ICA. We can note that the fuzzy system for Xi also managed to overcome the original ICA algorithm.

In the sphere and quartic functions, it is where our proposal behaves better, because it manages to converge quickly with a low number of decades. For the remaining functions are not very good if its results cannot be ruled without taking into account other aspects such as the search space, the number of variables, and a more extensive study.

Finally after analyzing the proposals and the results, this leads us to the conclusion that the FICA is better than the original algorithm.

Acknowledgment We would like to express our gratitude to the CONACYT and Tijuana Institute of Technology for the facilities and resources granted for the development of this research.

References

1. E. Atashpaz-Gargari, F. Hashemzadeh, R. Rajabioun and C. Lucas, «Colonial competitive algorithm: A novel approach for PID controller design in MIMO distillation column process,» International Journal of Intelligent Computing and Cybernetics, vol. 1, nº 3, pp. 337-355, 2008.
2. E. Atashpaz-Gargari and C. Lucas, «Imperialist competitive algorithm for minimum bit error rate beam forming,» International Journal Bio-Inspired Computation, vol. 1, nº 1/2, pp. 125-133, 2009.

3. E. Atashpaz-Gargari and C. Lucas, «Imperialist competitive algorithm: An algorithm for optimization inspired by imperialistic competition,» Evolutionary Computation, pp. 4661-4667, 2007.
4. B. Bontoux and D. Feillet, «Ant colony optimization for the traveling purchaser problem,» Computers & Operations Research, vol. 35, nº 2, pp. 628-637, 2008.
5. L. Dallegrave Afonso, V. Cocco Mariani and L. dos Santos Coelho, «Modified imperialist competitive algorithm based on attraction and repulsion concepts for reliability-redundancy optimization,» Expert Systems with Applications, vol. 40, nº 9, pp. 3794-3802, 2013.
6. M. Dorigo and C. Blum, «Ant Colony Optimization theory: A Survey,» Theoretical Computer Science, vol. 344, nº 2, pp. 243-278, 2005.
7. H. Duan and L. z. Huang, «Imperialist competitive algorithm optimized artificial neural networks for UCAV global path planning,» Neurocomputing, vol. 125, pp. 166-171, 2013.
8. H. Duan, C. Xu, S. Liu and S. Chao, «Template matching using chaotic imperialist competitive algorithm,» Pattern Recognition Letters, pp. 1968-1975, 2010.
9. S. M. Goldansaz, J. Fariborz and A. H. Zahedi Anaraki, «A hybrid imperialist competitive algorithm for minimizing makespan in a multi-processor open shop,» Applied Mathematical Modelling, vol. 37, nº 23, pp. 9603-9616, 2013.
10. D. Greenhalgh and S. Marshall, «Convergence Criteria for Genetic Algorithms,» SIAM Journal on Computing, vol. 30, nº 1, pp. 269-282, 2000.
11. R. L. Haunpt and S. E. Haunpt, Practical Genetic Algorithms, New Jersey: John Wiley and Sons, 2004.
12. S. m. Hosseini and A. Al Khaled, «A survey on the Imperialist Competitive Algorithm Metaheuristic: Implementation in Engineering Domain and Directions for Future Research.,» Applied Soft Computing Journal, p. 55, 2014.
13. J.-S. Jang, C.-T. Sun and E. Mizutani, Neuro-Fuzzy and Soft Computing A Computational Approach To Learning and Machine Intelligence, United States of America: Prentice Hall, 1997.
14. A. Jula, Z. Othman and E. Sundararajan, «Imperialist competitive algorithm with PROCLUS classifier for service time optimization in cloud computing service composition,» Expert Systems with Applications, vol. 42, nº 1, pp. 135-145, 2014.
15. C. Lucas, Z. Nasiri-Gheidari y F. Tootoonchian, «Application of an imperialist competitive algorithm to the design of a linear induction motor,» Energy Conversion and Management, vol. 51, nº 7, pp. 1407-1411, 2010.
16. P. Melin, F. Olivas, O. Castillo, F. Valdez, J. Soria and M. Valdez, «Optimal design of fuzzy classification systems using PSO with dynamic parameter adaptation through fuzzy logic,» Expert Systems with Applications, vol. 40, nº 8, p. 3196–3206, 2012.
17. A. Nourmohammadia, M. Zandiehb and R. Tavakkoli-Moghaddamca, «An imperialist competitive algorithm for multi-objective U-type assembly line design,» Journal of Computational Science, vol. 4, nº 5, pp. 393-400, 2012.
18 E. Rasul, H. Javedani Sadaei, A. H. Abdullah and A. Gani, «Imperialist competitive algorithm combined with refined high-order weighted fuzzy time series (RHWFTS–ICA) for short term load forecasting,» Energy Conversion and Management, vol. 76, pp. 1104-1116, 2013.
19. S. Shamshirband, A. Amini, A. Nor Badrul, M. L. Mat Kiah, W. T. Ying and S. Furnell, «D-FICCA: A density-based fuzzy imperialist competitive clustering algorithm for intrusion detection in wireless sensor networks.,» Journal of the International Measurement Confederation, vol. 55, pp. 212-226, 2014.
20. F. Valdez, P. Melin and O. Castillo, «An improved evolutionary method with fuzzy logic for combining particle swarm optimization and genetic algorithms,» Soft Computing, vol. 11, nº 2, pp. 2625-2632, 2011.
21. J. Woddis, An introduction to neo-colonialism, UK: Lawrence & Wishart, 1967.

Modification of the Bat Algorithm Using Type-2 Fuzzy Logic for Dynamical Parameter Adaptation

Jonathan Pérez, Fevrier Valdez and Oscar Castillo

Abstract We describe in this paper the Bat Algorithm and a new proposed approach using interval type-2 fuzzy systems to dynamically adapt its parameters. The Bat Algorithm (denoted in the literature as BA) is a metaheuristic inspired in micro bats based on echolocation. We analyze in detail the behavior of this proposed modification using interval type-2 fuzzy logic and compare it with type-1 fuzzy logic to compare the performance of the proposed new algorithm based on the behavior of the mega bat.

Keywords Bat algorithm · Mathematical functions · Fuzzy system

1 Introduction

The bat algorithm (BA) is a metaheuristic based on the behavior of the micro bat and proposed by Xin-She Yang in 2010 and has been applied to the solution of different types of optimization problems. In this paper we focus on the study of the BA analyzing the variants of the algorithm and the proposed modification using interval type-2 fuzzy logic system with the aim of dynamically setting some of the parameters of the BA and analyzing the results compared with a type-1 fuzzy logic system for the effects in the performance in the algorithm. In this case the modification in the BA corresponds to two parameters "Pulse Rate" and "Beta" integrated in the method. This algorithm is applied to benchmark mathematical functions and comparative tables between the two modifications are presented. The

J. Pérez · F. Valdez · O. Castillo (✉)
Tijuana Institute of Technology, Tijuana, B.C., Mexico
e-mail: ocastillo@tectijuana.mx

J. Pérez
e-mail: tecjonathan@gmail.com

F. Valdez
e-mail: fevrier@tectijuana.mx

P. Melin et al. (eds.), *Nature-Inspired Design of Hybrid Intelligent Systems*,
Studies in Computational Intelligence 667, DOI 10.1007/978-3-319-47054-2_23

main variants of the BA currently in the literature are the following in order of significance for the current work:

- BBA: A binary BA for feature selection is presented in [5], where they use the BA for feature selection using a binary version of the algorithm.
- CBAs: the Chaotic bat algorithm is presented in [2], where the proposed approach here is for global optimization, which used thirteen different chaotic maps are utilized to replace with the main parameters of the CBAs.
- HBA: A Hybrid bat algorithm is presented in [3], where it has been hybridized with differential evolution strategies.
- DLBA: A Novel Bat Algorithm Based on Differential Operator and Lévy Flights Trajectory is presented in [9], where a differential operator is introduced to accelerate the convergence speed of proposed algorithm.

This paper is organized as follows: Sect. 2 describes the original BA, Sect. 3 describes the fuzzy logic system, in Sect. 4 we describe the benchmark mathematical functions, in Sect. 5 we describe the results between type-1 fuzzy system and the type-2 fuzzy system, and Sect. 6 describes the conclusions and the future work.

2 Bat Algorithm

The BA is a metaheuristic based on the behavior of the micro bats, they use echolocation to find prey, for simplicity, we now use the following approximate or idealized rules:

1. All bats use echolocation to sense distance, ant they also 'know' the difference between food/prey and background barriers in some magical way.
2. Bats fly randomly witch velocity v_i at position x_i with a fixed frequency $f_{\min}$, varying wavelength λ and loudness A_0 to search for prey. They can automatically adjust the wavelength (or frequency) of their emitted pulses and adjust the rate of pulse emission $r \in [0, 1]$, depending on the proximity of their target.
3. Although loudness can vary in many ways, we assume that the loudness varies from a large (positive) A_0 to a minimum constant value $A_{\min}$.

For simplicity, the frequency $f \in [0, f_{\max}]$, the new solutions x_i^t and velocity v_i^t at a specific time step t are represented by a random vector drawn from a uniform distribution [3, 9].

2.1 Pseudocode for the Bat Algorithm

The basic steps of the BA, can be summarized as the pseudo code shown in Fig. 1.

```
Initialize the bat population xi(i=1, 2,..., n) and vi
Initialize frequency fi, pulse rates ri and the loudness Ai
While (t<Max numbers of iterations)
   Generate new solutions by adjusting frequency
   and updating velocities and locations/solutions [equations (1) to (3)]
   if(rand>ri)
     Select a solution among the best solutions
     Generate a local solution around the selected best solution
   end if
   Generate a new solutions by flying randomly
   if (rand <Ai& f(xi) < f(x*))
     Accept the new solutions
     Increase ri and reduce Ai
   end if
     Rank the bats and find the current best x*
end while
```

Fig. 1 Pseudo code of the bat algorithm

2.2 Movements in the Bat Algorithm

Each bat is associated with a velocity v_i^t and location x_i^t, at iteration t, in a dimensional search or solution space. Among all the bats, there exist a current best solution x_*. Therefore, the above three rules can be translated into the updating equations for x_i^t and velocities v_i^t:

$$f_i = f_{\min} + (f_{\max} - f_{\min})\beta, \tag{1}$$

$$v_i^t = v_i^{t-1} + (x_i^{t-1} - x_*)f_i, \tag{2}$$

$$x_i^t = x_i^{t-1} + v_i^t, \tag{3}$$

where $\beta \in [0, 1]$ is a random vector selected from a uniform distribution.

As mentioned earlier, we can either use wavelengths or frequencies for implementation, we will use $f_{\min} = 0$ and $f_{\max} = 1$, depending on the domain size of the problem of interest. Initially, each bat is randomly assigned a frequency which is drawn uniformly from $[f_{\min} - f_{\max}]$. The loudness and pulse emission rates essentially provide a mechanism for automatic control and auto zooming into the region with promising solutions [10].

2.3 Loudness and Pulse Rates

In order to provide an effective mechanism to control the exploration and exploitation and switch to the exploitation stage when necessary, we have to vary the loudness A_i and the rate r_i of pulse emission during the iterations. Since the loudness usually decreases once a bat has found its prey, while the rate of pulse emission increases, the loudness can be chosen as any value of convenience, between $A_{\min}$ and $A_{\max}$, assuming $A_{\min} = 0$ means that a bat has just found the prey and temporarily stop emitting any sound. With these assumptions, we have

$$A_i^{t+1} = \alpha A_i^t,\ r_i^{t+1} = r_i^0[1 - \exp(-\gamma^t)], \tag{4}$$

where α and γ are constants. In essence, here α is similar to the cooling factor of a cooling schedule in simulated annealing. For any $0 < \alpha < 1$ and $\gamma > 0$, we have

$$A_i^t \to 0, r_i^t \to r_i^0, \text{ as } \quad t \to \infty. \tag{5}$$

In the simplest case, we can use $\alpha = \gamma$, and we have used $\alpha = \gamma = 0.9$ to 0.98 in our simulations [8].

3 Fuzzy Logic System

In this paper the optimization of the parameter values of the membership functions was made with the implementation of fuzzy systems obtained using Type-1 and Type-2 fuzzy logic in the BA for the benchmark mathematical functions. In another area the optimization using fuzzy system in [4] has been applied in text optimization.

In the BA the selected parameters are: "Beta" and "Pulse Rate", the "Beta" parameters has valued between [0,1] which are increasing with the step iterations and the variable "Pulse Rate" has values between [0,1], which are decreasing with the step iterations, which are those integrated into the type-1 fuzzy system and Interval type-2 fuzzy system. In the type-1 and type-2 system the parameters "Iteration" and "Diversity" are defined as:

$$\text{Iteration} = \frac{\text{Current Iteration}}{\text{Maximun of Iterations}} \tag{6}$$

$$\text{Diversity}(S(t)) = \frac{1}{n_s}\sum_{i=1}^{n_x}\sqrt{\sum\nolimits_{j=1}^{n_x}(X_{ij}(t) - \overline{X}_j(t))^2} \tag{7}$$

The main difference between a Type-1 Fuzzy System (T1FS) and an Interval Type-2 Fuzzy System (IT2FS), is that the degree of membership in the later is also fuzzy, and is represented by the footprint of uncertainty (FOU), so if we shift from type-1 to type-2, theoretically we need a degree of footprint uncertainty, so that this degree was manually modified until the best possible FOU is obtained. The type-1 fuzzy system is shown in Fig. 2 and the type 2 fuzzy system shown in Fig. 3 based on the ideas building of fuzzy systems in the previous works [1, 6].

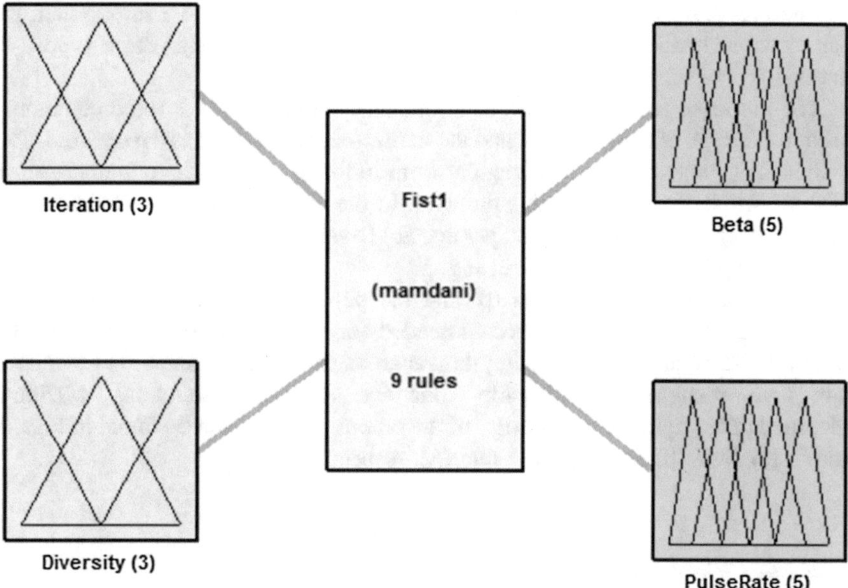

Fig. 2 Type-1 fuzzy system

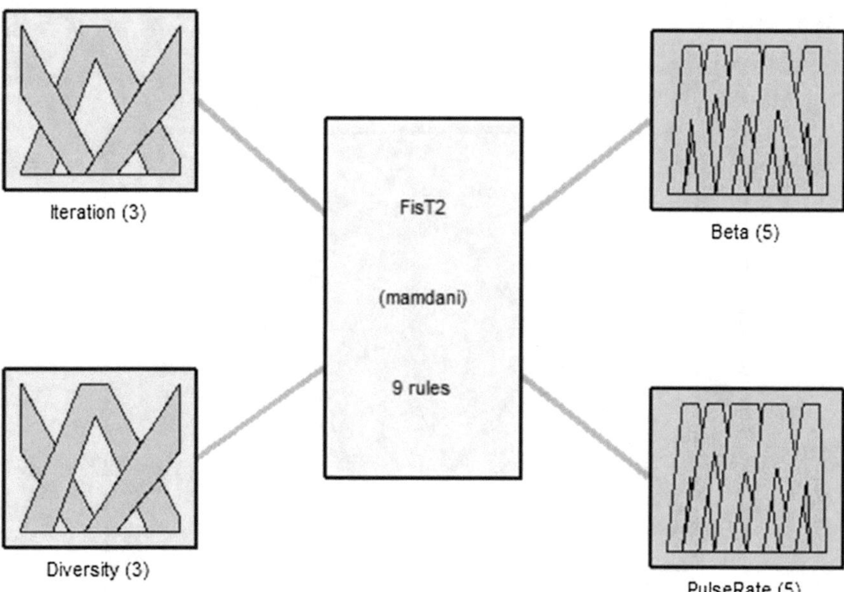

Fig. 3 Interval Type-2 fuzzy system

The Fig. 4 shows the fuzzy rule set from the original type-1 fuzzy system for parameter adaptation. These rules stay the same in the change from type-1 to interval type-2 fuzzy logic.

The proposed fuzzy system is of Mamdani type because it is more commonly used in this type of fuzzy control and the defuzzification method is the centroid. The membership functions are of triangular form in the inputs and the outputs because of their simple definition for this problem. In the "input1" variable "Iterations" as shown in Fig. 5 and the "input2" parameter "Diversity" the membership functions are of triangular form as shown in Fig. 6.

In the output variables, Beta (β) and the pulse rate (r_i), the literature values between the range of 0–1 are recommended for each of the output variables by which the same are designed using this range of values. Each output is granulated into five triangular membership functions (Low, MediumLow, Medium, MediumHigh, High), and the design of the output variables can be found in Figs. 7 and 8, for Beta (β) and the pulse rate (r_i), respectively.

```
1. If (Iteration is Low) and (Diversity is Low) then (Beta is Low)(PulseRate is High) (1)
2. If (Iteration is Low) and (Diversity is Medium) then (Beta is Low)(PulseRate is High) (1)
3. If (Iteration is Low) and (Diversity is High) then (Beta is MediumLow)(PulseRate is MediumHigh) (1)
4. If (Iteration is Medium) and (Diversity is Low) then (Beta is MediumLow)(PulseRate is MediumHigh) (1)
5. If (Iteration is Medium) and (Diversity is Medium) then (Beta is Medium)(PulseRate is Medium) (1)
6. If (Iteration is Medium) and (Diversity is High) then (Beta is MediumHigh)(PulseRate is MediumLow) (1)
7. If (Iteration is High) and (Diversity is Low) then (Beta is MediumHigh)(PulseRate is MediumLow) (1)
8. If (Iteration is High) and (Diversity is Medium) then (Beta is MediumHigh)(PulseRate is MediumLow) (1)
9. If (Iteration is High) and (Diversity is High) then (Beta is High)(PulseRate is Low) (1)
```

Fig. 4 Rule set from type-1 and interval type-2 fuzzy system

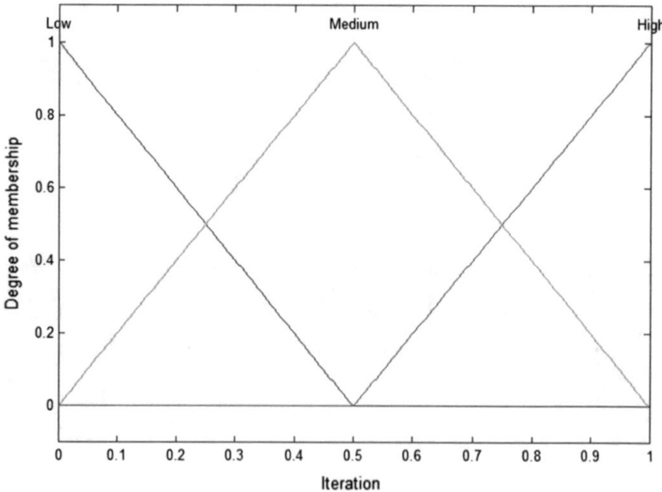

Fig. 5 Type-1 membership function for the iteration variable

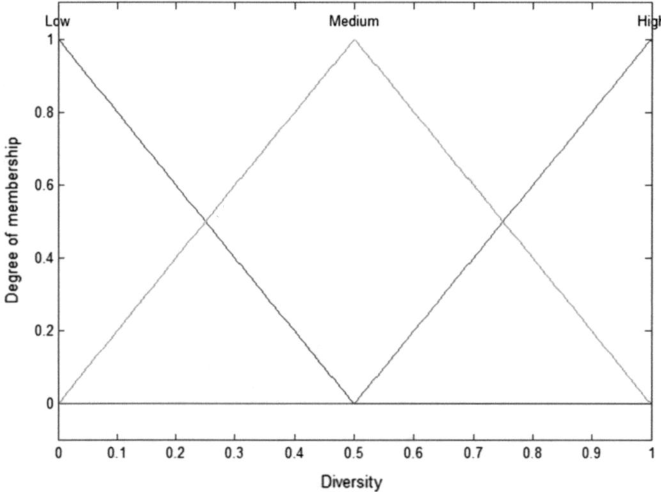

Fig. 6 Type-1 membership function for the iteration variable

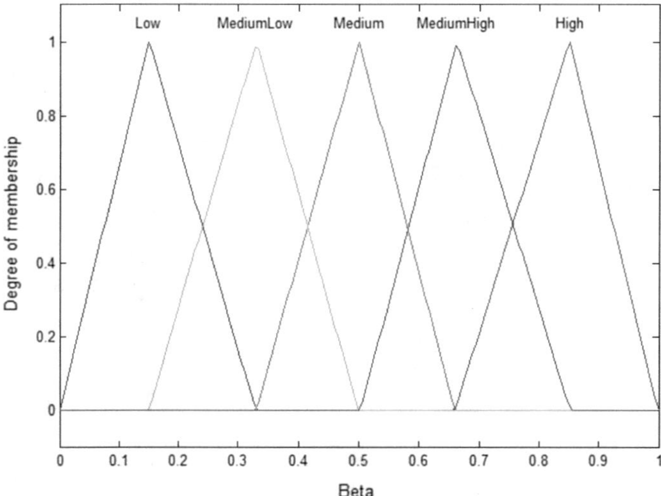

Fig. 7 Type-1 membership function for the beta parameter

The interval type-2 fuzzy system is designed similarly to the one in [7], for the parameter adaptation and is shown in Fig. 2, and we develop this system manually, this is, we change the levels of FOU of each point of each membership function, but each point has the same level of FOU, also for the input and output variables we have only interval type-2 triangular membership functions.

The membership functions are of trapezoidal form in the inputs and the outputs because of their simple definition for this problem. In the "input1" for the

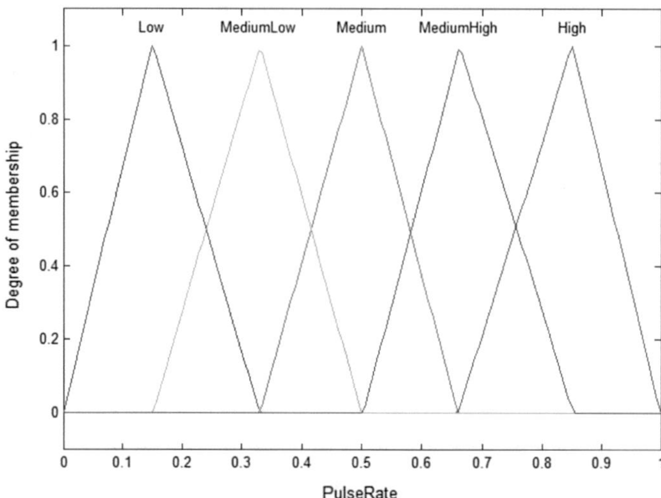

Fig. 8 Type-1 membership function for the pulse rate parameter

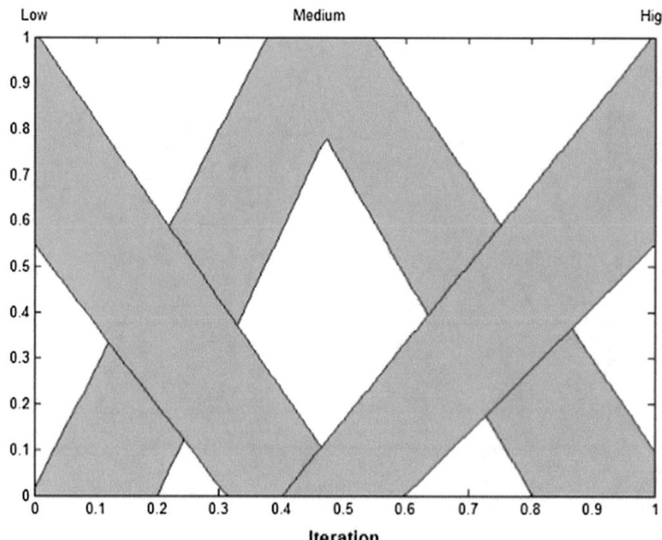

Fig. 9 Interval type-2 membership functions of the iteration variable

"Iterations" variable as shown in Fig. 9 and the "input2" for the "Diversity" the membership functions are of triangular form as shown in Fig. 10.

In the output variables, Beta (β) and the pulse rate (r_i) are granulated into five triangular membership functions (Low, LowMedium, Medium, MediumHigh, High), and the design of the output variables can be found in Figs. 11 and 12, for Beta (β) and the pulse rate (r_i), respectively.

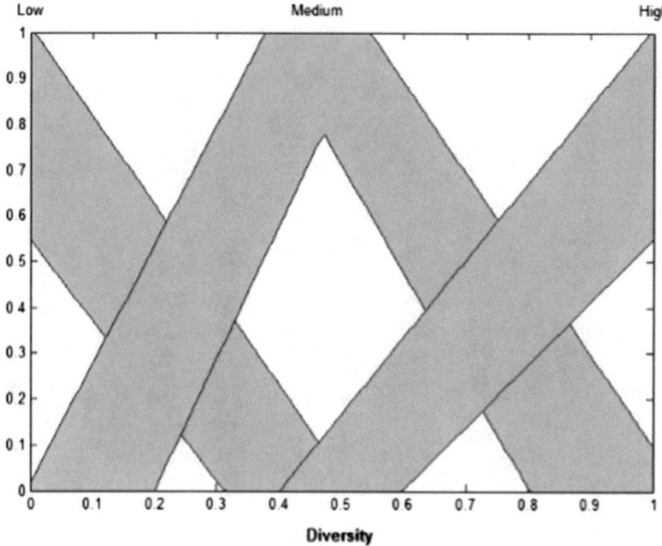

Fig. 10 Interval type-2 membership function of the diversity variable

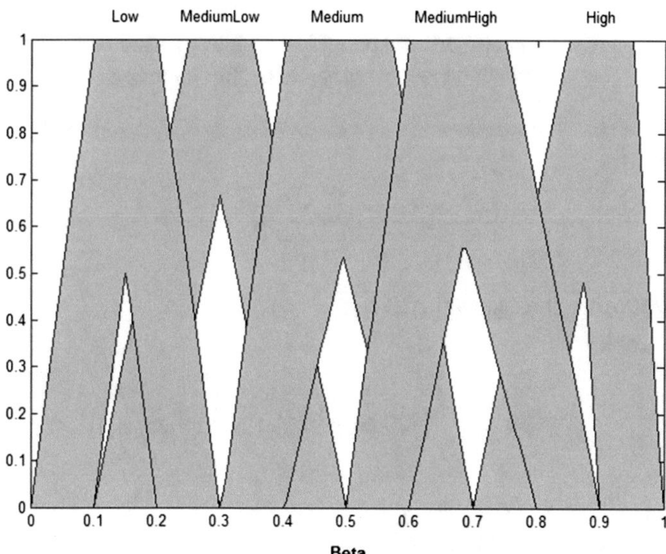

Fig. 11 Interval type-2 membership functions of the beta parameter

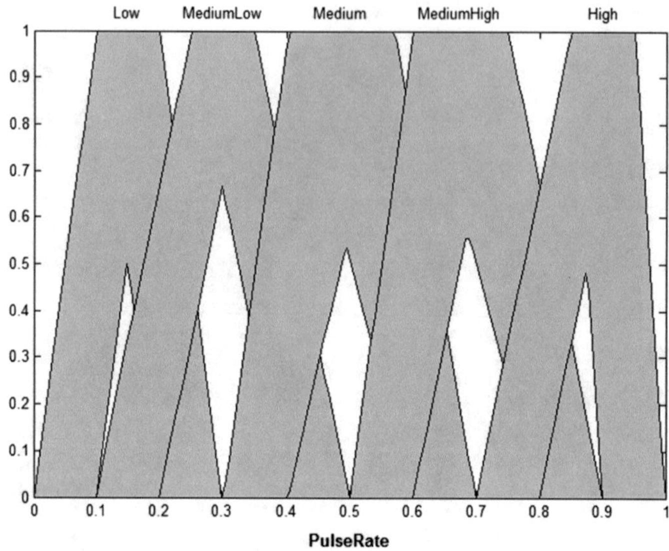

Fig. 12 Interval type-2 membership functions of the pulse rate parameter

4 Benchmark Mathematical Functions

This section lists the benchmark mathematical functions used to evaluate the performance of the optimization algorithms, and the mathematical functions are defined below:

- **Sphere**

$$f(x) = \sum_{j=1}^{n_x} x_j^2 \tag{8}$$

with $x_j \in [-100, 100]$ and $f^*(x) = 0.0$

- **Rosenbrock**

$$f(x) = \sum_{j=1}^{n_z/2} [100(x_{2j} - x_{2j-1}^2)^2 + (1 - x_{2j-1})^2] \tag{9}$$

with $x_j \in [-2.048, 2.048]$ and $f^*(x) = 0.0$

- **Axis Parallel Hyper-Ellipsoid**

$$f(x) = \sum_{i=1}^{n} i \cdot x_i^2 \tag{10}$$

with $x_j \in [-5.12, 5.12]$ and $f^*(x) = 0.0$

- **Griewank**

$$f(x) = \sum_{i=1}^{d} \frac{x_i^2}{4000} - \prod_{i=1}^{d} \cos\left(\frac{x_i}{\sqrt{i}}\right) + 1 \tag{11}$$

with $x_j \in [-600, 600]$ and $f^*(x) = 0.0$

The mathematical functions were integrated directly into the code of the BA

5 Results Between Type-1 and the Interval Type-2 for the Bat Algorithm

In this section the BA is applied and compare against the Type-1 and Interval Type-2Fuzzy System. In each of the algorithms, 4 Benchmark math functions were used separately for dimensions of 100, 150, 200 and 256 variables, and the parameters in the BA are obtained based on previous works [8].

The results of the tests made with the five functions for the BA with a type-1 fuzzy system are shown in Table 1.

The results of the tests made with the five functions for the BA with an interval type-2 fuzzy system are shown in Table 2.

In the above table it can be noted in the means that when we increase the number of dimensions the problem is more complex to reach the best value this shows the effectiveness.

The application of type 2 fuzzy systems in the BA as shown in the above tables could obtain the best values, in this case Type-1 and type-2 get the bests values.

Table 1 Simulation results for the type-1 fuzzy system

Function		Dimensions			
		100	150	200	256
Sphere	Best	0.00012913	0.00030837	0.00056627	0.00095417
	Mean	0.67533244	2.36897875	1.67918428	6.71745662
	Worst	20.2556499	67.5344513	49.6947439	161.569895
Rosenbrock	Best	5.211E−11	2.0279E−10	8.5742E−12	5.1827E−11
	Mean	1.344E−10	5.7476E−10	8.9226E−11	4.4564E−10
	Worst	4.4177E−10	1.2511E−09	5.4853E−10	2.6897E−09
Axis parallel hyper-ellipsoid	Best	1.0753E−11	2.2491E−12	2.1031E−11	7.5661E−12
	Mean	4.2956E−11	3.3677E−11	2.8646E−11	2.627E−11
	Worst	4.4865E−10	9.4507E−10	1.4546E−10	1.8604E−10
Griewank	Best	1.9482E−06	4.6829E−06	11.4855214	28.4378093
	Mean	3.21452715	160.909937	408.770743	486.592714
	Worst	37.4759221	579.126662	1092.01259	1231.51904

Table 2 Simulation results for the interval type−2 fuzzy system

Function		Dimensions			
		100	150	200	256
Sphere	Best	0.00013919	0.00033171	0.00055865	0.00094694
	Mean	82.6940987	161.762658	443.937531	56.020309
	Worst	1119.5465	1819.41804	3689.18143	839.697861
Rosenbrock	Best	9.1337E−11	1.026E−10	1.6376E−10	2.7268E−11
	Mean	2.7781E−10	3.6283E−10	3.7201E−10	7.8791E−11
	Worst	4.2513E−09	3.0177E−09	1.6215E−09	1.5224E−10
Axis parallel hyper-ellipsoid	Best	9.6233E−12	2.6257E−13	7.2596E−12	5.7609E−13
	Mean	2.842E−11	1.5235E−11	9.9284E−12	2.4434E−11
	Worst	1.2426E−10	1.7555E−10	3.3947E−11	3.6971E−10
Griewank	Best	0.00985948	2.229E−06	2.4232E−06	2.3291E−06
	Mean	55.9165128	95.3839741	90.4612914	79.3224917
	Worst	284.027694	409.752717	395.097708	360.824881

This does not mean that one is better than the other, if not this shows that the application of type 2 in the changing parameters in the BA is effective. Type-2 is better when we increase the complexity of the problem with the increase in the dimensions where we can attack more complex problems certainty skilled to find good solutions integrating in this case a higher level uncertainty.

6 Conclusion and Future Work

The application of the BA to various problems has been a very wide field, where its effectiveness is demonstrated in various applications, and their use can be mentioned in the processing digital pictures, search for optimal values, neural networks, and many applications.

Future work analyzing the results obtained here consists in the construction of a new algorithm based on the behavior of the mega bat, which as main feature has the application of vision that the bat uses for orientation in space search.

Acknowledgment We would like to express our gratitude to the CONACYT and Tijuana Institute of Technology for the facilities and resources granted for the development of this research.

References

1. Amador-Angulo L., Castillo O., Statistical Analysis of Type-1 and Interval Type-2 Fuzzy Logic in dynamic parameter adaptation of the BCO. IFSA-EUSFLAT 2015.

2. Gandomi A., Yang X. S., Chaotic bat algorithm, Department of Civil Engineering, The University of Akron, USA, 2013.
3. Goel N., Gupta D., Goel S., Performance of Firefly and Bat Algorithm for Unconstrained Optimization Problems, Department of Computer Science Maharaja Surajmal Institute of Technology GGSIP university C-4, Janakpuri, New Delhi, India, 2013.
4. Madera Q., García-Valdez M., Castillo O., Fuzzy Logic for Improving Interactive Evolutionary Computation Techniques for Ad Text Optimization." in Novel Developments in Uncertainty Representation and Processing, Advances in Intelligent Systems and Computing, pages. Springer, 291-300.
5. Nakamura R., Pereira L., Costa K., Rodrigues D., Papa J., BBA: A Binary Bat Algorithm for Feature Selection, Department of Computing Sao Paulo State University Bauru, Brazil, 2012.
6. Olivas F., Valdez F., Castillo O., Patricia Melin, Dynamic parameter adaptation in particle swarm optimization using interval type-2 fuzzy logic. Soft Comput. 20(3): 1057-1070 (2016).
7. Olivas F., Valdez F., Castillo O., Dynamic parameter adaptation in Ant Colony Optimization using a fuzzy system for TSP problems, IFSA-EUSFLAT 2015, Gijón, Asturias (Spain).
8. Pérez J., Valdez F., Castillo O., Bat Algorithm Comparison with Genetic Algorithm Using Benchmark Functions. Recent Advances on Hybrid Approaches for Designing Intelligent Systems 2014: 225-237.
9. Yang X. S., A New Metaheuristic Bat-Inspired Algorithm, Department of Engineering, University of Cambridge, Trumpington Street, Cambridge CB2 1PZ, UK, 2010.
10. Yang X. S., Bat Algorithm: Literature Review and Applications, School of Science and Technology, Middlesex University, The Burroughs, London NW4 4BT, United Kingdom, 2013.

Flower Pollination Algorithm with Fuzzy Approach for Solving Optimization Problems

Luis Valenzuela, Fevrier Valdez and Patricia Melin

Abstract In this paper, we present a new hybrid approach of flower pollination algorithm (FPA). This is a Bio-Inspired technique based on the pollination process carried out by the flowers. We used a Fuzzy inference system to adapt the probability of switching and this is the mechanism by which there is a change of global and local pollination; thus, the algorithm can explore and exploit in a different way to the original method. To validate in the best way the proposed method we present a comparison results among different optimization algorithms to evaluate the performance using a set of benchmark mathematical functions.

Keywords Flower pollination algorithm · Fuzzy inference systems · Mathematical functions · Nature-Inspired algorithms

1 Introduction

Today optimization problems face the fact that these are increasingly more complex and this is why researchers continually develop new methods for solving these issues. And these new methodologies should cover more variables that allow solving the constraints that arise. In recent years, researchers have proposed new methodologies which improve the results of traditional optimization algorithms. Some of these methodologies are the nature-inspired algorithms whose main characteristic is the ability of biological systems to adapt to constant changes in the environment.

Optimization can be found everywhere and, therefore, represents an important paradigm for a lot of applications. We can find applications in industry and engineering which are often trying to optimize something, either minimize cost, energy consumption or to maximize profit, production, performance, and efficiency.

L. Valenzuela · F. Valdez (✉) · P. Melin
Tijuana Institute of Technology, Tijuana, B.C., Mexico
e-mail: fevrier@tectijuana.mx

P. Melin et al. (eds.), *Nature-Inspired Design of Hybrid Intelligent Systems*,
Studies in Computational Intelligence 667, DOI 10.1007/978-3-319-47054-2_24

Resources such as time and money are limited; therefore, optimization plays a crucial role in practice [15].

Fuzzy logic is based on the theory of fuzzy sets proposed by Zadeh [16]. Recently, the use of fuzzy logic in nature-inspired algorithms has been used as an approach to improve the performance of these algorithms [7].

2 Flower Pollination Algorithm

The flower pollination algorithm (FPA) was developed by Xin-She Yang in 2012 [14].

2.1 Pollination

Inspired by the pollination process carried out by flowering plants, in this process we can distinguish two types of pollination.

2.1.1 Self-pollination

When the pollen from a flower pollinates the same flower or flowers of the same plant, the process is called self-pollination. It occurs when a flower contains both the male and the female gametes [5] (Fig. 1).

2.1.2 Cross-pollination

Cross-Pollination occurs when pollen grains are moved to a flower from another plant. The process of cross-pollination happens with the help of abiotic or biotic agents, such as insects, birds, snails, bats, and other animals as pollinators. Abiotic pollination is a process where the pollination happens without involvement of external agent [5].

2.1.3 Pollination Process in the Algorithm

Inspired by the flow pollination process of flowering plants obeying the following rules:

Rule 1: Biotic cross-pollination can be considered as a process of global pollination, and pollen carrying pollinators move in a way that obeys Lévy flights.

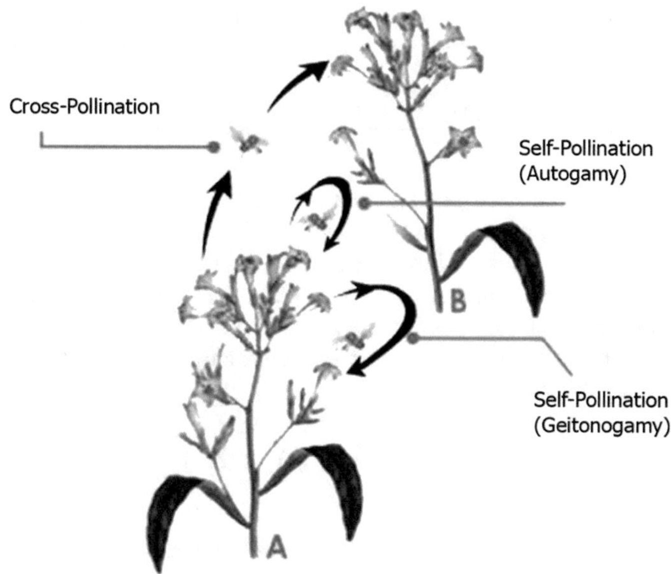

Fig. 1 Pollination [12]

Rule 2: For local pollination, abiotic pollination and self-pollination are used.

Rule 3: Pollinators such as insects can develop flower constancy, which is equivalent to a production probability that is proportional to the similarity of two flowers involved.

Rule 4: The interaction or switching of local pollination and global pollination can be controlled by a switch probability $p \in [0,1]$, slightly biased toward local pollination.

In order to formulate the updating formulas for the algorithm, the above rules have to be converted into proper updating equations. For example, in the global pollination step, flower pollen gametes are carried by pollinators, such as insects, and pollen can travel over along distance because insects can often fly and move over a much longer range. Therefore, Rule 1 and flower constancy (Rule 3) can be represented mathematically as

$$x_i^{t+1} = x_i^t + L(\lambda)(G_* - x_i^t) \tag{1}$$

where x_i^t is the pollen i or solution vector x_i at iteration t, and g_* is the current best solution found among all solutions at the current generation/iteration. Here γ is a scaling factor to control the step size.

Here $L(\lambda)$ is the parameter, more specifically the Lévy flights-based step size, that corresponds to the strength of the pollination. Since insects may move over a long distance with various distance steps, a Lévy flight can be used to mimic this characteristic efficiently. That is, $L > 0$ is drawn from a Lévy distribution.

$$L \sim \frac{\lambda \Gamma(\lambda) \sin(\pi/2)}{\pi} \frac{1}{s^{1+\lambda}}, \quad (s \gg s_0 > 0). \tag{2}$$

Here $\Gamma(\lambda)$ is the standard gamma function, and this distribution is valid for large steps s > 0. In theory, it is required that $|s_0| \gg 0$, but in practice s0 can be as small as 0.1. However, it is not trivial to generate pseudorandom step sizes that correctly obey this Lévy distribution Eq. (2). There are a few methods for drawing such random numbers, and the most efficient one from our studies is the so-called Mantegna algorithm for drawing step size s by using two Gaussian distributions U and V by the following transformation

$$s = \frac{U}{|V|^{1/\lambda}}, \quad U \sim N\left(0, \sigma^2\right), \quad V \sim N(0, 1) \tag{3}$$

Here $U \sim N(0, \sigma^2)$ means that the samples are drawn from a Gaussian normal distribution with a zero mean and a variance of σ^2. The variance can be calculated by

$$\sigma^2 = \left[\frac{\Gamma(1+\lambda)}{\lambda \Gamma\left(\frac{1+\lambda}{2}\right)} \cdot \frac{\sin(\pi\lambda/2)}{2^{(\lambda-1)/2}}\right]^{1-\lambda}, \tag{4}$$

For local pollination, both Rules 2 and 3 can be represented as follows

$$x_i^t = x_i^t + \in (x_j^t - x_k^t), \tag{5}$$

where x_j^t and x_k^t are pollen from different flowers of the same plants species. This essentially mimics flower constancy in a limited neighborhood. Mathematically, if x_j^t and x_k^t come from the same species or are selected from the same population; this equivalently becomes a local random walk if $\in$ is drawn from a uniform distribution in [0, 1].

In principle, flower pollination activities can occur at all scales, both local and global. But in reality, adjacent flower patches or flowers in the not-so-far-away neighborhood are more likely to be pollinated by local flower pollen tan those far away. In order to mimic this feature, a switch probability (Rule 4) or proximity probability p can be effectively used to switch between common global pollination to intensive local pollination. To start with, a naive value of $p = 0.5$ may be used as an initial value. A preliminary parametric study showed that $p = 0.8$ may work better for most applications [14].

2.2 Pseudocode Flower Pollination Algorithm

The pseudocode for FPA is presented below:

Define Objective function f (x), x = (x1, x2,…, xd)
Initialize a population of n flowers/pollen gametes with random solutions Find the best solution B in the initial population
Define a switch probability p ∈ [0, 1]
Define a stopping criterion (either a fixed number of generations/iterations or accuracy)
while (t < MaxGeneration)
for i = 1: n (all n flowers in the population)
if rand < p,
Draw a (d-dimensional) step vector L which obeys a Lévy distribution
Global pollination via $x_i^{t+1} = x_i^t + L(B - x_i^t)$
Else
Draw U from a uniform distribution in [0,1]
Do local pollination via $x_i^{t+1} = x_i^t + U(x_j^t - x_i^t)$
End if
Evaluate new solutions
If new solutions are better, update them in the population
End for
Find the current best solution B
End while
Output the best solution found

2.3 Applications

Based on characteristics of flower pollination described above, Xin-She Yang developed the Flower pollination algorithm (FPA) in 2012, and this is why in the literature we can find few applications among which we highlight “Hybrid Flower Pollination Algorithm with Chaotic Harmony Search for Solving Sudoku Puzzles” [1],“Training Feedforward Neural Network using Hybrid Flower Pollination-Gravitational Search Algorithm” [3], “A Study on Flower Pollination Algorithm and Its Applications” [5], “Nature Inspired Flower Pollen Algorithm For WSN Localization Problem” [4], “A Comparative Study of Flower Pollination Algorithm and Bat Algorithm on Continuous Optimization Problems” [8], “Implementing Flower Multi-objective Algorithm for selection of university academic credits” [9], DE-FPA: A Hybrid Differential Evolution-Flower Pollination Algorithm for Function Minimization [2].

3 Proposed Method FFPA

For this paper, we propose a fuzzy system that allows dynamically update the *p*-value which is responsible for switching between local and pollination global. It was shown that $p = 0.8$ might work better for most applications [14].

3.1 Flowchart

As we can see in Fig. 2, in the first step we define the search space, population size, number of iterations, switching commutation, and find de current best. If a random number is less than *p* we execute global pollination else local pollination, evaluate new solutions, and if these are best we update population. Then, we added our fuzzy system after evaluate the entire population in each iteration. We can see in Fig. 3 the input of the fuzzy system. In this case, we use the iterations to represent the input variable. In Fig. 4 we show the output variable to represent the change in the adjusted parameter.

3.2 Input and Output of the Fuzzy System

In this case, we propose a fuzzy system for the FFPA of Mamdani type with 1-input, 1-output. To define our method, previously, we tested by increasing and decreasing the value *p*; this experiment showed that by decreasing *p* we obtained better performance.

So we have used four rules described in Fig. 5 and granulated in *low*, *medium-low*, *medium-high,* and *high.*

We use Eq. 6 to calculate the percentage of iterations and this is the input to the fuzzy system (Figs. 3 and 4).

$$\text{iteration} = \frac{\text{current\_iteration}}{\text{max\_iterations}} \tag{6}$$

4 Benchmark Mathematical Functions

For the experiments we used eight benchmark mathematical functions. For this, we have relied on [11, 6, 13].

In this section, we list some the classical benchmark functions used to validate optimization algorithms.

To validate the method we used a set of eight benchmark mathematical functions, called Spherical, Ackley, Rastrigin, Zakharov, Griewank, Sum of different powers, Michalewitz, and Rosenbrock (*only second experiment*).

The mathematical functions are shown in Table 1.

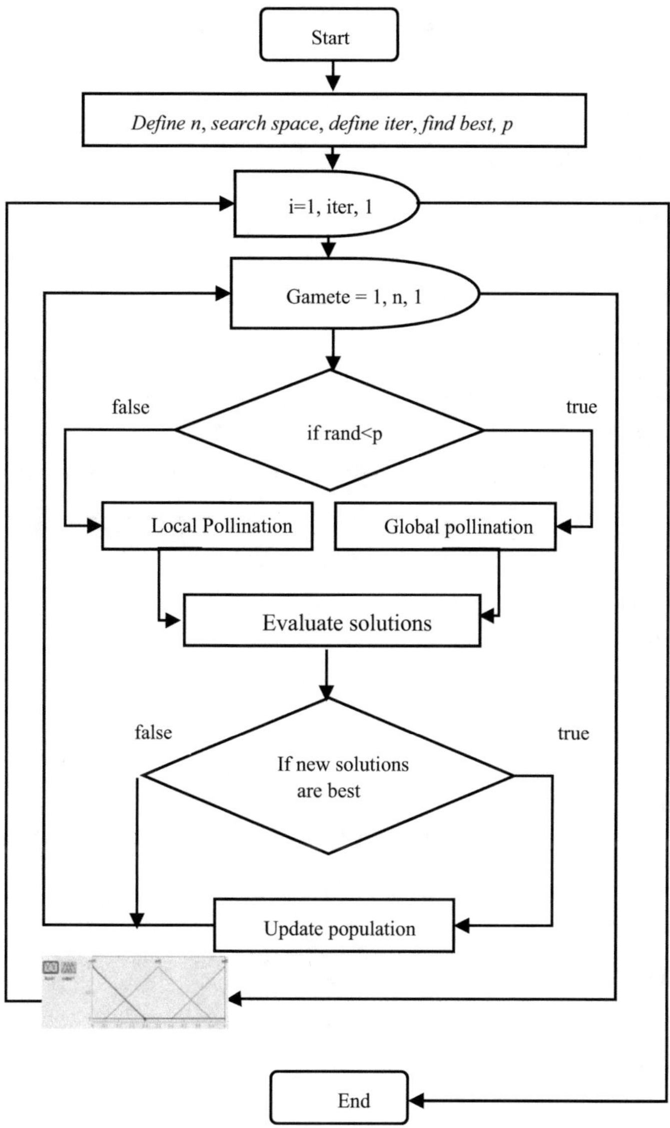

Fig. 2 Flowchart

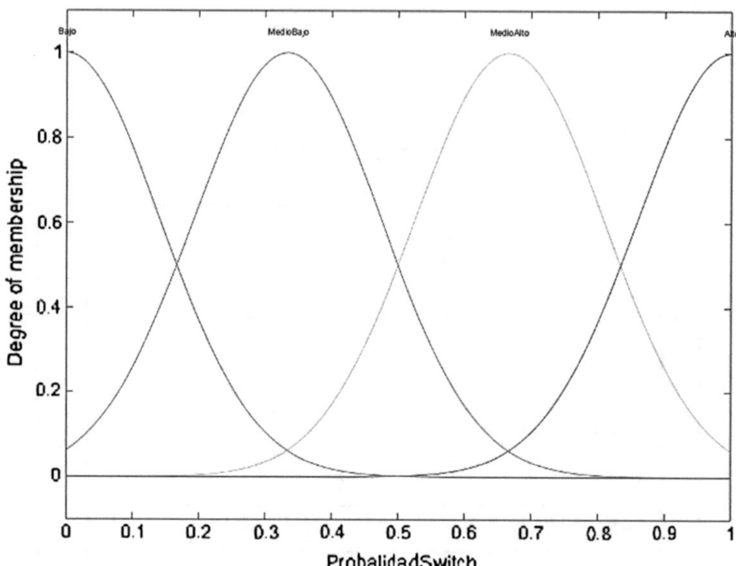

Fig. 3 Input-iteration (%)

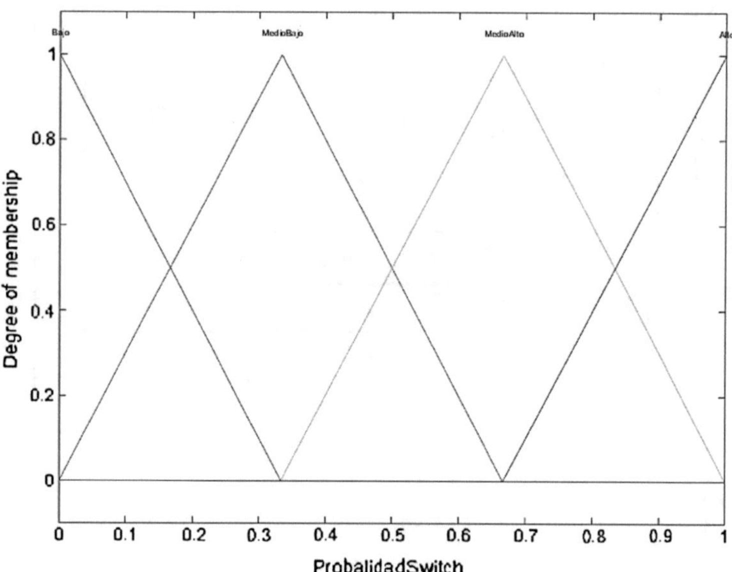

Fig. 4 Output-*p*

Fig. 5 Rules set

1. If (Iteration is low) then (p is high)
2. If (Iteration is medLow) then (p is medHigh)
3. If (Iteration is medHigh) then (p is medLow)
4. If (Iteration is high) then (p is low)

Table 1 8 Mathematical functions used to test the method

No.	Function	Range [MIN, MAX]	Act. Opt.
1.	Sphere, $f1(x)$	[−5.12, 5.12]	0
2.	Ackley, $f2(x)$	[−30, 30]	0
3.	Rastrigin, $f3(x)$	[−5.12, 5.12]	0
4.	Zakharov, $f4(x)$	[−5, 10]	0
5.	Griewank, $f5(x)$	[−600, 600]	0
6.	Sum of different powers, $f6(x)$	[−1, 1]	0
7.	Michalewitz, $f7(x)$	[0, π]	−0.966d
8.	Rosenbrock, $f8(x)$	[−5, 10]	0

5 Experiment Results

In this section, we present the results obtained of the FPA and FFPA, compared with Differential Evolution, GA, and PSO. For the first comparison we take the results from [6], and for the second one we performed the corresponding experiments.

5.1 First Experiment

In the first experiment, we used the parameters of the paper mentioned above and performed the corresponding tests for the FPA and the proposed FFPA with the following results. The parameters are shown in Table 2.

Table 2 Test parameters

No.	Parameters	Value
1.	Iterations	2000
2.	Population size, n	40
3.	Dimensions number, d	30
4.	Crossover probability GA	0.7
5.	Mutation probability GA	0.01
6.	Crossover probability DE	0.8
7.	Differential weight DE	0.3
8.	Inertia weight PSO	0.7
9.	Acceleration constant PSO	2
10.	Random weights PSO	[0, 1]
11.	Maximum speed PSO	Midrange
12.	Maximum position PSO	Pool range
13.	Experiments number	100

Table 3 In this first test we use 30 dimensions

Func.	Act. Opt.	Method	Best	Worst	Ave.
$f1(x)$	0	GA	0.0008	0.1978	0.0535
		DE	0.004	0.3654	0.0932
		PSO	0.0099	0.0719	0.0237
		FPA	0.0302	0.3213	0.1255
		Fuzzy FPA	0.0073	0.0224	0.0152
$f2(x)$	0	GA	0.0169	3.6098	0.6513
		DE	0.5545	4.5842	2.2593
		PSO	0.8398	2.3545	1.7527
		FPA	2.6333	4.4619	3.6088
		Fuzzy FPA	0.0010	4.5938	2.6244
$f3(x)$	0	GA	0.0563	19.1326	4.6809
		DE	5.0154	12.3525	9.7992
		PSO	46.1047	116.193	79.3988
		FPA	64.1695	95.6657	85.2551
		Fuzzy FPA	45.8711	85.7563	63.6051
$f4(x)$	0	GA	0.2972	1.6587	0.7395
		DE	0.0649	1.9861	0.6328
		PSO	0.0613	0.6920	0.2835
		FPA	3.28738	23.2394	13.4698
		Fuzzy FPA	1.5580	7.5484	4.8033
$f5(x)$	0	GA	2414.34	2420.59	2416.512
		DE	91.0157	206.649	164.3646
		PSO	361.0020	1081	621.9848
		FPA	1.1570	2.1337	1.6363
		Fuzzy FPA	0.8348	0.8348	1.0556
$f6(x)$	0	GA	1.00E−06	5.41E−04	1.03E−04
		DE	8.94E−11	1.97E−06	3.12E−07
		PSO	1.65E−16	2.45E−14	5.24E−15
		FPA	3.72E−15	4.84E−12	8.64E−13
		Fuzzy FPA	1.23E−15	9.28E−13	1.4594E−13
$f7(x)$	−28.98	GA	−24.9448	−22.3162	−23.9686
		DE	−23.7439	−17.0515	−20.0908
		PSO	−22.9425	−14.9033	19.0958
		FPA	−15.7955	−13.7761	−14.6617
		Fuzzy FPA	−18.7786	−15.6142	−16.7907

As we can note in Table 3, the GA obtained better results by winning in three of the seven Benchmark functions, on the hand PSO and FFPA only got better results in 2 of them.

5.2 Second Experiment

In this case in Table 4, we used 100 dimensions and compared ourselves to GA, PSO, FPA, and FFPA. These are the results, as we can see the FFPA achieved better results in half of the functions (Sphere, Griewank, Michalewicz, Rosenbrock). On the other hand FPA and PSO with only two functions.

Table 4 In the second experiment we increased the complexity of the problem with 100 dimensions

Func.	Act. Opt.	Method	Best	Worst	Ave.
$f1(x)$	0	GA	32.5424	53.6624	47.1804
		PSO	13.7761	14.5472	−14.3083
		FPA	5.9610	15.0828	9.2921
		Fuzzy FPA	0.0668	4.1207	2.0055
$f2(x)$	0	GA	5.0154	12.3525	9.7992
		PSO	1.9744	4.2044	2.1508
		FPA	3.6392	6.8716	5.0354
		Fuzzy FPA	2.7539	5.0853	3.9620
$f3(x)$	0	GA	271.0810	317.2890	296.7937
		PSO	45.8711	85.7563	66.0546
		FPA	225.0241	421.9527	317.2890
		Fuzzy FPA	226.9238	391.0345	321.9301
$f4(x)$	0	GA	1089.3483	1100.4062	1096.3845
		PSO	361.0020	621.9848	583.8823
		FPA	302.9564	880.3748	543.8928
		Fuzzy FPA	323.6544	1030.5314	553.6868
$f5(x)$	0	GA	43.4633	55.9846	47.6626
		PSO	164.3646	206.6490	361.0020
		FPA	7.4012	20.7380	13.3170
		Fuzzy FPA	6.8286	22.1806	13.4716
$f6(x)$	0	GA	93.6477	240.5874	205.0356
		PSO	472.5683	501.4673	480.3562
		FPA	3.515E−17	7.7808E−10	1.8078E−11
		Fuzzy FPA	5.9453E−17	8.487E−11	2.853E−12
$f7(x)$	−96.60	GA	−23.5645	−10.5634	−20.1547
		PSO	−44.3251	15.4451	−40.8757
		FPA	−40.3673	−29.1310	−33.4871
		Fuzzy FPA	−64.0829	−27.7103	−53.7921
$f8(x)$	0	GA	20473.7437	25041.6674	21377.8401
		PSO	62488.8044	84673.3572	71974.6825
		FPA	18140.031	63521.2689	33108.9655
		Fuzzy FPA	15662.2387	52351.933	32232.2923

6 Conclusions

In this paper, we present a modification to the flower pollination algorithm through a fuzzy inference system that enables to dynamically set the value of p. This strategy allows the algorithm to have good results as it helps the algorithm to jump off local minima in the last iterations, as there is greater possibility of performing global pollination.

By analyzing this information and the previous study, we can argue that the fuzzy flower pollination algorithm is a method easy to implement, and based on the results shown in Tables 3 and 4 with Benchmark mathematical functions, compared to other methods we can realize works best in more complex problems.

We are confident that our results can be further improved by adding diversity as input to our fuzzy inference system.

As future work we shall improve the performance of fuzzy flowers pollination algorithm, this by calculating diversity. This strategy has already been used in [10]. Test our method in time series prediction with neural networks and other problems, and publish at least one article in a national or international journal.

References

1. Abdel-Raouf, O. et al.: A Novel Hybrid Flower Pollination Algorithm with Chaotic Harmony Search for Solving Sudoku Puzzles. Int. J. Eng. Trends Technol. 7, 3, 126–132 (2014).
2. Chakraborty, D. et al.: DE-FPA : A Hybrid Differential Evolution-Flower Pollination Algorithm for Function Minimization. In: High Performance Computing and Applications. pp. 1–6 IEEE, Bhubaneswar, (2014).
3. Chakraborty, D. et al.: Training Feedforward Neural Networks using Hybrid Flower Pollination-Gravitational Search Algorithm. Int. Conf. Futur. trend Comput. Anal. Knowl. Manag. 261–266 (2015).
4. Harikrishnan, R. et al.: Nature Inspired Flower Pollen Algorithm For WSN Localization Problem. ARPN J. Eng. Appl. Sci. 10, 5, 2122–2125 (2015).
5. Kamalam, B., Karnan, M.: A Study on Flower Pollination Algorithm and Its Applications. Int. J. Appl. or Innov. Eng. Manag. 3, 11, 230–235 (2014).
6. Lim, S.P., Haron, H.: Performance Comparison of Genetic Algorithm, Differential Evolution and Particle Swarm Optimization Towards Benchmark Functions. Open Syst. 41–46 (2013).
7. Melín, P. et al.: Optimal design of fuzzy classification systems using PSO with dynamic parameter adaptation through fuzzy logic. (2013).
8. Nazmus, S. et al.: A Comparative Study of Flower Pollination Algorithm and Bat Algorithm on Continuous Optimization Problems. Int. J. Appl. Inf. Syst. 7, 9, 13–19 (2014).
9. Ochoa, A. et al.: Implementing Flower Multi-objective Algorithm for selection of university academic credits. In: Nature and Biologically Inspired Computing. pp. 7–11 IEEE, Porto, (2014).
10. Olivas, F. et al.: Ant Colony Optimization with Parameter Adaptation Using Fuzzy Logic for TSP Problems. In: Design of Intelligent Systems Based on Fuzzy Logic, Neural Networks and Nature-Inspired Optimization. pp. 593–603 Springer, Gewerbestrasse, (2015).
11. Olivas, F., Castillo, O.: Particle Swarm Optimization with Dynamic Parameter Adaptation Using Fuzzy Logic for Benchmark Mathematical Functions. In: Recent Advances on Hybrid Intelligent Systems. pp. 247–258 Springer (2013).

12. Pearson: Pollination, http://biology.tutorvista.com/plant-kingdom/pollination.html.
13. Solano-Aragón, C., Castillo, O.: Optimization of Benchmark Mathematical Functions Using the Firefly Algorithm with Dynamic Parameters. In: Fuzzy Logic Augmentation of Nature-Inspired Optimization Metaheuristics: Theory and Applications. pp. 81–89 Springer (2015).
14. Yang, X.-S. et al.: Flower pollination algorithm: A novel approach for multiobjective optimization. Eng. Optim. DOI:10.1080/0305215X.2013.832237, (2013).
15. Yang, X.-S. et al.: Swarm Intelligence and Bio-Inspired Computation: Theory and applications. Elsevier, London, UK (2013).
16. Zadeh, L.A.: Fuzzy sets information and control. Elsevier, California, USA (1965).

A Study of Parameters of the Grey Wolf Optimizer Algorithm for Dynamic Adaptation with Fuzzy Logic

Luis Rodríguez, Oscar Castillo and José Soria

Abstract The main goal of this paper is to present a general study of the Grey Wolf Optimizer algorithm. We perform tests to determine in the first part which parameters are candidates to be dynamically adjusted and in the second stage to determine which are the parameters that have the greatest effect in the algorithm. We also present a justification and results of experiments as well as the benchmark functions that were used for the tests that are shown.

Keywords Performance · GWO · Benchmark functions · Adaptation of parameters

1 Introduction

Nowadays computer science is solving problems through their different areas, depending on the complexity of the problem and the results expected from the idea to optimize (minimize or maximize) the solution to a problem.

Researchers lately have turned to metaheuristics can have superior abilities than conventional optimization techniques, and this is mainly due to their ability to avoid a local optimum result.

Optimization techniques based on metaheuristics have become very popular over the past two decades for example Genetic Algorithm (GA) [1], Particle Swarm Optimization (PSO) [2], Ant Colony Optimization (ACO) [3], and Artificial Bee Colony (ABC) [4]. In addition to the huge amount of theoretical work, these optimization techniques have been used in various areas of application. The answer to why they have become so popular can be summarized into four main reasons: simplicity, flexibility, derivative free mechanism, and because they have more ability to avoid local optima.

L. Rodríguez · O. Castillo (✉) · J. Soria
Tijuana Institute of Technology, Tijuana, Mexico
e-mail: ocastillo@tectijuana.mx

P. Melin et al. (eds.), *Nature-Inspired Design of Hybrid Intelligent Systems*,
Studies in Computational Intelligence 667, DOI 10.1007/978-3-319-47054-2_25

This paper is organized as follows: Sect. 2 shows the Grey Wolf Optimizer algorithm. Section 3 describes the Benchmark Functions, Sect. 4 shows experiments and Results and Sect. 5 shows the Conclusions.

2 Grey Wolf Optimizer

The Grey Wolf Optimizer algorithm (GWO) [5] is a metaheuristic that was originated in 2014 created by Seyedali Mirjalili, inspired basically because there exists in the literature a Swarm Intelligence (SI) [6] technique that mimics the hierarchy of leadership of the Grey Wolf.

Metaheuristics techniques can be classified as follows:

- Evolutionary (based on concepts of evolution in nature) [7]: Genetic Algorithms (GA), Evolutionary Programming (EP), and Genetic Programming (GP).
- Based on Physics (imitates the physical rules) [8]: Big-Bang Big Crunch (BBBC), Gravitational Search Algorithm (GSA), and Artificial Chemical Reactions Optimization Algorithm (ACROA).
- Swarm Intelligence (Social Behavior of swarms, herds, flocks or schools of creatures in nature) [9]: Particle Swarm Optimization (PSO), Ant Colony Optimization (ACO), and Bat-inspired Algorithm (BA).

NFL Theorem [10] (No Free Lunch) has logically proved that there is no metaheuristic appropriately suited for solving all optimization problems (Wolpert and Macready in 1997).

For example, a particular metaheuristic can give very promising results for a set of problems, but the same algorithm can show poor performance in a set of different problems. Therefore exits a diversity of techniques and are born new metaheuristic for solving optimization problems.

2.1 *Grey Wolf Optimizer*

The hierarchy of leaders and mechanism of the grey wolf hunting is illustrated in the next pyramid (Fig. 1).

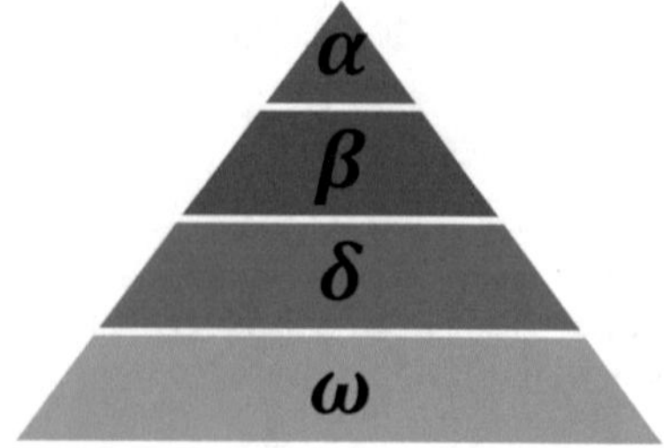

Fig. 1 Hierarchy pyramid

In addition, we can say that the social hierarchy of the grey wolf in the group of hunters has a very interesting social behavior.

According to C. Muro [11] the main phases of the grey wolf hunting are as follows:

- Tracking, chasing, and approaching the prey.
- Pursuing, encircling, and harassing the prey until it stops moving.
- Attack towards the prey

2.2 *Social Hierarchy*

In order to mathematically model the social hierarchy of wolves when designing the GWO, we consider the fittest solutions as the alpha (α) wolves. Consequently, the second and third best solutions are named beta (β) and delta (δ), respectively. The rest of the candidate solutions are assumed to be omega (ω). In the GWO algorithm the hunting (optimization) is guided by α, β, and δ. The ω wolves follow these three wolves.

2.3 *Encircling Prey*

As mentioned above, grey wolves encircle prey during the hunt. In order to mathematically model encircling behavior the following equations are proposed:

$$D = \left\| C \cdot \mathbf{X_p}(t) - \mathbf{X}(t) \right\| \tag{1}$$

$$X(t+1) = \mathbf{X_p}(\mathrm{t}) - A\mathbf{D} \tag{2}$$

where t indicates the current iteration, A and C are coefficients, $\boldsymbol{X_p}$ is the position vector of the prey, and $\boldsymbol{X}$ indicates the position vector of a grey wolf.

The coefficients A and C are calculated as follows:

$$A = 2a \cdot r_1 - a \tag{3}$$

$$C = 2 \cdot r_2 \tag{4}$$

where the components of $\boldsymbol{a}$ are linearly decreased from 2 to 0 over the course of iterations and r_1, r_2 are random numbers in [0, 1].

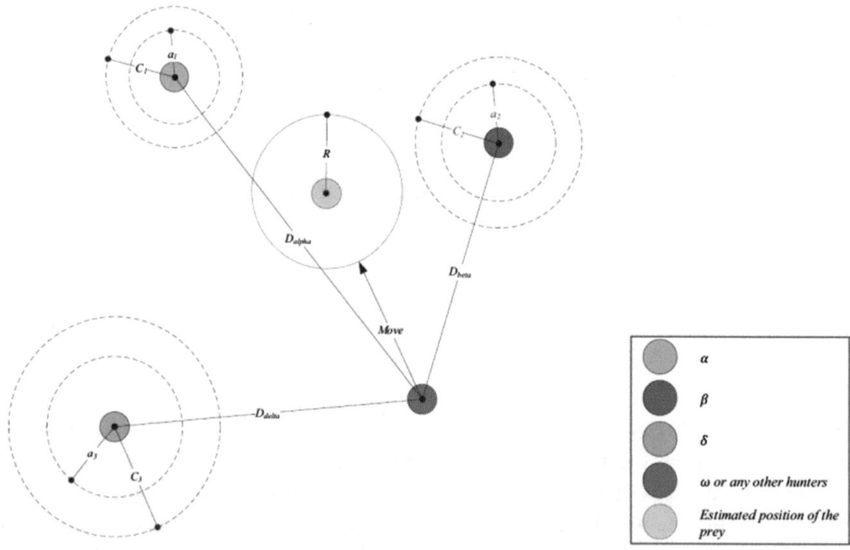

Fig. 2 Updating position in the algorithm

2.4 Hunting

Grey wolves have the ability to recognize the location of prey and encircle them. The hunt is usually guided by the alpha wolf.

In order to mathematically simulate the hunting behavior of grey wolves, we suppose that the alpha (best candidate solution), beta, and delta have better knowledge about the potential location of prey. Therefore, we save the first three best solutions obtained so far and oblige the other search agents (including the omegas) to update their positions according to the position of the best search agents. The following Equations have been proposed in this regard (Fig. 2).

$$\mathbf{D}_\alpha = \|C_1 \cdot \mathbf{X}_\alpha - \mathbf{X}\|, \quad \mathbf{D}_\beta = \|C_2 \cdot \mathbf{X}_\beta - \mathbf{X}\|, \quad \mathbf{D}_\delta = \|C_3 \cdot \mathbf{X}_\delta - \mathbf{X}\| \tag{5}$$

$$\mathbf{X}_1 = \mathbf{X}_\alpha - A_1 \cdot (\mathbf{D}_\alpha),\ \mathbf{X}_2 = \mathbf{X}_\beta - A_2 \cdot (\mathbf{D}_\beta),\ \mathbf{X}_3 = \mathbf{X}_\delta - A_3 \cdot (\mathbf{D}_\delta) \tag{6}$$

$$\mathbf{X}(t+1) = \frac{\mathbf{X}_1 + \mathbf{X}_2 + \mathbf{X}_3}{3} \tag{7}$$

2.5 Attacking the Prey

As mentioned above, the grey wolves finish the hunt by attacking the prey when it stops moving. In order to mathematically model the process of approaching the

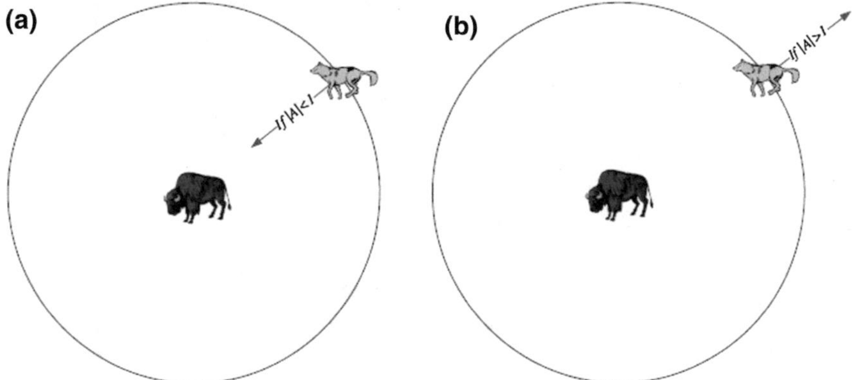

Fig. 3 Attacking prey versus searching for prey. **a** If $|A| < 1$ then attacking prey (exploitation). **b** If $|A| > 1$ then searching for prey (exploration)

prey, we decrease the value of "*a*." Note that the fluctuation range of *A* is also decreased by "*a*."

When random values of *A*are in $[-1, 1]$, the next position of a search agent can be in any position between its current position and the position of the prey. Figure 3a shows that $|A| < 1$ forces the wolves to attack towards the prey.

2.6 Searching for Prey

Grey wolves mostly search according to the position of the alpha, beta, and delta wolves. They diverge from each other to search for prey and converge to attack prey. In order to mathematically model divergence, we utilize *A* with random values greater than 1 or less than -1 to force the search agent to diverge from the prey. This emphasizes exploration and allows the GWO algorithm to search globally. Figure 3b also shows that $|A| > 1$ forces the grey wolves to diverge from the prey to hopefully find a fitter prey. Figure 4 shows the pseudocode of the algorithm.

3 Benchmark Functions

In this paper, we consider seven Benchmark functions, which are briefly explained below [12–15].

- **Sphere Function**
  Number of variables: *n* variables.
  Definition

```
Initialize the grey wolf population X_i(i = 1, 2, …, n)
Initialize a, A and C
Calculate the fitness of each search agent
X_α = the best search agent
X_β = the second best agent
X_δ = the third best search agent
while (t < Max number of iterations)
      for each search agent
              Update the position of the current
              search agent by equation (7)
      end for
      Update a, A and C
      Calculate the fitness of all search agents
      Update X_α , X_β and X_δ
      t = t+ 1
end while
return Xα
```

Fig. 4 Pseudocode of the algorithm

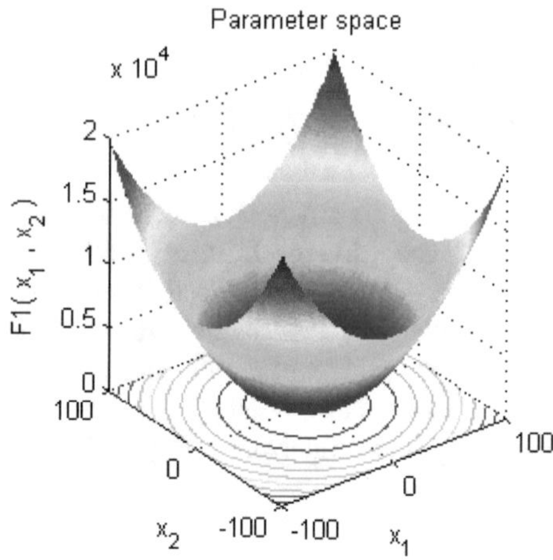

Fig. 5 Sphere function

$$f(x) = \sum_{i=1}^{n} x_i^2 \tag{8}$$

Search domain: $-100 \leq x_i \leq 100$, $i = 1, 2, \ldots, n$.
Number of local minima: no local minimum except the global one.
The global minima: $x^* = (0, \ldots, 0)$, $f(x^*) = 0$
Function graph: for $n = 2$ is presented in Fig. 5.

- **Griewank Function**
  Number of variables: n variables.
  Definition

$$f(x) = \frac{1}{4000}\sum_{i=1}^{n} x_1^2 - \prod_{i=1}^{n} \cos\left(\frac{x_i}{\sqrt{i}}\right) + 1 \tag{9}$$

  Search domain: $-600 \leq x_i \leq 600$, $i = 1, 2, \ldots, n$.
  Number of local minima: no local minimum except the global one.
  The global minima: $x^* = (0, \ldots, 0)$, $f(x^*) = 0$
  Function graph: for $n = 2$ is presented in Fig. 6.
- **Schwefel Function**
  Number of variables: n variables.
  Definition

$$f(x) = \sum_{i=1}^{n} -x_i \sin\left(\sqrt{|x_i|}\right) \tag{10}$$

  Search domain: $-500 \leq x_i \leq 500$, $i = 1, 2, \ldots, n$.
  Number of local minima: no local minimum except the global one.
  The global minima: $x^* = (0, \ldots, 0)$, $f(x^*) = 0$
  Function graph: for $n = 2$ is presented in Fig. 7.
- **Rastringin Function**
  Number of variables: n variables.
  Definition

$$f(x) = \sum_{i=1}^{n} \left[x_i^2 - 10\cos(2\pi x_i) + 10\right] \tag{11}$$

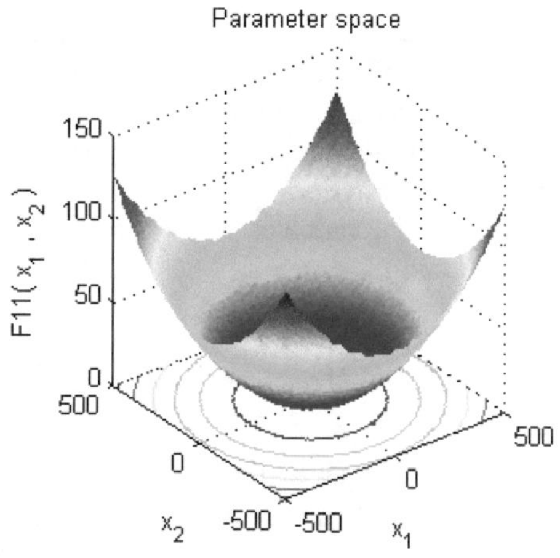

Fig. 6 Griewank function

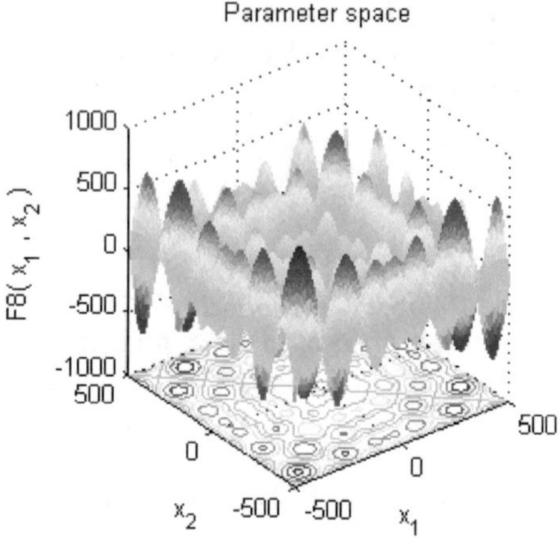

Fig. 7 Schwefel function

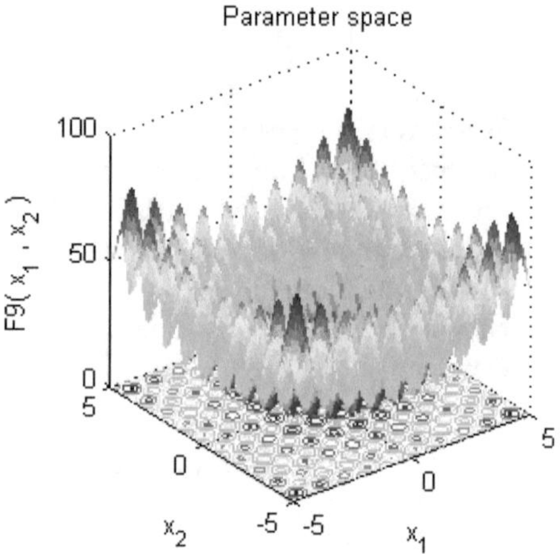

Fig. 8 Rastrigin function

Search domain: $-5.12 \leq x_i \leq 5.12$, $i = 1, 2, \ldots, n$.
Number of local minima: no local minimum except the global one.
The global minima: $x^* = (0, \ldots, 0)$, $f(x^*) = 0$
Function graph: for $n = 2$ is presented in Fig. 8.

- **Ackley Function**
  Number of variables: n variables.
  Definition

$$f(x) = -20\exp\left(-0.2\sqrt{\frac{1}{n}\sum_{i=1}^{n} x_i^2}\right) - \exp\left(\frac{1}{n}\sum_{i=1}^{n}\cos(2\pi x_i)\right) + 20 + e \quad (12)$$

Search domain: $-32 \leq x_i \leq 32$, $i = 1, 2, \ldots, n$.
Number of local minima: no local minimum except the global one.

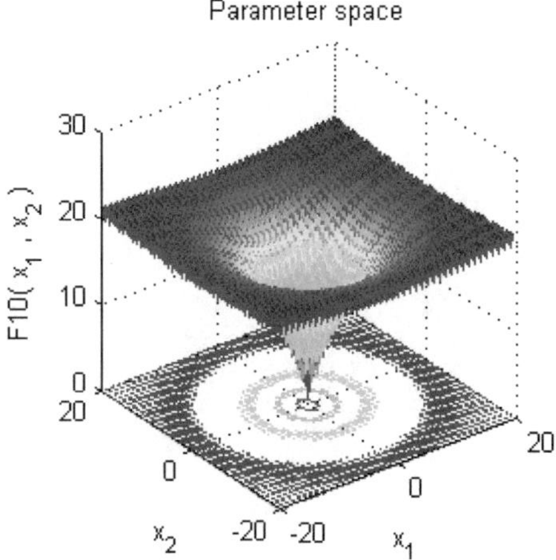

Fig. 9 Ackley function

The global minima: $x^* = (0, \ldots, 0)$, $f(x^*) = 0$
Function graph: for $n = 2$ is presented in Fig. 9.

- **Rosenbrock Function**
  Number of variables: n variables.
  Definition

$$f(x) = \sum_{i=1}^{n-1} \left[100(x_{i+1} - x_i^2)^2 + (x_1 - 1)^2 \right] \tag{13}$$

  Search domain: $-30 \leq x_i \leq 30$, i = 1, 2, … , n.
  Number of local minima: no local minimum except the global one.
  The global minima: $x^* = (0, \ldots, 0)$, $f(x^*) = 0$
  Function graph: for $n = 2$ is presented in Fig. 10.

- **Quartic Function**
  Number of variables: n variables.
  Definition

$$f(x) = \sum_{i=1}^{n} ix_i^4 + \text{random}\,[0, 1] \tag{14}$$

  Search domain: $-1.28 \leq x_i \leq 1.28$, $i = 1, 2, \ldots, n$.
  Number of local minima: no local minimum except the global one.
  The global minima: $x^* = (0, \ldots, 0)$, $f(x^*) = 0$
  Function graph: for $n = 2$ is presented in Fig. 11.

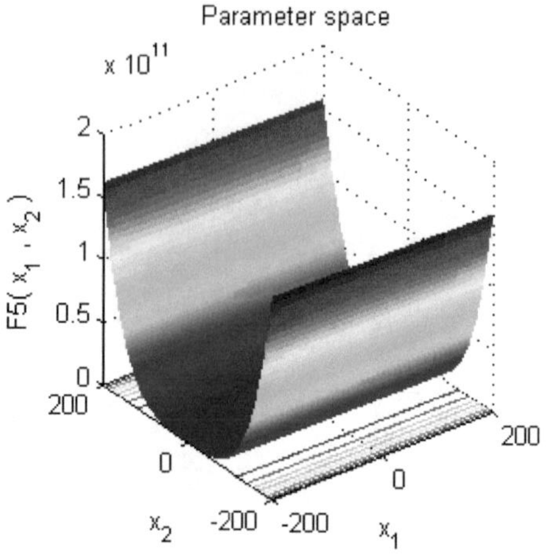

Fig. 10 Rosenbrock function

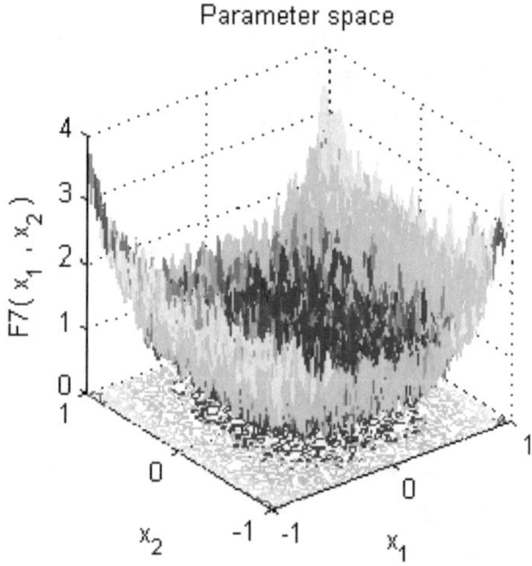

Fig. 11 Quartic function

4 Experiments and Results

The parameters where we can work in the area of optimization are the coefficients, which are the values of *A* and *C*.

The parameters for the experiments are presented by "*a*," and *C*.

4.1 Results, Experiments, and Analysis of Parameter "a"

As shown in the previous equations, the parameter "*a*" affects directly the coefficient *A*, this is the parameter that affects the two main Eqs. (1) and (2).

The parameter *A* simulates the act of attacking their prey (exploitation), when the values of *A* are between [−1, 1], the next position of a search agent (wolf) can be in any position between its current position and the position of the prey.

Parameter "*a*" is key factor in the algorithm, not just for the exploitation, also explicitly affects decision-making to explore, the result for the experiments the parameter "*a*" are presented.

Giving new values of "*a,*" to observe the algorithm with different values of "*a,*" for example if the values are high, medium, and low (with respect to the values of the author where a decrease of 2–0 is recommended). The algorithm has values that decrease from 2 to 0 along iterations.

The first part is the exploration with manual values for "*a,*" the values are from 0 to 3, with an increase of 0.1 is presented, and the results are shown. For each value of "*a,*" we presented 30 runs and the results that are shown are the average of them.

4.2 *Result, Experiments, and Analysis of Parameter* C

In the equations, the parameter *C* is a random value between 0 and 2, which directly affects the position of the prey.

This component provides for the prey random weights in order to emphasize stochastically ($C > 1$) or deemphasize ($C < 1$) the effect of the prey (Eq. 1). This helps the GWO algorithm to show a random behavior over optimization, which encourages exploration and avoids local results.

It is important to say that *C* is not linearly decreasing like "*a*." *C* provides random values at all times in order to emphasize the exploration not only during initial iterations, also in the final iterations. This feature is very useful when local optima stagnation occurs, especially in the final iterations. The parameter *C* can be also considered the effect of obstacles to approach prey in nature.

The algorithm GWO has *C* values between 0 and 2 randomly along iterations. As the first part of the experiments are manual values for *C* from 0 to 3, with an increase of 0.1, and the results are shown.

For each value of *C*, we presented 30 runs and the results shown are the average of them.

We show in Table 1 the results that were obtained in analysis of parameter "*a*" in the sphere function, increasing the original range of "*a*" (from 0 to 2) from 0 to 3 with an increase of 0.1 between each experiment.

Each experiments show the average of 30 runs for each increase, for example when $a = 0$, the average of 30 runs is 25124.5184, when $a = 0.1$, the average of 30 runs is 60.7062, etc.

The same way is like the parameter C which is represented in the column Average of *C*, for example when $C = 0$, the average of 30 runs is 1.3560E−43, when $C = 0.1$, the average of 30 runs is 2.2439E−39, etc.

Table 1 Simulation results of the Sphere function

Sphere function		
Original average: 6.59E−28	Max iterations: 500	
Value of a and C	Average of a	Average of C
0	25124.5184	1.3560E−43
0.1	60.7062	2.2439E−39
0.2	0.2519	1.0187E−36
0.3	0.0011	2.5584E−33
0.4	3.3366E−06	3.0897E−29
0.5	8.9334E−09	3.0251E−25
0.6	1.7444E−11	1.0480E−20
0.7	1.7378E−14	8.1066E−16
0.8	2.0714E−17	3.5977E−10
0.9	6.6541E−21	4.1666E−04
1	6.0795E−25	4.6100E+03
1.1	1.9949E−29	4.3400E−02
1.2	6.8560E−35	1.1600E−07
1.3	3.1482E−41	3.0400E−13
1.4	4.9586E−48	1.0800E−17
1.5	3.7108E−54	2.9300E−21
1.6	5.7559E−58	1.8800E−23
1.7	2.1617E−56	2.2300E−25
1.8	2.6385E−54	1.7500E−26
1.9	4.5515E−48	1.8300E−26
2	2.5433E−41	2.3000E−27
2.1	3.4014E−33	3.2600E−27
2.2	4.6882E−26	3.9600E−27
2.3	3.1137E−19	5.3600E−27
2.4	2.0968E−13	8.8000E−27
2.5	1.2566E−08	1.3300E−26
2.6	1.4655E−04	1.1300E−25
2.7	0.1296	1.2700E−25
2.8	29.6186	4.1100E−25
2.9	665.1587	8.0500E−25
3	4222.6477	1.8900E−24

It is important to say that when the value of "a" was experimented, the value of C was obtained with the original formula that was proposed by the author, and vice versa with the parameter C.

Table 2 Simulation results of the Rosenbrock function

Rosenbrock function		
Original average: 26.8126		Max iterations: 500
Value of *a* and *C*	Average of *a*	Average of *C*
0	46235583.63	27.4989
0.1	743.2122	27.2596
0.2	50.4988	27.5281
0.3	37.4896	27.4245
0.4	35.5024	27.3892
0.5	34.6096	27.0373
0.6	28.4861	27.3600
0.7	28.5327	27.3573
0.8	28.6514	27.8414
0.9	28.4485	33.9965
1	28.3160	2803103.3310
1.1	28.2197	39.6635
1.2	27.9439	28.6720
1.3	28.0584	29.5859
1.4	28.1365	27.1489
1.5	27.4881	26.9709
1.6	27.2782	26.8633
1.7	27.3860	26.8743
1.8	27.1124	27.0907
1.9	27.0989	26.9911
2	27.0881	27.2755
2.1	26.9945	27.0176
2.2	27.0637	27.0486
2.3	27.2281	26.8839
2.4	27.3600	27.1547
2.5	27.7725	27.2925
2.6	29.6583	27.3907
2.7	981.8355	27.2043
2.8	68422.97246	27.4405
2.9	1375919.189	27.3958
3	9947203.443	27.4632

We show in Table 2 the results of the parameters the "*a*" and *C* for the Rosenbrock function, although we experiment with a large range which is notable that this function is difficult for GWO algorithm.

Table 3 Simulation results of the Quartic function

Quartic function		
Original average: 2.2130E−03	Max iterations: 500	
Value of a and C	Average of a	Average of C
0	22.7004	6.8677E−04
0.1	0.53051	9.0543E−04
0.2	0.16345	1.1809E−03
0.3	0.07317	1.3961E−03
0.4	0.04208	1.8320E−03
0.5	0.02600	2.9122E−03
0.6	0.01706	4.6489E−03
0.7	0.01284	8.5442E−03
0.8	9.5829E−03	0.0207
0.9	7.9161E−03	0.0765
1	5.3744E−03	4.3148
1.1	4.0344E−03	0.0755
1.2	3.4281E−03	0.0157
1.3	3.0185E−03	6.6635E−03
1.4	2.4490E−03	4.4531E−03
1.5	1.7767E−03	3.2043E−03
1.6	1.5296E−03	2.4836E−03
1.7	1.3481E−03	2.1526E−03
1.8	1.4453E−03	2.0670E−03
1.9	1.3713E−03	1.6335E−03
2	1.1016E−03	1.6453E−03
2.1	1.4415E−03	2.1304E−03
2.2	2.4441E−03	1.5360E−03
2.3	3.6614E−03	1.4875E−03
2.4	8.9087E−03	1.6733E−03
2.5	0.0182	2.0917E−03
2.6	0.0296	1.8925E−03
2.7	0.0894	1.8921E−03
2.8	0.2803	1.7002E−03
2.9	1.1013	1.8749E−03
3	4.4280	1.9008E−03

We show in Table 3 the results in Quartic function which are only better if the C parameter is with lower value for example between 0 and 0.4, in this range the algorithm have a good performance with respect the author.

We show in Table 4 the results that were obtained in analysis of parameter "a" also the analysis of parameter C for the Schwefel function.

Table 4 Simulation results of the Schwefel function

Schwefel function		
Original average: −6123.10	Max iterations: 500	
Value of a and C	Average of a	Average of C
0	−2344.148477	−5010.3696
0.1	−2340.168715	−5014.4897
0.2	−3065.00585	−5301.9475
0.3	−3485.956032	−5220.6817
0.4	−4413.376354	−5311.6921
0.5	−4628.105893	−5514.4644
0.6	−4699.960723	−5656.2490
0.7	−4604.297534	−6225.5404
0.8	−4387.300483	−6266.4229
0.9	−4254.729856	−6687.2381
1	−4155.170942	−6508.9224
1.1	−3833.088683	−7194.7190
1.2	−3628.844783	−7177.3934
1.3	−3661.409077	−6717.1709
1.4	−3606.25134	−6820.4830
1.5	−3569.150188	−6638.7364
1.6	−3476.074396	−6588.0930
1.7	−3397.647409	−6206.9022
1.8	−3373.379504	−5828.3448
1.9	−3406.556174	−5795.6171
2	−3430.886702	−5051.9349
2.1	−3506.31382	−4547.2465
2.2	−3612.682924	−4193.5002
2.3	−3523.425052	−4614.9171
2.4	−3626.836179	−4285.8446
2.5	−3828.2345	−4121.7761
2.6	−3889.978723	−4413.1229
2.7	−3954.50416	−4114.1458
2.8	−3972.69278	−4120.1431
2.9	−3949.110596	−4087.7812
3	−4056.931437	−4144.0507

We show in Table 5 the results of the parameters "a" and C for the Rastrigin function, although we experimented with a large range, we can say that the best value for the parameter C is between the last values, for example [2.6, 3].

Table 6 shows that the results in the Ackley function are only better if the C parameter is with a value between 0 and 0.4, and for the parameter "a" the best

Table 5 Simulation results of the Rastrigin function

Rastrigin function		
Original average: 0.3105		Max iterations: 500
Value of a and C	Average of a	Average of C
0	356.7057	0.1438
0.1	103.7436	7.3896E−14
0.2	101.5066	0.2237
0.3	117.8934	5.1334
0.4	128.2469	6.0594
0.5	104.2054	14.0440
0.6	109.3073	19.9834
0.7	106.2402	27.7009
0.8	92.8126	39.8306
0.9	94.8878	70.3188
1	86.3684	130.7050
1.1	95.4060	67.0678
1.2	90.0946	38.3061
1.3	79.0978	22.7429
1.4	52.1075	14.9657
1.5	27.1762	6.3668
1.6	10.1409	3.5884
1.7	3.4672	1.1350
1.8	7.2002E−14	0.8350
1.9	3.5323	0.5670
2	0	0.8444
2.1	0.3367	0.1705
2.2	0.2735	0.3174
2.3	1.0234E−07	0.1714
2.4	2.0056	0.1957
2.5	37.5093	0.4392
2.6	111.6454	7.5791E−14
2.7	166.6798	2.4253E−13
2.8	210.6517	2.6338E−13
2.9	240.9622	0.1172
3	269.7178	9.7032E−12

performance is between [1.1, 2.2] in these ranges the algorithm has a good performance with respect ranges of the author [0–2].

In Table 7, we find the results for the Griewank function that were inconsistent but is important to say that in three values of the parameters, the average was 0 (zero) and this is the optimal.

Table 6 Simulation results of the Ackley function

Ackley function		
Original average: 1.06E−13	Max iterations: 500	
Value of a and C	Average of a	Average of C
0	18.5966	1.3915E−14
0.1	16.9974	2.0191E−14
0.2	13.5467	2.7652E−14
0.3	9.3749	3.9494E−14
0.4	6.6771	7.1232E−14
0.5	5.5308	2.7374E−13
0.6	3.7108	2.3765E−11
0.7	3.7652	8.2559E−09
0.8	3.7467	0.4971
0.9	3.2684	3.3045
1	0.8283	13.5448
1.1	3.7126E−14	4.3760
1.2	2.8599E−14	0.5542
1.3	2.2560E−14	1.1997E−07
1.4	1.5810E−14	6.9117E−10
1.5	1.3323E−14	1.4707E−11
1.6	1.0244E−14	7.4293E−13
1.7	9.4147E−15	2.5100E−13
1.8	8.4673E−15	1.2997E−13
1.9	8.8226E−15	1.0794E−13
2	8.9410E−15	0.6796
2.1	9.6515E−15	5.5499
2.2	4.7902E−14	4.7456
2.3	1.1023E−10	10.8470
2.4	1.0365E−07	14.2245
2.5	2.2030E−05	13.5158
2.6	0.0020	18.8763
2.7	0.0898	19.5178
2.8	2.4654	20.1800
2.9	9.3836	18.8065
3	15.8409	19.4487

4.3 Summary of the Results

Tables 8 and 9 contain general information on the results obtained by studying each of the parameters in the algorithm.

Table 7 Simulation results of the Griewank function

Griewank function		
Original average: 6.59E−28	Max iterations: 500	
Value of a and C	Average of a	Average of C
0	223.25044	6.8525E−04
0.1	1.51228	1.1247E−03
0.2	0.37056	1.5709E−03
0.3	0.02928	4.4261E−03
0.4	0.01441	3.9091E−03
0.5	0.01027	4.9412E−03
0.6	9.8862E−03	6.5429E−03
0.7	8.9718E−03	0.0111
0.8	7.4841E−03	0.0109
0.9	8.7353E−03	0.0189
1	8.6446E−03	41.8165
1.1	6.8745E−03	0.0866
1.2	6.4324E−03	0.0147
1.3	5.4714E−03	0.0071
1.4	4.5249E−03	4.8015E−03
1.5	1.5083E−02	4.2360E−03
1.6	3.1753E−03	8.7963E−04
1.7	5.2585E−03	1.8190E−03
1.8	1.3133E−03	3.8830E−03
1.9	9.6964E−05	8.3722E−04
2	4.5438E−04	1.1578E−03
2.1	4.3597E−06	1.6554E−03
2.2	0	3.4449E−04
2.3	1.0785E−12	7.8263E−04
2.4	4.7815E−08	0
2.5	0.0304	5.3289E−04
2.6	0.1813	3.3307E−17
2.7	0.7163	1.4949E−03
2.8	1.3164	6.8147E−04
2.9	7.2187	1.0045E−03
3	41.5016	0

In Table 8 we can find that the algorithm performed better with the range values being proposed in this paper, which was the range [0.5, 2.5] that is obtained after analyzing the results of previous experiments.

Experiments performed so far are independently made, only changing one parameter at time, and the other parameter values obtained by the author proposed in his work.

Table 8 Summary of results with parameter "*a*"

Function	Mean	Mean	Mean
Sphere	6.5900E−28	7.1605E−23	1.1527E−30
Rosenbrock	26.8126	26.7090	27.0824
Quartic	2.2130E−03	2.0822E−03	1.6376E−03
Schwefel	−6123.1000	−5453.3390	−3972.06268
Rastrigin	0.3105	2.6729	1.4438E−12
Ackley	1.0600E−13	3.3964	6.1064
Griewank	4.4850E−03	3.6077E−03	1.3771E−03
Values of *a*	*a* [0,2]	*a* [0,3]	*a* [0.5,2.5]

Table 9 Summary of results with parameter *C*

Function	Mean	Mean	Mean	Mean
Sphere	6.59E−28	1.90E−32	1.17E−35	2.03E−25
Rosenbrock	26.81	27.0841	27.4340	27.2901
Quartic	2.21E−03	1.43E−03	1.53E−03	1.58E−03
Schwefel	−6123.10	−5270.75	−4997.29	−4274.15
Rastrigin	0.3105	2.46E−14	6.74E−01	4.79E−01
Ackley	1.06E−13	3.61E−14	3.05E−14	18.8594
Griewank	4.49E−03	0	2.03E−03	4.79E−04
Values of *C*	*C* [0,2]	*C* [0,3]	*C* [0, 0.5]	*C* [2.5, 3]

Table 9 shows the results of the *C* parameter, similarly this was increased range from [0, 2] to [0, 3] for analyzing the results.

From the results of the parameter it can be noted that there are two possible ranges for the optimization algorithm. When the function has one optimal solution (for example 0) values that work best for that problem is a range of [0, 0.5] as the function has more than one target point, it can be observed in experiments that the best values for *C*, also can be at higher values, in the range of [2.5, 3].

In this time the range that has a best result is in the range of [0, 3].

The main objective of these tables (Tables 8 and 9) is to analyze which of the two parameters have a greater impact within the algorithm, and we can see that the parameter with the best results in both values (author and this study) is the value of *C* with five functions of the seven analyzed, so we can conclude that this is the parameter that most affects the evolution of the algorithm.

5 Conclusions

We conclude that in a study of the algorithm one can find parameters which can be optimized using a dynamic method, which is intended as the next part of the research, making a fuzzy system with ranges based on a best result obtained in these experiments.

Analyzing the experiments, we observed that the parameter having the greatest impact on the algorithm is the C parameter, so in the next part of the research is explorer with this parameter and also with "a" only for prove and unexpected some change that cannot been observed in experiments.

We expected that with fuzzy logic algorithm improves significantly, and the study of a modification to the algorithm, which basically is adding a weight of greater impact to the alpha wolf, after the beta and delta finally, to simulate the hierarchy of these three wolves.

References

1. Bonabeau E, Dorigo M, Theraulaz G. Swarm intelligence: from natural to artificial systems: OUP USA; 1999.
2. Kennedy J, Eberhart R. Particle swarm optimization, in Neural Networks, 1995: Proceedings, IEEE international conference on; 1995. p. 1942–1948.
3. Dorigo M, Birattari M, Stutzle T. Ant colony optimization. Comput Itell Magaz, IEEE 2006;1:28–39.
4. Basturk B, Karaboga D. An artificial bee colony (ABC) algorithm for numeric function optimization, IEEE swarm intelligence symposium; 2006. p. 12–4.
5. Mirjalili S., Mirjalili M., Lewis A: Grey Wolf Optimizer. Advances in Engineering Software69 (2014) 46-61.
6. Beni G, Wang J. Swarm intelligence in cellular robotic systems. In: Robots and biological systems: towards a new bionics? Springer; 1993. p. 703–12.
7. Maier H.R., Kapelan Z: Evolutionary algorithms and other metaheuritics in water resources: Current status, research challenges and future directions. Environmental Modelling and Software 62 (2014) 271-299.
8. Can U., Alatas B: Physics Based Metaheuristic Algorithms for Global Optimization, American Journal of Information Science and Computer Engineering 1(2015) 94-106.
9. Yang X., Karamanoglu M: Swarm Intelligence and Bio-Inspired Computation: An Overview, Swarm Intelligence and Bio-Inspired Computation (2013) 3-23.
10. Wolpert DH, Macready WG. No free lunch theorems for optimization. EvolutComput, IEEE Trans 1997;1:67–82.
11. Muro C, Escobedo R, Spector L, Coppinger R. Wolfpack (Canis lupus) hunting strategies emerge from simple rules in computational simulations. BehavProcess 2011;88:192–7.
12. Yao X, Liu Y, Lin G. Evolutionary programming made faster. Evolut Comput, IEEE Trans 1999;3:82–102.
13. Digalakis J, Margaritis K. On benchmarking functions for genetic algorithms.Int J Comput Math 2001;77:481–506.
14. Molga M, Smutnicki C. Test functions for optimization needs. Test functions for optimization needs; 2005.
15. Yang X-S. Test problems in optimization, arXiv, preprint arXiv: 1008.0549; 2010.

Gravitational Search Algorithm with Parameter Adaptation Through a Fuzzy Logic System

Frumen Olivas, Fevrier Valdez and Oscar Castillo

Abstract The contribution of this paper is to provide an analysis of the parameters of Gravitational Search Algorithm (GSA), to include a fuzzy logic system for dynamic parameter adaptation through the execution of the algorithm, in order to control the behavior of GSA based on some metrics like the iterations and the diversity of the agents in an specific moment of its execution.

Keywords Fuzzy logic · Gravitational search algorithm · Dynamic parameter adaptation · GSA

1 Introduction

Most optimization algorithms lack the ability to set parameters that dynamically adjust conforming to the problem in which is applied, our methodology bring this ability with the use of a fuzzy system, this is based on some metrics about the algorithm (i.e., number of iterations elapsed, average error, diversity of population, etc.), we can control one or more parameters, and with the help of fuzzy logic model this problem.

Fuzzy logic and fuzzy set theory was introduced by Zadeh in [14] and [15], and provides to us an easy way of modeling problems, with the use of membership functions and fuzzy rules [16]. With membership functions we can represent fuzzy sets and the knowledge of an expert on the problem can be represented with fuzzy rules, in our particular case the knowledge of the expert can be acquired by man-

F. Olivas · F. Valdez · O. Castillo (✉)
Tijuana Institute of Technology, Tijuana, Mexico
e-mail: ocastillo@tectijuana.mx

F. Olivas
e-mail: frumen@msn.com

F. Valdez
e-mail: fevrier@tectijuana.mx

P. Melin et al. (eds.), *Nature-Inspired Design of Hybrid Intelligent Systems*,
Studies in Computational Intelligence 667, DOI 10.1007/978-3-319-47054-2_26

ually set kbest parameter in GSA and experiment with several variations of this parameter, this is to decide in which case is better a level of kbest low or high.

GSA is a method based on populations of agents, in which search agents are considered a whole of masses obeying the Newtonian laws of gravitation and motion [11].

The contribution of this paper is the inclusion of a fuzzy system in GSA that help in the control of the kbest parameter, also the contribution is the analysis of kbest and the impact that it has in GSA. This was done by analyzing the algorithm behavior and performing several experiments to conclude that changing the value of kbest helps in GSA to improve the quality of the results. We also review several applications of it [4, 10, 13] and approaches as in [1, 3, 7] so that in contradistinction to others methods, where kbest parameter is fixed [8], we have a fuzzy system that dynamically change this parameter along iterations.

2 The Law of Gravity and Second Motion Law

Isaac Newton proposed the law of gravity in 1685 and stated that "The gravitational force between two particles is directly proportional to the product of their masses and inversely proportional to the square of the distance between them" [5], it is one of the four fundamental interactions in nature, together with the electromagnetic force, the weak nuclear force, and the strong nuclear force. The gravity force is present in each object in the universe and it's behave is called "action at a distance", this means gravity acts between separated particles without any intermediary and without any delay. The gravity law is represented by the following equation

$$F = G\frac{M_1 M_2}{R^2}, \tag{1}$$

where

F	is the magnitude of the gravitational force,
G	is gravitational constant,
M_1 and M_2	are the mass of the first and second particles, respectively, and,
R	is the distance between the two particles

The gravitational search algorithm furthermore to be based on Newtonian gravity law it is also based on Newton's second motion law, which says "The acceleration of an object is directly proportional to the net force acting on it and inversely proportional to its mass" [5]. The second motion law is represented by the following equation

$$a = \frac{F}{M}, \tag{2}$$

where

a is the magnitude of acceleration,
F is the magnitude of the gravitational force and,
M is the mass of the object

We can see that in Eq. 1 appears the gravitational constant G, this is a physic constant which determines the intensity of the gravitational force between the objects and it is defined as a very small value. The constant G is defined by the following equation

$$G(t) = G(t_0) \times \left(\frac{t_0}{t}\right)^{\beta}, \quad \beta < 1, \tag{3}$$

where

$G(t)$ is the value of the gravitational constant at time t and,
$G(t_0)$ is the value of the gravitational constant at the first cosmic quantum-interval of time t_0

The Eqs. (1) and (2) have been modified using concepts as active gravitational mass (M_a), passive gravitational mass (M_p) and inertial mass (M_i), it was done in order to rewrite Newton's laws. For example

$$F_{ij} = G\frac{M_{aj} \times M_{pi}}{R^2} \tag{4}$$

Now we can say "The gravitational force, F_{ij}, that acts on mass i by mass j, is proportional to the product of the active gravitational of mass j and passive gravitational of mass i, and inversely proportional to the square distance between them" [11]. Moreover, the acceleration has the following way

$$a_{ij} = \frac{F_{ij}}{M_{ii}} \tag{5}$$

and expresses, "a_i is proportional to F_{ij} and inversely proportional to inertia mass of i".

The theory of general relativity rests on the assumption that inertial and passive gravitational mass are equivalent [6]. Although inertial mass, passive gravitational mass, and active gravitational mass are conceptually distinct, no experiment has ever unambiguously demonstrated any difference between them [11].

3 Gravitational Search Algorithm

The gravitational search algorithm was introduced by E. Rashedi et al. [11], this algorithm is based on populations and at the same time it takes as fundamental principles the law of gravity and second motion law, its principal features consist that agents are considered as objects and their performance is measured by their masses, all these objects are attract each other by the gravity force, and this force causes a global movement of all objects, the masses cooperate using a direct form of communication, through gravitational force, an agent with heavy mass correspond to good solution therefore its move more slowly than lighter ones, finally its gravitational and inertial masses are determined using a fitness function. In this algorithm each mass is a solution and the algorithm is navigated by properly adjusting the gravitational and inertia masses, through iterations the masses are attracted by the heaviest mass. This mass will represent the best solution found in the search space [11].

The way in which the position of number N of agents is represented by

$$X_i = (x_i^1, \ldots, x_i^d, \ldots, x_i^n) \quad \text{for} \quad i = 1, 2, \ldots, N, \tag{6}$$

where x_i^d presents the position of ith agent in the dth dimension.

Now Eq. (1) with new concepts of masses is defined as following: the force acting on mass i from mass j in a specific time t, is

$$F_{ij}^d(t) = G(t) \frac{M_{pi}(t) \times M_{aj}(t)}{R_{ij}(t) + \varepsilon} (x_j^d(t) - x_i^d(t)) \tag{7}$$

where M_{aj} is the active gravitational mass related to agent j, M_{pi} is the passive gravitational mass related to agent i, $G(t)$ is gravitational constant at time t, ε is a small constant, and $R_{ij}(t)$ is the Euclidian distance between two agents i and j

$$R_{ij}(t) = \left\| X_i(t), X_j(t) \right\|_2 \tag{8}$$

The stochastic characteristic of this algorithm is based on the idea of the total force that acts on agent i in a dimension d be a randomly weighted sum of dth components of the forces exerted from other agents

$$F_i^d(t) = \sum_{j=i, j \neq i}^{N} \text{rand}_j F_{ij}^d(t), \tag{9}$$

where rand_j is a random number in the interval [0, 1]. The acceleration now is showed as,

$$a_i^d(t) = \frac{F_i^d(t)}{M_{ii}(t)} \tag{10}$$

where M_{ii} is the inertial mass of ith agent. For determine the velocity of an agent we considered as a fraction of its current velocity added to its acceleration.

$$v_i^d(t+1) = \text{rand}_i \times v_i^d(t) + a_i^d(t) \tag{11}$$

The position of agents could be calculated as the position in a specific time t added to its velocity in a time $t + 1$ as follows

$$x_i^d(t+1) = x_i^d(t) + v_i^d(t+1) \tag{12}$$

In this case the gravitational constant G, is initialized at the beginning and will be reduced with time to control the search accuracy. Its equation is

$$G(t) = G(G_0, t) \tag{13}$$

This is because G is a function of the initial value G_0 and time t.

The equation used to update G is

$$G(t) = G_0 e - \alpha t/T, \tag{14}$$

where α is a very small negative value (user-defined), t is the actual iteration and T is the maximum number of iterations.

As mentioned previously gravitational and inertia masses are simply calculated by the fitness evaluation and a heavier mass means a more efficient agent. The update the gravitational and inertial masses by the following equations

$$M_{ai} = M_{pi} = M_{ii} = M_i, i = 1, 2, \ldots, N, \tag{15}$$

$$m_i(t) = \frac{\text{fit}_i(t) - \text{worst}(t)}{\text{best}(t) - \text{worst}(t)} \tag{16}$$

$$M_i(t) = \frac{m_i(t)}{\sum_{j=1}^{N} m_j(t)} \tag{17}$$

the fitness value of the agent i at time t is defined by $\text{fit}_i(t)$, and best(t) and worst (t) are represented as

$$\text{best}(t) = \min_{j \in \{1,\ldots,N\}} \text{fit}_j(t) \tag{18}$$

$$\text{worst}(t) = \max_{j \in \{1,\ldots,N\}} \text{fit}_j(t) \tag{19}$$

If you want to use GSA for a maximization problem you only have to change Eqs. (18) and (19) as following

$$\text{best}(t) = \max_{j \in \{1,\ldots,N\}} \text{fit}_j(t) \tag{20}$$

$$\text{worst}(t) = \min_{j \in \{1,\ldots,N\}} \text{fit}_j(t) \tag{21}$$

The gravitational search algorithm has a mechanism of elitism in order to only allow the agents with bigger mass apply their force to the other. This is with objective to have a balance between exploration and exploitation with lapse of time it is achieved by the only the *kbest* agents will attract the others *kbest* is a function of time, with the initial value K_0 at the beginning and decreasing with time. In such a way, at the beginning, all agents apply the force, and as time passes, *kbest* is decreased linearly and at the end there will be just one agent applying force to the others. [11] For that reason the Eq. (9), can be modified as follows

$$F_i^d(t) = \sum_{j \in K\text{best}, j \neq 1} \text{rand}_i F_{ij}^d(t), \tag{22}$$

where *kbest* is the set of first *K* agents with the best fitness value and biggest mass. A better representation of GSA process is showing it next, it is the principle of this algorithm.

In Fig. 1 is shown the flowchart of GSA, in which we can see the main steps in the algorithm.

First an initial population is generated, next fitness of each agent is evaluated, thereafter update the gravitational constant *G*, best and worst of the population; next step is calculating mass and acceleration of each agent, if meeting end of iterations,

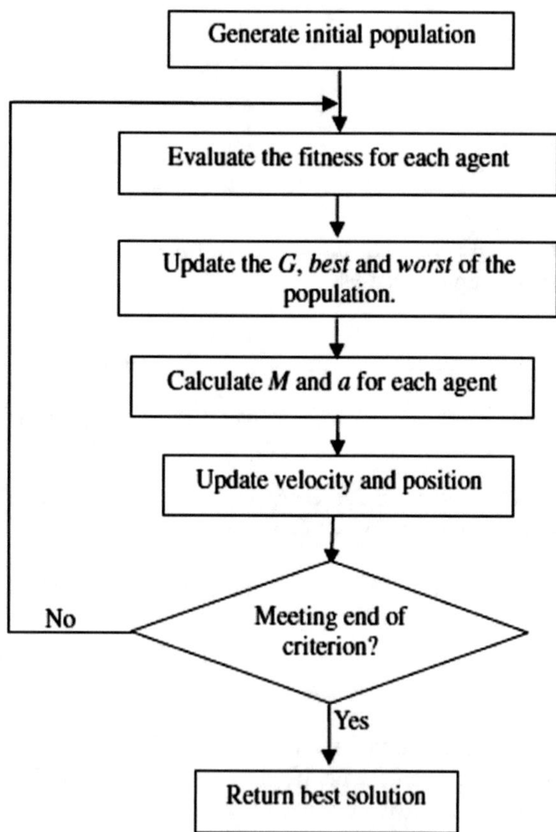

Fig. 1 General principle of GSA, taken from [11]

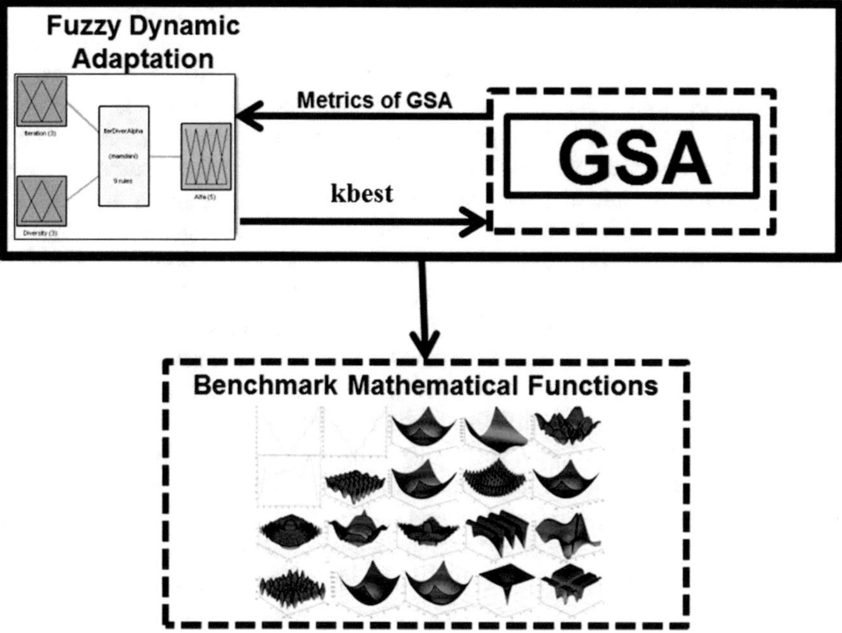

Fig. 2 Proposal for change kbest parameter using fuzzy logic

in this case maximum of iterations then returns the best solution, else executes the same steps starting from fitness evaluation. In step five, where we apply the modification in this algorithm, we propose changing kbest parameter to update how many agents apply their force to other agents and help GSA to obtain better results.

Our methodology is centered in the fifth block, because is in this block that kbest is used to upgrade the positions of the agents based on the mass of the best agents, Fig. 2 is a representation of the proposal, where we use a fuzzy system to dynamically adjust the values of kbest.

4 Methodology

In order to obtain some kind of knowledge about the behavior of GSA, we perform several experiments, in each experiment the parameters are change and with this we can see which parameter has the biggest impact in the performance and quality of the results of GSA, this give us an idea of the best way to control the abilities of GSA to perform a local or global search. In this case, kbest parameter from Eq. (22) has the biggest impact in GSA because this parameter influence the level the number of best agents that are allowed to affects the other agents by its gravitational force providing an opportunity to agents explore others good solutions in the search space and improve the final result.

Our methodology uses a fuzzy system to control the values of kbest in order to control how many agents apply their gravitational force to the other agents. After doing an exhaustive search by experiencing with different kbest values, we observed better results than the original approach showing that it is possible to make changes in the value of this parameter in order to get a better performance of the algorithm throughout its iterations. We have chosen the kbest parameter to be modified through the iterations of the algorithm because is the manner that GSA has to apply some kind of elitism, this means that only the best agents can affect the other agents, in other words, if all the agents apply their gravitational force to all of the other agents, then all agents will move with no destination, but if only the best agent apply its gravitational force then all of the other agent will move towards the best agent and for this reason move to the best region of the search space.

The main metrics about GSA that we use are iteration and diversity, for this purpose iteration is defined as the percentage of iteration elapsed given by Eq. (23), and diversity is defined as the level of dispersion of the agents given by Eq. (24), the reason we use this metrics is because with the percentage of iteration elapsed we can decide if GSA has to perform a global or a local search (i.e., in early iterations we want that GSA perform a global search), and with the level of dispersion of the agents we ensure that the agents are separately from each other or together (i.e., if the agents are close together in early iterations we want that the agents search by their own).

$$\text{Iteration} = \frac{\text{Current Iteration}}{\text{Maximum of Iterations}}, \tag{23}$$

where in Eq. (23), *current iteration* is the number of elapsed iterations, and *maximum of iterations* is the number iterations established for GSA to find the best solution.

$$\text{Diversity}\,(S(t)) = \frac{1}{n_s}\sum_{i=1}^{n_s}\sqrt{\sum_{j=1}^{n_s}\left(x_{ij}(t) - \bar{x}_j(t)\right)^2} \tag{24}$$

In Eq. (24), S is the population of GSA; t is the current iteration or time, n_s is the size of the population, i is the number of agent, n_x is the total number of dimensions, j is the number of the dimension, x_{ij} is the j dimension of the agent i, $\bar{x}j$ is the j dimension of the current best agent of the population.

In the fuzzy system the metrics about GSA are considered as inputs, and the output is kbest, Fig. 3 shows the input variable iteration, with a granularity of three triangular membership functions, with a range from 0 to 1.

Iteration as input variable has a range from 0 to 1 because the Eq. (23), given to us a percentage that only can be 0–100 % in other words from 0 to 1. The granularity onto three triangular membership functions is because with the increase in the number of membership functions in the inputs can cause that the number of possible rules increase exponential, and with three membership functions we

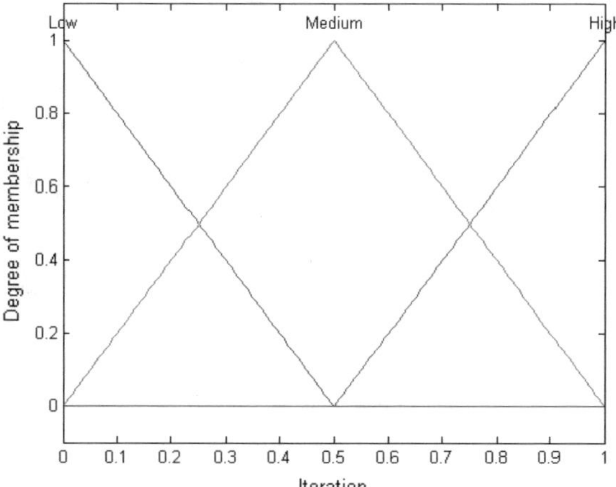

Fig. 3 Iteration as input variable for the fuzzy system

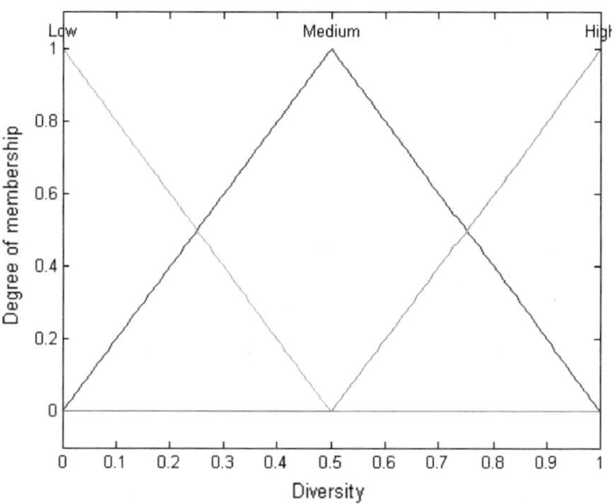

Fig. 4 Iteration as input variable for the fuzzy system

considered enough, the triangular type of membership functions was chosen based on an investigation reported on [9], which says that in this kind of problem the type of membership function does not have a big difference, in addition the membership functions is easier to configure compared with the other types of membership functions.

The input variable diversity with a range from 0 to 1 has been granulated into three triangular membership functions, and is represented in Fig. 4.

Diversity as input variable has a range from 0 to 1, because the results from Eq. (24) are between 0 and 1, and can be interpreted as if the population is close together, this means there is no diversity represented by 0, any other result means that there is diversity and with the result of 1 or more means that there is enough diversity.

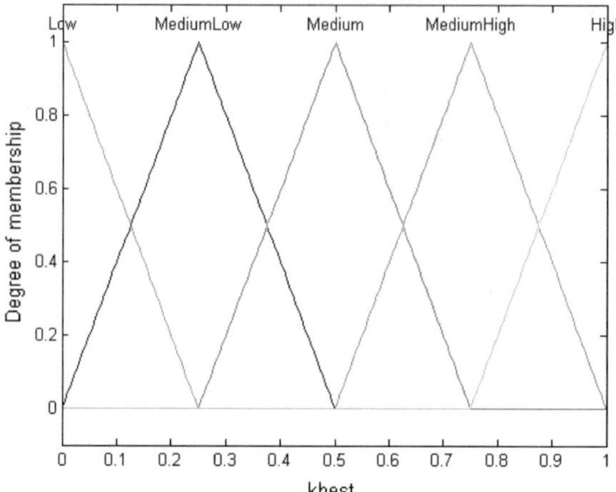

Fig. 5 kbest as output variable for the fuzzy system

1. If (Iteration is Low) and (Diversity is Low) then (kbest is High)
2. If (Iteration is Low) and (Diversity is Medium) then (kbest is MediumHigh)
3. If (Iteration is Low) and (Diversity is High) then (kbest is Medium)
4. If (Iteration is Medium) and (Diversity is Low) then (kbest is MediumHigh)
5. If (Iteration is Medium) and (Diversity is Medium) then (kbest is Medium)
6. If (Iteration is Medium) and (Diversity is High) then (kbest is MediumLow)
7. If (Iteration is High) and (Diversity is Low) then (kbest is Medium)
8. If (Iteration is High) and (Diversity is Medium) then (kbest is MediumLow)
9. If (Iteration is High) and (Diversity is High) then (kbest is Low)

Fig. 6 Rule set used for the fuzzy system

The output variable kbest has a range from 0 to 1 and has been granulated into five triangular membership functions, and is shown in Fig. 5.

kbest parameter as output has been defined with a range from 0 to 1, because this means the percentage of the best agents that can apply their gravity force to all other agents, also has been granulated into five triangular membership functions, and this is because with the two input variables with three membership functions each resulting into nine possible rules when combined all with the "and" operator, with this many rules we need to granulated more to allow a better performance of the fuzzy system.

The rule set used in this system is illustrated in Fig. 6 and there are a total of nine rules.

The rule set were created based on some principles about the desired behavior that we want for GSA, for example: in early iterations GSA must use exploration, in final iterations GSA must use exploitation, and with a low diversity GSA must use exploration and with a high diversity GSA must use exploitation, in other words,

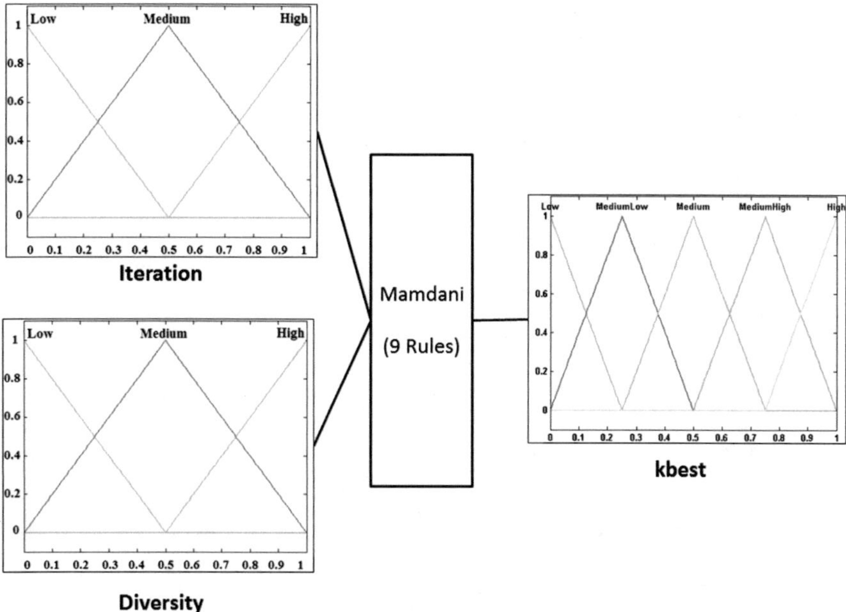

Fig. 7 Fuzzy system used to control kbest parameter

when GSA start the iterations must perform a global search to find the best regions of the search space, and in final iterations GSA must perform a local search in the best region found so far, with the diversity is all about trying to avoid local minimum because if the diversity is low this means that all the agents are in a very little area, and if is not the final iterations these agents must separate. And by the results of the experiments with the modification of kbest we can accomplish the task of control the abilities of GSA of perform a global or a local search.

The complete fuzzy system is shown in Fig. 7, where it can be seen the inputs and output, is a Mamdani type and has the number of fuzzy rules used.

We also created other fuzzy systems that have only one input and all of them to control kbest, but the fuzzy system that is shown in Fig. 7 is the best of all so our proposal is only this system, with iteration and diversity as inputs and kbest as output.

5 Experiments and Results

To test the proposal we apply GSA with dynamic parameter adaptation to 15 mathematical functions with 30 dimensions as applied in the original proposal and other improvements to GSA that we use to make a comparison with our proposal. We perform 30 experiments for each of the 15 benchmark mathematical functions that are shown below.

$$F_1(X) = \sum_{i=1}^{n} = x_i^2$$

$$F_2(X) = \sum_{i=1}^{n} |x_i| + \prod_{i=1}^{n} |x_i|$$

$$F_3(X) = \sum_{i=1}^{n} \left(\sum_{j=1}^{i} x_j\right)^2$$

$$F_4(X) = \max_i \{|x_i|, 1 \leq i \leq n\}$$

$$F_5(X) = \sum_{i=1}^{n-1} \left[100\left(x_{i+1} - x_i^2\right)^2 + (x_i - 1)^2\right]$$

$$F_6(X) = \sum_{i=1}^{n} ([x_i + 0.5])^2$$

$$F_7(X) = \sum_{i=1}^{n} i x_i^4 + \text{random}(0, 1)$$

$$F_8(X) = \sum_{i}^{n} -x_i \sin\left(\sqrt{|x_i|}\right)$$

$$F_9(X) = \sum_{i=1}^{n} \left[x_i^2 - 10\cos(2\pi x_i) + 10\right]$$

$$F_{10}(X) = -20\exp\left(-0.2\sqrt{\frac{1}{n}\sum_{i=1}^{n} x_i^2}\right) - \exp\left(\frac{1}{n}\sum_{i=1}^{n}\cos(2\pi x_i)\right) + 20 + e$$

$$F_{11}(X) = \frac{1}{4000}\sum_{i=1}^{n} x_i^2 - \Pi_{i=1}^{n} \cos\left(\frac{x_i}{\sqrt{i}}\right) + 1$$

$$F_{12}(X) = \frac{\pi}{n}\left\{10\sin(\pi y_1) + \sum_{i=1}^{n-1} (y_i - 1)^2\left[1 + 10\sin^2(\pi y_{i+1})\right] + (y_n - 1)^2\right\} + \sum_{i=1}^{n} u(x_i, 10, 100, 4) \quad y_i = 1 + \frac{x_i + 1}{4}$$

$$u(x_i, a, k, m) = \begin{cases} k(x_i - a)^m & x_i > a \\ 0 & -a < x_i < a \\ k(-x_i - a)^m & x_i < -a \end{cases}$$

$$F_{13}(X) = 0.1\left\{\sin^2(3\pi x_1) + \sum_{i=1}^{n} (x_i - 1)^2\left[1 + \sin^2(3\pi x_i + 1)\right] + (x_n - 1)^2\left[1 + \sin^2(2\pi x_n)\right]\right\} + \sum_{i=1}^{n} u(x_i, 5, 100, 4)$$

$$F_{14}(X) = \left(\frac{1}{500} + \sum_{j=1}^{25} \frac{1}{j + \sum_{i=1}^{2} \left(x_i - a_{ij}\right)^6}\right)^{-1}$$

$$F_{15}(X) = \sum_{i=1}^{11} \left[a_i - \frac{x_1\left(b_i^2 + b_i x_2\right)}{b_i^2 + b_i x_3 + x_4}\right]^2$$

Table 1 Comparison GSA versus our proposal

Function	GSA	Proposal
*F*1	7.3e−11	1.7276e−35
*F*2	4.03e−5	1.5117e−17
*F*3	0.16e−3	3.1201e−4
*F*4	3.7e−6	4.4420e−7
*F*5	25.16	24.0340
*F*6	8.3e−11	0
*F*7	0.018	1.2035e−2
*F*8	−2.8e+3	−2.9547e+3
*F*9	15.32	13.2337
*F*10	6.9e−6	6.6909e−15
*F*11	0.29	0.2362
*F*12	0.01	1.0509e−3
*F*13	3.2e−32	1.5e−33
*F*14	3.70	2.0921
*F*15	8.0e−3	4.1273e−3

We made 30 experiments of each function on the same conditions as the original proposal, with number of *agents* = 50, *maximum of iterations* = 1000, *dimensions* = 30, G_0 = 100 and α = 20, kbest is linear decreasing from 100 to 2 %, but in our proposal kbest is dynamic, that is the only difference. In Table 1 is shown the results of each method with each membership function, please note that each of the results in Table 1 is the average of 30 experiments, and the results in bold are the best when compared our proposal with the original GSA.

From the results in Table 1, is easy to see that our proposal outperform the quality of the results when compared with the original GSA method, and with the use of iteration and diversity as inputs the fuzzy system using the rule set provided can control the behavior of kbest and thereby control the behavior of GSA.

We also perform a comparison against other GSA improvement, in this case from Sombra et al. [12] in which they used a fuzzy system to control alpha, in fact this paper is continuation of that research. Table 2 shows the results of the comparison between our proposal and the modification from Sombra et al., these results are obtained using the same conditions as in the Table 1, so each result is the average of 30 experiments, and the results in bold are the best when compared our proposal with Sombra et al.

The comparison in Table 2 in some cases the results are close and only in function F10 our proposal is not better than the results from Sombra et al., but in all of the other benchmark mathematical functions our proposed approach can achieve better results, also remember that each result is the average of 30 experiments, this is our proposal can have a better control of the kbest parameter thereby a better control of the abilities of GSA to perform exploration or exploitation when is needed.

Table 2 Comparison Sombra et al. [12] versus our proposal

Function	Sombra et al. [12]	Proposal
$F1$	8.8518e−34	1.7276e−35
$F2$	1.1564e−10	1.5117e−17
$F3$	468.4431	3.1201e−4
$F4$	0.0912	4.4420e−7
$F5$	61.2473	24.0340
$F6$	0.1000	0
$F7$	0.0262	1.2035e−2
$F8$	−2.6792e+3	−2.9547e+3
$F9$	17.1796	13.2337
$F10$	6.3357e−15	6.6909e−15
$F11$	4.9343	0.2362
$F12$	0.1103	1.0509e−3
$F13$	0.0581	1.5e−33
$F14$	2.8274	2.0921
$F15$	0.0042	4.1273e−3

Besides of the approach presented in this paper we look forward to find an even better way to improve the quality of the results of GSA by using another inputs or controlling other parameter than kbest.

6 Conclusions

From the results in the previous section we can see that our proposed approach using the given methodology, can help us to achieve a better quality results, and with the use of fuzzy logic can be an easy way to model a complex problem, in this case trying to find the best kbest parameter depending on the situation given by the inputs.

With the inclusion of a fuzzy system, GSA can be applied to a wide variety of problem without the need of find the best values for kbest because the fuzzy system will do that for you.

From the results we can conclude that the modification of the kbest parameter using a fuzzy system is a good way to obtain better results when compared with the original method that use a fixed value of kbest. As we mentioned before we look forward to find a better way to improve GSA in the quality of the result, by so far this is the best fuzzy system we found, but we will continuum in the search of a better fuzzy system.

References

1. Bahrololoum, A.; Nezamabadi-pour, Bahrololoum, H.; Saeed, M. "A prototype classifier based on gravitational search algorithm", in ELSEVIER: Applied Soft Computing, Volume 12, Issue 2, Iran, 2012, pp. 819–825.
2. Engelbrecht, Andries P. "Fundamentals Of Computational Swarm Intelligence", University Of Pretoria, South Africa.
3. Hassanzadeh, H.R.; Rouhani, M. "A Multi-objective Gravitational Search Algorithm", in IEEE: Second International Conference on Computational Intelligence, Communication Systems and Networks (CICSyN), Liverpool, 2010, pp. 7–12.
4. Hatamlou, A.; Abdullah, S.; Othman, Z. "Gravitational search algorithm with heuristic search for clustering problems", in IEEE: 3rd Conference on Data Mining and Optimization (DMO), Putrajaya, 2011, pp. 190–193.
5. Holliday, D., Resnick, R., Walker, J., Fundamental of physic, John Wiley & Son, 1993.
6. Kennedy, J., and R. C. Eberhart. 2001. Swarm Intelligence. San Francisco: Morgan Kaufmann.
7. Mirjalili, S.; Hashim, S.Z.M. "A new hybrid PSOGSA algorithm for function optimization", in IEEE: International Conference on Computer and Information Application (ICCIA), Tianjin, 2010, pp. 374–377.
8. Mirjalili, S.; Hashim, S.; Sardroudi, H. "Training feedforward neural networks using hybrid particle swarm optimization and gravitational search algorithm", in ELSEVIER: Applied Mathematics and Computation, Volume 218, Issue 22, Malaysia, 2012, pp. 11125–11137.
9. Olivas, F., Valdez, F., & Castillo, O. (2014). A Comparative Study of Membership Functions for an Interval Type-2 Fuzzy System used to Dynamic Parameter Adaptation in Particle Swarm Optimization. In Recent Advances on Hybrid Approaches for Designing Intelligent Systems (pp. 67–78). Springer International Publishing.
10. Pagnin,A.; Schellini,S.A.; Spadotto,A.; Guido,R.C.; Ponti,M.; Chiachia,G.; Falcao,A.X. "Feature selection through gravitational search algorithm", in IEEE: IEEE International Conference on Acoustics, Speech and Signal Processing (ICASSP), Prague, 2011, pp. 2052–2055.
11. Rashedi, E.; Nezamabadi-pour, H.; Saryazdi, S. "GSA: A Gravitational Search Algorithm", in ELSEVIER: Information Sciences, Volume 179, Issue 13, Iran, 2009, pp. 2232–2248.
12. Sombra, A., Valdez, F., Melin, P., & Castillo, O. (2013, June). A new gravitational search algorithm using fuzzy logic to parameter adaptation. In Evolutionary Computation (CEC), 2013 IEEE Congress on (pp. 1068–1074). IEEE.
13. Verma, O.P.,Sharma, R. "Newtonian Gravitational Edge Detection Using Gravitational Search Algorithm", in IEEE: International Conference on Communication Systems and Network Technologies (CSNT), Rajkot, 2012, pp. 184–188.
14. Zadeh L. (1965) "Fuzzy sets". Information & Control, 8, 338–353.
15. Zadeh L. (1988) "Fuzzy logic". IEEE Computer Mag., vol. 1, pp. 83–93.
16. Zadeh L. (1975) "The concept of a linguistic variable and its application to approximate reasoning—I," Inform. Sci., vol. 8, pp. 199–249.

Part IV
Metaheuristic Applications

Particle Swarm Optimization of Ensemble Neural Networks with Type-1 and Type-2 Fuzzy Integration for the Taiwan Stock Exchange

Martha Pulido, Patricia Melin and Olivia Mendoza

Abstract This paper describes an optimization method based on particle swarm optimization (PSO) for ensemble neural networks with type-1 and type-2 fuzzy aggregation for forecasting complex time series. The time series that was considered in this paper to compare the hybrid approach with traditional methods is the Taiwan Stock Exchange (TAIEX), and the results shown are for the optimization of the structure of the ensemble neural network with type-1 and type-2 fuzzy integration. Simulation results show that ensemble approach produces good prediction of the Taiwan Stock Exchange.

Keywords Ensemble neural networks · Time series · Particle swarm · Fuzzy system

1 Introduction

Time series are usually analyzed to understand the past and to predict the future, enabling managers, or policy makers to make properly informed decisions. Time series analysis quantifies the main features in data, like the random variation. These facts, combined with improved computing power, have made time series methods widely applicable in government, industry, and commerce. In most branches of science, engineering, and commerce, there are variables measured sequentially in time. Reserve banks record interest rates and exchange rates each day. The government statistics department will compute the country's gross domestic product on a yearly basis. Newspapers publish yesterday's noon temperatures for capital cities from around the world. Meteorological offices record rainfall at many different sites with differing resolutions. When a variable is measured sequentially in time over or at a fixed interval, known as the sampling interval, the resulting data form a time series [1].

M. Pulido · P. Melin (✉) · O. Mendoza
Tijuana Institute of Technology, Tijuana, Mexico
e-mail: pmelin@tectijuana.mx

P. Melin et al. (eds.), *Nature-Inspired Design of Hybrid Intelligent Systems*,
Studies in Computational Intelligence 667, DOI 10.1007/978-3-319-47054-2_27

Time series predictions are very important because based on them we can analyze past events to know the possible behavior of futures events and thus can take preventive or corrective decisions to help avoid unwanted circumstances.

The choice and implementation of an appropriate method for prediction has always been a major issue for enterprises that seek to ensure the profitability and survival of business. The predictions give the company the ability to make decisions in the medium and long term, and due to the accuracy or inaccuracy of data this could mean predicted growth or profits and financial losses. It is very important for companies to know the behavior that will be the future development of their business, and thus be able to make decisions that improve the company's activities, and avoid unwanted situations, which in some cases can lead to the company's failure. In this paper, we propose a hybrid approach for time series prediction by using an ensemble neural network and its optimization with particle swarm optimization (PSO). In the literature, there have been recent produced work of time series [2–10].

2 Preliminaries

In this section we present basic concepts that are used in this proposed method:

2.1 *Time Series and Prediction*

The word "prediction" comes from the Latin prognosticum, which means I know in advance. Prediction is to issue a statement about what is likely to happen in the future, based on analysis and considerations of experiments. Making a forecast is to obtain knowledge about uncertain events that are important in decision-making [6]. Time series prediction tries to predict the future based on past data, it take a series of real data $x_t - n, \ldots, x_t - 2, 0\, x_t - 1, x_t$ and then obtains the prediction of the data $x_t + 1$, $x_t + 2 \ldots x_n + \mathrm{n}$. The goal of time series prediction or a model is to observe the series of real data, so that future data may be accurately predicted [1, 11, 12].

2.2 *Neural Networks*

Neural networks (NNs) are composed of many elements (Artificial Neurons), grouped into layers and are highly interconnected (with the synapses), this structure has several inputs and outputs, which are trained to react (or give values) in a way you want to input stimuli. These systems emulate in some way, the human brain. Neural networks are required to learn to behave (Learning) and someone should be responsible for the teaching or training (Training), based on prior knowledge of the environment problem [13, 14].

2.3 *Ensemble Neural Networks*

An ensemble neural network is a learning paradigm where many NNs are jointly used to solve a problem [15]. A neural network ensemble is a learning paradigm where a collection of a finite number of NNs is trained for a task [16]. It originates from Hansen and Salamon's work [17], which shows that the generalization ability of a neural network system can be significantly improved through ensembling a number of NNs, i.e., training many NNs and then combining their predictions. Since this technology behaves remarkably well, recently it has become a very hot topic in both NNs and machine-learning communities [18], and has already been successfully applied to diverse areas, such as face recognition [19, 20], optical character recognition [21–23], scientific image analysis [24], medical diagnosis [25, 26], seismic signals classification [27], etc.

In general, a neural network ensemble is constructed in two steps, i.e., training a number of component NNs and then combining the component predictions.

There are also many other approaches for training the component NNs. Examples are as follows. Hampshire and Waibel [23] utilize different object functions to train distinct component NNs.

2.4 *Fuzzy Systems as Methods of Integration*

There exists a diversity of methods of integration or aggregation of information, and we mention some of these methods below.

Fuzzy logic was proposed for the first time in the mid-sixties at the University of California Berkeley by the brilliant engineer Lofty A. Zadeh., who proposed what it is called the principle of incompatibility: "As the complexity of system increases, our ability to be precise instructions and build on their behavior decreases to the threshold beyond which the accuracy and meaning are mutually exclusive characteristics." Then introduced the concept of a fuzzy set, under which lies the idea that the elements on which to build human thinking are not numbers but linguistic labels. Fuzzy logic can represent the common knowledge as a form of language that is mostly qualitative and not necessarily a quantity in a mathematical language that means of fuzzy set theory and function characteristics associated with them [13].

2.5 Optimization

The process of optimization is the process of obtaining the 'best,' if it is possible to measure and change what is 'good' or 'bad.' In practice, one wishes the 'most' or 'maximum' (e.g., salary) or the 'least' or 'minimum' (e.g., expenses). Therefore, the word 'optimum' is takes the meaning of 'maximum' or 'minimum' depending on the circumstances; 'optimum' is a technical term which implies quantitative measurement and is a stronger word than 'best' which is more appropriate for everyday use. Likewise, the word 'optimize,' which means to achieve an optimum, is a stronger word than 'improve.' Optimization theory is the branch of mathematics encompassing the quantitative study of optima and methods for finding them. Optimization practice, on the other hand, is the collection of techniques, methods, procedures, and algorithms that can be used to find the optima [28].

2.6 Particle Swarm Optimization

PSO is a bio-inspired optimization method proposed by Eberhart and Kennedy [29] in 1995. PSO is a search algorithm based on the behavior of biological communities that exhibits individual and social behavior [30], and examples of these communities are groups of birds, schools of fish, and swarms of bees [30].

A PSO algorithm maintains a swarm of particles, where each particle represents a potential solution. In analogy with the paradigms of evolutionary computation, a swarm is similar to a population, while a particle is similar to an individual. In simple terms, the particles are "flown" through a multidimensional search space, where the position of each particle is adjusted according to its own experience and that of its neighbors. Let x_i denote the position i in the search space at time step t, unless otherwise stated, t denotes discrete time steps. The position of the particle is changed by adding a velocity, $v_i(t)$, to the current position, i.e.,

$$x_i(t+1) = x_i\ (t) + v_i(t+1) \quad (1)$$

with $x_i(0) \sim U(X_{\min}, X_{\max})$.

3 Problem Statement and Proposed Method

The goal of this work was to implement PSO to optimize the ensemble neural network architectures. In these cases the optimization is for each of the modules, and thus to find a neural network architecture that yields optimum results in each of the time series to be considered. In Fig. 1 we have the historical data of each time series prediction, then the data is provided to the modules that will be optimized

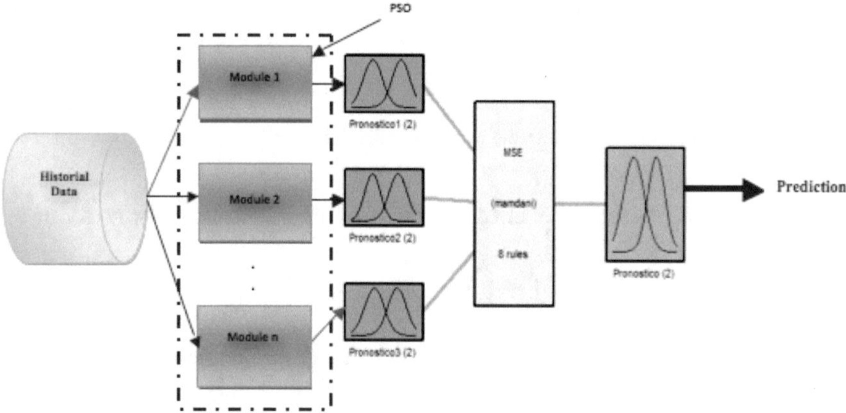

Fig. 1 General architecture of the proposed ensemble model

with the PSO for the ensemble network, and then these modules are integrated with integration based on type-1 and type-2 Fuzzy Integration.

Historical data of the Taiwan Stock Exchange time series was used for the ensemble neural network trainings, where each module was fed with the same information, unlike the modular networks, where each module is fed with different data, which leads to architectures that are not uniform.

The Taiwan Stock Exchange (Taiwan Stock Exchange Corporation) is a financial institution that was founded in 1961 in Taipei and began to operate as stock exchange on 9 February 1962. The Financial Supervisory Commission regulates it. The index of the Taiwan Stock Exchange is the TWSE [31].

Data of the Taiwan Stock Exchange time series: We are using 800 points that correspond to a period from 03/04/2011 to 05/07/2014(as shown in Fig. 2). We used 70 % of the data for the ensemble neural network trainings and 30 % to test the network [31].

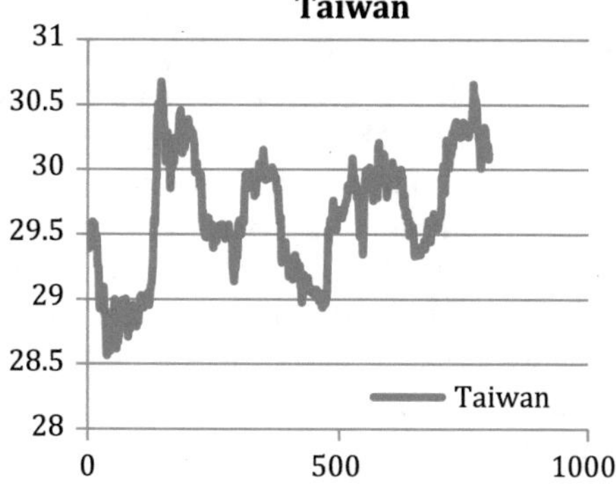

Fig. 2 Taiwan stock exchange

Number of Modules	Number of Layers	Neurons 1	…	Neurons n

Fig. 3 Particle structure to optimize the ensemble neural network

The objective function is defined to minimize the prediction error as follows:

$$\mathrm{EM} = \left(\sum\nolimits_{i=1}^{D} |a_i - x_i|\right)/D, \tag{2}$$

where *a,* corresponds to the predicted data depending on the output of the network modules, *X* represents real data, *D* the Number of Data points, and *EM* is the total prediction error.

The corresponding particle structure is shown in Fig. 3.

Figure 3 represents the particle structure to optimize the ensemble neural network, where the parameters that are optimized are the number of modules, number of layers, and number of neurons of the ensemble neural network. PSO determines the number of modules, number of layers, and number of neurons per layer that the neural network ensemble should have, to meet the objective of achieving the better prediction error.

The parameters for the PSO algorithm are: 100 Particles, 100 iterations, Cognitive Component (C1) = 2, Social Component (C2) = 2, Constriction coefficient of linear increase (C) = (0–0.9) and Inertia weight with linear decrease (W) = (0.9–0). We consider a number of 1–5 modules, number of layers of 1–3 and neurons number from 1 to 30.

The aggregation of the responses of the optimized ensemble neural network is performed with type-1 and type-2 fuzzy systems. In this work, the fuzzy system consists of five inputs depending on the number of modules of the neural network ensemble and one output is used. Each input and output linguistic variable of the fuzzy system uses two Gaussian membership functions. The performance of the type-2 fuzzy aggregators is analyzed under different levels of uncertainty to find out the best design of the membership functions for the 32 rules of the fuzzy system. Previous tests have been performed only with a three input fuzzy system and the fuzzy system changes according to the responses of the neural network to give us better prediction error. In the type-2 fuzzy system, we also change the levels of uncertainty to obtain the best prediction error.

Figure 4 shows a fuzzy system consisting of five inputs depending on the number of modules of the neural network ensemble and one output. Each input and output linguistic variable of the fuzzy system uses two Gaussian membership functions. The performance of the type-2 fuzzy aggregators is analyzed under different levels of uncertainty to find out the best design of the membership functions for the 32 rules of the fuzzy system. Previous experiments were performed with triangular, and Gaussian and the Gaussian produced the best results of the prediction.

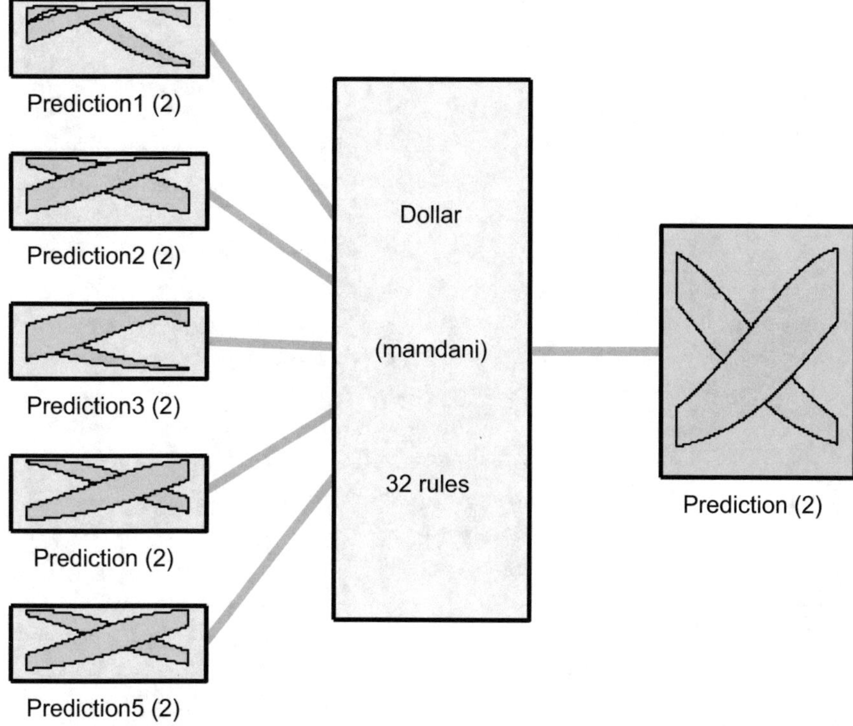

Fig. 4 Fuzzy inference system for integration of the ensemble neural network

Figure 5 represents the 32 possible rules of the fuzzy system; we have five inputs in the fuzzy system with two membership functions, and the outputs with two membership functions. These fuzzy rules are used for both the type-1 and type-2 fuzzy systems. In previous work, several tests were performed with three inputs, and the prediction error obtained was significant and the number of rules was greater, and this is why we changed to two inputs.

4 Simulation Results

In this section, we present the simulation results obtained with the genetic algorithm and PSO for the Taiwan Stock Exchange.

We consider working with a genetic algorithm to optimize the structure of an ensemble neural network and the best architecture obtained was the following (shown in Fig. 6).

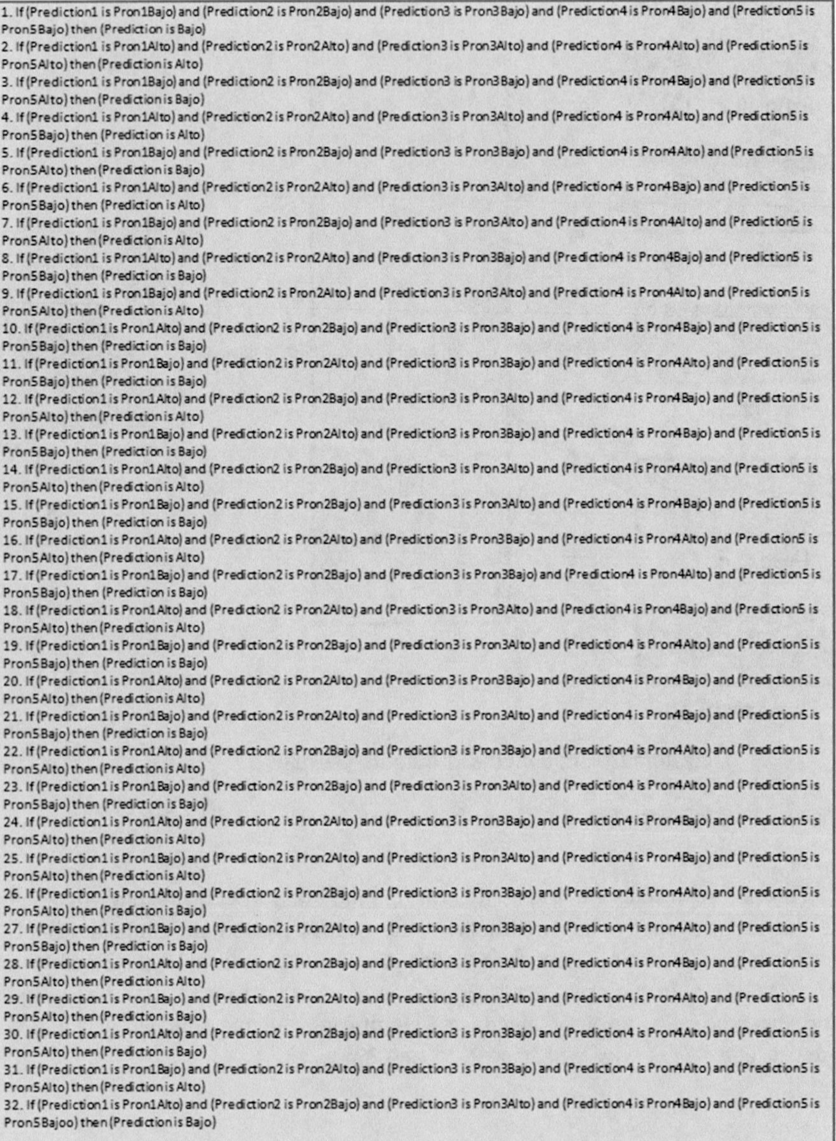

1. If (Prediction1 is Pron1Bajo) and (Prediction2 is Pron2Bajo) and (Prediction3 is Pron3Bajo) and (Prediction4 is Pron4Bajo) and (Prediction5 is Pron5Bajo) then (Prediction is Bajo)
2. If (Prediction1 is Pron1Alto) and (Prediction2 is Pron2Alto) and (Prediction3 is Pron3Alto) and (Prediction4 is Pron4Alto) and (Prediction5 is Pron5Alto) then (Prediction is Alto)
3. If (Prediction1 is Pron1Bajo) and (Prediction2 is Pron2Bajo) and (Prediction3 is Pron3Bajo) and (Prediction4 is Pron4Bajo) and (Prediction5 is Pron5Alto) then (Prediction is Bajo)
4. If (Prediction1 is Pron1Alto) and (Prediction2 is Pron2Alto) and (Prediction3 is Pron3Alto) and (Prediction4 is Pron4Alto) and (Prediction5 is Pron5Bajo) then (Prediction is Alto)
5. If (Prediction1 is Pron1Bajo) and (Prediction2 is Pron2Bajo) and (Prediction3 is Pron3Bajo) and (Prediction4 is Pron4Alto) and (Prediction5 is Pron5Alto) then (Prediction is Bajo)
6. If (Prediction1 is Pron1Alto) and (Prediction2 is Pron2Alto) and (Prediction3 is Pron3Alto) and (Prediction4 is Pron4Bajo) and (Prediction5 is Pron5Bajo) then (Prediction is Alto)
7. If (Prediction1 is Pron1Bajo) and (Prediction2 is Pron2Bajo) and (Prediction3 is Pron3Alto) and (Prediction4 is Pron4Alto) and (Prediction5 is Pron5Alto) then (Prediction is Alto)
8. If (Prediction1 is Pron1Alto) and (Prediction2 is Pron2Alto) and (Prediction3 is Pron3Bajo) and (Prediction4 is Pron4Bajo) and (Prediction5 is Pron5Bajo) then (Prediction is Bajo)
9. If (Prediction1 is Pron1Bajo) and (Prediction2 is Pron2Alto) and (Prediction3 is Pron3Alto) and (Prediction4 is Pron4Alto) and (Prediction5 is Pron5Alto) then (Prediction is Alto)
10. If (Prediction1 is Pron1Alto) and (Prediction2 is Pron2Bajo) and (Prediction3 is Pron3Bajo) and (Prediction4 is Pron4Bajo) and (Prediction5 is Pron5Bajo) then (Prediction is Bajo)
11. If (Prediction1 is Pron1Bajo) and (Prediction2 is Pron2Alto) and (Prediction3 is Pron3Bajo) and (Prediction4 is Pron4Alto) and (Prediction5 is Pron5Bajo) then (Prediction is Bajo)
12. If (Prediction1 is Pron1Alto) and (Prediction2 is Pron2Bajo) and (Prediction3 is Pron3Alto) and (Prediction4 is Pron4Bajo) and (Prediction5 is Pron5Alto) then (Prediction is Alto)
13. If (Prediction1 is Pron1Bajo) and (Prediction2 is Pron2Alto) and (Prediction3 is Pron3Bajo) and (Prediction4 is Pron4Bajo) and (Prediction5 is Pron5Bajo) then (Prediction is Bajo)
14. If (Prediction1 is Pron1Alto) and (Prediction2 is Pron2Bajo) and (Prediction3 is Pron3Alto) and (Prediction4 is Pron4Alto) and (Prediction5 is Pron5Alto) then (Prediction is Alto)
15. If (Prediction1 is Pron1Bajo) and (Prediction2 is Pron2Bajo) and (Prediction3 is Pron3Alto) and (Prediction4 is Pron4Bajo) and (Prediction5 is Pron5Bajo) then (Prediction is Bajo)
16. If (Prediction1 is Pron1Alto) and (Prediction2 is Pron2Alto) and (Prediction3 is Pron3Bajo) and (Prediction4 is Pron4Alto) and (Prediction5 is Pron5Alto) then (Prediction is Alto)
17. If (Prediction1 is Pron1Bajo) and (Prediction2 is Pron2Bajo) and (Prediction3 is Pron3Bajo) and (Prediction4 is Pron4Alto) and (Prediction5 is Pron5Bajo) then (Prediction is Bajo)
18. If (Prediction1 is Pron1Alto) and (Prediction2 is Pron2Alto) and (Prediction3 is Pron3Alto) and (Prediction4 is Pron4Bajo) and (Prediction5 is Pron5Alto) then (Prediction is Alto)
19. If (Prediction1 is Pron1Bajo) and (Prediction2 is Pron2Bajo) and (Prediction3 is Pron3Alto) and (Prediction4 is Pron4Alto) and (Prediction5 is Pron5Bajo) then (Prediction is Bajo)
20. If (Prediction1 is Pron1Alto) and (Prediction2 is Pron2Alto) and (Prediction3 is Pron3Bajo) and (Prediction4 is Pron4Bajo) and (Prediction5 is Pron5Alto) then (Prediction is Alto)
21. If (Prediction1 is Pron1Bajo) and (Prediction2 is Pron2Alto) and (Prediction3 is Pron3Alto) and (Prediction4 is Pron4Bajo) and (Prediction5 is Pron5Bajo) then (Prediction is Bajo)
22. If (Prediction1 is Pron1Alto) and (Prediction2 is Pron2Bajo) and (Prediction3 is Pron3Bajo) and (Prediction4 is Pron4Alto) and (Prediction5 is Pron5Alto) then (Prediction is Alto)
23. If (Prediction1 is Pron1Bajo) and (Prediction2 is Pron2Bajo) and (Prediction3 is Pron3Alto) and (Prediction4 is Pron4Alto) and (Prediction5 is Pron5Bajo) then (Prediction is Bajo)
24. If (Prediction1 is Pron1Alto) and (Prediction2 is Pron2Alto) and (Prediction3 is Pron3Bajo) and (Prediction4 is Pron4Bajo) and (Prediction5 is Pron5Alto) then (Prediction is Alto)
25. If (Prediction1 is Pron1Bajo) and (Prediction2 is Pron2Alto) and (Prediction3 is Pron3Alto) and (Prediction4 is Pron4Bajo) and (Prediction5 is Pron5Alto) then (Prediction is Alto)
26. If (Prediction1 is Pron1Alto) and (Prediction2 is Pron2Bajo) and (Prediction3 is Pron3Bajo) and (Prediction4 is Pron4Alto) and (Prediction5 is Pron5Alto) then (Prediction is Bajo)
27. If (Prediction1 is Pron1Bajo) and (Prediction2 is Pron2Alto) and (Prediction3 is Pron3Bajo) and (Prediction4 is Pron4Alto) and (Prediction5 is Pron5Bajo) then (Prediction is Bajo)
28. If (Prediction1 is Pron1Alto) and (Prediction2 is Pron2Bajo) and (Prediction3 is Pron3Alto) and (Prediction4 is Pron4Bajo) and (Prediction5 is Pron5Alto) then (Prediction is Alto)
29. If (Prediction1 is Pron1Bajo) and (Prediction2 is Pron2Alto) and (Prediction3 is Pron3Alto) and (Prediction4 is Pron4Alto) and (Prediction5 is Pron5Alto) then (Prediction is Bajo)
30. If (Prediction1 is Pron1Alto) and (Prediction2 is Pron2Bajo) and (Prediction3 is Pron3Bajo) and (Prediction4 is Pron4Alto) and (Prediction5 is Pron5Alto) then (Prediction is Bajo)
31. If (Prediction1 is Pron1Bajo) and (Prediction2 is Pron2Alto) and (Prediction3 is Pron3Bajo) and (Prediction4 is Pron4Alto) and (Prediction5 is Pron5Alto) then (Prediction is Alto)
32. If (Prediction1 is Pron1Alto) and (Prediction2 is Pron2Bajo) and (Prediction3 is Pron3Alto) and (Prediction4 is Pron4Bajo) and (Prediction5 is Pron5Bajoo) then (Prediction is Bajo)

Fig. 5 Rules of the type-2 fuzzy system

In this architecture, we have two layers in each module. In module 1, in the first layer we have 23 neurons and the second layer we have 9 neurons, and in module 2 we used 9 neurons in the first layer and the second layer we have 15 neurons the Levenberg–Marquardt (LM) training method was used; three delays for the network were considered.

Table 1 shows the PSO results (as shown in Fig. 6) where the prediction error is of 0.0013066.

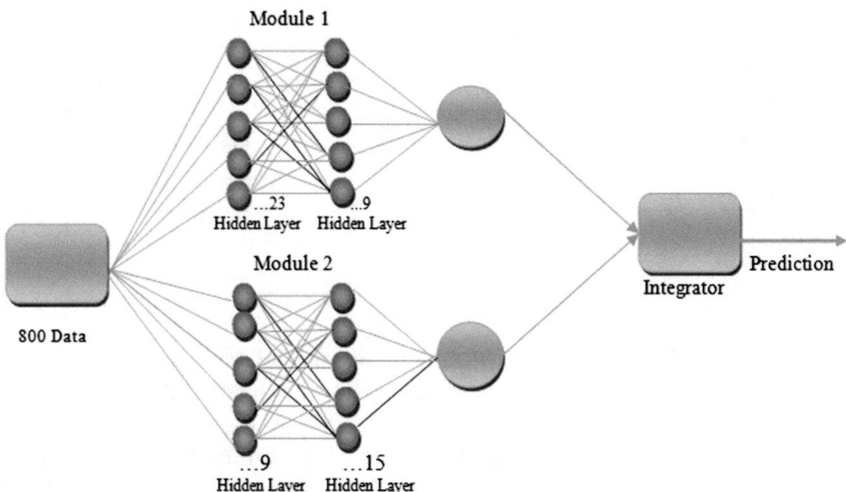

Fig. 6 Prediction with the optimized ensemble neural network with GA of the TAIEX

Table 1 Particle swarm optimization result for the ensemble neural network

No.	Iterations	Particles	Number of modules	Number of layers	Number of neurons	Duration	Prediction error
1	100	100	2	3	13, 16, 2 18,20,18	01:48:30	0.002147
2	100	100	2	2	3, 9 14, 19	01:03:09	0.0021653
3	100	100	2	2	20, 4 10, 7	01:21:02	0.0024006
4	100	100	2	2	16, 19 3,12	01:29:02	0.0019454
5	100	100	2	2	19, 19 24, 17	02:20:22	0.0024575
6	100	100	2	3	21, 14, 23 14, 24, 20	01:21:07	0.0018404
7	100	100	2	2	23, 9 9, 15	01:19:08	0.0013065
3	*100*	*100*	2	2	15, 17 9, 22	01:13:20	0.0018956
9	*100*	*100*	2	2	22, 15 20, 16	01:13:35	0.0023377
10	*100*	*100*	2	2	23, 8 10, 17	01:04:23	0.0023204

Table 2 Result PSO for the type-1 fuzzy integration of the TAIEX

Experiments	Prediction error with fuzzy integration type-1
Experiment 1	0.0473
Experiment 2	0.0422
Experiment 3	0.0442
Experiment 4	0.0981
Experiment 5	0.0253
Experiment 6	0.0253
Experiment 7	0.0253
Experiment 8	0.0235
Experiment 9	0.0253
Experiment 10	0.0253

Fuzzy integration is performed initially by implementing a type-1 fuzzy system in which the best result is in experiment of row number 8 of Table 2 with an error of: 0.0235.

As a second phase, to integrate the results of the optimized ensemble neural network a type-2 fuzzy system is implemented, where the best results that are obtained are as follows: with a degree uncertainty of 0.3 a forecast error of 0.01098 is obtained, with a degree of uncertainty of 0.4 the error is of 0.01122 and with a degree of uncertainty of 0.5 the error is of 0.001244, as shown in Table 3.

Table 3 Result PSO for the type-2 fuzzy integration of the TAIEX

Experiment	Prediction error 0.3 uncertainty	Prediction error 0.4 uncertainty	Prediction error 0.5 uncertainty
Experiment 1	0.0335	0.033	0.0372
Experiment 2	0.0299	0.5494	0.01968
Experiment 3	0.0382	0.0382	0.0387
Experiment 4	0.0197	0.0222	0.0243
Experiment 5	0.0433	0.0435	0.0488
Experiment6	0.0121	0.0119	0.0131
Experiment 7	0.01098	0.01122	0.01244
Experiment 8	0.0387	0.0277	0.0368
Experiment 9	0.0435	0.0499	0.0485
Experiment 10	0.0227	0.0229	0.0239

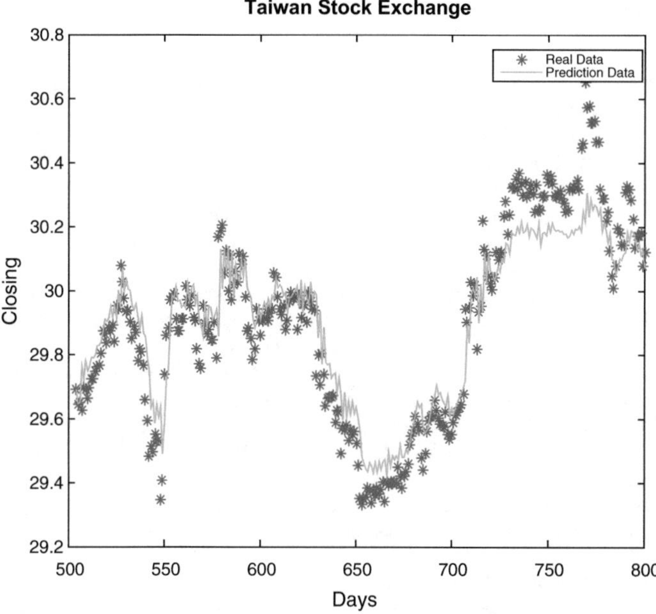

Fig. 7 Prediction with the optimized ensemble neural network with PSO of the TAIEX

Figure 7 shows the plot of real data against the predicted data generated by the ensemble neural network optimized with the PSO.

5 Conclusions

The best result when applying the particle swarm to optimize the ensemble neural network was: 0.0013066 (as shown in Fig. 6 and Table 1). Implemented a type-2 fuzzy system for ensemble neural network, in which the results where for the best evolution as obtained a degree of uncertainty of 0.3 yielded a forecast error of 0.01098, with an 0.4 uncertainty error: 0.01122, and 0.5 uncertainty error of 0.01244, as shown in Table 3. After achieving these results, we have verified efficiency of the algorithms applied to optimize the neural network ensemble architecture. In this case, the method was efficient but it also has certain disadvantages, sometimes the results are not as good, but genetic algorithms can be considered as good technique a for solving search and optimization problems.

Acknowledgment We would like to express our gratitude to the CONACYT, Tijuana Institute of Technology for the facilities and resources granted for the development of this research.

References

1. P. Cowpertwait, A. Metcalfe, Time Series, "Introductory Time Series with R.", Springer Dordrecht Heidelberg London New York, 2009, pp. 2–5.
2. O. Castillo, P. Melin. "Hybrid intelligent systems for time series prediction using neural networks, fuzzy logic, and fractal theory" Neural Networks, IEEE Transactions on Volume 13, Issue 6, Nov. 2002, pp. 1395–1408.
3. O. Castillo, P. Melin. "Simulation and Forecasting Complex Economic Time Series Using Neural Networks and Fuzzy Logic", Proceeding of the International Neural Networks Conference 3, 2001, pp. 1805–1810.
4. O. Castillo, P. Melin. "Simulation and Forecasting Complex Financial Time Series Using Neural Networks and Fuzzy Logic", Proceedings the IEEE the International Conference on Systems, Man and Cybernetics 4, 2001, pp. 2664–2669.
5. N. Karnikand M. Mendel, "Applications of type-2 fuzzy logic systems to forecasting of time-series", Information Sciences, Volume 120, Issues 1–4, November 1999, pp. 89-111.
6. A. Kehagias and V. Petridis., "Predictive Modular Neural Networks for Time Series Classification", Neural Networks,, January 1997, Volume 10, Issue 1 pp. 31–49.2000, pp.245–250.
7. L. P. Maguire., B. Roche, T.M. McGinnity and L. J. McDaid., Predicting a chaotic time series using a fuzzy neural network, Information Sciences, December 1998, Volume 112, Issues 1–4, pp. 125–136.
8. P. Melin., O. Castillo., S. Gonzalez., J Cota., W., Trujillo., and P. Osuna., "Design of Modular Neural Networks with Fuzzy Integration Applied to Time Series Prediction", Springer Berlin / Heidelberg, 2007, Volume 41/2007, pp. 265–273.
9. R. N. Yadav. Kalra P.K. and J. John., "Time series prediction with single multiplicative neuron model", Soft Computing for Time Series Prediction,, Applied Soft Computing, Volume 7, Issue 4, August 2007, pp. 1157–1163.
10. L. Zhao and Y. Yang., "PSO-based single multiplicative neuron model for time series prediction", Expert Systems with Applications, March 2009, Volume 36, Issue 2, Part 2, pp. 2805–2812.
11. P. T. Brockwell, & R. A. Davis., (2002). "Introduction to Time Series and Forecasting", Springer-Verlag New York, pp 1–219.
12. N. Davey, S. Hunt, R. Frank., "Time Series Prediction and Neural Networks", University of Hertfordshire, Hatfield, UK, 1999.
13. Neuro-Fuzzy and Soft Computing. J.S.R. Jang, C.T. Sun, E. Mizutani, Prentice Hall 1996.
14. I .M. Multaba, M A. Hussain., Application of Neural Networks and Other Learning. Technologies in Process Engineering. Imperial College Press. 2001.
15. A. Sharkey., Combining artificial neural nets: ensemble and modular multi-net systems, Springer- Verlag, London, 1999.
16. P. Sollich., and A. Krogh., Learning with ensembles: however-fitting can be useful, in: D.S. Touretzky, M.C. Mozer, M. E. Hasselmo (Eds.), *Advances in Neura lInformation Processing Systems 8*, Denver, CO, MIT Press, Cambridge, MA, 1996, pp.190–196.
17. L. K. Hansen., and P. Salomon., Neural network ensembles, *IEEE Trans. Pattern Analysis and Machine Intelligence* 12 (10) 1990 pp. 993-1001.
18. Sharkey A., "One combining Artificial of Neural Nets", Department of Computer Science University of Sheffield, U.K., 1996.
19. S. Gutta., H. Wechsler., Face recognition using hybrid classifier systems, in: *Proc. ICNN-96*, Washington, DC, IEEE Computer Society Press, Los Alamitos, CA, 1996, pp.1017–1022.
20. F. J. Huang., Z. Huang, H-J. Zhang, and T. H. Chen., Poseinvariantface recognition, in: *Proc. 4th IEEE International Conference on Automatic Face and Gesture Recognition*, Grenoble, France, IEEE Computer Society Press, Los Alamitos, CA.

21. H. . Drucker.,, R. Schapire., P .Simard P., Improving performance in neural networks using a boosting algorithm, in:S.J. Hanson, J.D. Cowan Giles (Eds.), *Advancesin Neural Information Processing Systems 5*, Denver, CO, Morgan Kaufmann, San Mateo, CA, 1993, pp.42–49.
22. J. Hampshire., A. Waibel., A novel objective function for improved phoneme recognition using time-delay neural networks, *IEEE Transactions on Neural Networks* 1(2), 1990, pp. 216–228.
23. J. Mao., A case study on bagging, boosting and basic ensembles of neural networks for OCR, in:*Proc. IJCNN-98*, vol. 3, Anchorage, AK, IEEE Computer Society Press, Los Alamitos, CA,1998,pp.1828–1833.
24. K. J. Cherkauer., Human expert level performance on a scientific image analysis task by a system using combine dartificial neural networks, in:P. Chan, S. Stolfo, D. Wolpert (Eds.), *Proc. AAAI-96 Workshop on Integrating Multiple Learned Models for Improving and Scaling Machine Learning Algorithms*, Portland, OR, AAAI Press, Menlo Park, CA, 1996, pp.15–21.
25. P. Cunningham., J. Carney, S. Jacob., Stability problems with artificial neural networks and the ensemble solution, *Artificial Intelligence in Medicine* 20(3) (2000) pp. 217–225.
26. Z.-H Zhou, Y. Jiang,,Y.-B. Yang., and S.-F. Chen., Lung cancer cell identification based on artificial neural network ensembles, *Artificial Intelligence in Medicine* 24(1) (2002) pp. 25–36.
27. Y.N. Shimshon., Intrator Classification of seismic signal by integrating ensemble of neural networks, IEEE Transactions Signal Processing 461 (5) (1998) 1194–1201.
28. Practical Optimization Algorithms and Engineering Applications "Introduction Optimization", Antoniou A. and Sheng W., Ed. Springer 2007, pp. 1–4.
29. R. Eberhart and J. Kennedy, "A new optimizer using swarm theory", in proc. 6th Int. Symp. Micro Machine and Human Science (MHS), Oct.1995, pp. 39–43.
30. J. Kennedy and R. Eberhart R., "Particle Swarm Optimization", in Proc. IEEE Int. Conf. Neural Network (ICNN), Nov. 1995, vol.4, pp. 1942–1948.
31. Taiwan Bank Database: www.twse.com.tw/en (April 03, 2011).

A New Hybrid PSO Method Applied to Benchmark Functions

Alfonso Uriarte, Patricia Melin and Fevrier Valdez

Abstract According to the literature of particle swarm optimization (PSO), there are problems of local minima and premature convergence with this algorithm. A new algorithm is presented called the improved particle swarm optimization using the gradient descent method as operator of particle swarm incorporated into the Algorithm, as a function to test the improvement. The gradient descent method (BP Algorithm) helps not only to increase the global optimization ability, but also to avoid the premature convergence problem. The improved PSO algorithm IPSO is applied to Benchmark Functions. The results show that there is an improvement with respect to using the conventional PSO algorithm.

Keywords Particle swarm optimization · Gradient descent method · Benchmark functions

1 Introduction

The particle swarm optimization (PSO) algorithm is inspired by the movements of bird flocks and the mutual collaboration among themselves in seeking food within the shortest period of time [14]. PSO is one of the most popular optimization algorithms due to its extremely simple procedure, easy implementation, and very fast rate of convergence. Apart from all the advantages, the algorithm has its drawbacks too [1]. The PSO algorithm faces problems with premature convergence as it is easily trapped into local optima. It is known that it is almost impossible for the PSO algorithm to escape from the local optima once it has been trapped, causing the algorithm to fail in achieving the near optimum result. Many methods have been proposed throughout the years to counter this drawback of the PSO algorithm [2, 17, 18].

A. Uriarte · P. Melin (✉) · F. Valdez
Division of Graduate Studies and Research, Tijuana Institute of Technology, Tijuana, Mexico
e-mail: pmelin@tectijuana.mx

P. Melin et al. (eds.), *Nature-Inspired Design of Hybrid Intelligent Systems*, Studies in Computational Intelligence 667, DOI 10.1007/978-3-319-47054-2_28

Although some improvement measures have been proposed, such as increasing population scale, the dynamic adjustment coefficient of inertia, these to a certain extent improve the optimization algorithm performance, but the algorithm itself has some of the nature problems that are not solved.

2 Gradient Descent Algorithm

The gradient descent algorithm is a method used to find the local minimum of a function [3, 6]. It works by starting with an initial guess of the solution, and it takes the gradient of the function at that point. It moves the solution in the negative direction of the gradient and it repeat the process [9, 10]. The algorithm will eventually converge where the gradient is zero (which corresponds to a local minimum) [11, 12].

A similar algorithm, the gradient ascent, finds the local maximum nearer to the current solution by stepping it toward the positive direction of the gradient [8]. They are both first-order algorithms because they take only the first derivative of the function [15].

The BP algorithm is a kind of simple deterministic local search method. It uses local adjustments aspects and shows a strong performance, along the direction of gradient descent can quickly find local optimal solution, but the gradient descent method (BP algorithm) is very sensitive to the choice of the initial position, and does not ensure that the optimal solution is the global optimal [16, 21, 23]. Figure 1 shows the Gradient Descent Method Visualization.

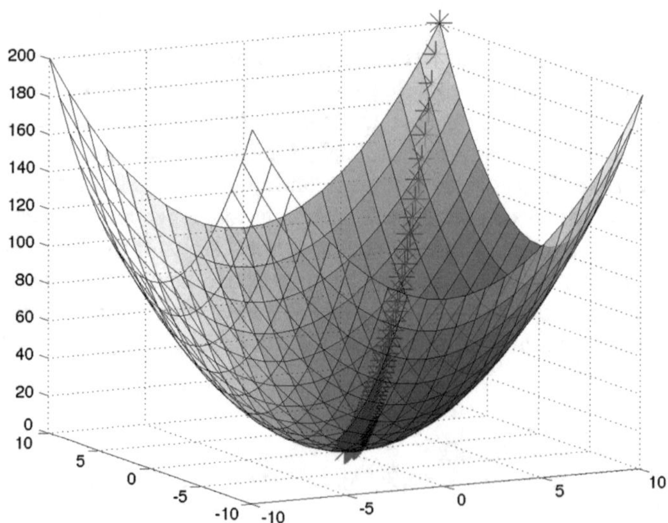

Fig. 1 Gradient descent method visualization

3 Particle Swarm Optimization

PSO was formulated originally by Eberhart and Kennedy in 1995. The thought process behind the algorithm was inspired by the social behavior of animals, such as bird flocking or fish schooling. PSO is similar to the continuous Genetic Algorithm (GA) in that it begins with a random population matrix [4]. Unlike the GA, PSO has no evolutionary operators such as crossover and mutation. The rows in the matrix are called particles (similar to the GA chromosome). They contain the variable values and are not binary encoded. Each particle moves about the cost surface with certain velocity [13, 19, 20]. The particles update their velocities and positions based on the local and global best solutions:

The equation to update the velocity is:

$$V_{ij}(t+1) = V_{ij}(t) + c_1 r_{1j}(t)[y_{ij}(t) - x_{ij}(t)] + c_2 r_2(t)[\hat{y}_j(t) - x_{ij}(t)] \tag{1}$$

The equation to update the position is:

$$x_i(t+1) = x_i(t) + v_i(t+1) \tag{2}$$

In Eqs. (1) and (2), x_i is particle i position, v_i is particle I velocity, y_i is best position of particle i, ŷ is the best global position, c_1 is cognitive coefficient, c_2 social coefficient, and r_{1j} and r_{2j} are random values.

The PSO algorithm updates the velocity vector for each particle then adds that velocity to the particle position or values. Velocity updates are influenced by both the best global solution associated with the lowest cost ever found by a particle and the best local solution associated with the lowest cost in the present population. If the best local solution has a cost less than the cost of the current global solution, then the best local solution replaces the best global solution [5].

The particle velocity is reminiscent of local minimizers that use derivative information, because velocity is the derivative of position. The constant C1 is called the cognitive parameter.

The constant C2 is called the social parameter. The advantages of PSO are that it is easy to implement and there are few parameters to adjust [7].

4 Improved Particle Swarm Optimization

The process is as follows: all particles first are improved by PSO in the group of every generation algorithm, according to the type (1) and (2) update each particle speed and position, and calculate for each particle the fitness function of value.

According to the particle fitness value choose one or several of the adaptable particles. These particles called elite individuals. The elite individual is not directly

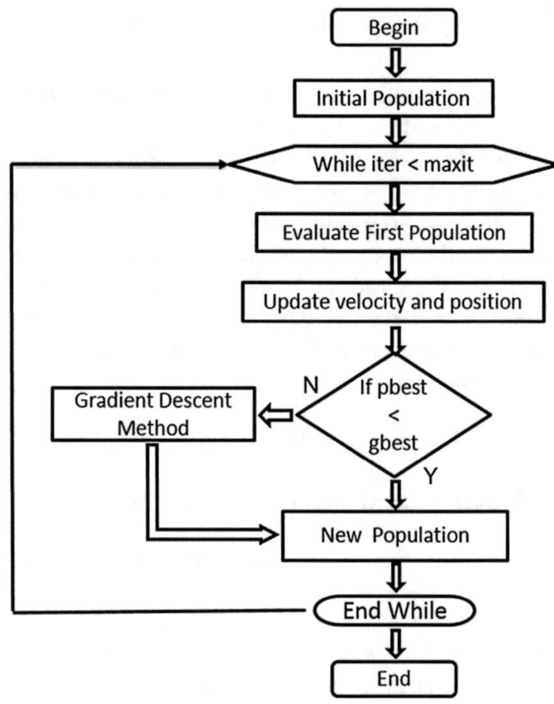

Fig. 2 Improved particle swarm optimization diagram

into the next generation of algorithm in the group, but rather through BP operator to improve performance [22, 24].

That is the BP operator in the area around to develop individual elite more excellent performance of particle position, and they lead the species particles rapid evolution in the next generation.

From the overall perspective, the improved particle swarm optimization algorithm is in the PSO algorithm of the next generation groups merge with BP algorithm (Gradient Descent Method) [25, 26]. Figure 2 shows a flow diagram of the Improved Particle Swarm Optimization.

5 Benchmark Functions

In order to verify the performance of the proposed algorithm, five benchmark functions are selected in the experiment. These functions are a set of nonlinear minimization problems which are described as follows:

1. Rosenbrock Function

$$f(x) = \sum_{i=1}^{d-1} \left[100\left(x_{i+1} - x_i^2\right)^2 + (x_i - 1)^2 \right]$$

2. Rastrigin Function

$$f(x) = 10d + \sum_{i+1}^{d} \left[x_i^2 - 10\cos(2\pi x_i) \right]$$

3. Sphere Function

$$f(x) = \sum_{i+1}^{d} x_i^2$$

4. Ackley Function

$$f(x) = -a\ \exp\left(-b\sqrt{\frac{1}{d}\sum_{i=1}^{d} x_i^2} \right) - \exp\left(\frac{1}{d}\sum_{i=1}^{d} \cos(cx_i) \right) + a + \exp(1)$$

5. Griewank Function

$$f(x) = \sum_{i=1}^{d} \frac{x_i^2}{4000} - \prod_{i=1}^{d} \cos\left(\frac{x_i}{\sqrt{i}} \right) + 1$$

6 Experimental Results

We made 30 experiments to test the algorithm performance with Benchmark mathematical functions: Rosenbrock, Rastrigin, Sphere, Ackley, and Griewank where the goal is approximate its value to zero. Table 1 shows the parameters of the Benchmark Functions.

Tables 2, 3, 4, 5, and 6 show the results of the benchmark functions.

Figure 3 shows the convergence graph of the IPSO.

Table 1 Parameters of the Benchmark functions

Function	n	Range of search
Rosenbrock	10	[−30, 30]
Rstrigin	10	[−5.12, 5.12]
Sphere	10	[−100, 100]
Ackley	10	[−15, 30]
Griewank	10	[−600, 600]

Table 2 Results of Rosenbrock function

Function	Algorithm	Minimum	Average
Rosenbrock	PSO	3.5132	4.2580
Rosenbrock	IPSO	0.1069	0.4261

Table 3 Results of Rastrigin function

Function	Algorithm	Minimum	Average
Rastrigin	PSO	1.99	3.34
Rastrigin	IPSO	0	0.18

Table 4 Results of Sphere function

Function	Algorithm	Minimum	Average
Sphere	PSO	1.25e−19	5.46e−15
Sphere	IPSO	2.95e−37	4.18e−21

Table 5 Results of Ackley function

Function	Algorithm	Minimum	Average
Ackley	PSO	7.04e−11	6.77e−09
Ackley	IPSO	0	1.27e−14

Table 6 Results of Griewank function

Function	Algorithm	Minimum	Average
Griewank	PSO	1.48e−02	1.17e−01
Griewank	IPSO	0	1.14e−02

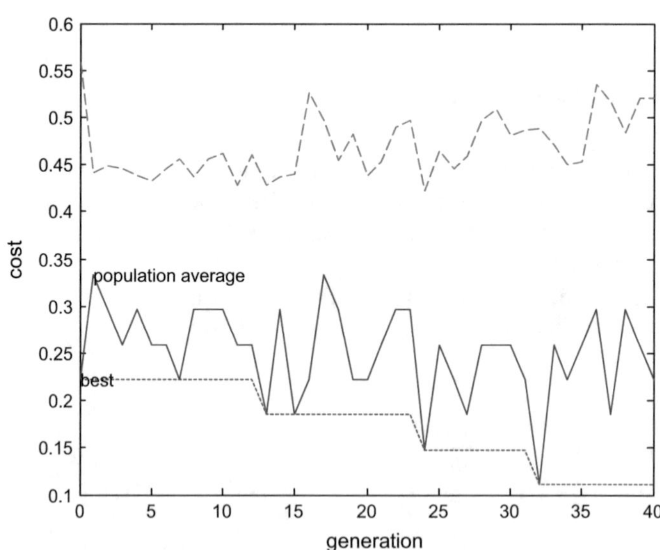

Fig. 3 Convergence graph of the IPSO

7 Conclusion

In order to reduce the search process for the particle swarm algorithm and the existing early convergence problem, a kind of ""variation" into the idea of the particle swarm algorithm is proposed, the gradient descent method (BP algorithm) as a particle swarm operator is embedded in particle swarm algorithm and helps to solve problems of local minimum and premature convergence with faster convergence velocity approaching the global optimal solution. Numerical experiments demonstrate that the ability to find the global optima; the convergent speed and the robustness of the proposed algorithm are better than those of the simple PSO algorithm.

References

1. Abraham, H. Guo, and H. Liu, "Swarm intelligence: Foundations, perspectives and applications, swarm intelligent systems", in Studies in Computational Intelligence. Berlin, Germany: Springer Verlag, pp. 3-25, 2006.
2. Afsahi Z, Javadzadeh R, "Hybrid Model of Particle Swarm Optimization and Cellular Learning Automata with New Structure of Neighborhood", International Conference on Machine Learning and Computing (1CMLC 2011), 2011.
3. Aibinun A.M., M.J.E. Salami, A.A. Shafie, "Artificial neural network based autoregressive modeling technique with application in voice activity detection", Engineering Applications of Artificial Intelligence, Volume 25, 2012, pp. 1265–1276.
4. Behnam A., Nader A., Detlef D. Nauck, 2000. Intelligent Systems and Soft Computing Prospects, Tools and Applications. Springer New York. 376p.
5. Engelbrecht A P., "Fundamentals of Computational Swarm Intelligence", University of Pretoria, South Africa, 2005, 1st edition.
6. Fei H., Qing Liu, "A diversity-guided hybrid particle swarm optimization based on gradient search", School of Computer Science and Communication Engineering, Jiangsu University, Zhenjiang, Jiangsu 212013, China.
7. GaoHai-bing, GAO Liang, "Particle Swarm Optimization Based Algorithm for Neural Network Learning", Acta Electronica Sinica, 32(9) (2004), 1572-1574.
8. Gevaert W., Tsenov G., Mladenov V., "Neural Networks used for Speech Recognition", Journal of Automatic Control, University of Belgrade, Vol. 20:1-7, 2010.
9. H. Shayeghi, H. A. Shayanfar, G. Azimi, "STLF Based on Optimized Neural Network Using PSO", International Journal of Electrical and Computer Engineering 4:10 2009.
10. Happel Bart L.M., Murre Jacob M.J., "Design and evolution of modular neural network architectures", Neural Networks, Volume 7, Issues 6–7, 1994, Pages 985–1004.
11. Ittichaichareon C., Siwat S. and Thaweesak Y., "Speech Recognition using MFCC", International Conference on Computer Graphics, Simulation and Modeling (ICGSM'2012) July 28-29, 2012 Pattaya (Thailand).
12. Jang J. Sun, J. Mitzutani E., "Neuro-Fuzzy and soft computing: a computational approach to learning and machine Intelligence", Prentice Hall, 1997.
13. Jun-qing, Quan-ke Pan, Sheng Xie, Bao-xian Jia and Yu-ting Wang. 2009. A Hybrid Particle Swarm Optimization and Tabu Search algorithm for Flexible Job-Shop Scheduling Problem. IACSIT.
14. Kennedy J. and Eberhart R., 1995. Particle swarm optimization, Proc. IEEE Int. Conf. on Neural Networks, IV. Piscataway, NJ: IEEE Service Center, pp. 1942–1948.

15. Mathew M. N., A new gradient based particle swarm optimization algorithm for accurate computation of global minimum, Appl. Soft Comput. 12(1) (2012) 353–359.
16. Mirjalili S., Siti Z., Mohd H., Hossein M. S., "Training feed forward neural networks using hybrid particle swarm optimization and gravitational search algorithm", Applied Mathematics and Computation, Volume 218, 2012, pp. 11125–11137.
17. Palupi R. D. and M. S. Siti, "Particle swarm optimization: technique, system and challenges," International Journal of Applied Information Systems, vol. 1, pp. 19-27, October 2011.
18. Q. H. Bai. "Analysis of particle swarm optimization algorithm," Computer and Information Science, vol. 3, pp. 180-184, February 2010.
19. Rehab F., Abdel-Kader, "Hybrid discrete PSO with GA operators for efficient QoS-multicast routing", Ain Shams Engineering Journal, Volume 2, Issue 1, March 2011, Pages 21–31.
20. RlKuo,C.C.Huang. Application of particle swam optimization algorithm for solving bi-level linear programming Problem [J]. Computers and Mathematics with Applications, 58 (2009), 678-685.
21. Sirko M., Michael P., Ralf S., and Hermann N., "Computing Mel-Frequency Cepstral Coefficients on the Power Spectrum", Lehrstuhlfür Informatik VI, Computer Science Department, RWTH Aachen – University of Technology, 52056 Aachen, Germany, 2001.
22. Weidong J., Keqi Wang, "An Improved Particle Swarm Optimization Algorithm", 2011 International Conference on Computer Science and Network Technology, (2011).
23. X.Yao. Evolving Artificial Networks Proceedings of the IEEE 87 (1999), 1423-1447.
24. Y Shi, R C Eberhart, "A modified particle swarm optimizer", Proc. IEEE Int Conf Evol Computing, Anchorage, Alaska, May (1998), 69-73.
25. Ying-ping Chen, Pei Jiang, "Analysis of particle interaction in particle swarm optimization", Theoretical Computer Science, 411 (2010), 2101-2115.
26. Ying-ping Chen, Pei Jiang. Analysis of particle interaction in particle swarm optimization [J]. Theoretical Computer Science, 411 (2010), 2101-2115.

On the Use of Parallel Genetic Algorithms for Improving the Efficiency of a Monte Carlo-Digital Image Based Approximation of Eelgrass Leaf Area I: Comparing the Performances of Simple and Master-Slaves Structures

Cecilia Leal-Ramírez, Héctor Echavarría-Heras, Oscar Castillo and Elia Montiel-Arzate

Abstract Eelgrass is a relevant sea grass species that provides important ecological services in near shore environments. The overall contribution of this species to human welfare is so important that upon threats to its permanence that associate to deleterious anthropogenic influences, a vigorous conservation effort has been recently enforced worldwide. Among restoration strategies transplanting plays a key role and the monitoring of the development of related plots is crucial to assess the restoration of the ecological features observed in donor populations. Since traditional eelgrass assessment methods are destructive their use in transplants could lead to undesirable effects such as alterations of shoot density and recruitment. Allometric methods can provide accurate proxies that sustain nondestructive estimations of variables required in the pertinent assessments. These constructs rely on extensive data sets for precise estimations of the involved parameters and also depend on precise estimations of the incumbent leaf area. The use of electronic scanning technologies for eelgrass leaf area estimation can enhance the nondestructive nature of associated allometric methods, because the necessary leaf area assessments could be obtained from digital images. But when a costly automatic leaf area meter is not available, we must rely on direct image processing, usually achieved through computationally costly Monte Carlo procedures. Previous results

C. Leal-Ramírez (✉) · H. Echavarría-Heras
Centro de Investigación Científica y de Educación Superior de Ensenada, Carretera Ensenada-Tijuana, No 3918, Zona Playitas, 22860 Ensenada, BC, Mexico
e-mail: cleal@cicese.mx

O. Castillo
Tijuana Institute of Technology, Tijuana, Mexico
e-mail: ocastillo@tectijuana.mx

E.M. Arzate
Instituto Tecnológico Nacional de México, Sede Ensenada,
Boulevard Tecnológico, No. 1, 22890 Ensenada, BC, Mexico

P. Melin et al. (eds.), *Nature-Inspired Design of Hybrid Intelligent Systems*,
Studies in Computational Intelligence 667, DOI 10.1007/978-3-319-47054-2_29

show that the amendment of simple genetic algorithms could drastically reduce the time required by regular Monte Carlo methods to achieve the estimation of the areas of individual eelgrass leaves. But even though this amendment, the completion of the task of measuring the areas of the leaves of a data set with an extension, as required for precise parameter estimation, still leads to a burdensome computational time. In this paper, we have explored the benefits that the addition of a master-slave parallel genetic algorithm to a Monte Carlo based estimation routine conveys in the aforementioned estimation task. We conclude that unless a suitable number of processors are involved, and also the proper mutation and crossover rates are contemplated the efficiency of the overall procedure will not be noticeably improved.

Keywords Nondestructive eelgrass monitoring · Leaf area estimation · Monte Carlo methods · Parallel genetic algorithms

1 Introduction

Eelgrass (*Zostera marina L.*) is a prevalent and ecologically important seagrass species that distributes in temperate estuaries worldwide. Eelgrass is the dominant seagrass species along the coasts of both the North Pacific and North Atlantic [1]. This seagrass forms dense meadows that play an important trophic role by provisioning the shallow-water food web with significant stocks of organic material. Eelgrass also bears important structural benefits by supplying habitat and shelter for associated algae and epifaunas, as well as, for waterfowl and many commercially important fish species and their larvae [2]. But beyond these fundamental trophic and structural roles, eelgrass beds contribute to the remediation of contaminated sediments [3], filter and retain nutrients from the water column [4], help in the stabilization of sediments [5] and reduce erosion forces by stumping wave energy, thus promoting the stabilization of adjacent shorelines [6]. These ecological services represent relevant eelgrass benefits for human welfare that are nowadays at the verge of disappearing due to deleterious anthropogenic, influences that threaten the permanence of this important species. The resulting concern has made eelgrass recipient of a vigorous research trust aimed to provide efficient restoration strategies. Among eelgrass reestablishing efforts, transplants play a fundamental role [7–9]. The monitoring of the associated plots provides key information of the success of reinstatement of the ecological functions and values of natural populations. These endeavors ordinarily require monitoring leaf area, rhizome diameter, individual shoot height, density, and weight, and overall productivity of transplanted plants and comparing with similar assessments taken on plants of the reference population, which usually settles nearby [10].

The variability in the production of eelgrass biomass constitutes a dynamic linkage between its structural and trophic roles. Therefore, the assessment of the productivity and resulting standing stock is fundamental for the valuation of the overall status of a given population. In eelgrass shoots are the fundamental biomass production units and include roots, rhizomes, and the leaves. A rhizome is a flat stem that propagates along the sediment, and as it elongates produces roots. Root appearance befalls at leaf scars that are also identified as rhizome nodes. Leaves appear from a meristem lying within a protecting sheath. This structure also grips the leaves together and is attached at the actively growing end of the rhizome. The production of leaves and rhizome nodes is connected in such a way that every leaf formed is associated with a rhizome node. Hence, the overall production of shoots can be estimated by measuring the production of the leaves [11]; this makes knowing the rate of the production of leaves fundamental to the assessment of eelgrass populations.

In an attempt to make this presentation self-contained we first provide formal descriptions of the average biomass of the leaves in eelgrass shoots and the corresponding average rate of leaf growth per shoot-day. For that aim in what follows t will label time and $w(t)$ eelgrass leaf biomass. A letter s used as a subscript will label a generic eelgrass shoot, that will be assumed to hold a number $nl(s)$ of leaves, with combined biomass symbolized by means of $w_s(t)$. Then, observed biomass in shoots can be formally represented by means of

$$w_s(t) = \sum_{nl(s)} w(t), \tag{1}$$

here summation of the leaves that the shoot s holds is indicated by means of $\sum_{nl(s)}$. Meanwhile, the corresponding average for the leaf biomass of a number $ns(t)$ of shoots collected at a time t is denoted using the symbol $w_{ms}(t)$ and calculated through,

$$w_{ms}(t) = \frac{\sum_{ns(t)} w_s(t)}{ns(t)} \tag{2}$$

where $\sum_{ns(t)}$ stands for summation over the number $ns(t)$ of collected shoots. Correspondingly, if $\Delta w(t, \Delta t)$ represents the increase in biomass produced by an individual leaf over an interval $[t, t + \Delta t]$, then denoting by means of $L_{sg}(t, \Delta t)$ the consequential average rate of growth of the leaves on the shoot s, we have,

$$L_{sg}(t, \Delta t) = \frac{\sum_{nl(s)} \Delta w(t, \Delta t)}{\Delta t}, \tag{3}$$

and consequently if $L_g(t, \Delta t)$ stands for the associated average rate of leaf growth per shoot and per day, we then have,

$$L_g(t, \Delta t) = \frac{\sum_{ns(t,\Delta t)} L_{sg}(t, \Delta t)}{ns(t, \Delta t)}, \tag{4}$$

where $\sum_{ns(t,\Delta t)}$ is used to indicate the sum of the shoots retrieved over the marking interval $[t, t+\Delta t]$ and $ns(t, \Delta t)$ denoting their number.

Due to seasonal effects, monitoring average leaf biomass and mean rate of leaf growth per shoot-day in eelgrass demands extensive sampling during the entire annual cycle. Many methods for the estimation of these variables have required sampling procedures that include the removal of a representative number of shoots, followed by time-consuming dry-weight determinations. Particularly, the destruction of shoots during the early stages of growth might seriously disturb transplanted populations; therefore in avoiding undesirable effects, sampling endeavors required to characterize the dynamics of the variables described by Eqs. (1) and (4) should be preferably achieved using nondestructive methods. It is at this stage of eelgrass assessment, that allometric methods become fundamental [12–17]. Indeed, as explained in the appendix the scaling relationship

$$w(t) = \beta a(t)^{\alpha}, \tag{5}$$

where $a(t)$ stand for eelgrass leaf area, and α and β are parameters, substantiates the derivation of a proxy $w_{ms}(\alpha, \beta, t)$ for the average leaf biomass in shoots $w_{ms}(t)$,and another $L_g(\alpha, \beta, \Delta t)$ corresponding to the average rate of leaf growth per shoot and per day $L_g(t, \Delta t)$. Indeed, the allometric alternate $w_{ms}(\alpha, \beta, t)$ becomes

$$w_{ms}(\alpha, \beta, t) = \frac{\sum_{ns(t)} w_s(\alpha, \beta, t)}{ns(t)}, \tag{6}$$

where

$$w_s(\alpha, \beta, t) = \sum_{nl(s)} \beta a(t)^{\alpha}. \tag{7}$$

Correspondingly the $L_g(\alpha, \beta, t, \Delta t)$ proxy for $L_g(t, \Delta t)$ is given by,

$$L_g(\alpha, \beta, \Delta t) = \frac{\sum_{ns(t,\Delta t)} L_{sg}(\alpha, \beta, t, \Delta t)}{ns(t, \Delta t)}, \tag{8}$$

where

$$L_{sg}(\alpha, \beta, \Delta t) = \frac{\sum_{nl(s)} \Delta w(\alpha, \beta, t, \Delta t)}{\Delta t}, \tag{9}$$

and

$$\Delta w(\alpha, \beta, t, \Delta t) = \beta a(t + \Delta t)^{\alpha}\left(1 - \frac{\beta a(t)^{\alpha}}{a(t + \Delta t)^{\alpha}}\right) \tag{10}$$

The appropriateness of the allometric surrogates $w_{ms}(\alpha, \beta, t)$ and $L_{sg}(\alpha, \beta, \Delta t)$ to furnish truly nondestructive assessments, mainly depends on the time invariance of the parameters α and β, because estimates previously fitted at a site could be used to readily obtain substitutes for currently observed values of $w_{ms}(t)$ and $L_g(t, \Delta t)$. Solana-Arellano et al. [18] demonstrated that within a given geographical region the parameters α and β can be in fact considered as time invariant. But even though α and β are statistically invariant environmental influences are expected to induce a relative extent of variability on local estimates of α and β and as Solana-Arellano et al. [18] and Echavarria-Heras et al. [16] warned inherent imprecision could propagate significant uncertainties on the values of $w_{ms}(\alpha, \beta, t)$ and $L_{gs}(\alpha, \beta, \Delta t)$. Particularly, Echavarría-Heras et al. [16] pointed out that data quality, analysis method, as well as, sample size are factors that could sensibly affect the precision of estimates of the parameters α and β in Eq. (5) thereby influencing the suitability of $w_{ms}(\alpha, \beta, t)$ and $L_{gs}(\alpha, \beta, t, \Delta t)$. Moreover, if the fitting of the model of Eq. (5) is performed on raw data, due to data quality issues, in order to assure consistent model identification a data set including over 10,000 leaves will be necessary [16]. On addition to precision issues related to the reliability of fittings of Eq. (5) to available data, by noticing from Eq. (5) that both $w_{ms}(\alpha, \beta, t)$ and $L_{gs}(\alpha, \beta, t, \Delta t)$ depend on leaf area $a(t)$, then the accuracy of the procedure used to estimate this variable is another factor spreading uncertainty over these allometric surrogates. And by recalling the large number of leaves that is necessary for accurate estimation of the parameters of Eq. (5) surely an additional efficiency concern arises. This relates to the processing time required for the estimation of the areas of the leaves in the extensive data set that is required.

The leaves of *Z. marina* present a strip-like appearance and this feature facilitates obtaining direct estimations of observed blade length $l(t)$ and width $h(t)$. As explained in the appendix, allometric methods also substantiate that these variables provide opportune estimations of the corresponding blade area $a(t)$ through an approximation $a(t)_p$ given by,

$$a(t)_p = l(t)h(t). \tag{11}$$

For regularly shaped eelgrass leaves the substitute given by Eq. (11) will yield handy approximations to $a(t)$ provided $h(t)$ is measured at a point halfway the leaf base and its tip [19]. When both $h(t)$ and $l(t)$ can be measured in situ without removing the leaves Eq. (11) could produce convenient nondestructive leaf area assessments. Moreover, the contents of the appendix also demonstrate that the $a(t)_p$ approximation can be used in place of real leaf area $a(t)$ in order to adapt nondestructive proxies similar to those given by Eqs. (7) and (8). But sampling conditions could impair the possibility of taking direct measurements of leaf length $l(t)$

and width $h(t)$, so besides Eq. (11), the use of an automatic leaf area meter could be recommended for producing nondestructive estimations of leaf area. An affordable alternative comprises the use of direct image processing techniques that could sustain a convenient cost effective and noninvasive approach for eelgrass leaf area estimation [20, 21]. Certainly, by counting the number of pixels contained in the image of a leaf and knowing as well, the image resolution, that is the number of pixels in a squared centimeter; leaf area estimation could reduce to calculating a simple proportion. Nevertheless, in practice the amount of pixels that an image can enclose will be limited by the technical features of the generating gauge [22]. Most commercial devices used to produce leaf images withstand a maximum resolution which seems unsuitable for highly precise estimation, and as a consequence counting pixels in an image could undervalue the actual area in the leaf. Consequently, improving precision entails increasing the resolution of the leaf image which could eventually require the use of an expensive automatic leaf area meter. In addition, leaf area estimates obtained by portable commercial electronic scanning devices depend sensibly on leaf shape. Moreover for eelgrass the closer the leaf contour is to a rectangle the higher the accuracy of digital image based assessments will be. But data examinations show that we can expect environmental effects bearing severe damage to eelgrass leaves, rendering the associated digital image processing tasks problematic. Indeed, observations demonstrate that the effects of herbivory or drag forces can induce noticeable changes in the shapes of eelgrass leaves, and that these changes could be of such an extent that it might be difficult to identify the peripheral contour of the associated digital images [20].

The Monte Carlo method is a widespread procedure for the approximation of the area of a plane region. It can also be adapted to produce estimates of the area of images of eelgrass leaves [22]. While in the direct image processing method enhancing precision implies increasing the resolution of the leaf image, pertaining to the Monte Carlo method, an enhancement in precision will entail increasing the number of points randomly placed over the leaf image. But, this could boost processing time in such a way that the procedure would become computationally burdensome. Certainly while evaluating the performance of a direct Monte Carlo routine in producing estimates of eelgrass leaf area we found that this procedure requires a processing time t_{test} of 83.11 min to calculate the area $a(t)_{test}$ of a regularly shaped eelgrass leaf amounting to 0.257 cm^2. Assuming that a leaf has a regular shape, so that the total processing time for calculating its area $a(t)$ could be obtained by multiplying the ratio $a(t)/a(t)_{test}$ by t_{test} we can obtain a lower bound for the time required by the Monte Carlo routine in estimating the incumbent area. Then assuming that all available leaves for parameter identification have regular shapes, it will require a relative total processing time t_{tot}, of about 4.0×10^7 h to accomplish the task of measuring the areas of a batch as large as one containing $n_{tot} = 10{,}412$ sampled leaves reported by Echavarria-Heras et al. [16]. Nevertheless, we must recall that most sampled leaves display irregularities caused by herbivory or drag forces that pose contour identification problems. This will make the computational time required to achieve the task of measuring the areas of a sampled batch as large as n_{tot} considerably larger than t_{tot}. In other words using the regular Monte Carlo method for producing accurate estimations of the areas of the leaves of a sampling bunch as reported in [16] could involve a quite onerous processing time.

Genetic algorithms are adaptive methods that can be used to solve search and optimization problems. A genetic algorithm added to regular Monte Carlo technique can be considered a valued device that permits that digital images of eelgrass leaves produced by commercially available scanners could provide reliable and efficient estimations of the associated area. This could reduce the uncertainty in allometric estimations of the pertinent aboveground biomass and leaf growth rates, which can greatly contribute to the conservation of the considered seagrass species. Leal-Ramirez et al. [22] evaluated the prospective of a genetic algorithm for the improvement of the efficiency of the Monte Carlo method for approximating the area of a *Z. marina* leaf. Specifically, they focused in the problem of optimizing the time of execution of the Monte Carlo method. These authors report both a markedly increased accuracy, as well as, a noticeably boosting on computational efficiency to estimate the sample leaf area $a(t)_{\mathrm{test}}$. But, results reported by Leal-Ramirez et al. [22] imply that calculating the area of an average sized eelgrass leaf $a(t)_{av} = 690.05$ mm^2 could require an approximated processing time of $t_{tgsav} = 8.95$ h, which determines lower bound $(t_{tgsav} \cdot n_{\mathrm{tot}})$ of about 9×10^4 h for the computational time required in estimating the areas of a batch of n_{tot} leaves, which is still quite onerous. Therefore, the quest for a more efficient technique for estimation of the addressed leaf area is justified. And here we have considered that since genetic algorithms allows the execution of parallel processing, the benefits in computational time reduction derived from introducing these techniques could be expected to be manifest.

The basic idea behind most parallel programs is to divide a task into parts and solving these parts simultaneously using multiple processors. This divide-and-conquer approach can be applied to simple genetic algorithms in many different ways. Moreover, parallel genetic algorithms (PGAs) are capable of combining speed, efficacy, and efficiency (e.g., [23–27]). There are many examples in the literature showing why PGAs have become versatile calculation tools [28]. Some examples of PGAs applications to biological problems are: racing biological weapons, the coevolution of parasite-host interactions, symbiosis, and the flow of resources. For an overview of the applications of PGAs the reader is referred to [24, 29–31]. In this paper we add a parallel genetic algorithm to a regular Monte Carlo routine in order to explore the capabilities of PGAs in finding optimal solutions for the problem of estimating eelgrass leaf areas. We also compare the solutions obtained by using PGAs with those obtained by means of a simple genetic algorithm (SGA). Our fundamental aim is the search for values of parameters characterizing a parallel genetic algorithm which could reduce the processing time involved in calculating the area of a representative eelgrass leaf by means of Monte Carlo techniques. The key motivation in performing this task is the need to reduce the oversized processing time required by the simple genetic algorithm and Monte Carlo method approach in obtaining a reasonably accurate estimation of the areas of a sampling batch of hundreds of leaves such as n_{tot}.

2 Methods

2.1 Monte Carlo Approximation of Eelgrass Leaf Area

We take some of the fundamental ideas of the Leal-Ramirez et al. [22] method to explain the Monte Carlo estimation of eelgrass leaf area. We first consider a plane region L_{fg} contained in the Cartesian plane R^2, bounded above and below by the plots of two nonintersecting smooth real valued and continuous functions $f(x)$ and $g(x)$ which depend on an independent variable x defined over a domain $D = \{x|c \leq x \leq d\}$ that belongs to the set of real numbers R (see Fig. 1), with the inequality $f(x) > g(x)$ holding through D. We also assume that the domain L_{fg} is contained in a rectangle $\Omega = \{(x, y)|x_a \leq x \leq x_b, y_a \leq y \leq y_b\}$.

The length and width of the rectangle Ω are known (see Fig. 1), therefore the involved area $A\Omega$ is given by

$$A\Omega = (x_b - x_a) * (y_b - y_a). \tag{12}$$

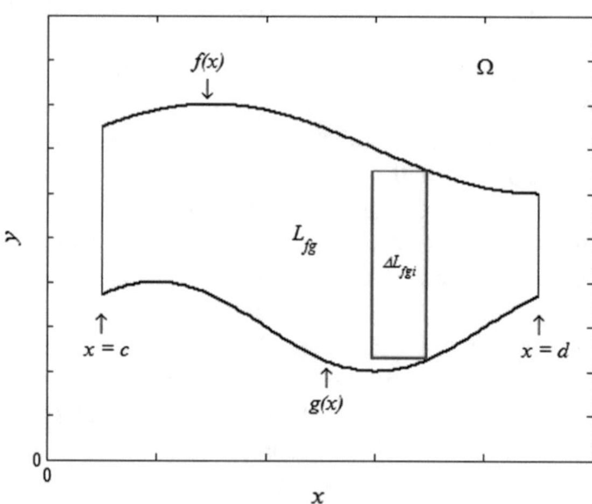

Fig. 1 The domain L_{fg} bounded *above* and *below* by the plots of the nonintersecting functions $f(x)$ and $g(x)$ extending from $x = c$ through $x = d$, and enclosed by a *rectangle* $\Omega = \{(x, y)|x_a \leq x \leq x_b, y_a \leq y \leq y_{b,}\}$. The area AL_{fg} of L_{fg} can be approximated by means of a collection of partition cells ΔL_{fgi}, $1 \leq i \leq k$ [22]

If AL_{fg} stands for the area of the domain L_{fg}, then a Monte Carlo procedure for the estimation of AL_{fg} is formally expressed through,

$$AL_{fg} = AL_{fgm} + e_m, \tag{13}$$

where,

$$AL_{fgm} = \left\langle \frac{np(L_{fg})}{np(\Omega)} A\Omega \right\rangle_m \tag{14}$$

stands for the average of the ratio $np(L_{fg})A\Omega/np(\Omega)$, taken over a large number m of trials of a random experiment that distributes a total number $np(\Omega)$ of points over Ω, with $np(L_{fg})$ of these points striking the region L_{fg} and with $|e_m|$ denoting the involved average approximation error, defined through

$$|e_m| = |AL_{fg} - AL_{fgm}| \tag{15}$$

In estimating eelgrass leaf area by using the Monte Carlo method it will be first required to adapt the dimensions of the reference rectangle Ω, surrounding the leaf image which is characterized as a region L_{fg} contained in Ω. Then we need to randomly place an appropriately large amount $np(\Omega)$ of points in the reference rectangle Ω. Afterwards we must directly count the number $np(L_{fg})$ of these points falling within the leaf image L_{fg} itself, which becomes a difficult task. Moreover, completion of this chore will entail slicing the leaf image in order to form a partition of disjoint rectangles ΔL_{fgi} (Fig. 1), with $1 \leq i \leq k$. And in estimating the leaf area, a procedure must recursively count the number of points randomly placed on each one the partition cells covering the leaf image L_{fg}.

An appropriate number k of screen rectangles, is a prime requirement for accuracy, nevertheless on spite of an optimal partition has been chosen, given the amount of points that must be placed in the interior of a leaf for a truthful estimation of its area, whenever a leaf has a long length and presents irregularities, the processing time required by the Monte Carlo method could be onerous. It is worthwhile to stress, that on addition to the computational load of testing the placement of points inside the image in every single cycle, the Monte Carlo method requires as well to perform a large number m of replicates for averaging purposes, which could rise the overall processing time in a significant way.

2.2 Structure of a Simple Genetic Algorithm

Simple genetic algorithms (SAGs) deal with a population n_{ind} of individuals, each one representing a feasible solution to a given problem. Each individual is assigned a value that relates to the suitability of the solution it provides. The greater the adaptation of an individual to the problem, the larger the likelihood that it will be

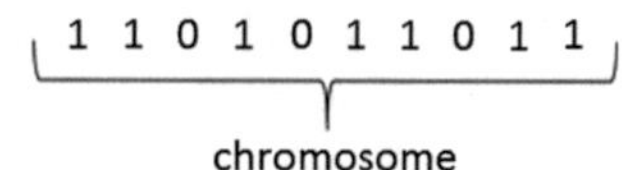

Fig. 2 Chromosome formed by binary entries

selected for breeding and then crossing its genetic material with another individual selected in the same way. This crossing will produce new individuals called children, which share some of the characteristics of their parents but include a higher proportion of good features entailing a better solution to the problem. In this way a new population of possible solutions is produced, that replaces the previous one, and along many generations good features spread through the remaining population. While it is true that the ideal size of a population cannot be exactly specified, we will choose the size of populations provided the search space is adequately covered. By favoring the crossing of the fittest individuals, the most promising areas of the search space are explored, achieving through this process convergence toward an optimal solution of the problem.

Individuals are represented by a set of parameters, which grouped form a chromosome (Fig. 2).

Adaptation of an individual to the problem depends on its assessment, which can be obtained from the chromosome using the evaluation function (13). The adaptation or objective function [cf. Eq. (15)] assigns to the individual represented by a chromosome a real number that sets its level of adaptation to the problem. Fittest individuals are those who in accordance with their assessments show the minimum value of the error specified by Eq. (15). During the execution of the algorithm, the fittest individuals should be selected for breeding. Each individual has probability $p_{i,t}$ of selection as a parent, being this likelihood proportional to the value of the objective function

$$p_{i,t} = \frac{|e_m(i,t)|}{\sum_{i=1}^{n} |e_m(i,t)|}, \tag{16}$$

where $|e_m(i,t)|$ stands for absolute approximation error linked to the ith individual in the population [cf. Eq. (15)].

Selected individuals are paired using a point based crossover operator, which cuts two chromosomes at a randomly chosen position to produce two new whole chromosomes (Fig. 3).

The crossover operator does not apply to all parents who have been selected to mate, but it is applied randomly, usually with a probability ranging between 0.5 and 1.0. A mutation operator acts on each of the children. The mutation operator is applied to each child individually, inducing a random alteration (usually with a small probability) on each gene composing the chromosome (Fig. 4). The mutation operator safeguards that no point in the search space has zero probability of being examined, which is very important to ensure convergence of the genetic algorithm.

The combination of the crossover and mutation functions will produce a set of individuals (possible solutions), which as the genetic algorithm evolves will be part

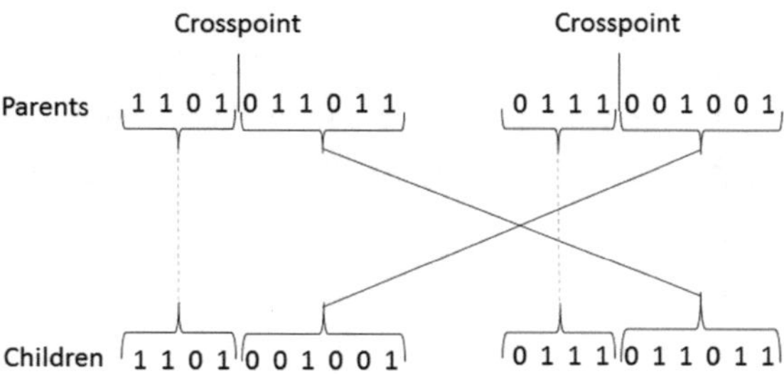

Fig. 3 Crossbreeding based on a point

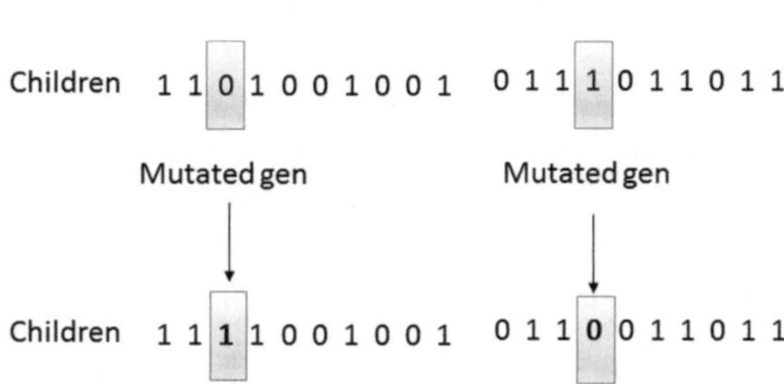

Fig. 4 Mutation

of the succeeding population. The population develops over a number n_{gen} of successive generations so that the extended average adaptation of all individuals in the population, as well as the adaptation of the best individuals will increase allowing the procedure to progress toward the global optimum. When at least 95 % of individuals in the population share the same fitness value [cf. Eq. (15)] then the genetic algorithm has attained convergence.

According to the aforesaid ideas sustaining the development of the theory of genetic algorithms the implementation of the related methods to the present settings will be achieved by the ensuing procedure

- Create an initial population
- Compute the function of evaluation of each individual
  WHILE NOT FINISHED = TRUE DO
  BEGIN /* create a new generation */
  - Select individuals with better fitness of the prior generation for breeding
    FOR selected population size /2 DO
    BEGIN
    - Select two individuals of the selected population for breeding.
    - Cross the two individuals to obtain two descendants
    - Mute descendants
    - Insert descendants into new generation

END

- Insert not selected population for breeding into new generation
- Compute the evaluation function of descendants
- IF population has produced convergence THEN FINISHED = TRUE

END

2.3 Forms of Parallel Genetic Algorithms

The basic idea of the parallel genetic algorithm is to divide a task into chunks and to solve them simultaneously using multiple processors. There are three commonly used types of PGAs: (1) master-slave, (2) fine-grained and (3) coarse-grained. In a master-slave parallel genetic algorithm there is a single population, selection and crossover involves the entire population, nevertheless, fitness assessment is performed by several processors (Fig. 5).

Fine-grained PGAs are suited for massively parallel computers and consist of one spatially structured population. The processes of selection mating occur only around a small neighborhood, nevertheless, the neighborhoods can intersect allowing a certain degree of interaction among the complete set of individuals. In an ideal arrangement there is only one individual for every processing element available.

Multiple-population PGAs consist on several subpopulations which interchange individuals occasionally. This trading of individuals is named migration, and it is ruled by several parameters making this process intricate. Since the size of the population is smaller than that used by a single genetic algorithm, we would expect the PGA converging faster. However, we must assess the overall suitability of the associated solutions because the involved error could increase.

An important characteristic of the master-slave parallelization method is that this does not alter the performance of the algorithm while the last two methods change the operating mode of the PGA. For example, in master-slave PGAs, selection is carried over the whole population, but in the other two PGAs considered here, selection takes into account only a share of the total number of individuals. Also, in the master-slave any two individuals in the population can randomly mate, nonetheless in the former methods mating is restricted to a subgroup of individuals. For the present aims we adopted a master-slave parallelization method, where the

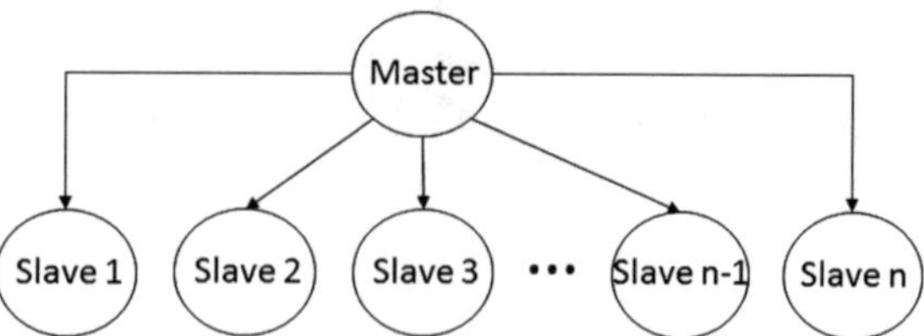

Fig. 5 Structure of a master-slave parallel genetic algorithm

evaluation of fitness is distributed among several processors (see Fig. 5) because the fitness of an individual is independent from that of the other members of the population and slaves are not intended to communicate during this phase, but the crossover involves only a selected subset of individuals. Additionally, the evaluation of individuals [cf. Eq. (15)] is parallelized by assigning a fraction of the number m of Monte Carlo trials to each of the available slaves. Communication between master and slaves occurs only as each slave receives the whole population and a fraction of m to evaluate. Communication also occurs when the slaves return the fitness average values. The master receives the average fitness values produces by each one of the slaves to obtain the average fitness of all the individuals of the population. Moreover, the master applies the genetic operators of selection, crossover and mutation to produce the next generation. A probability P_c determines whether two selected individuals are crossed or not. The crossing of two individuals is done at a point randomly chosen extracted from a uniform probability distribution to form two descendants. If crossing does not occur, then exact copies of parents are taken as the descendants. Likewise, a probability P_m determines whether or not descendants mutate. The mutation of the descendants occurs at a randomly chosen point extracted from a uniform probability distribution.

3 Results

We adapted a parallel genetic algorithm of master-slaves type with the aim of optimizing the execution time required by a standard Monte Carlo method in counting the number of points randomly distributed over a collection of rectangles or cells approximating the target area L_{fg} (Fig. 1). Our motivation here is the exploration of the capabilities of PAGs to find optimal solutions and also on performing a comparison of the results with the those obtained by using a simple genetic algorithm (e.g., [22]). Therefore, by setting $x_a = 0.0$, $x_b = 1.29$, $y_a = 0.0$, $y_b = 0.25$ we consider the reference region $\Omega = \{(x, y)|0.0 \leq x \leq 1.29, 0.0 \leq y \leq 0.25\}$, with x and y measured in cm. In order to characterize L_{fgo} we chose $f(x) = 0.254$ and $g(x) = 0.0254$, that is, we have $L_{fgo} = \{(x, y)|0.0254 \leq x \leq 1.2954,\ 0.0254 \leq y \leq 0.254\}$. And, to determine if a point (x, y) lies within the target area AL_{fg} we verified the following conditions

$$0.0254 \leq i \leq 1.2954 \quad \text{and} \quad 0.0254 \leq j \leq 0.254 \tag{17}$$

The exact area to be estimated is $AL_{fgo} = 0.290322\ \text{cm}^2$. For the purposes of this study, it will be only necessary that our chromosome represents the number of point $n_p(\Omega)$ distributed in Ω. Nevertheless, in order to increase precision of estimates of areas of leaves, it will be necessary that the chromosome includes as well the number of rectangles ΔL_{fgi} and the number m of iterations that the Monte Carlo method requires to approximate the target area L_{fg}. In this work, we fixed $i = 1$

rectangles while the number of iterations required by the Monte Carlo method was set at $m = 40$.

For the involved task, we depended on a computer with a Quad Core AMD processor, with a velocity of 2.4 HRZ and including 8 Gbytes RAM memory. In order to test the efficiency of the parallel genetic algorithm, we performed several simulation runs using one master and three slaves and setting a value of $|e_m| = 1.0 \times 10^{-3}$ as a search stopping condition, as well as, a fixed number n_{gen} of generations required to estimate AL_{fgo}. That is, the simulation runs stopped when an error value of $|e_m| = 1.0 \times 10^{-3}$ or n_{gen} generations were reached.

In Tables 1, 2 and 3, we provide the results of the simulations, this allows a comparison of efficiencies between simple and parallel genetic algorithms, and also the identification of the effects of mutation rate P_m, crossover rate P_c, number of individuals n_{ind} and number of generations n_{gen} on optimizing the processing time required to estimate the area of the blade.

Table 1 shows the number of generations that minimize the average approximation error $|e_m|$. The greater the number of generations is, the smaller the attained average approximation error $|e_m|$ becomes (Fig. 6).

Our results also show, that the average approximation error $|e_m| = 0.00039$ cm^2 achieved by the parallel genetic algorithm when $P_c = 0.1$, $P_m = 0.3$, $n_{\text{ind}} = 50$ and $n_{\text{gen}} = 50$, amounts to 35 % of the average approximation error of $|e_m| = 0.0011$ cm^2 accomplished by the parallel genetic algorithm when $P_c = 0.1$, $P_m = 0.3$, $n_{\text{ind}} = 50$, but now setting $n_{\text{gen}} = 30$. On the other hand, from Table 1 it can be ascertained that for most cases mutation rate minimizes average processing time, that is, the parallel genetic algorithm requires less processing time when the mutation rate is larger. As a matter of fact, the average processing time $\tilde{t} = 9.1274$ s achieved by the parallel genetic algorithm, when $P_c = 0.1$, $P_m = 0.3$, $n_{\text{ind}} = 50$ and $n_{\text{gen}} = 50$, corresponds to the 45 % of the average processing time of $\tilde{t} = 19.9707$ s accomplished by the simple genetic algorithm under the same parametric conditions.

Table 1 Comparison of average processing times $\tilde{t}$, and differences between average approximation errors $|e_m|$ obtained by the simple and the parallel genetic algorithms

Parameters				Simple genetic algorithm		Parallel genetic algorithm	
P_c	P_m	n_{ind}	n_{gen}	$\tilde{t}$ [s]	$\|e_m\|[cm^2]$	$\tilde{t}$ (s)	$\|e_m\|[cm^2]$
0.1	0.1	50	30	2.7680	0.0049	3.3079	0.0050
0.1	0.1	50	50	4.6868	0.0040	5.2211	0.0039
0.1	0.2	50	30	2.9507	0.0028	3.0503	0.0028
0.1	0.2	50	50	6.1572	0.0014	5.5511	0.0015
0.1	0.3	50	30	4.0147	0.0011	3.3774	0.0011
0.1	0.3	50	50	19.9707	0.00036	9.1274	0.00039

Calculations were obtained by setting $P_c = 0.1$ and $n_{\text{ind}} = 50$, and also by considering the variation ranges $0.1 \leq P_m \leq 0.3$ and $30 \leq n_{\text{gen}} \leq 50$

Table 2 Comparison of average processing times $\tilde{t}$, and of average approximation error $|e_m|$ obtained by the simple and parallel genetic algorithms, by setting $P_c = 0.2$ and $n_{\text{ind}} = 50$, and also by considering the variation ranges $0.1 \leq P_m \leq 0.3$ and $30 \leq n_{\text{gen}} \leq 50$

Parameters				Simple genetic algorithm		Parallel genetic algorithm	
P_c	P_m	n_{ind}	n_{gen}	$\tilde{t}$ [s]	$\|e_m\|[cm^2]$	$\tilde{t}$ [s]	$\|e_m\|[cm^2]$
0.2	0.1	50	30	2.7841	0.0037	2.7801	0.0040
0.2	0.1	50	50	4.9200	0.0025	4.7713	0.0026
0.2	0.2	50	30	3.5644	0.0021	2.8953	0.0019
0.2	0.2	50	50	8.3909	0.00087	5.9469	0.00094
0.2	0.3	50	30	5.1308	0.00081	3.6387	0.00076
0.2	0.3	50	50	49.7957	0.00023	9.7461	0.00035

Table 3 Comparison of average processing times $\tilde{t}$ and also of average approximation errors $|e_m|$ obtained by the simple and parallel genetic algorithms by setting $P_c = 0.3$ and $n_{\text{ind}} = 50$, and letting P_m and n_{gen} satisfy: $0.1 \leq P_m \leq 0.3$ and $30 \leq n_{\text{gen}} \leq 50$

Parameters				Simple genetic algorithm		Parallel genetic algorithm	
P_c	P_m	n_{ind}	n_{gen}	$\tilde{t}$ [s]	$\|e_m\|[cm^2]$	$\tilde{t}$ [s]	$\|e_m\|[cm^2]$
0.3	0.1	30	30	3.3571	0.0026	2.9209	0.0025
0.3	0.1	50	50	7.0889	0.0012	5.2032	0.0012
0.3	0.2	30	30	3.6104	0.0014	2.9740	0.0014
0.3	0.2	50	50	16.2421	0.00047	8.2554	0.00048
0.3	0.3	30	30	7.3970	0.00057	4.1402	0.00060
0.3	0.3	50	50	183.2114	0.00010	57.6975	0.00010

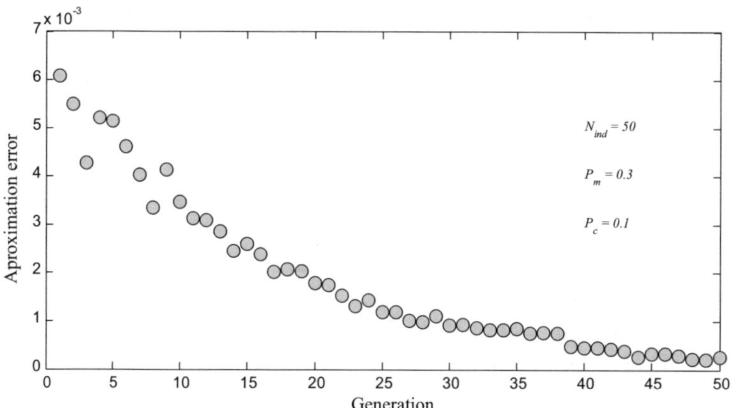

Fig. 6 The parallel genetic algorithm required a total processing average time of $\tilde{t} = 9.1274$ s to achieve the leaf area approximation task. The stopping condition was attained for $n_{\text{gen}} = 50$. The procedure required a total of $np(\Omega) = 37{,}310$, randomly selected points

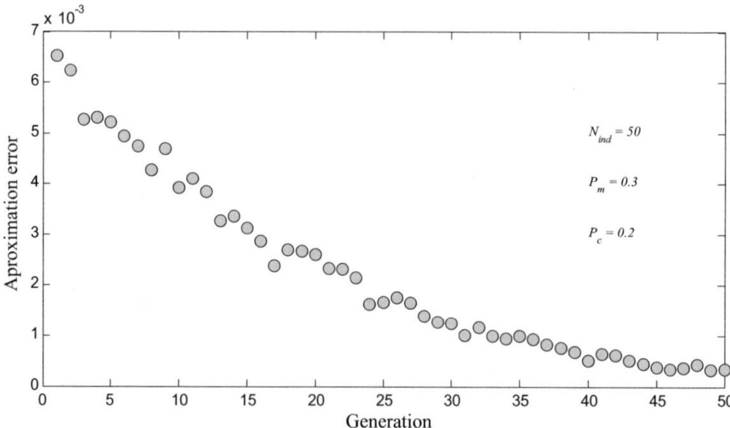

Fig. 7 The parallel genetic algorithm required a total processing average time of $\tilde{t} = 9.1274$ s to complete the leaf area approximation task. The stopping condition was attained for $n_{\text{gen}} = 50$. The chore required a total of $np(\Omega) = 58{,}401$, randomly selected points

Table 2 shows that the number of generations minimizes the average approximation error $|e_m|$. The greater the number of generations becomes, the smaller the attained average approximation error $|e_m|$ is (Fig. 7).

Moreover, the average approximation error of $|e_m| = 0.00035\ \text{cm}^2$ attained by the parallel genetic algorithm when $P_c = 0.2$, $P_m = 0.3$, $n_{\text{ind}} = 50$ and $n_{\text{gen}} = 50$, amounts to only 46 % of the average approximation error of $|e_m| = 0.00076\ \text{cm}^2$ achieved by the parallel genetic algorithm when $P_c = 0.2$, $P_m = 0.3$, $n_{\text{ind}} = 50$, but setting $n_{\text{gen}} = 30$. From Table 2 we can assess that unlike the results presented in Table 1, in all cases, mutation rate minimizes the average processing time. In other words, the parallel genetic algorithm requires less processing time when the mutation rate is larger. We also observed that the average processing time $\tilde{t} = 9.7461$ s achieved by the parallel genetic algorithm, when $P_c = 0.1$, $P_m = 0.3$ and $n_{\text{ind}} = 50$, corresponds to the amounts to only 19 % of the average processing time of $\tilde{t} = 49.7957$ s requiered by the simple genetic algorithm under the same parametric conditions. Also, we can verify that the average approximation errors presented in Table 2 are lower than those presented in Table 1, at least for half the cases. Furthermore, average processing times obtained by the parallel genetic algorithm are lower in Table 1 that these presented in Table 2, which means that the combination of $P_c = 0.2$ and $0.1 \leq P_m \leq 0.3$ induces a slight reduction in the average processing time achieved by the parallel genetic algorithm (Table 2).

From Table 3 we once more observe that the number of generations minimizes the average approximation error $|e_m|$, that is, the greater the number of generations, the smaller is the average accuracy error $|e_m|$ achieved (Fig. 8). Indeed, the average approximation error $|e_m| = 0.00010\ \text{cm}^2$ reached by the parallel genetic algorithm when $P_c = 0.3$, $P_m = 0.3$, $n_{\text{ind}} = 50$ and $n_{\text{gen}} = 50$, matches the 16 % of the

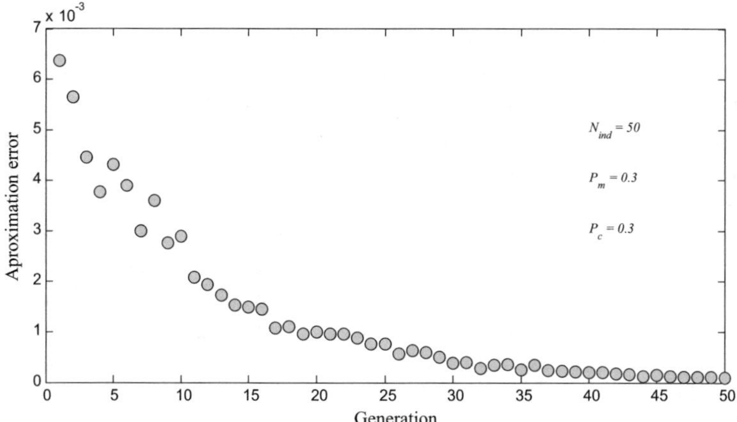

Fig. 8 The parallel genetic algorithm required a total processing average time of $\tilde{t} = 57.6975$ s to accomplish the leaf area approximation chore. The stopping condition was attained for $n_{\text{gen}} = 50$. The routine required a total of $np(\Omega) = 831{,}458$, randomly selected points

average approximation error of $|e_m| = 0.00060\ \text{cm}^2$ accomplished by the parallel genetic algorithm with $P_c = 0.3$, $P_m = 0.3$, $n_{\text{ind}} = 50$, but setting $n_{\text{gen}} = 30$.

We can also verify from Table 3 that for all cases, mutation rate minimizes average processing time $\tilde{t}$, in other words, the parallel genetic algorithm requires less processing time when mutation rate becomes larger. It turns out, that the average processing time of $\tilde{t} = 57.6975$ s achieved by the parallel genetic algorithm, when $P_c = 0.3$, $P_m = 0.3$, $n_{\text{ind}} = 50$ and $n_{\text{gen}} = 50$, amounts to 31 % of the average processing time $\tilde{t} = 183.2114$ s accomplished by the simple genetic algorithm under the same parametric conditions. Finally, from Table 3 it can be also established that an average approximation error $|e_m| = 0.00010\ \text{cm}^2$ can be attained when $P_c = 0.3$, $P_m = 0.3$, $n_{\text{ind}} = 50$ and $n_{\text{gen}} = 50$ with an average processing time of $\tilde{t} = 57.6975$ s, which corresponds to 31 % of the average processing time it takes for the simple genetic algorithm to complete the task.

4 Discussion

The benefits granted by cosmopolitan eelgrass on behalf of human welfare are relevant. Nevertheless, at a global scale, current anthropogenic influences are seriously threatening the permanence of this valuable seagrass species. Among remediation efforts aimed to eelgrass conservation, transplanting plays an important role. The monitoring of the average biomass of leaves in shoots [cf. Eq. (2)] and corresponding leaf growth rates [cf. Eq. (4)] in an eelgrass transplant and comparison with corresponding assessments taken at the donor population is fundamental to assess the success of the restoration endeavor. Traditional assessments of eelgrass relevant variables like those described by Eqs. (2) and (4) involve shoot destruction, and this practice can seriously alter both density and recruitment of shots in a transplant. As it is explained in the appendix, allometric methods convey important tools aimed to indirect monitoring like those formally exhibited by Eqs. (6) and (8) that could provide cost-effective, nondestructive estimations of average biomass of leaves in shoots and corresponding leaf growth rates. But results show [16], that factors like analysis method sample size and data quality could seriously reduce the precision of estimates of the allometric parameters α and β in Eq. (5), which as we have pointed out, in turn, influences the reproducibility capabilities of the proxies of Eqs. (6) and (8). But it is worth to recall that even though the involved methods guaranty highly precise estimates of α and in β leaf area estimations, provide the basis of the allometric projections of Eqs. (6) and (8). Therefore, it is mandatory to evaluate the primary influences that both the precision and efficiency of leaf area estimations could bear in the overall adequacy of these allometric devices. And since the noninvasive attribute of the aforementioned allometric surrogates could be greatly strengthened by involving digital imagery for the related leaf area estimations, an evaluation of pertinence and precision, as well as, the efficiency of this amendment is mandatory. Leal-Ramírez and Echavarría-Heras [20] and Echavarría-Heras et al. [21] performed the related precision study and Leal-Ramirez et al. [22] partially addressed the issue of improving the efficiency of Monte Carlo methods aimed to estimate the area of an eelgrass leaf from digital images. The present study extended the results in [22] by adding parallel processing to the Monte Carlo and genetic algorithm procedure considered there. Our arrangement that includes a master-slaves genetic algorithm structure can reduce processing time to a 25 % of that required by the procedure considered in [22] to complete the task of obtaining the areas of the leaves associated to the data set reported in Echavarría-Heras et al. [16]. Therefore, compared with a simple genetic algorithm and Monte Carlo procedure performing the task of estimating the area of a given eelgrass leaf, we can observe that the efficiency of a master-slaves parallel genetic algorithm and Monte Carlo approach is more clearly seen when mutation and crossover rates are greater than or equal to 0.3 (Tables 1, 2 and 3). Besides, we conjecture that the efficiency of the present arrangement could be improved (1) by setting $P_m = 0.3$ and $P_c = 0.3$, as well as, a greater number of slaves and (2) by contemplating a genetic algorithm with either a fine grain or a multiple-population structure. These provisions in our view could reduce the processing time required in achieving the calculation of the areas of the leaves in the extensive data set contemplated by Echavarria-Heras et al. [16] to few hours while also reducing the associated approximation error. The exploration of these alternatives will be the subject of further research. In the meantime, we must take into account that the suitability of the allometric methods of Eqs. (6) and (8) depends sensibly on appropriate sample size because this is a factor that determines the precision of estimates of the parameters α and β in Eq. (5). We must also recall that a data set of an extent as contemplated in [16] is required for highly precise estimates for these parameters. Therefore, in light of the present results, we can ascertain that the quest for the suitable structure of a parallel genetic algorithm is a must if we want to expand beyond the present findings the efficiency of Monte Carlo methods aimed to the estimation of eelgrass leaf area from digital imagery. In conclusion, within the overall strategy that leads to an effective eelgrass remediation effort, the use of genetic algorithms in leaf area assessment techniques is highly recommended.

Appendix

Allometric methods have provided convenient tools that describe biomass or other traits of seagrasses [15, 32, 33]. With the aim of making this contribution self-contained, in this appendix we illustrate how allometric models can substantiate indirect assessment tools commonly used in eelgrass monitoring and that relate to the associated leaf area. Particularly, we will review the substantiation of the leaf length times width proxy of Eq. (11), which provides handy approximations to the involved leaf area, and also the verification of the formulae in Eqs. (6) and (8) that yield allometric proxies based on leaf area for the average of the biomass of leaves in shoots and corresponding leaf growth rates, respectively. In order to achieve our goals we will keep here the notation presented in the introduction and will let $l(t)$ denote the length of a representative eelgrass leaf at time t, this defined as the extent between the base and the tip. Likewise, we will let $h(t)$ stand for a typical width measurement taken at a fixed point over the span of leaf length. We will also let $a(t)$ and $w(t)$, respectively, denote the linked area and dry weight of leaves.

By taking into account the architecture and growth form of eelgrass leaves, we can use allometric models to scale leaf dry weight in terms of the matching length or area [16, 17]. These allometric assumptions can also be represented through the differential equations

$$\frac{1}{w}\frac{\mathrm{d}w}{\mathrm{d}t} = \frac{\theta}{l}\frac{\mathrm{d}l}{\mathrm{d}t} \tag{18}$$

$$\frac{1}{w}\frac{\mathrm{d}w}{\mathrm{d}t} = \frac{k}{a}\frac{\mathrm{d}a}{\mathrm{d}t}, \tag{19}$$

where k and θ are positive constants [34]. Now from Eqs. (18) to (19), we can write

$$\frac{\mathrm{d}a}{\mathrm{d}t} = \frac{ca}{l}\frac{\mathrm{d}l}{\mathrm{d}t}, \tag{20}$$

where $c = \theta/k$.

Since it has been observed, that through elongation, leaf architecture and growth form in eelgrass lead to slight values for width increments, it is sound to assume that for a characteristic leaf $h(t)$ remains fairly constant, and thus

$$\frac{\mathrm{d}h}{\mathrm{d}t} = 0 \tag{21}$$

Hence using this assumption and Eq. (20) we may write

$$\frac{\mathrm{d}a}{\mathrm{d}t} = \frac{ca}{l}\frac{\mathrm{d}l}{\mathrm{d}t} + \frac{ca}{l}\frac{\mathrm{d}h}{\mathrm{d}t} \tag{22}$$

Moreover, if we consider the function $U(t)$ defined through

$$U(t) = ch(t)l(t) + c\int e(t)\frac{\mathrm{d}\Phi}{\mathrm{d}t}, \tag{23}$$

where

$$e(t) = a(t) - h(t)l(t), \tag{24}$$

and

$$\Phi(t) = ln(h(t)l(t)), \tag{25}$$

then it is straightforward to show that

$$\frac{\mathrm{d}a}{\mathrm{d}t} = \frac{\mathrm{d}U}{\mathrm{d}t}. \tag{26}$$

It turns out that there exists a constant p such that we have

$$a(t) = ch(t)l(t) + p + z(t), \tag{27}$$

where

$$z(t) = c\int e(t)\frac{\mathrm{d}\Phi}{\mathrm{d}t}. \tag{28}$$

We can then assume that if $h(t)$ has been selected in a way that $e(t)$ in Equation (24) is negligible, then $z(t)$ will also be negligible, so we can consider the linear regression model criterion

$$a(t) = ca_p(t) + p, \tag{29}$$

where

$$a_p(t) = h(t)l(t). \tag{30}$$

This result can provide criteria to test the assumption that $h(t)l(t)$ provides a reliable proxy for $a(t)$. This can indeed be ascertained if after considering data on eelgrass leaf area $a(t)$, lengths $l(t)$ and suitably taken width $h(t)$ measurements, the linear regression model of Eq. (29) was reliably identified as having a slope near 1 and an intercept 0. As it is explained in Echavarría-Heras et al. [19] who tested the criterion of Eq. (29) for eelgrass leaf width and length data the proxy $h(t)l(t)$ will provide accurate assessments of leaf area $a(t)$ only in case width is measured half way between the base and tip of the leaf. This establishes the result

$$a(t) = a_p(t) + \varepsilon_p(t), \tag{31}$$

where $\varepsilon_p(t)$ is the involved approximation error.

By assuming that Eqs. (18) and (19) hold, Echavarría-Heras et al. [35] generalized the model of Hamburg and Homann [15] and established that $w(t)$ can be scaled by using an allometric model of the form

$$w(t) = \rho l(t)^{\alpha} h(t)^{\beta}, \tag{32}$$

where $\rho > 0$, α and β are constants. We will first show that we can similarly represent the above scaling for $w(t)$ only in terms of the considered proxy for leaf area, that is, $a_p(t) = h(t)l(t)$. To be precise, rearranging Eq. (32) we can obtain the alike form

$$w(t) = \rho a_p(t)^{\alpha} h(t)^{\theta}, \tag{33}$$

where $\theta = \beta - \alpha$. Now, let's consider a time increment $\Delta t \geq 0$. Then direct algebraic manipulation of Eq. (33) yields

$$\frac{w(t+\Delta t)}{w(t)} = \left(\frac{a_p(t+\Delta t)}{a_p(t)}\right)^{\alpha} \left(1 + \frac{\Delta h}{h(t)}\right)^{\theta} \tag{34}$$

where $\Delta h = h(t+\Delta t) - h(t)$. Since we observed that during a growing interval of negligible length Δt eelgrass leaf architecture and growth form yield small values for Δh, we may expect the ratio involving $a_p(t)$ in Eq. (34) to be dominant. On the other hand, if we assume that $w(t)$ can be allometrically scaled in terms of $a_p(t)$, then as it is stated by Eq. (19), there will be a constant m [34] such that

$$\frac{1}{w}\frac{\mathrm{d}w}{\mathrm{d}t} = \frac{m}{a_p}\frac{\mathrm{d}a_p}{\mathrm{d}t} \tag{35}$$

$$\frac{w(t+\Delta t)}{w(t)} = \frac{a_p(t+\Delta t)^m}{a_p(t)^m}. \tag{36}$$

Therefore, the dominance of the term containing $a_p(t)$ in Eq. (34) is consistent with the assumption of an allometric scaling of $w(t)$ in terms of $a_p(t)$. Besides, supposing that Eq. (36) is satisfied for all values of t, then both sides must be equal to a constant. Let c be such a constant. Then we will have

$$w(t) = ca_p(t)^k, \tag{37}$$

The result for $w(t)$ in the form of Eq. (34) can be also obtained for real leaf area values $a(t)$. In fact, by recalling that $a_p(t)$ is an estimator for the real area $a(t)$ then in accordance with Eq. (31) there exist an approximation error $\varepsilon_p(t)$. Moreover $a_p(t)$ would provide a reliable approximation for $a(t)$ if for the ratio

$$s(t) = \frac{\varepsilon_p(t)}{a(t)}, \tag{38}$$

we set the consistency condition

$$s(t) \ll 1, \tag{39}$$

Solving for $a_p(t)$ in Eq. (31) and replacing in Eq. (34), after few steps we can obtain

$$\frac{w(t+\Delta t)}{w(t)} = \left(\frac{a(t+\Delta t)}{a(t)}\right)^m \left(1 - \frac{s(t+\Delta t) - s(t)}{1 - s(t)}\right)^m. \tag{40}$$

By virtue of the consistency condition (38), the right-hand side of Eq. (40) will be dominated by the ratio $a(t+\Delta t)^m a(t)^{-m}$. Therefore, for α and β constants can also propose an allometric model, for eelgrass leaf dry weight $w(t)$ and corresponding leaf area values $a(t)$, e.g.

$$w(t) = \beta a(t)^\alpha. \tag{41}$$

Readily, using Eq. (41) from Eq. (1) we can obtain an allometric proxy for $w_s(t)$, this is denoted through the symbol $w_s(\alpha, \beta, t)$, and it turns out to be,

$$w_s(\alpha, \beta, t) = \sum_{nl(s)} \beta a(t)^\alpha. \tag{42}$$

Since the parameters α and β in Eq. (41) are identified by means of regression methods, the uncertainty ranges for their estimates imply that $w_s(t)$ and $w_s(\alpha, \beta, t)$ are linked through the expression

$$w_s(t) = w_s(\alpha, \beta, t) + \in_s (\alpha, \beta, t), \tag{43}$$

that is, $w_s(\alpha, \beta, t)$ provides only an approximation to the true value of $w_s(t)$ being $\varepsilon_s(\alpha, \beta, t)$ the associated approximation error. Similarly from Eq. (42), we can obtain an allometric proxy for $w_{ms}(t)$, the average leaf biomass in shoots at a time t. This surrogate which we denote here through the symbol $w_{ms}(\alpha, \beta, t)$ is given by

$$w_{ms}(\alpha, \beta, t) = \frac{\sum_{ns(t)} w_s(\alpha, \beta, t)}{ns(t)}. \tag{44}$$

Similarly, we have

$$w_{ms}(t) = w_{ms}(\alpha, \beta, t) + \varepsilon_{ms}(\alpha, \beta, t), \tag{45}$$

where $\varepsilon_{ms}(\alpha, \beta, t)$ is the resulting approximating error.

Moreover, as it has been explained by Echavarría-Heras et al. [17], we can use Eqs. (3) and (4) in order to obtain an allometric approximation for $L_{sg}(t, \Delta t)$, which here we symbolize by means of $L_{sg}(\alpha, \beta, t, \Delta t)$ and that we express through

$$L_{sg}(\alpha, \beta, t, \Delta t) = \frac{\sum_{nl(s)} \Delta w(\alpha, \beta, t, \Delta t)}{\Delta t}, \tag{46}$$

where $\Delta w(\alpha, \beta, t, \Delta t)$ is an allometric surrogate for $\Delta w(t, \Delta t)$.

Therefore, Eq. (41) yields

$$\Delta w(\alpha, \beta, t, \Delta t) = \beta a(t + \Delta t)^{\alpha} - \beta a(t)^{\alpha}.$$

Factoring $a(t + \Delta t)^{\alpha}$ then considering that $a(t + \Delta t) = a(t) + \Delta a$ where Δa stands for the increment in leaf area attained over the interval $[t, t + \Delta t]$ we get,

$$\Delta w(\alpha, \beta, t, \Delta t) = \beta a(t + \Delta t)^{\alpha}(1 - (1 - \rho(t, \Delta t))^{\alpha})$$

where

$$\rho(t, \Delta t) = \frac{\Delta l}{a(t + \Delta t)},$$

and by letting

$$\delta(t, \Delta t) = (1 - (1 - \rho(t, \Delta t))^{\alpha})$$

we obtain

$$\Delta w(\alpha, \beta, t, \Delta t) = \beta a(t + \Delta t)^{\alpha} \Delta(t, \Delta t).$$

Therefore, from (46) we have

$$L_{sg}(\alpha, \beta, t, \Delta t) = \frac{\sum_{nl(s)} \beta a(t + \Delta t)^{\alpha} \Delta(t, \Delta t)}{\Delta t}. \tag{47}$$

Similarly, if $L_g(\alpha, \beta, t, \Delta t)$ stands for the allometric proxy for $L_g(t, \Delta t)$, then according to Eq. (4), this is given by

$$L_g(\alpha, \beta, t, \Delta t) = \frac{\sum_{ns(t,\Delta t)} L_{sg}(\alpha, \beta, t, \Delta t)}{ns(t, \Delta t)}, \tag{48}$$

this explains Eq. (8) and in turn we have

$$L_g(t, \Delta t) = L_g(\alpha, \beta, t, \Delta t) + \varepsilon_g(\alpha, \beta, t, \Delta t) \tag{49}$$

being $\varepsilon_g(\alpha, \beta, t, \Delta t)$ the involved approximation error.

References

1. Short, F.T., Coles, R.G., Pergent-Martini, C.: Global seagrass distribution. In: Short, F.T., Coles, R.G. (eds.) Global Seagrass Research Methods, pp. 5-30. Elsevier Science B.V., Amsterdam, TheNetherlands (2001).
2. McRoy, C.P.: Standing stock and ecology of eelgrass (*Zostera marina L.*) in zembek Lagoon, Alaska. MS.D., University of Washington, Seattle, WA, USA (1966).
3. Williams, T.P., Bubb, J.M., Lester, J. N.: Metal accumulation within salt marsh environments. Marine Pollution Bulletin, **28**(5), 277-289 (1994).
4. Short, F.T., Short, C.A.: The seagrass filter: purification of coastal water. In: Kennedy, V.S. (ed.) The Estuary as a Filter, pp. 395-413. Academic Press, Massachusetts (1984).
5. Ward, L.G., Kemp, W.M., Boynton, W.R.: The influence of waves and seagrass communities on suspended particulates in an estuarine embayment. Mar. Geol. **59**(1-4), 85-103 (1984).
6. Fonseca, M.S., Fisher, J.S.: A comparison of canopy friction and sediment movement between four species of seagrass with reference to their ecology and restoration. Mar. Ecol. Prog. Ser. **29**, 15-22 (1986).
7. Orth, R.J., Harwell, M.C., Fishman, J.R.: A rapid and simple method for transplanting eelgrass using single, unanchored shoots. Aquat. Bot. **64**, 77-85 (1999).
8. Campbell, M.L., Paling, E.I.: Evaluating vegetative transplanting success in *Posidonia Australis*: a field trial with habitat enhancement. Mar. Pollut. Bull. **46**, 828-834 (2003).
9. Fishman, J.R., Orth, R.J., Marion, S., Bieri, J.: A comparative test of mechanized and manual transplanting of eelgrass, *Zostera marina*, in Chesapeake Bay. Restoration Ecol. **12**, 214–219 (2004).
10. Li, W.T., Kim, J.H., Park, J.I., Lee, K.S.: Assessing establishment success of *Zostera marina* transplants through measurements of shoot morphology and growth. Estuarine, Coastal and Shelf Science. **88**(3), 377-384 (2010).
11. Dennison, W.C.: Shoot density. In: Philips, R.C., McRoy, C.P. (eds.) Seagrass Research Methods, pp. 77-79. UNESCO, (1961).
12. MacRoy, C.P.: Standing stock and other features of eelgrass (*Zostera marina*) populations on the coast of Alaska. J. Fish. Res. Bd. Canada. **27**, 1811-1812 (1970).
13. Patriquin, D.G.: Estimation of growth rate, production and age of the marine angiosperm, Thalassiatestudinum. Konig. Carib. J. Sci. **13**, 111-123 (1973).
14. Jacobs, R.P.W.M.: Distribution and aspects of the production and biomass of eelgrass, *Zostera marina L.* at Roscoff, France. Aquat. Bot. **7**, 151 (1979).
15. Hamburg, S.P., Homman, P.S.: Utilization of Growth parameters of eelgrass *Zostera marina* for productivity estimation under laboratory and insitu conditions. Mar. Biol. **93**, 299-303 (1986).
16. Echavarría-Heras, H.A., Leal-Ramírez, C., Villa-Diharce, E., Cazarez-Castro, N.R.: The effect of parameter variability in the allometric projection of leaf growth rates for eelgrass (*Zostera*

marina L.) II: the importance of data quality control procedures in bias reduction. Theoretical Biology and Medical Modelling. **12**(30), 1-12 (2015).

17. Echavarria-Heras, H., Solana-Arellano, E., Franco-Vizcaino, E.: An allometric method for the projection of eelgrass leaf biomass production rates. Math. Biosci. **223**, 58-65 (2009).
18. Solana-Arellano, M.E., Echavarría-Heras, H.A., Leal-Ramírez, C., Kun-Seop, L.: The effect of parameter variability in the allometric projection of leaf growth rates for eelgrass (*Zostera marina L.*). Latin American Journal of Aquatic Research. **42**(5), 1099-1108 (2014).
19. Echavarria-Heras, H., Solana-Arellano, E., Leal-Ramirez, C., Franco-Vizcaino, E.: The length-times-width proxy for leaf area of eelgrass: criteria for evaluating the representativeness of leaf-width measurements. Aquat. Conserv. Mar. Freshw. Ecosyst. **21**(7), 604-613 (2011).
20. Leal-Ramirez, C., Echavarria-Heras, H.: A method for calculating the area of *Zostera marina* leaves from digital images with noise induced by humidity content. Sci. World J. **2014**, 11 (2014).
21. Echavarría-Heras, H., Leal-Ramírez, C., Villa-Diharce, E. and Castillo O.: Using the Value of Lin's Concordance Correlation Coefficient, as a criterion for efficient estimation of areas of eelgrass *Zostera marina* leaves from noisy digital images. Source Code For Biology and Medicine. **9**, 1-29 (2014).
22. Leal-Ramírez, C., Echavarría-Heras, H.A., Castillo, O.: Exploring the suitability of a genetic algorithm as tool for boosting efficiency in Monte Carlo estimation of leaf area of eelgrass. In: Melin, P., Castillo, O., Kacprzyk, J. (eds.) Design of Intelligent Systems Based on Fuzzy Logic, Neural Networks and Nature-Inspired Optimization. Studies in Computational Intelligence, vol. 601, pp. 291-303. Springer, Edition 17746-5 (2015).
23. Adamidis, P.: Review of Parallel Genetic Algorithms Bibliography. Internal Technical Report, Aristotle University of Thessaloniki, (1994).
24. Alba, E., Cotta, C.: Evolution of Complex Data Structures. Informática y Automática, **30**(3), 42-60 (1997).
25. Cantú-Paz, E.: A Summary of Research on Parallel Genetic Algorithms. R. 95007, (1995).
26. Pettey, C.C., Leuze, M.R., Grefenstette, J.: A Parallel Genetic Algorithm. Proceedings of the 2nd ICGA, J. Grefenstette (ed.), Lawrence Erlbraum Associates, pp. 155-161, (1987).
27. Ribeiro-Filho J.L., Alippi, C., Treleaven, P.: Genetic algorithm programming environments. In: Stender, J. (ed.), Parallel Genetic Algorithms: Theory & Applications. IOS Press., (1993).
28. Cantú-Paz, E.: A survey of parallel genetic algorithms. CalculateursParalleles, Reseauxet Systems Repartis. 10(2), 141-171 (1998).
29. Alba, E., Aldana, J.F., Troya, J.M.: Full automatic ANN design: a genetic approach. In: Mira, J., Cabestany, J., Prieto, A. (eds.) New Trends in Neural Computation. Lecture Notes in Computer Science, vol. 686, pp. 399-404. Springer-Verlag, IWANN'93 (1993).
30. Alba, E., Aldana, J.F., Troya. J.M.: A genetic algorithm for load balancing in parallel query evaluation for deductive relational databases. In: Pearson, D.W., Steele, N.C., Albrecht, R.F. (eds.) Procs. of the I. C. on ANNs and Gas. pp. 479-482. Springer-Verlag, (1995).
31. Stender, J. (ed.): Parallel Genetic Algorithms: Theory and Applications. IOS Press. (1993).
32. Duarte, C.M.: Allometric scaling of seagrass form and productivity. Mar. Ecol. Prog. Ser. **77**, 289-300 (1991).
33. Brun, F.G., Cummaudo, F., Olivé, I., Vergara, J.J, Pérez-Lloréns, J.L.: Clonal extent, apical dominance and networking features in the phalanx angiosperm *Zostera noltii Hornem.* Marine Biology. **151**, 1917-1927 (2007).
34. Batschelet, E. (ed.): Introduction to Mathematics for Life Scientists. vol. XV, number edition 3, pp. 646. Springer-Verlag Berlin Heidelberg, (1979).
35. Echavarria-Heras, H., Lee, K.S., Solana-Arellano E. and Franco-Vizcaíno E.: Formal analysis and evaluation of allometric methods for estimating above-ground biomass of eelgrass. Annals of Applied Biology. **159**(3), 503-515 (2011).

Social Spider Algorithm to Improve Intelligent Drones Used in Humanitarian Disasters Related to Floods

Alberto Ochoa, Karina Juárez-Casimiro, Tannya Olivier, Raymundo Camarena and Irving Vázquez

Abstract The aim of this study was to implement an optimal arrangement of equipment, instrumentation and medical personnel based on the weight and balance of the aircraft and to transfer humanitarian aid in a drone, by implementing artificial intelligence algorithms. This is due to the problems presented by the geographical layout of human settlements in southeast of the state of Chihuahua. The importance of this research is to understand the multivariable optimization associated with the path of a group of airplanes associated with different kinds of aerial in order to improve the evaluation of flooding and to send medical support and goods; to determine the optimal flight route, including speed, storage and travel resources. To determine the cost–benefit, this has been partnered with a travel plan to rescue people, which has as its principal basis the orography airstrip restriction, although this problem has been studied on several occasions by the literature failed to establish by supporting ubiquitous computing for interacting with the various values associated with the achievement of the group of drones and their cost–benefit of each issue of the company and comparing their individual trips for the rest of group. There are several factors that can influence in the achievement of a group of drones for our research. We propose the use of a bioinspired algorithm.

Keywords Aeronautics logistics · Social spider algorithm · Drones and mobile devices

1 Introduction

The Cessna 208 Caravan, also known as Cargo master, is a regional jet/turboprop short-range utility manufactured in the USA by the company Cessna. The standard version has 10 places (nine passengers and a pilot), although, according to new regulations of the Federal Aviation Administration (FAA), a subsequent design can

A. Ochoa (✉) · K. Juárez-Casimiro · T. Olivier · R. Camarena · I. Vázquez
Maestría en Cómputo Aplicado, Juarez City University, Ciudad Juárez, Mexico
e-mail: alberto.ochoa@uacj.mx

P. Melin et al. (eds.), *Nature-Inspired Design of Hybrid Intelligent Systems*,
Studies in Computational Intelligence 667, DOI 10.1007/978-3-319-47054-2_30

Fig. 1 A Cessna 208 used for flight with passengers and goods in Southwestern Chihuahua

carry up to 14 passengers. The aircraft is also widely used to make connections in freight services, so that goods that arrive at smaller airports are transported to major hubs for distribution, as in Fig. 1.

The concept of the Cessna 208 arose in early 1980, and the first prototype flew on 8 December 1982. After two years of testing and review, in October 1984, the FAA certified the model for flight. Since then, the Caravan has undergone many evolutions. Working in hand with international logistics company FedEx, Cessna produced first the Cargo master. This was followed by an improved and extended version called the Super Cargo master and a passenger model called the Grand Caravan. The Practitioners will then be free fall after boarding a Cessna 208 in the Dutch island of Texel. Currently, the Cessna 208 offers different configurations to meet the varied market demand. The core 208 can be supplemented with different types of landing gear and can operate in a variety of terrains. Some adaptations include skis, larger tires for unprepared runways, or floats with wheels in the case of the Caravan Amphibian. Cabin seats can be included or room can be left for cargo in various configurations. The standard configuration of the airline consists of four rows of one to two seats, with two seats in the cockpit. This variant is capable of carrying up to 13 passengers, albeit only lead to four longer an operation rentable one. The cabin can also be configured to a low density of passengers, in combination or alone as a freighter. Some versions include an additional compartment at the bottom to increase the capacity or luggage. In the cockpit, the 208 has standard analog gauges with a modern digital avionics equipped with autopilot and GPS, modern radio and transponder. Cessna currently offers two different packages with avionics manufacturers, one from Garmin and the other from Bendix/King, a subsidiary of Honeywell. Vehicle Routing Problems (VRPs) are actually a broad range of variants and customization problems, from those that are simple to those that remain the subject of research, such as in [1]. They were trying to figure out the routes of a transportation fleet in order to service a customer, which nowadays includes aerial transportation. This type of problem is a combinatorial optimization problem. In the scientific literature, Dantzig and Ramser were the first authors in 1959, when they studied the actual application of the distribution of gasoline for fuel stations. The objective function depends on the type and characteristics of the problem. The most common is to try to minimize the total cost of ownership, minimize total transportation time, minimize the total distance traveled, minimize waiting time, maximize profit, maximize customer service, minimize the use of vehicles, and balance the resource utilization.

Other considerations to this problem

- ***Optimization of spaces associated with medical emergencies for patients without mobility in a Cessna 208***

The southeastern state of Chihuahua is an area whose geographical characteristics confer greater difficulty in moving the terrestrial inter hospitalary. For that reason, it is necessary is to develop an intelligent system that can consider the transfer of patients using specially adapted airplanes.

The Cessna 208 is a short-range regional aircraft turboprop that meets the expectations for routes to cover the landing strips of wood, Social Spideropilas, Temoris, Balleza Morelos and all these with connection to the airport in Hidalgo del Parral, estimated to last approximately 45 min.

Dates and times of the planes that travel the state of Chihuahua are an important consideration, so that no problems regarding the accumulation of air traffic in that area are caused. The departures and arrivals of each Cessna aircraft are determined through logistics (Fig. 2).

- ***Identifying and optimizing response times for aerial medical emergencies for climbers in southwestern Chihuahua***

Today, technological advances have allowed new devices to be at our disposal and it is possible make use of them for commercial or personal purposes. Among these new devices is the drone, which is able to fly and reach heights that humans need for multimedia tasks, such as taking pictures and videos.

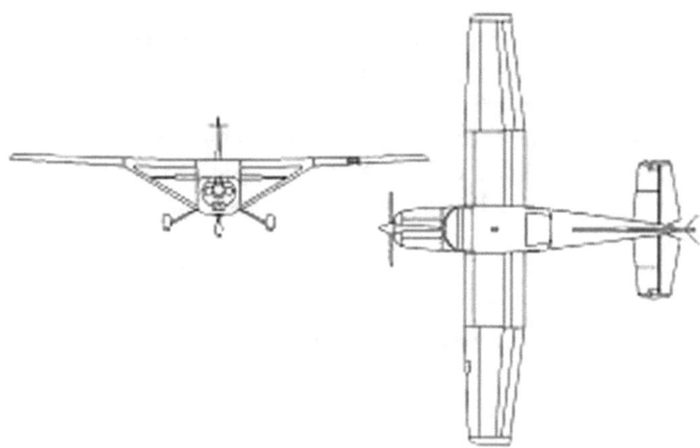

Fig. 2 Views of fuselage dimensions associated with Cessna 208

Fig. 3 The study regions in southwestern Chihuahua

The proposal is focused on how we can use these devices to search for people in need or in distress and can get to various support centers more quickly and effectively (Fig. 3).

As part of the Cisco initiative, under the competition drone, we developed applications for technological purposes. The main idea behind this project was based on searches for people in situations of natural disasters or accidents in inaccessible areas, such as the mountains of the country, forests and jungles where access by any means of transport is difficult and it takes longer.

Statistics in southern Chihuahua show that in an accident of this type, the transfer of the person to an urban area takes about an hour. This length of time is problematic if the injury is life threatening.

Originality

Currently, the use of drones is primarily for panoramic photographs and personal recreational use.

The originality of this project lies in taking the features and capabilities of drones as a starting point, and taking them to an area that is difficult to access and where current means of transportation are limited.

Thus, the use of these technologies would impact significantly on search methods, supporting brigades such as civil protection and the relevant state government, considering that this type of environment complicates its location and signal decreases (Fig. 4).

Benefit

The main objective of this proposal is to meet quickly and efficiently in accidents in natural areas in order to:

Fig. 4 One of the areas where the proposed aeronautical logistics can be applied to medical emergencies

- Saving lives
- Improve time search and identification of people at risk
- Optimize search and rescue operations.
- Recognition of land
- Minimize as much as possible the use of rescue personnel

The priority is to safeguard the integrity and welfare of a person in a situation of risk as quickly as possible.

The viability lies mainly in the fact that this technology is within reach and tested, and is currently in use, which is why it is such artifacts are intended for implementation, focused on communities prone to suffer any type of natural disaster (Fig. 5).

The fact that this initiative is implemented as such involves limited financial resources and the main concerns are the device and the operating personnel.

Analysis of data

The process of data analysis is first to launch the drone in the area near to where the person was last seen, taking that as a starting point. Then the operator can install the drone to search, starting with the closest areas, gradually moving forward walking along the area. The drone activates the associated camera at all times, wirelessly transmitting video, which will be seen by the operator in progress. If the main social spider search is exhausted, in seconds the spare social spider search can enhance the autonomy travel drone.

Once the person is located, the current coordinates are taken to rescue personnel in order to approach and establish presence of the person.

Fig. 5 The priority is to safeguard human lives in alpine rescues

Likewise, the drone provides guidance to the location of the person to be rescued.

The drone increases the organization, monitoring and response times for rescue either at air or ground level and is therefore an indispensible tool.

Simply saving and secure something as valuable as human life gives this project great importance in terms of implementing drones in risk areas.

The cost of drones rises above 60,000 pesos, and this is considering the implementation of the best technologies that mind current account market.

This project would provide a great benefit in terms of technology and customer growth, considering that the loss of a rescuer represents a major negative economic impact, which can be avoided with the implementation of this technology.

Determine charges

The charges are the distances that require convoy travel from the airstrip, to the point where there is demand and return to the airstrip of origin. The rise, which will cause the vehicle height change and help (with decreasing elevation) or give more load (with increasing elevation), is a factor that has to be taken into account (Fig. 6).

To simplify the determination of the load with altitude, we consider the altitude at the start, the final altitude and also the way between the two points. A simple formulation was used to take the load:

$$c = r_{\text{horizontal}} * \left(1 + \frac{h_f - h_i}{r_{\text{horizontal}}}\right)$$

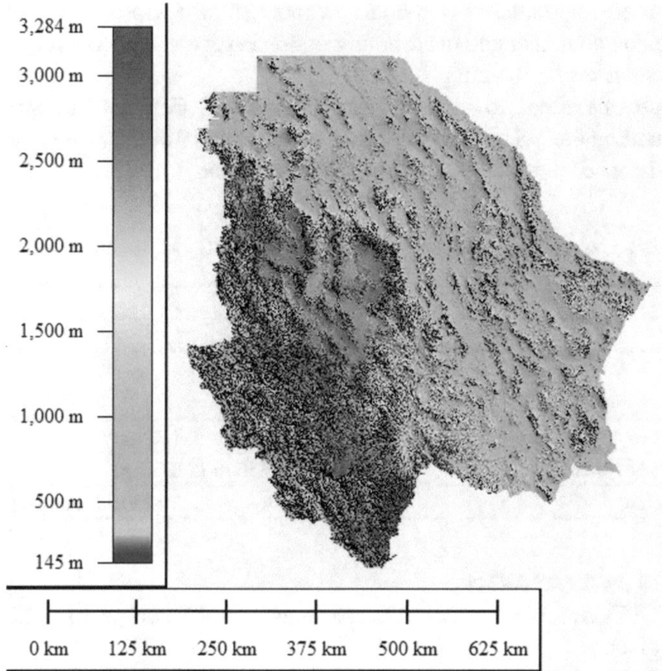

Fig. 6 Topography of Chihuahua

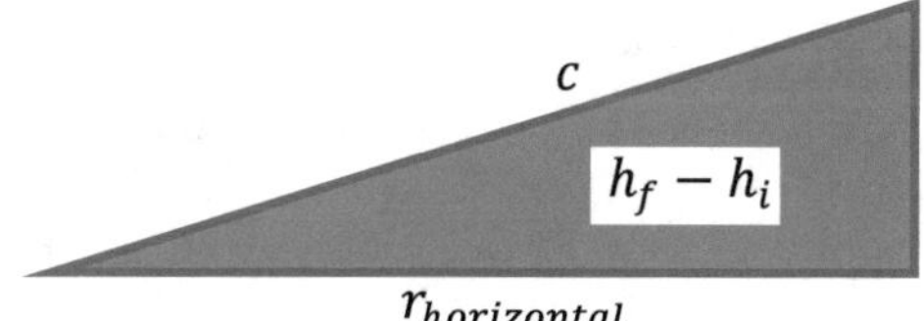

Fig. 7 Graphical representation associated with charge

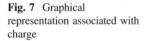

where $r_{\text{horizontal}}$ is the horizontal path, h_f is the final altitude and h_i is the initial altitude. This formulation is shown in Fig. 7. The final altitude is increased if the load increases and the load decrease is smaller.

- ***Implementation of an optimization model for the transfer of organs in a non-commercial aircraft***

The importance of this research is the implementation organ in a non commercial aircraft. It is necessary because in cities of Chihuahua, distances are very long or the roads are very damaged for the transportation of organs and some have very little time preservation, such as the heart and lung. To preserve the organs, associated hypothermia at 4 °C for short distances is commonly used in ambulances, while the aircraft is used for routes of two hours or more. It will approach the Cessna 208 Caravan, using container and determining optimal space and energy required for the cooling system inside the aircraft.

The type of aircraft to be used is a Cessna 208 Caravan turboprop aircraft manufactured by the USA. The traditional version has 10 seats, nine passengers and one pilot. Later designs can take up to 14 passengers.

Dimensions

Wingspan	15.88 m
Height	4.71 m
Width	12.68 m

Limitations

Maximum takeoff weight (with cargo)	9.062 lb (4.110 kg)
MTOW in icy conditions	8.550 lb (3.878 kg)
Minimum crew	Two pilots for freight transport

Electric Power (ATA-24)

Diesel GPU 28.5 (28.5 VDC, 800 continuous, 2000 amp peak) (CE)

Power plant

The PT6A-114A turboprop Pratt & Whitney Canada engine is flat-rated 675 horsepower at 1900 rpm shaft. Motor operation is controlled by the engine indication system, showing numerical readouts of engine fuel and critical electric signs for the pair, propeller speed, fuel flow and others. Torque meter was also installed pending wet type.

Manufacturer: Pratt & Whitney Canada
Model: (1) PT6A-114A
Output Power: 675 hps
Propeller Manufacturer: McCauley
Description: 3 sheet metal, constant speed, full plumage

For the transportation of bodies, hypothermic perfusion machines shall be used. These are new tools for the conservation of the organ: they allow organs to be perfused constantly and at the same time, monitor and add perfusion liquid drug substances that will improve their viability.

Storage organ preservation and transport

The organ preservation techniques used to reduce the damage, improve function and survival of the transplanted organ. Conditions may vary according to the type of organ. To preserve solid organs associated hypothermia at 4 °C.

Hypothermic preservation techniques most commonly used today are cold storage and preservation in hypothermic machine perfusion.

Preservation by cold storage (CF)

The most common preservation is also the least expensive method, and is to perfuse the body internally or wash with cold preservation solution immediately after removal in the operating room itself. Later, that preservation solution, or the like to Social Spiderhe, and keep it stored in a refrigerator at 4 °C until it is time for implantation. For its extreme simplicity, cold storage has a number of advantages, such as its almost universal availability and ease of transport, and it is the method most used in preservation.

Preservation hypothermic machine perfusion (HMP)

The concern within transplantation science to improve the quality of donated organs and the need for expanded criteria donors given the shortage of organs and the inability to assess the viability of organs before transplantation, has promoted in recent years development of methods that are better than simple cold storage preservation. The combination of continuous perfusion and hypothermic storage used by Belzer in 1967 represented a new paradigm in organ preservation, as it managed to successfully preserve canine kidneys for 72 h. According to this technique, after the initial washing is performed in the operating room during perfusion, the organ is introduced into a device that maintains a controlled flow continuously or pulsatile preservation solution of cold (0–4 °C). This flow allows full organ perfusion and clean micro thrombi the bloodstream and facilitates the removal of metabolic end products. Its beneficial effects are a lower incidence of delayed initial graft function, the possibility of evaluating viability in real time and the ability to provide metabolic support (oxygen and substrates) or drugs during the infusion. The continuous flow MPH has shown advantages over the pulsatile flow. Due to the size that a hypothermic machine perfusion and the energy required, it is more feasible in a Cessna 208 Caravan carry out a cooler where

Transfer process for the transfer of organs.

When a transplant over long distances is required and a donor in the hospital has become available, the National Organization of Transportation is responsible for arranging a short ambulance journey and, for a superior ride, 2 h in the plane.

If an organ for a transplant is going to be transported using a cooler, this requires special labeling to indicate the type of organ and that patient should receive

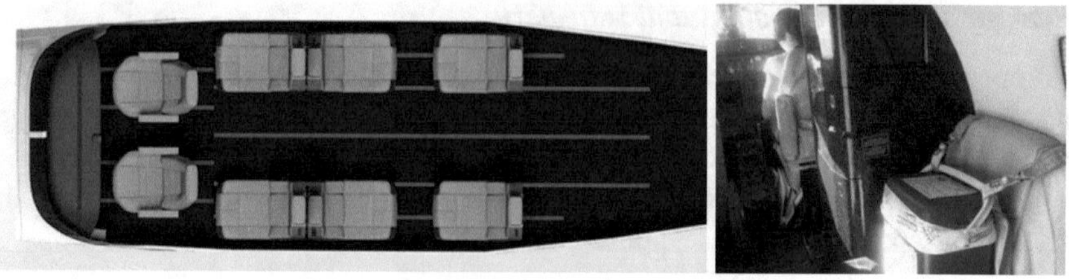

Fig. 8 This image shows the distribution of passengers by special missions

Large distances

In these cases, given the short cold ischemia time (time since being placed in transport solution time and the start of disinfection) that tolerate bodies, private aircraft aviation companies are hired and occasionally uses to aircraft of the Air Force. The preparation of a flight needs no less than 2 h (in order to check the aircraft, notify the crew, conduct flight plan preparation, etc.), so it is important to communicate the existence of the donor to the ONT to most as soon as possible. Once hired, the flight and the scheduled times, the coordinators of the hospitals involved is notified. Most domestic airports do not operate on a 24 h basis, so the coordinator of the ONT must be taken into account because if it is not 24 h, they will implement the mechanisms necessary for opening or to keeping it operating after hours.

Possible incidents when transporting organs
Improper transport temperature

Misidentification or reading the temperature during processing, storage or distribution. Failure to maintain storage facilities at risk of loss or damage to tissues or cells stored. Leakiness in packaging. Misidentification or discrepancy between labeling and shows (Fig. 8; Table 1).

1.1 Project Development

To do this research, the project was developed by outlining three sections, the modules of application development, implementation of the server, and the

Table 1 Body and preservation time

Liver	24 h
Kidneys	48–72 h
Hearth	3–5 h
Lung	3–5 h
Pancreas	12–24 h
Corneas	7–10 days
Marrow	Until 3 years
Skin	Until 5 years

intelligent module associated with social spider algorithm and data mining. Android is the operating system that is growing into Streak 5 from Dell; for this reason, we select this mobile dispositive along with other manufacturers that are propelling the Latin American landing on Android with inexpensive equipment. On the other hand, some complain about the fragmentation of the platform due to the different versions. Android is free software, so any developer can download the SDK (development kit) that contains the API [2]. This research tries to improve group travel related to recreational vehicles in Chihuahua, where 7500 people form the Caravan Range Community.

1.2 Components of the Application

Social spider algorithm

In social spider algorithm (SSA), we formulate the search space of the optimization problem as a hyper-dimensional spider web. Each position on the web represents a feasible solution to the optimization problem and all feasible solutions to the problem have corresponding positions on this web. The web also serves as the transmission media of the vibrations generated by the spiders. Each spider on the web holds position and the quality (or fitness) of the solution is based on the objective function, and represented by the potential of finding a food source at the position. The spiders can move freely on the web. However, they cannot leave the web as the positions off the web represent infeasible solutions to the optimization problem. When a spider moves to a new position, it generates a vibration which is propagated over the web. Each vibration holds the information of one spider and other spiders can get the information upon receiving the vibration.

Spider

The spiders are the agents of SSA to perform optimization. At the beginning of the algorithm, a pre-defined number of spiders are put on the web. Each spider, s, holds a memory, storing the following individual information:

The position of s on the web.
The fitness of the current position of s.
The target vibration of s in the previous iteration.
The number of iterations since s has last changed its target vibration.
The movement that s performed in the previous iteration.
The dimension mask that s employed to guide movement in the previous iteration.

The first two types of information describe the individual situation of s, while all others are involved in directing s to new positions. The detailed scheme of movement will be elaborated on in the last section. Based on observations, spiders are found to have a very accurate sense of vibration. Furthermore, they can separate different vibrations propagated on the same web and sense their respective intensities [11]. In SSA, a spider will generate a vibration when it reaches a new position different from the previous one. The intensity of the vibration is correlated with the fitness of the position. The vibration will propagate over the web and it can be sensed by other spiders. In such a way, the spiders on the same web share their personal information with others to form a collective social knowledge.

Vibration

Vibration is a very important concept in SSA. It is one of the main characteristics that distinguish SSA from other metaheuristics. In SSA, we use two properties to define a vibration, namely, the source position and the source intensity of the vibration. The source position is defined by the search space of the optimization problem, and we define the intensity of a vibration in the range [0, +∞). Whenever a spider moves to a new position, it generates a vibration at its current position. We define the position of spider at time t as Pa(t), or simply as Pa if the time argument is t. We further use I(Pa, Pb, t) to represent the vibration intensity sensed by a spider at position Pb at time t and the source of the vibrations at position Pa. With these notations, we can thus use I(Ps, Ps, t) to represent the intensity of the vibration generated by spider s at 1. The dimension mask is a 0–1 binary vector of length D, where D is the dimension of the optimization problem the source position. This vibration intensity at the source positions correlated with the fitness of its position f (Ps), and we define the intensity value as follows: (1) where C is a confidently small constant such that all possible fitness values are larger than C. Please note that we consider minimization problems in this paper. The design of (1) takes the following issues into consideration.

All possible vibration intensities of the optimization problem are positive. The positions with better fitness values, i.e. smaller values for minimization problems, have larger vibration intensities than those with worse fitness values. When a solution approaches the global optimum, the vibration intensity does not increase excessively, and cause malfunctioning of the vibration attenuation scheme. As a form of energy, vibration attenuates over distance. This physical phenomenon is accounted for in the design of SSA. We define the distance between spider a and b as D(Pa, Pb) and we use 1-norm (Manhattan distance) to calculate the distance; i.e., (2) The standard deviation of all spider positions along each dimension is represented by _. With these definitions, we further define the vibration attenuation over distance as follows: (3) In the above formula, we introduce a user-controlled parameter $ra \in (0, \infty)$. This parameter controls the attenuation rate of the vibration intensity over distance. The larger ra is, the weaker the attenuation imposed on the vibration.

Search pattern

Here we demonstrate the above ideas in terms of an algorithm. There are three phases in SSA: initialization, iteration, and final. These three phases are executed sequentially. In each run of SSA, we start with the initialization phase, then

perform searching in an iterative manner, and finally terminate the algorithm and output the solutions found. In the initialization phase, the algorithm defines the objective function and its solution space. The value for the parameter used in SSA is also assigned. After setting the values, the algorithm proceeds to create an initial population of spiders for optimization. As the total number of spiders remains unchanged during the simulation of SSA, a fixed size memory is allocated to store their information. The positions of spiders are randomly generated in the search space, with their fitness values calculated and stored. The initial target vibration of each spider in the population is set at its current position, and the vibration intensity is zero. All other attributes stored by each spider are also initialized with zeros. This finishes the initialization phase and the algorithm starts the iteration phase, which performs the search with the artificial spiders created. In the iteration phase, the algorithm performs a number of iterations. In each iteration, all spiders on the web move to a new position and evaluate their fitness values. Each iteration can be further divided into the following sub-steps: fitness evaluation, vibration generation, mask changing, random walk, and constraint handling. The algorithm first calculates the fitness values of all the artificial spiders on different positions on the web, and updates the global

$$
\begin{aligned}
(\mathrm{Ps}, \mathrm{Ps}, t) &= \log\_1 f(\mathrm{Ps}) - C + 1\_D(\mathrm{Pa}, \mathrm{Pb}) = ||\mathrm{Pa} - \mathrm{Pb}||1\,.\,I(\mathrm{Pa}, \mathrm{Pb}, t) \\
&= I(\mathrm{Pa}, \mathrm{Pa}, t) \times \exp\_ - D(\mathrm{Pa}, \mathrm{Pb})\_ \times ra\_.
\end{aligned}
$$

optimum value if possible. The fitness values are evaluated once for each spider during each iteration. Then these spiders generate vibrations at their positions using (1). After all the vibrations are generated, the algorithm simulates the propagation process of these vibrations using (3). In this process, each spider, s, will receive pop| different vibrations generated by other spiders where pop are the spider population. The received information of these vibrations includes the source position of the vibration and its attenuated intensity. We use *V* to represent these |pop| vibrations. Upon the receipt of *V*, *s* will select the strongest vibration *v*bests from *V* and compare its intensity with the intensity of the target vibration *v*tars stored in its memory. *s* will store *v*bests as *v*tars if the intensity of *v*bests is larger, and *cs*, or the number of iterations since *s* last changed its target vibration, is reset to zero; otherwise, the original *v*tar is retained and cs is incremented by one. We use Pisv and Ptarsb to represent the source positions of *V* and *v*tar, respectively, and $i = \{1, 2, \ldots, |\mathrm{pop}|\}$. The algorithm then manipulates s to perform a random walk towards *v*tars. Here, we utilize a dimension mask to guide the movement. Each spider holds a dimension mask, m, which is a 0–1 binary vector of length *D* and *D* is the dimension of the optimization problem. Initially, all values in the mask are zero. In each iteration, each spider has a probability of 1 − pccs to change its mask where pc ∈ (0, 1) is a user-defined attribute that describes the probability of changing mask. If the mask is decided to be changed, each bit of the vector has a probability of pm to be assigned with a one, and 1 − pm to be a zero. Pm is also a user-controlled parameter defined in (0, 1). Each bit of a mask is changed independently and does not have any correlation with the previous mask. In case all bits are zeros, one random value of the mask is changed to one. Similarly, one random bit is assigned to zero if all values are ones. After the dimension mask is determined, a new following position, Pfos, is generated based on the mask for s. The value of the dimension of the following position, Pfos, is generated as follows: (4) Where r is a random integer value generated in $[1, |pop|]$, and ms, I stands for the ith dimension of the dimension mask, m, of spiders. Here, the random number, r, for two different dimensions with ms, $i = 1$ is generated independently. With the generated Pfos, s performs a random walk to the position. This random walk is conducted using the following equation: (5) Where _ denotes element-wise multiplication and R is a vector of random float-point numbers generated from zero to one uniformly. Before following Pfos, s first moves along its previous direction, which is the direction of movement in the previous iteration. The distance along this direction is a random portion of the previous movement. Then s approaches Pfos along each dimension with random factors generated in (0, 1). This random factor for different dimensions is generated independently. After these random walks are performed then we need to store the movements in the current iteration for using them in the next iteration. This ends the random walk sub-step. The final sub-step of the iteration phase is the constraint handling. The spiders may move out of the web during the random walk step, which causes the constraints of the optimization problem to be violated. There are many methods to handle the boundary constraints in the previous literature, and the random approach, absorbing approach, and the reflecting approach are three most widely adopted methods [11]. In this paper, we adopt the reflecting approach for constraint handling and produce a boundary-constraint-free position Ps(t + 1) by (6) where $\bar{x}\ i$ is the upper bound of the search space in the ith dimension, and $x\ i$ is the lower bound of the corresponding dimension. r is a random floating point number generated in (0, 1). The iteration phase loops until the stopping criteria is matched. The stopping criteria can be defined as the maximum iteration number reached, the maximum CPU time used, the error rate reached, the maximum number of iterations with no improvement in the best fitness value, or any other appropriate criteria. After the iteration phase, the algorithm outputs the best solution with the best fitness found. The above three phases constitute the complete algorithm of SSA and its pseudo-code can be found in Algorithm 1.

Algorithm 1 Social spider algorithm

```
Assign values to the parameters of SSA.
Create the population of spiders pop and assign memory for them.
Initialize vtars for each spider.
while stopping criteria not met do
for each spider s in pop do
Evaluate the fitness value of s.
Generate a vibration at the position of s.
end for
for each spider s in pop do
Calculate the intensity of the vibrations V generated by all spiders.
Select the strongest vibration vbests from V.
if The intensity of vbestsis larger than vtars then
Store vbests as vtars.
```

```
end if
Update cs.
Generate a random number r from [0,1).
if r > pccs then
Update the dimension mask ms.
end if
Generate Pfos.
Perform a random walk.
Address any violated constraints.
end for
end while
Output the best solution found.
```

3.4. There exist some differences between SSA and other evolutionary computation algorithms in the literature. A number of swarm intelligence algorithms have been proposed in the past few decades. Among them, PSO and ACO are the two most widely employed and studied algorithms. SSA may also be classified as a swarm intelligence algorithm, but it has many differences from PSO and ACO, which are elaborated on below. PSO, like SSA, was originally proposed for solving continuous optimization problems. It was also inspired by animal behavior. However, the first crucial difference between SSA and PSO is in individual following patterns. In PSO, all particles follow a common global best position and their own personal best position. However in SSA, all spiders follow positions constructed by others' current positions and their own historical positions. These following positions are not guaranteed to be visited by the population before, and different spiders can have different following positions. Since the global best position and spiders' current positions differ greatly during most of the time of the optimization process, these two following patterns lead to different searching behaviors.

2 Implementation of an Intelligent Application

When designing an interface for mobile devices, it has to be taken into account that the space is very small screen. Plus, there are many resolutions and screen sizes so it is necessary to design an interface that suits most devices. This module explains how to work with different layouts provided by the Android API. The programming interface is through XML. Obtaining the geographical position of a device can be made by different suppliers; the most commonly used in this project are through GPS and using access points (Wi-Fi) nearby, which perform the same action but differ in some accuracy, speed and resource consumption. Data Server Communication is the module most important because it allows communication with the server, allowing you to send the GPS position obtained by receiving the processed image and map of our location, thus showing the outcome of your

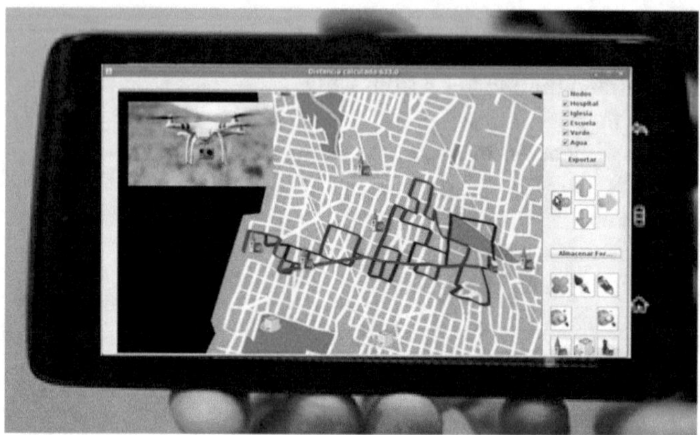

Fig. 9 Intelligent tool recommend a group travel associated with limited resources and optimize energy (oil), time and comfort

application that is the indicator of insecurity. To communicate to a server requires a HTTP client, which can send parameters and to establish a connection using TCP/IP, client for HTTP, can access to any server or service as this is able to get response from the server and interpreted by a stream of data. The Android SDK has two classes with which we can achieve this, HttpClient and HttpPost. With the class HttpClient is done to connect to a remote server, it needs HttpPost class will have the URI or URL of the remote server. This method receives a URL as a parameter and uses classes HttpPost HttpClient and the result is obtained and received from the server. In this specific case, it is only text, which can be JSON or XML format. Here, the server responds with a JSON object, which would give the indicator and is then used to create the map, as is shown in the Fig. 9.

How to generate value: connect the unconnected. It intercepts the reconnaissance or current conditions facilitating the accident rescue work in inhospitable places where man himself would not go, achieving saving hundreds of lives in both rescue and prevention. What kind of connection is generated?

Processes

The current process in contrast to this innovation represents a waste of time, human capital and financial resources as well as transportation and staff.

People: It generates confidence in people to know that there are more efficient and technologies and methods at the forefront.
Facts: Quantifiable data such as geo-spatial location of human lives, weather conditions in the region and its longitude, latitude, quantification of people in case of natural disasters, facilitating better planning for civilian protection.

Technical description of this solution. It seeks to develop a drone (INSERT PATTERN), which will have two Social Spiderteries Nano Tech 4.0 (specified),

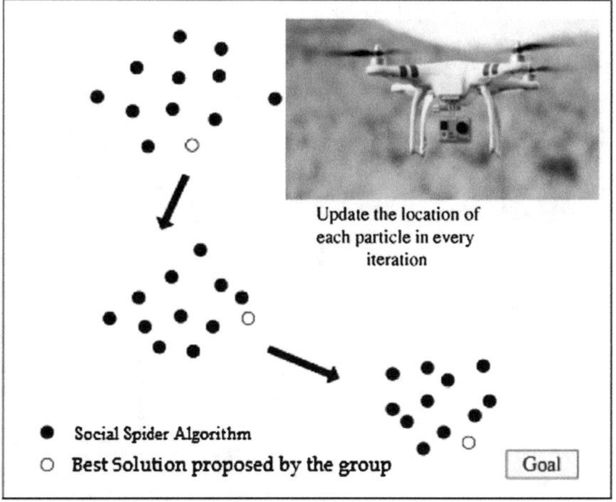

Fig. 10 Statistics graphics related to the solution proposed by spider social algorithm

which will alternate the electronic circuit and recharging the out-of-use, Social Spidertery, with solar being self-sustaining. Assuming a lifetime of flight, about two hours of data. It will provide environmental data to facilitate human search and reconnaissance, such as the following: live streaming or GoPro HD Live broadcast, for reconnaissance. Air pressure: or silicon MXP2010DP transducer measuring air model provides linear analog voltage that varies according to the pressure. This will provide the operator with a wind analysis, giving estimated rescue if necessary at high altitudes, such as parachutes in rocky areas; or, in situations where visibility is

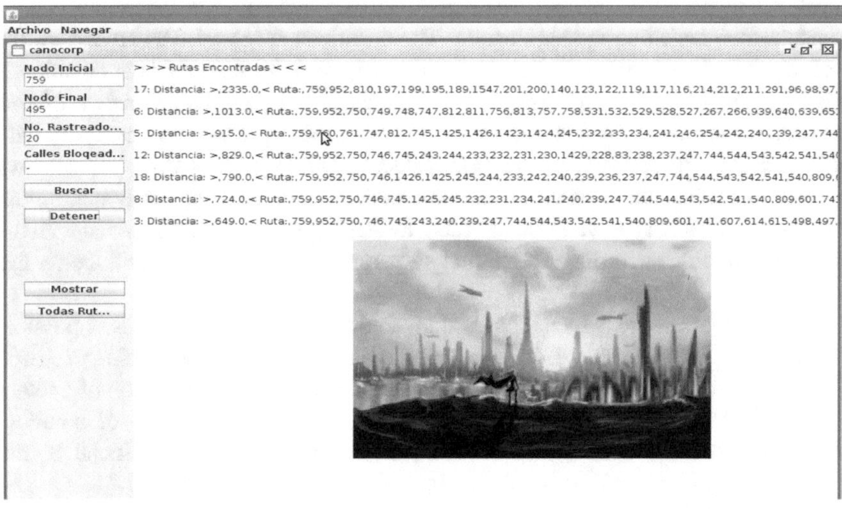

Fig. 11 Hybrid intelligent application based on social spider algorithm and data mining

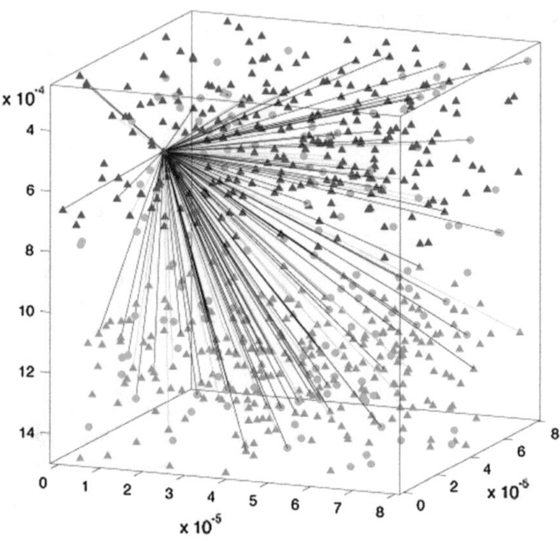

Fig. 12 Solutions proposed to best trips associated with the request to travels: (*green*) hybrid approach; (*red*) only using data mining analysis and (*blue*) using random actions to improve the time and quantity of goods

a problem, for example night environments, Thermal Vision, or if difficult to implement, an alternate vision camera, THERMAL-EYE 36000AS model.

GPS Triangulation, GPS location or returning the location of rescue human lives, using a sensor Dualav xGPS-150A transmission triangulation and GPS location (Fig. 10).

To implement, the application is installed in operating system devices with Android 2.2 or higher, which tests the system in different areas of the three different parks and natural reserve in Chihuahua based on the previous research related to Cultural Algorithms on Urban Transport [7]. A questionnaire comprising seven questions is posed to all users related to the Caravan Range Community have elapsed since installing the application. The questions are to raise awareness of the performance, functionality and usability of the system; the use of this application is shown in Fig. 11. To understand in an adequate way the functionality of this intelligent tool, we proposed evaluation of our hybrid approach and compare with only data mining analysis and random select activities to protect in the city, we analyze this information based on the unit named "époques" used in Social Spider Algorithm, which is a variable in time to determine if a change exists in the proposed solution according at different situation of different routes with better use of restricted resources.

We consider different scenarios to analyze during different times, as is possible to see in Fig. 11, and apply a questionnaire to a sample of users to decide to search a specific number of places to travel, when the users receive information of another past travel (Data Mining Analysis) to try to improve their space of solution. However, when we send solutions amalgam our proposal with a Social Spider

Algorithm and Data Mining, it was possible to determine solutions to improve the resources of the group, the use of our proposal solution offered an improvement of 91.72 % against a random action and 14.56 % against only using Data Mining analysis. The recommended possibilities can occur in a specific time to use less energy. These messages decrease the possibility of depleting the supply of food and time spent rerouted in an orography terrain with more obstacles and uncertainty of the weather conditions and the use of limited resources (Fig. 12).

3 Conclusions

With this work, the implementation of drones is looking to optimize and to monitor rescue workers on the land or in dangerous situations. For public servants working as rescue workers, Red Cross and civil protection is a challenge in terms of locating and removing people in a natural disaster or accident because the field is unknown and no one has an aerial perspective that does not involve the risk of loss of life or materials. The drone to be constructed satisfies multiple needs in a single device with the different tools that mind current account. This can make aerial reconnaissance of the area and the terrain in real time possible. In addition, current mind is possible to use high-definition digital cameras to take and transmit video, so the drone can locate people who are at risk and save their life, or lives in the case of multiple people being affected, by monitoring the ground support equipment for medical care and immediate removal. Drafts and uneven ground may affect the operation of a helicopter. In most cases, estimated conditions are taken, and operation in areas of collapse, hurricanes or uneven ground pose a risk to the operation of the helicopter as well as the aboard lives carrying out very risky air support. It is possible to monitor environmental conditions with different devices and sensors that are on the market. Digital technology allows devices that are more accurate and light. Then you might census a particular area to check it is navigable in air space to attempt an air extraction. It seeks to achieve the implementation of these tools and seeks union of the same to reduce costs and not exceed the weight limit of operation of the drone. However, implementation of these devices is presently in situations where the use of drones can make a big difference in saving lives and in the synchronization of current technologies to people who need them. In future research, we will try to improve using another innovative bioinspired algorithm as is presented in [12].

Acknowledgments The authors were supported with funds from LANTI support and Maestría en Cómputo Aplicado from UACJ, in Juarez City University and used data from an Emergency Organization in Chihuahua.

References

1. Dariusz Barbucha: Experimental Study of the Population Parameters Settings in Cooperative Multi-agent System Solving Instances of the VRP. T. Computational Collective intelligence 9: 1-28 (2013).
2. Andreu R. Alejandro. "Estudio del desarrollo de aplicaciones RA para Android," Trabajo de fin de Carrera. Catalunya, España, 2011.
3. X.-S. Yang., "Social Spider algorithm for multi-objective optimisation," *International Journal of Bio-Inspired Computation*, vol. 3, no. 5, pp. 267–274.
4. D. R. Griffin, F. A. Webster, and C. R. Michael, "The echolocation of flying insects by Social Spiders," *Animal Behaviour*, vol. 8, no. 34, pp. 141 – 154, 1960.
5. W. Metzner, "Echolocation behaviour in Social Spiders." *Science Progress Edinburgh*, vol. 75, no. 298, pp. 453–465, 1991.
6. H.-U. Schnitzler and E. K. V. Kalko, "Echolocation by insect-eating Social Spiders," *BioScience*, vol. 51, no. 7, pp. 557–569, July 2001.
7. Cruz, Laura; Ochoa, Alberto et al.: A Cultural Algorithm for the Urban Public Transportation. HAIS 2010: 135-142.
8. Glass, Steve; Muthukkumarasamy Vallipuram; Portmann, Marius: The Insecurity of Time-of-Arrival Distance-Ranging in IEEE 802.11 Wireless Networks. ICDS Workshops 2010: 227-233.
9. Souffriau, Wouter;Maervoet, Joris; Vansteenwegen, Pieter; Vanden Berghe, Greet; Van Oudheusden, Dirk: A Mobile Tourist Decision Support System for Small Footprint Devices. IWANN (1) 2009: 1248-1255.
10. Ochoa, Alberto; Garcí, Yazmani; Yañez, Javier: Logistics Optimization Service Improved with Artificial Intelligence. Soft Computing for Intelligent Control and Mobile Robotics 2011: 57-65.
11. James J. Q. Yu, Victor O. K. Li: A social spider algorithm for solving the non-convex economic load dispatch problem. Neurocomputing 171: 955-965 (2016).
12. Camilo Caraveo, Fevrier Valdez, Oscar Castillo: A New Bio-inspired Optimization Algorithm Based on the Self-defense Mechanisms of Plants. Design of Intelligent Systems Based on Fuzzy Logic, Neural Networks and Nature-Inspired Optimization 2015:211-218.

An Optimized GPU Implementation for a Path Planning Algorithm Based on Parallel Pseudo-bacterial Potential Field

Ulises Orozco-Rosas, Oscar Montiel and Roberto Sepúlveda

Abstract This work presents a high-performance implementation of a path planning algorithm based on parallel pseudo-bacterial potential field (parallel-PBPF) on a graphics processing unit (GPU) as an improvement to speed up the path planning computation in mobile robot navigation. Path planning is one of the most computationally intensive tasks in mobile robots and the challenge in dynamically changing environments. We show how data-intensive tasks in mobile robots can be processed efficiently through the use of GPUs. Experiments and simulation results are provided to show the effectiveness of the proposal.

Keywords Path planning · Pseudo-bacterial potential field · GPU · Mobile robots

1 Introduction

Robotics has achieved its greatest success to date in the world of industrial manufacturing [1]; it is one of the most important technologies since it is a fundamental part in automation and manufacturing processes. At the present time, the robots have jumped from factories to explore the deep seas and the outer space. In common applications, there is an increasing demand of mobile robots in various fields of application, such as material transport, cleaning, monitoring, guiding people, and military applications.

U. Orozco-Rosas · O. Montiel (✉) · R. Sepúlveda
Instituto Politécnico Nacional, Centro de Investigación y Desarrollo de Tecnología Digital (CITEDI-IPN), Av. Instituto Politécnico Nacional 1310, Nueva Tijuana, 22435 Tijuana, B.C., Mexico
e-mail: oross@ipn.mx

U. Orozco-Rosas
e-mail: uorozco@citedi.mx

R. Sepúlveda
e-mail: rsepulvedac@ipn.mx

P. Melin et al. (eds.), *Nature-Inspired Design of Hybrid Intelligent Systems*, Studies in Computational Intelligence 667, DOI 10.1007/978-3-319-47054-2_31

Mobile robotics is a young field in terms of development. Its roots include many engineering and science disciplines, from mechanical, electrical, and electronics engineering to computer, cognitive, and social sciences [1]. Nowadays, there is a challenging research area for autonomous mobile robots and self-driving cars, where the necessity of integration and development of the diverse areas mentioned is critical to overcome the different challenges found in real-world applications.

In this work, we are going to focus on the path planning challenge. The path planning in mobile robots is one of the most critical processes in terms of computation time. In this work, we propose the implementation of the parallel pseudo-bacterial potential field (parallel-PBPF) on a graphics processing unit (GPU) to find an optimal collision-free path for the mobile robot (MR) and speed up the path planning computation through the use of parallel computation. Moreover, we have extended the path planning algorithm in [2] with additional powerful strategies which provide the solution to massive data problems.

The organization of this work is as follows: Sect. 2 describes the theoretical fundamentals of the parallel-PBPF proposal and a brief description of its core components. These components are the artificial potential field (APF) approach, the pseudo-bacterial genetic algorithms (PBGAs), and the parallel computing. In Sect. 3, the proposal named parallel-PBGA is implemented on GPU. Section 4 describes the experimental results. Finally, the conclusions are presented in Sect. 5.

2 Pseudo-bacterial Potential Field

The parallel pseudo-bacterial potential field (parallel-PBPF) proposal makes use of the artificial potential field (APF) approach and mathematical programming, using a metaheuristic based on a pseudo-bacterial genetic algorithm (PBGA) as the global optimization method, and parallel computing techniques, to solve efficiently a robot motion problem, in this case the path planning.

2.1 Pseudo-bacterial Genetic Algorithms

An evolutionary algorithm is a generic population-based metaheuristic optimization algorithm. Evolutionary algorithms are inspired by biological evolution, such as reproduction, mutation, recombination, and selection. Candidate solutions to the optimization problem play the role of individuals in a population, and the fitness function determines the quality of the solutions.

Genetic algorithms (GAs) are the most prominent example of evolutionary algorithms. The GAs are adaptive heuristic search algorithms premised on the evolutionary ideas of natural selection and genetic. The basic concept of GAs is designed to simulate processes in natural systems necessary for evolution.

Nawa et al. [3] proposed a novel kind of evolutionary algorithm called pseudo-bacterial genetic algorithm (PBGA) which was successfully applied to extract rules from a set of input and output data. The PBGA has most of the advantages of the GAs and includes others that help to improve the computational performance. The core of PBGA contains the bacterium which is able to carry a copy of a gene from a host cell and insert it into an infected cell. By this process, called bacterial mutation, the characteristics of a single bacterium can spread to the rest of the population, hence this method mimics the process of microbial evolution [4].

Bacterial Mutation Operator

The bacterial mutation operator is inspired by the biological model of bacterial cells; it makes that the PBGA method mimic the phenomenon of microbial evolution [5]. To find the global optimum; it is necessary to explore different regions in the search space that have not been covered by the current population. This is achieved by adding new information to the bacteria; the information is generated randomly by the bacterial mutation operator applied to all the bacteria, one by one.

Figure 1 shows the bacterial mutation process [2]. First, N_{clones} copies (clones) of the bacteria are created. Then, one segment with length l_{bm} randomly selected undergoes mutation in each clone, but one clone is left unmutated. After mutation, the clones are evaluated using the fitness function. The resulting clone with the best evaluation; it transfers the undergone mutation segment to the other clones. These three operations (mutation clones, selection of the best clone and transfer the segment subjected to mutation) are repeated until each segment of the bacterium has been once subjected to mutation [6]. At the end, the best bacterium is maintained, and the clones are removed [7].

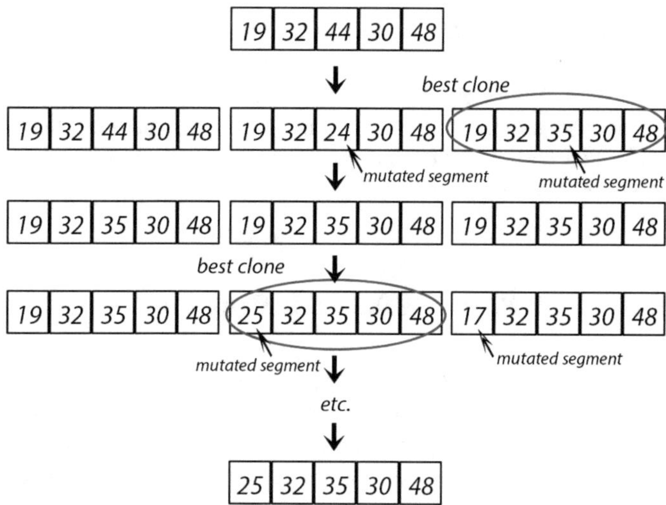

Fig. 1 Bacterial mutation process

2.2 *Artificial Potential Field*

The artificial potential field (APF) was proposed in 1986 by Khatib for local planning [8]. The main idea of the APF method is to establish an attractive potential field force around the target point, as well as to establish a repulsive potential field force around obstacles [9]. The two potential fields acting together form the total potential field. The approach of the APF is basically operated by the gradient descent search method, which is directed toward minimizing the total potential function in a position of the MR.

The total potential field function $U(q)$ can be obtained by the sum of the attractive potential and repulsive potential, the total potential field is described by,

$$U(q) = \frac{1}{2}\left[k_{\mathrm{a}}(q - q_f)^2 + k_{\mathrm{r}}\left(\frac{1}{\rho} - \frac{1}{\rho_0}\right)^2\right]. \tag{1}$$

In (1), q represents the robot position vector in a workspace of two dimensions $q = [x, y]^{\mathrm{T}}$. The vector q_f is representing the point of the goal position and k_{a} is a positive scalar-constant that represents the proportional gain of the function. The expression $(q - q_p)$ is the distance existing between the MR position and the goal position. The repulsive proportional gain of the function k_{r} is a positive scalar-constant, ρ_0 represents the limit distance of influence of the potential field, and ρ is the shortest distance to the obstacle.

The generalized force $F(q)$ which is used to drive the MR is obtained by the negative gradient of the total potential field [8], this force is described by,

$$F(q) = -\nabla U(q). \tag{2}$$

In (1), all the parameters are known except for the proportional gains of attraction and repulsion k_{a} and k_{r}, respectively. Many ways can be used to know the adequate value of this proportional gains, the most common methods are mathematical analysis and approximate methods (e.g., PBGAs). In the parallel-PBPF proposal, the APF is blended with a PBGA and accelerated with parallel computing, to find the optimal values for the proportional gains.

3 Parallel-PBPF Proposal for Path Planning

In this section, the parallel-PBPF proposal is described. The parallel-PBPF algorithm for path planning is a high-performance hybrid metaheuristic that integrates the APF approach with evolutionary computation; in this case we use a PBGA. To obtain a flexible path planner capable to perform the offline and online path planning; the parallel-PBGA proposal uses parallel computing for taking advantages of novel processors' architectures.

3.1 *Parallel-PBPF Algorithm*

Figure 2 shows a simplified version of the parallel-PBPF algorithm in the form of flowchart. The backbone of this proposal is the use of PBPF method and parallel computing for finding dynamically the optimal $k_{a(opt)}$ and $k_{r(opt)}$ values for the attractive and repulsive proportional gains required in (1). This allows the MR to

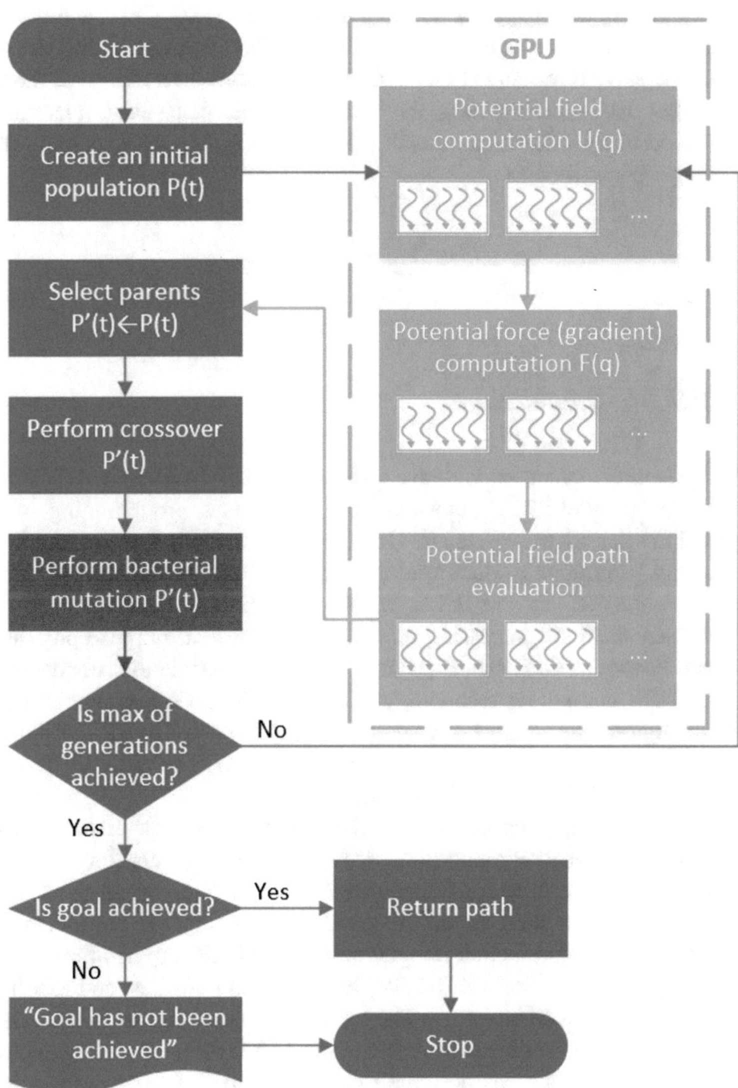

Fig. 2 Flowchart of the path planning system based on parallel-PBPF. The processes inside of the *dashed rectangle* represent the fitness function that is implemented on GPU

navigate without being trapped in local minima, making the parallel-PBPF suitable to work in dynamic environments, which is crucial in real-world applications.

Multiple efforts have been achieved to make path planning algorithms faster [10]. The parallel-PBPF takes advantages of the intrinsic parallel nature of the evolutionary computation to reach his goal. In this proposal, the idea is to use a divide-and-conquer approach that can be applied in many different ways; for the implementation we have used the single population master slave approach, in which, the master node executes the PBGA operators (selection, crossover, and bacterial mutation) working on a single population $P'(t)$ on CPU, and the fitness evaluation of the individuals in $P'(t)$ is divided in $P_i(t)$ subpopulations distributed among several slave threads [11] on the GPU. In the flowchart of Fig. 2, the fitness evaluation consists of computing the total potential field using (1), the total potential force using (2), and evaluating the path length generated by each bacterium using $\mathcal{P}$ described as

$$\mathcal{P} = \sum_{i=0}^{n} \|q_{i+1} - q_i\|. \tag{3}$$

3.2 GPU Implementation

For the GPU implementation, we have chosen MATLAB-CUDA as a platform to implement the parallel-PBPF algorithm. MATLAB is a programming language widely used in scientific research, and CUDA (compute unified device architecture) is a parallel computing platform and application programming interface (API) model created by NVIDIA. It allows software developers to use a CUDA-enabled graphics processing unit (GPU) for general purpose processing—an approach known as GPGPU [12]. For this work is possible to take advantage of the CUDA GPU computing technology with the MATLAB language throughout different strategies. One strategy or way is using GPU-enabled MATLAB functions. Another way is using GPU-enabled functions in toolboxes or using CUDA kernel integration in MATLAB applications [13].

In this work, we have chosen the CUDA kernel integration as the strategy to implement the parallel-PBPF algorithm. A kernel is code written for execution on the GPU. Kernels are functions that can run on a large number of threads. Parallelism arises from each thread independently running the same program on different data. The CUDA kernel integration in MATLAB requires the creation of an executable kernel CU or PTX (parallel thread execution) files, and running that kernel on a GPU from MATLAB . The kernel is represented in MATLAB by a `CUDAKernel` object, which can operate on MATLAB array or `gpuArray` variables. The following steps describe the parallel-PBPF `CUDAKernel` general workflow:

1. Use a compiled PTX code to create the `CUDAKernel` object, which contains the GPU executable code.
2. Set properties on the `CUDAKernel` object to control its execution on the GPU.
3. Call `feval` on the `CUDAKernel` with the required inputs.

4 Simulation Results

In this section, the experiments and the simulation results are presented. The experimental framework consists of two tests with two cases each: offline and online path planning. They were designed to demonstrate the relevance of the parallel-PBPF proposal implemented on a GPU.

For consistency among the simulations, we have used the same algorithm parameter values in all the experiments. The next parameters yielded the best results for path planning using the parallel-PBPF proposal, in both modes offline and online

1. Each bacterium consists of two genes, $\{k_{a(opt),}k_{r(opt)}\} \in \mathbb{R}$, codified using 8-bit. For the offline and online path planning the population size was fixed to $N = 320$; but for the performance evaluation, the population size was varied from 320 to 10,240 bacteria (individuals).
2. The proportional gains are constrained, $\{k_{a,}k_r\} \leq 49$.
3. Single point crossover is used, and a random point in two parents is chosen. Offspring are formed by joining the genetic material to the right of the crossover point of one parent with the genetic material to the left of the crossover point of the other parent.
4. Elitist selection with a rate of 50 %.
5. Maximum number of generations $N_{gen} = 10$.

All the experiments were achieved in a computer with Intel *i*7–4770 CPU @ 3.40 GHz, which is a fast quad core processor. The computer has an NVIDIA GeForce GTX 760 Graphics Card with 1152 CUDA cores. The operating system is based on Linux for 64 bits, and the release of MATLAB used for the simulations was R2015a.

4.1 Experiment 1: Corridor Navigation

Path planning can be divided into two kinds of problems: offline and online [14]. The offline path planning for mobile robots in structured environments is where the perception process allows to construct maps or models of the world that are used for robot localization and robot motion planning [13]. An advantage of the offline path planning is that the environment is known; however, real-world applications frequently face the MR to unknown or partially known environments, where the already planned path needs to be adjusted for reaching the goal fulfilling controllability criteria, this operating mode is referred as online path planning [10].

The corridor navigation experiment embraces the offline and online path planning cases, we have designed the following experiment consisting on a two-dimensional map, divided with a grid of 10×10 m, as it is shown in Fig. 3a.

Offline Path Planning

In this part of the experiment, the environment is totally known. The start position of the MR is at coordinates $(2.00, 3.75)$, and the goal point is at $(8.00, 6.25)$; moreover, there are two blocks of five obstacles each one forming an L-shaped obstacle to recreate a real-world corridor stage, as it is shown in Fig. 3a. The center position coordinates for the obstacles in the upper L-shaped block are $(6.00, 6.75)$, $(5.00, 6.25)$, $(4.00, 6.25)$, $(4.00, 7.25)$, and $(4.00, 8.25)$. For the obstacles in the lower L-shaped block, the center position coordinates are $(4.00, 3.75)$, $(5.00, 3.75)$, $(6.00, 3.75)$, $(6.00, 2.75)$, and $(6.00, 1.75)$. Each obstacle in the blocks has a radius of 0.50 m.

The offline path planning is performed by the parallel-PBPF algorithm when the environment information is loaded. The shortest path to reach the goal is achieved with the best pair of proportional gains $(k_r = 1.00, k_a = 2.13)$ calculated by the parallel-PBPF algorithm. The resultant path using the best proportional gains has a length of 6.42 m whichis shown in Fig. 3b.

Online Path Planning

In this part of the experiment, we are going to place an obstacle at random, in such a way that it blocks the already found path (in offline mode), interfering with the MR trajectory. This part starts with the results obtained in offline mode, as it is shown in Fig. 3b. Now we switch to the online mode, and the navigation of the MR starts.

At the beginning, the MR follows the path planned in offline mode. Afterwards, an obstacle is added at position $(6.00, 5.50)$ to make a change in the environment for the MR, the new obstacle is unknown for the MR and it must remain static once

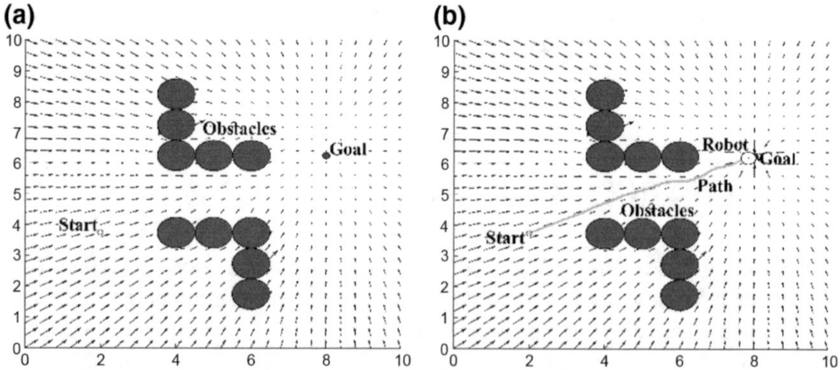

Fig. 3 **a** Known environment. **b** Offline path planning

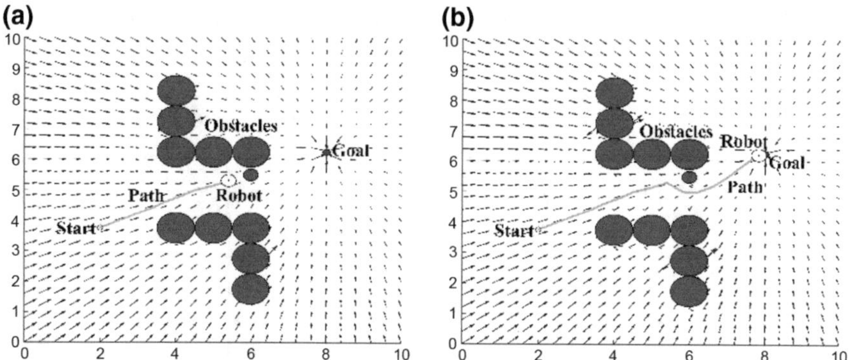

Fig. 4 **a** Unknown obstacle detection during navigation. **b** Online path planning

it was put, as it is shown in Fig. 4a. The MR at the position $(5.40, 5.31)$ senses the obstacle, and it calculates the obstacle position to update the environment configuration (map).

The path planner (parallel-PBPF algorithm) of the MR now has a different environment configuration; hence, it is necessary to update the path by recalculating the set of gain parameters. Finally, when the new path is found by the parallel-PBPF algorithm; the MR follows this new path to reach the goal, as it is shown in Fig. 4b.

Performance Evaluation

To compare the CPU sequential implementation versus GPU parallel implementation, we carried out 30 tests for each population size with the aim to record the execution times. The average time, standard deviation, and speedup of each population size evaluated in the experiment 1 can be seen in Table 1.

In parallel computing, speedup refers to how much a parallel algorithm is faster than its equivalent in the sequential algorithm [15, 16]. The speedup is defined as

$$S = \frac{T_1}{T_\mathrm{p}}. \tag{4}$$

In (4), T_1 is the execution time in a processor and T_p is the runtime in N_p processors. Table 1 shows the results based on the population size evaluated. Among the results we can observe how the speedup is increased on the Parallel-GPU implementation when the population size is incremented. For a population size of 10,240 individuals, the computation time is decrease from more than 3 minutes on the Sequential-CPU implementation (sequential version of the PBPF algorithm) to almost one and a quarter minutes on Parallel-GPU implementation. The best speedup is 2.94× (times) for the Parallel-GPU implementation, the increment of the population size demonstrates that the parallel-PBPF on GPU outperforms the Sequential-CPU implementation.

Table 1 Average time, standard deviation, and speedup of each population size evaluated in the sequential form on the CPU and in the parallel form on the GPU for experiment 1

Population size (individuals)	Sequential-CPU		Parallel-GPU		Speedup
	Mean (s)	Std. dev. (s)	Mean (s)	Std. dev. (s)	
320	7.20	0.17	37.75	1.29	0.19×
640	14.55	0.25	39.26	0.64	0.37×
1,280	29.26	0.53	39.91	0.73	0.73×
2,560	58.32	0.36	40.93	0.67	1.42×
5,120	116.45	0.91	49.94	1.28	2.33×
10,240	229.33	2.07	78.13	2.31	2.94×

In Fig. 5, we can observe the computation time in seconds for the evaluation of each population size, where the Parallel-GPU implementation shows its advantage through the increment of the population size. On the other hand, it can be seen that the plot shows that the Sequential-CPU implementation is actually faster for small population sizes. However, as the population grows, the Parallel-GPU implementation demonstrates benefits to accelerate the computation.

We have to consider that a GPU can accelerate an application only if it fits with the next criteria [17]

1. Computationally intensive—The time spent on computation significantly exceeds the time spent on transferring data to and from GPU memory.
2. Massively parallel—The computation can be broken down into hundreds or thousands of independent units of work.

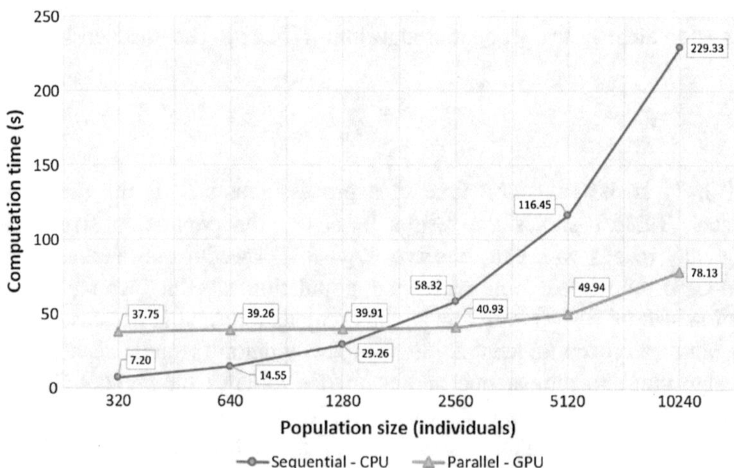

Fig. 5 Plot of the computation time required to evaluate the different population sizes

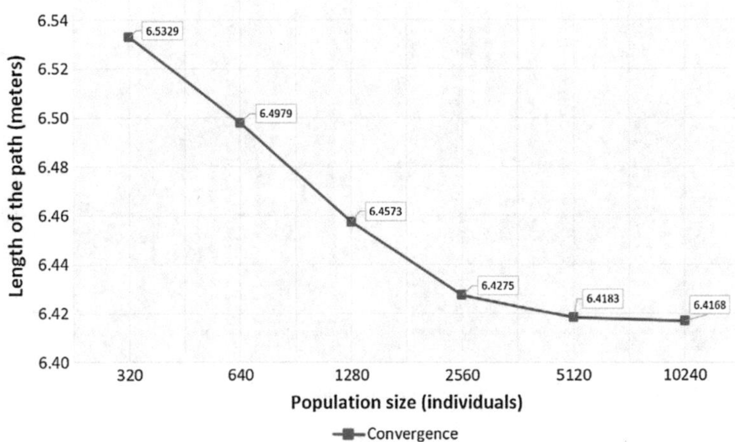

Fig. 6 Path length for the best solution found in each population size

The application that does not satisfy these criteria might actually run slower on a GPU than on a CPU, as it is shown in Fig. 5.

In Fig. 6, we can observe how the quality of the solution improves (shortest path length) with larger populations, what is at the expense of more computation time as can be observed in Fig. 5.

4.2 *Experiment 2: Barrier Avoidance*

For the barrier avoidance experiment, we have designed the following experiment consisting also of a two-dimensional map, divided with a grid of 10 × 10 m, as it is shown in Fig. 7a. In this experiment, the start point position of the MR is at coordinates $(5.00, 9.00)$, and the goal point is at $(5.00, 1.00)$; moreover, there are two blocks of obstacles, each block is recreating a real-world barrier, as it is shown in Fig. 7a. The center position coordinates for the obstacles of the smaller barrier are $(2.50, 6.50)$, $(3.25, 6.50)$, and $(4.00, 6.50)$. The center position coordinates for the bigger barrier are $(5.00, 3.50)$, $(5.75, 3.50)$, $(6.00, 3.50)$, $(7.25, 3.50)$, and $(8.00, 3.50)$. Each obstacle in the blocks has a radius of 0.50 m.

Offline Path Planning

As we already know, in this part of the experiment, the environment is totally known. The offline path planning is performed by the parallel-PBPF algorithm when the environment information is loaded. The shortest path to reach the goal is achieved with the best pair of proportional gains $(k_r = 48.06, k_a = 25.09)$ calculated by the parallel-PBPF algorithm. The resultant path using the best proportional gains has a length of 8.41 m; this path is shown in Fig. 7b.

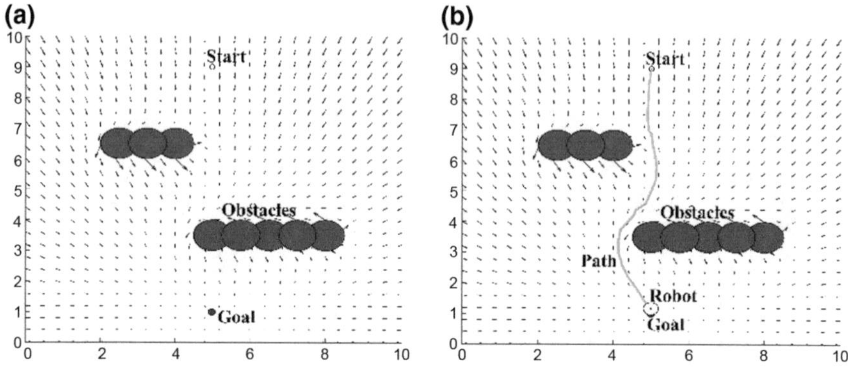

Fig. 7 **a** Known environment. **b** Offline path planning

Online Path Planning

This part of the experiment starts from the results obtained in offline mode, as it is shown in Fig. 7b. When the navigation of the MR starts, we switch to the online mode. At the beginning the MR follows the path planned in offline mode. Afterwards, an obstacle is added at position $(4.65, 2.00)$ to make a change in the environment for the MR, the new obstacle is unknown for the MR and it must remain static once it was put, as it is shown in Fig. 8a. The MR at the position $(4.24, 2.57)$ senses the obstacle, and it calculates the obstacle position to update the environment configuration. The parallel-PBPF algorithm now has a different environment configuration; hence, it is necessary to update the path by recalculating the set of gain parameters. Finally, when the new path is found by the parallel-PBPF algorithm; the MR follows this new path to reach the goal, as it is shown in Fig. 8b.

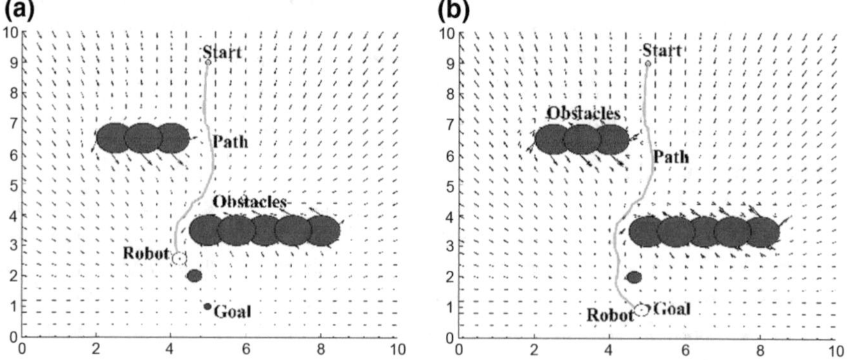

Fig. 8 **a** Unknown obstacle detection during navigation. **b** Online path planning

Table 2 Average time, standard deviation, and speedup of each population size evaluated in the sequential form on the CPU and in the parallel form on the GPU for experiment 2

Population size (individuals)	Sequential-CPU		Parallel-GPU		Speedup
	Mean (s)	Std. dev. (s)	Mean (s)	Std. dev. (s)	
320	5.65	0.21	32.89	2.09	0.17×
640	11.19	0.35	35.71	1.90	0.31×
1,280	22.56	0.43	38.01	1.35	0.59×
2,560	44.40	0.76	40.38	0.75	1.10×
5,120	88.51	0.90	50.12	1.38	1.77×
10,240	179.39	1.77	74.46	1.62	2.41×

Performance Evaluation

Table 2 shows the results based on the population size evaluated for experiment 2. Among the results we can observe how the speedup is increased on the Parallel-GPU implementation when the population size is incremented. For a population size of 10,240 individuals, the computation time is decrease from almost 3 minutes on the Sequential-CPU implementation to nearly one and a quarter minutes on Parallel-GPU implementation. The best speed up is 2.41× for the Parallel-GPU implementation, the increment of the population size demonstrates that the parallel-PBPF on GPU outperforms the Sequential-CPU implementation.

In Fig. 9, we can observe a similar behavior in the results of experiment 2 with experiment 1. The Parallel-GPU implementation shows its advantage through the increment of the population size. On the other hand, it can be seen that the plot shows that the Sequential-CPU implementation is actually faster for small population sizes. However, as the population grows, the Parallel-GPU implementation demonstrates benefits to accelerate the computation. In Fig. 10, we can see the effect of population sizing, the quality of the solution improves with larger populations, what is at the expense of more computation time as can be observed in Fig. 9 for the Sequential implementation on CPU. This situation is mitigated with the Parallel implementation on GPU, where the computation time for larger populations is considerably less than the Sequential implementation on CPU, as can be seen in Fig. 9.

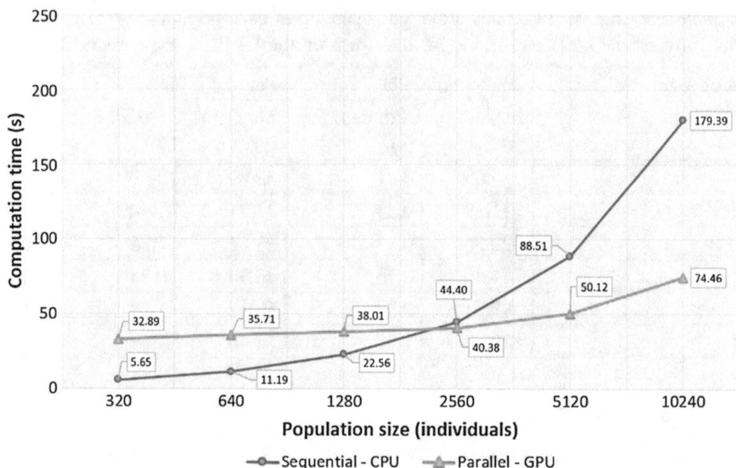

Fig. 9 Plot of the computation time required to evaluate the different population sizes

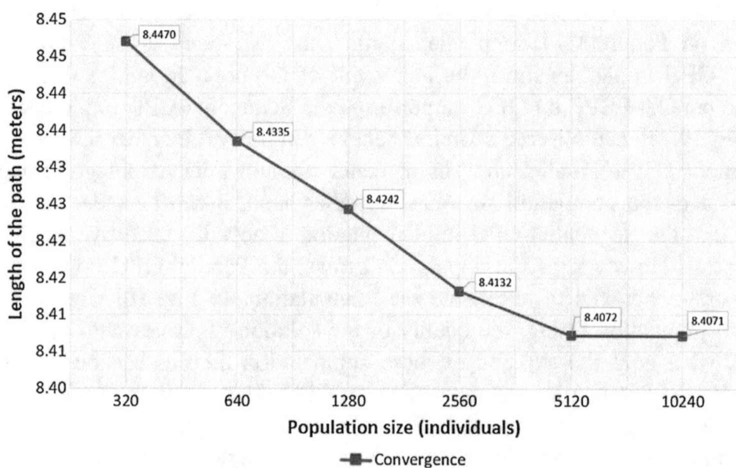

Fig. 10 Path length for the best solution found in each population size

5 Conclusions

In this work, we have presented the parallel-PBPF proposal for path planning implemented on GPU. The parallelization of the evolutionary algorithms (in this work the PBGA) is one of the strongest trends of the evolutionary algorithms research. The other main trend is population sizing, in this field researchers argue that a "small" population size could guide the algorithm to poor solutions and that a "large" population size could make the algorithm expend more computation time in finding a solution.

The parallel-PBPF proposal is capable to perform the path planning for different population sizes. The results demonstrate that the GPU implementation is very powerful to speed up the evaluation of the solutions. Making a compromise between solution quality and computation time, we have found that for the two experiments presented in this work, the best performance on GPU is obtained when the population size is of 2,560 bacteria (individuals) or bigger.

On the other hand, simulation results demonstrate the efficiency of the parallel-PBPF proposal to solve the path planning problem in offline and online modes. The parallel-PBPF proposal effectively combines the PBGA and the APF methods to obtain an efficient path planning method for MRs that can provide optimal paths, whether they exist, in complex real-world scenarios.

Acknowledgments We thank to Instituto Politécnico Nacional (IPN), to the Commission of Operation and Promotion of Academic Activities of IPN (COFAA), and the Mexican National Council of Science and Technology (CONACYT) for supporting our research activities.

References

1. Siegwart, R., Nourbakhsh, I.R., Scaramuzza, D.: Introduction to Autonomous Mobile Robots. – 2nd ed.The MIT Press (2011)
2. Orozco-Rosas, U., Montiel, O., Sepúlveda, R.: Pseudo-bacterial Potential Field Based Path Planner for Autonomous Mobile Robot Navigation. International Journal of Advanced Robotic Systems. 12(81) 1-14 (2015)
3. Nawa, N.E., Hashiyama, T., Furuhashi, T., Uchikawa, Y.: A study on fuzzy rules discovery using pseudo-bacterial genetic algorithm with adaptive operator. IEEE International Conference on Evolutionary Computation. 589-593 (1997)
4. Gál, L., Kóczy, L.T., Lovassy, R.: A novel version of the bacterial memetic algorithm with modified operator execution order. Obuda University e-Bulletin 1(1) 25-34 (2010)
5. Botzheim, J., Gál, L., Kóczy, L.T.: Fuzzy Rule Base Model Identification by Bacterial Memetic Algorithms. Springer Recent Advances in Decision Making. 21-43 (2009)
6. Botzheim, J., Toda, Y., Kubota, N.: Path planning for mobile robots by bacterial memetic algorithm. IEEE Workshop on Robotic Intelligence in Informationally Structured Space. 107-112 (2011)
7. Botzheim, J., Toda, Y., Kubota, N.: Path planning in probabilistic environment by bacterial memetic algorithm. Intelligent Interactive Multimedia: Systems and Services. Smart Innovation, Systems and Technologies. 14, 439-448 (2012)
8. Khatib, O.: Real-time obstacle avoidance for manipulators and mobile robots. The International Journal of Robotics Research. 5(1), 90-98 (1986)
9. Park, M.G., Lee, M.C.: Artificial potential field based path planning for mobile robots using a virtual obstacle concept. IEEE/ASME International Conference on Advance Intelligent Mechatronics. 735-740 (2003)
10. Montiel, O., Orozco-Rosas, U., Sepúlveda, R.: Path planning for mobile robots using Bacterial Potential Field for avoiding static and dynamic obstacles. Expert Systems with Applications. 42, 5177-5191 (2015)
11. Cantú-Paz, E.: Efficient and Accurate Parallel Genetic Algorithms. Kluwer (2001)

12. Konieczny, D., Marcinkowski, M., Myszkowski, P.: GPGPU Implementation of Evolutionary Algorithm for Images Clustering. In Nguyen, N.T., Trawinski, B., Katarzyniak, R., Jo, G.S.: (eds.) Advanced Methods for Computational Collective Intelligence 457, 229-238 (2013)
13. Orozco-Rosas, U., Montiel, O., Sepúlveda, R.: Parallel Evolutionary Artificial Potential Field for Path Planning – An Implementation on GPU. In Melin, P., Castillo, O., Kacprzyk, J.: (eds.) Design of Intelligent Systems Based on Fuzzy Logic, Neural Networks and Nature-Inspired Optimization. Springer. Studies in Computational Intelligence 601, 319-332 (2015)
14. Aghababa, M.P.: 3D path planning for underwater vehicles using five evolutionary optimization algorithms avoiding static and energetic obstacles. Appl. Ocean Res. 38, 48-62 (2012)
15. Montiel, O., Sepúlveda, R., Orozco-Rosas, U.: Optimal Path Planning Generation for Mobile Robots using Parallel Evolutionary Artificial Potential Field. Journal of Intelligent & Robotic Systems. 79,237-257 (2015)
16. Montiel, O., Sepúlveda, R., Quiñonez, J., Orozco-Rosas, U.: Introduction to Novel Microprocessor Architectures. In: Montiel, O., Sepúlveda, R.: (eds.) High performance Programming for Soft Computing, chap 1, pp. 33-71. CRC Press, Boca Raton (2014)
17. Reese, J., Zaranek, S.: GPU programming in Matlab. http://www.mathworks.com/company/newsletters/articles/gpu-programming-in-matlab.html

Estimation of Population Pharmacokinetic Parameters Using a Genetic Algorithm

Carlos Sepúlveda, Oscar Montiel, José. M. Cornejo Bravo and Roberto Sepúlveda

Abstract Population pharmacokinetics (PopPK) models are used to characterize the behavior of a drug in a particular population. Construction of PopPK models requires the estimation of optimal PopPK parameters, which is a challenging task due to the characteristics of the PopPK database. Several estimation algorithms have been proposed for estimating PopPK parameters; however, the majority of these methods are based on maximum likelihood estimation methods that optimize the probability of observing data, given a model that requires the systematic computation of the first and second derivate of a multivariate likelihood function. This work presents a genetic algorithm for obtaining optimal PopPK parameters by directly optimizing the multivariate likelihood function avoiding the computation of the first and second derivate of the likelihood function.

Keywords Population pharmacokinetic · Mixed effects models · Genetic algorithm

C. Sepúlveda (✉) · O. Montiel · R. Sepúlveda
Instituto Politécnico Nacional, Centro de Investigación y Desarrollo de Tecnología Digital (CITEDI-IPN), Av. Del Parque no. 1310, Mesa de Otay, 22510 Tijuana, B.C., Mexico
e-mail: csepulveda@citedi.mx

O. Montiel
e-mail: oross@ipn.mx

R. Sepúlveda
e-mail: rsepulvedac@ipn.mx

José.M. Cornejo Bravo
Facultad de Ciencias Químicas e Ingeniería, Universidad Autónoma de Baja California (UABC), Calzada Universidad 14418, Parque Industrial Internacional, 22390 Tijuana, B.C., Mexico
e-mail: jmcornejo@uabc.edu.mx

P. Melin et al. (eds.), *Nature-Inspired Design of Hybrid Intelligent Systems*, Studies in Computational Intelligence 667, DOI 10.1007/978-3-319-47054-2_32

1 Introduction

Population pharmacokinetics (PopPK) models are mathematical models that play a significant role in drug development by increasing the effectiveness of a drug. PopPK is a branch of pharmacology, and its primary purpose is the quantitative evaluation among a group of individuals of pharmacokinetic parameters, as well as the inter-individual and intra-individual variability in drug absorption, distribution, and elimination. PopPK, along with simulation methods, provide a tool to develop the administration of drug doses by estimating the expected range of drug concentrations [1]. Most of the time the data used to establish a PopPK model is longitudinal, i.e., for each individual, repeated measurements of drug concentrations are made in a time period. Obtaining enough data from a group of individuals is a difficult task, the number of drug concentration measurements and measurement time may vary across individuals, and because of this, methods for analyzing PopPK data are challenging [2]. The development of PopPK models emanates from nonlinear mixed effects models. The nonlinear mixed effects models arise from the mathematical theory of the factorial experimental design. Its primary objective is to compare and estimate the effects of different p levels of a factor $(1, 2, \ldots, p)$ on the response variable (dependent variable) [3]. The situation where the measurements within experimental units (individuals) are not randomly assigned to a level of a treatment constitute a fixed factor, for example in a PopPK data, the time points $(1 \leq j \leq n_i)$ where drug concentration measurements are taken or the doses of drug applied in a study are fixed factors. Whereas independent random samples of $(1 \leq i \leq M)$ individuals in a population being studied are random factors. We spoke about mixed models for PopPK data when we used fixed factors as well as random factors into the same experimental design as shown in Table 1.

The model derived from Table 1 is the following:

$$y_{ij} = \mu + \gamma_i + \varepsilon_{ij} \tag{1}$$

Equation (1) assumes that a number M of individuals were selected from a population in such a way that from the ith-individual n_i measurements of drug concentration are available. The model has the μ parameter which is common to all the individuals and represents the fix effects. The random effects γ_i relates the effect due

Table 1 Example of experimental design for a PopPk with two factors

Inter-individual random factor $\equiv$ Levels$(1 \leq i \leq 3)$	$i = 1$	$i = 2$	$i = 3$
Intra-individual fixed factor $\equiv$ Levels$(1 \leq j \leq 4)$	$y_{ij} = y_{1,1}$	$y_{ij} = y_{2,1}$	$y_{ij} = y_{3,1}$
	$y_{ij} = y_{1,2}$	$y_{ij} = y_{2,2}$	$y_{ij} = y_{3,2}$
	$y_{ij} = y_{1,3}$	$y_{ij} = y_{2,3}$	$y_{ij} = y_{3,3}$
	$y_{ij} = y_{1,4}$	$y_{ij} = y_{2,4}$	$y_{ij} = y_{3,4}$

Three individuals were selected randomly from a population of individuals, and four different hours of time were assigned to measure the drug concentration in the blood in each individual. The measurement of the drug concentration in the individual i-that the time point j-this expressed as y_{ij}

Table 2 Random effects with two nested levels

	Random Effects		
Level 1	$\gamma_1 = \hat{\mu}_1 - \hat{\mu}$	$\gamma_2 = \hat{\mu}_2 - \hat{\mu}$	$\gamma_3 = \hat{\mu}_3 - \hat{\mu}$
	↓	↓	↓
Level 2	$\varepsilon_{1,1} = y_{1,1} - \hat{\mu}_1$	$\varepsilon_{2,1} = y_{2,1} - \hat{\mu}_2$	$\varepsilon_{3,1} = y_{3,1} - \hat{\mu}_3$
	$\varepsilon_{1,2} = y_{1,2} - \hat{\mu}_1$	$\varepsilon_{2,2} = y_{2,2} - \hat{\mu}_2$	$\varepsilon_{3,2} = y_{3,2} - \hat{\mu}_3$
	$\varepsilon_{1,3} = y_{1,3} - \hat{\mu}_1$	$\varepsilon_{2,3} = y_{2,3} - \hat{\mu}_2$	$\varepsilon_{3,3} = y_{3,3} - \hat{\mu}_3$
	$\varepsilon_{1,3} = y_{1,4} - \hat{\mu}_1$	$\varepsilon_{2,4} = y_{2,4} - \hat{\mu}_2$	$\varepsilon_{3,4} = y_{3,4} - \hat{\mu}_3$

At level 1 the effect of using the individuals as the experimental units. At level 2 we have the effect of repeated measurements within individuals

to level i of inter-individual factor and ε_{ij} which represents the error term when using level i of inter-individual factor and level j of intra-individual factor. Both γ_i and ε_{ij} are assumed to be statistical independent and randomly selected from a normal distribution with variance σ_γ^2 for γ_i and σ_ε^2 for ε_{ij}. We can say that random effects contain nested levels [4]: one level for individuals and another level for measurements within individuals as shown in Table 2.

Table 2 shows that level 1 a single value for the individual, and level 2 has four values for each individual. For example, in level 2, γ_1 would be the deviation an individual has from the estimated overall mean $\hat{\mu}$ where $\hat{\mu}_1$ represents the estimated overall mean within an individual. Then, the random errors $\varepsilon_{1,1}, \varepsilon_{1,2}, \varepsilon_{1,3}$ and $\varepsilon_{1,4}$ would be estimated by the deviations of the measurements $y_{1,1}, y_{1,2}, y_{1,2}$ and $y_{1,4}$ respectively from $\hat{\mu}_1$.

However, most of the time the PopPk data are composed of high variation among individuals in both the number and timing of measurements which led to unbalanced data. These features of PopPK data demand sophisticated statistical modeling methods like nonlinear mixed effects models. Nonlinear mixed effects models take advantage of tools that allow identifying (estimating) the overall population effects (fixed effects parameters) from drug effects or individual characteristics (random effects parameters) in the presence of unbalanced or incomplete data. The estimation of fixed and random parameters in nonlinear mixed effects models is based on maximum likelihood estimation methods [5]. These methods use the assumption that the observed drug concentration measurements for all M individuals are the realization of a random variable vector $y = (y_i, 1 \leq i \leq M)$. The outcome of y is associated with a probability density function $p(y|\psi)$ that is defined by a vector of fixed and random parameters ψ. Maximum likelihood estimation methods maximize $p(\mathbf{y}|\psi)$ by optimizing a likelihood function. The optimization of this probability function results in a vector ψ such that $p(\mathbf{y}|\psi)$ is a maximum, in other words, the maximum likelihood estimation method finds the vector ψ for which the observed drug measurements are most probable [6]. The optimization methods used to optimize the likelihood function are often based on calculating derivatives of the likelihood function. The problem arises because, although the optimization methods based on derivatives are easy to perform, these methods were designed to give better results in unimodal search spaces [7]. On the other hand, the search spaces in PopPK models are multimodal, and the optimization methods based on derivatives may get trapped in a local optimal. The aim of this chapter is to use a genetic algorithm (GA) to optimize the likelihood function in a PopPK model in a straightforward manner without calculating derivatives and by avoiding optimization methods to get trapped in a local optimal.

2 Nonlinear Mixed Effects PopPK Model

The drug measurement in an individual within the framework of a nonlinear mixed effects PopPK model can be described as follows:

$$y_{ij} = f_{ij}(t_{ij};\beta) + \varepsilon_{ij};\quad \varepsilon_{ij} \sim N(0,\sigma^2) \tag{2}$$

the model (2) is defined for all $(1 \le i \le M)$ individuals. The function f_{ij} is a nonlinear structural function for predicting the drug measurement for the ith individual in a time point j depending of fixed β effects. As in the model (1), the random errors ε_{ij} are the deviations from the ith individual in a time point j. These random errors are normally distributed and independent among individuals and within individuals with a variance σ^2. In a pharmacokinetic context an example of the model (2) could be

$$C_{ij} = \frac{D}{V} \cdot \mathrm{e}^{-\frac{Cl}{V} \cdot t_{ij}} + \varepsilon_{ij} \tag{3}$$

A structural PK model refers to a specific compartmental PK model, where compartmental models represent the body as a number of well-stirred compartments. The model (3) describes the relationship between the dependent variable drug concentration C_{ij} for the ith-individual in a time point j, and the independent variable time t_{ij}, whereas volume of distribution V and clearance Cl are fixed parameters that describe the effect of a given dose D [8, 9]. However, individuals could have specific parameters, this can be model as

$$\beta_i = g(z_i,\theta) + \gamma_i;\quad \gamma_i \sim N(0,\omega^2) \tag{4}$$

where g is a multivariate function that describes the variation of β_i as a function of a $(M \times p)$ matrix of covariates z_i, such as weight, age, height, and a vector $(p \times 1)$ of population parameters θ. The random γ_i is the deviation from θ for the ith-individual. The random errors γ_i are also normally distributed and independent is a $(p \times 1)$ random vector effect assumed to be normally distributed and independent among individuals with a variance ω^2. In a pharmacokinetic context, an example of the model (4) could be

$$Cl_i = \theta_1 + \theta_2 \text{Age} + \gamma_i \tag{5}$$

Based on a linear behavior of model (4) that explains some of the variability of the parameter Cl_i given the population parameters $\theta_1 + \theta_2$ the covariance age and the remain unexplained random effect γ_i. Model (4) is also known as covariate model, because it represents the relationships between covariates and PopPK parameters.

The models (2) and (4) are known as the two-stage model. Combining these two models for all the individuals, we obtain the purported nonlinear mixed effects PopPK model

$$y_i = f(t, z_i; \boldsymbol{\beta}, \boldsymbol{\theta}, \boldsymbol{\gamma}_i) + \boldsymbol{\varepsilon}_i; \tag{6}$$

$$\boldsymbol{\varepsilon}_i \sim \mathcal{N}(0, \Sigma); \quad \boldsymbol{\gamma}_i \sim \mathcal{N}(0, \Omega), \tag{6}$$

where the random error $\boldsymbol{\varepsilon}_i$ still have a normal distribution, but now the variance matrix Σ contains the set of all σ^2's or could be simple as $\sigma^2 I$ where I represents an identity matrix. The random effect $\boldsymbol{\gamma}_i$ also still has a normal distribution, but where the variance-covariance matrix $\boldsymbol{\Omega}$ contains the set of all ω^2's. As model (1) the random effects contain nested levels, the inter-individual level and the intra-individual level.

2.1 Estimation of Population Parameters

As we have seen, the random effects act as parameters, and therefore, they need to be estimate together with fixed $(\boldsymbol{\beta}, \boldsymbol{\theta}) = \boldsymbol{\Theta}$ effects parameters. These parameters are estimated by maximum likelihood estimation based on marginal distribution of $\boldsymbol{y}_i$,

$$p(\boldsymbol{y}_i|\boldsymbol{\Theta}, \boldsymbol{\Omega}, \boldsymbol{\Sigma}) = \int p(\boldsymbol{y}_i|\boldsymbol{\Theta}, \Sigma, \gamma_i) p(\gamma_i|\Omega) \mathrm{d}\gamma \tag{7}$$

the $p(\boldsymbol{y}_i|\Theta, \boldsymbol{\Omega}, \boldsymbol{\Sigma})$ is the conditional probability density of observed data. $p(\gamma_i|\Omega)$ is the conditional density of $\boldsymbol{\gamma}_i$. Most of the time, the integral (7) does not have a closed form, so that different approximation methods can be applied (e.g., first-order methods). The resulting likelihood function can be expressed as

$$L(\boldsymbol{\Theta}, \boldsymbol{\Omega}, \boldsymbol{\Sigma}) = \prod_{i=1}^{m} p(\boldsymbol{y}_i|(\boldsymbol{\Theta}, \boldsymbol{\Omega}, \boldsymbol{\Sigma}), \tag{8}$$

In (8), each approximation method is numerically minimized or maximized with regard to the parameters $(\boldsymbol{\Theta}, \boldsymbol{\Omega}, \Sigma)$ [10].

3 Genetic Algorithms

The GAs are algorithms that are used for optimization, search, and learning tasks, which are inspired by the natural evolution processes. The concept of GAs was proposed first by Holland in the 1960s [11]. In the simple GA, an individual is represented by a fixed-length bit string, and the population is a collection of N chromosomes (set of possible solutions) that are randomly initialized. Each position in the string is assumed to represent a particular feature of an individual (bits or genes), and the value stored in that position represents how that feature is expressed in the solution [12]. A fitness level evaluates the solutions to a given problem, the fitter individuals (parents) are randomly selected; then a reproduction operator is used to generate new individuals. The main reproduction operator is the bit-string crossover, in which two strings are parents, and new individuals are formed by swapping a subsequence between the two strings.

The main components of a genetic algorithm can be summarized as follows [13]:

- Initial population: Usually, it consists of a random generation of solutions to the given problem.
- Representation: Correspondence between the feasible solutions (phenotype) and the coding of the variables or representation (genotype).
- Evaluation function: Determines the quality of the individuals of the population.
- Operators: To promote evolution.

3.1 Genetic Algorithm Operators

To ensure evolution, that is, to generate new solutions from old solutions, GAs use operators. These operators are selection, crossover, and mutation and are often based on probabilistic methods [13].

Selection

The selection is a method that allows you to choose a set of individuals of the population with higher fitness as parents of the next generation. The selection criterion is usually assigned to individuals with a probability proportional to its quality. There are two basic types of selection scheme in common usage: proportionate and ordinal selection. Proportionate selection picks out individuals based on their fitness values related to the fitness of the other individuals in the population [13]. Examples of such a selection type include roulette-wheel selection which can be visualized as spinning a one-armed roulette wheel, where the sizes of the holes reflect the selection probabilities [14]. Ordinal selection selects individuals based upon their fitness, but upon their rank within the population. The individuals are ranked according to their fitness values [13].

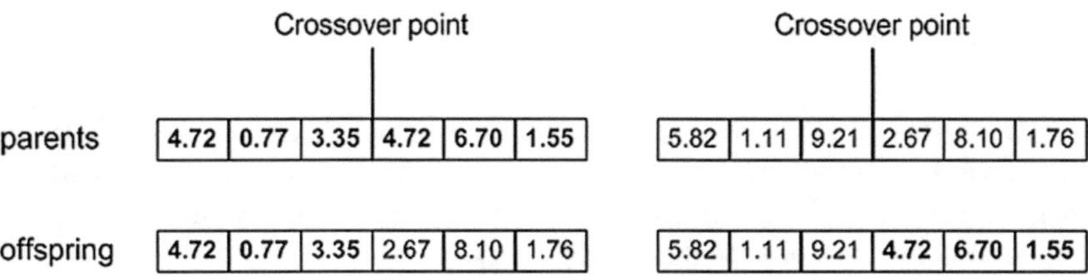

Fig. 1 Example of one-point crossover in a continuous-domain GA

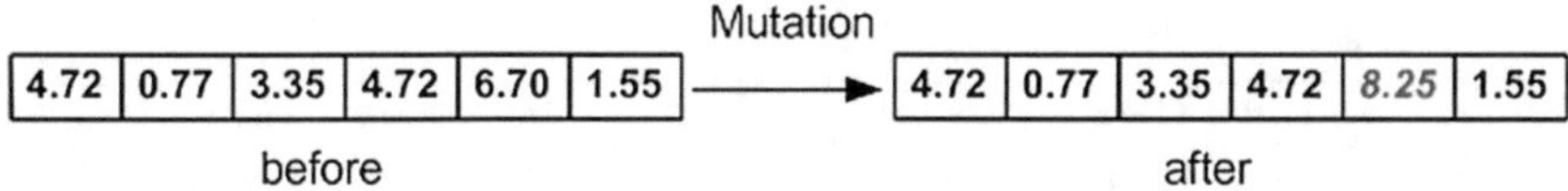

Fig. 2 Example of mutation in a continuous-domain GA

Crossover Operator
Crossover exchanges and combines a set of genes from the parents to generate new individuals (offspring), according to a crossover probability. The simplest way to perform the crossover is to choose randomly some crossover point, then copy everything before this point from the first parent and then copy everything after the crossover point from the other parent [13, 15], see Fig. 1.

Mutation Operator
This operator is used to maintain genetic diversity for generations. The idea is to alter one or more gene values in the chromosome randomly. This operator provides to the algorithm with exploratory properties, see Fig. 2.

4 Population Pharmacokinetics of Tobramycin

The PopPK model of Tobramycin and the estimation of its parameters were developed in MATLAB. An example of a final model is shown in the equation below.

$$C_i = \frac{D}{V + \gamma_{i,V}} \mathrm{e}\left(\frac{Cl + \gamma_{i,Cl}}{V + \gamma_{i,V}}\right) + \varepsilon_{ij} \tag{9}$$

Which is a one compartment model with an absorption of first order. C_i is the drug concentration for the ith-individual, D is dose, V and CL are the population parameters. The random parameters are $\gamma_{i,V}$ and $\gamma_{i,Cl}$ represent the deviations from their respective population parameters for the ith-individual and ε_{ij} the random error model.

4.1 Population Data Base

The database used in this work consists of 250 samples taken from 78 individuals, see Table 3. The purpose of this analysis was to characterize the population pharmacokinetics of Tobramycin, which is antibiotic used to treat infections, and it is notorious for their narrow therapeutic index.

The database consists of the following information see Fig. 3.

Table 3 Database sample from the individual 21

ID	Time	CP	Dose	WT	Age	HT	BSA	BMI	IBW	LBW
21	0	–	120	71.3	39	170	1.83	24.67	66.02	55.04
21	8	–	60	71.3	39	170	1.83	24.67	66.02	55.04
21	16	–	60	71.3	39	170	1.83	24.67	66.02	55.04
21	24	–	60	71.3	39	170	1.83	24.67	66.02	55.04
21	32	–	60	71.3	39	170	1.83	24.67	66.02	55.04
21	40	–	40	71.3	39	170	1.83	24.67	66.02	55.04
21	48	–	40	71.3	39	170	1.83	24.67	66.02	55.04
21	56	–	40	71.3	39	170	1.83	24.67	66.02	55.04
21	64	–	40	71.3	39	170	1.83	24.67	66.02	55.04
21	72	–	40	71.3	39	170	1.83	24.67	66.02	55.04
21	75	–	–	71.3	39	170	1.83	24.67	66.02	55.04
21	80	1.1	40	71.3	39	170	1.83	24.67	66.02	55.04

After importing the data for the individual 21, only one concentration was obtained. Note that we have written a dash where no information was available

Fig. 3 Database information

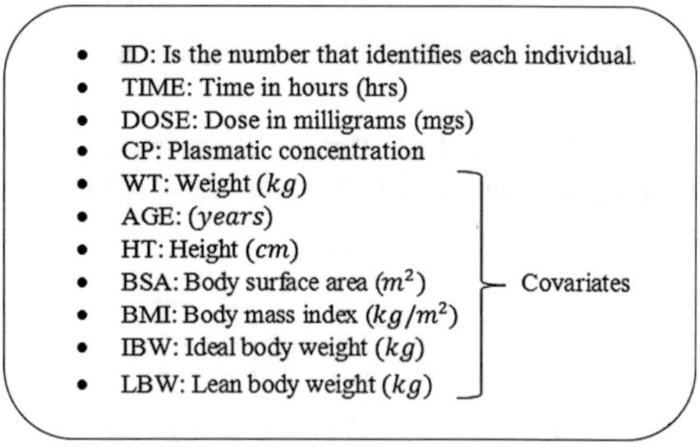

• ID: Is the number that identifies each individual.
• TIME: Time in hours (hrs)
• DOSE: Dose in milligrams (mgs)
• CP: Plasmatic concentration
• WT: Weight (kg)
• AGE: $(years)$
• HT: Height (cm)
• BSA: Body surface area (m^2)
• BMI: Body mass index (kg/m^2)
• IBW: Ideal body weight (kg)
• LBW: Lean body weight (kg)
Covariates

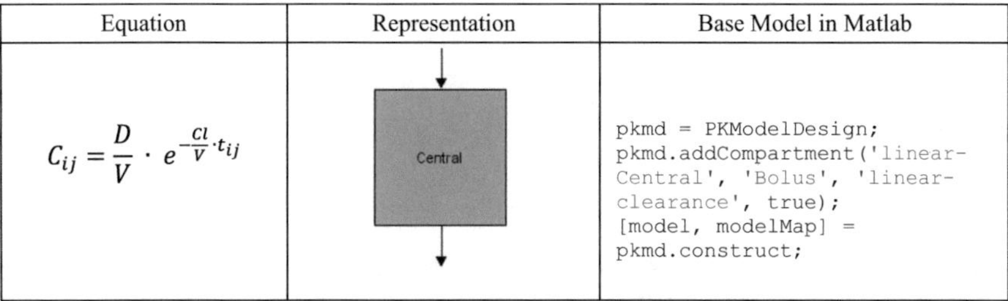

Equation
Representation
Base Model in Matlab
$C_{ij} = \frac{D}{V} \cdot e^{-\frac{Cl}{V} \cdot t_{ij}}$
Central
pkmd = PKModelDesign;
pkmd.addCompartment('linear-Central', 'Bolus', 'linear-clearance', true);
[model, modelMap] = pkmd.construct;

Fig. 4 One compartment model, and to the right side the mathematical expression representing the processes of absorption, distribution, and elimination of a drug from the body

4.2 Structural Model Development

The single compartment model was used to develop the PopPK model in MATLAB, as it is shown in Fig. 4.

The structural model for parameter *Cl* and *V* were defined as linear. The initial values of the fixed effects were 0.5 for *Cl*, and 0.5 for *V*, with a maximum number of iterations of 1×10^{-4}. Moreover, to derive the likelihood function in a straightforward manner, we used the assumption that the random effects $\gamma_{i,V}$, $\gamma_{i,Cl}$ and the individual errors ε_{ij} enter in the model in an additive form. Then a Taylor series approximation about $\gamma_{i,V}$ and $\gamma_{i,Cl}$ was employed. All this gives us the opportunity to adapt inferential methods for linear models as the restricted maximum likelihood estimation method (REML) to estimate the variance components in the context of unbalanced data. The likelihood function $L_{REML}(\boldsymbol{\theta})$ was minimized using the Newton-Raphson (N-R) algorithm for the optimization process.

4.3 Genetic Algorithm for Maximum Likelihood Estimation

A continuous GA can be implemented to maximize directly the likelihood function $L_{REML}(\theta)$ for the covariance parameters Ω without calculating derivatives. To achieve this satisfactory, we generate a population of *N* individuals in the context of our GA. Each individual *i*th is denoted as ω_i for $i \in [1, 2, \ldots, N]$, where *E*th element of ω_i is expressed as

$$\omega_i = [\omega_i(1), \omega_i(2), \ldots, \omega_i(n)]. \tag{10}$$

Assuming that the inter-individual variance terms have a normal distribution between $[\omega_{\min}(E), \omega_{\max}(E)]$, the random initial population is generated as follows:

for i **= 1:**N

for E **= 1:**n

$\omega_{(iE)} \leftarrow$ random value between $[\omega_{min}(0.1), \omega_{max}(1.1)]$
E **+ 1**
i**+1**

From here, $L_{REML}(\theta)$ is used as the fitness function, and roulette wheel selection is used for selecting potentially useful solutions for recombination. Only a single crossover point in the third position was used with a probability of 0.8 and a Gaussian mutation operator.

5 Results

We used a continuous GA for 50 generations with a population size equal to the number of individuals in the PopPK data base. For the REML method, we used 1000 iterations. Finally, the results of both N-R optimization method and the continuous GA are shown in Table 4.

Considering the square root of the estimated error variance (rmse) and the accomplishment of the normal assumption for the random errors ε_{ij}, see Fig. 5, the proposed GA can be a good method for obtaining PopPk parameter estimates than the N-R optimization.

Table 4 Comparative results between the stepwise method and the genetic algorithm

Description	V	Cl	$\gamma_{i,V}$	$\gamma_{i,Cl}$	rmse
N-R	3.1993	1.3923	0.18711	0.352	0.2412
Continuos GA	3.4578	1.254	0.24214	0.40004	0.0931

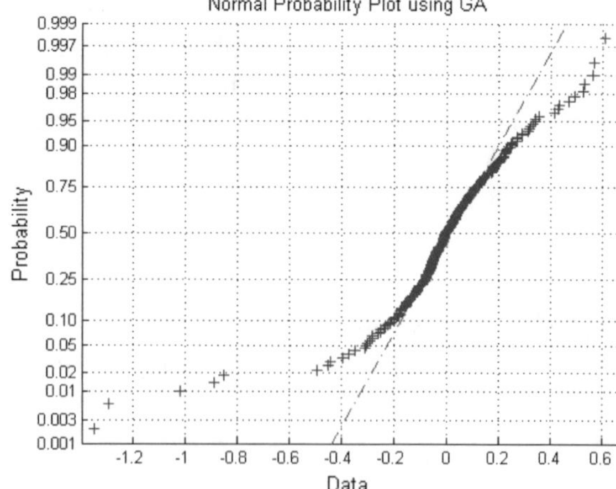

Fig. 5 The assumptions of a normal distribution of the random errors are still accomplished by the implemented GA

6 Conclusions

The main purpose of this work was to show that using genetic algorithms is possible to develop PopPK models as an alternative to the statistical methods; particularly, for obtaining PopPK parameter estimations. The comparison of the obtained results in this work exposes the possibility of incorporating methods of intelligent computation for the development of PopPK models.

Acknowledgments We thank to Instituto Politécnico Nacional (IPN), to the Commission of Operation and Promotion of Academic Activities of IPN (COFAA), and the Mexican National Council of Science and Technology (CONACYT) for supporting our research activities.

References

1. Paul J. Williams, Ene I. Ette. The Role of Population Pharmacokinetics in Drug Development in Light of the Food and Drug Administration's 'Guidance for Industry: Population Pharmacokinetics'. Springer, (2000)
2. Lang Wu. Mixed Effects Models for Complex Data. CRC Press, (2010)
3. Bruce L. Bowerman, RichardT. O'Conell, Emily S. Murpheree, Experimental Design: Unified Concepts, Practical Applications, and Computer Implementation. Business Expert Press. (2015)
4. Joel S. Owen and Jill Fieldler-Kelly. Introduction to Population Pharmacokinetic/ Pharmacodynamic Analysis with Nonlinear Mixed Effects Models. Wiley, (2014)
5. Ene I.Ette, Paul J. Williams. Pharmacometrics: the science of quantitative pharmacology (2007)
6. Marc Lavielle, Kevin Bleakley. Mixed Effects Models for the Population Approach: Models, Task, Methods and Tools. CRC Press, (2015)
7. Dan Simon. Evolutionary Optimization Algorithms: Biologically-Inspired and Population-Based Approaches to Computer Intelligence. Wiley. (2013)
8. Marie Davidian and David M. Giltinan. Nonlinear Models for Repeated Measurement Data, Chapman & Hall, (1995)
9. Johan Gabrielsson, Dan Weiner. Pharmacokinetic & Pharmacodynamic Data Analysis: Concepts and Applications. Apotekarsocieteten, (2006)
10. Seongho Kim, Lang Li. A novel global search algorithm for nonlinear mixed-effects models Springer Science+BusinessMedia, (2011)
11. Holland, J.H. Adaptation in Natural and Artificial Systems. Ann Arbor: University of Michigan Press, (1975)
12. S.N. Sivanandam,S.N. Deepa. Introduction to Genetic Algorithms. Springer, (2008)
13. Chang Wook Ahn. Advances in Evolutionary Algorithms. Theory, Design and Practice. Springer, (2006)
14. A.E. Eiben, J.E. Smith, Introduction to Evolutionary Computing. Springer, (2003)
15. Robert Lowen and Alain Verschoren. Foundations of Generic Optimization. Springer, (2008)

Optimization of Reactive Control for Mobile Robots Based on the CRA Using Type-2 Fuzzy Logic

David de la O, Oscar Castillo and Jose Soria

Abstract This paper describes the optimization of a reactive controller system for a mobile autonomous robot using the CRA algorithm to adjust the parameters of each fuzzy controller. A comparison with the results obtained with genetic algorithms is also performed.

Keywords Fuzzy logic · CRA · Control

1 Introduction

In this paper, we describe the optimization of Reaction Fuzzy Controller for mobile robots, using the chemical reaction algorithm that was originally proposed by Leslie Astudillo, the algorithm mimics the behavior of chemical elements around us and how they react to generate a new compound or product. We applied it to optimize the parameters of the membership functions of the type-2 fuzzy controller, both in the input and output.

We will limit what in this paper will be described as ability of reaction, this is applied in the navigation concept, so what this means is a forward moving robot that at some point of its journey it encounters and unexpected obstacle it will react to this simulation avoiding and continuing forward. One of the applications of fuzzy logic is the design of fuzzy control systems. The success of this control lies in the correct selection of the parameters and rules of the fuzzy controller; it is here where the Chemical Reaction Algorithm (CRA) metaheuristics is applied, and this approach is based on a static population of metaheuristic which applies an abstraction of chemical reactions. Recent research on optimized membership functions has shown good results [1–25].

D. de la O · O. Castillo (✉) · J. Soria
Tijuana Institute of Technology, Tijuana, Mexico
e-mail: ocastillo@tectijuana.mx

D. de la O
e-mail: ddsh@live.com

P. Melin et al. (eds.), *Nature-Inspired Design of Hybrid Intelligent Systems*,
Studies in Computational Intelligence 667, DOI 10.1007/978-3-319-47054-2_33

2 Mobile Robots

The particular mobile robot considered in this work is based on the description of the Simulation toolbox for mobile robots [28], which assumes a wheeled mobile robot consisting of one conventional, steered, unactuated and not-sensed wheel, and two conventional, actuated, and sensed wheels (conventional wheel chair model). This type of chassis provides two DOF (degrees of freedom) locomotion by two actuated conventional non-steered wheels and one unactuated steered wheel. The Robot has two degrees of freedom (DOFs): y-translation and either x-translation or z-rotation [5, 16–20, 26]. Figure 1 shows the robot's configuration, it has 2 independent motors located on each side of the robot and one castor wheel for support located at the front of the robot.

The kinematic equations of the mobile robot are as follows:

Equation 1 shows the sensed forward velocity solution [31]

$$\begin{pmatrix} V_{B_x} \\ V_{B_y} \\ \omega_{B_z} \end{pmatrix} = \frac{R}{2l_a} \begin{bmatrix} -l_b & l_b \\ -l_a & -l_a \\ -1 & -1 \end{bmatrix} \begin{pmatrix} \omega_{W_1} \\ \omega_{W_2} \end{pmatrix} \tag{1}$$

Equation 2 shows the Actuated Inverse Velocity Solution [6]

$$\begin{pmatrix} \omega_{W_1} \\ \omega_{W_2} \end{pmatrix} = \frac{1}{R(l_b^2+1)} \begin{bmatrix} -l_a l_b & -l_b^2-1 & -l_a \\ l_a l_b & -l_b^2-1 & l_a \end{bmatrix} \begin{pmatrix} V_{B_x} \\ V_{B_y} \\ \omega_{B_z} \end{pmatrix} \tag{2}$$

Under the Metric system are defined as:

V_{B_x}, V_{B_y} Translational velocities $\left[\frac{\text{m}}{\text{s}}\right]$,

ω_{B_z} Robot z-rotational velocity $\left[\frac{\text{rad}}{\text{s}}\right]$,

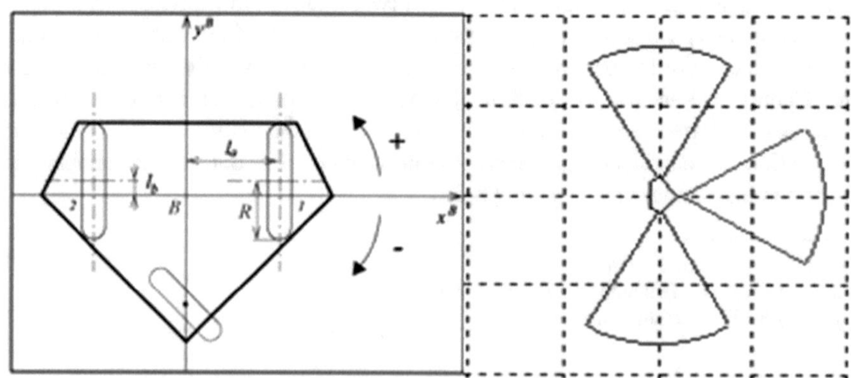

Fig. 1 Kinematic coordinate system assignments [31]

$\omega_{W_1}, \omega_{W_2}$ Wheel rotational velocities $\left[\frac{\text{rad}}{\text{s}}\right]$,
R Actuated wheel radius [m],
l_a, l_b Distances of wheels from robot's axes [m].

3 The Chemical Optimization Paradigm

The proposed chemical reaction algorithm is a metaheuristic strategy that performs a stochastic search for optimal solutions within a defined search space. In this optimization strategy, an element (or compound) represented every solution, and the fitness or performance of the element is evaluated in accordance with the objective function.

This algorithm has the advantage of not having external parameters (kinetic/potential energies, mass conservation, thermodynamic characteristics, etc.) as occurs in other optimization algorithms, this is a very straight forward methodology that takes the characteristics of the chemical reactions (synthesis, decomposition, substitution and double-substitution) to find the optimal solution [5, 16–20, 26].

This approach is based on a static population metaheuristic, which applies an abstraction of chemical reactions as intensifiers (substitution, double substitution reactions) and diversifying (synthesis, decomposition reactions) mechanisms. The elitist reinsertion strategy allows the permanence of the best elements and thus the average fitness of the entire element pool increases with each iteration. Figure 2 shows the flowchart of the algorithm.

The main components of this chemical reaction algorithm are described below.

The synthesis and decomposition reactions are used to diversify the resulting solutions; these procedures prove to be highly effective and rapidly lead the results to a desired value.

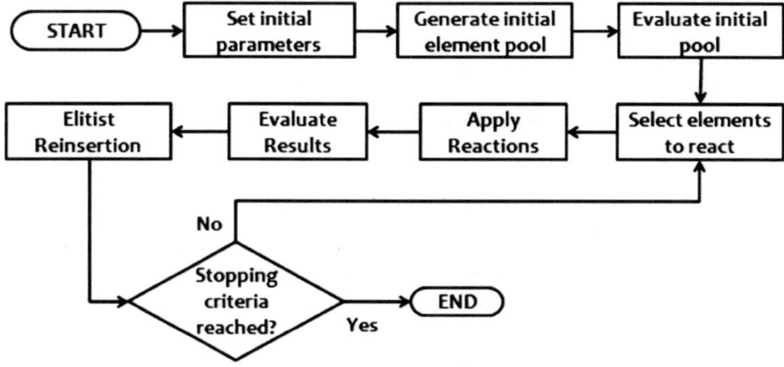

Fig. 2 Flowchart of the CRA

The single and double substitution reactions allow the algorithm to search for optimal values around a good previously found solution and they're described as follows.

The algorithm works mainly with chemical reactions with a change in at least one substance (element or possible solution), that changes its composition and property sets. Such reactions are classified into 4 types, which are described below.

3.1 Type 1: Combination Reactions

$$B + X \rightarrow BX \tag{3}$$

A combination reaction is a reaction of two reactants to produce one product. The simplest combination reactions are the reactions of two elements to form a compound. After all, if two elements are treated with each other, they can either react or not.

3.2 Type 2: Decomposition Reactions

$$BX \rightarrow B + X \tag{4}$$

The second type of simple reaction is decomposition. This reaction is also easy to recognize. Typically, only one reactant is given. A type of energy, such as heat or electricity, may also be indicated. The reactant usually decomposes to its elements, to an element and a simpler compound, or to two simpler compounds.

Binary compounds may yield two elements or an element and a simpler compound. Ternary (three-element) compounds may yield an element and a compound or two simpler compounds.

3.3 Type 3: Substitution Reactions

$$X + AB \rightarrow AX + B \tag{5}$$

Elements have varying abilities to combine. Among the most reactive metals are the alkali metals and the alkaline earth metals. On the opposite end of the scale of relativities, among the least active metals or the most stable metals are silver and gold, prized for their lack of reactivity. Reactive means the opposite of stable, but means the same as active.

When a free element reacts with a compound of different elements, the free element will replace one of the elements in the compound if the free element is more reactive than the element it replaces. In general, a free metal will replace the metal in the compound, or a free nonmetal will replace the nonmetal in the compound. A new compound and a new free element are produced [2, 21, 27, 28].

3.4 Type 4: Double-Substitution Reactions

Double-substitution or double-replacement reactions, also called double-decomposition reactions or metathesis reactions, involve two ionic compounds, most often in aqueous solution [4].

In this type of reaction, the cations simply swap anions. The reaction proceeds if a solid or a covalent compound is formed from ions in solution. All gases at room temperature are covalent. Some reactions of ionic solids plus ions in solution also occur. Otherwise, no reaction takes place.

Just as with replacement reactions, double-replacement reactions may or may not proceed. They need a driving force.

In replacement reactions the driving force is reactivity; here it is insolubility or co-valence.

At first sight, chemical theory and definitions may seem complex and none or few are related to optimization theory, but only the general schema will be considered here in order to focus on the control application.

3.5 Element Encoding

The Element consists of 25 real-valued vectors, representing the parameters for the triangular membership function; we use five membership functions for each variable. This encoding is shown Fig. 3 [29].

3.6 Reactive Controller Objective Function

The criteria used to measure the Reactive controller performance are the following

- Covered Distance
- Time used to cover the distance
- Battery life.

A Fitness FIS will provide the desired fitness value, adding very basic rules that reward the controller that provided the longer trajectories and smaller times and

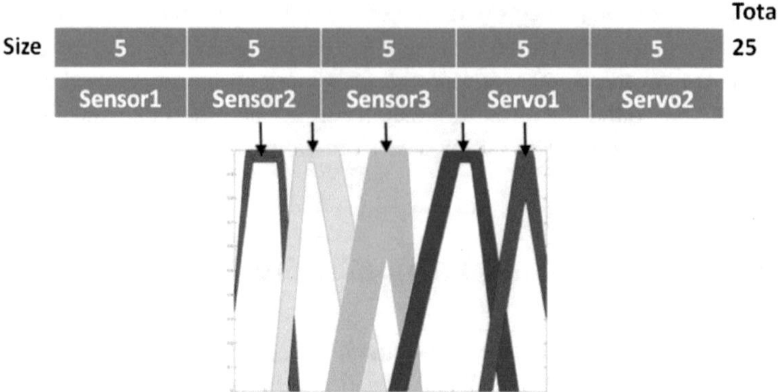

Fig. 3 Element encoding

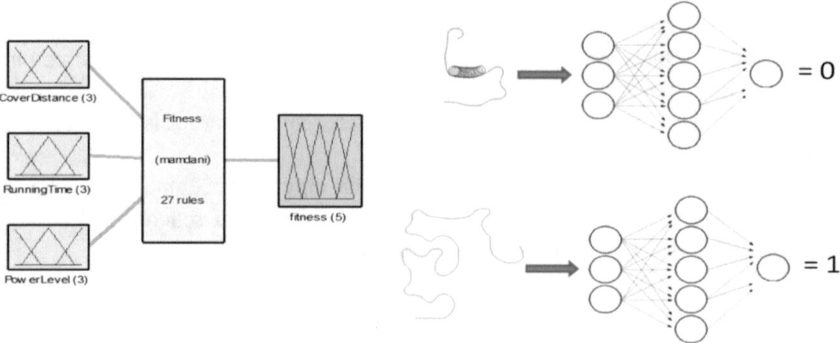

Fig. 4 Fitness function for the reactive controller

higher battery life. This seems like a good strategy that will guide the control population into evolving and provide the optimal control, but this strategy on its own is not capable of doing just that: it needs to have a supervisor on the robots trajectory to make sure it is a forward moving trajectory and that they does not contain any looping parts. For this, a Neural Network (NN), is used to detect cycle trajectories that do not have the desired forward moving behavior by giving low activation value and higher activation values for the ones that are cycle free. The NN has two inputs and one output, and 2 hidden layers, see Fig. 4.

The evaluation method for the reactive controller has integrated both parts of the FIS and the NN where the fitness value for each individual is calculated with Eq. 6. Based on the response of the NN the peak activation value is set to 0.35, this meaning that any activation lower than 0.35 will penalize the fitness given by the FIS [5, 16–20].

Equation 6 expresses how to calculate the fitness value of each individual

$$f(t) = \begin{cases} fv * nnv, & nnv < 0.35 \\ fv, & nnv \geq 0.35 \end{cases} \quad (6)$$

where

fi Fitness value of the *i*th individual,
fv Crisp value out of the fitness FIS,
nnv Looping trajectory activation value.

At the point of evaluation of the particle, we measure the effectiveness of the tracking controller FIS (Fuzzy Inference System) in the toolbox, which will be in a closed circuit with a given reference by a straight line [29–32] environment.

4 Simulation Results

The tools that were used to conduct the experiments are Matlab and the "Simrobot" simulation tool.

4.1 Reactive Controller

Table 1 shows the configuration of the CRA and the results, in which we have the fitness value obtained in each experiment. It also shows the mean, variance, as well as the best and worst value obtained.

Table 1 Summary of tracking results

Element	Iteration
20	1000
	Fitness
1	0.3299
2	0.3226
3	0.3143
4	0.3126
5	0.33
6	0.3607
7	0.3304
8	0.3299

(continued)

Table 1 (continued)

Element	Iteration
9	0.3179
10	0.33
Average	0.32783
Best	0.3607
Poor	0.3126
Std dev	0.01352

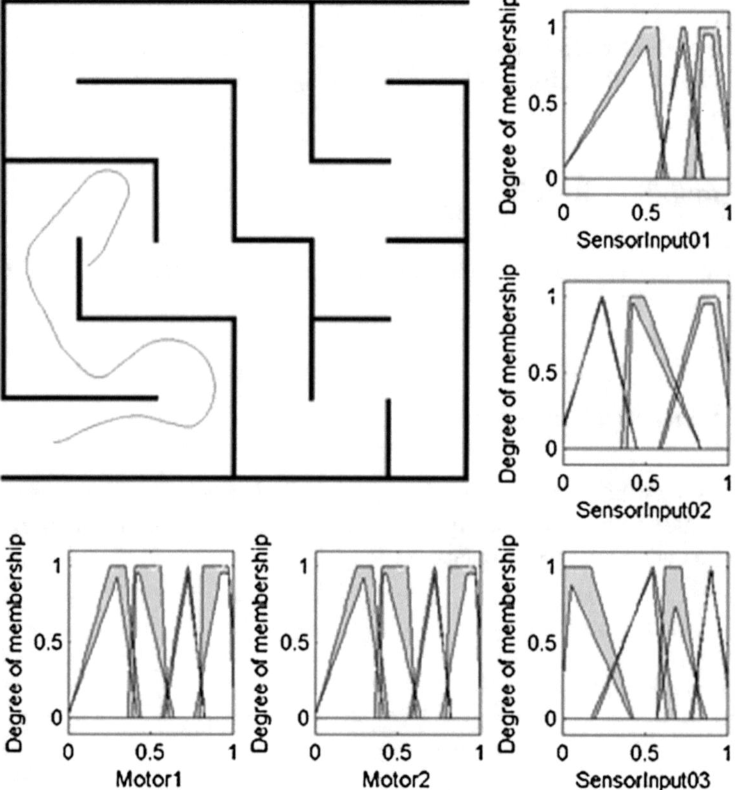

Fig. 5 Reactive controller results

Figure 5 shows the best path simulation during cycle CRA obtained for the Reactive controller.

5 Conclusions

In this paper, the improved CRA algorithm was used to tune the parameters of the fuzzy controller for the Reactive Controller.

Based on our experiments we observed that the algorithm converges slowly and therefore the expected results are not obtained; we propose the following improvements to the algorithm:

1. Optimizing both parameters and the rules.
2. Running the CRA with more iterations, the past work (CRA) we used 8000 iterations used Typ-1 fuzzy logic.

Acknowledgment We would like to express our gratitude to CONACYT, and Tijuana Institute of Technology for the facilities and resources granted for the development of this research.

References

1. Astudillo L., Castillo O., L. Aguilar and Martínez R., Hybrid Control for an Autonomous Wheeled Mobile Robot Under Perturbed Torques, vol. 4529, Melin P., Castillo O., L. Aguilar, J. Kacprzyk and W. Pedrycz, Edits., Springer Berlin Heidelberg, 2007, pp. 594-603.
2. Astudillo L., Melin P. and Castillo O., « Nature optimization applied to design a type-2 fuzzy controller for an autonomous mobile robot, » de Nature and Biologically Inspired Computing (NaBIC), 2012 Fourth World Congress on, 2012.
3. Bingül Z. and O. Karahan, « A Fuzzy Logic Controller Tuned with PSO for 2 DOF Robot Trajectory Control, » Expert Syst. Appl., vol. 38, n° 1, pp. 1017-1031, January 2011.
4. Cervantes L. and Castillo O., Design of a Fuzzy System for the Longitudinal Control of an F-14 Airplane, vol. 318, Castillo O., J. Kacprzyk and W. Pedrycz, Edits., Springer Berlin Heidelberg, 2011, pp. 213-224.
5. de la O D., Castillo O. and Meléndez A., « Optimization of Fuzzy Control Systems for Mobile Robots Based on PSO, » of Recent Advances on Hybrid Approaches for Designing Intelligent Systems, vol. 547, Castillo O., Melin P. and J. Kacprzyk, Edits., Springer Berlin Heidelberg, 2014, pp. 191-208.
6. De Santis E., A. Rizzi, A. Sadeghiany and F. Mascioli, « Genetic optimization of a fuzzy control system for energy flow management in microgrids, » de IFSA World Congress and NAFIPS Annual Meeting (IFSA/NAFIPS), 2013 Joint, 2013.
7. Esmin A., A. R. Aoki and G. Lambert-Torres, « Particle swarm optimization for fuzzy membership functions optimization, » de Systems, Man and Cybernetics, 2002 IEEE International Conference on, 2002.
8. Fang G., N. M. Kwok and Q. Ha, « Automatic Fuzzy Membership Function Tuning Using the Particle Swarm Optimization, » de Computational Intelligence and Industrial Application, 2008. PACIIA '08. Pacific-Asia Workshop on, 2008.
9. Fang G., N. M. Kwok and Q. Ha, « Automatic Fuzzy Membership Function Tuning Using the Particle Swarm Optimization, » de Computational Intelligence and Industrial Application, 2008. PACIIA '08. Pacific-Asia Workshop on, 2008.
10. García M., O. Montiel, Castillo O. and R. Sepúlveda, Optimal Path Planning for Autonomous Mobile Robot Navigation Using Ant Colony Optimization and a Fuzzy Cost Function Evaluation, vol. 41, Melin P., Castillo O., E. Ramírez, J. Kacprzyk and W. Pedrycz, Edits., Springer Berlin Heidelberg, 2007, pp. 790-798.

11. Garcia M., O. Montiel, Castillo O., R. Sepúlveda and Melin P., « Path planning for autonomous mobile robot navigation with ant colony optimization and fuzzy cost function evaluation, » Applied Soft Computing, vol. 9, nº 3, pp. 1102-1110, 2009.
12. Martínez R., Castillo O. and L. T. Aguilar, « Optimization of interval type-2 fuzzy logic controllers for a perturbed autonomous wheeled mobile robot using genetic algorithms, » Information Sciences, vol. 179, nº 13, pp. 2158-2174, 2009.
13. Martínez R.-Marroquín, Castillo O. and J. Soria, Particle Swarm Optimization Applied to the Design of Type-1 and Type-2 Fuzzy Controllers for an Autonomous Mobile Robot, vol. 256, Melin P., J. Kacprzyk and W. Pedrycz, Edits., Springer Berlin Heidelberg, 2009, pp. 247-262.
14. Martinez-Soto R., Castillo O., L. Aguilar and Melin P., Fuzzy Logic Controllers Optimization Using Genetic Algorithms and Particle Swarm Optimization, vol. 6438, G. Sidorov, A. Hernández Aguirre and C. Reyes García, Edits., Springer Berlin Heidelberg, 2010, pp. 475-486.
15. Martinez-Soto R., Castillo O., L. T. Aguilar and I. S. Baruch, « Bio-inspired optimization of fuzzy logic controllers for autonomous mobile robots, » de Fuzzy Information Processing Society (NAFIPS), 2012 Annual Meeting of the North American, 2012.
16. Melendez A. and Castillo O., « Optimization of type-2 fuzzy reactive controllers for an autonomous mobile robot, » de Nature and Biologically Inspired Computing (NaBIC), 2012 Fourth World Congress on, 2012.
17. Meléndez A. and Castillo O., Evolutionary Optimization of the Fuzzy Integrator in a Navigation System for a Mobile Robot, vol. 451, Castillo O., Melin P. and J. Kacprzyk, Edits., Springer Berlin Heidelberg, 2013, pp. 21-31.
18. Meléndez A. and Castillo O., Hierarchical Genetic Optimization of the Fuzzy Integrator for Navigation of a Mobile Robot, vol. 294, Melin P. and Castillo O., Edits., Springer Berlin Heidelberg, 2013, pp. 77-96.
19. Meléndez A., Castillo O. and J. Soria, « Reactive control of a mobile robot in a distributed environment using fuzzy logic, » de Fuzzy Information Processing Society, 2008. NAFIPS 2008. Annual Meeting of the North American, 2008.
20. Meléndez A., Castillo O., A. A. Garza and J. Soria, « Reactive and tracking control of a mobile robot in a distributed environment using fuzzy logic, » de FUZZ-IEEE, 2010.
21. Melin P., Astudillo L., Castillo O., Valdez F. and M. Garcia, « Optimal design of type-2 and type-1 fuzzy tracking controllers for autonomous mobile robots under perturbed torques using a new chemical optimization paradigm, » Expert Systems with Applications, vol. 40, nº 8, pp. 3185-3195, 2013.
22. Rajeswari K. and P. Lakshmi, « PSO Optimized Fuzzy Logic Controller for Active Suspension System, » de Advances in Recent Technologies in Communication and Computing (ARTCom), 2010 International Conference on, 2010.
23. Ross O., J. Camacho, R. Sepúlveda and Castillo O., Fuzzy System to Control the Movement of a Wheeled Mobile Robot, vol. 318, Castillo O., J. Kacprzyk and W. Pedrycz, Edits., Springer Berlin Heidelberg, 2011, pp. 445-463.
24. S. Amin and A. Adriansyah, « Particle Swarm Fuzzy Controller for Behavior-based Mobile Robot, » de Control, Automation, Robotics and Vision, 2006. ICARCV '06. 9th International Conference on, 2006.
25. Singh R., M. Hanumandlu, S. Khatoon and I. Ibraheem, « An Adaptive Particle Swarm Optimization based fuzzy logic controller for line of sight stabilization tracking and pointing application, » de Information and Communication Technologies (WICT), 2011 World Congress on, 2011.
26. Measurement and Instrumentation, Faculty of Electrical Engineering and Computer Science, Brno University of Technology, Czech Republic Department of Control. Autonomous Mobile Robotics Toolbox For Matlab 5. Online, 2001.
27. Sanchez C.. Melin P. and Astudillo L., « Chemical Optimization Method for Modular Neural Networks Applied in Emotion Classification, » of Recent Advances on Hybrid Approaches for Designing Intelligent Systems, vol. 547, Castillo O., Melin P. and J. Kacprzyk, Edits., Springer Berlin Heidelberg, 2014, pp. 381-390.

28. John Y. and L. Reza, Fuzzy Logic: Intelligence, Control, and Information, Prentice Hall, 1999.
29. Chen J. and L. Xu, « Road-Junction Traffic Signal Timing Optimization by an adaptive Particle Swarm Algorithm, » de Control, Automation, Robotics and Vision, 2006. ICARCV '06. 9th International Conference on, 2006.
30. Dorrah H. T., A. M. El-Garhy and M. E. El-Shimy, « PSO-BELBIC scheme for two-coupled distillation column process, » Journal of Advanced Research, vol. 2, n° 1, pp. 73-83, 2011.
31. Sanchez C.. Melin P. and Astudillo L., « Chemical Optimization Method for Modular Neural Networks Applied in Emotion Classification, » of Recent Advances on Hybrid Approaches for Designing Intelligent Systems, vol. 547, Castillo O., Melin P. and J. Kacprzyk, Edits., Springer Berlin Heidelberg, 2014, pp. 381-390.
32. Singh R., M. Hanumandlu, S. Khatoon and I. Ibraheem, « An Adaptive Particle Swarm Optimization based fuzzy logic controller for line of sight stabilization tracking and pointing application, » de Information and Communication Technologies (WICT), 2011 World Congress on, 2011.

Part V
Fuzzy Logic Applications

A FPGA-Based Hardware Architecture Approach for Real-Time Fuzzy Edge Detection

Emanuel Ontiveros-Robles, José González Vázquez, Juan R. Castro and Oscar Castillo

Abstract Edge detection is used on most pattern recognition algorithms for image processing, however, its main drawbacks are the detection of unreal edges and its computational cost; fuzzy edge detection is used to reduce false edges but at even higher computational cost. This paper presents a Field Programmable Gate Array (FPGA)-based hardware architecture that performs a real-time edge detection using fuzzy logic algorithms achieving a decrease in the amount of unreal edges detected while compensating the computational cost by using parallel and pipelining hardware design strategies. For image processing testing, image resolution is set to 480 × 640 pixels at 24 fps (frames per second), thus real-time processing requires 7,372,800 fuzzy logic inference per second (FLIPS). The proposed fuzzy logic edge detector is based on the morphological gradient; this algorithm performs the edge detection based in the gradient operator, getting vectors of edge direction, were the magnitude of these vectors determines if the pixel is edge or not. The hardware architecture processes each frame pixel by pixel with grayscale partial image inputs, at 8 bits resolution, represented with a 3 × 3 pixels matrix; subsequently the architecture executes the stages of the fuzzy logic system: fuzzification, inference, and defuzzification, however, taking advantage of the FPGAs versatility, the dedicated hardware-based processing is executed in parallel within a pipeline structure to achieve edge detection in real time. The real-time fuzzy edge detector is compared with several classic edge detectors to evaluate the performance in terms of quality of the edges and the processing rate in FLIPS.

Keywords Fuzzy logic · FPGA · Edge detection · Real time

E. Ontiveros-Robles (✉) · J.G. Vázquez · J.R. Castro
UABC University, Tijuana, Mexico
e-mail: emanuel.ontiveros@uabc.edu.mx

O. Castillo
Tijuana Institute of Technology, Tijuana, Mexico

P. Melin et al. (eds.), *Nature-Inspired Design of Hybrid Intelligent Systems*,
Studies in Computational Intelligence 667, DOI 10.1007/978-3-319-47054-2_34

1 Introduction

Many papers prove the efficacy of fuzzy logic edge detection, some of them use Type-1 Fuzzy Inference Systems (FIS) or are optimized inspired on: gradient measure [1, 2], Ant Colony Optimization (ACO) [3], Wavelets transform [4], Canny edge detector [5]; others use Interval Type-2 Fuzzy Inference Systems (IT2-FIS) [1], or Generalized Type-2 Fuzzy Inference Systems [6]. Other works emphasized their performance under noisy environments [7], or are computed on Graphic Processing Units (GPU) [8]. Applications of Fuzzy Edge Detectors have been implemented on different applications for example: medical [9, 10], biometrics systems [11], and others.

Most all of cited papers [1–11] document the better performance of the fuzzy logic edge detectors against the conventional edge detectors. This paper is focused on implementing real-time Fuzzy Edge Detection by means of a hardware-based dedicated architecture embedded on a Field Programmable Gate Array (FPGA), addressing the mayor limitation of the high computational cost of the fuzzy edge detectors.

FPGA is a reconfigurable digital device that allows the design and synthesis of dedicated hardware architectures, which in conjunction with diverse hardware design strategies allows for the reduction of fuzzy edge detection computational cost as it is presented on this paper; some examples of FPGA-based implementations are as follows: power optimization applications [12], photovoltaic systems [13], robotic [14], and finally real-time digital image processing [15–17].

The FPGA-based dedicated hardware architecture presented addresses the real-time high computational cost of Fuzzy Edge Detection; for test purposes, a target implementation of real-time fuzzy edge detection for a 480 $\times$ 640 pixels of resolution, at 24 frames per second (fps) was selected; this requires 7.32 millions of FLIPS.

The proposed architecture is evaluated for its tolerance to perturbations, its fidelity to software-based implementations and its processing rate measured in FLIPS.

2 Hardware Architecture Design and Edge Detection

2.1 Field Programmable Gate Array

Field Programmable Gate Array (FPGA) is a digital device that is composed of millions of logic resources embedded in a matrix structure and interconnection resources that allows development of complex architectures through Hardware Descriptor Languages as VHDL. Figure 1 shows a FPGA Schematic.

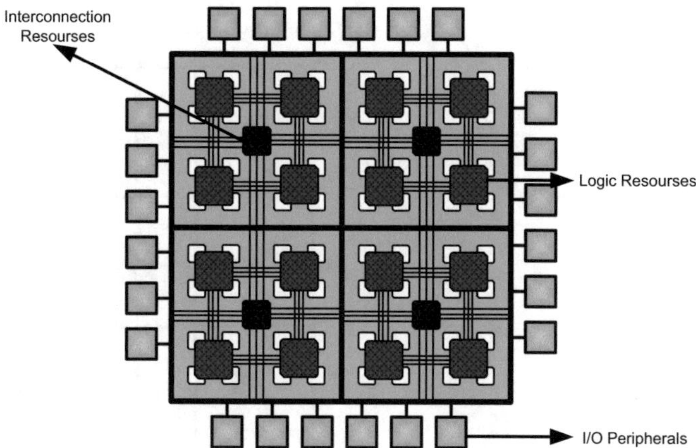

Fig. 1 FPGA schematic

Within FPGA versatility several design strategies can be used to optimize architecture performance; next are presented how the strategies used in the architecture design of this work:

- Parallelism: Characteristic of an architecture segmented in individual blocks with current computing operations, resulting on a system's capacity to execute two or more computations at the same time.
- Pipelining: Architecture's capability to realize different stages of a computation process at the same time in order to accelerate throughput.
- Distributed memory: FPGAs capability to create individual registers or memory blocks, allowing individual read/write access at the same time with multiple dedicated data buses.
- Fixed-point notation: Mathematical operations with fractional numbers represent a challenge for discrete digital systems due to truncation errors; for fractional number representations, a fixed-point representation is used, where the quantity of fractional bits highly determines the fidelity of the discrete operations. Reference [18] is an example of fixed-point notation.

2.2 *Morphological Gradient Fuzzy Edge Detection*

The morphological gradient is based on the evaluation of four gradient distances; through these distances the absolute gradient direction is obtain thus inferring on edge direction. Figure 2 shows the gradient directions.

Several methods of gradient distances evaluations can be used; in this work gradient direction is approximated with the absolute value of the difference of the pixels intensity pairs as is expressed in Eq. (1).

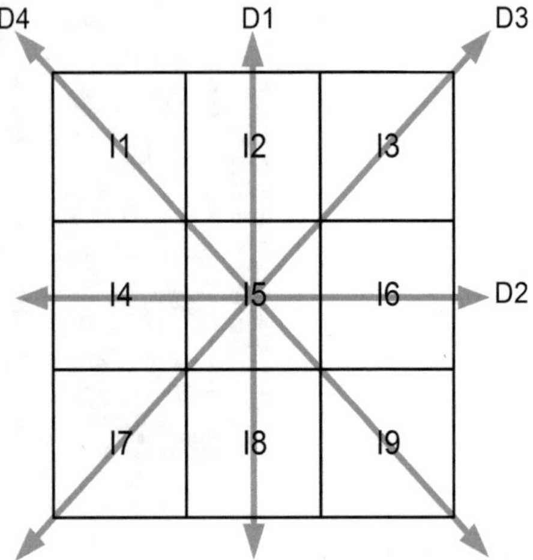

Fig. 2 Pixel's neighborhood

$$\begin{aligned} D1 &= |I2 - I8| \\ D2 &= |I4 - I6| \\ D3 &= |I3 - I7| \\ D4 &= |I1 - I9| \end{aligned} \tag{1}$$

Using these values as parameters it can be determined if the central pixel of the neighborhood is or not an edge, this is based on empirical knowledge about gradient evaluation; Fuzzy Inference processing of this four-scalar set can be synthesized to perform fuzzy edge detection. Knowledge representation can be synthesized with different Fuzzy Inference Systems (FIS), and the hardware system should be able to implement these FISs by adjusting parameters without modifications to its architecture.

3 Fuzzy Edge Detection Architecture Design

For the development of the Fuzzy Logic Edge Detection Architecture (FLED-Arch), first, the fuzzy inference system that performs the edge detection and the experimental platform is presented; second, the design of the architecture of the fuzzy system is realized stage by stage; and finally the aggregations of the designed modules, by taking advantage of the modularity and hierarchical design of the VHDL language.

3.1 FIS Properties

The proposed Fuzzy Inference System (FIS) is a Mamdani Inference System as shown in Fig. 3, this is selected because it allows the formulation based on expert knowledge and holds the interpretability of the decisions.

The Fuzzy Logic Edge Detector (FLED) is proposed to be adaptable, this means, the system parameters must be adaptable to different environments, i.e., illumination variations, different contrasts, etc.

Selected FIS properties are summarized in Table 1.

3.2 Fuzzifier

The fuzzifier evaluates all input MF's, at a computational cost proportional to the number of MF's used in the system.

In this work, the proposed distribution of MF's is shown in Fig. 4, where *D*1–*D*4 are gradient directions as calculated by (1), and horizontal axis is grayscale intensity, and vertical axis is degree of membership.

This distribution is based in two parameters; *C* represents the center and can be associated with the brightness of the image, and *A* represents the amplitude of the fuzzy set and can be associated with image contrast; these editable parameters allow the architecture to be adapted to different environments as needed. Figure 5 shows a proposed fuzzy set distribution.

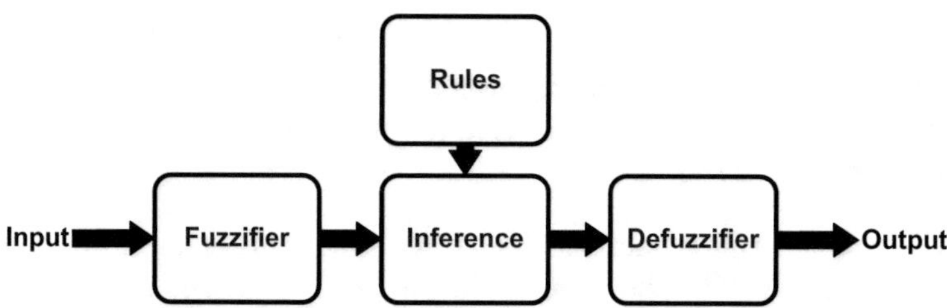

Fig. 3 Mamdani FIS

Table 1 FIS properties

Property	Selection
Inputs	*D*1, *D*2, *D*3, *D*4
Output	*Y* (*edge degree*)
MF	Trapezoidal MFs
Defuzzifier	First of Maxima defuzzifier (FoM)
Operators	T-Norm at minimum and S-Norm at maximum

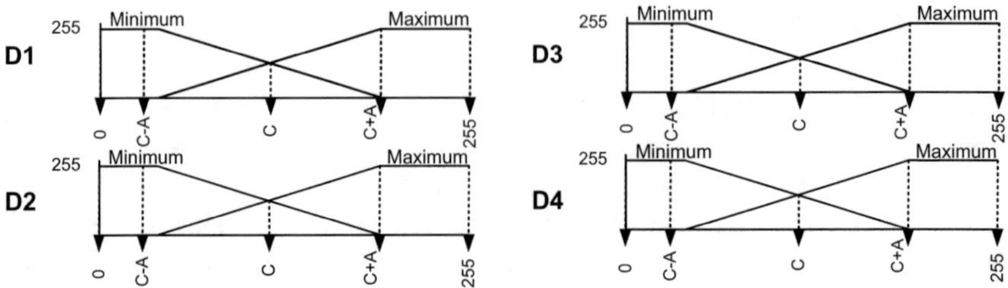

Fig. 4 Parametrized fuzzy sets distribution

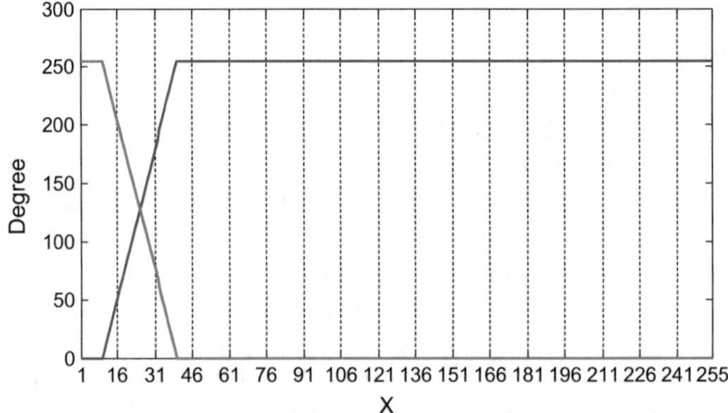

Fig. 5 Proposed fuzzy set distribution $C = 25$, $A = 30$

3.3 Rules and Inference

The inference and rules uses the min operator, such as the T-norm. Equation (2) shows the rules to perform edge detection, and these rules were proposed based as empirical expert knowledge.

$$\begin{array}{l}
\text{if}(D1 \text{ is min}) \text{ and } (D2 \text{ is max}) \text{ and } (D3 \text{ is max}) \text{ and } (D4 \text{ is max}) \text{ then } Y \text{ is Edge} \\
\text{if}(D1 \text{ is max}) \text{ and } (D2 \text{ is min}) \text{ and } (D3 \text{ is max}) \text{ and } (D4 \text{ is max}) \text{ then } Y \text{ is Edge} \\
\text{if}(D1 \text{ is max}) \text{ and } (D2 \text{ is max}) \text{ and } (D3 \text{ is min}) \text{ and } (D4 \text{ is min}) \text{ then } Y \text{ is Edge} \\
\text{if}(D1 \text{ is max}) \text{ and } (D2 \text{ is max}) \text{ and } (D3 \text{ is min}) \text{ and } (D4 \text{ is max}) \text{ then } Y \text{ is Edge} \\
\text{if}(D1 \text{ is max}) \text{ and } (D2 \text{ is max}) \text{ and } (D3 \text{ is max}) \text{ and } (D4 \text{ is min}) \text{ then } Y \text{ is Edge} \\
\text{if}(D1 \text{ is max}) \text{ and } (D2 \text{ is max}) \text{ and } (D3 \text{ is max}) \text{ and } (D4 \text{ is max}) \text{ then } Y \text{ is Edge}
\end{array} \tag{2}$$

All the rules are associated with the same consequent membership function, allowing a reduction of the complexity of the defuzzifier as follows.

3.4 Defuzzifier

The First of Maxima Defuzzifier (FoM) is used; the defuzzifier value is the first value of the core of the aggregated knowledge of the system, next is a mathematical reduction based on this defuzzifier approached on hardware implementation.

First, Eq. (3), aggregation knowledge, is the S-Norm of the individual knowledge associated to each rule.

$$B = \oplus Br_{r=1}^{6} \tag{3}$$

Each of the rule's individual knowledge is obtained through T-Norm implication between firing force Fr of the rule and output membership function Gr as express by Eq. (4).

$$B = \oplus(Gr \otimes Fr)_{r=1}^{6} \tag{4}$$

On proposed FIS in this work all of the rules are associated to the same output membership function, and this allows simplification of the aggregated knowledge G_{edge} as shown in Eq. (5).

$$B = G_{\text{edge}} \otimes \oplus Fr_{r=1}^{6} \tag{5}$$

Based on Eq. 5 and using the proposed T-Norm and S-Norm, aggregated knowledge is equivalent to the minimum operator applied to the consequent membership function and the maximum firing force of the rules, this is expressed in Eq. (6).

$$B = \min(G_{\text{edge}}, \text{Maximum } Fr) \tag{6}$$

Figure 6 shows the distribution of the consequent membership functions.

Based on this distribution of MF's and on Eq. (6), Fig. 7 shows the graphical interpretation of the FoM defuzzifier.

Where the shading space represents the aggregated knowledge that can be obtained through Eq. (6).

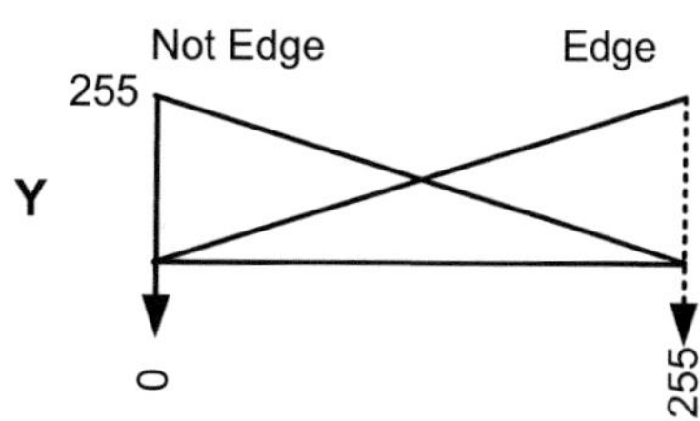

Fig. 6 Consequent fuzzy set

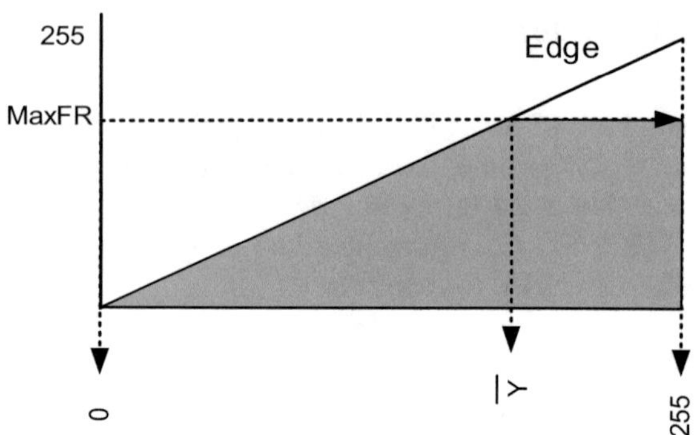

Fig. 7 Generalized system knowledge

Because of this particular consequent MF setup, the aggregated knowledge, and FoM defuzzifier two similar triangles can be observed, thus using basic geometry principles reduction is possible as showed in Eq. 7.

$$\frac{\bar{\tilde{y}}}{\text{Max } FR} = \frac{255}{255} \quad \therefore \quad \tilde{y} = \text{Max } FR \tag{7}$$

Based on this mathematical development, the output is approximated with only the max operator of the firing force of the rules.

3.5 Hardware Development Platform

For test proposes, the development platform selected is described in Table 2.

3.6 Architecture Design: TrapMF-Module

Defuzzifier block of Fig. 3 is composed of several input-MFs. Trapezoidal MFs were selected in this work as showed in Fig. 4.

Table 2 Development platform

Property
Atlys board
FPGA SPARTAN 6
Clock at 100 MHz
Conventional image sensor
8 bits RGB resolution
480 × 640 pixels
24 FPS

The trapezoidal-membership function is a nonlinear function showed in Fig. 8, and is described by intervals as showed in Eq. (8)

$$mf(x) = \begin{cases} \frac{x-a}{b-a}, & \text{for } a < x < b \\ 255 & \text{for } b < x < c \\ 1 - \frac{x-c}{d-c} & \text{para } c < x < d \end{cases} \quad (8)$$

The TrapMF-Module (Fig. 9) calculate the four intervals of the function at the same time, than selects the correct output depending input specific value; this is done concurrently using parallel dedicated hardware.

3.7 Architecture Design: Fuzzifier-Module

The Fuzzifier-Module performs evaluation of all the eight input-MFs; this module is implemented by eight TrapMF-Module's computing at parallel, allowing fuzzification evaluation at the same time. Figure 10 shows the Fuzzifier-Module schematic.

The fuzzification computation is executed in two stages, and it is controlled through the implementation of a Finite State Machine (FSM) as showed in Fig. 11; first stage evaluates all trapezoidal membership functions, and second stage stores the results on a local dedicated memory register.

3.8 Architecture Design: Inference Rules-Module

This module performs the evaluation of all firing force of the rules, in this case, six rules associated by the min operator. Figure 12 shows the module schematic.

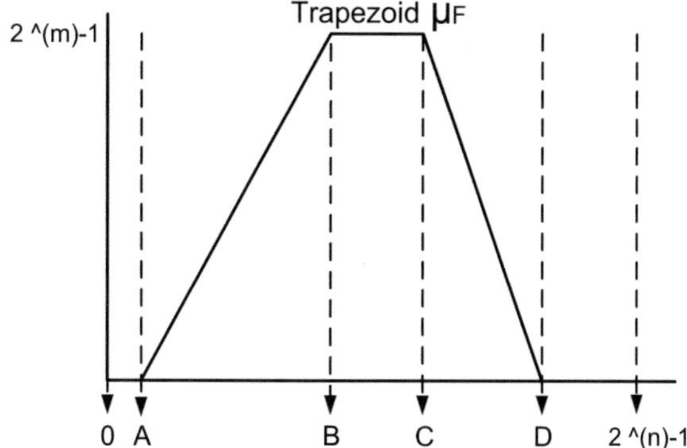

Fig. 8 TrapMF

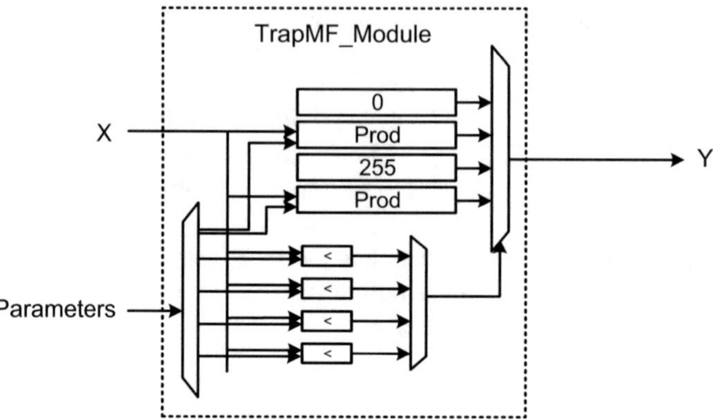

Fig. 9 TrapMF-Module

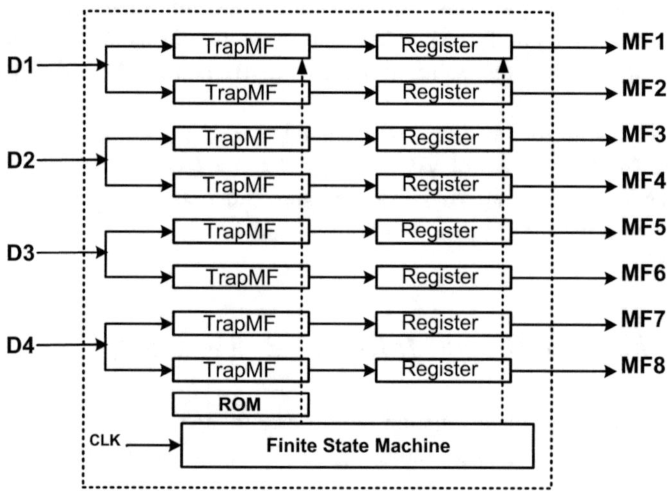

Fig. 10 Fuzzifier-Module

Fig. 11 Fuzzifier FSM

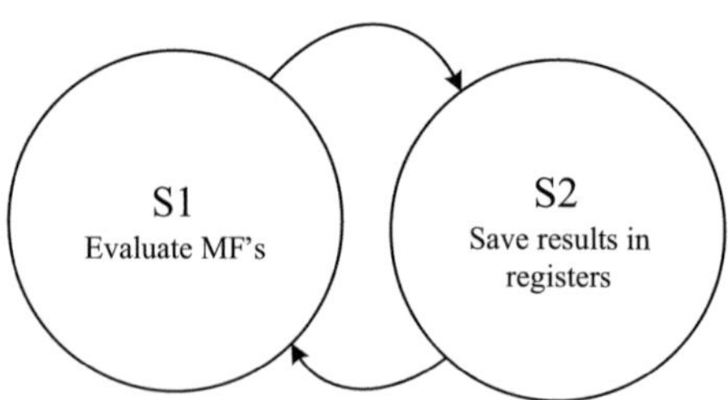

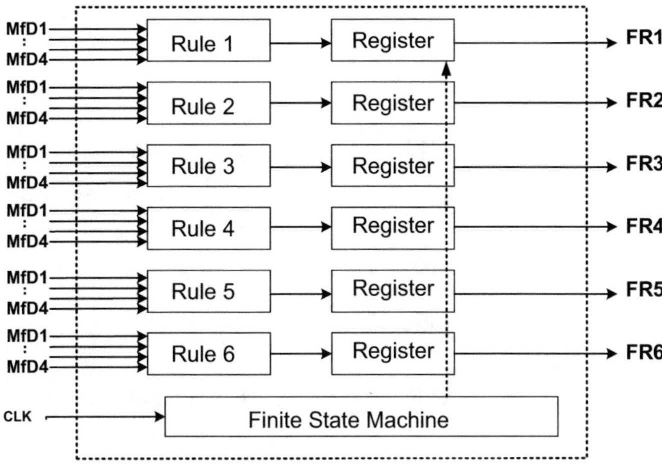

Fig. 12 Inference rules-module

The evaluation of the rules is done at the same time, and results are stored in local dedicated registers; a two-stage FSM controls data flow and execution, as showed in Fig. 13.

3.9 Architecture Design: FoM Defuzzifier

Based on Eq. (7), the FoM Defuzzifier-Module performs the defuzzification in one clock cycle evaluating the maximum value of the firing force of the rules. The schematic is showed in Fig. 14.

This module also contains a two-stage FSM, Fig. 15 resume the states.

Fig. 13 Inference rules FSM

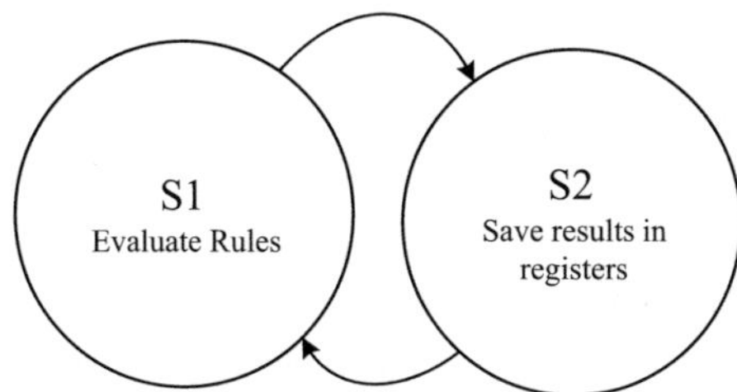

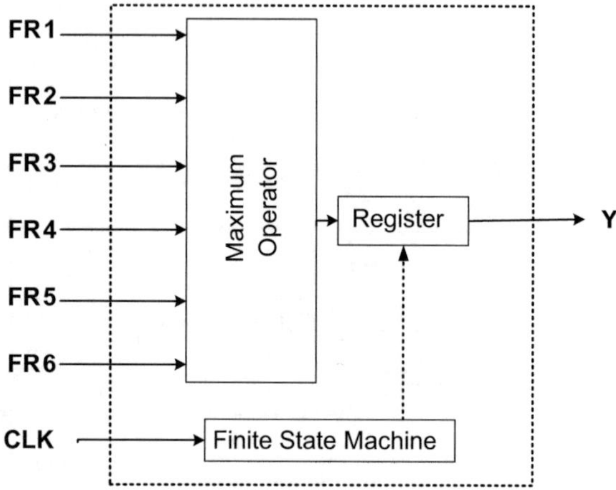

Fig. 14 Defuzzifier-Module

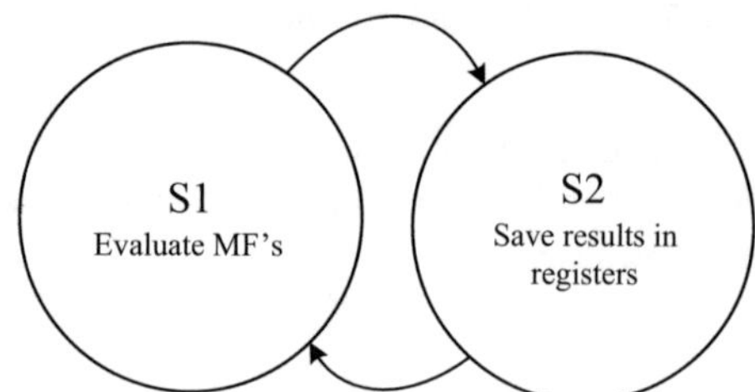

Fig. 15 Defuzzifier FSM

3.10 Fuzzy Edge Detector Architecture

Using the previous descripted modules, the FLED-Arch hierarchical design is synthesized; the low-level modules are controlled by a FSM that sequentially enables and disables the modules, in order to manage the data. Figure 16 shows the top-level architecture.

4 Test Results

In this section four experimental tests are presented; two tests demonstrate FPGA-based real-time execution of FLED-Arch and its fidelity to theoretical FIS under software-based processing; two other tests demonstrate FLED algorithm quality and robustness under image-noise conditions

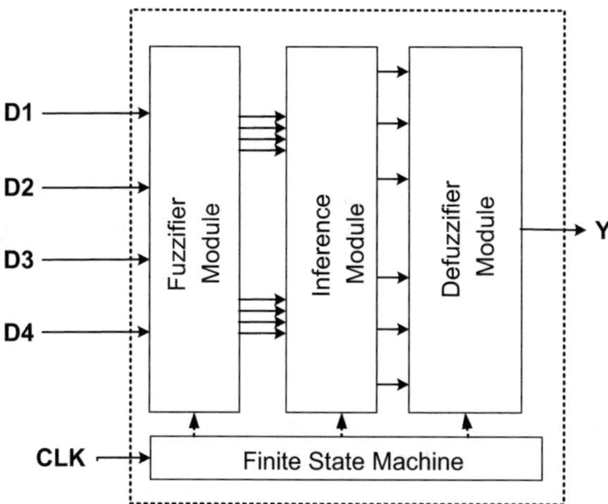

Fig. 16 FLED-Arch

4.1 Test 1: FPGA-Based Real-Time Execution

In this test, the FPGA-based dedicated hardware architecture is tested for its execution time in order to document possible real-time performance. Figure 16 FLED-Arch system's VHDL description is used in ISE Simulator 14.1 (Xilinx Inc.) with randomly selected input data that consisted of sets of four gradient values at 8-bit resolution each of them on their independent dedicated data bus as shown on Fig. 16.

Figure 17 shows input–output data sequence; top four data sequences are input gradient values, fifth is system clock signal at 100 MHz; fifth is output data where a value of is interpreted as "edge detected," and a zero value is "not an edge." Four data sets are input, respective values are displayed on respective data bus corresponding to the small images on the lower bottom of Fig. 17

Latency, measured as the time between the first input and the first available output of the system is of 40 ns or 4 clock cycles. Throughput time, representing the time between subsequently outputs, is of 20 ns or 2 clock cycles.

The execution time measurements demonstrate that at 100 MHz clock signal the hardware architecture is able to perform up to 50 million of fuzzy logic inference per second (MFLIPS), this are sufficient inferences to perform real-time edge detection as is evaluate in the conclusions section.

4.2 Test 2: Discrete Hardware Processing Fidelity

Because of discretization and limited resolution of all digital systems, fidelity is a metric of concern. In this test, input–output FPGA-based processing is compared to theoretical mathematical processing; the difference between them is hardware

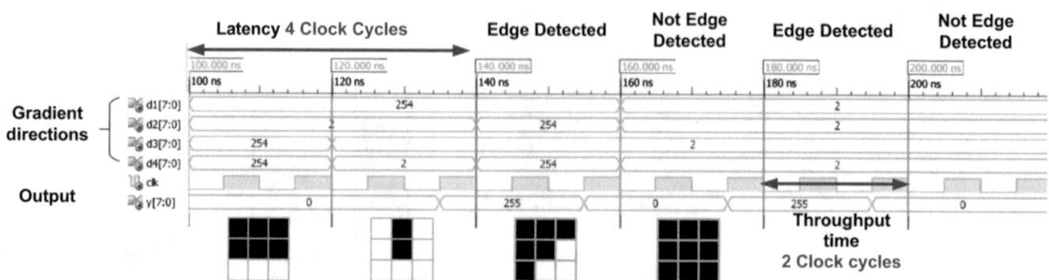

Fig. 17 System time plot

system's error, error is desired to be small enough that hardware discretization does not significantly compromises system's performance.

In FPGA and on all digital systems, data is managed bitwise, it means that scalar value representation have an inherent truncation error; truncation error is not of a constant value and depends on the amount represented by truncated bits.

In the presented system architecture, the major error source is the fuzzification hardware module, this is so because the evaluation of the TrapMF requires a product that uses an approximation of the slopes, i.e., a linear equation as stated in Eq. (9).

$$y = mx + b \tag{9}$$

Fidelity error in the difference between theoretical computation and discretized finite-resolution computation as expressed in Eq. (10).

$$e = m_r x - m_t x \tag{10}$$

where m_r is real-theoretical slope and m_t represents discretized-truncated slope, and x is MF-input, i.e., gradient value. Because of common factor x it could be expressed as in Eq. (11)

$$e = x(m_r - m_t) \tag{11}$$

Slope value m is specific to particular MF and is calculated as in Eq. (12).

$$m = \frac{\Delta y}{\Delta x} \tag{12}$$

The real-theoretical slope can be expressed as truncated slope function as shown in Eq. (13). and is replaced by 255 because in the present work the resolution in the output is 8 bits, the domain is limited depending on the value of the truncated slope, this is because the range is limited to 255, is important to remember that the values are binary data with 8 bits resolution, they are always entire numbers.

$$m_r = \frac{255 - m_t \Delta x}{\Delta x} + m_t, \quad x \leq \frac{255}{m_t} \tag{13}$$

Replacing the slope's expressions on (13), the error is expressed in (14).

$$e = x\left(\frac{255 - m_t \Delta x}{\Delta x} + m_t - m_t\right) \tag{14}$$

And simplifying the error is finally expressed in (15)

$$e = x\left(\frac{255 - m_t \Delta x}{\Delta x}\right), \quad x \leq \frac{255}{m_t} \tag{15}$$

Observing (15), the behavioral is not lineal, the error will be highest with high x value and this value is inversely proportional to m_t and this is inversely proportional to x, based on this appreciations, we infers the error have a nonlinear increase respecting Δx. Figure 18 shows the truncation error based on (15), respecting Δ.

The error shown is near 255 and this represents a near 100 % relative error and this is not desirable.

The fixed-point notation have a very important impact in the truncation error; the slope equation is expressed in (16).

$$m = \frac{2^b * 255}{\Delta x}, \quad x \in X, \quad (1, 255) \tag{16}$$

The slope are multiplied by a constant depending on the number of fractional bits used; the error equation changes as (17) expresses.

$$e = \frac{\left(x\left(\frac{2^b * 255 - m_t \Delta x}{\Delta x}\right)\right)}{2^b}, \quad x \leq \frac{255}{m_t} \tag{17}$$

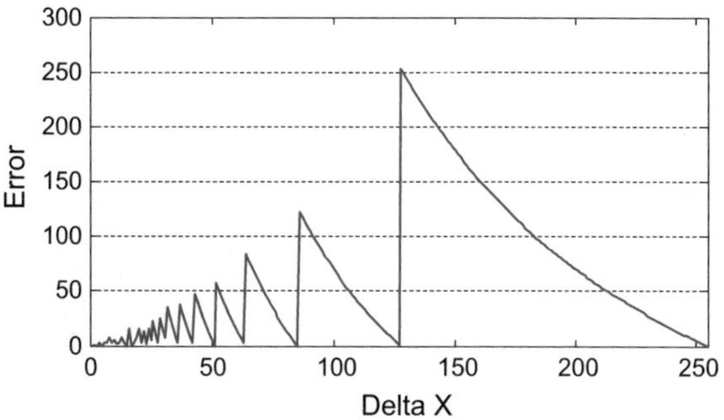

Fig. 18 Error behavioral

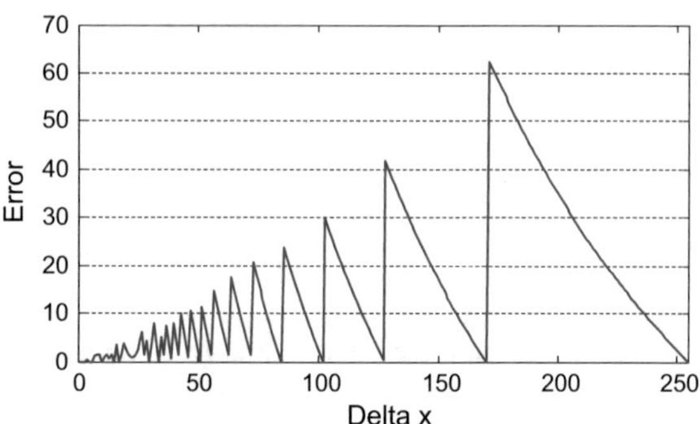

Fig. 19 Error behavioral with 1 fractional bit

Figures 19 and 20 show, respectively, the error behavioral with one fractional bit and three fractional bits.

The error is reduced at 63 and 4, respectively, this represents near 25 and 1.5 %.

The relation of these equations and the system proposed is because the variable Δx is the amplitude of the fuzzy sets in the proposed FIS.

The variable amplitude is limited to 127 and based on the error behavior, the maximum error is obtained for the maximum amplitude of the fuzzy set distribution.

Figure 21 shows the error in the edge detection evaluation with $C = 127$ and $A = 127$.

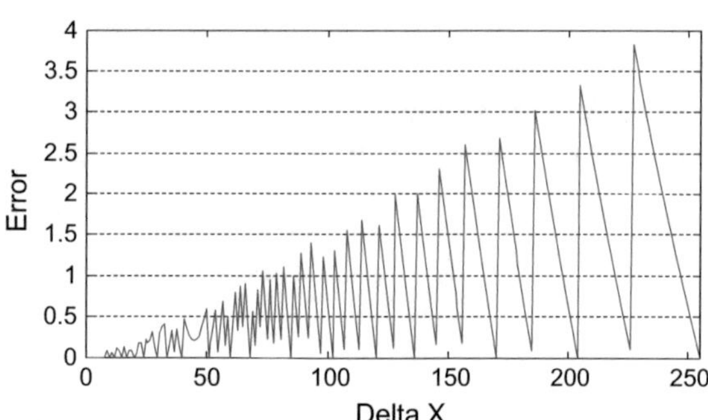

Fig. 20 Error behavioral with 3 fractional bits

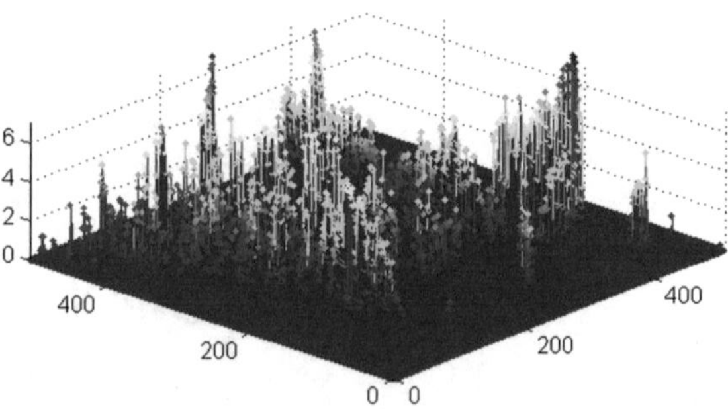

Fig. 21 Maximum error surface

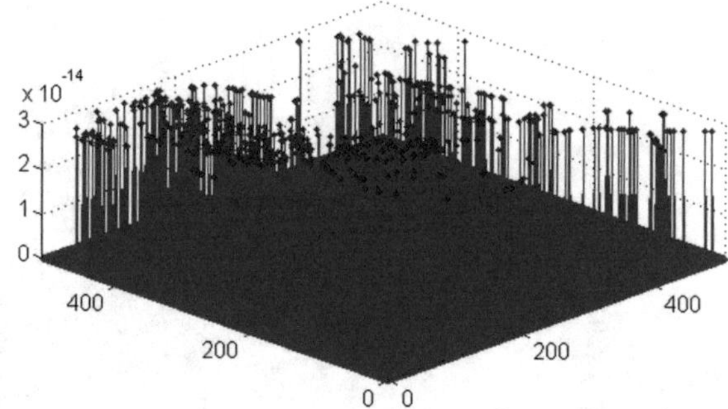

Fig. 22 Proposed FLED error surface

The maximum error showed is approximated seven per the whole FIS but this fuzzy set distribution does not perform a good edge detection, the proposed fuzzy set distribution achieve the error showed in Fig. 22

The error reported is very little and this represents a depreciated error.

4.3 *Test 3: Fuzzy Edge Detection*

In this experiment it is evaluated that the edge detection of a cameraman in order to demonstrate the results of the detection.

The quality of the detection is not explored in this work because this depends on the rules and the distribution of the parameters, this means, the architecture have the capability of perform better results if is adapted to the specific context of the input environment. Figure 23 shows the results of edge detection.

4.4 *Test 4: Performance Under Noisy-Image Environment*

The robustness in any image processing algorithm is very important, especially algorithms used as preprocessing as edge detection.

In order to evaluate this robustness, this experiment consists of the evaluations of the correlation between the edges detected in a clean image and the edge detected in a noisy image, this experiment compares the proposed architecture versus some conventional edge detectors. Equation (18) shows the setup of the experiment.

$$\begin{aligned} \text{Noisy image} = \text{Image} + \text{Noise edges}(\text{Noisy Image}) \approx \text{edges}(\text{Image}) \\ + \text{edges}(\text{Noise}) \text{ if } r \to 1 \text{ edges Noise} \to 0 \end{aligned} \tag{18}$$

Fig. 23 **a** Cameraman. **b** Fuzzy edge detection

Using the correlation coefficient between the edge detected for noise image and edge detected for clean image, it is possible to determine a degree of noise depreciation [19], because this coefficients measure the similarity of two images. Figure 24 shows the cameraman image plus a variance 0.005 of Gaussian noise.

Figure 25 shows the results of FLED-Arch edge detection and the canny edge detection for a clean image plus variance 0.005 of Gaussian noise.

Fig. 24 Noisy image

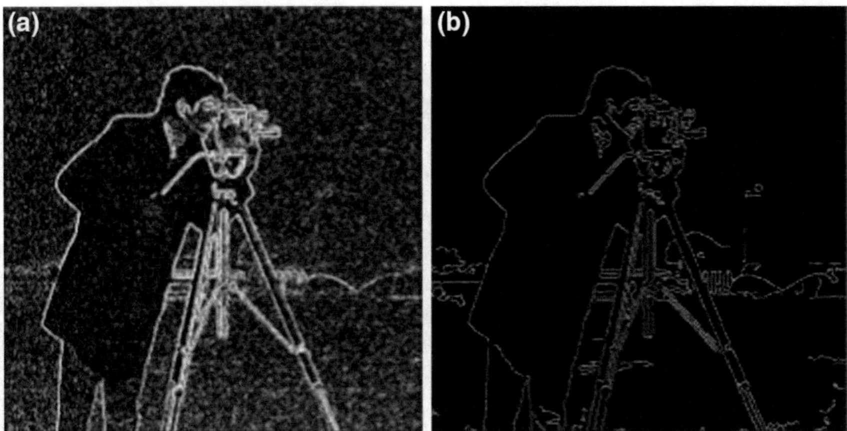

Fig. 25 FLED-Arch versus Canny edge detector

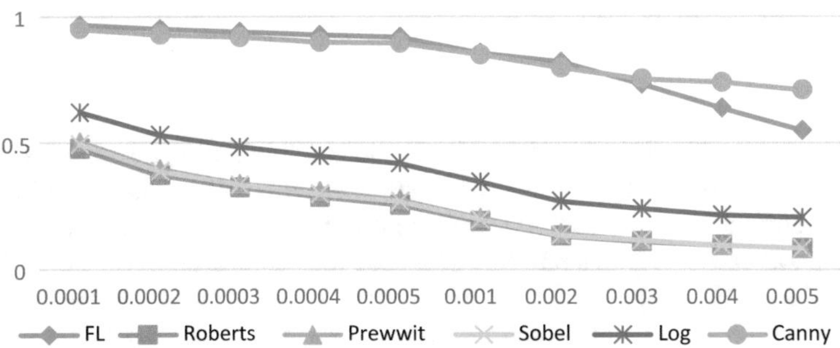

Fig. 26 Edges detectors robustness

These results can be appreciated better with the graphic showed in Fig. 26, the consistency of the proposed FIS evaluations in noisy environments.

For this noisy environments, it is interesting that the Canny edge detector shows better results than the FLED-Arch; this means our knowledge base or parameters distribution is not better in this noise environment but next is presented the results by adjusting the parameter of distribution of the inputs MF's. The improved detection is showed in Fig. 27.

Fig. 27 Improved FLED performance

5 Conclusion

The proposed architecture employs parallel and pipelining design structures that are described by VHDL language and synthesized on FPGA-based hardware platform demonstrating real-time fuzzy edge detection performance capability.

The experiments demonstrate that proposed architecture achieves 50 million of FLIPS, that at an image resolution of 480 × 640 pixels it can process over 162 FPS; this performance indicates that it can be used for higher resolution images.

Fidelity to theoretical software-based processing shown on Test 2 demonstrate that proposed architecture achieves a negligible error by means of the fixed-point notation strategy used, that balances integrity of discrete representation of fractional numbers and FPGA resources required for synthesis.

Test 3 shows tolerance to Gaussian noise, and the architecture performs better than conventional methods because of the FLED-Arch can be adapted to specific images context by editing it FIS-parameters without requiring changes to its hardware architecture.

In conclusion, the FLED-Arch is able to perform real-time edge detection, offers high fidelity to theoretical mathematical model, has inherent noise robustness, and can be adapted to different conditions of brightness and contrast without modifications to its hardware.

References

1. O. Mendoza y P. Melin, "Interval type-2 fuzzy integral to improve the performance of edge detectors based on the gradient measure", en *Fuzzy Information Processing Society (NAFIPS), 2012 Annual Meeting of the North American*, 2012, pp. 1–6.
2. W. Barkhoda, F. A. Tab, y O.-K. Shahryari, "Fuzzy edge detection based on pixel's gradient and standard deviation values", en *International Multiconference on Computer Science and Information Technology, 2009. IMCSIT '09*, 2009, pp. 7–10.
3. O. P. Verma, M. Hanmandlu, A. K. Sultania, y Dhruv, "A Novel Fuzzy Ant System for Edge Detection", en *2010 IEEE/ACIS 9th International Conference on Computer and Information Science (ICIS)*, 2010, pp. 228–233.
4. X.-P. Zong y W.-W. Liu, "Fuzzy edge detection based on wavelets transform", en *2008 International Conference on Machine Learning and Cybernetics*, 2008, vol. 5, pp. 2869–2873.
5. S. Sarangi y N. P. Rath, "Performance Analysis of Fuzzy-Based Canny Edge Detector", en *International Conference on Conference on Computational Intelligence and Multimedia Applications, 2007*, 2007, vol. 3, pp. 272–276.
6. P. Melin, C. I. Gonzalez, J. R. Castro, O. Mendoza, y O. Castillo, "Edge-Detection Method for Image Processing Based on Generalized Type-2 Fuzzy Logic", *IEEE Trans. Fuzzy Syst.*, vol. 22, núm. 6, pp. 1515–1525, dic. 2014.
7. X. Chen y Y. Chen, "An Improved Edge Detection in Noisy Image Using Fuzzy Enhancement", en *2010 International Conference on Biomedical Engineering and Computer Science (ICBECS)*, 2010, pp. 1–4.
8. F. Hoseini y A. Shahbahrami, "An efficient implementation of fuzzy edge detection using GPU in MATLAB", en *2015 International Conference on High Performance Computing Simulation (HPCS)*, 2015, pp. 605–610.
9. C. C. Leung, F. H. Y. Chan, K. Y. Lam, P. C. K. Kwok, y W. F. Chen, "Thyroid cancer cells boundary location by a fuzzy edge detection method", en *15th International Conference on Pattern Recognition, 2000. Proceedings*, 2000, vol. 4, pp. 360–363 vols.4.
10. Y. Zeng, C. Tu, y X. Zhang, "Fuzzy-Set Based Fast Edge Detection of Medical Image", en *Fifth International Conference on Fuzzy Systems and Knowledge Discovery, 2008. FSKD '08*, 2008, vol. 3, pp. 42–46.
11. V. K. Madasu y S. Vasikarla, "Fuzzy Edge Detection in Biometric Systems", en *36th IEEE Applied Imagery Pattern Recognition Workshop, 2007. AIPR 2007*, 2007, pp. 139–144.
12. D. Trevisan, W. Stefanutti, P. Mattavelli, y P. Tenti, "FPGA control of SIMO DC-DC converters using load current estimation", en *31st Annual Conference of IEEE Industrial Electronics Society, 2005. IECON 2005*, 2005, p. 6 pp.-pp.
13. J. Leuchter, P. Bauer, V. Rerucha, y P. Bojda, "Dc-Dc converters with FPGA control for photovoltaic system", en *Power Electronics and Motion Control Conference, 2008. EPE-PEMC 2008. 13th*, 2008, pp. 422–427.
14. B. Ding, R. M. Stanley, B. S. Cazzolato, y J. J. Costi, "Real-time FPGA control of a hexapod robot for 6-DOF biomechanical testing", en *IECON 2011 - 37th Annual Conference on IEEE Industrial Electronics Society*, 2011, pp. 252–257.
15. G. Chaple y R. D. Daruwala, "Design of Sobel operator based image edge detection algorithm on FPGA", en *2014 International Conference on Communications and Signal Processing (ICCSP)*, 2014, pp. 788–792.
16. P. K. Dash, S. Pujari, y S. Nayak, "Implementation of edge detection using FPGA amp; model based approach", en *2014 International Conference on Information Communication and Embedded Systems (ICICES)*, 2014, pp. 1–6.

17. A. Sanny y V. K. Prasanna, "Energy-efficient Median filter on FPGA", en *2013 International Conference on Reconfigurable Computing and FPGAs (ReConFig)*, 2013, pp. 1–8.
18. F. Cabello, J. Leon, Y. Iano, y R. Arthur, "Implementation of a fixed-point 2D Gaussian Filter for Image Processing based on FPGA", en *Signal Processing: Algorithms, Architectures, Arrangements, and Applications (SPA), 2015*, 2015, pp. 28–33.
19. A. M. Mahmood, H. H. Maras, y E. Elbasi, "Measurement of edge detection algorithms in clean and noisy environment", en *2014 IEEE 8th International Conference on Application of Information and Communication Technologies (AICT)*, 2014, pp. 1–6.

A Hybrid Intelligent System Model for Hypertension Diagnosis

Ivette Miramontes, Gabriela Martínez, Patricia Melin and German Prado-Arechiga

Abstract A hybrid intelligent system is made of a powerful combination of soft computing techniques for reducing the complexity in solving difficult problems. Nowadays hypertension (high blood pressure) has a high prevalence in the world population and is the number one cause of mortality in Mexico, and this is why it is called a silent killer because it often has no symptoms. We design in this paper a hybrid model using modular neural networks, and as response integrator we use fuzzy systems to provide an accurate diagnosis of hypertension, so we can prevent future diseases in people based on the systolic pressure, diastolic pressure, and pulse of patients with ages between 15 and 95 years.

Keywords BP (blood pressure) · Hypertension · ABPM (ambulatory blood pressure monitoring) · Fuzzy system · Modular neural network · Systolic pressure · Diastolic pressure · Pulse

1 Introduction

A hybrid intelligent system can be built from a prudent combination of two or more soft computing techniques for solving complex problems. The hybrid system is formed by the integration of different soft computing subsystems, each of which maintains its representation language and an inferring mechanism for obtaining its corresponding solutions. The main goal of hybrid systems is to improve the efficiency and the power of reasoning and expression of isolated intelligent systems [1, 2].

In this work, we used some recent soft computing techniques, such as modular neural networks and fuzzy inference systems.

I. Miramontes · G. Martínez · P. Melin (✉)
Tijuana Institute of Technology, Tijuana, Mexico
e-mail: pmelin@tectijuana.mx

G. Prado-Arechiga
Cardiodiagnostico, Tijuana, BC, Mexico

P. Melin et al. (eds.), *Nature-Inspired Design of Hybrid Intelligent Systems*,
Studies in Computational Intelligence 667, DOI 10.1007/978-3-319-47054-2_35

1.1 Blood Pressure and Hypertension

Blood pressure is the force exerted by the blood against the walls of blood vessels, especially the arteries [3]. Then we understand as high blood pressure, (called hypertension), as the sustained elevation of blood pressure (BP) above the normal limits [4, 5]. The regular blood pressure levels are those below 139 in systolic pressure (when the heart contracts and pushes the blood around the body) and over 89 in diastolic pressure (when the heart relaxes and refill with blood), and is measured in millimeters of mercury (mmHg) [3]. The heart rate is the number of times the heart beats per minute, and this is well-known to vary by ages, for example, the heart rate in a child is normal around 160 beats per minute, but in an adult at rest, the normal is between 50 and 90 beats per minute, also can change for some illness, in this case the change is abnormal.

For collecting the BP measurements of patients, we used a device called Ambulatory blood pressure monitoring, described below.

A 24-h blood pressure measurement is just the same as a standard blood pressure check: a digital machine measures the blood pressure by inflating a cuff around your upper arm and then slowly releasing the pressure. The device is small enough to be worn on a belt around the waist while the cuff stays on your upper arm for the full 24 h. The monitors are typically programmed to collect measurements every 15–20 min during the daytime and 30 min at night. At the end of the recording period, the readings are downloaded into a computer. This device can provide the following types of information, an estimate of the true or mean blood pressure level, the diurnal rhythm of blood pressure, and blood pressure variability [6, 7].

Studies with the Ambulatory blood pressure monitoring (ABPM) device have shown that when more than 50 % the readings of blood pressure are higher than 135/85 mmHg during the awake hours, and 120/80 mmHg for the sleep hours, there are signs of target organ damage (kidney, blood vessels, eyes, heart, and brain), so that this blood pressure level is already pathogenic and, therefore, has been concluded that the above-mentioned numbers should be considered abnormal [3].

The medical doctors recommend using the ABPM in different cases, for example if we suspect of having white-coat hypertension, in these patients, office blood pressures are substantially higher than ambulatory awake blood pressure averages [7].

Other causes of sporadic high blood pressure (in crisis), can be symptoms that are suggestive of sudden changes in blood pressure and if we are suspected of having masked hypertension, this means a normal clinic blood pressure and a high ambulatory blood pressure [3, 7].

1.2 Neural Network for Hypertension Diagnosis

Sumathi and Santhakumaran [8] used artificial neural networks for solving the problems of diagnosing hypertension using the Backpropagation learning algorithm. They constructed the model using eight risk factors, such as if the person consumes alcohol, smoking, is obese, if have stress [7] just to mention a few.

Hosseini et al. [9] presented a study to examine risk factors for hypertension and to develop a prediction model to estimate hypertension risk for rural residents over the age of 35 years. Also took some risk factors that were significantly associated with the illness, such as a high educational level, a predominantly sedentary job, a positive family history of hypertension, among others, they established predictive models using logistic regression model, and artificial neural network. The accuracy of the models was compared by receiver operating characteristic when the models applied to the validation set, and the artificial neural network model proved better than the logistic regression model.

For estimating Ambulatory Systolic Blood Pressure variations based on corporal acceleration and heart rate measurements, Hosseini et al. [10] used neural network models. The temporal correlation of the estimation residual, modeled by a first order autoregressive process, is used for training the neural network in a maximum likelihood framework, which yields a better estimation performance. As data are collected at irregular time intervals, the first order AR model is modified for taking into account this irregularity. The results are compared by those of a neural network trained using an ordinary least square method.

1.3 Fuzzy Logic and Hypertension

There are some recent works on using fuzzy logic in this area, for example for diagnosis of hypertension, Guzman et al. [11] have proposed a Mamdani fuzzy system model, based on the European Hypertension classification [14], which is shown in Table 1.

The model has two inputs, the first is the systolic blood pressure and the second is the diastolic blood pressure and this is done by taking into consideration all ranges of blood pressure, and the model has one output that is for the blood pressure level.

A fuzzy rule-based system for the diagnosis of heart diseases has been presented by Barman and Choudhury [12]. The inputs are the chest pain type, the resting

Table 1 Definitions and classification of office blood pressure levels (mmHg)

Category	Systolic		Diastolic
Optimal	<120	and	<80
Normal	120–129	and/or	80–84
High normal	130–139	and/or	85–89
Grade 1 hypertension	140–159	and/or	90–99
Grade 2 hypertension	160–179	and/or	100–109
Grade 3 hypertension	≥ 180	and/or	≥ 110
Isolated systolic hypertension	≥ 140	and	<90

blood pressure, serum cholesterol in mg, Numbers of years of being a smoker, fasting blood sugar, maximum heart rate achieved, and resting blood rate. The angiographic disease status of the heart of patients has been recorded as output. It is to state that the diagnosis of heart disease by angiographic disease status is assigned by a number between 0 and 1, and this number indicates whether the heart attack is mild or massive.

1.4 Fuzzy Logic and Pulse

For measuring health parameters of patients, Patil, and Mohsin [13] have proposed a wireless sensor network system for continuous monitors pulse and temperature of patients at remote or in the hospital, and it transmits the biosignals to the doctor and patient mobile phone. Data stored in a database is passed to the fuzzy logic controller to improve accuracy and amount of data to be sent to the remote user. The FLC system receives context information from the sensor as input (the patient age and pulse), and output is the status of the patient pulse.

We want to provide a risk diagnosis in patients with high blood pressure, for this, we used modular neural networks, where each module works independently. Each of the neural networks is built and trained for a specific task [2]. We used a response integrator of the modules and for giving the risk diagnosis, a traditional fuzzy rule system.

The paper is organized as follows: in Sect. 2 the proposed method is presented, in Sect. 3 a methodology description is presented, in Sect. 4 the results and discussions are presented, and in Sect. 5 the conclusions obtained after finishing the work with the modular neural network are offered.

2 Proposed Method

Measurements of the blood pressure are obtained by the ABPM for 100 people, and these data have been obtained from students of the master and doctorate in computer science from Tijuana Institute of Technology. In addition, the Cardio-Diagnostic Center of Tijuana has provided blood pressure data of his patients for this research, a databases with corresponding data to the systolic pressure, diastolic pressure, and pulse is created (Fig. 1).

The modular neural network is trained with 47 records of 27 patients (70 % for training phase and 30 % for testing phase) in the database, in other words, the first module was trained with the records of systolic pressure, the second with the diastolic pressure, and the third module with the pulse, the network is modeling the data for learning the blood pressure behavior.

The architecture for the modular neural network was changed in each experiment, showing the better results with the next parameters:

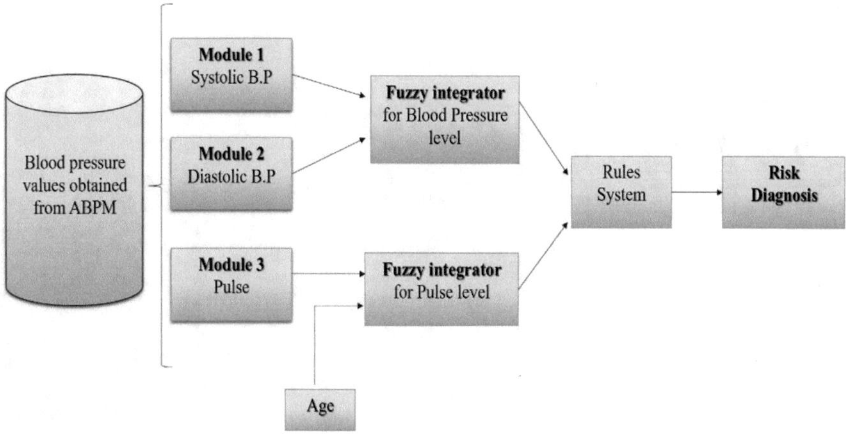

Fig. 1 Proposed method

Training method: Levenberg-Marquardt
Hidden Layers: 3
Neurons: 32, 18, 13
Error goal: 1.00 E^{-5}
Epochs: 300.

In Table 3 all experiments are shown.

We used fuzzy inference systems as integrators, the fuzzy model develop by Guzman et al. [11] is taken to obtain a blood pressure classification and the second fuzzy inference system to obtain the pulse level, this because there is no numerical relationship between blood pressure and pulse, but if there is a connection with some diseases. On the other hand, the age enters independently to the fuzzy system because it is an important variable to determinate the variation of the pulse, the output of the two integrators will be evaluated by traditional system rules to provide a risk diagnosis of a cardiovascular disease which the patient could have.

3 Methodology

In this work, we propose a Mamdani fuzzy inference system for finding the pulse level, and it was designed empirically; this has two inputs including the age and the pulse and has one output which corresponds to the Pulse level. The membership functions used on the age input are trapezoidal for "children" and "elder," and triangular for "young" and "middle," the membership functions used for the pulse input are trapezoidal for "low" and "very high," and triangular for "low," "normal" and "high" linguistic terms.

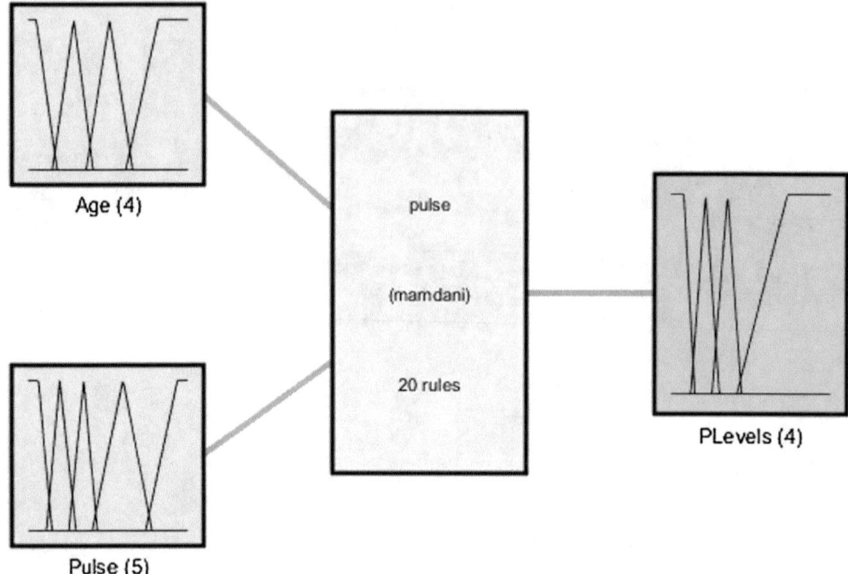

Fig. 2 Pulse fuzzy system

For the output trapezoidal membership functions are used for low and very low, and triangular for below normal, excellent, and above normal (Fig. 2).

Figure 3 shows the input and output variables; we can analyze the input for age and has a range of 0–100, and the pulse has a range of 0–220 because is the maximum level of the pulse in a person.

For the output of the fuzzy system this is considered in a range from 0 to 100 % because this is the range how well or how badly the patient is from low pulse to very high.

The rule set of the FLC contains 20 rules, which depends on the age and pulse for determining which pulse level the patient has. In Table 2 we present the rule set for this case (Fig. 4).

3.1 Graphic User Interface

A graphical user interface for the diagnosis of cardiovascular risk was designed, and is shown in Fig. 5, for which the final user can search for the appropriate file where they have saved the patient's record, the interface plots the behavior of the pressure and pulse obtained by ABPM and shows the patient information as name and age.

The medical doctor can make questions about risk factors, this means, which bad habits the patient has, this it will be evaluated together the records of ABPM and the fuzzy rules.

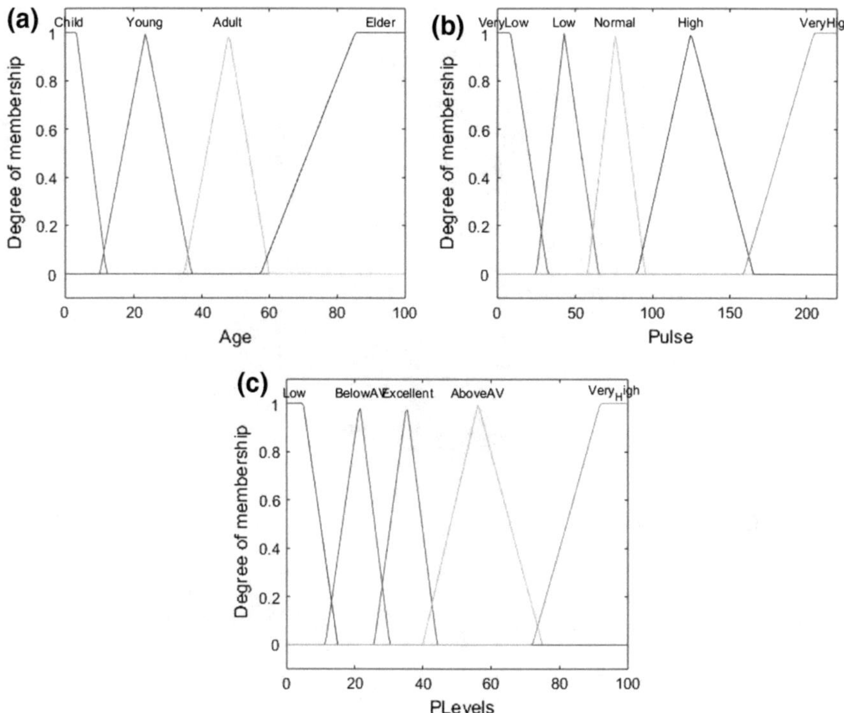

Fig. 3 **a** Input for age. **b** Input for pulse. **c** Output for pulse level

Table 2 Fuzzy rules set

Age/Pulse	Very low	Low	Normal	High	Very high
Child	Low	Low	Excellent	Excellent	Above AV
Young	Low	Below AV	Excellent	Above AV	Very high
Adult	Low	Below AV	Excellent	Above AV	Very high
Elder	Low	Below AV	Excellent	Very high	Very high

When the final user presses the evaluate button, it will display the results obtained by the fuzzy inference systems and the result of the fuzzy rules for cardiovascular risk diagnosis that the patient may have.

4 Results and Discussion

The modular neural network was trained with different architectures to observe the data behavior and find the better results. In Table 3, we show some experiments can be noted the training methods, which were the Levenberg–Marquardt (LM) and

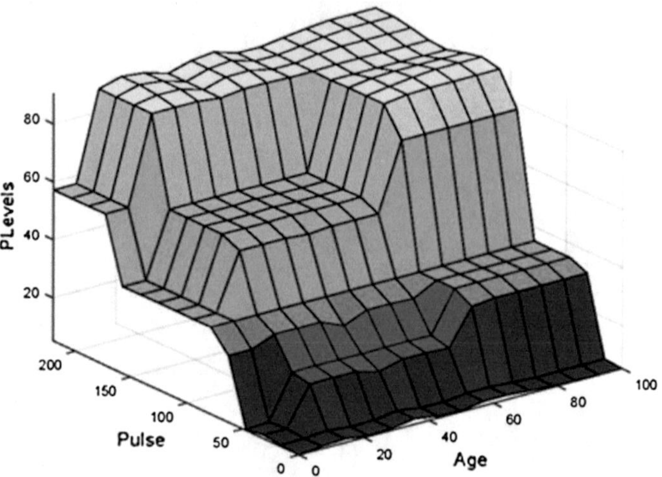

Fig. 4 Surface new of the fuzzy model

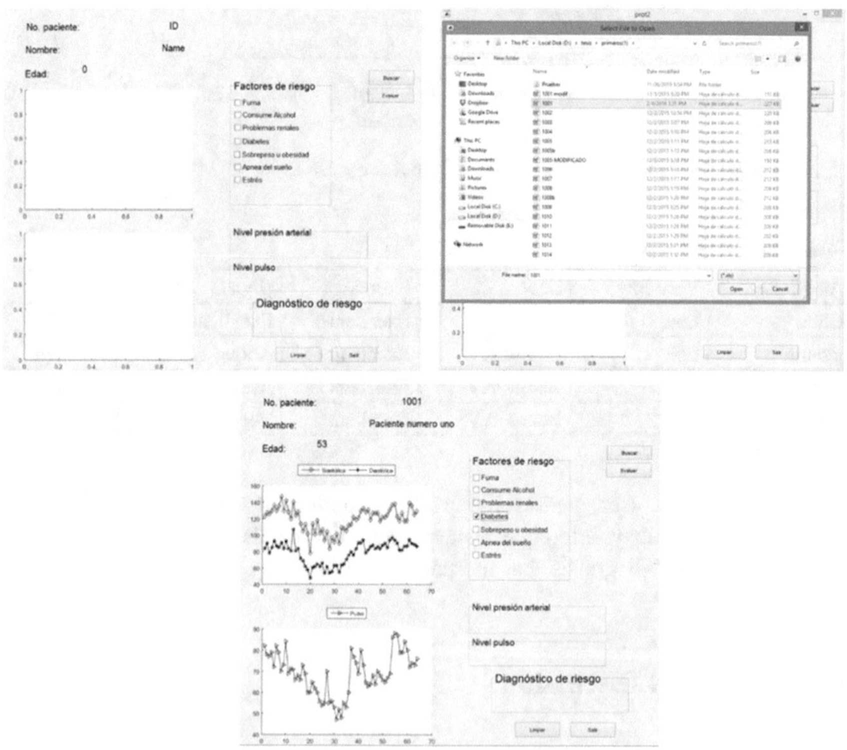

Fig. 5 Graphic user interface

Table 3 Neural network training

No	Train	Layers	Neurons	Errors					
				Systolic		Diastolic		Pulse	
				Train	Test	Train	Test	Train	Test
1	LM	1	25	0.0134	0.0101	0.0151	0.0159	0.009	0.0136
2	LM	1	30	0.013	0.0121	0.015	0.024	0.013	0.0121
3	LM	2	30,16	0.013	0.0119	0.0151	0.0261	0.009	0.0413
4	LM	2	25,12	0.0132	0.0337	0.0152	0.0024	0.009	0.4389
5	LM	1	18	0.0132	0.1036	0.0152	0.0142	0.009	0.0136
6	GDX	1	25	0.0104	0.0083	0.0166	0.0228	0.0057	0.0083
7	GDX	1	30	0.0118	0.0166	0.0131	0.0641	0.0079	0.0057
8	GDX	2	30,16	0.0099	0.0153	0.0139	0.0124	0.0062	0.0074
9	GDX	2	25,12	0.0122	0.0069	0.0138	0.0093	0.0065	0.0042
10	GDX	1	18	0.0105	0.0036	0.012	0.0344	0.008	0.0115
11	LM	1	35	0.0132	0.026	0.0151	0.1798	0.0091	0.0063
12	LM	1	12	0.013	0.0164	0.0152	0.0218	0.091	0.0108
13	LM	2	20,12	0.0133	0.0153	0.0152	0.247	0.0089	0.2576
14	LM	3	28,16,12	0.0131	0.151	0.0151	0.0649	0.0065	0.009
15	LM	3	32,18,13	0.0131	0.0128	0.0152	0.0315	0.0078	0.0073
16	GDX	1	35	0.0109	0.0045	0.0131	0.0484	0.0071	0.0151
17	GDX	1	12	0.0107	0.0045	0.0136	0.0138	0.0071	0.0095
18	GDX	2	20,12	0.0114	0.0032	0.0142	0.0154	0.0074	0.0078
19	GDX	3	28,16,12	0.0112	0.0104	0.0147	0.0058	0.006	0.0092
20	GDX	3	32,18,13	0.0113	0.0099	0.0155	0.0249	0.0081	0.0148
21	LM	1	33	0.0131	0.0245	0.015	0.0296	0.09	0.0081
22	LM	2	25,15	0.0131	0.0039	0.0151	0.3142	0.009	0.0273
23	LM	2	28,14	0.0132	0.0134	0.0152	0.3132	0.009	0.1519
24	LM	3	32,18,14	0.0131	0.018	0.0154	0.0746	0.0089	0.0486
25	LM	3	33,21,17	0.0132	0.0224	0.0157	0.0732	0.009	0.0864
26	GDX	1	33	0.0109	0.0149	0.014	0.0152	0.064	0.0226
27	GDX	2	25,15	0.0123	0.0109	0.0132	0.0182	0.0087	0.0136
28	GDX	2	28,14	0.0112	0.0083	0.0162	0.0336	0.0068	0.0048
29	GDX	3	32,18,14	0.012	0.0139	0.013	0.0244	0.0074	0.0042
30	GDX	3	33,21,17	0.0104	0.0152	0.0142	0.0169	0.0077	0.0067

Gradient descent with momentum and adaptive learning rate backpropagation (GDX), with different layers and neurons. There is a slight improvement in the training of LM method but is not significant.

Each training time was between 1 and 3 s and in the first 20 experiments we used 300 epochs, and in the last 10, we used 500 epochs. The error goal is 1.00 E^{-5}. We need to perform more tests to the neural network, try with another hidden layer, and maybe use delays for improving and obtain better results.

5 Conclusion and Future Work

This paper has presented a hybrid intelligent system for providing a risk diagnosis in patients with hypertension, and this type of system can be helpful for reducing the complexity of the problem to be solved.

We used a modular neural network with a fuzzy response integrator for being able to give an accurate result.

So far, we have good results but more experiments will be conducted and other factors will be added to the model to make improvements, such as more risk factors to give a final risk diagnosis more accurate.

Acknowledgment We would like to express our gratitude to the CONACYT and Tijuana Institute of Technology for the facilities and resources granted for the development of this research.

References

1. F. Fdez Riverola and J. M. Corchado, "Sistemas híbridos neuro-simbólicos: Una revisión," *Inteligencia Artificial. Revista Iberoamericana de Inteligencia Artificial,* vol. 4, pp. 12 - 26, 2000.
2. P. Melin and O. Castillo, Hybrid Intelligent Systems for Pattern Recognition Using Soft Computing, Springer, Ed., Germany: Springer, 2005.
3. J. F. Guadalajara Boo, Cardiología, Mexico, D.F: Méndez Editores, 2006.
4. G. Mancia, "2013 ESH/ESC Guidelines for the management of arterial hypertension," *Journal of Hypertension,* vol. 31, no. 7, pp. 1281-1357, 2013.
5. J. Narro Robles, O. Rivero Serrano and J. López Bárcena, Diagnóstico y tratamiento en la práctica médica, México: Manual moderno, 2010.
6. T. Pickering, P. Daichi Shimbo and D. Haas, "Ambulatory Blood-Pressure Monitoring," *New England Journal of Medicine,* vol. 354, no. 22, pp. 2368-2374, 2006.
7. W. White, Blood pressure monitoring in cardiovascular medicine and therapeuctis, Totowa, New Jersey: Humana Press, 2007.
8. B. Sumathi and A. Santhakumaran, "Pre-Diagnosis of Hypertension Using Artifical Neural Network," *Global Journal of Computer Scinece and Technology,* vol. 11, pp. 43-47, 2011.
9. H. Shuqiong, X. Yihua, Y. Li and W. Sheng, "Evaluating the risk of hypertension using an artificial neural network method in rural residents over the age of 35 years in a Chinese area," *Hypertension Research,* vol. 33, pp. 722-726, 2010.
10. S. Hosseini, C. Jutten and S. Charbonnier, "Neural network modeling of ambulatory systolic blood pressure for hypertension diagnosis," in *Artificial Neural Nets Problem Solving Methods*, Maó, Menorca, Spain, Springer, 2003, pp. 599-606.
11. J. C. Guzman, P. Melin and G. Prado-Arechiga, "Design of a Fuzzy System for Diagnosis of Hypertension," *Design of Intelligent Systems Based on Fuzzy Logic, Neural Networks and Nature-Inspired Optimization,* vol. 601, pp. 517- 526, 2015.
12. M. Barman and J. Choudhury, "A Fuzzy Rule Base System for the Diagnosis of Heart Disease," *International Journal of Computer Applications,* pp. 46-53, 2012.
13. P. Patil and S. Mohsin, "Fuzzy Logic based Health Care System using Wireless Body Area Network," *International Journal of Computer Applications,* vol. 80, no. 12, pp. 46-51, 2013.
14. I. Morsi and Z. Abd El Gawad, "Fuzzy logic in heart rate and blood pressure measuring system," in *Sensors Applications Symposium (SAS), 2013 IEEE*, Galveston, TX, IEEE, 2013, pp. 113-117.

Comparative Analysis of Designing Differents Types of Membership Functions Using Bee Colony Optimization in the Stabilization of Fuzzy Controllers

Leticia Amador-Angulo and Oscar Castillo

Abstract A study of the optimization of different types of membership functions (MF) using Bee Colony Optimization (BCO) for the stabilization of fuzzy controllers is presented. The main objective of the work is based on the main reasons for the comparative analysis of BCO as an optimization technique for the design of the Mamdani fuzzy controllers, specifically in tuning membership functions for two problems in fuzzy control. Simulations results confirmed that using the BCO to optimize the membership functions and the scaling gains of the fuzzy system improved the controller performance. The six metrics of the ITAE, ITSE, IAE, ISE, RMSE and MSE for the errors in control are implemented.

Keywords Fuzzy controller · Bee colony optimization · Membership function · Convergence · Type-1 fuzzy logic system

1 Introduction

In recent years, extensive work has been carried out on control system stabilization. However, all these control design methods require the exact mathematical models of the physical systems, which may not be available in practice. On the other hand, fuzzy control has been successfully applied to solve many nonlinear control problems, for example the works presented in [1–3, 6–8, 11].

The fuzzy logic system is one of the most used methods of computational intelligence, possibly thanks to the efficiency and simplicity of fuzzy systems, since they use linguistic terms that are similar to those used by humans [25]. A fuzzy controller can be considered to be a fuzzy rule-based controller, which consists of

L. Amador-Angulo (✉) · O. Castillo
Division of Graduate Studies, Tijuana Institute of Technology, Tijuana, Mexico
e-mail: leticia.amadorangulo@yahoo.com.mx

O. Castillo
e-mail: ocastillo@tectijuana.mx

P. Melin et al. (eds.), *Nature-Inspired Design of Hybrid Intelligent Systems*,
Studies in Computational Intelligence 667, DOI 10.1007/978-3-319-47054-2_36

input and output variables with membership functions, a set of (IF…THEN) rules and an inference system [26, 27].

It is empirically necessary to carry out various experiments to find the optimal types of the membership functions for a specific problem. This research focuses on the idea of demonstrating that BCO is a good technique in the optimization of parameters for the design of fuzzy controllers. Three types of MFs are presented which are: Gaussian, trapezoidal and triangular.

BCO has recently undergone many improvements and applications. The BCO algorithm mimics the food foraging behavior of swarms of honey bees [18]. Honey bees use several mechanisms such as waggle dance to optimally locate food sources and search for new ones. It is a very simple, robust and population-based stochastic optimization algorithm [5, 17]. The approach of BCO as a technique for solving problems of optimization is shown in the following research: [2, 3, 7, 9, 16, 18–20, 22].

This paper focuses on the comparative analysis for tuning two Mamdani fuzzy controllers using three type of MFs using BCO. Section 2 presents the theoretical basis of this work. Section 3 describes cases studies, indicating the two analyzed problems. Section 4 includes a description of BCO and its implementation in this paper. Section 5 shows the results of the simulation of the model. Section 6 includes the comparative analysis. Finally, Sect. 7 offers the conclusion and suggestions for future work.

2 Theoretical Basis

2.1 Fuzzy Logic System

A fuzzy logic system (FLS) that is defined entirely in terms of type-1 fuzzy sets, is known as a type-1 fuzzy logic system (type-1 FLS); its elements are defined in the following Fig. 1 [1–3, 6–8, 11, 23].

A type-1 fuzzy set in the universe X is characterized by a membership function $u_A(x)$ taking values in the interval [0,1], and can be represented as a set of ordered pairs of an element. The membership degree of an element to the set is defined by the following Eq. (1) [15, 24]:

$$A = \{(x, \mu_A(x))|x \in X\} \quad (1)$$

where $\mu_A : X \rightarrow [0, 1]$.

In this definition, $\mu_A(x)$ represents the membership degree of the element $x \in X$ to the set A. In this work, we are going the use the following notation: $A(x) = \mu_A(x)$ for all $x \in X$.

A variety of types of membership functions exist, and for this work three are used: Trapezoidal, Triangular and Gaussian. Figure 2 shows the three FMs.

The design of each of the fuzzy logic controllers implemented in this work is described in Sect. 3.

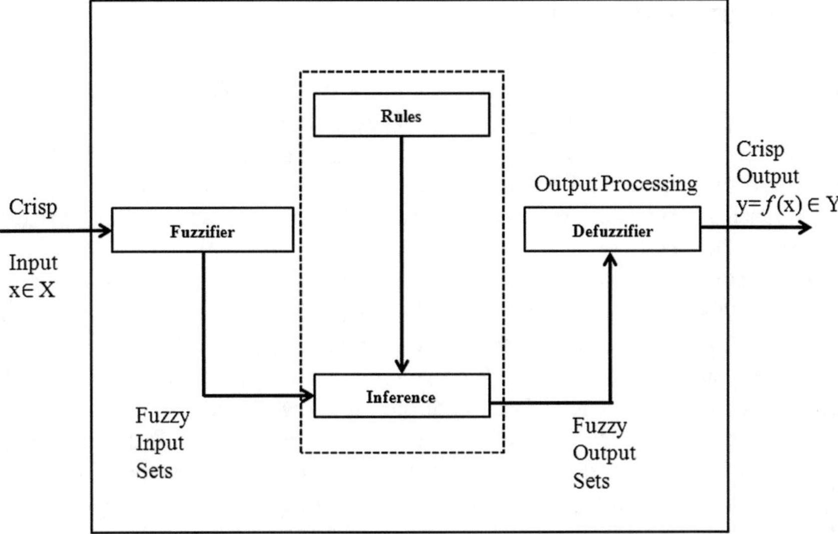

Fig. 1 Architecture of a type-1 fuzzy logic system

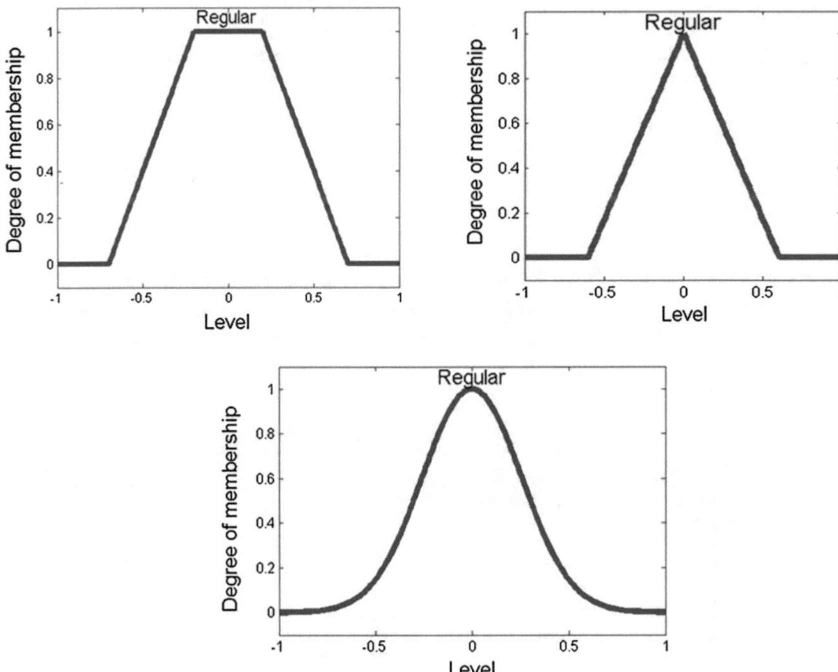

Fig. 2 Types of membership functions optimized by Bee Colony Optimization

2.2 *Fuzzy Logic Controller*

Mamdani's method was among the first control systems built using fuzzy set theory. It was proposed in 1975 by Ebrahim Mamdani as an attempt to control a steam engine and boiler combination by synthesizing a set of linguistic control rules obtained from experienced human operators [15]. Mamdani's effort was based on Lotfi Zadeh's 1973 paper on fuzzy algorithms for complex systems and decision processes [24–27].

The main principle of the automatic control system is the application of the concept of "feedback" shown in Fig. 3, which is the output of the adder we have the error signal, which is applied to the fuzzy controller together with a signal derived from this, which is the change in the error signal over time [1, 3, 8, 23]. A representation of a generic fuzzy controller is shown in Fig. 3.

The Bee Colony Optimization (BCO) metaheuristic [12–14, 18] has been introduced by Lučić and Teodorović as a new direction in the field of swarm intelligence, and has not been previously applied in type-1 fuzzy logic controller (T1FLC) design for these benchmark problems. The BCO is an algorithm implementation that adapts the behavior of real bees to solutions of minimum cost path problems on graphs. Figure 4 shows a representation of the BCO algorithm.

In this paper, the optimization of the parameter values of MFs in inputs and outputs was made with the implementation of fuzzy systems obtained with the fuzzy logic system with BCO for two benchmark problems using three types of MFs.

3 Case Studies

For the evaluation of T1FLC, we used two control problems, which in fuzzy control are mathematical models that allow us to analyze the behavior of the generated control. We describe each of the problems and the implemented fuzzy controller.

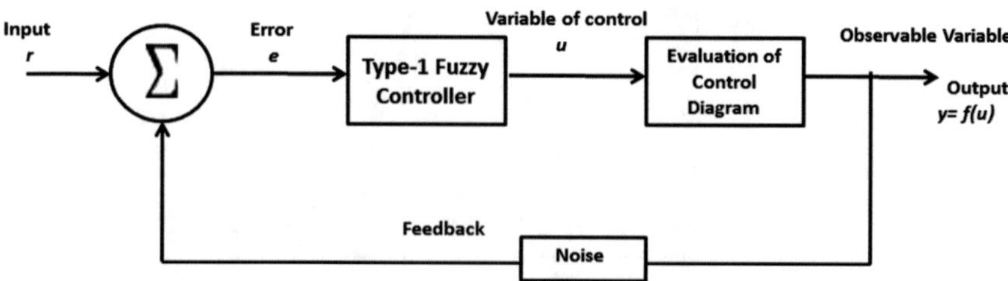

Fig. 3 Graphical representation of fuzzy controller

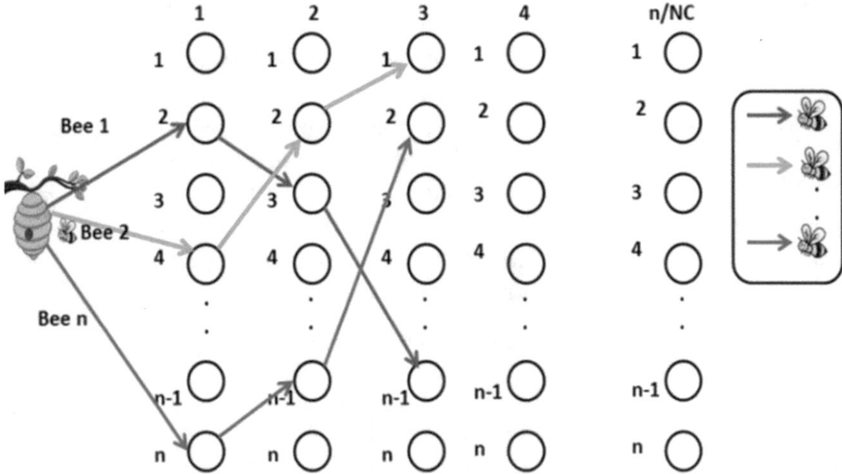

Fig. 4 Representation of Bee Colony Optimization

3.1 *Autonomous Mobile Robot Controller*

3.1.1 General Description of the Problem

The model used was a unicycle mobile robot [2, 3, 8], consisting of two driving wheels located on the same axis and a front free wheel. Figure 5 shows a graphical description of the robot model.

The robot model assumes that the motion of the free wheel can be ignored in its dynamics, as shown in Eqs. (2) and (3).

$$M(q)\dot{v} = C(q,\dot{q})v + Dv = \tau + P(t) \tag{2}$$

Fig. 5 Mobile robot model

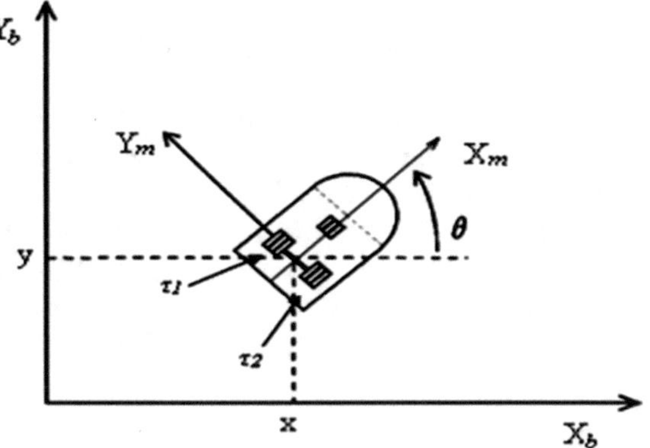

where

$q = (x, y, \theta)^T$	is the vector of the configuration coordinates,
$\upsilon = (v, w)^T$	is the vector of velocities,
$\tau = (\tau_1, \tau_2)$	is the vector of torques applied to the wheels of the robot where τ_1 and τ_2 denote the torques of the right and left wheel,
$P \in R^2$	is the uniformly bounded disturbance vector,
$M(q) \in R^{2\times2}$	is the positive-definite inertia matrix,
$C(q, \dot{q})\vartheta$	is the vector of centripetal and Coriolis forces, and
$D \in R^{2\times2}$	is a diagonal positive-definite damping matrix.

The kinematic system is represented by Eq. (3).

$$\dot{q} = \underbrace{\begin{bmatrix} \cos\theta & 0 \\ \sin\theta & 0 \\ 0 & 1 \end{bmatrix}}_{J(q)} \underbrace{\begin{bmatrix} v \\ w \end{bmatrix}}_{\upsilon} \tag{3}$$

where

(x, y)	is the position in the X–Y (world) reference frame,
θ	is the angle between the heading direction and the x-axis,
v and w	are the linear and angular velocities.

Furthermore, Eq. (4) shows the non-holonomic constraint, which this system has, which corresponds to a no-slip wheel condition preventing the robot from moving sideways.

$$\dot{y}\cos\theta - \dot{x}\sin\theta = 0 \tag{4}$$

The system fails to meet Brockett's necessary condition for feedback stabilization, which implies that no continuous static state feedback controller exists that can stabilize the closed-loop system around the equilibrium point.

3.1.2 Design of the Fuzzy Logic Controller

The main problem to study is controlling the stability of the trajectory in a mobile robot. The MFs are two inputs to the fuzzy system: the first is called $\mathbf{e_v}$ **(angular velocity)**, which has three MFs with linguistic values of *N*, *Z* and *P*. The second input variable is called $\mathbf{e_w}$ **(linear velocity)** with three MFs with the same linguistic values. The T1FLC has two outputs called **T1 (Torque 1)**, and **T2 (Torque 2)**, which are composed of three trapezoidal MFs with the following linguistic values, respectively: *N*, *Z*, *P*. Figure 6 shows the representation of the input and output variables.

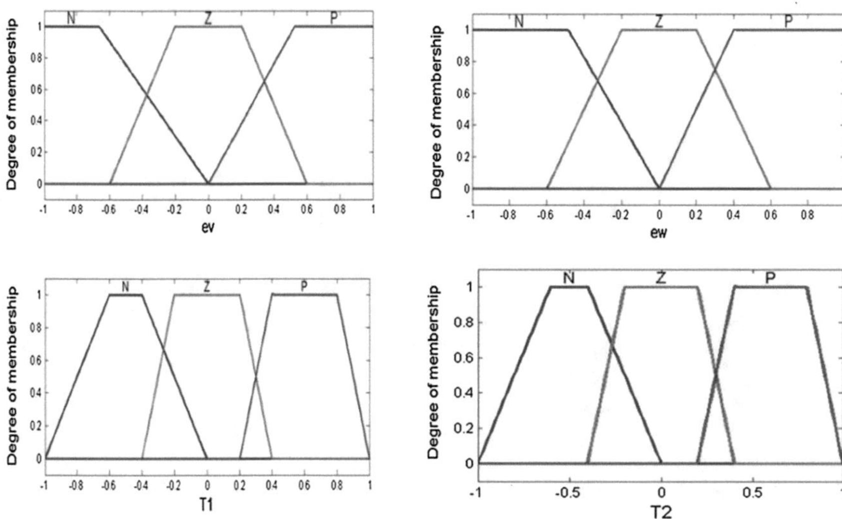

Fig. 6 Inputs and outputs of type-1 fuzzy logic controller for the mobile robot controller

Table 1 Fuzzy rules used by the fuzzy controller

# rules	Input 1 e_v	Input 2 e_w	Output 1 $T1$	Output 2 $T2$
1	N	N	N	N
2	N	Z	N	Z
3	N	P	N	P
4	Z	N	Z	N
5	Z	Z	Z	Z
6	Z	P	Z	P
7	P	N	P	N
8	P	Z	P	Z
9	P	P	P	P

The knowledge about the problem provides us with nine fuzzy rules for control. The combination of the rules is shown in Table 1 and the model of the T1FLC can be found in Fig. 7.

The test criteria are a series of Performance Indices, where the Integral Square Error (ISE), Integral Absolute Error (IAE), Integral Time Squared Error (ITSE), Integral Time Absolute Error (ITAE) and Root Mean Square Error (RMSE) are used, respectively, shown in Eqs. (5)–(9);

$$\mathrm{ISE} = \int_0^\infty e^2(t)\mathrm{d}t \tag{5}$$

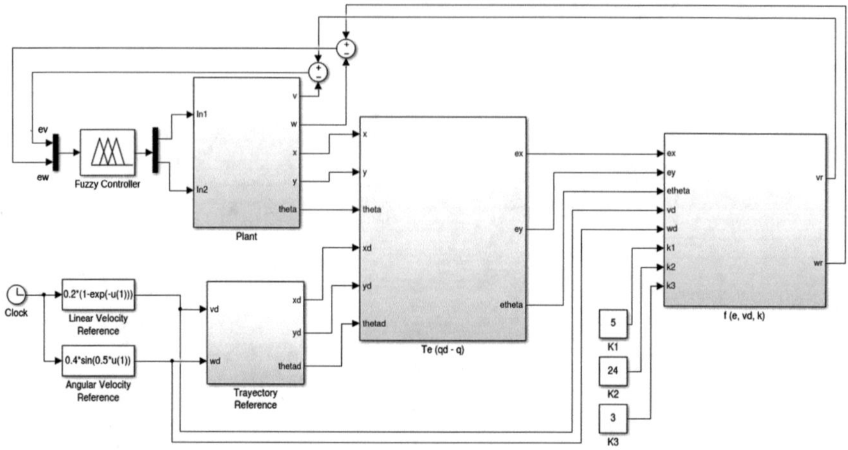

Fig. 7 Representation in model of type-1 fuzzy logic controller of the autonomous mobile robot

$$\mathrm{IAE} = \int_0^\infty |e(t)|\mathrm{d}t \tag{6}$$

$$\mathrm{ITSE} = \int_0^\infty e^2(t)t\mathrm{d}t \tag{7}$$

$$\mathrm{ITAE} = \int_0^\infty |e(t)|t\mathrm{d}t \tag{8}$$

$$\varepsilon = \sqrt{\frac{1}{N}\sum_{t=1}^{N}\left(X_t - \hat{X}_t\right)^2} \tag{9}$$

3.2 Water Tank Controller

The second problem is aimed at controlling the water level in a tank. Therefore, based on the actual water level in the tank, the controller has to be able to provide the proper activation of the valve. To evaluate the valve opening in a precise way, we rely on fuzzy logic, which is implemented as a fuzzy controller that performs the control on the valve that determines how fast the water enters the tank to maintain the level of water in a better way [1]. The process of filling the water tank is

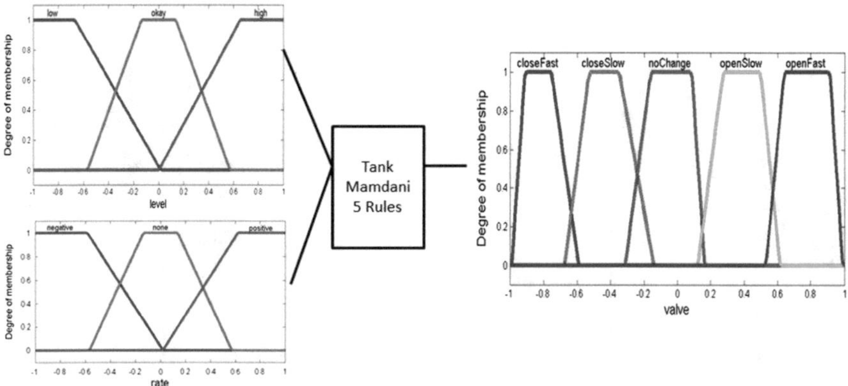

Fig. 8 Characteristics of the type-1 fuzzy logic controller

presented as a differential equation for the height of the water in the tank, H, and is given by Eq. (10).

$$\frac{\mathrm{d}}{\mathrm{d}t}\mathrm{Vol} = A\frac{\mathrm{d}H}{\mathrm{d}t} = bV - a\sqrt{H} \tag{10}$$

where **Vol** is the volume of water in the tank, **A** is the cross-sectional area of the tank, **b** is a constant related to the flow rate into the tank, and **a** is a constant related to the flow rate out of the tank. The equation describes the height of water, H, as a function of time, due to the difference between flow rates into and out of the tank. The MFs are for the two inputs to the fuzzy system: the first is called **level,** which has three trapezoidal MF with linguistic values of *low*, *okay* and *high*. The second input variable is called **rate** with three MFs corresponding to the linguistic values of *negative*, *none* and *positive*. The T1FLC has an output called **valve**, which is composed of five trapezoidal MFs with the following linguistic values: *closefast*, *closeslow*, *nochange*, *openslow* and *openfast*, and we show in Fig. 8 representations of the input and output variables. Figure 9 represents the T1FLC in the model.

All the combinations of rules were taken from experimental knowledge according to how the process is performed in a tank filled with water, which is detailed in Table 2.

4 Bee Colony Optimization Algorithm

The population of agents (artificial bees) consisting of bees collaboratively searches for the optimal solution. Every artificial bee generates one solution to the problem. The BCO is a population-based algorithm. Population of *artificial bees* searches for the optimal solution. *Artificial bees* represent agents, which collaboratively solve

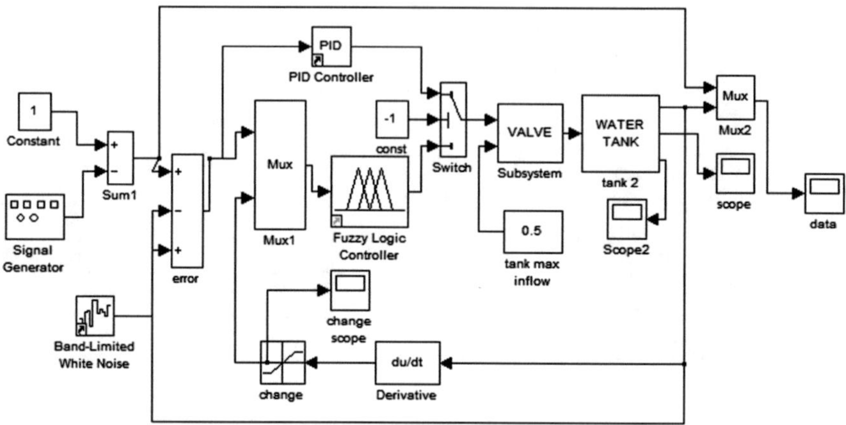

Fig. 9 Block diagram for the simulation of the type-1 fuzzy logic controller

Table 2 Rules for the second studied case

# rules	Input 1 level	Input 2 rate	Output valve
1	Okay	–	Nochange
2	Low	–	Openfast
3	High	–	Closefast
4	Okay	Positive	Closeslow
5	Okay	Negative	Openslow

complex combinatorial optimization problems. The foraging behavior in a bee colony remained mysterious for many years until Von Frisch translated the language embedded in a bee waggle dance [21]. The waggle dance operates as a communication tool among bees. Suppose a bee found a rich food source. Upon its return to the hive, it starts to dance in a figure-eight pattern. This figure-eight dance consists of a straight waggle run, followed by a turn to the right, back to the starting point, and then another straight waggle point again [4, 10].

The general steps and flowchart of BCO are indicated in Fig. 10, where "ScoutBees (n)" indicates the size of the population, "NC" represents the number of constructive moves during one forward pass and the "FollowerBees (m)" represents each bee exploring the possible solutions.

The dynamics of BCO are defined in Eqs. (11)–(14):

$$P_{ij,n} = \frac{[\rho_{ij,n}]^{\alpha} \cdot [\frac{1}{d_{ij}}]^{\beta}}{\sum\limits_{j \in A_{i,n}} [\rho_{ij,n}]^{\alpha} \cdot [\frac{1}{d_{ij}}]^{\beta}} \quad (11)$$

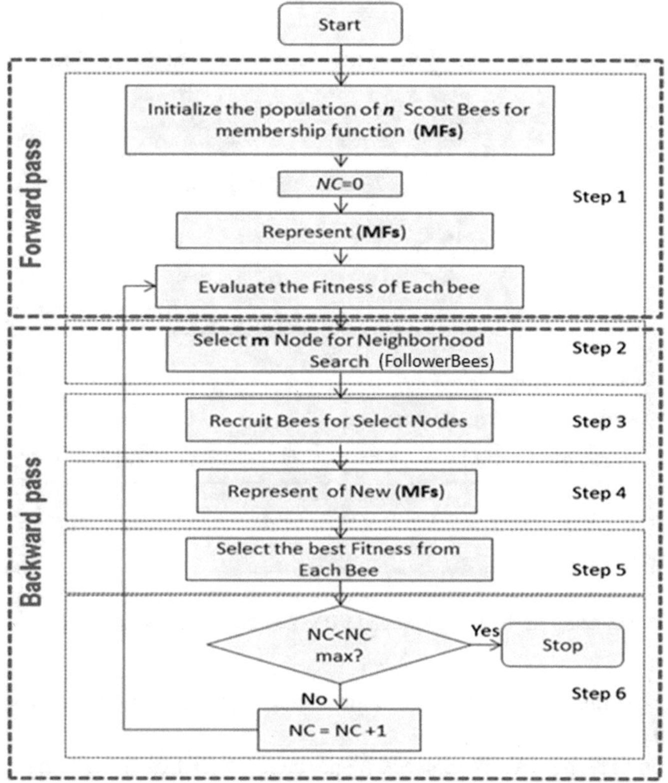

Fig. 10 Flowchart of the Bee Colony Optimization algorithm

$$D_i = K.\frac{Pf_i}{Pf_{\text{colony}}} \tag{12}$$

$$Pf_i = \frac{1}{L_I}, L_i = \text{Tour Length} \tag{13}$$

$$Pf_{\text{colony}} = \frac{1}{N_{\text{Bee}}}\sum_{i=1}^{N_{\text{Bee}}} Pf_i \tag{14}$$

Equation (11) represents the probability of a bee, k, located on a node, i, selecting the next node, denoted by j. β is the probability of visiting the following node. Note that the ρ_{ij} is inversely proportional to the city distance. d_{ij} represents the distance of node i until node j, for this algorithm indicates the total the dance that a bee have in this moment. Finally $\propto$ is a binary variable that is used for to find better solutions in the algorithm. Alpha and beta are variables that determine the

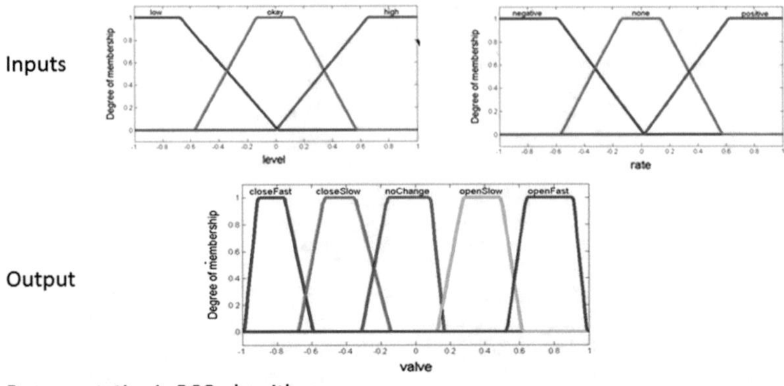

Representation in BCO algorithm

Input 1 (level)			Input 2 (rate)			Output (rate)				
Low	Okay	High	Negative	none	positive	CloseFast	CloseSlow	NoChange	OpenSlow	OpenFast
1 2 3 4	5 6 7 8	....	....	....	....	....	....	....	....	 44

Bee1 = [Pos1, Pos2...,...,...,...,...,..Posn]

Fig. 11 Representation of the fuzzy system in Bee Colony Optimization

heuristics of the algorithm. The alpha and beta values were 0.5 and 2.5, respectively.

Equation (12) represents that a waggle dance will last for a certain duration, determined by a linear function, where K denotes the waggle dance scaling factor [4], Pf_i denotes the profitability scores of bee i as defined in Eq. (13) and Pf_{colony} denotes the bee colony's average profitability, as in Eq. (14) and is updated after each bee completes its tour. For this research, the waggle dance is represented for the mean square error (MSE) that the model to find once that is done the simulation in the iteration of the algorithm [4, 10, 21]. Figure 11 indicates the representation of the BCO in the design of the T1FLC for the second studied case, which is the water tank controller.

The fitness function in the BCO algorithm is represented by the Mean Square Error (MSE) shown in Eq. (15). For each follower bee for N Cycles, the type-1 FLS design for the BCO algorithm is evaluated and the objective is to minimize the error.

$$\text{MSE} = \frac{1}{n}\sum_{i=1}^{n}(\bar{Y}_i - Y_i)^2 \tag{15}$$

In the BCO algorithm, a bee represents the values of the distribution of each of the MFs. Table 3 represents the size for each type of the MFs optimized by BCO.

Table 3 Size of the parameters of each membership function (MF) optimized by Bee Colony Optimization

Studied case	Type of MF	Size of the Bee
Autonomous mobile robot	Trapezoidal	48
	Gaussian	24
	Triangular	36
Water tank controller	Trapezoidal	44
	Gaussian	22
	Triangular	33

5 Simulation Results

Experimentation was performed with perturbations. We used the specific noise generators with band-limited noise with a value of 0.5 and delay of 1000. The configuration of the experiment was with the following parameters; *Population* of 50, *Follower Bees* of 30, *MaxCycle* of 20, *Alpha* and *Beta* of 0.5 and 2.5 for the BCO algorithm. Simulation results in Table 4 show the best experiment without perturbation in the fuzzy logic controller for each type of MF optimized by the BCO algorithm.

Table 4 shows that when the triangular MF is used, stability in the model is presented with a simulation error of **0.016**.

Simulation results in Table 5 show the best experiment with perturbation in the T1FLC for each type of MF optimized by the BCO algorithm.

Table 5 shows that when we used perturbation in the model with triangular FM, we continue with minimal errors with a simulation error of **0.155** and an average of **3.650**, which is an improvement compared with Gaussian and trapezoidal MFs.

For the second studied case, the simulations results without perturbation are shown in Table 6. Table 7 shows simulation results with perturbation in the model.

Table 6 shows that when the Gaussian MF is used, stability in the model is presented with a simulation error of **0.015**.

Simulation results in Table 7 show the best experiment with perturbation in the T1FLC for each type of MF optimized by the BCO algorithm.

Table 4 Average of 30 experiments for the autonomous mobile robot controller without perturbation

Index	Type of membership functions		
	Gaussian	Trapezoidal	Triangular
ISE	15.188	12.325	12.862
ITSE	603.833	476.002	504.023
IAE	32.614	30.285	31.575
ITAE	1295.375	1175.850	1241.770
MSE	6.453	11.430	1.351
RMSE	3.815	5.169	1.997
σ	5.873	13.840	1.999
Best	0.052	0.058	**0.016**
Worst	19.880	41.687	8.995

Table 5 Average of 30 experiments for the autonomous mobile robot controller with perturbation

Index	Type of membership functions		
	Gaussian	Trapezoidal	Triangular
ISE	16.352	13.609	13.547
ITSE	639.86	525.579	534.799
IAE	33.437	32.122	32.269
ITAE	1318.428	1270.120	1281.450
MSE	9.741	11.628	3.650
RMSE	3.832	5.755	2.611
σ	11.119	9.396	5.093
Best	0.340	0.762	**0.155**
Worst	39.725	37.554	25.963

Table 6 Average of 30 experiments for the water tank controller

Index	Type of membership functions		
	Gaussian	Trapezoidal	Triangular
ISE	4.364	3.176	4.117
ITSE	181.594	157.177	204.326
IAE	13.135	7.786	9.593
ITAE	529.943	397.513	494.95
MSE	**0.050**	0.052	0.058
RMSE	0.0236	0.0235	0.2489
σ	0.015	0.008	0.015
Best	**0.015**	0.034	0.027
Worst	0.081	0.068	0.093

Table 7 Average of 30 experiments for the water tank controller with perturbation

Index	Type of membership functions		
	Gaussian	Trapezoidal	Triangular
ISE	15.931	11.962	11.012
ITSE	636.377	477.761	441.413
IAE	32.605	18.873	18.186
ITAE	1300.274	753.602	726.281
MSE	0.069	0.013	0.014
RMSE	0.285	0.120	0.116
σ	0.029	0.003	0.007
Best	0.009	0.007	**0.005**
Worst	0.124	0.019	0.033

Table 7 shows that when perturbation was used in the model, Triangular MF obtained best results with a simulation error of **0.005** and an average of **0.014**, which was better compared with Gaussian and Trapezoidal MFs. It is important to mention that Trapezoidal MF obtained better standard deviation with a value of **0.003**, which indicates that the results of the 30 experiments are similar.

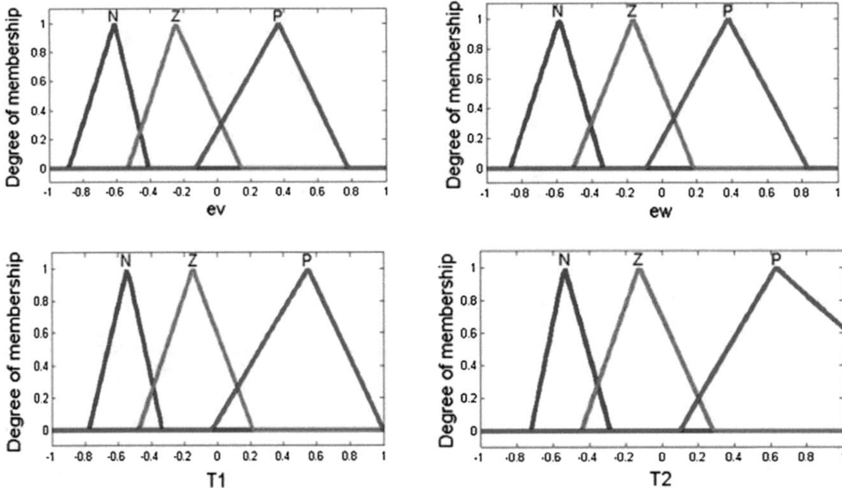

Fig. 12 Best type-1 fuzzy logic controller with triangular membership function of the autonomous mobile robot

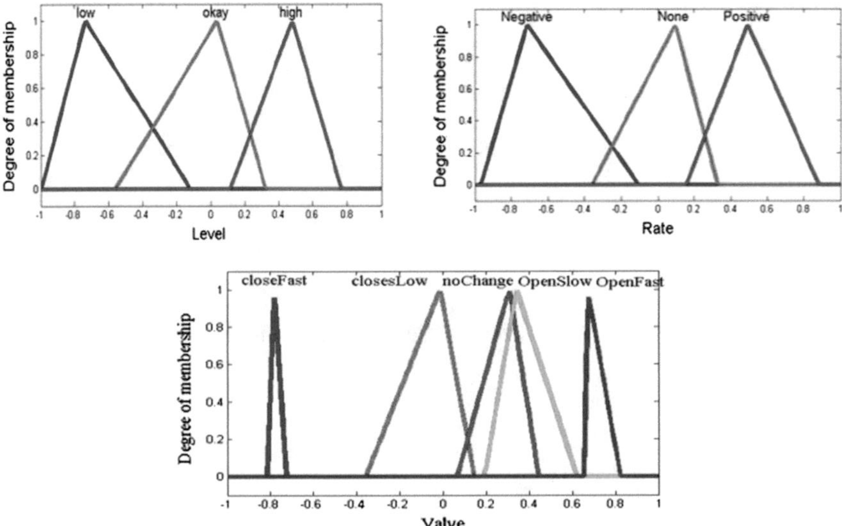

Fig. 13 Best type-1 fuzzy logic controller with trapezoidal membership function of the water tank controller

In both case studies, the best results were obtained using triangular MF. Figures 12 and 13 show the best design of type-1 FLS found by the BCO algorithm.

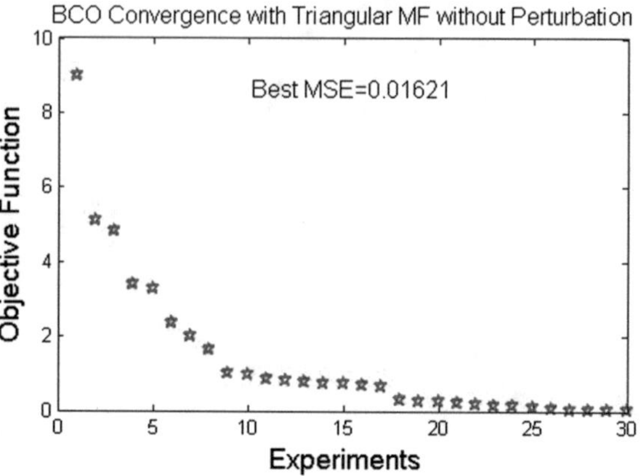

Fig. 14 Behavior of the best simulation results of the autonomous mobile robot

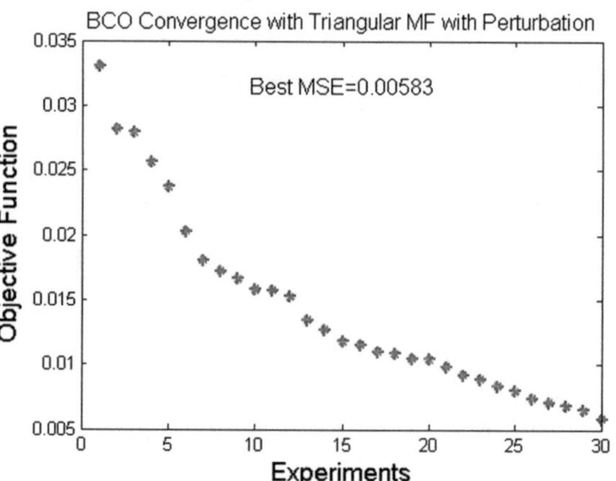

Fig. 15 Behavior of the best simulation results of the water tank controller

The behavior and the stabilization of the BCO algorithm in the model are shown in Figs. 14 and 15.

6 Comparative Analysis

The results mentioned in the previous section show that triangular MFs better allow for stability in the model for the two case studies analyzed in this work.

Figure 16 shows a comparative results of the average of 30 experiments for the first studied case without perturbation in the model, showing that with the use of triangular MFs the simulation error found by BCO is better, the MSE was **0.016** compared with **0.058** with Gaussian MFs and **0.053** with trapezoidal MFs.

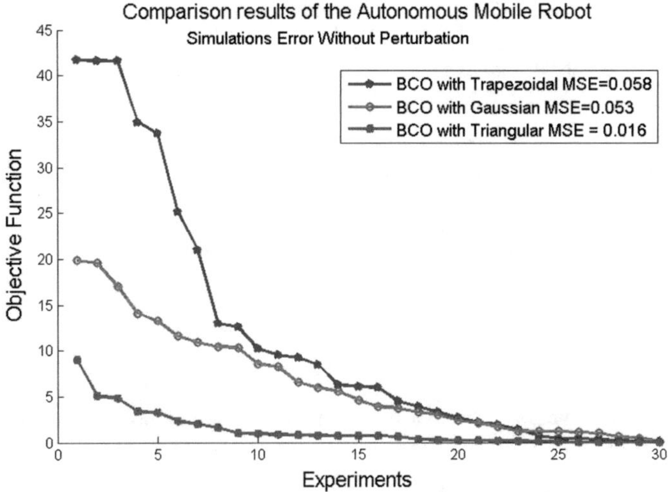

Fig. 16 Comparative results of the autonomous mobile robot without perturbation

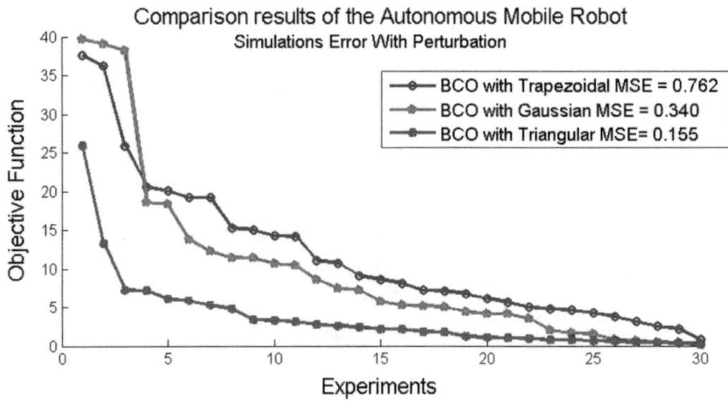

Fig. 17 Comparative results of the autonomous mobile robot with perturbation

Figure 17 shows comparative results with the applied level of noise in the model; it can be seen that triangular MF presents the best errors.

The behavior of the BCO with the best Gaussian type-1 FLS, trapezoidal type-1 FLS and triangular type-1 FLS are presented in Figs. 18, 19 and 20 for the mobile robot controller.

Simulation results for the second studied case are presented in Figs. 21 and 22. Figure 21 shows results without perturbation in the model; the Gaussian MF found better results compared with triangular and trapezoidal MF. When levels of noise are implemented in the model, stabilization is best seen with triangular and trapezoidal MF.

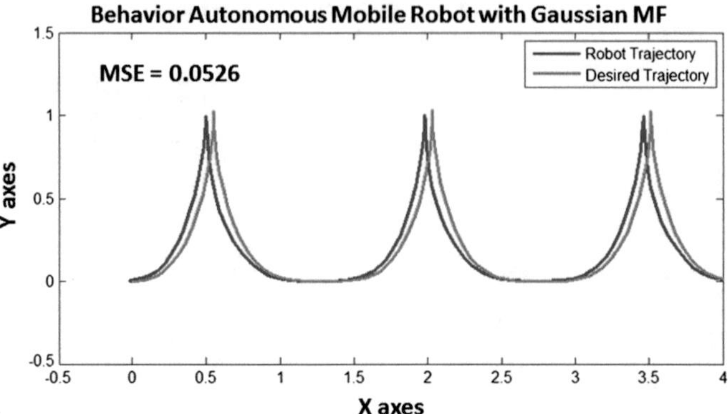

Fig. 18 Behavior of the Bee Colony Optimization optimized with Gaussian membership function

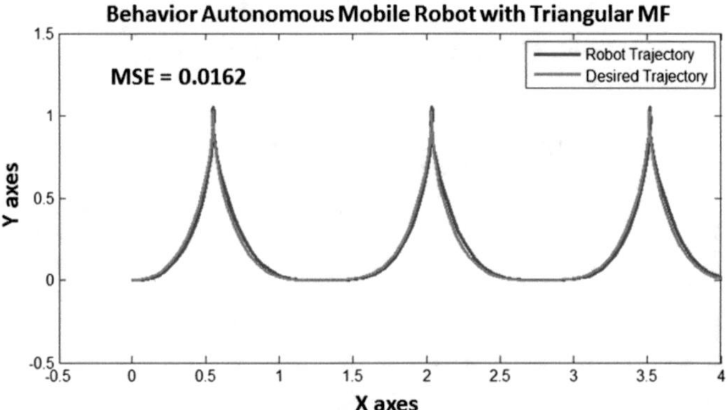

Fig. 19 Behavior of the Bee Colony Optimization optimized with triangular membership function

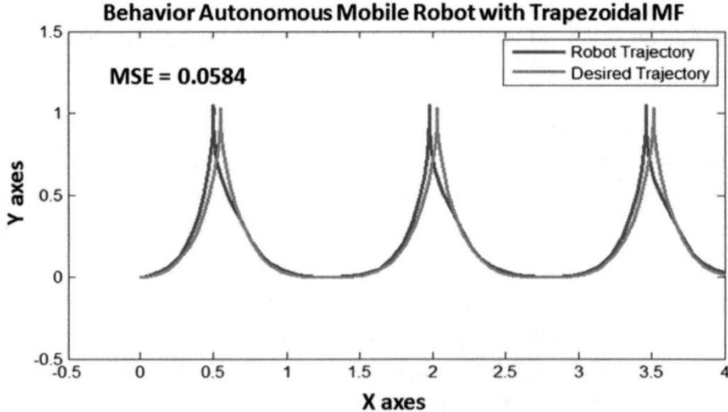

Fig. 20 Behavior of the Bee Colony Optimization optimized with trapezoidal membership function

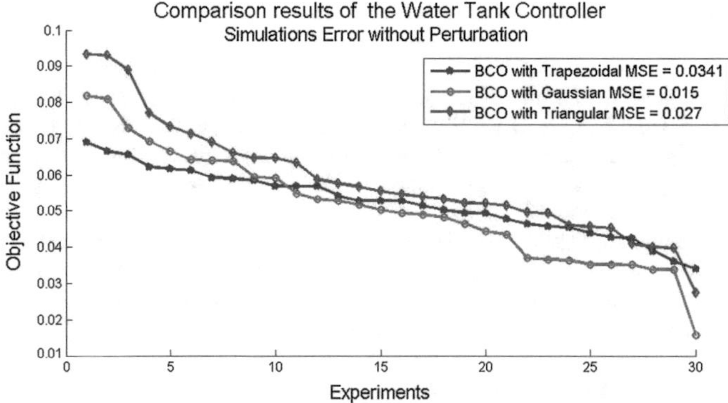

Fig. 21 Comparative results of the water tank controller without perturbation

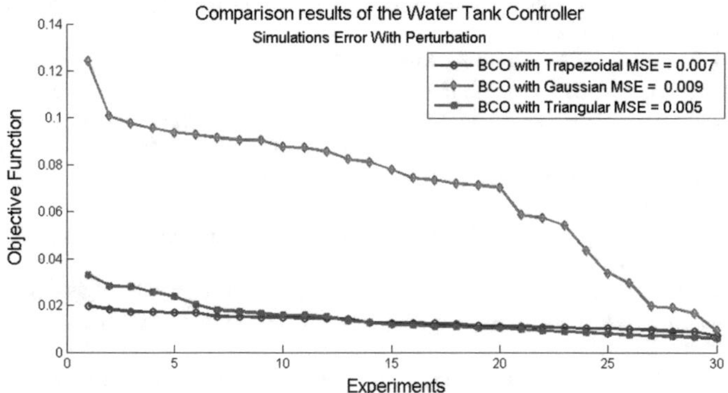

Fig. 22 Comparative results of the water tank controller with perturbation

7 Conclusions and Future Work

With the results obtained, we can conclude that the BCO algorithm is a good technique for the optimization for the design of fuzzy controllers applied to several types of MFs. Triangular MFs presented the better simulation errors in the model. BCO shows that it can obtain a better performance when dealing with external perturbation in the model.

Future work will include the optimization of BCO for an interval type-2 FLS with the goal of obtaining more precision in the simulations in the model, as well as better handling the uncertainty that is a characteristic of interval type-2 FLS.

References

1. Amador-Angulo, O. Castillo, "Comparison of Fuzzy Controllers for the Water Tank with Type-1 and Type-2 Fuzzy Logic", NAFIPS 2013, Edmonton, Canada, pp. 1 -6, 2013.
2. Amador-Angulo L., Castillo O., "*A Fuzzy Bee Colony Optimization Algorithm Using an Interval Type-2 Fuzzy Logic System for Trajectory Control of a Mobile Robot*". MICAI pp. 460-471, 2015.
3. Amador-Angulo A., Castillo O.: "Statistical Analysis of Type-1 and Interval Type-2 Fuzzy Logic in dynamic parameter adaptation of the BCO". IFSA-EUSFLAT, pp. 776-783, 2015.
4. Biesmeijer, J. C. and Seeley T. D., "*The use of waggle dance information by honey bees throughout their foraging careers*", Behavioral Ecology and Sociobiology, Vol. 59, Vo. 1, pp. 133-142, 2005.
5. Bonabeau E., Dorigo M., Theraulaz G., "*Swarm Intelligence*". Oxford University Press, Oxford, 1997.
6. Brown S.C. and Passino K. M., "Intelligence Control for an Acrobat", Journal of Intelligence and Robotic Systems, pp. 209-248, 1997.
7. Caraveo C., Valdez F. and Castillo O., "Optimization of fuzzy controller design using a new bee colony algorithm with fuzzy dynamic parameter adaptation", Applied Soft Computing, Vol. 43, pp. 131-142, 2016.
8. Castillo O., Martinez-Marroquin R., Melin P., Valdez P, and Soria J.: "Comparative study of bio-inspired algorithms applied to the optimization of type-1 and type-2 fuzzy controllers for an autonomous mobile robot", Information Sciences, Vol. 192, No. 1, pp. 19-38, 2010.
9. Chong C. S., Low M. Y. H., Sivakumar A. I., and Gay K. L.: "*A bee colony optimization algorithm to job shop scheduling*", in Proceedings of the 2006 Winter Simulation Conference, pp. 13-25, 2006.
10. Dyler F. C.,"*The biology of the dance language*", Annual Review of Entomology, Vol. 47, pp. 917-949, 2002.
11. Karakuzu C., "Parameters tuning of fuzzy sliding mode controller using particle swarm optimization", International Journal Innovative Computing Information Control, Vol. 6, pp. 4755-4770, 2010.
12. Lučić P. and Teodorović D., "*Computing with Bees: Attacking Complex Transportation Engineering Problems*", International Journal on Artificial Intelligence Tools, Vol. 12, No. 3, pp. 2003.
13. Lučić, P. and Teodorović, D., "*Vehicle routing problem with uncertain demand at nodes: the bee system and fuzzy logic approach*". In Verdegay, J.L. (Ed.): Fuzzy Sets in Optimization Springer-Verlag. Heidelberg, Berlin, pp. 67-82, 2003.
14. Lučić, P. and Teodorović, D., "*Transportation modeling: an artificial life approach*", In: Proceedings of the 14th IEEE International Conference on Tools with Artificial Intelligence, Washington, DC, pp. 216-223, 2002.
15. Mamdani, E.H. and Assilian S.: "*An experiment in linguistic synthesis with fuzzy logic controller"*. International Journal of Man-Machine Studies, pp. 1-13, 1975.
16. Pham, D. T., Darwish A. Haj., Eldukhri E. E. and Otri, S., "*Using the Bees Algorithm to tune a fuzzy logic controller for a robot gymnast*". *Innovative Production Machines and Systems*, online, pp. 1-2, 2007.
17. Teodorović D, "*Swarm Intelligence Systems for Transportation Engineering: Principles and Applications*", Transp. Res. Pt. C-Emerg. Technol., Vol. 116, pp. 651-782, 2008.
18. Teodorović, D., Lućić P., Marcković G., Orco, M. Dell', "*Bee colony optimization: principles and applications*". In: Reljin, B. Stanković, S. (Eds.): Proceedings of the Eight Seminar on Neural Network Applications in Electrical Engineering- NEUREL, University of Belgrade, Belgrade pp. 151-156, 2006.
19. Teodorović D., Selmić M., "*The BCO algorithm For The p Median Problem*", In: Proceedings of the XXXIV Serbian Operations Research Conference, Zlatibor, Serbia, 2007.

20. Tiacharoen S. and Chatchanayuenyong T.: "*Design and Development of an Intelligent Control by Using Bee Colony Optimization Technique*", American Journal of Applied Sciences, Vol. 9, No. 9, pp. 1464-1471, 2012.
21. Von Frisch K.,"*Decoding the language of the bee*", Science, Vol. 185, No. 4152, pp. 663-668, 1974.
22. Wong, L. P., Low, M. Y. H. and Chong, C. S., "*A bee colony optimization algorithm for traveling salesman problem*", in proceedings of Second Asia International Conference on Modeling & Simulation (AMS), pp. 818-823, 2008.
23. Yen, J.,Langari R.: Fuzzy Logic: Intelligence, Control and Information. Prentice Hall, 1999.
24. Zadeh, L.A., "Fuzzy sets", Information Control, Vol.8, pp. 338-353, 1965.
25. Zadeh, L. A.: The concept of a linguistic variable and its application to approximate reasoning, Part I. Information Sciences 8, pp. 199-249, 1975.
26. Zadeh, L. A.: The concept of a Linguistic Variable and its Application to Approximate Reasoning, Part II, Information Sciences 8, pp. 301-357, 1975.
27. Zadeh, L.A.: Toward a theory of fuzzy information granulation and its centrality in human reasoning and fuzzy logic, Fuzzy Sets and Systems, Vol. 90, Elsevier, pp. 117-117, 1997.

Neuro-Fuzzy Hybrid Model for the Diagnosis of Blood Pressure

Juan Carlos Guzmán, Patricia Melin and German Prado-Arechiga

Abstract We propose a neuro-fuzzy hybrid model for the diagnosis of blood pressure to provide a diagnosis as accurate as possible based on intelligent computing techniques, such as neural networks and fuzzy logic. The neuro-fuzzy model uses a modular architecture which works with different number of layers and different learning parameters so that we can have a more accurate modeling. So for the better diagnosis and treatment of hypertension patients, an intelligent and accurate system is needed. In this study, we also design a fuzzy expert system to diagnose blood pressure for different patients. The fuzzy expert system is based on a set of inputs and rules. The input variables for this system are the systolic and diastolic pressures and the output variable is the blood pressures level. It is expected that this proposed neuro-fuzzy hybrid model can provide a faster, cheaper, and more accurate result.

Keywords Fuzzy system · Blood pressure · Diagnosis

1 Introduction

Nowadays different techniques of artificial intelligence, such as fuzzy systems, neural networks, and evolutionary computation, are used in the areas of medicine. One of the most important problems in medicine is hypertension diagnosis. Hypertension is one of the most dangerous diseases that seriously threaten the health of people around the world. This type of disease often leads to fatal results, such as

J.C. Guzmán · P. Melin (✉)
Tijuana Institute of Technology, Tijuana, B.C., Mexico
e-mail: pmelin@tectijuana.mx

G. Prado-Arechiga
Excel Medical Center, Tijuana, B.C., Mexico

P. Melin et al. (eds.), *Nature-Inspired Design of Hybrid Intelligent Systems*, Studies in Computational Intelligence 667, DOI 10.1007/978-3-319-47054-2_37

- Heart attack,
- Cerebrovascular accident,
- Renal insufficiency.

One of the most dangerous aspects of hypertension is that people perhaps may not know they have it. In fact, nearly a third of the people with high blood pressure do not know. The only way to know if the blood pressure is high is through regular checkups.

Today in Medicine various modeling approaches have been applied to diagnose some future illness of patients and to treat them in time. This is why the implementation of an appropriate method of modeling has always been a major issue in Medicine, because it helps to deal with diseases in time and save lives. The modeling of future data for medicine provides the ability to make decisions in the medium and long term due to the accuracy or inaccuracy of modeled data and provide better control in the health of any patient.

For a doctor, it is important to know the future behavior of blood pressure for a patient and to make correct decisions that will improve the patient's health and avoid unwanted situations that in the worst case can lead to early death of a person for not having a proper treatment to control blood pressure.

Therefore in this work, blood pressure monitoring tests were performed to 20 people for 6 days with 4 daily intakes at the same time, with different activities in their daily lives to have data with different settings, so we can reach a more concrete conclusion on the fuzzy system when making an accurate diagnosis based on information previously obtained through daily neuro monitoring system. Experiments with different neural network architectures are performed, and thus it was determined, which provided the optimal behavior for prognosis.

The idea of using modular neural networks was to divide the information of the systolic and diastolic pressure, as they are complementary to provide a result, and they need to be analyzed separately to yield a better prognosis and thereby obtain better results. Designing an appropriate modular neural network to model the behavior of the blood pressure, using time series, it is with the aim that we can provide results to help decision-making necessary in the future regarding the diagnosis of blood pressure.

The paper is organized as follows: in Sect. 2 a methodology of hypertension is presented, in Sect. 3 the Development and final design of the neural-fuzzy hybrid model is presented, in Sect. 4 the conclusion obtained after tests the neuro-fuzzy hybrid model for blood pressure diagnosis is presented.

In the current literature, there are papers that try to achieve a good diagnosis of blood pressure, like in Hypertension Diagnosis Using Fuzzy Expert System [8]. In this study, a fuzzy expert system to diagnose hypertension for different patients was designed. In the paper Genetic Neuro Fuzzy System for Hypertension Diagnosis [7], the authors propose and evaluate genetic neuro fuzzy system for diagnosing hypertension risk systolic and diastolic blood pressure, body mass index, heart rate, cholesterol, glucose, blood urea, creatinine and uric acid have been taken as inputs to the system. In the paper Fuzzy Expert System for Fluid Management in General Anesthesia [12], anesthetists use rules of thumb for managing patients. In the article

A Neuro-Fuzzy Approach to FMOLP Problems [6], fuzzy neural networks for finding a good compromise solution to fuzzy multiple objective linear programs are proposed. In the paper Fuzzy Expert System for the Management of Hypertension [5], the authors focused on the use of information and communication technology (ICT) to design a web-based fuzzy expert system for the management of hypertension using the fuzzy logic approach. In the study Fuzzy Medical Diagnosis [9], often on the borderline between science and art, is an excellent exponent: vagueness, linguistic uncertainty, hesitation, measurement imprecision, natural diversity, subjectivity all these are prominently present in medical diagnosis. In the work entitled Can Fuzzy Logic Make Things More Clear [10], the authors propose that in clinical decision support and artificial intelligence using fuzzy logic and closed-loop techniques are methods that might help us to handle this complexity in a safe, effective and efficient way. In the work entitled Neuro-Fuzzy and Soft Computing —A Computational Approach to Learning and Machine Intelligence—JSR Jang et al. [11], in consistent notation, all the information on computational intelligence (CI), such as neural networks (NN), fuzzy logic (FL), and genetic algorithms (GA) can be found collected in one place,. In the work Experimental Study of Intelligent Computer Aided Diagnostic and Therapy [2], the limitations of the conventional methods for the diagnosis of diseases have been highlighted. In the paper Design and Development of Fuzzy Expert System for Diagnosis of Hypertension [3], the aim is to design a fuzzy expert system (FES) for diagnosis of hypertension risk for patients aged between 20, 30 and 40 years and is divided into male and female genders. In the paper A Note on Hypertension Classification Scheme and Soft Computing Decision Making System [13], a soft computing diagnostic support system for the risk assessment of hypertension is proposed. In another work the authors propose a pre-diagnosis of hypertension using artificial neural networks [14], which deals with artificial neural networks solving the problems of diagnosing hypertension using backpropagation learning algorithm. The network is constructed based on various factors, which are classified into some categories, to be trained tested and validated using the respective data sets.

2 Methodology

2.1 Blood Pressure

Blood pressure is needed to deliver oxygen and nutrients to body organs. In the human body, the blood circulates through the blood vessels. They are mainly arteries and veins. The blood flowing through the vessels constantly exerts pressure on the vessel walls. The pressure is determined by the heart's pumping strength and elasticity of the vessels.

In general, the heart contracts and expands again, on average, 60–80 times per minute. This pressure pumps blood into the arteries to deliver oxygen and nutrients to body organs. The blood vessels branch more and more to become capillary blood vessels (capillaries). This offers more or less resistance to blood stream, if you have enough pressure.

The pressure is the highest at the time of the heartbeat, when the heart contracts. This pressure is known as systolic blood pressure. The contraction phase of the heart which increases blood pressure is called systolic. Blood pressure is low between two heartbeats, that is, when the heart muscle relaxes. Blood pressure at this point is called diastolic blood pressure. The phase in which the heart relaxes and blood pressure decreases is called diastole.

Blood pressure is measured in mmHg. For example, 120/80 mmHg means that the systolic blood pressure is 120 mmHg and diastolic blood pressure of 80.

2.2 *Type of Blood Pressure Diseases*

Hypertension is the most common disease and it markedly increases both morbidity and mortality from cardiovascular and many other diseases. Different types of hypertension are observed when the disease is sub-categorized. These types are shown in Table 1.

2.3 *Low Blood Pressure (Hypotension)*

Unlike hypertension, hypotension is not life threatening and does not cause other potentially serious diseases. It helps to protect against many cardiovascular diseases, such as heart attack or stroke.

Table 1 Definitions and classification of the blood pressure levels (mmHg)[a]

Category	Systolic		Diastolic
Hypotension	<90	and/or	<60
Optimal	<120	and	<80
Normal	120–129	and/or	80–84
High normal	130–139	and/or	85–89
Grade 1 hypertension	140–159	and/or	90–99
Grade 2 hypertension	160–179	and/or	100–109
Grade 3 hypertension	≥180	and/or	≥110
Isolated systolic hypertension	≥140	and	<90

[a]The blood pressure (BP) category is defined by the highest BP level, whether systolic or diastolic. Isolated systolic hypertension should be graded as 1, 2, or 3 according to the systolic BP value in the indicated ranges

However, people with low blood pressure (hypotension) may also exhibit symptoms, which can make them suffer sometimes of dizziness; impaired concentration and fatigue are also possible symptoms. In addition, mental performance can be affected. Healthy people who have low blood pressure may have trouble concentrating and can be reacting more slowly.

The so-called primary hypotension (essential) is the most common form of low blood pressure and is not classified as a disease. It occurs mainly in young women.

Basically, low blood pressure can be considered as a simple measured value and not a disease. The World Health Organization (WHO) has defined as less than 100/60 mmHg in women and less than 110/70 mmHg in men as low blood pressure (hypotension). However, the appearance of these symptoms with these values depends on the individuals. Particularly sensitive individuals may also experience dizziness and lightheadedness with higher values.

2.4 High Blood Pressure (Hypertension)

Blood pressure increases with the increasing pumping power of the heart or when blood vessels contract. High blood pressure (hypertension) is a disease of the cardiovascular system. Increased blood pressure is widespread, particularly in industrialized countries.

The risk of hypertension increases with age. However, hypertension can also be suffered by young people. Hypertension can also be caused by hormones, such as adrenaline and noradrenaline, but also because of kidney disease or drugs.

However, in 95 % of the cases, the hypertension has no obvious organic cause. Physical inactivity, obesity, excessive alcohol or salt, and stress are the most common causes of hypertension; initially, hypertension has no symptoms. Often, people affected do not perceive it. More than half of those affected do not know they are part of the group of hypertensive patients. This is dangerous, since a permanently high blood pressure increases the risk of damage to vital organs such as heart, brain, kidneys, and eyes. Some possible consequences are myocardial infarction, heart failure, stroke, kidney failure, and vision loss.

2.5 Risk Factors

Some of the primary risk factors for hypertension include the following [1]:

- Obesity
- Lack of exercise
- Smoking
- Consumption of salt
- Consumption of alcohol

- Stress level
- Age
- Sex
- Genetic factors

3 Development and Final Design of the Neuro-Fuzzy Hybrid Model

In Fig. 1, the neuro-fuzzy hybrid model in which we have as input the values of the blood pressure of a person, which are divided into systolic and diastolic pressure, and are for 24 samples. These values enter the neural network as two inputs, the first is the systolic and the diastolic is the second, once they enter the network and the process of learning and prediction is performed two outputs are obtained, which are introduced as inputs to the fuzzy system are obtained. Once the systolic and diastolic blood pressures are used in the fuzzy system, this in turn classifies the information and produces a final diagnosis, and this is finally sent to the GUI.

Finally once we have the fuzzy system and a well-defined architecture of the modular neural network, we can continue with the GUI in which we work combining the above mentioned techniques of intelligent computing.

Figure 2 shows the GUI of the final neuro-fuzzy hybrid model, where we are able to add the Excel file with the data of the patients, and this shows the information in a graphic as it is shown in Figs. 2 and 3. Figure 4 shows the modeling of

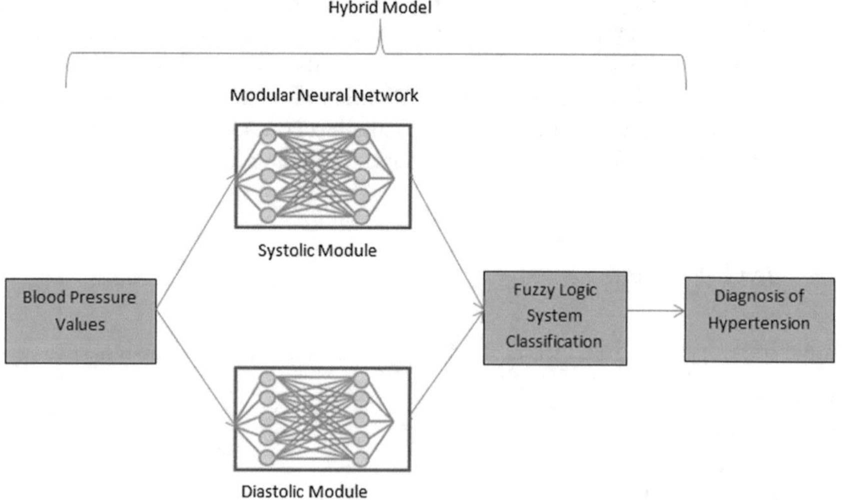

Fig. 1 Neuro-fuzzy hybrid model

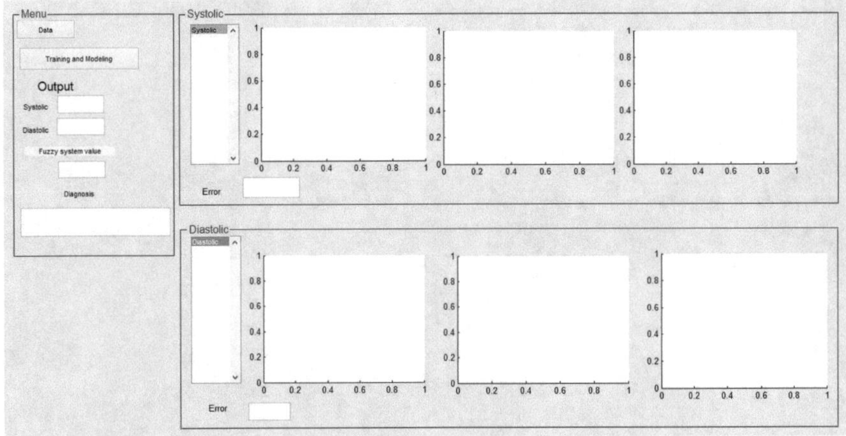

Fig. 2 Final GUI of the neuro-fuzzy hybrid model

the systolic and diastolic pressures and, finally, Fig. 5 shows the final results, such as the fuzzy system output values and diagnosis.

Figure 3 shows the graphic interface and select file window, which has the information to be simulated and used for modeling of blood pressure.

Figure 4 shows the graphic interface with the models of the systolic and diastolic pressures.

Figure 5 shows the graphic interface with the modeling of the monitoring of blood pressure of a patient, which is based on the provided information.

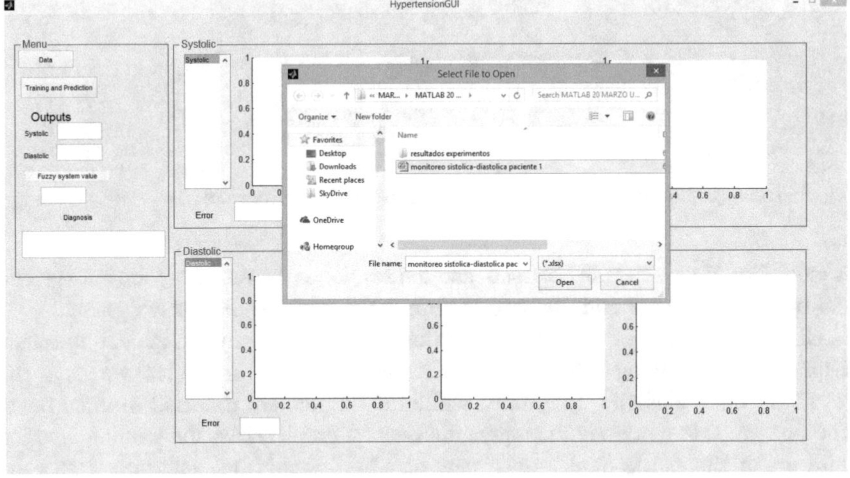

Fig. 3 Graphic interface and select file window

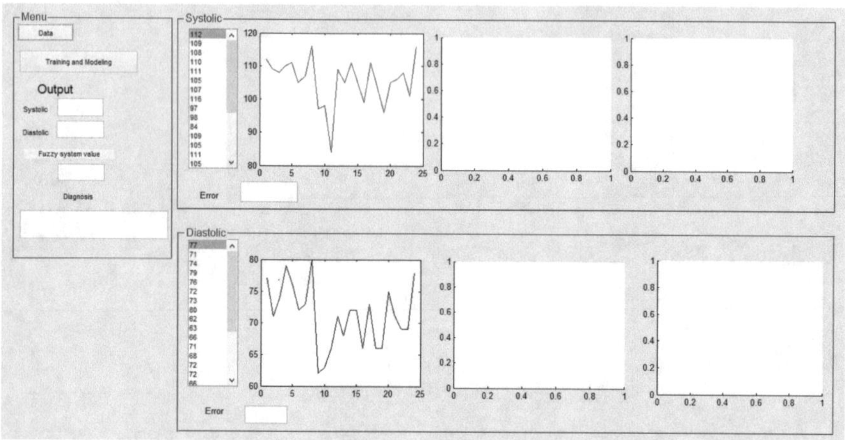

Fig. 4 Modeling the systolic and diastolic inputs to the modular neural network

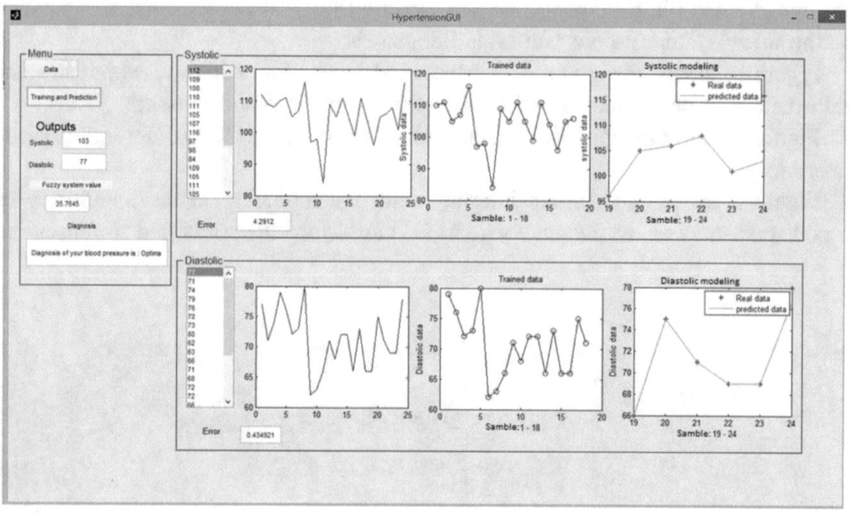

Fig. 5 Training, modeling, and result of the diagnosis of blood pressure

A modeling of new data is obtained, and the results of the modeling are the systolic and diastolic pressure and this information is the input to the fuzzy system.

Once the training process and modeling are done, the graphical interface displays the following information, as shown in Fig. 5, in which we have the "systolic" and "diastolic" pressures, while the outputs are modeled in three parts. The first graph is modeling the inputs, the second graph shows the training, and the third graph shows the modeling of new data and, finally, the interface shows the fuzzy output result and diagnosis.

4 Conclusions

This type of fuzzy systems actually implements the human intelligence and reasoning. Using a set of decision rules, we can provide different suggestions for diagnosing diseases, in this case hypertension. This is a very efficient, less time consuming and more accurate method to calculate the risk of hypertension. Finally, we can note that it is a very effective method for a diagnosis of hypertension, which can help a physician to achieve a better accuracy when giving a diagnosis to the patient.

Acknowledgments We would like to express our gratitude to the CONACYT and Tijuana Institute of Technology for the facilities and resources granted for the development of this research.

References

1. Abrishami, Z. and Tabatabaee, H. "Design of a Fuzzy Expert System and a Multi-layer Neural Network System for Diagnosis of Hypertension", MAGNT Research Report (ISSN. 1444-8939) Vol.2 (5). PP: 913-926, 2014.
2. Akinyokun, O.C. and Adeniji, O.A. 1991. "Experimental Study of Intelligent Computer Aided Diagnostic and Therapy". AMSE Journal of Modeling Simulation and Control. 27 (3):9-20.
3. Azamimi, A., Abdullah, and Zulkarnay, Z. and Mohammad Nur Farahiyah. "Design and development of Fuzzy Expert System for diagnosis of hypertension" International Conference on Intelligent Systems, Modelling and Simulation, IEEE 2011.
4. Das, S., Ghosh, P.K and Kar, S. "Hypertension diagnosis: a comparative study using fuzzy expert system and neuro fuzzy system."Fuzzy Systems, IEEE International Conference on. IEEE, 2013.
5. Djam,X.Y and Kimbi, Y.H."Fuzzy expert system for the management of hypertension." The Pacific Journal of Science and Technology, Volume 12. Number 1. May 2011 (Spring).
6. Fuller, R. and Giove, S. 1994. "A Neuro-Fuzzy Approach to FMOLP Problems". Proceedings of CIFT'94. Trento, Italy. 97-101.
7. Kaur, A. and Bhardwaj, A. "Genetic Neuro Fuzzy System for Hypertension Diagnosis", International Journal of Computer Science and Information Technologies, Vol. 5 (4), 2014, 4986-4989.
8. Kaur, R. and Kaur, A. "Hypertension Diagnosis Using Fuzzy Expert System", International Journal of Engineering Research and Applications (IJERA) ISSN: 2248-9622 National Conference on Advances in Engineering and Technology, AET- 29th March 2014.
9. Ludmila, I.K. and Steimann F. 2008. Fuzzy Medical Diagnosis. School of Mathematics, University of Wales: Bangor, UK.
10. Merouani, M., Guignard, B., Vincent, F., Borron, S.W., Karoubi, P., Fosse, J.P., Cohen, Y., Clec'h, C., Vicaut, E., Marbeuf-Gueye, C., Lapostolle, F., and Adnet, F. 2009. "Can Fuzzy Logic Make Things More Clear?". Critical Care. 13:116.
11. Rahim, F., Deshpande, A., and Hosseini, A.2007. "Fuzzy Expert System for Fluid Management in General Anesthesia". Journal of Clinical and Diagnostic Research. 256-267.
12. Srivastava, P. "A Note on Hypertension Classification Scheme and Soft Computing Decision Making System." ISRN Biomathematics 2013.

13. Sumathi, B. and Santhakumaran, A. "Pre-diagnosis of hypertension using artificial neural network." Global Journal of Computer Science and Technology, Volume 11 Issue 2 Version 1.0 February 2011.
14. Zadeh, L.A. 1965. "Fuzzy Sets and Systems". In Proceedings Symposium on System Theory. Fox, J. (editor). Polytechnic Institute of Brooklyn: New York, NY. April 1965. 29-37.

Microcalcification Detection in Mammograms Based on Fuzzy Logic and Cellular Automata

Yoshio Rubio, Oscar Montiel and Roberto Sepúlveda

Abstract In the early diagnosis of breast cancer, computer-aided diagnosis (CAD) systems help in the detection of abnormal tissue. Microcalcifications can be an early indication of breast cancer. This work describes the implementation of a new method for the detection of microcalcifications in mammographies. The images were obtained from the mini-MIAS database. In the proposed method, the images are preprocessed using an x and y gradient operators, the output of each filter is the input of a fuzzy system that will detect areas with high-tone variation. The next step consists of a cellular automaton that uses a set of local rules to eliminate noise and keep the pixels with higher probabilities of belonging to a microcalcification region. Comparative results are presented.

Keywords Breast cancer · Microcalcification · Mammography image Image enhancement · Fuzzy system · Cellular automata

1 Introduction

In 2012, the International Agency for Research on Cancer of the World Health Organization reported 1.67 million cases of breast cancer and an estimated of 552 thousand deaths. Breast cancer is the most frequent cancer in women and is the second most common cancer in the world and it is a major health problem in both developed countries and least developed regions [1]. Breast cancer develops from the different tissues of the breast.

Y. Rubio (✉) · O. Montiel · R. Sepúlveda
Instituto Politécnico Nacional, Centro de Investigación y Desarrollo de Tecnología Digital (CITEDI-IPN), Av. Del Parque no. 1310, Mesa de Otay, 22510 Tijuana, B.C., Mexico
e-mail: rrubio@citedi.mx

O. Montiel
e-mail: oross@ipn.mx

R. Sepúlveda
e-mail: rsepulvedac@ipn.mx

P. Melin et al. (eds.), *Nature-Inspired Design of Hybrid Intelligent Systems*, Studies in Computational Intelligence 667, DOI 10.1007/978-3-319-47054-2_38

In the United States, 234 thousand cases of breast cancer were recorded in 2015, with a mortality rate of 17 %. It represents the 29 % of the cancer-related cases, and it is the second deadliest type of cancer in the US, just below lung cancer [2]. In Mexico, of the 38,000 cancer-related deaths in women, 15.8 % are from breast cancer. The mortality rate of breast cancer cases in this country is around 30 % [3].

A mammography is a low-dose X-ray image of the breasts; it helps in the detection and diagnosis of breast cancer [4]. The computer-aided diagnosis (CAD) systems applied in breast cancer imaging are mainly focused on the detection of microcalcifications or abnormal masses [5]. As Oliver et al. [6] described CAD is a useful tool to detect positive cancer cases that the specialist does not detect at first glance.

Microcalcifications are calcium deposits that appear as small and white flecks on mammograms, their dimensions go from the 0.01–1 mm, with an average of 0.3 mm, making them easily confused with artifacts. Microcalcifications are brighter than the tissue around them, this variation of brightness helps to segment the microcalcification from the breast tissue, there is a high correlation between breast cancers and microcalcifications [4, 7, 8].

Because microcalcifications nature and dimensions are irregular, in some cases the contrast between them and the breast tissue is low, making the analysis harder in dense mammograms [9]. The drawback of the actual methods and algorithms for mammographic analysis is the high number of false positive results [8, 10]. There are a significant number of microcalcification clusters that are not spotted by the radiologist because the size of the microcalcification is too small or the contrast in the region that includes the calcification very low [8].

The development of breast cancer detection algorithms is not recent. Pereira et al. [11] propose a set of computational tools for segmentation of mammographies and detection of masses in mammograms. As they point out, modern detection algorithms have a high success rate for detecting microcalcifications, but with a high false positive rate. They developed an algorithm that uses artifacts elimination, noise reduction, and gray level enhancement based on a Wiener filter and Wavelet transform.

Abnormal breast mass detection also has been studied by Berber et al. [12]. In their research, they propose a new segmentation algorithm based on classical seed region growth that enhances the contour of masses of images that have been filtered, and then classifies the segmented regions.

1.1 Cellular Automata for Microcalcification Detection

At present, cellular automata (CA, singular *Cellular Automaton*) are applied to different tasks in medical imaging. Viher made one of the earliest research, Dobnikar and Zazula [13] use a two-dimensional CA for follicle recognition. Qadir et al. [14] developed a method to reduce noise in medical images using a CA with Moore neighborhood.

Wongthanavasu and Tangvoraphonkchai [15] use a CA to identify areas with suspected cancer cells. Anitha and Peter [16] developed a method that detects abnormal masses in mammograms, their method uses a CA that modifies it local rules to segment suspicious areas. The method is compared with others using a data base of 70 mammograms.

Benmazou et al. [5] used a four-state CA to classify suspicious regions on mammographies with 89 % of sensitivity in the 79 real case images studied. Cordeiro et al. [17] created an interactive platform using cellular automata to delimitate regions of masses in mammograms. Kumar and Sahoo [18] use a CA for edge detection based on the pixel intensity of the neighborhood of each pixel.

Hadi et al. [19] proposed a based method to detect microcalcifications in mammograms. They obtained an 87.5 % success rate on identifying cases that were evaluated as malignant cases, and 100 % when identifying benign cases.

1.2 Fuzzy Systems for Microcalcification Detection

Another approach for detecting microcalcifications in mammograms is the use of fuzzy inference systems (FIS). Cheng et al. [20, 21] presented one of the early works, were they used fuzzy set theory and geometrical statistics to increase the contrast of microcalcifications in mammograms. They implemented this technique in a five-step algorithm to detect microcalcifications in mammograms, even in those with high density.

Cheng et al. [22] use fuzzy entropy and fuzzy set theory to process and fuzzify an image and then enhance it. After that, they use Gaussian filtering to detect the microcalcification and measure them. Fuzzy enhancement on the contrast of an image for microcalcification detection has been the center of other research studies with slight variations in the type of membership functions and the adjustment of their parameters [23–25].

Pandey et al. [26] use a set of rules to identify if a pixel is a microcalcification or not, based in their neighbor pixels without using image enhancement and they implemented their algorithm in a fuzzy logic application specific processor. Chumklin et al. [27] apply a type-2 fuzzy system with automatic membership function generation with a high percentage of correct classification, 89.47 %, the main problem was the average detection of false positives, six per image.

The other branch of microcalcification detection using fuzzy systems is classification. Bhattacharya and Das [28] classify the pixels in mammograms as microcalcifications using fuzzy c-means and get the original size using morphological erosion. A similar approach is the one developed by Quintanilla-Dominguez et al. [29], where the use a k-means and fuzzy c-means algorithm but instead of erosion the method uses dilatation.

The most significant problem in microcalcification detection is the high number of false positive, detecting cases that do not have regions with calcifications as positive, in the computer evaluation. The aforementioned gives the opportunity of developing methods that lower the amount of false positive results for a better diagnosis.

Fuzzy logic systems are a useful tool in the enhancing of the characteristics of regions of interest in mammograms, and also, as pixels classifiers indicating if they are microcalcification or not. The variability in gray intensity, and other characteristics as shape and density of the breast, is well adapted by class adjustments of the fuzzy system applied to detect these regions of interest.

Many of the mentioned fuzzy systems use neighbor pixel operations, similar to cellular automaton models that have been used independently to detect microcalcifications. The aforementioned techniques serve as inspiration to the methodology presented in this work that implements both tools to identify regions of interest, which are the ones with a high probability of microcalcifications.

The proposed methodology uses a fuzzy inference system (FIS) to enhance zones with high-tone variation in the x and y-axes, which is accomplished by giving them a high membership value, and a small membership value to the low variation zones. The output of the fuzzy system is then evaluated using a cellular automaton with a Moore neighborhood that verifies if each pixel corresponds to a region with microcalcifications.

In Sect. 2, the concepts of computer vision, CA, and fuzzy system are described; Sect. 3 explains the steps of the methodology proposed; in Sect. 4, the results of the test using the mini-MIAS database are presented; and finally, in Sect. 5, the conclusions are presented.

2 Computer Vision, Fuzzy Logic, and Cellular Automata

The proposed methodology takes advantage of three main subjects: computer vision theory, fuzzy logic (FL), and CA. The next sections describe the relevant theoretical framework of these topics.

2.1 Computer Vision

Computer vision integrates image processing and pattern recognition. Image processing involves the manipulation and analysis of image information, from one or a group of them. The main focus of image processing is to enhance the image information for human interpretation or its numeric data processing for machine analysis [30].

Pattern recognition task consists in the identification of objects and forms in images. This process has a wide range of practical applications: robot vision, character recognition, voice recognition, military recognition applications, hand signature recognition applications, and medical imaging.

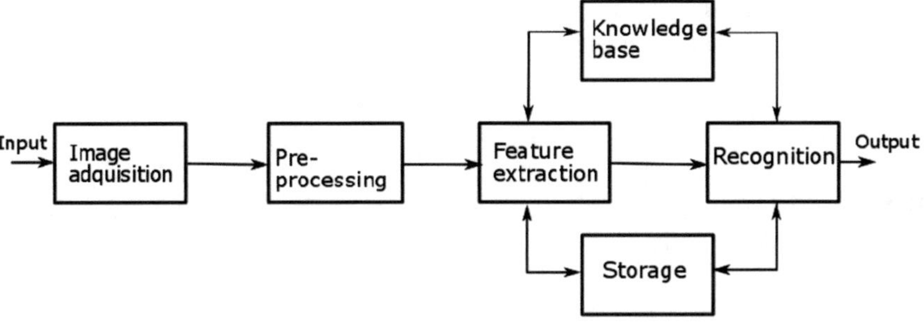

Fig. 1 Computer vision system description

A computer vision system can be summarized in the next stages: images acquisition, preprocessing, feature extraction, storage, a base of knowledge, and recognition (Fig. 1).

The first stage of the computer vision system is image acquisition, where the image is converted to digital format. An image is an intensity of light function in two dimensions $f(x, y)$, where x and y are spatial coordinates, and the value of f in any point (x, y) is proportional to the brightness or intensity of the gray tone at that particular point. A digital image is a numeric representation of an image, and every element that forms part of the image is called picture element or pixel.

The second stage is preprocessing, that includes a variety of mathematical operations, being the most typical grayscale manipulation, noise filtering, region isolation, geometry correction, image reconstruction, and segmentation. The aim of feature extraction is to obtain entities with significant data for their analysis; this helps to reduce the amount of data analyzed.

The three last stages—storage, knowledge base, and recognition—use high-end processing. The data is stored in memory addresses that will be allocated and accessed depending on the system requirements. Recognition classifies the information giving to the data a label based on the knowledge provided by an expert. This classification can be supervised or unsupervised. The classification rules are stored in a knowledge base; this base interacts with the other stages.

2.2 *Fuzzy Logic Inference System*

Fuzzy logic imitates human-like decisions based on a group of rules and memberships values to fuzzy sets with a linguistic label. It is an alternative to classical logic, where being part of a set is Boolean, and the only alternative of an object is to be a member of a set or not. Fuzzy logic is based on fuzzy set theory, where the object has a membership value in the closed interval [0, 1].

Definition 1 Classic set. Given a collection of objects *B*, an object *x* has a membership value $\mu_B(x)$ defined by

$$\mu_B(x) = \begin{cases} 1 & \text{if } x \in B; \\ 0 & \text{if } x \notin B. \end{cases} \quad (1)$$

Definition 2 Fuzzy set. Given a collection of objects *X*, a fuzzy set *A* in *X* is defined as a set of ordered pairs:

$$A = \{(x, \mu_A(x)) \,|\, x \in X\}\} \quad (2)$$

where μ_A, is the membership function of the fuzzy set *A*, $\mu_A(x)$ is the degree of membership of element *x* in the fuzzy set *A*, a value that is in the whole interval of real numbers between zero and one. In fuzzy sets, *X* is defined as the universe of discourse, or universe, and it can be a discrete or continuous set [31, 32].

A fuzzy inferences system (FIS) converts crisp numerical information (inputs) to fuzzy numerical values though linguistic terms and variables using a process called *fuzzification*. The linguistic information is processed using a fuzzy inference engine and fuzzy rule base to generate conclusions, then output numerical values (crisp) can be obtained using a process called *defuzzyfication* [34]. A very common FIS is the Mandani model [33].

2.3 Cellular Automata

Cellular automata were introduced by John Von Newman and Stanislaw Ulam with the name of cellular spaces.

John Von Newman and Stanislaw Ulam introduced the CA with the name of cellular spaces. They were conceived as an idealization of biological systems, as a model of biological self-reproduction [34]. CA have been applied and reintroduced in a variety of names: tessellation automata, homogeneous structures, cellular structures, tessellation structures, and iterative arrays [35].

CA is mathematical idealizations of physical systems, where time and space are discrete, and physical quantities have a finite set with a discrete range of values. The CA consists of a regular uniform lattice, or array, with a discrete value at each site that is called *cell* [35].

This array of cells is represented by $\mathbb{Z}^d$, where $\mathbb{Z}$ is the geometry of the automaton, and *d* refers to the dimensionality of the automaton. The geometry establishes the amount of elements of the CA and the shape of the array. The states of each cell come from the finite set *S*, where all the possible number the states are included.

A cellular automaton evolves in discrete time intervals, the states of each cell updates at every discrete time iteration, or generation. At each generation, the states of the whole automaton are represented by the mapping c: $\mathbb{Z}^d \rightarrow$ S [36]. The state

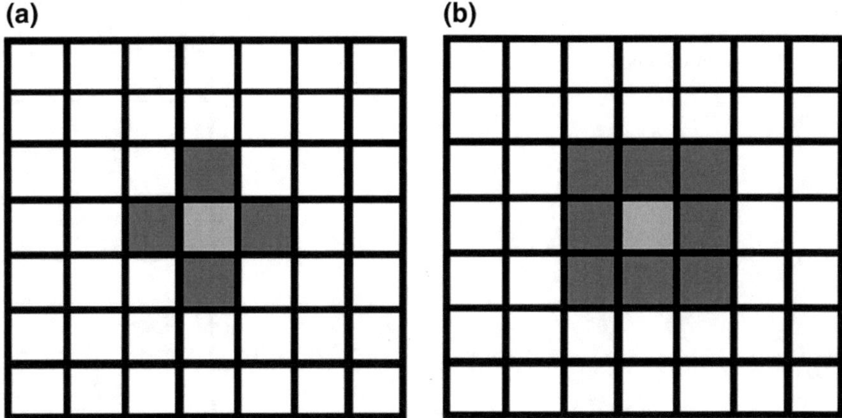

Fig. 2 **a** Von Neumann neighborhood; **b** Moore neighborhood

of each cell at a t discrete time is affected by the states of the cells in their surrounding neighborhood N, in a time $t - 1$.

The neighborhood can be described as a function that contains the coordinates of the elements that are part of it. For a cell in the α position, the neighborhood function can be defined as $g(\alpha) = \{\alpha, \alpha + \delta_1, \alpha + \delta_2, \ldots, \alpha + \delta_n\}$, where $\alpha \in \mathbb{Z}^d$ and $\delta_n(i = 1, 2, \ldots, n) \in \mathbb{Z}^d$ [36]. Every cell is synchronously updated based on the states of their neighborhood at $t - 1$, and a set of local rules denoted by L [35].

Two typical neighborhoods are Moore and Von Neumann neighborhoods, see Fig. 2. The first one includes all the elements that surround the cell in a defined radius r. For a two dimension CA, with a $r = 1$, the neighborhood function is defined as $g(\alpha) = \{\alpha, \alpha + (1, 0), \alpha + (0, 1), \alpha + (-1, 0), \alpha + (0, -1), \alpha + (1, 1), \alpha + (1, -1), \alpha + (-1, 1), \alpha + (-1, -1)\}$. The Von Neumann neighborhood dismiss the diagonal elements of the Moore neighborhood, the function associated for a two dimension CA with $r = 1$ is $g(\alpha) = \{\alpha, \alpha + (1, 0), \alpha + (0, 1), \alpha + (-1, 0), \alpha + (0, -1)\}$.

The characteristics of a CA can be resume in the 4-tuple:

$$\{\mathbb{Z}^d, N, L, S\} \tag{3}$$

3 Methodology

Microcalcifications in mammograms are small white flecks that appear as high-tone variation. At first glance, edge detection appears as a candidate to segment the regions with microcalcifications, but the different tissues of the breast, the variation in density, the small dimensions of the calcifications, and their irregular pattern and dimensions make the traditional edge detectors non-efficient to this task.

Fig. 3 Proposed methodology for microcalcification detection

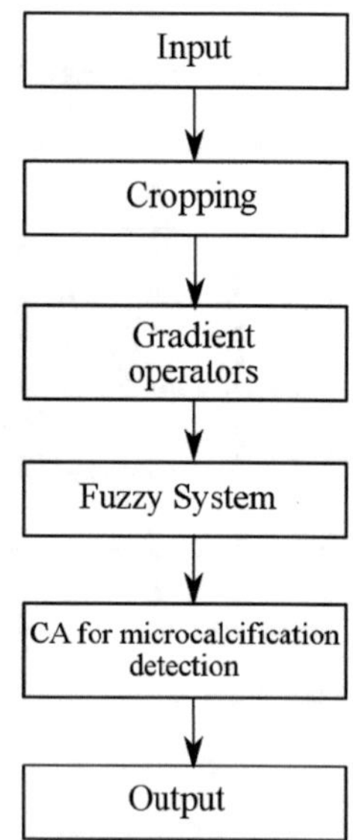

Contrast enhancement in mammograms for microcalcification detection lowers the intensity of the background (fatty tissue, muscle, and tissue near the skin), intensifies the tone of the microcalcifications, and makes the difference in tonality between the pixels that constitute the calcification regions and their closest pixels as high as possible.

The proposed methodology is shown in Fig. 3. The first step is to input the image in grayscale. Next, the cropping removes most of the background, labels, medical annotations, and breast muscle. Every pixel is normalized from zero to one.

In the third step, the gradient of each pixel on the x and y-axis is calculated. Microcalcifications have high-tone variation and irregular shape, gradient operators that measure these differences in tonalities in the immediate neighbors help to identify those regions that are suspicious of being calcifications. Equation (4) describes the use mask for the x-axis, and (5) the y-axis, it is the transpose of x.

$$\begin{bmatrix} 0 & 0 & 0 \\ 0 & 1 & -1 \\ 0 & 0 & 0 \end{bmatrix} \tag{4}$$

$$\begin{bmatrix} 0 & 0 & 0 \\ 0 & 1 & 0 \\ 0 & -1 & 0 \end{bmatrix} \tag{5}$$

Instead of applying the spatial filter using the masks and convoluting it with the image, both filters were applied using the concept of two dimensions CA, this helps reduce the number of operations. Two CA are needed, one for each axis, each one

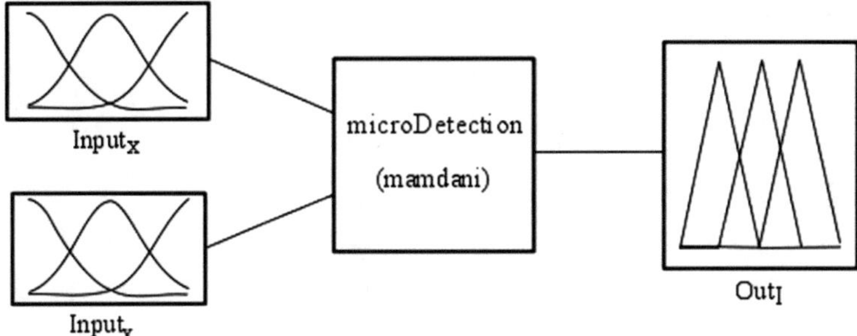

Fig. 4 Fuzzy system implemented

applied only for one generation. For the x-axis the neighborhood function is $g_x(\alpha) = \{\alpha, \alpha + (1, 0)\}$, and for the y-axis the function is $g_y(\alpha) = \{\alpha, \alpha + (0, 1)\}$. The local rule for both is to multiply each neighborhood by the weight is given by the mask for each axis at time t, the state of cell α at $t + 1$ will be the sum of the cells that are part of the neighborhood.

The output of the gradient filters will be inputs of a Mamdani fuzzy system is described in Fig. 4. It has two inputs (Input$_x$ and Input$_y$) and one output (Out$_I$). The inputs are the result of the gradients filters—Input$_x$ for the x-axis, and Input$_y$ for the y-axis—both inputs have identical membership functions (Fig. 5). The output membership functions are shown in Fig. 6.

The membership functions for each image are different and are tuned manually, making the zones with high intensity a higher value than the zones that are part of the background.

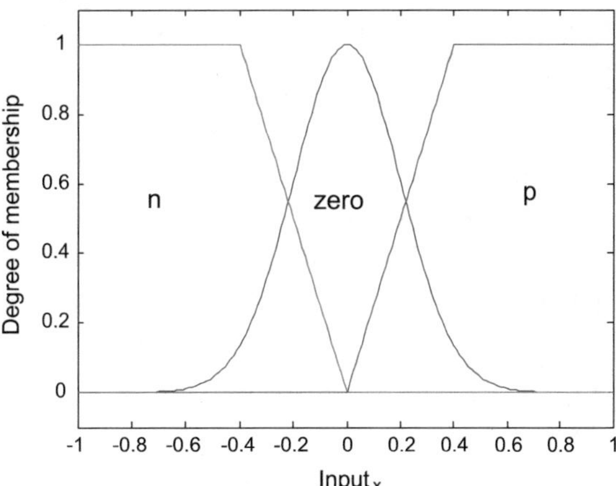

Fig. 5 Input membership functions for the input x. *Note* Input y uses the same function shapes and parameters values

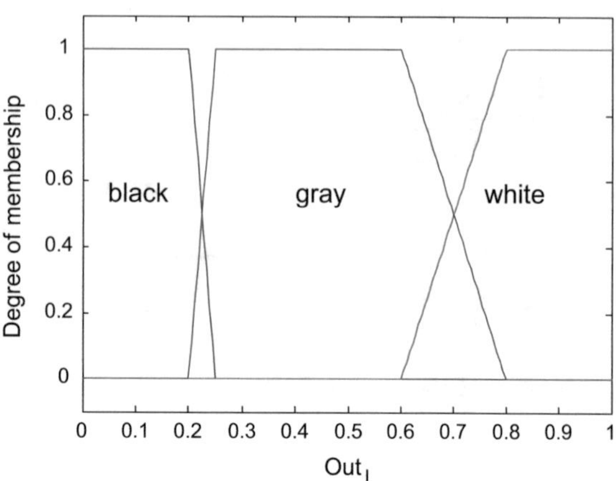

Fig. 6 Output membership functions

The rules of the fuzzy system are:

- If I_x is zero and I_y is zero then I_{out} is black.
- If I_x is zero and I_y is n then I_{out} is gray.
- If I_x is zero and I_y is p then I_{out} is gray.
- If I_x is n and I_y is zero then I_{out} is gray.
- If I_x is p and I_y is zero then I_{out} is gray.
- If I_x is n and I_y is n then I_{out} is white.
- If I_x is n and I_y is p then I_{out} is white.
- If I_x is p and I_y is n then I_{out} is white.
- If I_x is p and I_y is p then I_{out} is white.

The output of the fuzzy system will be the input of the CA. In this stage, the CA designed for microcalcification detection is applied. The CA is two-dimensional, with the same size and geometry as the input image; it is a 3×3 Moore neighborhood because this size provided the best results. The rules of the CA were designed to preserve the states of the cells that have high value and with a neighborhood in which every cell have high value. The local rules of the CA are the following:

1. At $t = 0$, the state of the cell $\alpha(x, y)$ is the same as the corresponding state of the output of the fuzzy system in (x, y). Go to Step 2.
2. If the state of $\alpha(x, y)$ is lower than the established cell threshold, the state of $\alpha(x, y)$ at $t = t + 1$ will be zero, go to step 5; else go to Step 3.
3. If the sum of the states in the neighborhood of $\alpha(x, y)$ is lower than the established neighborhood threshold, the state of $\alpha(x, y)$ at $t = t + 1$ will be zero, go to Step 5; else go to Step 4.
4. The state of the cell $\alpha(x, y)$ at $t = t + 1$ is preserved; go to Step 5.
5. Repeat from step two until the number of generations is reached.

4 Results

To test the proposed methodology, the mini-MIAS database was used [37], a total of 4 images of the database that were classified as containing microcalcifications where selected. The mammograms original sizes are 1024 × 1024 in grayscale. The images have the center and the radius of the microcalcification zones identified by the expert. We implemented the test system in MATLAB.

Because the images had labels, annotations made by the professional, salt and pepper noise; the images had to be cropped trying to maintain most of the breast as possible. The part of the breast that was cropped was most of the breast muscle, the zones with annotations and the zones with a clear indication of an artifact. The characteristics of the input images are described in Table 1.

For every image, it was needed to tune the output membership functions and the two thresholds of the CA. The membership functions were tuned manually, with the main objective of incrementing the tone difference of the small bright regions and the background (Fig. 7).

The histogram of the output of the fuzzy system was used to adjust the thresholds of the CA. The threshold of the cell is based on the values that are around the 90–95 % of the distribution in the histogram. The threshold for the neighborhood is a value that depends on the distribution of those cells that have preserved their state once evaluated by the cell threshold. It is usually between one Gaussian distribution of the cell threshold. The automaton only keeps the state of the pixels with a high probability of belonging to a calcification zone; the other pixels are given the state zero. The output of the system provides an image display of the zones that are suspicious Fig. 8.

With the locations of the possible microcalcifications zones, a comparison of the regions that were detected as calcifications by the system and those that were indicated by the database are shown in Figs. 9, 10, 11, and 12. The blue circles in the images presented were manually put in based on the data given by the system. Images displayed have been enhanced for a better perception of the calcifications.

Table 2 shows that all the regions with microcalcifications were identified, according to the information provided by the mini-MIAS database. Three of the mammograms processed gave false positive results, Table 3 describes them. The main reason of these false positives is the salt and pepper noise. Tiny black spots are also high-tone variation, and the gradient operators do not discriminate white spots from black spots. The other possible reason is that in some mammograms

Table 1 Description of the images used for the test stage

Image	Background	Severity	Coordinates	Ratio
A	Fatty-Glandular	Malign	547,520	45
B	Fatty-Glandular	Benign	546,756	29
C	Fatty-Glandular	Malign	No central point	
D	Fatty	Malign	No central point	

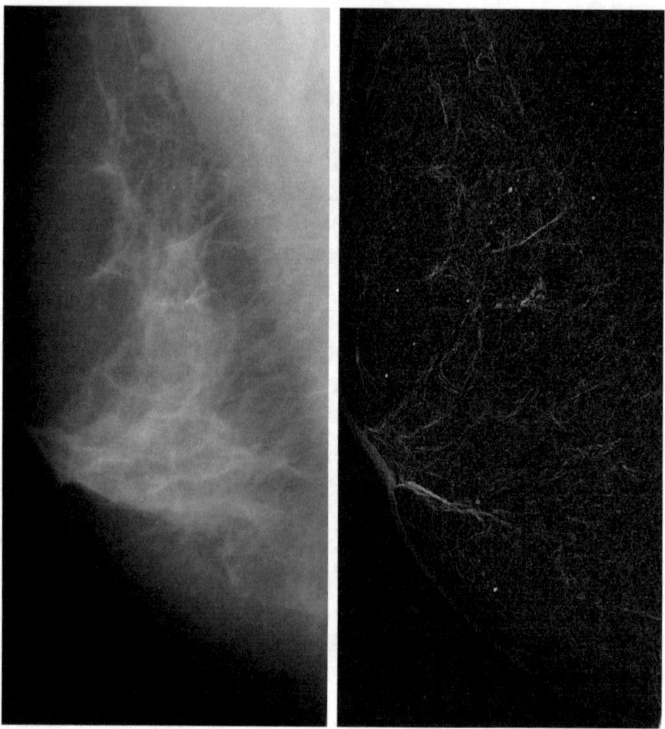

Fig. 7 (*left*) Mammogram A; (*right*) output of the fuzzy system, the images is enhanced for a better display

zones with solitary calcifications are present, and the system detected them, but the professional did not value them as important due to its isolation.

Other regions that were detected as false positive are zones with high density, which appear as bright white spots, and are irregular in size but many times larger than calcifications. These high-density zones can be part of the anatomy of the breast (glandular or fatty tissue) or an abnormal mass. Detecting the perimeter or segments of these areas is what is giving the false positives.

The main problem is the pixels that belong to the outer regions of these high-density zones. The gradient operators detect a high-tone variation on both axis, these pixels are evaluated by the fuzzy system given them a high value. The pixels are part of the perimeter of the high-density zone, and most of these pixels will be evaluated by the fuzzy system with a high value, consequentially, they are preserved by the CA.

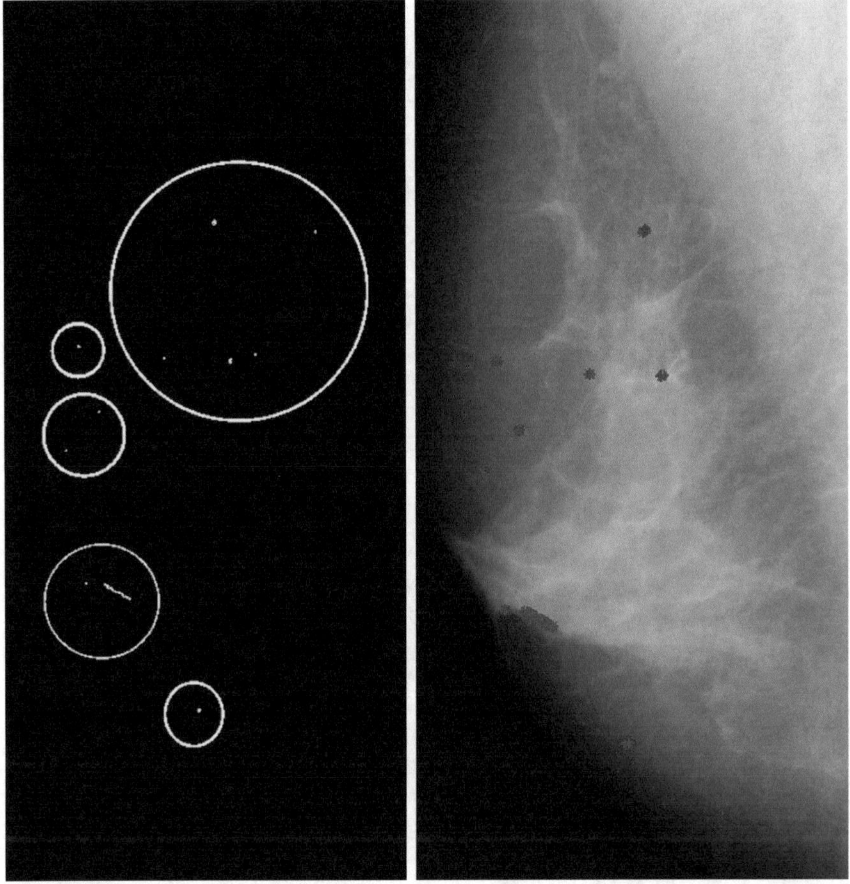

Fig. 8 (*left*) Output of the CA, the *white circles* where manually inserted; (*right*) zones that where identified in the A mammogram as regions with possible calcifications

5 Discussion and Future Work

The results of the test achieved using the proposed methodology, detected the zones that were identified by the database as microcalcifications in the selected mammograms. The nature of the gradient operator makes a pixel shift in both directions, x and y, so a pixel position correction is needed for fine-precision. Three out of four of the mammograms analyzed had false positives, but none had false negatives. Most of the false positives come from isolated calcifications, salt and pepper noise, artifacts, and high-density zones.

Identifying the isolated calcifications does not play an important role in the efficiency of this method because it was developed to detect microcalcifications and not to evaluate the clusters. Cluster evaluation is a valuable tool to identify

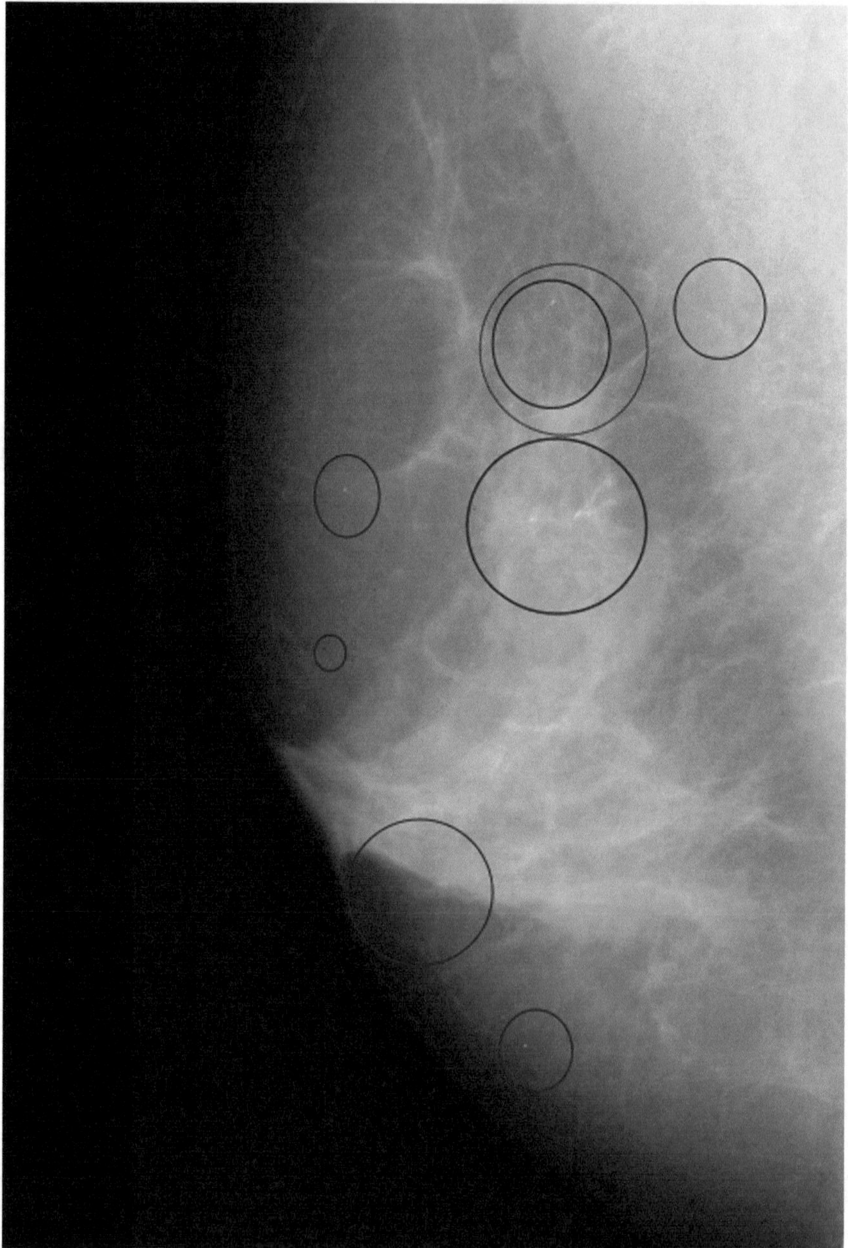

Fig. 9 Comparisons of the zones detected in the A mammography, *blue circles* where the zones identified by the system, *red zones* where detected by the specialist

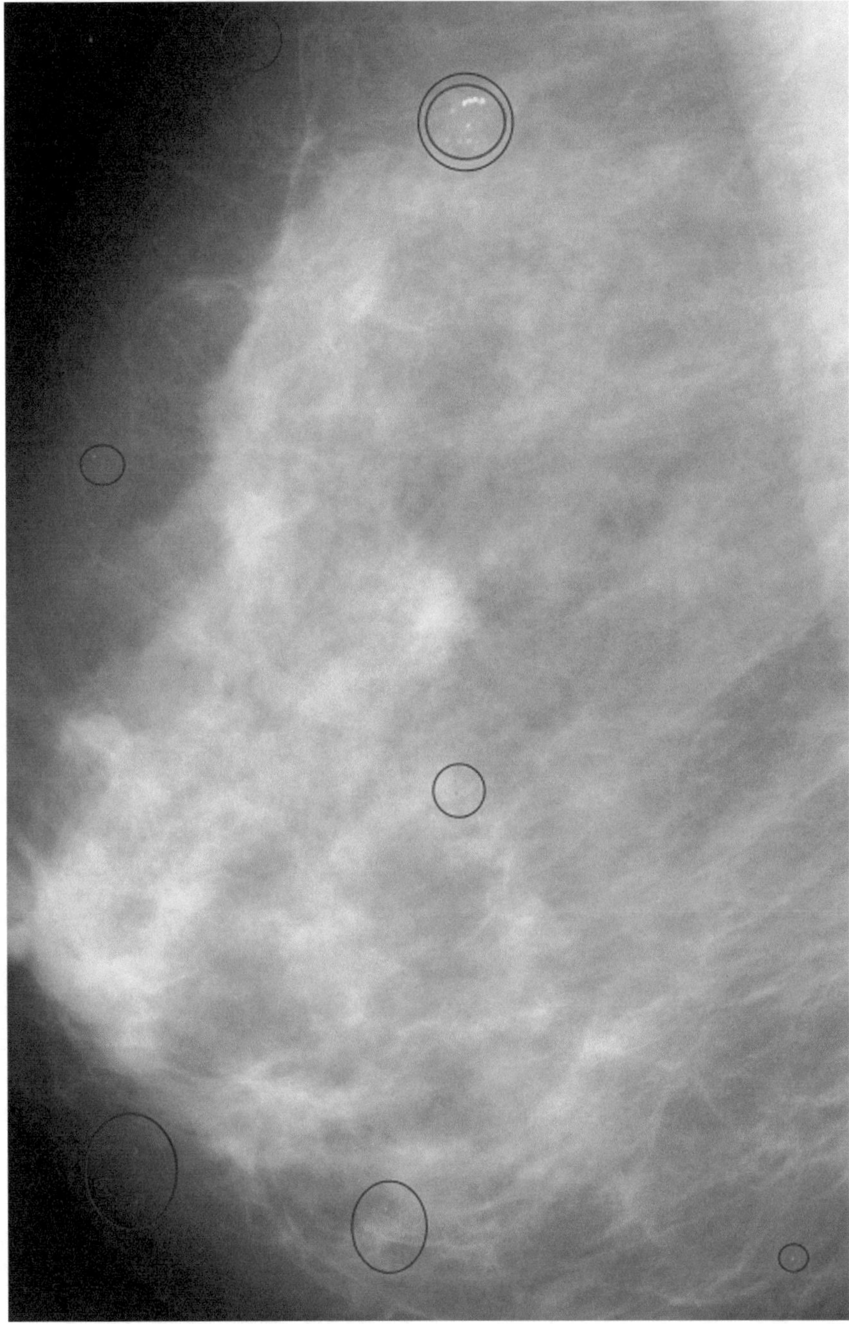

Fig. 10 Comparisons of the zones detected in the B mammography, *blue circles* where the zones identified by the system, *red zones* where detected by the specialist

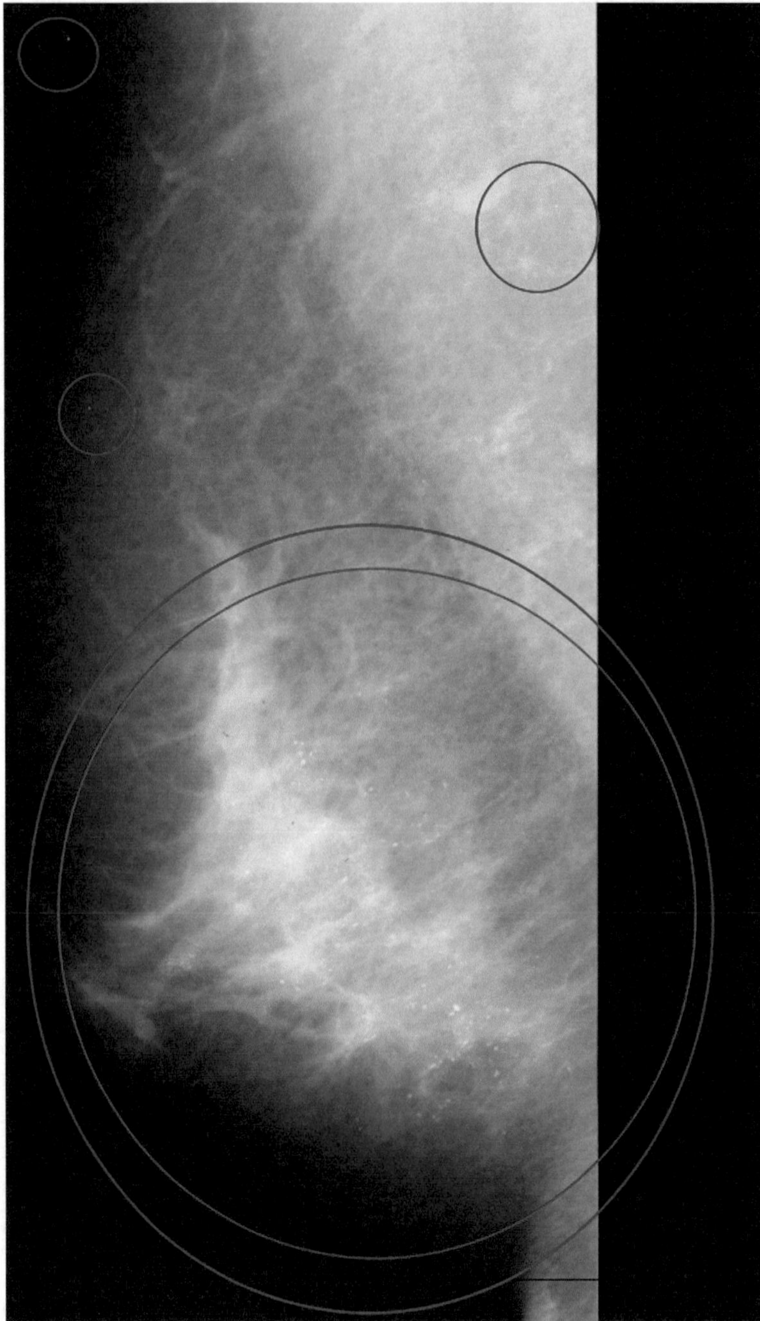

Fig. 11 Comparisons of the zones detected in the C mammography, *blue circles* where the zones identified by the system, *red zones* where detected by the specialist

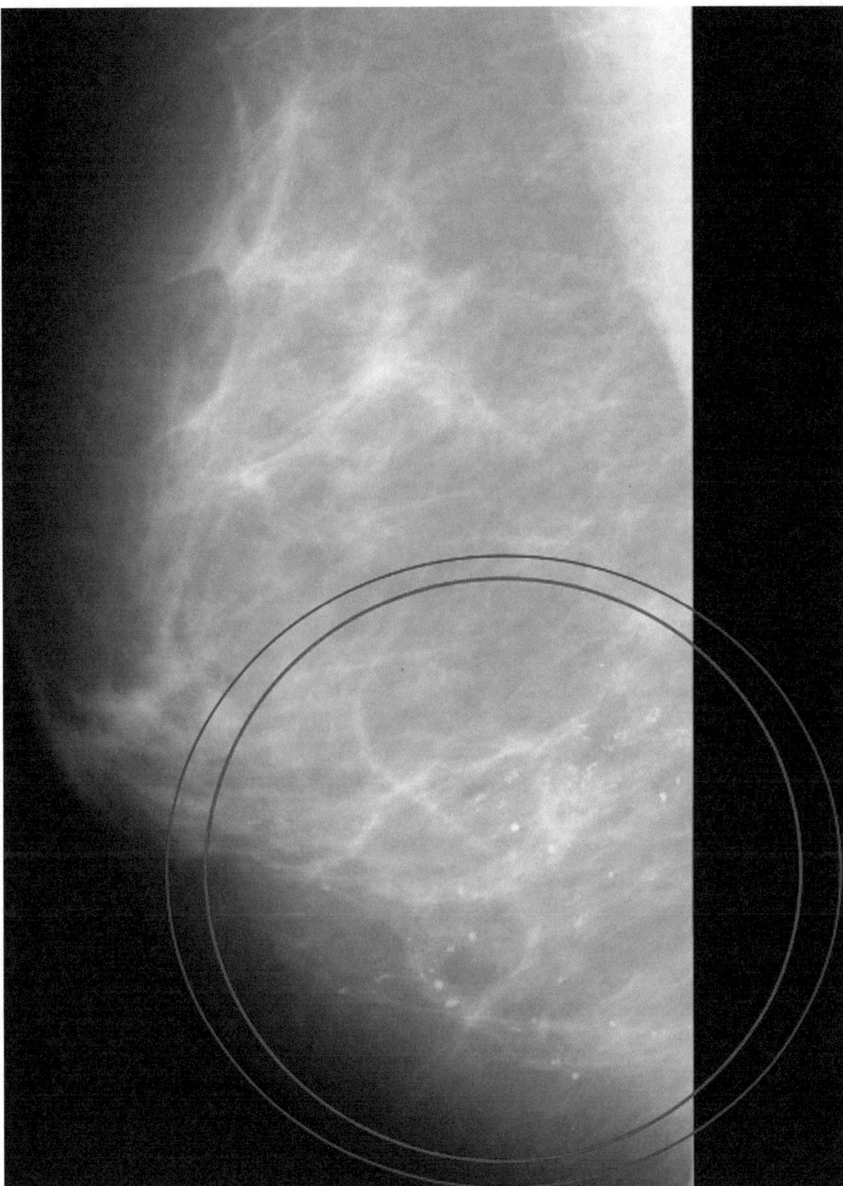

Fig. 12 Comparisons of the zones detected in the D mammography, *blue circles* where the zones identified by the system, *red zones* where detected by the specialist

Table 2 Results of the processed mammograms

Mammogram	Regions detected	False negatives	False positives
A	6	0	5
B	7	0	6
C	4	0	3
D	1	0	0

Table 3 False positives description

Mammogram	Isolated calcifications/salt noise	Pepper noise	Salt noise out of the breast area	Pectoral muscle region	High-density fibro glandular tissue
A	2	1	0	1	2
B	4	1	1	0	0
C	1	0	1	1	0
D	0	0	0	0	0

malignancy which is not a goal of the proposed algorithm. Artificial neural networks can be applied to assist in the stage of cluster evaluation.

Salt noise is easily confused with isolated calcifications and it is hard for the system to differentiate both. False positive as a result of pepper noise can be reduced with a preprocessing stage that puts to zero the regions in the output of the fuzzy system belonging to pixels in the original mammography that are near to zero can be applied to correct this feature. As mentioned, high-density zones induce false positives, mainly in the pixels that are part of its perimeter, being this topic a relevant case of study for future research based on the proposed method.

Although the tested mammograms have different characteristics, none of the results had false negatives, which is a promising feature of the proposed method. The feature that allows evaluating the regions even if the images have different features is the tuning phase of the output membership functions. Automatic tuning of the membership functions using soft computing can be a future approach to this issue.

Acknowledgments We thank Instituto Politécnico Nacional (IPN), the Commission of Operation and Promotion of Academic Activities of IPN (COFAA), and the Mexican National Council of Science and Technology (CONACYT) for supporting our research activities.

References

1. International Agency for Research on Cancer (2012). GLOBOCAN 2012: Estimated Cancer Mortality and Prevalence Worldwide in 2012. http://globocan.iarc.fr/Pages/fact_sheets_cancer.aspx. Accessed 1 Dec 2015
2. American Cancer Society Inc. Cancer facts and figures 2015. http://www.cancer.org/acs/groups/content/@editorial/documents/document/acspc-044552.pdf. Accessed 6 Sep 2015.

3. World Health Organization (2015) Cancer Country Profiles (2014). http://www.who.int/cancer/country-profiles/mex_en.pdf?ua=1. Accessed 1 Sep 2015
4. Paredes ES (2007) Atlas of mammography. Lippincott Williams and Wilkins
5. Benmazou S, Merouani HF, Layachi S, Nedjmeddine B (2014). Classification of mammography images based on cellular automata and Haralick parameters. In: Evolving Systems 5 (3):209-216
6. Oliver A, Freixenet J, Martí J, Pérez E, Pont J, Denton E, Zwiggelaar R (2009) A review of automatic mass detection and segmentation in mammographic images. Medical Image Analysis, vol. 14. Elsevier, pp. 87-110
7. Nawalade Y (2009) Evaluation of Breast Calcification. The Indian Journal of Radiology and Imaging, 19:282-286
8. Balakumaran T, Vennila I, Shankar CG (2010) Detection of Microcalcification in Mammograms Using Wavelet Transform and Fuzzy Shell Clustering. International Journal of Computer Science and Information Security 7(1):121-125
9. Mohanalin J, Kalra PK, Kumar N (2008) Fuzzy based micro calcification segmentation. In: International Conference on Electrical and Computer Engineering, 2008. ICECE 2008. Dec 2008. pp 49-52
10. American Cancer Society Inc. (2015) Cancer Facts and Figures (2015). http://www.cancer.org/acs/groups/content/@editorial/documents/document/acspc-044552.pdf. Accessed 6 Sep 2015
11. Pereira DC, Ramos RP, do Nascimento MZ (2014). Segmentation and detection of breast cancer in mammograms combining wavelet analysis and genetic algorithm. Computer Methods and Programs in Biomedicine, 114:88-101
12. Berber T, Alpkocak A, Balci P, Dicle O (2012) Breast mass contour segmentation algorithm in digital mammograms. Computer Methods and Programs in Biomedicine 110:150-159
13. Viher B, Dobnikar A, Zazula D (1998) Cellular automata and follicle recognition problem and possibilities of using cellular automata for image recognition purposes. International Journal of Medical Informatics 49:231-241
14. Qadir F, Peer MA, Khan KA (2012) Efficient edge detection methods for diagnosis of lung cancer based on two-dimensional cellular automata. Advances in Applied Science Research 3 (4):2050-2058
15. Wongthanavasu S, Tangvoraphonkcha V (2007) Cellular Automata-Based Algorithm and its Application in Medical Image Processing. In: ICIP 2007. IEEE International Conference on Image Processing 50:11-13
16. Anitha J, Peter JD (2015) Mammogram segmentation using maximal cell strength updation in cellular automata. Medical & Biological Engineering & Computing vol. 53(8):737-749
17. Cordeiro FR, Santos WP, and Silva-Filho AG (2014) Segmentation of mammography by applying growcut for mass detection. In: Liu J, Doi K, Fenster A (ed) MEDINFO 2013: Studies in Health Technology and Informatics, vol. 192. IOS Press, pp. 87-91
18. Kumar T, Sahoo G (2010) A novel method of edge detection using cellular automata. International Journal of Computer Applications 9:38-44
19. Hadi R, Saeed S, Hamid A (2013) A modern approach to the diagnosis of breast cancer in women based on using Cellular Automata In: First Iranian Conference on Pattern Recognition and Image Analysis (PRIA) 2013, pp 1-5
20. Cheng H, Lui YM, Freimanis RI (1996) A new approach to microcalcification detection in digital mammograms. In: Nuclear Science Symposium, Nov 1996. Conference Record, vol. 2. IEEE, pp 1094-1098
21. Cheng H, Lui YM, Freimanis RI (1998) A novel approach to microcalcification detection using fuzzy logic technique. IEEE Transactions on Medical Imaging. 17(3):442-450
22. Cheng HD, Wang J, Shi X (2004) Microcalcification detection using fuzzy logic and scale space approaches. Pattern Recognition, vol 37. Elsevier, pp 363-375
23. Cheng HD, Wang J (2003) Fuzzy logic and scale space approach to microcalcification detection. In: 2003 IEEE International Acoustics, Speech, and Signal Processing, 2003. Proceedings. (ICASSP '03). April 2003, 2:345-348

24. Chen X, Chen Y (2010) An Improved Edge Detection in Noisy Image Using Fuzzy Enhancement. In: International Conference on Biomedical Engineering and Computer Science (ICBECS), Apr 2010. pp 1-4
25. Begum S, Devi O (2011) Fuzzy Algorithms for Pattern Recognition in Medical Diagnosis. Physical Sciences and Technology 7(2):1-12
26. Pandey N, Salcic Z, Sivaswamy J (2000) Fuzzy logic based microcalcification detection. In: Proceedings of the 2000 IEEE Signal Processing Society Neural Networks for Signal Processing X. 2:662-671
27. Chumklin S, Auephanwiriyakul S, Theera-Umpon, N (2010) Microcalcification detection in mammograms using interval type-2 fuzzy logic system with automatic membership function generation. In: IEEE International Conference on Fuzzy Systems, July 2010. pp 1-7
28. Bhattacharya M, Das A (2007) Fuzzy Logic Based Segmentation of Microcalcification in Breast Using Digital Mammograms Considering Multiresolution. In: International Machine Vision and Image Processing Conference, 2007. IMVIP 2007. pp 98-105
29. Quintanilla-Dominguez J, Ojeda-Magaña B, Cortina-Januchs MG, Ruelas R, Vega-Corona A, Andina D (2011) Image segmentation by fuzzy and possibilistic clustering algorithms for the identification of microcalcifications. Scientia Iranica Transactions: Computer Science & Engineering and Electrical Engineering 18:580–589
30. Kulkarni A (2001) Computer vision and fuzzy-neural systems. Prentice Hall
31. Jantzen J (2013) Foundations of Fuzzy Control: A practical Approach. John Wiley & Sons, Second Edition.
32. Jang J, Sun C, Mizutani E (1997) Neuro-fuzzy and Soft Computing: A Computational Approach to Learning and Machine Intelligence. Prentice Hall
33. Espinosa J, Vandewalle J, Wetz V (2004) Fuzzy Logic, Identification and Predictive Control. Springer-Verlag
34. Von Newmann J, Burks A (1966) Theory of Self-Reproducing Automata. University of Illinois Press, 1966.
35. Wolfram S (1983) Reviews of Modern Physics, Statistical Mechanics of Cellular Automata. The American Physical Society 55: 601-643
36. Kari J (2005) Theory of cellular automata: A survey. Theoretical Computer Science 334:3-333
37. Suckling J (1994) The mini-MIAS database of mammograms. http://peipa.essex.ac.uk/info/mias.html. Acceded 5 Nov 2015

Sensor Less Fuzzy Logic Tracking Control for a Servo System with Friction and Backlash

Nataly Duarte, Luis T. Aguilar and Oscar Castillo

Abstract The tracking problem for an electrical actuator consisting of a DC motor and a reducer part (load) operating under uncertainty conditions due to friction and backlash is addressed. The Mamdani type fuzzy logic control will be designed to enforce the load position to track a prespecified reference trajectory. Since it is assumed that the dynamic model is not available, Lyapunov stability theory coupled together with the comparison principle will be used to conclude stability of the closed-loop system.

Keywords Fuzzy control · Backlash · Fuzzy logic

1 Introduction

The tracking control problem of a DC motor operating under uncertain conditions due the friction and backlash and driven by a Mamdani type fuzzy logic control is addressed. The DC motor under consideration is governed by the following dynamic equation [11]

$$J\ddot{q} = u - F(\dot{q}) - F_1(t) + \omega(t) \tag{1}$$

N. Duarte (✉) · L.T. Aguilar
Instituto Politécnico Nacional—CITEDI, Avenida Instituto Politécnico Nacional no. 1310, Colonia Nueva Tijuana, 22435 Tijuana, Mexico
e-mail: nduarte@citedi.mx

L.T. Aguilar
e-mail: laguilarb@ipn.mx

O. Castillo
Instituto Tecnológico de Tijuana, Calzada del Tecnológico S/N, 22379 Tijuana, Mexico
e-mail: ocastillo@tectijuana.mx

P. Melin et al. (eds.), *Nature-Inspired Design of Hybrid Intelligent Systems*, Studies in Computational Intelligence 667, DOI 10.1007/978-3-319-47054-2_39

where $q(t) \in \mathbb{R}$ is the rotor angular position, $\dot{q}(t) \in \mathbb{R}$ is the rotor angular velocity, $u(t) \in \mathbb{R}$ is the control input, $t \in \mathbb{R}$ is the time variable, and $\omega(t)$ is the external disturbance assumed unknown but uniformly bounded, that is

$$\|\omega(t)\| \leq W, \quad \text{for all } t. \tag{2}$$

The parameter $J > 0$ denotes the inertia mass of the motor, K is the amplifier constant, and $F(\dot{q})$ is the friction force governed by the following static model

$$F(\dot{q}) = f_v \dot{q} + f_c \operatorname{sign}(\dot{q}). \tag{3}$$

Here, the first term defines the viscous friction with positive coefficient f_v and the last term is the Coulomb friction model, with positive coefficient f_c, which is an ideal relay model, multi-valued for zero velocity, that is

$$\operatorname{sign}(\dot{q}) = \begin{cases} 1 & \text{if } \dot{q} > 0 \\ [-1, 1] & \text{if } \dot{q} = 0 \\ -1 & \text{if } \dot{q} < 0. \end{cases} \tag{4}$$

Thus, stiction, describing the Coulomb friction force at rest, can take any value in the segment $[-f_c, f_c]$ been a source of uncertainty the plant. In other words steady-state position error is caused by stiction.

Finally, $F_1(t)$ given by

$$F_1(t) \begin{cases} -K(\Delta\theta - j_0) & \text{if } \Delta\theta \geq j_0 \\ 0 & \text{if } \|\Delta\theta\| < j_0 \\ -K(\Delta\theta + j_0) & \text{if } \Delta\theta \leq -j_0 \end{cases} \tag{5}$$

describes the backlash as it was in [14] in terms of the maximal backlash magnitude $j_0 > 0$, the stiffness coefficient $K > 0$, and the deviation $\Delta\theta = q - \theta$ of the motor position q from the load position θ governed by

$$J\ddot{\theta} = -F_1(t). \tag{6}$$

Throughout, the precise meaning of the differential Eqs. (1) and (3) with a piecewise continuous right-hand side is defined in the sense of Filippov [2].

Systems with actuator non-idealities, such as backlash, dead-zone, and Coulomb friction produce steady-steady errors and limit cycles at the output, and in the worst case could cause instability. Therefore, the motivation behind this paper is that a number of industrial systems are characterized by such non-idealities degrading severally the system performance [12, 13] thus leading to a challenging problem in terms of position and velocity stabilization.

The *control problem* is formulated as follows: given the desired load position $\theta_d(t) \in \mathbb{R}$, the control objective is to drive the load output $\theta(t)$ to the desired trajectory $\theta_d(t)$, namely,

$$\lim_{t\to\infty}\|\theta_d(t) - \theta(t)\| = 0 \tag{7}$$

for an arbitrary initial condition $\theta(0) \in \mathbb{R}$ despite the presence of external disturbances.

Fuzzy control is long-recognized for its robustness features against plant uncertainties and unmodeled dynamics (see, e.g., [1, 5]). Lyapunov stability analysis, on the other hand, has proven to be an essential tool for studying the stability of model-based control systems including model-based fuzzy control systems [6]. In contrast, Mamdani type fuzzy control is a heuristic and model free method where Lyapunov stability analysis is nontrivial. An additional contribution of the paper is in studying the stability of the latter control systems in the sense of Lyapunov, based on the comparison principle [4, p. 354], and its application in the tracking control of DC motor where the presence of friction is under consideration as unknown disturbance.

The paper is organized as follows. We present instrumental tools for stability analysis of non-smooth systems and the comparison principle as well in Sect. 2. A fuzzy controller for a DC motor is presented in Sect. 3. Verification of the fuzzy logic control synthesis and stability analysis of the closed-loop system are discussed in Sect. 4. Finally, Sect. 5 presents the conclusions.

2 Tools for Stability Analysis of Non-smooth Systems

The following definitions of stability for non-smooth systems, taken from Orlov [10], will be essential to conclude stability.

Consider a non-autonomous differential equation of the form

$$\dot{x} = \varphi(x,t) \tag{8}$$

with the state vector $x = (x_1, \ldots, x_n)^T$, with the time variable $t \in \mathbb{R}$, and with a piece-wise continuous right-hand side $\varphi(x,t) = (\varphi_1(x,t), \ldots, \varphi_n(x,t))^T$.

The differential Eq. (8) with a piecewise continuous right-hand side is defined in the sense of Filippov [2] as that of the differential inclusion

$$\dot{x} \in \Phi(x,t) \tag{9}$$

with $\Phi(x,t)$ being the smallest convex closed set containing all the limit values of $\varphi(x^*,t)$ for $(x^*,t) \in \mathbb{R}^{n+1}\backslash N$, $x^* \to x$, $t = \text{const}$.

Definition 1 The equilibrium point $x = 0$ of the differential inclusion (9) is stable (uniformly stable) if and only if for each $t_0 \in \mathbb{R}$, $\varepsilon > 0$, there is $\delta = \delta(\varepsilon, t_0) > 0$, dependent on ϵ and possibly dependent on t_0 (respectively, independent on t_0) such that each solution $x(t, t_0, x_0)$ of (8) with the initial data $x(t_0) = x_0 \in B_\delta$ within the ball B_δ, centered at the origin with radius δ, exists for all $t \leq t_0$ and satisfies the inequality

$$\|x(t, t_0, x_0)\| < \epsilon, \quad t_0 \leq t < \infty. \tag{10}$$

The following lemma, extracted from [6], will be useful to verify stability of the closed-loop system.

Lemma 1 (Comparison Lemma) *Consider the scalar differential equation*

$$\dot{u} = f(t, u), \quad u(t_0) = u_0, \tag{11}$$

where $f(t, u)$ is continuous in t and locally Lipschitz in u, for all $t \geq 0$ and all $u \in B_u \subset \mathbb{R}$. Let $[t_0, \infty)$ be the maximal interval of existence of the solution $u(t)$, and suppose $u(t) \in B_u$ for all $t \in [t_0, \infty)$. Let $v(t)$ be a continuous function whose upper right-hand derivative $\dot{v}(t)$ satisfies the differential inequality

$$\dot{v}(t) \leq f(t, v(t)), \quad v(t_0) = u_0 \tag{12}$$

with $v(t) \in B_u$ for all $t \in [t_0, \infty)$. Then, $v(t) \leq u(t)$ for all $t \in [t_0, \infty)$.

3 Fuzzy Logic Control Synthesis

3.1 Fuzzy Logic

The aim of the fuzzy logic is represent the human expertise (linguistic data) into mathematical formulas represented by rules with format *if-then* (fuzzy rules). A fuzzy set A in W is defined as a set of ordered pairs

$$A = \{(w, \mu_A(w)) \mid w \in W\}, \tag{13}$$

where $\mu_A(w)$ is called the membership function (or MF for short) for the fuzzy set A. The MF maps each element of W to a membership grade (or membership value) between 0 and 1 [3].

The fuzzy logic system consists of four basic elements see Fig. 1: the *fuzzifier*, the *fuzzy rules-base*, the *inference engine*, and *the defuzzifier*.

According to Mamdani and Assilian [7, 8], the fuzzy rules has the following form:

$$\text{IF } y_1 \text{ is } A_1^l \text{ AND } y_2 \text{ is } A_2^l \text{ THEN } y_3 \text{ is } B^l, \tag{14}$$

where $y = (y_1, y_2)^T \in U = U_1 \times U_2 \subset \mathbb{R}$ where y_1 is the fuzzified value of a measured variable, y_2 is the fuzzified value of the derivative of the measured variable, and $y_3 \in V \subset \mathbb{R}$. For each input fuzzy set A_l^k in $y_k \subset U_k$ with output

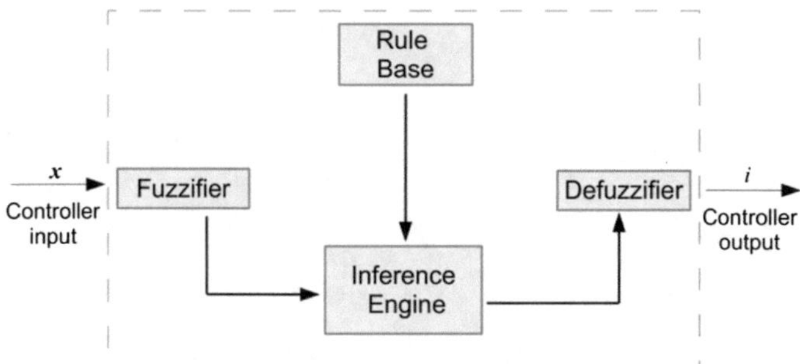

Fig. 1 General structure for a fuzzy inference system where $x = (e, \dot{e})^T$

membership function $\mu_{B^l} \in y_3 \subset V$, respectively; being l the number of membership functions associated to the input k. The particular choice of each $\mu_{B^l}(y_3)$ will depend on the heuristic knowledge of the experts over the plant. The fuzzifier maps the crisp input into fuzzy sets (FS), which are subsequently used as inputs to the inference engine, whereas the defuzzifier maps the FS produced by the inference engine into crisp numbers [9].

3.2 Fuzzy Logic Controller

The design of the fuzzy logic controller (FLC) was based on ideology proposed by Mamdani [7, 8], where the basic assumption of this idea is that in the absence of an explicit plant model, informal knowledge of the operation of the plant can be codified in terms of if then rules (14), which form the basis for a linguistic control strategy. Two-inputs and one-output rules in the formulation of the knowledge base were selected. The position error $e\,(t) = \theta\,(t) - \theta_d(t)$ and the change of error $\dot{e}\,(t) = \dot{\theta}\,(t) - \dot{\theta}_d(t)$ were chosen as inputs, both granulating in three fuzzy sets. The applied current to the motor was chosen as output, granulated in five fuzzy sets. These can be seen in Figs. 2 and 3.

The fuzzy rules are summarized in Table 1. The elements in the first row and column are the input linguistic variables, while the others correspond to the output linguistic variables.

For the inference process, we implement the Mamdani type of fuzzy inference, with minimum as disjunction operator, maximum as conjunction operator, minimum as implication operator, maximum as aggregation operator and centroid (COA) as our defuzzification method, which is

$$Z_{\mathrm{COA}} = \frac{\int \mu_A(y_3)\, y_3\, dy_3}{\int \mu_A(y_3)\, dy_3}, \tag{15}$$

(a)

Degree of membership
NL NS Z PL PS
1.2 1 0.8 0.6 0.4 0.2 0
-1 -0.8 -0.6 -0.4 -0.2 0 0.2 0.4 0.6 0.8 1
e

(b)

Degree of membership
NL NS Z PL PS
1.2 1 0.8 0.6 0.4 0.2 0
-1 -0.8 -0.6 -0.4 -0.2 0 0.2 0.4 0.6 0.8 1
De

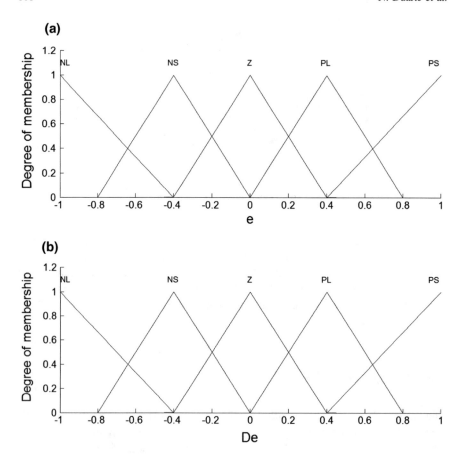

Fig. 2 Input variables: error and change in error

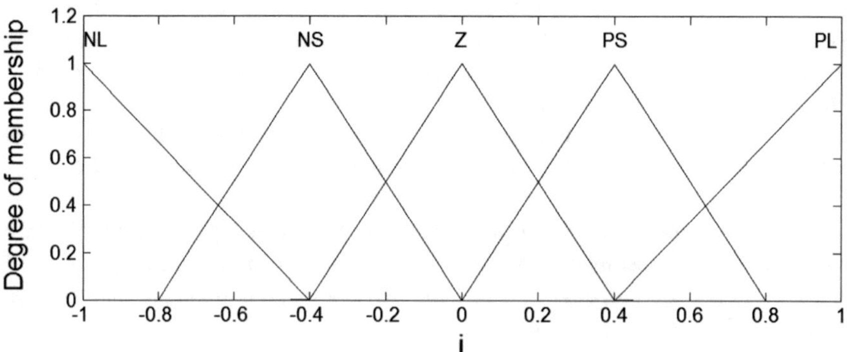

Fig. 3 Output variable

Table 1 Set of rules

$\dot{e}$	e				
	NL	NS	Z	PS	PL
NL	PL	PL	PL	PS	Z
NS	PL	PL	PS	Z	NS
Z	×	×	Z	×	×
PS	PS	Z	NL	NL	NL
PL	Z	NS	NL	NL	NL

where $\mu_A(z)$ is the aggregated output of MF and, R denotes the union of $(y_3, \mu(y_3))$ pairs. Figure 4 shows the input–output curves.

4 Stability Analysis for the DC Motor with Backlash and Friction

First, we used the test-bed installed in the robotics and control laboratory of CITEDI-IPN which involves a DC motor manufactured by *LEADSHINE*. The system is governed by the dynamical model (1)–(6) whose parameters are provided in Table 2. The feedback system with fuzzy logic control and the plant with backlash is shown in Fig. 5. For the simulation, the load was required to move from the initial static position $\theta(0) = 0$ rad/s rad to the desired trajectory. The initial velocity was set to $\dot{\theta}(0) = 0$ rad/s.

Figure 6 shows output responses and the control signal for the DC motor with friction and backlash, driven by the Mamdani-type fuzzy logic controller. Figure 6a shows that the load position reaches the desired position. The motor position depicted in Fig. 6b shows a slightly steady-state error allowing the load reaches the desired position. Figure 6c shows the load position error where the settling time

Fig. 4 Control surface

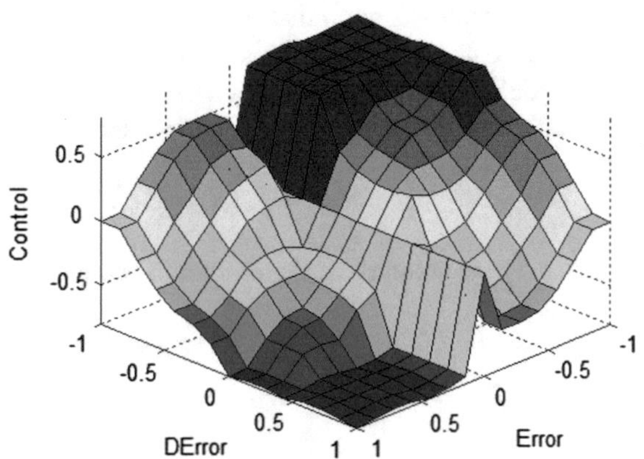

Table 2 Parameters

Parameter	Value	Unit
Rotor inertia (J)	3.11×10^{-5}	Kg m^2
Viscous friction coefficient (f_v)	2.25×10^{-5}	Nm s/rad
Coulomb friction coefficient (f_c)	7.745×10^{-3}	Nm
Amplifier constant (K_a)	4.2	A / V
Stiffness coefficient (K)	10000	
Dead-zone level (j_0)	0.0025	

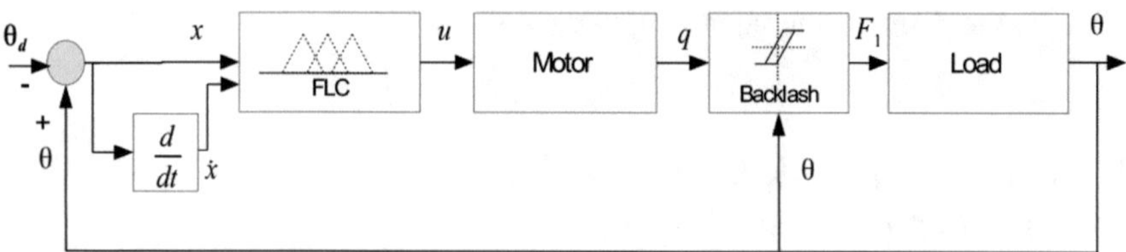

Fig. 5 The feedback system with fuzzy logic control and the plant with backlash

was T $\approx$ 5 s. Finally, Fig. 6d shows the control input u. These simulation results corroborates that the control objective is achieved.

4.1 Comparison Principle

Let us take into account the load position error, depicted in Fig. 7, as the trajectory to be analyzed. Although the trajectory may appear to be converging to the origin in finite-time, we will assume that the equilibrium point of the closed-loop system is stable according to Definition 1. The heuristic condition of the controller makes no possible to predict analytically the stability of the closed-loop system. This is the reason on the use of the comparison principle. Then, consider that there exist a fuzzy logic control input u that drives the position and velocity errors to the origin and there exists a positive definite function $V(t) = x^T x$. Assume that there exists a positive definite and symmetrical matrix Q such that the time derivative of V along the solution of the closed-loop system should satisfy

$$\dot{V} \leq -\lambda_{\min}\{Q\}\|x\|^2 = -\lambda_{\min}\{Q\}V. \tag{16}$$

Let $u\,(t)$ the solution of the differential equation

$$\dot{u} = -\lambda_{\min}\{Q\}u \tag{17}$$

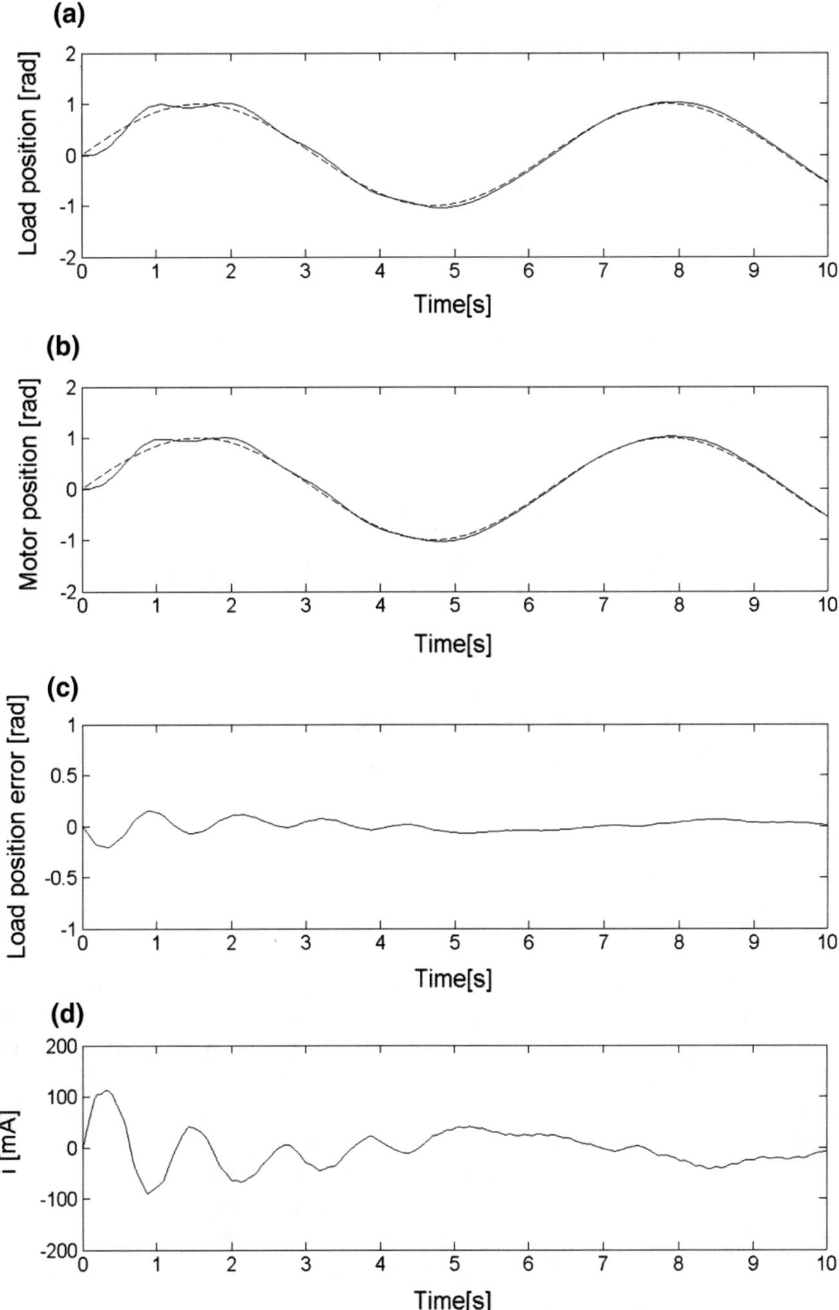

Fig. 6 **a** Load position, **b** motor position, **c** load position error, and **d** control input

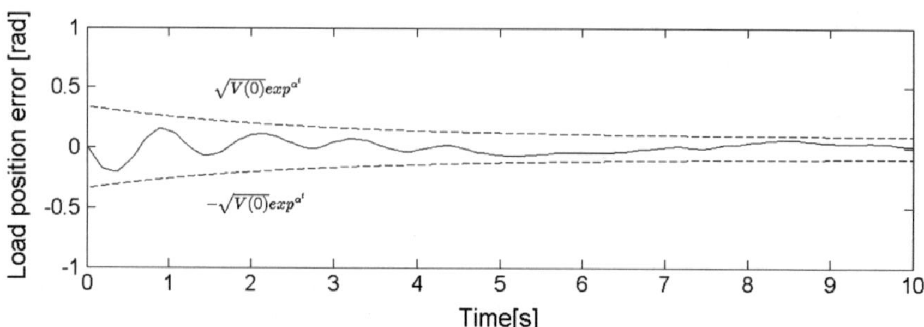

Fig. 7 Load position error (*solid line*) and the upper and lower solution of the differential inequality (*dotted line*) where $\alpha = \lambda_{\min}\{Q\}/2 = 0.38$

which is given by

$$u(t) = u(0)\, \exp^{-\lambda_{\min}\{Q\}t}. \tag{18}$$

By the comparison lemma, the solution $x(t)$ has the following bound:

$$\|x(t)\| = \sqrt{V(t)} \leq \sqrt{V(0)}\, \exp\left\{-\frac{\lambda_{\min}\{Q\}}{2}t\right\} \tag{19}$$

for all $t \geq 0$. Figure 7 shows that the solution of the differential inequality (17), under $V(0) = e(0)^2 + \dot{e}(0)^2 = \pi^2/4$ and $\lambda_{\min}\{Q\} = 0.76$, is a bound of the time response of the system.

5 Conclusions

The solution to the tracking control problem of a DC motor with friction and backlash using motor position measurements has been addressed. Mamdani fuzzy logic control proves to solve the problem. Due to the heuristic condition of the Mamdani type fuzzy controller, added to a free-model control design, makes it not possible to predict analytically the stability of the closed-loop system straightforwardly. The latter motivates the use of the comparison principle for a DC motor coupled together with the Mamdani type fuzzy controller. Since the linear nature of the plant, it is possible to propose an exponential solution of a linear differential inequality that can be a storage function also called Lyapunov function. Generalization of the results for a class of linear systems deserves a new paper. Simulation result corroborates the proposal.

References

1. Feng, G: A survey on analysis and design of model-based fuzzy control systems. IEEE Trans. Fuzzy Systems 14, 676–697 (2006)
2. Filippov A (1988). Differential equations with discontinuous right-hand sides. Kluwer Academic Publisher, Dordrecht.
3. Jang J, Sun C, and Mizutani E (1997). Neuro-fuzzy and Soft Computing: A Computational Approach to Learning and Machine Intelligence. MATLAB curriculum series. Prentice Hall, New Jersey.
4. Khalil H (2002). Nonlinear systems. Prentice Hall, New Jersey, third edition.
5. Kovacic, Z, Bogdan, S: Fuzzy Controller Design: Theory and Applications. CRC, Boca Raton (2006)
6. Lee, D., Joo, T., Tak, M.: Local stability analysis of continuous-time Takagi-Sugeno fuzzy systems: A fuzzy Lyapunov function approach. Information Sciences 257, 163–175 (2014)
7. Mamdani E. (1977). Application of fuzzy logic to approximate reasoning using linguistic synthesis. IEEE Transactions in Computers, 26(12), p 1182–1191.
8. Mamdani E and Assilian S (1975). An experiment in linguistic synthesis with a fuzzy logic controller. International Journal of Man-Machine Studies, 7(1), 1–13.
9. Melin P and Castillo O (2008). Type-2 Fuzzy Logic: Theory and Applications, volume 223. Springer, Berlin, Heidelberg.
10. Orlov Y (2009). Discontinuous Systems—Lyapunov Analysis and Robust Synthesis under Uncertainty Conditions. Springer-Verlag, London.
11. Orlov Y, Aguilar L, and Cadiou J (2003). Switched chattering control vs. backlash/friction phenomena in electrical servo-motors. International Journal of Control, 76(9/10), p 959–967.
12. Ponce I, Bentsman J, Orlov Y, and Aguilar L (2015). Generic nonsmooth $\mathrm{H}_{¥}$ output synthesis: application to a coal-fired boiler/turbine unit with actuator dead zone. IEEE Transactions on Control Systems Technology, 23(6), p 2117–2128.
13. Rascón R, Alvarez J, and Aguilar L (2016). Discontinuous $H_{¥}$ control for underactuated mechanical systems with friction and backlash. International Journal of Control, Automation and Systems.
14. Tao G and Kokotovic P (1996). Adaptive control of systems with actuator and sensor non-linearities. Wiley, New York.

Part VI
Optimization: Theory and Applications

Differential Evolution with Self-adaptive Gaussian Perturbation

M.A. Sotelo-Figueroa, Arturo Hernández-Aguirre, Andrés Espinal and J.A. Soria-Alcaraz

Abstract Differential evolution is a population-based metaheuristic that is widely used in Black-Box Optimization. The mutation is the main search operator and there are different implementation schemes reported in state of art literature. Nonetheless, such schemes lack mechanisms for an intensification stage, which can enable better search and avoid local optima. This article proposes a way to adapt the Covariance Matrix parameter of a Gaussian distribution that is used to generate a disturbance that improves the performance of two well-known mutation schemes. This disturbance allows working with problems with correlated variables. The test was performed over the CEC 2013 instances and the results were compared through the Friedman nonparametric test.

Keywords Differential evolution · Black-Box Optimization · Metaheuristics

M.A. Sotelo-Figueroa (✉) · J.A. Soria-Alcaraz
División de Ciencias Económico Administrativas, Departamento de Estudios Organizacionales, Universidad de Guanajuato, Fraccionamiento I El Establo, 36250 Guanajuato, Gto, Mexico
e-mail: masotelo@ugto.mx

J.A. Soria-Alcaraz
e-mail: jorge.soria@ugto.mx

A. Hernández-Aguirre
Departamento de Ciencias de la Computación, Centro de Investigación en Matemáticas, Jalisco S/N, 36240 Guanajuato, Gto, Mexico
e-mail: artha@cimat.mx

A. Espinal
Departamento de Estudios de Posgrado e Investigación, Tecnologico Nacional de México, Instituto Tecnológico de León, Avenida Tecnológico S/N, 37290 León, Gto, Mexico
e-mail: andres.espinal@itleon.edu.mx

P. Melin et al. (eds.), *Nature-Inspired Design of Hybrid Intelligent Systems*, Studies in Computational Intelligence 667, DOI 10.1007/978-3-319-47054-2_40

1 Introduction

Differential Evolution (DE) [23] is a population-based algorithm that has been widely studied. The research based on Differential Evolution basically has two lines [5]: the search for new mutation schemes with the purpose to improve their performance [1, 2, 9, 12, 17], and the study of the parameters for better results [3, 4, 6, 15, 20].

To improve the results obtained by DE it has been hybridized with some other techniques such as Local Search [19, 22], Particle Swarm Optimization [27], among others. They have also tried to propose new schemes of mutation as a Gaussian mutation [17]. However, those proposals seek to improve the DE performance but it makes the DE more complex than it was originally.

The black-box problems [8, 10, 11, 25] are problems that are not known a priori objective function with which it is working and cannot be used gradient-based algorithms.

In this article a self-adaptive Gaussian mutation is proposed. This mutation can be used with any mutation scheme to improve the results obtained by DE. To test, the mutation used the 28 functions from the CEC 2013 Special Session on Real-Parameter Optimization.[1] The results obtained by the self-adaptive Gaussian mutation, the Gaussian mutation were compared using the nonparametric Friedman Test [7].

2 Differential Evolution

Differential Evolution (*DE*) [23] was developed by R. Storn and K. Price in 1996. It is a vector-based evolutionary algorithm, and it can be considered as a further development to a Genetic Algorithm (*GA*) [13]. It is a stochastic search algorithm with self-organizing tendency and does not use the information of derivatives [26].

For a d-dimensional problem with d-parameters, a population of n solution are initially generated, so we have x_i solution vectors where $i = 1, 2, \ldots, n$. For each solution x_i, at any generation t we use the conventional notation as

$$x_i^t = \left(x_1^t, x_2^t, \ldots, x_d^t\right) \tag{1}$$

which consists of d-components in the d-dimensional space. This vector can be considered as the chromosomes or genomes.

This metaheuristic consists of three main steps: mutations, crossover, and selection.

---

[1] http://www.ntu.edu.sg/home/EPNSugan/index_files/CEC2013/CEC2013.htm.

Mutation is carried out by the mutation scheme. For each vector x_i at any time or generation t, first randomly choose three distinct vector x_p, x_q and x_r at t, and then generate a so-called donor vector by the mutation scheme.

$$v_i^{t+1} = x_p^t + F\left(x_q^t - x_r^t\right), \tag{2}$$

where F is in [0, 2] is a parameter, often referred to as the scale factor. This requires that the minimum number of population size is $n \geq 4$. We can see that the perturbation $\delta = F\left(x_q - x_r\right)$ to the vector x_p is used to generate a donor vector v_i, and such perturbation is directed and self-organized.

The crossover is controlled by a probability $C_r \in [0, 1]$ and actual crossover can be carried out in two ways: binomial and exponential. The binomial scheme performs crossover on each of the d-components or variables/parameters. By generating a uniformly distributed random number $r_i \in [0, 1]$, the jth components of v_i is manipulated as

$$u_{j,i}^{t+1} = \begin{cases} v_{j,i} & \text{if } r_i \leq C_r \\ x_{j,i} & \text{otherwise} \end{cases} \quad j = 1, 2, \ldots, d \tag{3}$$

This way, each component can be decided randomly whether to exchange with donor vector or not.

In the exponential scheme, a segment of the donor vector is selected and this segment starts with a random k with a random length L, which can include many components. Mathematically, this is to choose $k \in [0,\ d - 1]$ and $L \in [1,\ d]$ randomly, and we have

$$u_{j,i}^{t+1} = \begin{cases} v_{j,i}^t & \text{for } j = k, \ldots, k - L \in [1, d] \\ x_{j,i} & \text{otherwise} \end{cases} \tag{4}$$

The binomial crossover, due to its popularity in many DE literatures [23], is utilized in our implementation.

Selection is essentially the same as that used in genetic algorithms. It is to select the fittest, and for minimization problem, the minimum objective value. Therefore, we have

$$x_i^{t+1} = \begin{cases} u_i^{t+1} & \text{if } f(u_i^{t+1}) \leq f(x_i^t) \\ x_i^t & \text{otherwise} \end{cases} \tag{5}$$

All the three components can be seen in the pseudocode as shown in Algorithm 1. It is worth pointing out there that the use of J_r is to ensure that $v_i^{t+1} \neq x_i^t$ which may increase the evolutionary or exploratory efficiency. The overall search efficiency is controlled by two parameters: the differential weight F and the crossover probability C_r.

Algorithm 1 Differential Evolution Algorithm

Require: F differential weight, C_r crossover probability, n population size
Initializate the initial population.
while stopping criterion not met **do**
  for $i = 1$ to n **do**
    For each x_i randomly choose 3 distinct vector x_p, x_q and x_r.
    Generate a new vector v by DE scheme (2).
    Generate a random index $J_r \in \{1, 2, \ldots, d\}$ by permutation.
    Generate a randomly distributed number $r_i \in [0, 1]$
    for $j = 1$ to n **do**
      For each parameter $v_{j,i}$ (jth component of v_i), update
      $u_{j,i}^{t+1} = \begin{cases} v_{j,i}^{t+1} & \text{if } r_i \leq C_r \text{ or } j = J_r \\ x_{j,i}^{t} & \text{if } r_i > C_r \text{ or } j \neq J_r \end{cases}$
    end for
    Select and update the solution by (5).
  end for
  Update the counters such as $t = t + 1$
end while

In [18], five strategies to mutate the v vector are shown

DE/rand/1: $V_i = X_{r1} + F(X_{r2} - X_{r3})$.
DE/best/1: $V_i = X_{\text{best}} + F(X_{r1} - X_{r2})$.
DE/current to best/1: $V_i = X_i + F(X_{\text{best}} - X_i) + F(X_{r1} - X_{r2})$.
DE/best/2: $V_i = X_{\text{best}} + F(X_{r1} - X_{r2}) + F(X_{r3} - X_{r4})$.
DE/rand/2: $V_i = X_{r1} + F(X_{r2} - X_{r3}) + F(X_{r4} - X_{r5})$.

where r_1, r_2, r_3, r_4 are random and mutually different indices, witch should also be different from the trial vector's index i and X_{best} is the individual vector with best fitness.

2.1 *Gaussian Perturbation*

The Gaussian perturbation lets the DE to evolve using a scheme, if the vector is not improved then a perturbation for each dimension is applied. It is used to try to leave a local minimum, adding a randomly generated number by a normal distribution with zero mean and variance of one. An example of the Gaussian perturbation is shown in Fig. 1 and the Algorithm 2 explains the perturbation.

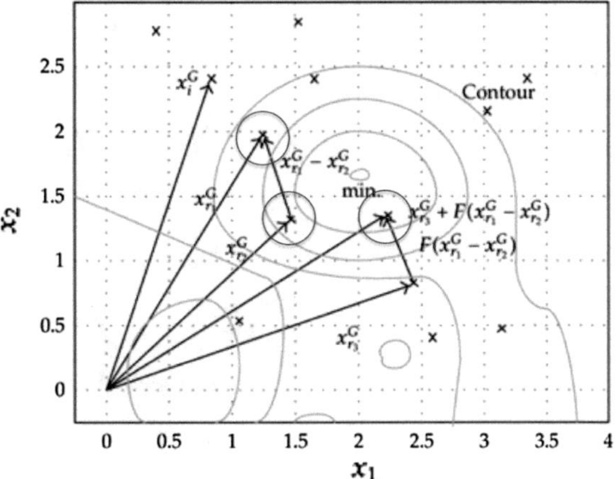

Fig. 1 Differential Evolution with Gaussian perturbation, the function contours was taken by Taher and Afsari [24]

Algorithm 2 Gaussian Perturbation

---

for $j = 1$ to n **do**
  if Individual j doesn't be updated **then**
    $u_j^{t+1} = N(x_j^t, 1)$
    Select and update the solution by (5).
  end if
end for

2.2 *Self-adaptive Gaussian Perturbation*

The Self-Adaptive Gaussian Perturbation lets the DE to start generating individuals oriented to the best position located. The perturbations are based on a Covariance Matrix, which is updated using a Rank-One Update [14], using the Eq. 6. This update is based on the best element found through the iterations. If the vector is not improved, then a perturbation for each dimension is applied. It tries to leave a local minimum, adding a randomly generated number by a normal distribution with zero mean and the covariance matrix is updated. An example of the Self-Adaptive Gaussian perturbation is shown in Fig. 2 and Algorithm 3 explains the perturbation.

$$\mathrm{Cov}_j^{t+1} = (1 - C_1) \times \mathrm{Cov}_j^t + C_1 \times (x_{\mathrm{best}} \times x_{\mathrm{best}}^T) \tag{6}$$

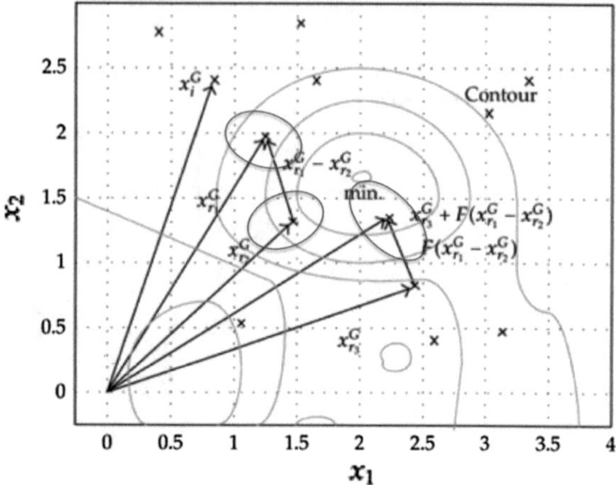

Fig. 2 Differential evolution with self-adaptive Gaussian perturbation, the function contours was taken by Taher and Afsari [24]

Algorithm 3 Self adaptive Gaussian Perturbation

for $j = 1$ to n **do**
  if Individual j doesn't be updated **then**
    $Cov_j^{t+1} = (1 - C_1) \times Cov_j^t + C_1 \times \left(x_{best} \times x_{best}^T\right)$
    $u_j^{t+1} = N(x_j^t, Cov_j^{t+1})$
    Select and update the solution by (5).
  end if
end for

3 Experiments

The 28 functions, which are used, are shown in Table 1, which were taken from the special session optimization CEC 2013. These functions are 2, 10, 30, and 50 dimensions. Fifty-one independent tests were performed to determine the average and standard deviation for each function. The function calls used are based on the dimension, 10,000*D we have 2D = 20,000, 10D = 100,000, 30D = 300,000, and 50D = 500,000. The search space for each function is $[-100, 100]^D$, the population must initialize randomly uniformly distributed evenly and must use the same method of optimization for all functions.

The parameters used by the DE were $F = 0.9$ and $C_r = 0.8$. These parameters were obtained by a process parameter optimization based on Covering Arrays (CA) [21]. The CAs were generated by the Covering Array Library (CAS) [16] the National Institute of Standards and Technology (NIST).[2]

[2]http://csrc.nist.gov/groups/SNS/acts/index.html.

Table 1 CEC2013 Functions

	No.	Function	$f_i^* = f_i(x^*)$
Unimodal functions	1	Sphere function	−1400
	2	Rotated high conditioned elliptic function	−1300
	3	Rotated Bent Cigar Function	−1200
	4	Rotated discus function	−1100
	5	Different powers function	−1000
Basic multimodal functions	6	Rotated Rosenbrock's function	−900
	7	Rotated Schaffers F7 function	−800
	8	Rotated Ackley's function	−700
	9	Rotated Weierstrass function	−600
	10	Rotated Griewank's function	−500
	11	Rastrigin's function	−400
	12	Rotated Rastrigin's function	−300
	13	Non-continuous rotated Rastrigin's function	−200
	14	Schwefel's function	−100
	15	Rotated Schwefel's function	100
	16	Rotated Katsuura function	200
	17	LunacekBi_Rastrigin function	300
	18	Rotated LunacekBi_Rastrigin function	400
	19	Expanded Griewank's plus Rosenbrock's function	500
	20	Expanded Scaffer's F6 function	600
Composition functions	21	Composition function 1	700
	22	Composition function 2	800
	23	Composition function 3	900
	24	Composition function 4	1000
	25	Composition function 5	1100
	26	Composition function 6	1200
	27	Composition function 7	1300
	28	Composition function 8	1400

To compare the results obtained by different schemes, mutation was performed by a nonparametric Friedman test [7] with a significance level of 99 % or p-value less than 0.1.

4 Results

Tables 2, 3, 4, and 5 show the results for each of the different dimensions tested. The results shown are the median and standard deviation obtained by applying each perturbation with the schemes mutation to different test function.

Table 2 Results obtained in two dimensions

Function	Rand1	Rand1 GP	Rand1 SAGP	Best1	Best1 GP	Best SAGP
1	−1.40E+03	−1.40E+03	−1.40E+03	−1.40E+03	−1.40E+03	−1.40E+03
2	−1.30E+03	−1.30E+03	−1.30E+03	−1.30E+03	−1.30E+03	−1.30E+03
3	−1.20E+03	−1.20E+03	−1.20E+03	−1.20E+03	−1.20E+03	−1.20E+03
4	−1.10E+03	−1.10E+03	−1.10E+03	−1.10E+03	−1.10E+03	−1.10E+03
5	−1.00E+03	−1.00E+03	−1.00E+03	−1.00E+03	−1.00E+03	−1.00E+03
6	−9.00E+02	−9.00E+02	−9.00E+02	−9.00E+02	−9.00E+02	−9.00E+02
7	−8.00E+02	−8.00E+02	−8.00E+02	−8.00E+02	−8.00E+02	−8.00E+02
8	−6.89E+02	−7.00E+02	−7.00E+02	−7.00E+02	−7.00E+02	−7.00E+02
9	−6.00E+02	−6.00E+02	−6.00E+02	−6.00E+02	−6.00E+02	−6.00E+02
10	−5.00E+02	−5.00E+02	−5.00E+02	−5.00E+02	−5.00E+02	−5.00E+02
11	−4.00E+02	−4.00E+02	−4.00E+02	−4.00E+02	−4.00E+02	−4.00E+02
12	−3.00E+02	−3.00E+02	−3.00E+02	−3.00E+02	−3.00E+02	−3.00E+02
13	−2.00E+02	−2.00E+02	−2.00E+02	−2.00E+02	−2.00E+02	−2.00E+02
14	−1.00E+02	−1.00E+02	−1.00E+02	−9.97E+01	−9.97E+01	−9.97E+01
15	1.00E+02	1.00E+02	1.00E+02	1.00E+02	1.00E+02	1.01E+02
16	2.00E+02	2.00E+02	2.00E+02	2.00E+02	2.00E+02	2.00E+02
17	3.00E+02	3.00E+02	3.00E+02	3.00E+02	3.02E+02	3.00E+02
18	4.00E+02	4.00E+02	4.00E+02	4.00E+02	4.02E+02	4.00E+02
19	5.00E+02	5.00E+02	5.00E+02	5.00E+02	5.00E+02	5.00E+02
20	6.00E+02	6.00E+02	6.00E+02	6.00E+02	6.00E+02	6.00E+02
21	7.00E+02	7.00E+02	7.00E+02	7.00E+02	7.00E+02	7.00E+02
22	8.00E+02	8.00E+02	8.00E+02	8.00E+02	8.00E+02	8.00E+02
23	9.00E+02	9.00E+02	9.00E+02	9.00E+02	9.00E+02	9.00E+02
24	1.00E+03	1.00E+03	1.00E+03	1.00E+03	1.00E+03	1.00E+03
25	1.10E+03	1.10E+03	1.10E+03	1.10E+03	1.10E+03	1.10E+03
26	1.20E+03	1.20E+03	1.20E+03	1.20E+03	1.20E+03	1.20E+03
27	1.40E+03	1.30E+03	1.30E+03	1.40E+03	1.30E+03	1.30E+03
28	1.40E+03	1.40E+03	1.40E+03	1.40E+03	1.40E+03	1.40E+03

The results for two-dimensions are in Table 2, we can see that the median reported is practically the best for each of the test functions, and the standard deviation almost equals to zero, which means that each one of the 51 independent tests almost reached the same result. For other dimensions, the results obtained with the different schemes for each objective function mutation vary.

Table 6 shows the values of the nonparametric Friedman test applied to each of the dimensions, the dimension 2 is the only one with p-value less than 0.1 thus you cannot determine statistically if there is a scheme that has a different mutation others performance. The other dimensions have p-values less than 0.1 and a post-hoc procedure was performed [7] to determine which mutation schemes gave better results.

Table 3 Results obtained in ten dimensions

Function	Rand1	Rand1 GP	Rand1 SAGP	Best1	Best1 GP	Best SAGP
1	−1.40E+03	−1.40E+03	−1.40E+03	−1.40E+03	−1.40E+03	−1.40E+03
2	4.16E+05	3.42E+05	1.59E+04	−1.30E+03	−9.23E+02	−1.30E+03
3	3.43E+05	2.54E+06	3.82E+04	−1.20E+03	−1.19E+03	−1.20E+03
4	9.85E+02	−5.55E+02	−8.50E+02	−1.10E+03	−1.10E+03	−1.10E+03
5	−1.00E+03	−1.00E+03	−1.00E+03	−1.00E+03	−1.00E+03	−1.00E+03
6	−9.00E+02	−9.00E+02	−9.00E+02	−9.00E+02	−9.00E+02	−9.00E+02
7	−7.79E+02	−7.87E+02	−7.91E+02	−8.00E+02	−8.00E+02	−8.00E+02
8	−6.80E+02	−6.80E+02	−6.80E+02	−6.80E+02	−6.80E+02	−6.80E+02
9	−5.91E+02	−5.92E+02	−5.91E+02	−5.91E+02	−5.99E+02	−5.96E+02
10	−4.99E+02	−4.99E+02	−4.99E+02	−5.00E+02	−5.00E+02	−5.00E+02
11	−3.89E+02	−3.80E+02	−3.91E+02	−3.96E+02	−3.97E+02	−3.97E+02
12	−2.64E+02	−2.57E+02	−2.68E+02	−2.87E+02	−2.85E+02	−2.91E+02
13	−1.60E+02	−1.58E+02	−1.66E+02	−1.80E+02	−1.70E+02	−1.77E+02
14	1.35E+03	1.03E+03	1.27E+03	1.69E+02	8.71E+01	1.37E+02
15	2.09E+03	1.11E+03	1.83E+03	1.98E+03	4.79E+02	1.46E+03
16	2.01E+02	2.01E+02	2.01E+02	2.01E+02	2.01E+02	2.01E+02
17	3.33E+02	3.38E+02	3.34E+02	3.17E+02	3.16E+02	3.16E+02
18	4.25E+02	4.31E+02	4.23E+02	4.15E+02	4.14E+02	4.13E+02
19	5.03E+02	5.03E+02	5.03E+02	5.01E+02	5.01E+02	5.01E+02
20	6.04E+02	6.02E+02	6.04E+02	6.04E+02	6.02E+02	6.03E+02
21	1.10E+03	1.10E+03	1.10E+03	1.10E+03	1.10E+03	1.10E+03
22	2.55E+03	2.24E+03	2.50E+03	1.23E+03	1.08E+03	1.14E+03
23	3.05E+03	2.54E+03	2.79E+03	2.91E+03	2.03E+03	2.54E+03
24	1.22E+03	1.22E+03	1.22E+03	1.22E+03	1.22E+03	1.21E+03
25	1.32E+03	1.32E+03	1.32E+03	1.32E+03	1.32E+03	1.31E+03
26	1.43E+03	1.40E+03	1.40E+03	1.40E+03	1.40E+03	1.40E+03
27	1.85E+03	1.84E+03	1.84E+03	1.83E+03	1.84E+03	1.81E+03
28	1.70E+03	1.70E+03	1.70E+03	1.70E+03	1.70E+03	1.70E+03

Table 7 contains the results of post hoc procedure for the dimensions 20, 30, and 50. This table contains the position in each dimension that occupies each of the schemes used; the results are displayed in ascending order with the smallest number that scheme had as the best performance.

Table 4 Results obtained in 30 dimensions

Function	Rand1	Rand1 GP	Rand1 SAGP	Best1	Best1 GP	Best SAGP
1	−1.35E+03	−1.12E+03	−1.40E+03	−1.40E+03	−1.40E+03	−1.40E+03
2	1.12E+09	9.79E+07	6.12E+06	6.59E+07	3.32E+07	3.63E+06
3	2.07E+12	1.97E+10	1.38E+10	1.55E+09	7.16E+08	8.88E+07
4	3.43E+05	2.92E+04	9.48E+04	1.83E+05	2.07E+04	4.98E+04
5	−6.70E+02	−6.54E+02	−9.99E+02	−1.00E+03	−1.00E+03	−1.00E+03
6	−8.52E+02	−8.10E+02	−8.78E+02	−8.83E+02	−8.77E+02	−8.84E+02
7	1.04E+03	−6.67E+02	3.54E+02	−4.51E+02	−7.13E+02	−2.92E+02
8	−6.79E+02	−6.79E+02	−6.79E+02	−6.79E+02	−6.79E+02	−6.79E+02
9	−5.60E+02	−5.60E+02	−5.71E+02	−5.60E+02	−5.60E+02	−5.71E+02
10	6.89E+02	1.09E+02	−4.97E+02	−5.00E+02	−4.99E+02	−5.00E+02
11	−1.89E+02	−1.66E+02	−2.41E+02	−3.63E+02	−3.64E+02	−3.33E+02
12	−1.26E+01	−1.33E+01	−7.92E+01	−7.18E+01	−4.95E+01	−1.64E+02
13	8.91E+01	9.11E+01	2.43E+01	3.16E+01	5.17E+01	−4.13E+01
14	6.75E+03	6.74E+03	4.23E+03	2.25E+03	1.65E+03	4.05E+03
15	9.00E+03	7.17E+03	4.39E+03	8.73E+03	6.98E+03	4.32E+03
16	2.03E+02	2.02E+02	2.02E+02	2.03E+02	2.02E+02	2.02E+02
17	5.78E+02	5.80E+02	5.65E+02	3.78E+02	3.87E+02	3.81E+02
18	6.66E+02	6.75E+02	6.44E+02	4.69E+02	4.69E+02	4.75E+02
19	5.40E+02	5.45E+02	5.21E+02	5.03E+02	5.08E+02	5.16E+02
20	6.14E+02	6.12E+02	6.13E+02	6.14E+02	6.11E+02	6.13E+02
21	1.24E+03	1.14E+03	9.26E+02	1.00E+03	1.00E+03	1.00E+03
22	8.73E+03	7.95E+03	5.32E+03	3.11E+03	2.53E+03	5.33E+03
23	1.02E+04	8.55E+03	5.44E+03	9.70E+03	8.52E+03	5.53E+03
24	1.31E+03	1.30E+03	1.28E+03	1.30E+03	1.30E+03	1.27E+03
25	1.41E+03	1.40E+03	1.37E+03	1.40E+03	1.40E+03	1.37E+03
26	1.61E+03	1.53E+03	1.57E+03	1.60E+03	1.41E+03	1.57E+03
27	2.68E+03	2.62E+03	2.34E+03	2.64E+03	2.60E+03	2.31E+03
28	2.48E+03	2.49E+03	1.72E+03	1.70E+03	1.70E+03	1.70E+03

Table 5 Results obtained in 50 dimensions

Function	Rand1	Rand1 GP	Rand1 SAGP	Best1	Best1 GP	Best SAGP
1	−1.04E+03	−3.35E+02	−1.40E+03	−1.40E+03	−1.40E+03	−1.40E+03
2	3.21E+09	3.05E+08	1.40E+07	2.00E+08	8.54E+07	1.54E+07
3	2.09E+13	4.61E+10	1.81E+09	3.11E+11	9.91E+09	1.52E+09
4	6.10E+05	4.91E+04	1.33E+05	4.05E+05	4.13E+04	1.11E+05
5	1.15E+02	−5.50E+02	−9.89E+02	−1.00E+03	−1.00E+03	−1.00E+03
6	−7.74E+02	−6.85E+02	−8.55E+02	−8.55E+02	−8.54E+02	−8.54E+02
7	2.24E+03	−6.60E+02	2.87E+02	−5.27E+01	−6.93E+02	−3.75E+02
8	−6.79E+02	−6.79E+02	−6.79E+02	−6.79E+02	−6.79E+02	−6.79E+02
9	−5.27E+02	−5.37E+02	−5.45E+02	−5.27E+02	−5.62E+02	−5.45E+02
10	2.23E+03	7.03E+02	−4.87E+02	−4.99E+02	−4.96E+02	−4.90E+02
11	1.23E+01	3.64E+01	−5.07E+01	−3.01E+02	−2.87E+02	−1.76E+02
12	2.33E+02	2.44E+02	2.23E+02	1.55E+02	1.88E+02	7.88E+01
13	3.29E+02	3.53E+02	3.32E+02	2.57E+02	3.16E+02	2.10E+02
14	1.17E+04	1.23E+04	8.22E+03	4.96E+03	3.95E+03	8.11E+03
15	1.65E+04	1.44E+04	8.75E+03	1.61E+04	1.42E+04	8.75E+03
16	2.03E+02	2.03E+02	2.03E+02	2.03E+02	2.03E+02	2.03E+02
17	8.28E+02	8.29E+02	8.53E+02	4.75E+02	4.91E+02	4.67E+02
18	9.03E+02	9.05E+02	9.16E+02	5.42E+02	5.57E+02	7.80E+02
19	1.34E+03	7.46E+02	5.45E+02	5.08E+02	5.19E+02	5.41E+02
20	6.24E+02	6.21E+02	6.22E+02	6.24E+02	6.21E+02	6.22E+02
21	1.57E+03	1.42E+03	1.02E+03	1.00E+03	1.10E+03	1.00E+03
22	1.42E+04	1.45E+04	9.90E+03	6.37E+03	5.08E+03	9.76E+03
23	1.74E+04	1.57E+04	1.04E+04	1.72E+04	1.55E+04	1.03E+04
24	1.40E+03	1.39E+03	1.34E+03	1.39E+03	1.39E+03	1.34E+03
25	1.49E+03	1.48E+03	1.44E+03	1.49E+03	1.48E+03	1.44E+03
26	1.70E+03	1.68E+03	1.64E+03	1.69E+03	1.69E+03	1.64E+03
27	3.56E+03	3.47E+03	3.06E+03	3.50E+03	3.45E+03	3.00E+03
28	2.38E+03	2.24E+03	1.81E+03	1.80E+03	1.80E+03	1.80E+03

Table 6 Friedman nonparametric test

	2D	10D	30D	50D
Value	4.4183	68.3418	66.7551	69.4234
p-value	0.4908	6.20E−11	5.09E−11	5.41E−11

Table 7 Post-hoc procedure to the Friedman nonparametric test

Scheme	10	30	50
Best1 SA Mutation	1.78	2	2.01
Best1 Mutation	2.46	2.67	2.6
Best1	3.12	2.35	3.01
Rand1 SA Mutation	3.69	2.92	3.21
Rand1 Mutation	4.78	4.53	4.6
Rand1	5.14	5.5	5.53

5 Conclusions

With the results obtained in Sect. 4 we can conclude the following:

Among the mutation schemes reported in the state of the art and tested the Best1 scheme is enabling better results.
The problems with low dimensionality do not need the application of perturbations, the statistical test does not show significant difference. However, high dimensionality problems have improved with the use of perturbations.
Self-adaptive Gaussian Perturbation orientates the individuals through mutations to the best result obtained, and it allows working with any kind of Black-Box problems like unimodal functions or multimodal functions.

Acknowledgments The authors want to thank to *Universidad de Guanajuato* (UG) for the support to this research.

References

1. Al-dabbagh, R., Botzheim, J., Al-dabbagh, M.: Comparative analysis of a modified differential evolution algorithm based on bacterial mutation scheme. In: Differential Evolution (SDE), 2014 IEEE Symposium on. pp. 1–8 (Dec 2014)
2. Bhowmik, P., Das, S., Konar, A., Das, S., Nagar, A.: A new differential evolution with improved mutation strategy. In: Evolutionary Computation (CEC), 2010 IEEE Congress on. pp. 1–8 (July 2010)
3. Brest, J., Greiner, S., Boskovic, B., Mernik, M., Zumer, V.: Self-adapting control parameters in differential evolution: A comparative study on numerical benchmark problems. Evolutionary Computation, IEEE Transactions on 10(6), 646–657 (Dec 2006)
4. Brest, J., Maučec, M.: Self-adaptive differential evolution algorithm using population size reduction and three strategies. Soft Computing 15(11), 2157–2174 (2011)
5. Das, S., Suganthan, P.: Differential evolution: A survey of the state-of-the-art. Evolutionary Computation, IEEE Transactions on 15(1), 4–31 (Feb 2011)
6. Das, S., Konar, A., Chakraborty, U.K.: Two improved differential evolution schemes for faster global search. In: Proceedings of the 7th Annual Conference on Genetic and Evolutionary Computation. pp. 991–998. GECCO '05, ACM, New York, NY, USA (2005)
7. Derrac, J., García, S., Molina, S., Herrera, F.: A practical tutorial on the use of nonparametric statistical tests as a methodology for comparing evolutionary and swarm intelligence algorithms. Swarm and Evolutionary Computation pp. 3–18 (2011)

8. Droste, S., Jansen, T., Wegener, I.: Upper and lower bounds for randomized search heuristics in black-box optimization. Theory of Computing Systems 39(4), 525–544 (2006)
9. Einarsson, G., Runarsson, T., Stefansson, G.: A competitive coevolution scheme inspired by de. In: Differential Evolution (SDE), 2014 IEEE Symposium on. pp. 1–8 (Dec 2014)
10. El-Abd, M.: Black-box optimization benchmarking for noiseless function testbed using artificial bee colony algorithm. In: Proceedings of the 12th Annual Conference Companion on Genetic and Evolutionary Computation. pp. 1719–1724. GECCO '10, ACM, New York, NY, USA (2010)
11. El-Abd, M., Kamel, M.S.: Black-box optimization benchmarking for noiseless function testbed using particle swarm optimization. In: Proceedings of the 11th Annual Conference Companion on Genetic and Evolutionary Computation Conference: Late Breaking Papers. pp. 2269–2274. GECCO '09, ACM, New York, NY, USA (2009)
12. Fan, Q., Yan, X.: Self-adaptive differential evolution algorithm with zoning evolution of control parameters and adaptive mutation strategies. Cybernetics, IEEE Transactions on PP (99), 1–1 (2015)
13. Holland, J.: Adaptation in natural and artificial systems. University of Michigan Press (1975)
14. Igel, C., Suttorp, T., Hansen, N.: A computational efficient covariance matrix update and a (1 +1)-cma for evolution strategies. In: Proceedings of the 8th Annual Conference on Genetic and Evolutionary Computation. pp. 453–460. GECCO '06, ACM, New York, NY, USA (2006), http://doi.acm.org/10.1145/1143997.1144082
15. Jin, W., Gao, L., Ge, Y., Zhang, Y.: An improved self-adapting differential evolution algorithm. In: Computer Design and Applications (ICCDA), 2010 International Conference on. vol. 3, pp. V3–341–V3–344 (June 2010)
16. Kacker, R.N., Kuhn, D.R., Lei, Y., Lawrence, J.F.: Combinatorial testing for software: An adaptation of design of experiments. Measurement 46(9), 3745 – 3752 (2013)
17. Li, D., Chen, J., Xin, B.: A novel differential evolution algorithm with gaussian mutation that balances exploration and exploitation. In: Differential Evolution (SDE), 2013 IEEE Symposium on. pp. 18–24 (April 2013)
18. Luke, S.: Essentials of Metaheuristics. Lulu (2009)
19. Noman, N., Iba, H.: Accelerating differential evolution using an adaptive local search. Evolutionary Computation, IEEE Transactions on 12(1), 107–125 (Feb 2008)
20. Omran, M., Salman, A., Engelbrecht, A.: Self-adaptive differential evolution. In: Hao, Y., Liu, J., Wang, Y., Cheung, Y.m., Yin, H., Jiao, L., Ma, J., Jiao, Y.C. (eds.) Computational Intelligence and Security, Lecture Notes in Computer Science, vol. 3801, pp. 192–199. Springer Berlin Heidelberg (2005)
21. Rodriguez-Cristerna, A., Torres-Jiménez, J., Rivera-Islas, I., Hernandez-Morales, C., Romero-Monsivais, H., Jose-Garcia, A.: A mutation-selection algorithm for the problem of minimum brauer chains. In: Batyrshin, I., Sidorov, G. (eds.) Advances in Soft Computing, Lecture Notes in Computer Science, vol. 7095, pp. 107–118. Springer Berlin Heidelberg (2011)
22. Sotelo-Figueroa, M.A., Hernández-Aguirre, A., Espinal, A., Soria-Alcaraz, J.A.: Evolución diferencial con perturbaciones gaussianas. Research in Computing Science 94, 111–122 (2015)
23. Storn, R., Price, K.: Differential evolution - a simple and efficient heuristic for global optimization over continuous spaces. J. of Global Optimization 11, 341–359 (December 1997)
24. Taher, S.A., Afsari, S.A.: Optimal location and sizing of upqc in distribution networks using differential evolution algorithm. Mathematical Problems in Engineering p. 20 (2012)
25. Wang, G., Goodman, E., Punch, W.: Toward the optimization of a class of black box optimization algorithms. In: Tools with Artificial Intelligence, 1997. Proceedings., Ninth IEEE International Conference on. pp. 348–356 (Nov 1997)
26. Yang, X.S.: Nature Inspired Metaheuristic Algorithms. Luniver Press, 2da edn. (2008)
27. Zavala, A., Aguirre, A., Diharce, E.: Particle evolutionary swarm optimization algorithm (peso). In: Computer Science, 2005. ENC 2005. Sixth Mexican International Conference on. pp. 282–289 (2005)

Optimization Mathematical Functions for Multiple Variables Using the Algorithm of Self-defense of the Plants

Camilo Caraveo, Fevrier Valdez and Oscar Castillo

Abstract In this work a new bio-inspired metaheuristic based on the self-defense mechanism of plants is presented. This new optimization algorithm is applied to solve optimization problems, in this case optimization of mathematical functions for multiple variables, other works related to this, where the same algorithm is used with some modifications and improvements is presented in (Caraveo et al. in Advances in artificial intelligence and soft computing, Springer International Publishing, pp. 227–237, 2015 [3]). Since its inception the planet has gone through changes, so plants have had to adapt to these changes and adopt new techniques to defend from natural predators in this case. Many works have shown that plants have mechanisms of self-defense to protect themselves from predators. When the plants detect the presence of invading organisms this triggers a series of chemical reactions that are released to air and attract natural predators of the invading organism (Bennett and Wallsgrove in New Phytol, 127(4):617–633, 1994 [1]; Neyoy et al. Recent Advances on hybrid intelligent systems, Springer, Berlin, pp. 259–271, 2013 [10]; Ordeñana in Costa Rica, 63:22–32, 2002 [11]). For the development of this algorithm we consider as a main idea the predator prey model of Lotka and Volterra, where two populations are considered and the objective is to maintain a balance between the two populations.

Keywords Predator prey model · Plants · Self-defense · Mechanism · Clon · Lévy flights

C. Caraveo (✉) · F. Valdez · O. Castillo
Division of Graduate Studies, Institute of Technology Tijuana,
Tijuana, Mexico
e-mail: Camilo.caraveo@Gmail.com

F. Valdez
e-mail: fevrier@tectijuana.mx

O. Castillo
e-mail: Ocastillo@tectijuana.mx

P. Melin et al. (eds.), *Nature-Inspired Design of Hybrid Intelligent Systems*,
Studies in Computational Intelligence 667, DOI 10.1007/978-3-319-47054-2_41

1 Introduction

Several metaheuristic algorithms have been developed to solve various combinatorial optimization problems. These can be classified into different groups based on the criteria that are, for example, deterministic, iterative, population based, stochastic, etc. While an algorithm that works with a set of solutions and the use of multiple iterations to approach the desired solution is called as iterative algorithm based on populations [3].

Recently, there have been proposed and developed multiple methods of search and optimization inspired by natural processes. This with the goal of solving particular problems in the area of computer science has tried different bio-inspired methods, such as ACO, PSO, BCO, GA, SGA, FPA, etc. [8, 9, 12, 17], in all cases trying to get the resolution of a specific problem with a smaller error [3].

In this work the application of a new optimization algorithm inspired in the self-defense mechanisms of plants is presented. This in order to compete against the existing optimization methods. In nature, plants are exposed to many different pathogens in the environment. However, only a few can affect them. If a particular pathogen is unable to successfully attack a plant, it is said that it is resistant to it, in other words, cannot be affected by the pathogen [3].

The proposed approach takes as its main basis the Lotka and Volterra predator-prey model, which is a system formed by a pair of first-order differential equations, nonlinear for moderating the growth of two populations that interact with each other (predator and prey) [3, 7].

2 Self-defense Mechanisms of the Plants

In nature, plants as well as animals are exposed to a large number of invading organisms, such as insects, fungi, bacteria, and viruses that can cause various types of diseases, and even death [3, 11, 16].

Defense mechanisms are automatic processes that protect the individual against external or internal threats. The plant is able to react to external stimuli. When it detects the presence or attack of an organism triggers a series of chemical reactions that are released into the air that attracts natural predator of the assailant or cause internal damage to the aggressor [10]. In Fig. 1 a general scheme is shown to illustrate the behavior of the plant when it detects the presence or attack by a predator [3].

The leaves normally release into the air small amounts of volatile chemicals, but when a plant is damaged by herbivorous insects, the amount of chemicals tends to grow. Volatile chemicals vary depending on the plant and species of herbivorous insects [13]. These chemicals attract both predators and parasitic insects that are natural enemies of herbivores, see Fig. 1. Such chemicals, which work in the communication between two or more species, as well as those who serve as

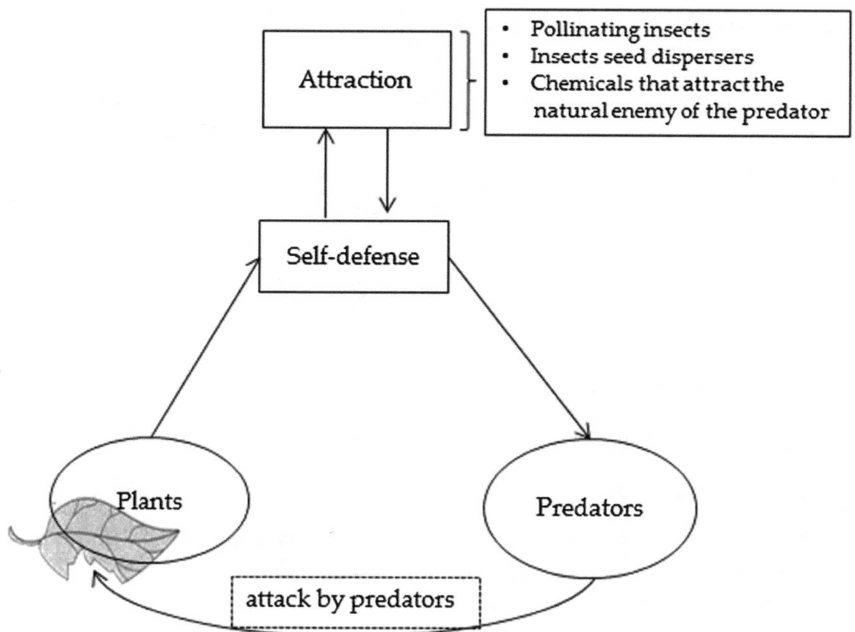

Fig. 1 Illustration of the process of self-defense of the plant

messengers between members of the same species are called semi-chemicals [1–3, 6, 11].

3 Predator-Prey Model

The organisms live in communities, forming intricate relationships of interaction, where each species directly or indirectly dependent on the presence of the other. One of the tasks of Ecology is to develop a theory of community organization for understanding the causes of diversity and mechanisms of interaction [14, 15, 18]. In this paper, we consider the interaction of two whose population size at time t is $x\ (t)$ and $y\ (t)$ species [2, 3]. Furthermore, we assume that the change in population size can be written as [3]:

$$\frac{\mathrm{d}y}{\mathrm{d}t} = I(x, y) \tag{1}$$

$$\frac{\mathrm{d}x}{\mathrm{d}t} = P(x, y) \tag{2}$$

3.1 Analysis of the Lotka and Volterra Model

This model is based on the following assumptions. We have a model of interaction between *x*(*t*) and *y*(*t*) is given by the following system: Eqs. (3) and (4)

$$\frac{dx}{dt} = \alpha x - \beta xy \tag{3}$$

$$\frac{dy}{dt} = -\delta xy + \lambda y \tag{4}$$

x is the number of prey
y is the number of predators
$\frac{dx}{dt}$ is the growth of the population of prey time t
$\frac{dy}{dt}$ is the growth of the population of predator at time t
α it represents the birth rate of prey in the absence of predator
β it represents the death rate of predators in the absence of prey
δ measures the susceptibility of prey
λ measures the ability of predation.

4 Proposed Optimization Algorithm Based on the Self-defense Mechanisms of Plants

The proposed approach takes as its main basis the Lotka and Volterra predator-prey model, which is a system formed by a pair of differential equations of first-order nonlinear moderating the growth of two populations that interact with each other (predator and prey) [4]. In Fig. 2 a general scheme of our proposal was designed as the based on the traditional model of predator prey shows, using the principle of the dynamics of both populations the evolutionary process of plants is generated to develop the techniques of self-defense [5].

In nature, plants have different methods of reproduction, in our approach we consider only the most common: clone, graft and pollen. **Clone**: the offspring identical to the parent plant. **Graft**: it takes a stem of a plant and is encrusted on another to generate an alteration in the structure of the plant nd inherit characteristics of other individual. **Pollen**: one plant pollinates other flowers and generates a seed and the descent is a plant with characteristics of both plants [3].

To generate the initial population of the algorithm we use the equations of the model of Lotka and Volterra, the mathematical representation is shown in Sect. 3.1 Eqs. (3) and (4). Equation (3) is used to generate the population of prey in this case (plants), and Eq. (4) is used to generate the population of predators in this case

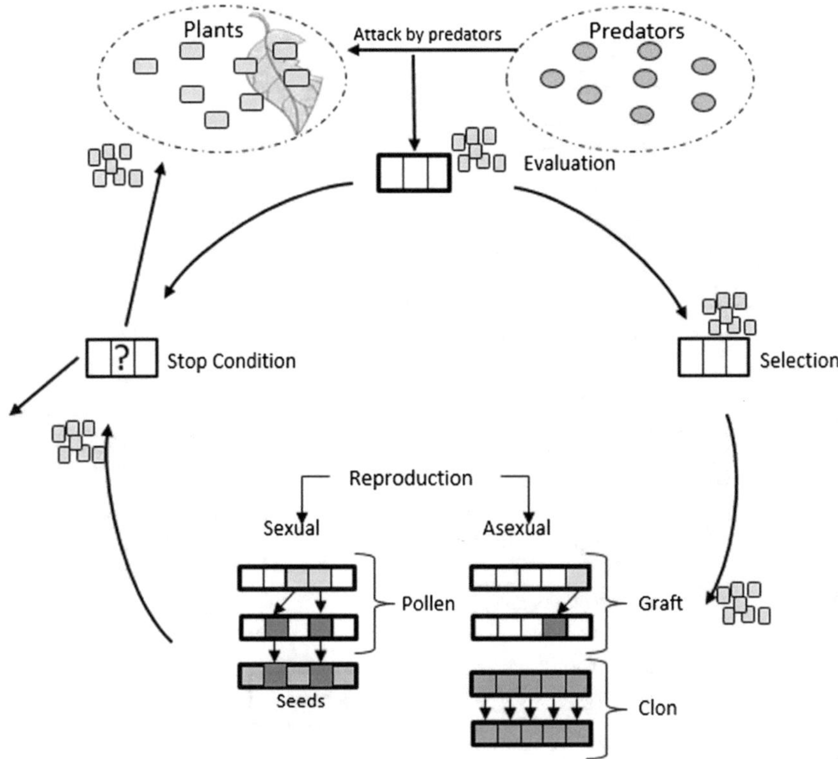

Fig. 2 General representation of our proposal

(herbivores), as mentioned above functions predator prey model is used to model our proposed model variables adapted to the proposal.

Figure 3 describes the stages of the optimization algorithm inspired by the defense mechanisms of plants.

The initial sizes of both populations (prey, predators) and parameters (α, β, δ, λ) are also defined by the user, the model of Lotka and Volterra recommended the following parameter values $\alpha = 0.4$, $\beta = 0.37$, $\delta = 0.3$, $\lambda = 0.05$. Both populations that initiated these populations interact with each other prey and predator, use this method to generate new offspring of plants, these plant reproduction in biological processes are applied. The population is reevaluated and if the stop criterion is not satisfied, return the iterative cycle of the algorithm.

In [3] the authors present a work using a clone reproduction method, in this paper we are presenting a new variant of the method using the pollination of plants as biological reproduction method, to emulate pollination by insect pollinators, we propose to use the method of Lévy flights, algorithms that use this method are able to improve the performance of the algorithm allowing them greater balance between exploration and exploitation in the search space of our proposal.

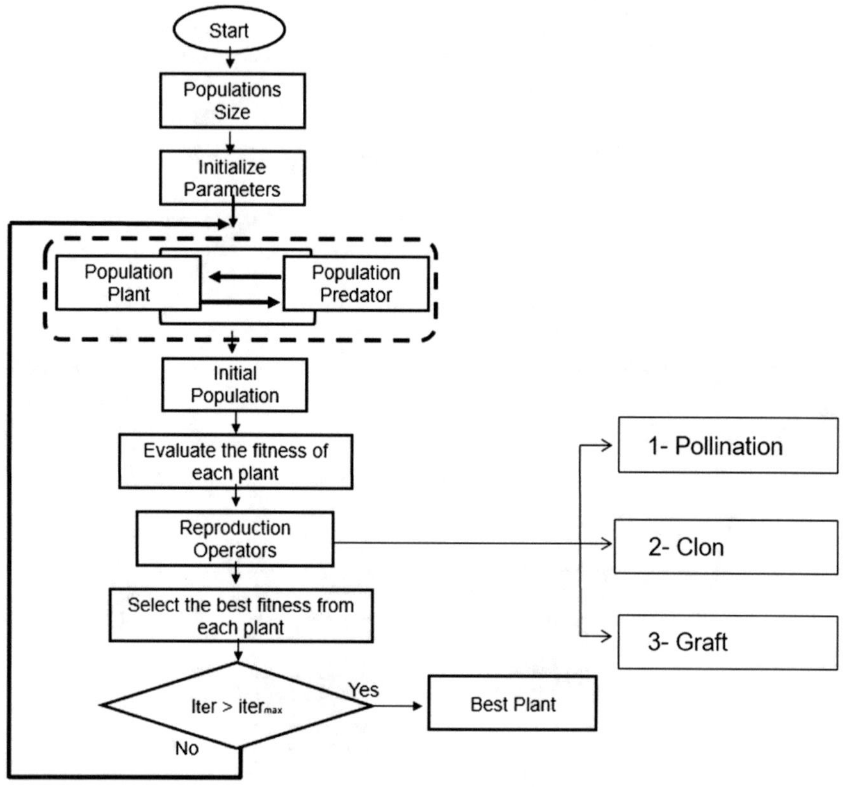

Fig. 3 Flowchart illustrating the proposed algorithm

5 Simulation Results

In this section, the results obtained are shown to the test the performance of the algorithm using the method of reproduction by clone and pollination with Lévy flights, we used a set of five benchmark functions, where the goal is to approximate the value of the function to zero. 30 experiments were performed for the following mathematical functions the evaluation is for 10, 30, 50 Variables. Table 1 shows definitions of the mathematical functions used in this work.

5.1 Parameters Settings for the Algorithm

The common parameters in the control of algorithms are the population size and the maximum number of iterations and some others, in this case the algorithm uses many parameters compared to other bio-inspired methods. We are planning to find

Table 1 Mathematical functions

Function	Representation mathematical
Sphere	$f(x) = \sum_{i=1}^{n} x_i^2$
Griewank	$f(x) = \sum_{i=1}^{d} \frac{x_i^2}{4000} - \prod_{i=1}^{d} \cos\left(\frac{x_i}{\sqrt{i}}\right) + 1$
Rastrigin	$f(x) = -a \cdot exp\left(-b \cdot \sqrt{\frac{1}{n}\sum_{i=1}^{n} x_i^2}\right) - exp\left(\frac{1}{2}\sum_{i=1}^{n} cos(cx_i)\right) + a + exp(1)$
Ackley	$f(x) = 10n + \sum_{i=1}^{n}\left[x_i^2 - 10\cos(2\pi x_i)\right]$
Rosenbrock	$f(x) = \sum_{i=1}^{n-1}\left[100\left(x_{i+1} - x_i^2\right)^2 + (1 - x_i)^2\right]$

a way to optimize the parameters setting of the algorithm using intelligent computing techniques.

The proposal uses more parameters than other traditional algorithms, it is difficult to find the optimal parameters to achieve acceptable results. The parameters of the algorithm are usually moved manually by trial and error to observe and determine the range of parameters used. The configuration parameters are defined below: the parameters (α, β, δ, λ), α = 0–1, β = 0–1, δ = 0–1, λ = 0.0–0.1. We also need to define the size of prey populations (plants) and predators (herbivores), the model does not recommend optimal values for populations, and we use plants = 350, Herbivores = 300, must also define the time and maximum number of iterations, and we use the iteration in the range of 200–500 to observe the results.

Tables 2, 3 and 4 show the experimental results for the benchmark mathematical functions used in this work. The Tables show the results of the evaluations for each function with 10, 30, 50 dimensions; where we can see the best, worse, standard deviation, average, values shown was obtained from 30 experiments performed by moving the parameters manually α, β, δ, λ in the above ranges established.

Table 2 30 Experimental results with 10 dimensions

Function	Reproduction operator	Algorithm performance with 10 dimensions			
		Best	Worse	σ	Average
Sphere	Clone	3.62E−308	4.994E−23	1.2628E−23	3.700E−24
Sphere	Pollination	1.63094E−84	1.66902E−37	3.04705E−38	5.57106E−39
Ackley	Clone	8.88E−16	3.21E−09	6.8865E−10	2.06E−10
Ackley	Pollination	8.88E−16	2.93E−14	6.96208E−15	2.02E−14
Rosenbrock	Clone	0.0339	8.9300	2.16266548	1.4421
Rosenbrock	Pollinitation	0.060235154	6.197935618	1.463038781	3.3679405
Rastrigin	Clone	3.94E−09	2.32E−05	4.2294E−06	8.42E−07
Rastrigin	Pollination	0	2.84217E−14	1.1563E−14	8.52651E−15
Griewank	Clone	1.90E−10	2.23E−09	3.8869E−10	8.23E−10
Griewank	Pollination	0	0	0	0

Table 3 30 Experimental results with 30 dimensions

Function	Reproduction operator	Algorithm performance with 30 dimensions			
		Best	Worse	σ	Average
Sphere	Clone	1.523E−252	3.427E−07	6.4224E−08	1.529E−08
Sphere	Pollination	1.7702E−36	6.73049E−15	1.23284E−15	2.9397E−16
Ackley	Clone	8.88E−16	3.78E−03	0.00069861	1.94E−04
Ackley	Pollination	8.88E−16	7.55E−14	1.67575E−14	5.28E−14
Rosenbrock	Clone	0.0494	37.2144	7.53849331	4.9807
Rosenbrock	Pollination	1.91872105	28.71735277	8.618153096	19.58295956
Rastrigin	Clone	1.57E−07	7.97E−03	0.00160841	4.22E−04
Rastrigin	Pollination	2.2774E−20	0.00048881	8.92437E−05	1.62962E−05
Griewank	Clone	5.07E−10	2.45E−08	4.0204E−09	4.92E−09
Griewank	Pollination	0	0.010177886	0.001858219	0.000339263

Table 4 30 Experimental results with 50 dimensions

Function	Reproduction operator	Algorithm performance with 50 dimensions			
		Best	Worse	σ	Average
Sphere	Clone	2.136E−214	3.938E−04	7.2512E−05	1.594E−05
Sphere	Pollination	3.15201E−76	5.92104E−13	1.08078E−13	1.9872E−14
Ackley	Clone	8.88E−16	3.27E−01	0.05971378	1.13E−02
Ackley	Pollination	8.88E−16	1.22E−13	2.03287E−14	9.15E−14
Rosenbrock	Clone	0.2656	41.0000	8.79412663	6.4178
Rosenbrock	Pollination	13.67637767	111.40382	13.81014363	44.694589
Rastrigin	Clone	1.86E−05	3.78E−02	0.0095950	3.01E−03
Rastrigin	Pollination	0	0.170530257	0.031133714	0.0056881
Griewank	Clone	8.10E−09	3.23E−04	5.8959E−05	1.09E−05
Griewank	Pollination	0	0.015510002	0.002831726	0.000517

The results shown in the previous tables show that we are on track in the development process of the bio-inspired optimization algorithm based on the self-defense mechanisms of plants, with all experiments we deduce a pattern of behavior data moving the parameters (α, β, δ, λ) manually by trial and error for optimization problems of mathematical functions. Shows that when the variables are in the following ranges α = 0.3–0.7, β = 0.1–0.4, δ = 0.2–0.3, λ = 0.01–0.05, we can maintain a best balance between the two populations managing to obtain less dispersion of data to model them in the mathematical functions used.

6 Conclusions

The proposal is to create, develop and test a new optimization algorithm inspired by the bio-defense mechanism of plants. The first challenge is to adapt the predator-prey model and test the algorithm in an optimization problem, in this case, we decided to test the performance in mathematical functions and we have achieved acceptable results. When we move the parameters manually we observe that the algorithm has better performance when the values are in a range of values, these observations are only for this problem, we need to apply it to other optimization problems to analyze the behavior. The approach shows that the results are good, but we think that it can be improved by adding other biological processes as a method of reproduction.

References

1. Bennett, R. N., & Wallsgrove, R. M. (1994). Secondary metabolites in plant defense mechanisms. New Phytologist, 127(4), 617-633
2. Berryman, A. A. (1992). The origins and evolution of predator-prey theory. Ecology, 1530-1535.
3. Caraveo, C., Valdez, F., & Castillo, O. (2015). Bio-Inspired Optimization Algorithm Based on the Self-defense Mechanism in Plants. In Advances in Artificial Intelligence and Soft Computing (pp. 227-237). Springer International Publishing.
4. Cruz, J. M. L., & González, G. B. (2008). Modelo Depredador-Presa. Revista de Ciencias Básicas UJAT, 7(2), 25-34.
5. García-Garrido, J. M., & Ocampo, J. A. (2002). Regulation of the plant defense response in arbuscular mycorrhizal symbiosis. Journal of experimental botany, 53(373), 1377-1386.
6. Karaboga, D., & Basturk, B. (2007). A powerful and efficient algorithm for numerical function optimization: artificial bee colony (ABC) algorithm. Journal of global optimization, 39(3), 459-471.
7. Law, J. H., & Regnier, F. E. (1971). Pheromones. Annual review of bio-chemistry, 40(1), 533-548.
8. Lez-Parra, G. G., Arenas, A. J., & Cogollo, M. R. (2013). Numerical-analytical solutions of predator-prey models. WSEAS Transactions on Biology & Biomedicine, 10(2).
9. Melin, P., Olivas, F., Castillo, O., Valdez, F., Soria, J., & Valdez, M. (2013). Optimal design of fuzzy classification systems using PSO with dynamic parameter adaptation through fuzzy logic. Expert Systems with Applications, 40(8), 3196-3206.
10. Neyoy, H., Castillo, O., & Soria, J. (2013). Dynamic fuzzy logic parameter tuning for ACO and its application in TSP problems. In Recent Advances on Hybrid Intelligent Systems (pp. 259-271). Springer Berlin Heidelberg.
11. Ordeñana, K. M. (2002). Mecanismos de defensa en las interacciones planta-patógeno. Revista Manejo Integrado de Plagas. Costa Rica, 63, 22-32.
12. Paré, P. W., & Tumlinson, J. H. (1999). Plant volatiles as a defense against insect herbivores. Plant physiology, 121(2), 325-332.
13. Teodorovic. Bee colony optimization (BCO). In C. P. Lim, L. C. Jain, and S. Dehuri, editors, Innovations in Swarm Intelligence, pages 39–60. Springer-Verlag, 2009. 65, 215
14. Tollsten L, Mu ̈ller PM (1996) Volatile organic compounds emitted from beech leaves. Phytochemistry 43: 759–762

15. Vivanco, J. M., Cosio, E., Loyola-Vargas, V. M., & Flores, H. E. (2005). Mecanismos químicos de defensa en las plantas. Investigación y ciencia, 341(2), 68-75.
16. Xiao, Y., & Chen, L. (2001). Modeling and analysis of a predator–prey model with disease in the prey. Mathematical Biosciences, 171(1), 59-82.
17. Yang, X. S. (2012). Flower pollination algorithm for global optimization. In Unconventional computation and natural computation (pp. 240-249). Springer Berlin Heidelberg.
18. Yoshida, T., Jones, L. E., Ellner, S. P., Fussmann, G. F., & Hairston, N. G. (2003). Rapid evolution drives ecological dynamics in a predator–prey system. Nature, 424(6946), 303-306.

Evaluation of the Evolutionary Algorithms Performance in Many-Objective Optimization Problems Using Quality Indicators

Daniel Martínez-Vega, Patricia Sanchez, Guadalupe Castilla, Eduardo Fernandez, Laura Cruz-Reyes, Claudia Gomez and Enith Martinez

Abstract The need to address more complex real-world problems gives rise to new research issues in many-objective optimization field. Recently, researchers have focused in developing algorithms able to solve optimization problems with more than three objectives known as many-objective optimization problems. Some methodologies have been developed into the context of this kind of problems, such as A^2-NSGA-III that is an adaptive extension of the well-known NSGA-II (Non-dominated Sorting Genetic Algorithm II). A^2-NSGA-III was developed for promoting a better spreading of the solutions in the Pareto front using an improved approach based on reference points. In this paper, a comparative study between NSGA-II and A^2-NSGA-III is presented. We examine the performance of both algorithms by applying them to the project portfolio problem with 9 and 16 objectives. Our purpose is to validate the effectiveness of A^2-NSGA-III to deal with

D. Martínez-Vega · P. Sanchez (✉) · G. Castilla · L. Cruz-Reyes (✉)
C. Gomez · E. Martinez
Tecnologico Nacional de Mexico, Instituto Tecnologico de Ciudad Madero, Tamaulipas, Mexico
e-mail: jpatricia.sanchez@gmail.com

L. Cruz-Reyes
e-mail: lauracruzreyes@itcm.edu.mx

D. Martínez-Vega
e-mail: danielmartinezvega@hotmail.com

G. Castilla
e-mail: gpe_cas@yahoo.com.mx

C. Gomez
e-mail: cggs71@hotmail.com

E. Martinez
e-mail: emc.enith@gmail.com

E. Fernandez
Universidad Autonoma de Sinaloa, Sinaloa, Mexico
e-mail: eddyf@uas.edu.mx

P. Melin et al. (eds.), *Nature-Inspired Design of Hybrid Intelligent Systems*,
Studies in Computational Intelligence 667, DOI 10.1007/978-3-319-47054-2_42

many-objective problems and increase the variety of problems that this method can solve. Several quality indicators were used to measure the performance of the two algorithms.

Keywords Many-objective problems · Project portfolio selection · Algorithm performance analysis

1 Introduction

Most real-world problems often involve the fulfillment of multiple objectives, which should be solved simultaneously [1]. These kinds of problems are known as multi-objective optimization problems (MOPs). Given the conflicting nature of objectives, there is usually no single optimal solution and, consequently, the ideal solution of MOPs cannot be reached. The optimal solution for a MOP is not a single solution but rather a set of alternative solutions called the *Pareto set*.

Solving this type of problems, it means obtaining a set of Pareto optimal solutions that satisfy the requirements of minimizing the distance of solutions to the optimal front (*convergence*) and maximizing the distribution of solutions over the optimal front (*diversity*) [2].

During the last two decades, multi-objective evolutionary algorithms (MOEAs) have been proposed and successfully used in solving MOPs since are capable of achieving an approximation of the Pareto set in a single run. MOEAs usually work very well on problems with two or three objectives, but their search ability severely deteriorate when more than three objectives are involved [3]. MOPs with more than three objectives are referred to as many-objective problems.

Due to the large number of objectives in many real-world applications and the loss of selective pressure toward the Pareto front when the number of objectives is increased have attracted increasing interest in the evolutionary multi-objective optimization community in recent years [3]. According to Yang et al. [4], some techniques have been developed in this area, such as, test functions scalable for higher number of objectives, quality indicators for a high-dimensional space, and visualization tools to show solutions with four or more objectives. These procedures have made it possible to examine the performance of algorithms on many-objective problems. Balancing the performance in convergence and diversity is not a simple task in many-objective optimization [4].

Most Pareto-based EMO algorithms lose selection pressure toward the Pareto front since the number of non-dominated solutions in a population rises quickly as the number of objectives increase. In these cases, the diversity mechanism plays a key role in determining the survival of solutions [3].

Recently, some efforts about new diversity-preservation operators have been reported in the literature. Such is the case of the A^2-NSGA-III proposed by Jain and

Deb [5], which is an adaptive extension of the algorithm NSGA-II [6]. In A^2-NSGA-III, the maintenance of diversity among population members is aided by a set of adaptive reference points. A^2-NSGA-III has demonstrated its efficacy in solving many-objective optimization problems.

In this paper, we present a comparative study between NSGA-II and its successor A^2-NSGA-III. The aim of this work is to validate the capability of A^2-NSGA-III in solving many-objectives problems and expand the range of problems that it can effectively handle. The study was carried out considering the project portfolio problem having 3, 9, and 16 objectives and to evaluate the performance of the algorithms several quality indicators were used.

2 Background

In this section, some definitions about multi-objective optimization and quality indicators are given, as well as a short description of the project portfolio problem addressed in this work. Finally, the NSGA-II and A^2-NSGA-III algorithms will be briefly described.

2.1 *Multi-objective Optimization*

2.1.1 Pareto Dominance

In multi-objective optimization, the Pareto dominance [7] is often used to compare two solutions (vectors). A vector $\vec{x}$ dominates to other vector $\vec{y}$ if and only if fulfills the following conditions (maximization):

$$f_i(\vec{x}) \geq f_i(\vec{y}), \quad \forall i \in [1, 2, \ldots, k] \quad \text{and} \tag{1}$$

$$f_j(\vec{x}) > f_j(\vec{y}), \quad \exists j \in [1, 2, \ldots, k] \tag{2}$$

where k is the number of objectives.

That is, a vector $\vec{x}$ is Pareto optimal if and only if no component of $\vec{y}$ is larger than the corresponding component of $\vec{x}$ and, at least, one component of $\vec{x}$ is strictly larger. The Pareto dominance almost always gives not one but several solutions called as the Pareto optimal set. The solutions contained in the Pareto optimal set are called non-dominated.

Between two solutions there can only be three possible results: $\vec{x}$ dominates $\vec{y}$, $\vec{y}$ dominates $\vec{x}$, or $\vec{x}$ and $\vec{y}$ are non-dominated.

2.1.2 Pareto Front

For a given multi-objective function $F(\vec{x})$ and a Pareto optimal set Ω, the Pareto front $\mathcal{PF}$ is defined as follow:

$$\mathcal{PF} = \{F(\vec{x}) = (f_1(\vec{x}), f_2(\vec{x}), \ldots, f_k(\vec{x}))|\vec{x} \in \mathcal{F}\} \tag{3}$$

where $\mathcal{F}$ is the feasible region.

2.2 *Quality Indicators*

To improve the performance of a multi-objective algorithm two issues must be considered: minimizing the distance between solutions set and the solutions in the optimal Pareto, and maximizing the uniform dispersion of the solutions on the front [8]. The performance of the studied algorithms is assessed using five quality indicators described in the following lines:

Dominance. This indicator receives two sets of solutions (Pareto's fronts of each algorithm) that joins in a single set, to which a nonfast dominated sorting method is applied in order to split the solutions in fronts, after that, calculates how many solutions from Algorithm 1 are in the zero front (f_0). The percentage of dominance for this algorithm (a_1) is obtained calculating the percentage that represents the number of solutions that are in (f_0) and belongs to a_1, in relation to the total number of solutions from Algorithm 1, see Eq. 4. The same calculation is done for the second Algorithm (a_2). Consequently, as higher percentage of dominance have an algorithm, better performance exhibits. Figure 1 shows an example of this indicator.

$$\left(\% \text{ of dominance for } a_1 = \frac{\text{number of sol. from } a_1 \in f_0}{\text{number of sol. from } \mathrm{a}_1} * 100\right) \tag{4}$$

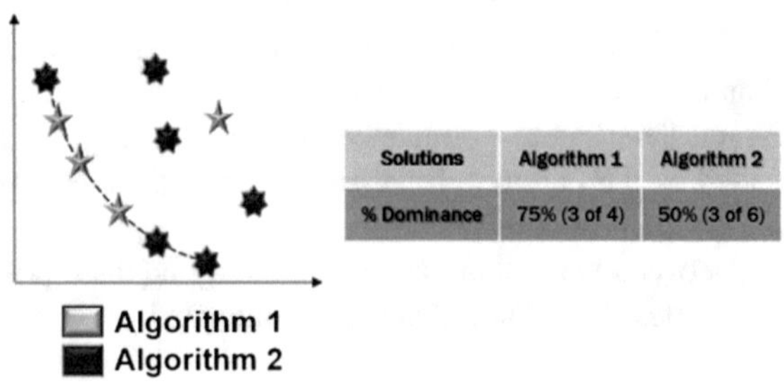

Solutions
Algorithm 1
Algorithm 2
% Dominance
75% (3 of 4)
50% (3 of 6)
Algorithm 1
Algorithm 2

Fig. 1 Dominance

Spread. This metric measures the dispersion degree of the solutions obtained. Equation 5 [8] obtains this value, wherein, instead of using two adjacent solutions it uses the distance between each solution and its nearest neighbor:

$$\Delta(X,S) = \frac{\sum_{i=1}^{m} d(e_i,S) + \sum_{X\in S}\left|d(X,S) - \bar{d}\right|}{\sum_{i=1}^{m} d(e_i,S) + |S| * \bar{d}} \quad (5)$$

S is the set of solutions, S^* is the optimal set of Pareto solutions, $(e_1,\ldots,e_m)$ are the m extreme solutions of S^*, m is the number of objectives, $d(X,S)$ is obtained by Eq. 6 and $\bar{d}$ is calculated by Eq. 7.

$$d(X,S) = \min_{Y\in S, Y\neq X} \|F(X) - F(Y)\|^2, \quad (6)$$

$$\bar{d} = \frac{1}{S^*}\sum_{X\in S^*} d(X,S) \quad (7)$$

This metric requires normalization of the objective function values. In the best case $\Delta(X,S) = 0$, which corresponds with the second curve in Fig. 2.

Hypervolume. Calculates the volume in the space of objectives covered by a set of non-dominated solutions or Pareto front found (FP_f). For each solution $s_i \in \mathrm{FP}_f$, a hypercube h_i is constructed from the diagonal defined by a reference point w and the solution s_i. The point w represents the anti-optimal and it is obtained using the worst values for the objective functions. The union operation of all hypercubes, considering the principle of inclusion–exclusion, defines hypervolume for the solutions front considered [8, 9]. This indicator is calculated by Eq. 8.

$$\mathrm{HV} = \mathrm{volume}\left(\bigcup_{i=1}^{|\mathrm{FP}_f|} h_i\right) \quad (8)$$

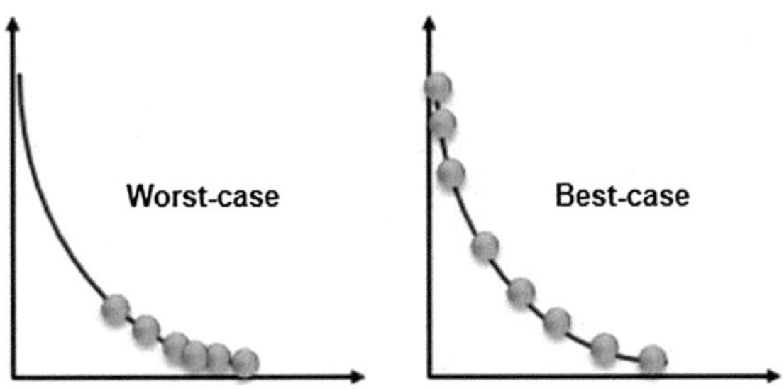

Fig. 2 Spread indicator

Generational distance. It measures the distance between the non-dominated solution set and the Pareto optimal set [8, 10]. It is calculated with the following equation:

$$\mathrm{GD} = \frac{\sqrt{\sum_{i=1}^{n} d_i^2}}{n} \tag{9}$$

where

n is the number of elements of non-dominated solution set.

d_i is the Euclidian distance between each non-dominated solution and the closest solution of front of Pareto optimal.

Just as that for Spread, this method requires a previous normalization to the front of Pareto solutions found and to the front of Pareto optimal.

Inverted Generational Distance (IGD). It consists in determining the average distance between each point in *true Pareto front* and the closest solution in the *found Pareto front*, this metric indicates the proximity between the true Pareto front (PF_t) and the Pareto front found. The mathematical expression [11] for this calculation is the following:

$$\mathrm{IGD} \triangleq \frac{\left(\sum_{i=1}^{n} d_i^p\right)^{\frac{1}{p}}}{n} \tag{10}$$

where n is the number of solutions in PF_t, p value is 2, and d_i is the Euclidian distance in the objective space between each vector of PF_t and the closest point of PF_f. IGD = 0 indicates that both fronts are equal; any other value points to a divergence between them; therefore, it is desirable to obtain low values of *IGD*.

2.3 Project Portfolio Problem

This problem is about proper selection of projects that will integrate a portfolio which will receive financial support to its realization.

A project is a process which pursues a set of objectives; its quality is measured as the number of beneficiaries for each of objectives previously established. Each objective is associated with some social class and a level of impact.

A feasible portfolio has to satisfy the following constraints: a total budget and a budget by geographic area and by region.

The decision-maker is responsible for choosing the portfolio that the organization will implement. Please refer to [12] for more details about the problem.

2.4 *Multi-objective Evolutionary Algorithms*

2.4.1 Non-dominated Sorting Genetic Algorithm II (NSGA-II)

NSGA-II [6] starts with parent population P_0 created randomly which is then sorted according to the level of non-domination. Genetic operations are performed on P_0 to create an offspring population Q_0. Both populations are of size N.

From the first generation forward, the procedure is different, let us describe the tth generation.

The P_t and Q_t populations are combined to form a population R_t of size $2N$. The first step is to select the best N members from R_t. To accomplish this, the population R_t is sorted according to different non-domination levels or fronts (F_1, F_2, and so on).

Then, a new parent population P_{t+1} is created with solutions of the best non-dominated fronts. Each front is selected, one at a time, starting from F_1, until the size of P_{t+1} is equal to N or when it is exceeded for the first time. Let us say the last front included is the lth front.

The solutions from front $(l + 1)$ onward are discarded. Usually, the number of solutions included, from F_1 to F_l, could be larger than the population size, doing that the solutions of the F_l are accepted partially to construct P_{t+1}. In this case, only those solutions that will maximize the diversity are selected.

In NSGA-II, this is achieved through the crowded-comparison operator which computes the crowding distance for every solution on the last front. Subsequently, these solutions are sorted in descending order according to the crowding distance values. The solutions having larger crowding distance values are chosen, then, the new parent population P_{t+1} is finally created. Genetic operators are used on P_{t+1} to create a new offspring population Q_{t+1}.

2.4.2 Efficiently Adaptive NSGA-III Procedure (A^2-NSGA-III)

A^2-NSGA-III, proposed by Jain and Deb in [5], is an improved extension of the NSGA-II algorithm. The main feature of this algorithm is that to promote diversity among solutions through a set of reference points which can be provided by the DM or calculated in a structured manner. Reference points lead the search such that the resulting Pareto front has a graphical shape very similar to the Pareto optimal front, avoiding falling into local optima, at the same time a greater search space is explored in the search for solutions. The procedure is shown in Algorithm 1 and is described below.

Algorithm 1. Generation t of A^2-NSGA-III procedure.

Input: Structurated reference points Z^s or supplied aspiration points Z^a, parents population P_t
Output: P_{t+1}

1. $S_t = \emptyset,\ i = 1$
2. $Q_t = Recombination + Mutation\ (P_t)$
3. $R_t = P_t \cup Q_t$
4. $(F_1, F_2, \ldots) = non-dominated\ sort\ (R_t)$
5. **Repeat**
6. $S_t = S_t \cup F_i$ and $i = i + 1$
7. **Until** $|S_t| \geq N$
8. Last front included: $F_l = F_i$
9. **If** $|S_t| = N$ **then**
10. $P_{t+1} = S_t,\ break\ cycle$
11. **Else**
12. $P_{t+1} = \cup_{j=1}^{l-1} F_j$
13.   Selected points of F_l: $K = N - |P_{t+1}|$
14.   Normalize objectives and create reference set Z^r:
$$Normalize\ (f^n, S_t, Z^r, Z^s, Z^a)$$
15.   Associate each member of $s \in S_t$ with a reference point:
$[\pi(s), d(s)] = Associate\ (S_t, Z^r)$ where:
$$\pi(s) = closest\ reference\ point\ and$$
$$d(s) = distance\ between\ s\ and\ \pi(s)$$
16.   Compute niche count of reference point $j \in Z^r$:
$$\rho_j = \sum_{s \in S_t/F_l} ((\pi(s) = j)\ ?\ 1 : 0)$$
17.   Select K members one at a time of F_l to build $P_{t+1}: Niching\ (K, \rho_j, \pi, d, Z^r, F_l, P_{t+1})$
18. **End if**
19. $Add\ Reference\ Points\ (Z^r, \rho, p, \lambda)$

Steps 1–4 are virtually the same as those used in the NSGA-II algorithm, where from an initial parent population (P_t) the crossover operations are carried out (recombination) and mutation, generating an offspring population (Q_t).

Subsequently Q_t and P_t are joined in the same set which is called R_t on which a non-dominated sorting is applied, after which, the next parent population (P_{t+1}) is extracted.

The set S_t incorporates fronts one by one until the number of solutions added is equal to or greater than the number of members in the first parent population of size N, the last front included is called F_l (steps 5–8). If the condition is fulfilled in step 9 (the magnitude of S_t is equal to N) S_t is set as P_{t+1} (step 10).

If is not satisfied the condition of step 9, the steps 11–18 will be done. The population P_{t+1} acquires all those fronts in S_t which were fully added while its magnitude does not exceed N (step 12). K is the number of solutions which is going to be added by diversity (step 13). Reference points are normalized (step 14). In step 15, each member in S_t is associated with the nearest reference point having the smallest orthogonal distance. The value of an indicator called niche in step 16 (ρ_j) which contains the number of solutions in $S_t \backslash F_l$ associated to each reference point is calculated. Finally in step 17 the process called Niching selects K solutions from F_l that will be part of P_{t+1}.

Step 19 is performed only if during m generations (in this experiment was tested with 6) the highest ρ_j remains unchanged.

3 Experimentation and Results

In this section, we present the results of the experimentation carried out.

3.1 Experimentation

The experiments were performed on a computer with the following characteristics:

- Hardware. Inside i3 processor 2.1 Ghz of 64 bits, 6 Gb in RAM, and HDD 7200 rpm.
- Software. O.S. Windows 8 × 64, language Java, JDK 1.8.0, IDE NetBeans 8.0.1.

A parameter setting to the metaheuristics NSGA-II and A^2-NSGA-III was performed, for each one of the experiments 30 runs of each metaheuristic were done. Table 1 shows the population size for each instance.

The NSGA-II implementation was made according to [6]. The algorithm configuration was crossover probability = 90 %, mutation probability = 1 %, number of generations = 500. The selection method is a binary tournament aided by crowding distance and one insertion movement for mutation is applied.

The A^2-NSGA-III implementation is based on [5]. The configuration used for this metaheuristic was crossover probability = 90 %, mutation probability = 1 %, and number of generations = 500. Genetic operators are a random selection of parents, one point crossover for recombination, and a mutation by insertion. The stop condition is reached at 500 iterations.

Table 1 Information of instances and experimentation parameters

Instances	No. of obj.	No. of projects	NSGA-II	A^2-NSGA-III		No. of iterations
			Pop. size	Pop. Size	No. ref. points	
1	3	100	105	105	105	500
2	9	100	165	165	165	500
3	16	500	136	136	136	500

3.2 Results

In this section, we provide the experimental results. We compared NSGA-II with A^2-NSGA-III to investigate their performance. We experimented with three instances, described in Table 1, addressing the project portfolio problem.

First, we compared the performance through the quality indicator of Pareto dominance, the results can be observed in Table 2. In right column, it is shown how A^2-NSGA-III has a better performance in all cases; it indicates that the A^2-NSGA-III's result sets provide generally better results. For example in the three objectives instance, NSGA-II has 3 %, it means that of 3146 results, 102 of them were at the same level that 636 solutions of 3150 provided by A^2-NSGA-III which is the 20 % of its own set, it means that 3 % of solutions provided by NSGA-II outranks to 2514 solutions obtained from A^2-NSGA-III.

Secondly, Hypervolume value for each solution set was computed. Table 3 summarizes the average of the 30 runs, respectively. It's clear that both metaheuristics have a very similar performance for this metric. It means that the space solution spanned by both of metaheuristics has the same size.

In Table 4, the Spread metric results are presented; we can see that A^2-NSGA-III shows a better performance in all cases, in fact the difference between the results increases in the same way that problem difficulty.

Table 2 Pareto dominance values for NSGA-II and A^2-NSGA-III approaches

			Average	
No. of objectives	Algorithm	Size of the final solution set	Solutions that remain non-dominated in $A \cup B$	Rate of solutions that remain non-dominated (%)
3	NSGA-II	3146	102	3
	A^2-NSGA-III	3150	636	20
9	NSGA-II	4950	1263	25
	A^2-NSGA-III	4950	1417	28
16	NSGA-II	4950	1866	37
	A^2-NSGA-III	4080	2063	50

Note A is the set of solutions obtained by NSGA-II; B is the set obtained by A^2-NSGA-III

Table 3 Hypervolume values for NSGA-II and A^2-NSGA-III approaches

No. of objectives	Algorithm	Average rate
3	NSGA-II	0.9642321
	A^2-NSGA-III	0.9719522
9	NSGA-II	0.8411267
	A^2-NSGA-III	0.7906261
16	NSGA-II	0.7903667
	A^2-NSGA-III	0.8161129

Table 4 Spread values for NSGA-II and A^2-NSGA-III approaches

No. of objectives	Algorithm	Average rate
3	NSGA-II	0.8011445
	A^2-NSGA-III	0.7631850
9	NSGA-II	0.5969997
	A^2-NSGA-III	0.5252820
16	NSGA-II	0.8319083
	A^2-NSGA-III	0.6988260

Table 5 GD values for NSGA-II and A^2-NSGA-III approaches

No. of objectives	Algorithm	Average rate
3	NSGA-II	0.0000178
	A^2-NSGA-III	0.00001463
9	NSGA-II	0.0003578
	A^2-NSGA-III	0.0003236
16	NSGA-II	0.0015496
	A^2-NSGA-III	0.0003197

Table 6 IGD values for NSGA-II and A^2-NSGA-III approaches

No. of objectives	Algorithm	Average rate
3	NSGA-II	0.0000980
	A^2-NSGA-III	0.0000254
9	NSGA-II	0.0004124
	A^2-NSGA-III	0.0003668
16	NSGA-II	0.0029294
	A^2-NSGA-III	0.0004465

Finally, the performance in terms of proximity between the true *PF* and the best front found was computed. In Tables 5 and 6, results corresponding to the GD and IGD indicators for instances with 3, 9, and 16 objectives, respectively, are presented. As we can see A^2-NSGA-III has a superior performance than NSGA-II, except for the instance with 9 objectives.

4 Conclusion and Future Work

In this paper, a comparative study about the performance achieved by the NSGA-II and A^2-NSGA-III algorithms for solving the public portfolio problem is performed; the first has proven to be successful to solve a wide range of multi-objective problems but, it has also exhibited difficulties when the number of objectives increases. In order to address this issues, some improved approaches of NSGA-II were proposed in the literature, from which A^2-NSGA-III is reported as the most successful version in relation to its capability to solve many-objective problems (more than 4) [3]. Its improvement is supported in a less expensive process for updating the reference points.

Regarding the indicator of dominance, superiority of the A^2-NSGA-III is remarkable, as can be seen in Table 1, where differences in the percentage in favor for the instances of 3, 9, and 16 objectives were 17, 3, and 13, respectively. These results indicate a stable performance regardless of the number of objectives to achieve.

In the same vein are the results corresponding to the DG and IGD indicators, showing an advantage in favor to A^2-NSGA-III as it is shown in Tables 5 and 6. For these indicators, it is possible to see that the difference in favor grows as grows the number of objectives of the problem to address.

Finally, the results for HV and spread indicators are analyzed, in order to assess the capacity of spatial coverage and the ability to produce solutions homogeneously distributed in the Pareto front, respectively. Table 3 shows that for the hypervolume A^2-NSGA-III is superior in the instances of 3 and 16 objectives, and NSGA for 9 objectives. Furthermore the results of Table 4 indicates that A^2-NSGA-III obtains a favorable difference in the spread indicator, which scale with the number of objectives of the addressed instance.

The capability of the algorithm A^2-NSGA-III to solve the project public portfolio problem on the scale of many-objective problems is confirmed according to the results above described; also they indicate that it is robust in terms of the number of objectives.

As future work, we will incorporate the reference solutions relocation strategy in other population metaheuristic as scatter search or PSO, in order to incorporate diversity in the search process.

Acknowledgments This work was partially financed by CONACYT, COTACYT, DGEST, TECNM, and ITCM.

References

1. Deb, K.: Multi-Objective Optimization using Evolutionary Algorithms (Vol. 16). John Wiley & Sons. (2001).
2. Talbi, E. G.: Metaheuristics: from design to implementation (Vol. 74). John Wiley & Sons. (2009).
3. Deb, K., Jain, H.: An Evolutionary Many-Objective Optimization Algorithm Using Reference-point Based Nondominated Sorting Approach, Part I: Solving Problems with Box Constraints. In Proceedings of IEEE Transactions on Evolutionary Computation. (2013).
4. Yang, S., Li, M., Liu, X., & Zheng, J. A grid-based evolutionary algorithm for many-objective optimization. Evolutionary Computation, IEEE Transactions on, 17(5), 721-736.(2013).
5. Jain, H., & Deb, K. An improved adaptive approach for elitist nondominated sorting genetic algorithm for many-objective optimization. In Evolutionary Multi-Criterion Optimization (pp. 307-321). Springer Berlin Heidelberg.(2013).

6. Deb, K., Agrawal, S., Pratap, A. and Meyarivan, T.: A fast elitist non-dominated sorting genetic algorithm for multi-objective optimization: NSGA-II. Lecture notes in computer science, 1917, pp. 849-858. (2000).
7. Pareto, V.: Politique, Cours D' economie. Rouge, Lausanne, Switzerland. (1896).
8. Nebro, A. J., Luna, F., Alba, E., Dorronsoro, B., & Durillo, J. J. Un algoritmo multiobjetivo basado en búsqueda dispersa.
9. Mirjalili, S., & Lewis, A. Novel performance metrics for robust multi-objective optimization algorithms. Swarm and Evolutionary Computation, 21, 1-23.(2015).
10. Yen, G. G., & He, Z. Performance metric ensemble for multiobjective evolutionary algorithms. Evolutionary Computation, IEEE Transactions on, 18(1), 131-144.(2014).
11. Fabre, M. G. *Optimización de problemas con más de tres objetivos mediante algoritmos evolutivos* (Doctoral dissertation, Master's thesis, Centro de Investigación y de Estudios Avanzados del Instituto Politécnico Nacional, Ciudad Victoria, Tamaulipas, México).(2009).
12. Cruz-Reyes, L., Fernandez, E., Gomez, C., Sanchez, P., Castilla, G., & Martinez, D. Verifying the Effectiveness of an Evolutionary Approach in Solving Many-Objective Optimization Problems. In Design of Intelligent Systems Based on Fuzzy Logic, Neural Networks and Nature-Inspired Optimization (pp. 455-464). Springer International Publishing (2015).

Generating Bin Packing Heuristic Through Grammatical Evolution Based on Bee Swarm Optimization

Marco Aurelio Sotelo-Figueroa, Héctor José Puga Soberanes, Juan Martín Carpio, Héctor J. Fraire Huacuja, Laura Cruz Reyes, Jorge Alberto Soria Alcaraz and Andrés Espinal

Abstract In the recent years, Grammatical Evolution (GE) has been used as a representation of Genetic Programming (GP). GE can use a diversity of search strategies including Swarm Intelligence (SI). Bee Swarm Optimization (BSO) is part of SI and it tries to solve the main problems of the Particle Swarm Optimization (PSO): the premature convergence and the poor diversity. In this paper we propose using BSO as part of GE as strategies to generate heuristics that solve the Bin Packing Problem (BPP). A comparison between BSO, PSO, and BPP heuristics is performed through the nonparametric Friedman test. The main contribution of this paper is to propose a way to implement different algorithms as search strategy in GE. In this paper, it is proposed that the BSO obtains better results than the ones obtained by PSO, also there is a grammar proposed to generate online and offline heuristics to improve the heuristics generated by other grammars and humans.

Keywords Genetic Programming · Grammatical Evolution · Bee Swarm Optimization · Bin Packing Problem · Heuristic · Metaheuristic

---

M.A. Sotelo-Figueroa (✉) · H.J.P. Soberanes · J.M. Carpio · J.A.S. Alcaraz · A. Espinal
División de Estudios de Posgrado e Investigación, Instituto Tecnológico de León, 37290 León, Gto, Mexico
e-mail: marco.sotelo@itleon.edu.mx

H.J.P. Soberanes
e-mail: pugahector@yahoo.com

J.M. Carpio
e-mail: jmcarpio61@hotmail.com

J.A.S. Alcaraz
e-mail: soajorgea@gmail.com

H.J. Fraire Huacuja · L.C. Reyes
División de Estudios de Posgrado e Investigación,
Instituto Tecnológico de Ciudad Madero, 89440 Tamaulipas, Tamp., Mexico
e-mail: hfraire@prodigy.net.mx

L.C. Reyes
e-mail: lcruzreyes@prodigy.net.mx

P. Melin et al. (eds.), *Nature-Inspired Design of Hybrid Intelligent Systems*, Studies in Computational Intelligence 667, DOI 10.1007/978-3-319-47054-2_43

1 Introduction

The methodology development to solve a specific problem is a process that entails the problem study and the analysis instances from such problem. There are many problems [1] for which there are no methodologies that can give the exact solution, because the size of the problem search space makes it intractable in time, and it is necessary to search and improve methodologies that can give a solution in a finite time. There are methodologies based on Artificial Intelligence that does not give exact solutions, however, those methodologies give an approximation, we can find the following methodologies:

- **Heuristics** are defined as "a type of strategy that dramatically limits the search for solutions" [2, 3]. One important characteristic of heuristics is that they can obtain a result for an instance problem in polynomial time [1], although heuristics are developed for a specific instance problem.
- **Metaheuristics** are defined as "a master strategy that guides and modifies other heuristics to obtain solutions, generally better that the ones obtained with a local search optimization" [4]. The metaheuristics can work over several instances of a given problem or various problems, but it is necessary to adapt the metaheuristics to work with each problem.

There are shown that the metaheuristic Genetic Programming [5] can generate a heuristic that can be applied to an instance problem [6]. Also, there exist metaheuristics that are based on Genetic Programming's paradigm [7] such as *Grammatical Differential Evolution* [8], *Grammatical Swarm* [9], *Particle Swarm Programming* [10], *Geometric Differential Evolution* [11], etc.

BSO [12, 13] is a hybrid metaheuristic based on Particle Swarm Optimization [14] and the Bee Algorithm [15]. Recently the BSO has been applied to solve the Knapsack Problem [16] to optimize the energy in Ambient Intelligence [17], to minimize the Cyclic Instability in Intelligent Environments [18] among other applications.

The Bin Packing Problem (BPP) has been widely studied because it has many Industrial Applications, like cutting wood or glass, packing into transportation and warehousing [19], job scheduling on uniform processors [20, 21], etc. There are many heuristics for this problem [22–25] because there does not exist an exact algorithm to solve it for its complexity. This is the reason to try to apply the metaheuristics [26–28], because it looks for a better result than the one obtained by heuristics.

In the present paper it is shown that it is possible to generate BPP heuristics by using BSO and PSO as a search strategy. It is also shown that the heuristics generated with the proposed grammar have better performance than the BPP heuristics, which were designed by an expert in operational research. Those results were obtained by comparing the results obtained by the GE and the BPP heuristics by means of Friedman nonparametric test [29].

The GE is described in Sect. 2, including the PSO and BSO. The Sect. 3 describe the Bin Packing Problem, the heuristics from the state of art, the instances used, and the fitness function. We describe the experiments performed in Sect. 4. Finally, general conclusions about the present work are presented in Sect. 5, including future perspectives of this work.

2 Grammatical Evolution

Grammatical Evolution (*GE*) [7] is a grammar-based form of Genetic Programming (*GP*) [30]. GE joins the principles from molecular biology, which are used by the GP, and the power of formal grammars. Unlike GP, the GE adopts a population of lineal genotypic integer strings, or binary strings, which are transformed into functional phenotypic through a genotype-to-phenotype mapping process [31], this process is also know as *Indirect Representation* [32]. The genotype strings are evolved with no knowledge of their phenotypic equivalent, only uses the fitness measure.

The transformation is governed through a Backus Naur Form grammar (*BNF*), which is made up of the tuple N, T, P, S; where N is the set of all nonterminal symbols, T is the set of terminals, P is the set of production rules that map $N \rightarrow T$, and S is the initial start symbol where $S \in N$. There are a number of production rules that can be applied to a nonterminal, a "|" (or) symbol separates the options.

Even though the GE uses the Genetic Algorithm (*GA*) [7, 31, 33] as search strategy, it is possible to use another search strategy like the Particle Swarm Optimization, called Grammatical Swarm (*GS*) [8].

In the GE, each individual is mapped into a program using the BNF, using (1) proposed in [31] to choose the next production based on the nonterminal symbol. An example of the mapping process employed by GE is shown in Fig. 1.

$$\text{Rule} = c \,\%\, r \tag{1}$$

where c is the codon value and r is the number of production rules available for the current nonterminal.

The GE can use different search strategies; our proposed model is shown in the Fig. 2. This model includes the problem instance and the search strategy as an input. In [31] the search strategy is part of the process, however, it can be seen as an additional element that can be chosen to work with GE. The GE through the search strategy selected will generate a solution, and it will be evaluated in the objective function using the problem instance.

2.1 Particle Swarm Optimization

Particle Swarm Optimization (PSO) [14, 35–38] is a Bio-inspired metaheuristic in flocks of birds or schools of fish. It was developed by J. Kennedy and R. Eberthart based on a concept called social metaphor. This metaheuristics simulates a society

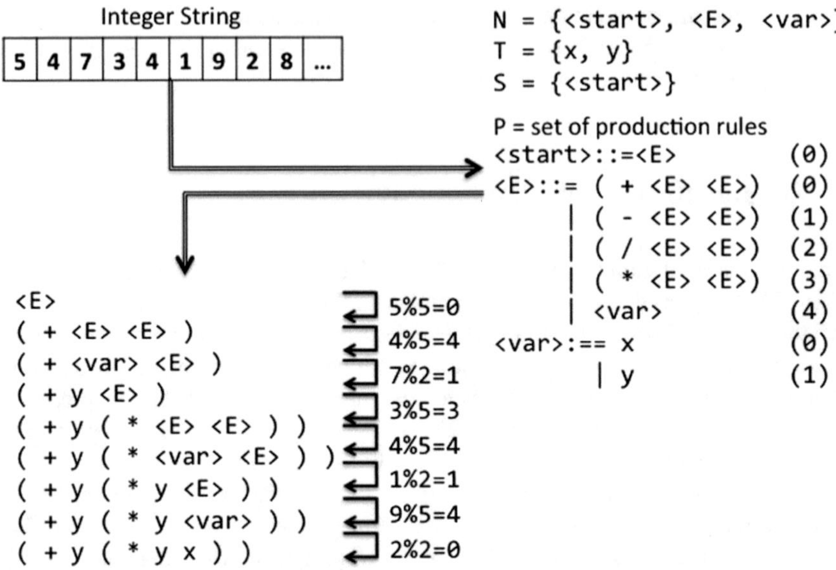

Fig. 1 An example of transformation from genotype to phenotype using a BNF Grammar. It begins with the start symbol, if the production rule from this symbol is only one rule, then the production rule gets instead of the start symbol, and the process begins to choose the productions rules base on the current genotype. It is taking each genotype and the nonterminal symbol from the left to realize the next production using (1) until all the genotypes are mapped or there are no more nonterminals in the phenotype

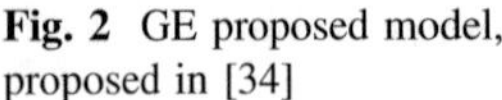

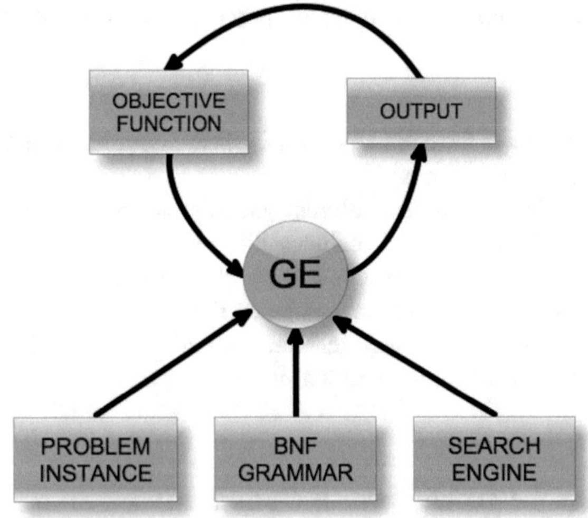

Fig. 2 GE proposed model, proposed in [34]

where all individuals contribute with their knowledge to obtain a better solution. There are three factors that influence for change in status or behavior of an individual:

- The knowledge of the environment or adaptation: it is related to the importance given to the experience of the individual.
- His Experience or local memory: it is related to the importance given to the best result found by the individual.
- The Experience of their neighbors or Global memory: this is related to how important it is the best result obtained by their neighbors or other individuals.

In this metaheuristic, each individual is considered as a particle, and moves through a multidimensional space that represents the social space or search space depends on the dimension of space which depends on the variables used to represent the problem.

For the update of each particle, we use the velocity vector which tells how fast it will move the particle in each of the dimensions, the method for updating the speed of PSO is given by Eq. (2), and it is updated by the Eq. (3). Algorithm 1 shows the complete PSO algorithm.

$$v_i = wv_i + \varphi_1(x_i - B_{\text{Global}}) + \varphi_2(x_i - B_{\text{Local}}) \tag{2}$$

$$x_i = x_i + v_i \tag{3}$$

where

v_i	is the velocity of the ith particle.
w	is adjustment factor to the environment.
φ_1	is the memory coefficient in the neighborhood.
φ_2	is the coefficient memory.
x_i	is the position of the ith particle.
B_{Global}	is the best position found so far by all particles.
B_{Local}	is the best position found by the ith particle.

Algorithm 1 PSO Algorithm

Require: w adaptation to environment coefficient, ϕ_1 neighborhood memory coefficient, ϕ_2 memory coefficient, n swarm size.

```
Start the swarm particles.
Start the velocity vector for each particle in the swarm.
while stopping criterion not met do
    for i = 1 to n do
        If the i-particle's fitness is better than the local best then replace the local best with the i-particle.
        If the i-particle's fitness is better than the global best then replace the global best with the i-particle.
        Update the velocity vector by (2).
        Update the particle's position with the velocity vector by (3).
    end for
end while
```

Fig. 3 Example of search radius based on binary operations

2.2 *Bee Swarm Optimization*

BSO [12, 13] is a hybrid metaheuristic based on Particle Swarm Optimization [14] and the Bee Algorithm [15]. The BSO uses the PSO and BA elements to try to avoid the problems observed in PSO. The premature convergence is avoided through the radius search and the poor diversity with the scout bees.

The BSO core is the PSO with its equations; speed Eq. (2) and updating Eq. (3). After this applies the exploration and a search radius from the BA to explore and exploit. The search radius is based on binary operations, adding and subtracting the radius number to the solution vector as shown in the Fig. 3.

Algorithm 2 BSO Algorithm

Require: w adaptation to environment coefficient, ϕ_1 neighborhood memory coefficient, ϕ_2 memory coefficient, n swarm size, sb scout bees, r search radius.

```
Start the bee swarm.
Start the velocity vector for each bee in the swarm.
while stopping criterion not met do
    for i = 1 to n do
        If the i-bee's fitness is better than the local best then replace the local best with the i-bee.
        If the i-bee's fitness is better than the global best then replace the global best with the i-bee.
        Update the velocity vector by (2).
        Update the bee's position with the velocity vector by (3).
        Apply the search radius r to the local best bee.
    end for
    Restart the worst sb bees.
end while
```

3 Bin Packing Problem

The Bin Packing Problem (BPP) [39] can be described as follows: we have n items and we want to pack them in the lowest possible number of bins, each item has a weight w_j, where j is the element; and we also have the max capacity of the bins c. The objective is to minimize the bins used to pack all the items, where each item is assigned only to one bin, and the sum of all the items in the bin cannot exceed the bin's size.

This problem has been widely studied, including the following:

- Proposing new theorems [40, 41].
- Developing new heuristics algorithms based on Operational Research concepts [25, 42].
- Characterizing the problem instances [43–45].
- Implementing metaheuristics [26, 46–48].

This problem has been shown to be an NP-Hard optimization problem [1]. A mathematical definition of the BPP is:

Minimize:

$$z = \sum_{i=1}^{n} y_i \tag{4}$$

Subject to the following constrains and conditions:

Subject to:

$$\sum_{j=1}^{n} w_j x_{ij} \leq c y_i \quad i \in N = \{1, \ldots, n\} \tag{5}$$

$$\sum_{i=1}^{n} x_{ij} = 1 \quad j \in N \tag{6}$$

$$y_i \in \{0, 1\} \quad i \in N \tag{7}$$

$$x_{ij} \in \{0, 1\} \quad i \in N, j \in N \tag{8}$$

where

w_j weight of item j.
y_j binary variable that shows if the bin i have items.
x_{ij} indicates whether the item j is in the bin i.
n number of available bins (also the number of items n).
c capacity of each bin.

The algorithms for the BPP instances can be classified as *online* or *offline* [45]. We have algorithms considered *online* if we do not know the items before starting the packing process, and *offline* if we know all the items before starting. In this research, we worked with both algorithms.

3.1 Tests Instances

Beasley [49] proposed a collection of test data sets, known as *OR-Library* and maintained by the Beasley University, which were studied by Falkenauer [27]. This collection contains a variety of test data sets for a variety of Operational Research problems, including the BPP in several dimensions. For the one dimensional BPP case, the collection contains eight data sets that can be classified in two classes:

- *Unifor* The data sets from binpack1 to binpack4 consist of items of sizes uniformly distributed in $(200, 100)$ to be packed into bins of size 150. The number of bins in the currently known solution was found by [27].
- *Triplets* The data sets from binpack5 to binpack8 consist of items from $(24, 50)$ to be packed into bins of size 100. The number of bins can be obtained dividing the size of the data set by three.

Scholl et al. [50] proposed another collection of data sets, only 1184 problems were solved optimally. Alvim et al. [51] reported the optimal solutions for the remaining 26 problems. The collection contains three data sets:

- *Set 1* It has 720 instances with items drawn from a uniform distribution on three intervals [1, 100], [20, 100], and [30, 100]. The bin capacity is C = 100, 120, and 150 and n = 50, 100, 200, and 500.
- *Set 2* It has 480 instances with C = 1000 and n = 50, 100, 200, and 500. Each bin has an average of 3–9 items.
- *Set 3* It has 10 instances with C = 100,000, n = 200, and items are drawn from a uniform distribution on [20,000, 35,000]. Set 3 is considered the most difficult of the three sets.

3.2 Heuristics

Heuristics have been used to solve the BPP, obtaining good results [25], because the BPP is an NP-Hard optimization problem [1] and an exact algorithm to solve it has not been found. The following heuristics are used in the *online BPP*:

- *Best Fit* [24] Puts the piece in the fullest bin that has room for it, and opens a new bin if the piece does not fit in any existing bin.
- *WorstFit* [25] Puts the piece in the emptiest bin that has room for it, and opens a new bin if the piece does not fit in any existing bin.
- *AlmostWorstFit* [25] Puts the piece in the second emptiest bin if that bin has room for it, and opens a new bin if the piece does not fit in any open bin.
- *NextFit* [22] Puts the piece in the right-most bin and opens a new bin if there is not enough room for it.
- *FirstFit* [22] Puts the piece in the left-most bin that has room for it and opens a new bin if it does not fit in any open bin.

The previous heuristics can be considered as offline algorithms if the items are sorted before to start the packing process.

3.3 Fitness Measure

There are many *Fitness Measure* used to discern the results obtained by heuristics and metaheuristics algorithms. In [52] are shown two Fitness Measures, the first measure [see Eq. (9)] try to find the difference between the used bins and the theoretical upper bound on the bins needed. The second [see Eq. (10)] was proposed in [46] and rewards full or almost full bins; the objective is to fill each bin, minimizing the free space.

$$\text{Fitness} = B - \left(\frac{\sum_{i=1}^{n} w_i}{c}\right) \tag{9}$$

$$\text{Fitness} = 1 - \left(\frac{\sum_{i=1}^{n}\left(\frac{\sum_{j=1}^{m} w_j x_{ij}}{c}\right)^2}{n}\right) \tag{10}$$

where

n number of bins.
m number of item.
w_j weight of item jth.
x_{ij} $x_{ij} = \begin{cases} 1 & \text{if the item } j \text{ is in the bin} \\ 0 & \text{otherwise} \end{cases}$
c bin capacity.

4 Grammar Design and Testing

To improve the Bin Packing heuristics was necessary to design a grammar that represents the Bin Packing problem. In [53] Grammar 1 is shown that is based on heuristics elements taken by Burke et al. [6]. That Grammar has been improved in the Grammar 2 [54] to obtain similar results to the one obtained by the Best Fit heuristic. However, this grammar cannot be applied to Bin Packing offline problems because it does not sort pieces. Grammar 3 is proposed to improve the results obtained by Grammar 2, it could be generating heuristics online and offline.

$$
\begin{aligned}
\langle \text{inicio} \rangle &\models ((\langle \text{expr} \rangle) <= (\langle \text{expr} \rangle)) \\
\langle \text{expr} \rangle &\models ((\langle \text{expr} \rangle \langle \text{op} \rangle \langle \text{expr} \rangle)) \mid \langle \text{var} \rangle \mid abs(\langle \text{expr2} \rangle)) \\
\langle \text{expr2} \rangle &\models ((\langle \text{expr2} \rangle \langle \text{op} \rangle \langle \text{expr2} \rangle)) \mid \langle \text{var} \rangle \\
\langle \text{var} \rangle &\models \text{F} \mid \text{C} \mid \text{S} \\
\langle \text{op} \rangle &\models + \mid * \mid - \mid /
\end{aligned}
$$

Grammar 1 Grammar based on FirstFit Heuristic, it was proposed in [53] and use the Heuristic Components shown in [6]

$$
\begin{aligned}
\langle \text{inicio} \rangle &\models \langle \text{exprs} \rangle.((\langle \text{expr} \rangle) <= (\langle \text{expr} \rangle)) \\
\langle \text{exprs} \rangle &\models Sort(\langle \text{exprk} \rangle, \langle \text{order} \rangle) \mid \lambda \\
\langle \text{exprk} \rangle &\models \text{Bin} \mid \text{Content} \\
\langle \text{order} \rangle &\models \text{Asc} \mid \text{Des} \\
\langle \text{expr} \rangle &\models ((\langle \text{expr} \rangle \langle \text{op} \rangle \langle \text{expr} \rangle)) \mid \langle \text{var} \rangle \mid abs(\langle \text{expr2} \rangle)) \\
\langle \text{expr2} \rangle &\models ((\langle \text{expr2} \rangle \langle \text{op} \rangle \langle \text{expr2} \rangle)) \mid \langle \text{var} \rangle \\
\langle \text{var} \rangle &\models \text{F} \mid \text{C} \mid \text{S} \\
\langle \text{op} \rangle &\models + \mid * \mid - \mid /
\end{aligned}
$$

Grammar 2 Grammar proposed in [54], it was based on BestFist Heuristic

$$
\begin{aligned}
\langle \text{begin} \rangle &\models \langle \text{exproff} \rangle \langle \text{exprsort} \rangle ((\langle \text{expr} \rangle) <= (\langle \text{expr} \rangle)) \\
\langle \text{exproff} \rangle &\models Sort(Elements, \langle \text{order} \rangle) \mid \lambda \\
\langle \text{exprsort} \rangle &\models Sort(\langle \text{exprkind} \rangle, \langle \text{order} \rangle) \mid \lambda \\
\langle \text{exprkind} \rangle &\models \text{Bins} \mid \text{SumElements} \\
\langle \text{order} \rangle &\models \text{Asc} \mid \text{Des} \\
\langle \text{expr} \rangle &\models ((\langle \text{expr} \rangle \langle \text{op} \rangle \langle \text{expr} \rangle)) \mid \langle \text{var} \rangle \mid abs(\langle \text{expr2} \rangle)) \\
\langle \text{expr2} \rangle &\models ((\langle \text{expr2} \rangle \langle \text{op} \rangle \langle \text{expr2} \rangle)) \mid \langle \text{var} \rangle \\
\langle \text{var} \rangle &\models \text{F} \mid \text{C} \mid \text{S} \\
\langle \text{op} \rangle &\models + \mid * \mid - \mid /
\end{aligned}
$$

Grammar 3 Grammar proposal to generate heuristics online and offline, this grammar is based on Grammar 2

where

S	size of the current piece.
C	bin capacity.
F	sum of the pieces already in the bin.
Elements	sort the elements.
Bin	sort the bins base on the bin number.
Cont	sort the bins base on the bin contents.
Asc	sort in ascendant order.
Des	sort in descendant order.

Table 1 PSO and BSO parameters

Parameter	Algorithm PSO BSO	
Population size	50	
w	1.0	
φ_2	0.8	
φ_2	0.5	
Sb	–	5
Search radius	–	2
Function calls	1500	

In order to generate the heuristics Grammar 3 was used. The search strategies applied to the GE were BSO and PSO. The number of function calls is only 10 % of the number of functions calls used by [6] (more details are available on [54]). To obtain the parameters shown in the Table 1 a fine-tuning process based on Covering Arrays (CA) was applied [55]. In this case, the CA was generated using the Covering Array Library (CAS) [56] from the National Institute of Standards and Technology (NIST).[1]

In order to generate the heuristics one instance from each set was used. Once this instance was obtained the heuristic for each instance set was applied to all the set to obtain the heuristic's fitness. The instance sets used were detailed in Sect. 3.1. It performed 33 experiments independently and used the median to compare the results against those obtained with the heuristics described in Sect. 3. The comparison was implemented through the nonparametric test of Friedman [29, 57]. This nonparametric test used a post hoc analysis to discern the performance between the experiments, and gives ranking to them.

5 Results

In Tables 2 and 3 the results obtained with the online and offline heuristics (described in Sect. 3.2) are shown. These results were obtained after applying the heuristics to each instance and all the results from an instance set were added.

In Table 4 examples of heuristics generated using the Grammar proposed with GE for each instance set are shown, some heuristics can be reduced but this is not part of the present work.

[1]http://csrc.nist.gov/groups/SNS/acts/index.html.

Table 2 Results obtained by each Online Heuristics

Instance	BestFit	FirstFit	NextFit	WorstFit	Almost WorstFit
bin1data	44.5601	44.5614	110.0537	47.6550	67.5970
bin2data	44.5601	44.5614	110.0537	47.6550	67.5970
bin3data	1.3902	1.3902	2.6996	1.3970	1.5188
binpack1	2.4258	2.6049	7.9410	5.0896	5.3657
binpack2	2.2591	2.3968	7.9891	4.9163	4.9920
binpack3	2.0143	2.1333	7.9946	4.7259	4.7696
binpack4	1.8386	1.9357	7.9612	4.5973	4.6221
binpack5	0.0	0.0	0.0	0.0	1.3243
binpack6	0.0	0.0	0.0	0.0	0.6820
binpack7	0.0	0.0	0.0	0.0	0.3349
binpack8	0.0	0.0	0.0	0.0	0.1671
hard28	0.6554	0.6553	13.1333	1.5473	1.9278

Table 3 Results obtained by each Offline Heuristics

Instance	BestFit	FirstFit	NextFit	WorstFit	Almost WorstFit
bin1data	44.5601	44.5614	110.0537	47.6550	67.5970
bin2data	44.5601	44.5614	110.0537	47.6550	67.5970
bin3data	1.3902	1.3902	2.6996	1.3970	1.5188
binpack1	0.9139	0.9140	9.5046	1.2334	1.4791
binpack2	0.7059	0.7060	9.4594	0.8464	1.0432
binpack3	0.5915	0.5915	9.4123	0.6664	0.7467
binpack4	0.4955	0.4955	9.4043	0.5722	0.6224
binpack5	4.8308	4.8308	6.4206	4.8487	5.4537
binpack6	4.5536	4.5536	6.1979	4.5573	4.9941
binpack7	4.5569	4.5569	6.0789	4.5586	4.7760
binpack8	4.4005	4.4005	6.0595	4.4007	4.5277
hard28	0.6554	0.6553	13.1333	1.5473	1.9278

The results obtained by the PSO and BSO with the Grammars are shown in Table 5, these results are the median from 33 individual experiments. With the results obtained by the heuristics and the GE with BSO and PSO, the Friedman nonparametric test was performed to discern the results. The value obtained by the Friedman nonparametric test is 83.023897 and the p-value 9.312428E−11, it means that the Heuristics tested have different performance. Due to this it was necessary to apply a post hoc procedure to obtain the Heuristics Ranking shown in Table 6.

Table 4 Example of Heuristics obtained for each instance set using the Grammar 3

Instance	Heuristic Generated
dasaset1	Sort(Elements, Des). Sort(Bin, Des). ((F + S)) $\leq$ (abs(C))
dataset2	Sort(Elements, Des). Sort(Cont, Des). (abs(S)) $\leq$ (abs((C − F)))
dataset3	Sort(Elements, Des). Sort(Cont, Des). (S) $\leq$ ((C − F))
binpack1	Sort(Content, Des). (abs(F)) $\leq$ ((C − abs(S)))
binpack2	Sort(Content, Des). ((F + S)) $\leq$ (C)
binpack3	Sort(Content, Des). (F) $\leq$ (abs((C − S)))
binpack4	Sort(Content, Asc). (S) $\leq$ ((C − F))
binpack5	((S+F)) $\leq$ (C)
binpack6	(F) $\leq$ ((abs(C) − S))
binpack7	(abs(F)) $\leq$ (abs((S − C)))
binpack8	(abs((S + F))) $\leq$ (C)
hard28	Sort(Cont, Des). (F) $\leq$ (abs((C − S)))

Table 5 Results obtained by each heuristics over the instance set

Instance	PSO			BSO		
	Grammar 1	Grammar 2	Grammar 3	Grammar 1	Grammar 2	Grammar 3
bin1data	44.56230	47.54510	44.56143	44.561455	44.560112	44.560112
bin2data	44.56230	44.56100	44.56143	44.561455	44.560112	44.560112
bin3data	1.39020	1.39020	1.39029	1.390289	1.390289	1.390289
binpack1	2.60470	2.60470	0.91403	2.604965	2.425894	0.913949
binpack2	2.39700	2.25910	0.70600	2.396851	2.259154	0.705970
binpack3	2.13330	2.01450	0.59154	2.133326	2.014334	0.591541
binpack4	1.93580	1.83870	0.49552	1.935710	1.838669	0.495512
binpack5	0.00000	0.00000	0.00000	0.000000	0.000000	0.000000
binpack6	0.00000	0.00000	0.00000	0.000000	0.000000	0.000000
binpack7	0.00000	0.00000	0.00000	0.000000	0.000000	0.000000
binpack8	0.00000	0.00000	0.00000	0.000000	0.000000	0.000000
hard28	0.65530	0.65550	0.65535	0.655350	0.655480	0.655480

Table 6 Rankings of the algorithms

Algorithm	Ranking Friedman
BSO—Grammar3	4.083333
PSO—Grammar3	4.583333
BestFit	6.166667
BSO—Grammar2	6.166667
BestFit—Offline	6.458333
PSO—Grammar2	6.916667
FirstFit—Offline	7.208333
PSO—Grammar1	7.250000
FirstFit	7.416667
BSO—Grammar1	7.416667
WorstFit	10.000000
WorstFit—Offline	10.166667
AlmostWorstFit—Offline	11.500000
NextFit	12.000000
AlmostWorstFit	12.833333
NextFit—Offline	15.833333

6 Conclusions and Future Works

In the present work, a model was proposed to use different metaheuristics as search strategy. This model uses the search strategy and the problem instance as inputs for the GE. It is presented as a Grammar to generate online and offline heuristics to improve the heuristics generated by other grammars and humans. Also, it was proposed to use BSO as search strategy based on Swarm Intelligence to avoid the problems observed in PSO.

Through the results obtained in Sect. 5, it was possible to conclude that it is possible to generate good heuristics with the Grammar proposed. Additionally it can be seen that the quality of these heuristics strongly depend on the grammar used to evolve.

The grammar proposed in the present work shows that it is possible to generate heuristics with better performance than the well-known heuristics BestFit, FirstFit, NextFit, WorstFit, and Almost WorstFit.

The results obtained by BSO are better than those obtained by PSO when using Grammar 2 and 3, but with the Grammar 1 PSO had better performance than BSO.

The current investigation is based on the one dimensional Bin Packing problem but this methodology can be used to solve other problems, due to the generality of the approach. It is necessary to apply the heuristic generation to another problem and investigate if the GE with PESO as search strategy gives better results than the GP or GE with another search strategy.

As a future work, we propose to find a methodology to choose the instance or instances for the training process as well as determine if the instances are the same or classify the instances in groups with the same features and to generate only one heuristic by group.

Also, it will be necessary to research another metaheuristics that does not need the parameter tuning because the metaheuristic shown in the present paper were tuned using Covering Arrays.

References

1. M.R. Garey, D.S. Johnson, *Computers and Intractability: A Guide to the Theory of NP-Completeness* (W. H. Freeman & Co., New York, NY, USA, 1979).
2. E.A. Feigenbaum, J. Feldman, *Computers and Thought* (AAAI Press, 1963).
3. M.H.J. Romanycia, F.J. Pelletier, Computational Intelligence **1**(1), 47 (1985).
4. F.W. Glover, Comput. Oper. Res. **13**, 533 (1986).
5. J.R. Koza, in *IJCAI* (1989), pp. 768–774.
6. E.K. Burke, M. Hyde, G. Kendall, in *Parallel Problem Solving from Nature - PPSN IX*, *Lecture Notes in Computer Science*, vol. 4193, ed. by T. Runarsson, H.G. Beyer, E. Burke, J. Merelo-Guervós, L. Whitley, X. Yao (Springer Berlin / Heidelberg, 2006), pp. 860–869.
7. C. Ryan, J. Collins, J. Collins, M. O'Neill, in *Lecture Notes in Computer Science 1391, Proceedings of the First European Workshop on Genetic Programming* (Springer-Verlag, 1998), pp. 83–95.
8. O. M., B. A, in *International Conference on Artificial Intelligence (ICAI'06)* (CSEA Press, Las Vegas, Nevada, 2006).
9. M. O'Neill, A. Brabazon, Natural Computing **5**(4), 443 (2006).
10. J. Togelius, R.D. Nardi, A. Moraglio, IEEE Congress on Evolutionary Computation pp. 3594–3600 (2008).
11. A. Moraglio, S. Silva, in *Genetic Programming*, *Lecture Notes in Computer Science*, vol. 6021, ed. by A. Esparcia-Alcázar, A. Ekárt, S. Silva, S. Dignum, A. Uyar (Springer Berlin / Heidelberg, 2010), pp. 171–183.
12. M.A. Sotelo-Figueroa, M. del Rosario Baltazar-Flores, J.M. Carpio, in *International Seminar on Computational Intelligence 2010* (Springer-Verlag, 2010).
13. B. Xing, W.J. Gao, in *Innovative Computational Intelligence: A Rough Guide to 134 Clever Algorithms*, *Intelligent Systems Reference Library*, vol. 62 (Springer International Publishing, 2014), pp. 45–80. 10.1007/978-3-319-03404-1\s\do5(4). URL http://dx.doi.org/10.1007/978-3-319-03404-1\s\do5(4).
14. J. Kennedy, R.C. Eberhart, IEEE Int. Conf. Neural Netw **4**, 1942 (1995).
15. D. Pham, A. Ghanbarzadeh, E. Koç, S. Otri, S. Rahim, M. Zaidi, in *Intelligent Production Machines and Systems* (Elsevier, 2006), pp. pp. 454–460.
16. M. Sotelo-Figueroa, R. Baltazar, M. Carpio, Journal of Automation, Mobile Robotics and Intelligent Systems (JAMRIS) **5** (2011).
17. W. Romero, V.M.Z. Rodríguez, R.B. Flores, M.A.S. Figueroa, J.A.S. Alcaraz, in *Artificial Intelligence (MICAI), 2011 10th Mexican International Conference on*, vol. 1 (Puebla, 2011), vol. 1.
18. L.A. Romero, V. Zamudio, R. Baltazar, E. Mezura, M. Sotelo, V. Callaghan, Sensors **12**(8), 10990 (2012). 10.3390/s120810990. URL http://www.mdpi.com/1424-8220/12/8/10990.
19. A. Lodi, S. Martello, D. Vigo, Discrete Applied Mathematics **123**(1–3), 379 (2002). http://dx.doi.org/10.1016/S0166-218X(01)00347-X.
20. H.V.D. Vel, S. Shijie, The Journal of the Operational Research Society **42**(2), 169 (1991).
21. B. Han, G. Diehr, J. Cook, Annals of Operations Research **50**(1), 239 (1994). 10.1007/BF02085642.

22. D.S. Johnson, A. Demers, J.D. Ullman, M.R. Garey, R.L. Graham, SIAM Journal on Computing **3**(4), 299 (1974).
23. A.C.C. Yao, J. ACM **27**, 207 (1980).
24. W.T. Rhee, M. Talagrand, Mathematics of Operations Research **18**(2), 438 (1993).
25. E. Coffman, Jr., G. Galambos, S. Martello, D. Vigo, *Bin Packing Approximation Algorithms: Combinatorial Analysis* (Kluwer Academic Publishers, 1998).
26. T. Kämpke, Annals of Operations Research **16**, 327 (1988).
27. E. Falkenauer, Journal of Heuristics **2**, 5 (1996).
28. A. Ponce-Pérez, A. Pérez-Garcia, V. Ayala-Ramirez, in *Proceedings of the 15th International Conference on Electronics, Communications and Computers (CONIELECOMP 2005)* (IEEE Computer Society, Los Alamitos, CA, USA, 2005), pp. 311–314.
29. J. Derrac, S. García, S. Molina, F. Herrera, Swarm and Evolutionary Computation pp. 3–18 (2011).
30. J.R. Koza, R. Poli, in *Search Methodologies: Introductory Tutorials in Optimization and Decision Support Techniques*, ed. by E.K. Burke, G. Kendall (Kluwer, Boston, 2005), pp. 127–164.
31. I. Dempsey, M. O'Neill, A. Brabazon, in *Foundations in Grammatical Evolution for Dynamic Environments*, vol. 194 (Springer-Verlag, New York, NY, USA, 2009).
32. H. lan Fang, H. lan Fang, P. Ross, P. Ross, D. Corne, D. Corne, in *Proceedings of the Fifth International Conference on Genetic Algorithms* (Morgan Kaufmann, 1993), pp. 375–382.
33. J. Holland, University of Michigan Press (1975).
34. M.A. Sotelo-Figueroa, H.J. Puga Soberanes, J.M. Carpio, H.J. Fraire Huacuja, L. Cruz Reyes, J.A. Soria-Alcaraz, Mathematical Problems in Engineering **2014** (2014).
35. C. Maurice, *Particle Swarm Optimization* (Wiley-ISTE, Estados Unidos, 2006).
36. R. Poli, J. Kennedy, T. Blackwell, Swarm Intelligence **1**(1), 33 (2007).
37. M.F. Tasgetiren, P.N. Suganthan, Q.Q. Pan, in *GECCO '07: Proceedings of the 9th annual conference on Genetic and evolutionary computation* (ACM, New York, NY, USA, 2007), pp. 158–167.
38. T. Gong, A.L. Tuson, in *GECCO '07: Proceedings of the 9th annual conference on Genetic and evolutionary computation* (ACM, New York, NY, USA, 2007), pp. 172–172.
39. S. Martello, P. Toth, *Knapsack Problems, Algorithms and and Computer Implementations* (John Wiley & Sons Ltd., New York, NY, USA, 1990).
40. E.G. Coffman, Jr., C. Courcoubetis, M.R. Garey, D.S. Johnson, P. Shor, R.R. Weber, M. Yannakakis, SIAM J. Disc. Math. **13**, 384 (2000).
41. T.G. Crainic, G. Perboli, M. Pezzuto, R. Tadei, Computers & Operations Research **34**(11), 3439 (2007). http://dx.doi.org/10.1016/j.cor.2006.02.007.
42. J. Edward G. Coffman, G. Galambos, S. Martello, D. Vigo, *Bin packing approximation algorithms: Combinatorial analysis* (Kluwer Academic Pub., 1999), pp. 151–207.
43. S.P. Fekete, J. Schepers, Mathematical Programming **91**(1), 11 (2001). 10.1007/s101070100243.
44. S.S. Seiden, R. van Stee, L. Epstein, SIAM Journal on Computing **32**, 2003 (2003).
45. E.G. Coffman Jr., J. Csirik, Acta Cybernetica **18**, 47 (2007).
46. E. Falkenauer, A. Delchambre, in *Robotics and Automation, 1992. Proceedings., 1992 IEEE International Conference on* (1992), pp. 1186 –1192 vol.2.
47. A. Lodi, S. Martello, D. Vigo, INFORMS Journal on Computing **11**(4), 345 (1999). 10.1287/ijoc.11.4.345.
48. E. Hopper, B. Turton, Artificial Intelligence Review **16**(4), 257 (2001). 10.1023/A:1012590107280.
49. J. Beasley, Journal of the Operational Research Society **41**(11), 1069 (1990).
50. A. Scholl, R. Klein, C. Jürgens, Computers & Operations Research **24**(7), 627 (1997).
51. A. Alvim, C. Ribeiro, F. Glover, D. Aloise, Journal of Heuristics **10**(2), 205 (2004).
52. M. Hyde, A genetic programming hyper-heuristic approach to automated packing. Ph.D. thesis, University of Nottingham (2010).

53. M.A. Sotelo-Figueroa, H.J. Puga Soberanes, J. Martín Carpio, H.J. Fraire Huacuja, C.L. Reyes, J.A. Soria-Alcaraz, in *Recent Advances on Hybrid Intelligent Systems*, *Studies in Computational Intelligence*, vol. 451, ed. by O. Castillo, P. Melin, J. Kacprzyk (Springer Berlin Heidelberg, 2013), pp. 349–359.
54. M. Sotelo-Figueroa, H. Puga Soberanes, J. Martin Carpio, H. Fraire Huacuja, L. Cruz Reyes, J. Soria-Alcaraz, in *Nature and Biologically Inspired Computing (NaBIC), 2013 World Congress on* (2013), pp. 92–98.
55. A. Rodriguez-Cristerna, J. Torres-Jiménez, I. Rivera-Islas, C. Hernandez-Morales, H. Romero-Monsivais, A. Jose-Garcia, in *Advances in Soft Computing*, *Lecture Notes in Computer Science*, vol. 7095, ed. by I. Batyrshin, G. Sidorov (Springer Berlin Heidelberg, 2011), pp. 107–118.
56. R.N. Kacker, D.R. Kuhn, Y. Lei, J.F. Lawrence, Measurement **46**(9), 3745 (2013).
57. D.J. Sheskin, *Handbook of Parametric and Nonparametric Statistical Procedures*, 2nd edn. (CRC, 2000).

Integer Linear Programming Formulation and Exact Algorithm for Computing Pathwidth

Héctor J. Fraire-Huacuja, Norberto Castillo-García, Mario C. López-Locés, José A. Martínez Flores, Rodolfo A. Pazos R., Juan Javier González Barbosa and Juan M. Carpio Valadez

Abstract Computing the Pathwidth of a graph is the problem of finding a linear ordering of the vertices such that the width of its corresponding path decomposition is minimized. This problem has been proven to be NP-hard. Currently, some of the best exact methods for generic graphs can be found in the mathematical software project called *SageMath*. This project provides an integer linear programming model (IPSAGE) and an enumerative algorithm (EASAGE), which is exponential in time and space. The algorithm EASAGE uses an array whose size grows exponentially with respect to the size of the problem. The purpose of this array is to improve the performance of the algorithm. In this chapter we propose two exact methods for computing pathwidth. More precisely, we propose a new integer linear programming formulation (IPPW) and a new enumerative algorithm (BBPW). The formulation IPPW generates a smaller number of variables and constraints than IPSAGE. The algorithm BBPW overcomes the exponential space requirement by using a *last-in-first-out* stack. The experimental results showed that, in average, IPPW reduced the number of variables by 33.3 % and the number of constraints by 64.3 % with respect to IPSAGE. This reduction of variables and constraints allowed IPPW to save approximately 14.9 % of the computing time of IPSAGE. The results also revealed that BBPW achieved a remarkable use of memory with respect to EASAGE. In average, BBPW required 2073 times less amount of memory than EASAGE for solving the same set of instances.

Keywords Pathwidth · Exact solution methods · Integer linear programming formulations · Enumerative algorithms

H.J. Fraire-Huacuja · N. Castillo-García (✉) · M.C. López-Locés
J.A. Martínez Flores · R.A. Pazos R. · J.J. González Barbosa
Tecnológico Nacional de México. Instituto Tecnológico de Ciudad Madero, Ciudad Madero, Mexico
e-mail: norberto_castillo15@hotmail.com

J.M. Carpio Valadez
Tecnológico Nacional de México. Instituto Tecnológico de León, León, Mexico

P. Melin et al. (eds.), *Nature-Inspired Design of Hybrid Intelligent Systems*,
Studies in Computational Intelligence 667, DOI 10.1007/978-3-319-47054-2_44

1 Introduction

The Pathwidth problem belongs to a family of optimization problems in which the goal is to find a linear ordering of the vertices of a generic graph such that a certain objective function is optimized. In particular, the objective function of the Pathwidth problem minimizes the number of vertices to the left of position t with at least one or more vertices adjacent to the right of t, for all positions of the linear ordering [1].

The Pathwidth problem is NP-hard [2] and has important practical applications in a variety of domains such that very large scale integration design [3–5], computer language compiler design [6], natural language processing [7], order processing of manufactured products [8], and computational biology [9].

Due to its practical applications, the Pathwidth problem has been widely studied and a lot of methods have been proposed to solve it. Thus, there are several exact and stochastic methods for both generic and structured graphs (see for example [1, 10–17]).

In the literature, we can find the mathematical software project called *SageMath* (visit www.sagemath.org/ for more details). In this project, there are two of the best exact methods for computing the vertex separation number of a generic graph. Recall that the Pathwidth problem is equivalent to finding the vertex separation number of a graph [18]. The *SageMath* project provides an integer linear programming formulation (IPSAGE) and an enumerative algorithm (EASAGE) [19].

IPSAGE is a very efficient integer linear programming (ILP) model that uses four types of variables: three groups of binary variables and one integer variable. The efficiency of IPSAGE is due to the small number of variables and constraints generated. In particular, IPSAGE generates only $3n^2+1$ variables and $3n^2+2mn$ constraints, where n and m represent, respectively, the numbers of vertices and edges of the graph.

EASAGE is an enumerative algorithm that computes the optimal objective value of small graphs very rapidly. The efficiency of EASAGE is due to an array whose size is exponential in the number of vertices, that is, an array of size 2^n. This array is used to store previous calculations that could be used later. Hence, this array avoids unnecessary computations and speeds up the global process. However, the spatial complexity of EASAGE is really high, i.e., $\theta(2^n)$. This considerably limits the size of the graphs that EASAGE can solve. Specifically, EASAGE can only deal with graphs with at most 32 vertices since it already requires 4 GB of memory. The requirement of memory doubles for each additional vertex, that is, EASAGE requires 8 GB for solving graphs with 33 vertices, 16 GB for graphs with 34 vertices, and so on. Obviously, the requirement of memory constitutes the main drawback of EASAGE.

In this chapter, we propose a new ILP formulation (IPPW) and a new branch and bound algorithm (BBPW). Our formulation IPPW is actually an improvement of IPSAGE. Specifically, IPPW reduces the number of variables by n^2 and the number of constraints by $n^2+2mn-n$ with respect to IPSAGE. Our branch and bound algorithm uses a *last-in-first-out* (LIFO) stack to overcome the excessive use of memory of EASAGE. Moreover, BBPW performs a binary search to find the optimal value of the instance rather than the sequential search performed by EASAGE. The use of the LIFO stack allows our algorithm to deal with larger instances. Specifically, BBPW solved graphs with 49 vertices. The amount of memory required by EASAGE to solve a graph with 49 vertices is about 512 *Terabytes*. This amount of memory exceeds by far the storage capacity of the computer where the experiment was conducted (256 GB).

We have conducted a computational experiment to assess the performance of our exact methods. The experimental results revealed that our formulation IPPW outperformed the formulation from the literature IPSAGE by 14.9 % in execution time. Moreover, IPPW generated 33.3 % less variables and 64.3 % less constraints than IPSAGE. Our exact algorithm BBPW spent 37 times more amount of time but required 2073 times less amount of memory than that required by EASAGE.

The remainder of this chapter is organized as follows. Section 2 presents the formal definition of the Pathwidth problem. Sect. 3 details our ILP formulation (IPPW) and our branch and bound algorithm (BBPW). In Sect. 4, we present the theoretical and experimental evaluation of the exact methods proposed. Finally, in Sect. 5, we discuss the major findings of this research.

2 Formal Definition of the Pathwidth Problem

The formal definition of the Pathwidth problem was taken from [6]. Let $G = (V, E)$ be a connected undirected and unweighted graph without loops. In the graph G, the sets V and E represent the sets of vertices and edges, respectively. In this research, we assume that the sets V and E are discrete and finite. Thus, the number of vertices in G is $n = |V|$ and the number of edges is $m = |E|$.

A *path decomposition* of G is a sequence of subsets of vertices $X_1, \ldots, X_r$ called *bags*, such that:

- $\bigcup_{1 \leq i \leq r} X_i = V$,
- $\forall (u, v) \in E$ there is an index i such that $1 \leq i \leq r$, with $u \in X_i$ and $v \in X_i$, and
- $\forall i \leq j \leq k, X_i \cap X_k \subseteq X_j$.

The width of some path decomposition is the cardinality of the largest bag minus one, i.e., $\max_{1 \leq i \leq r}\{|X_i| - 1\}$. The *pathwidth* of G is the minimum width over all possible path decompositions.

According to Suchan and Villanger [20], a path decomposition can be obtained from some linear ordering of vertices $(v_1, v_2, \ldots, v_n)$ as follows. Let v_p be the vertex placed at position p in the ordering (for all $1 \leq p \leq n$). The sequence of subsets $V_t = \bigcup_{p=1}^{t} v_p$ (with $1 \leq t \leq n$) represents a path with the following properties:

- $V_t \subset V_{t+1} \quad \forall t = 1, \ldots, n-1$, and
- $|V_t| = t \quad \forall t = 1, \ldots, n$

where index t is considered a *step* of the path $V_1, \ldots, V_n$. A path decomposition can be obtained from the path $V_1, \ldots, V_n$ by Eq. (1):

$$X_t = N(V_t) \backslash V_t \quad \forall t = 1, \ldots, n, \tag{1}$$

where $N(V_t) = \bigcup_{u \in V} \Gamma(u)$ is the set of all vertices adjacent to each vertex in V_t and $\Gamma(u) = \{v \in V | (u, v) \in E\}$ is the set of vertices adjacent to the vertex u. The cardinality of each bag is known as *cut value*. There are n cut values since there are n positions in the ordering. The width of the path decomposition induced by the path $V_1, \ldots, V_n$ is given by the largest cut value. This constitutes the objective value of the solution $V_1, \ldots, V_n$ and its formalization is presented in Eq. (2).

$$pw(V_1, \ldots, V_n, G) = \max_{1 \le t \le n} \{|X_t|\} \tag{2}$$

3 Solution Methods Proposed

This section presents our exact methods proposed for solving the Pathwidth problem. Specifically, we describe our ILP formulation (IPPW) in Sect. 3.1 and our branch and bound algorithm (BBPW) in Sect. 3.2.

3.1 Integer Linear Programming Formulation IPPW

Our formulation IPPW is an improvement of the integer linear programming formulation IPSAGE [19]. In particular, we redesign IPSAGE in such a way that only two groups of binary variables and one integer variable are needed. This leads to a new formulation that not only reduces the number of variables, but also the number of constraints. Our formulation IPPW only uses the following variables:

x_u^t is a binary variable whose value is 1 if the vertex u is in the set V_t and 0 otherwise. For all $u \in V, t = 1, \ldots, n$.

y_u^t is a binary variable whose value is 1 if $u \notin V_t$ and $\exists w \in V_t \bigcap \Gamma(u)$; and 0 otherwise. For all $u \in V, t = 1, \ldots, n$.

PW is an integer variable used to represent the objective value. This value is minimized.

We now present the complete mathematical formulation of IPPW:

$$\min PW \tag{3}$$

subject to:

$$x_u^t \leq x_u^{t+1}, \quad \forall u \in V, t = 1, \ldots, n-1, \tag{4}$$

$$\sum_{u \in V} x_u^t = t, \quad \forall t = 1, \cdots, n, \tag{5}$$

$$\sum_{w \in \Gamma(u)} x_w^t - deg(u) \cdot x_u^t \leq deg(u) \cdot y_u^t, \quad \forall u \in V, t = 1, \ldots, n, \tag{6}$$

$$\sum_{u \in V} y_u^t \leq PW, \quad \forall t = 1, \ldots, n, \tag{7}$$

$$x_u^t, y_u^t \in \{0, 1\}, \quad \forall u \in V, t = 1, \ldots, n, \tag{8}$$

$$PW \in \mathbb{Z}^+. \tag{9}$$

Constraints (4) and (5) are designed to accept feasible solutions only. Specifically, constraint (4) establishes that if vertex u is in the set V_t, then u must be also in the subsequent sets $V_{t+1}, V_{t+2}, \ldots, V_n$. Constraint (5) forces each set V_t to have exactly t elements, i.e., $|V_t| = t$.

The key constraint of our formulation is (6). In this constraint $deg(u) = |\Gamma(u)|$ is the adjacency degree of vertex u. This constraint forces the variable y_u^t to take the value 1 if and only if the vertex u is not in the set V_t, and it has one or more adjacent vertices in V_t. The number of vertices adjacent to vertex u which are in V_t (i.e., $|V_t \bigcap \Gamma(u)|$) is computed by summing the binary values of the variables associated to the vertices adjacent to u, i.e., $k = \sum_{w \in \Gamma(u)} x_w^t$. Thus, when u is not in V_t ($x_u^t = 0$) and u has at least one adjacent vertex in V_t ($k > 0$), the variable y_u^t is forced to take the value 1. The coefficient of the variables x_u^t and y_u^t is $deg(u)$. This is to consider the case in which all the vertices adjacent to u are in V_t, i.e., $k = deg(u)$.

Constraint (7) computes the number of elements in each bag X_t (with $1 \leq t \leq n$) and keeps the largest one. This constitutes the objective value PW of the current feasible solution. Finally, the objective function (3) minimizes the value of the variable PW. Constraints (8) and (9) formally define the variables used.

3.2 Branch and Bound Algorithm BBPW

As mentioned previously, in this chapter we will call our exact algorithm BBPW. The description of the algorithm is organized in the following way. Section 3.2.1

presents the complete description of the search tree. In Sect. 3.2.2, we describe the computation of the lower bound, the upper bound, and the cost of the nodes. The branching and pruning strategies are explained in Sect. 3.2.3. Finally, Sect. 3.2.4 shows the high-level pseudocode of BBPW.

3.2.1 The Search Tree

The algorithm BBPW models the instances by means of a search tree. The search tree is explored exhaustively in order to find the optimal value of the given instance. The nodes of the search tree are subsets of vertices. The cardinality of some node corresponds to the level in which the node is placed. More precisely, all the nodes (subsets of vertices) in the level t have exactly t elements. Thus, all the nodes in the first level have one element; all the nodes in the second level have two elements, and so on. The number of descendant nodes of some node placed in the level t is $n - t$. Moreover, the number of nodes in the level t of the search tree is given by:

$$\binom{n}{t} = \frac{n!}{t!(n-t)!},$$

which corresponds to the number of different ways to select t vertices from the set of vertices V. Therefore, the total number of nodes in the search tree for a graph with n vertices is

$$\sum_{t=1}^{n} \binom{n}{t} = 2^n - 1.$$

Figure 1 depicts an example of a search tree for a graph with $n = 4$ vertices.

From the search tree we can observe the following. The level at the top of the tree is the first one while the level at the bottom is the last one. There are four levels since the graph of the example has four vertices. The total number of nodes is

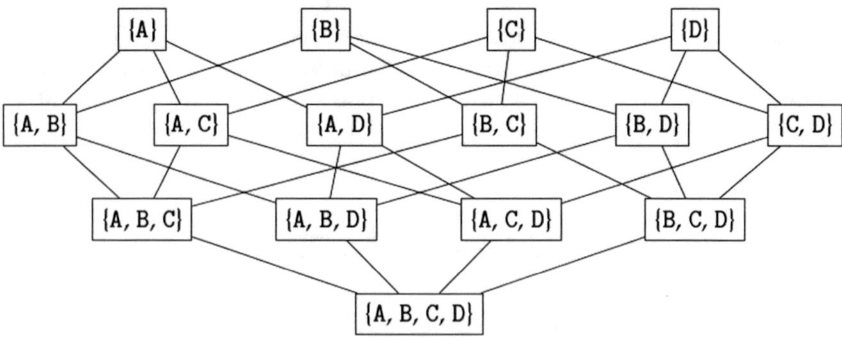

Fig. 1 Example of the search tree for a graph with $n = 4$ vertices

$2^4 - 1 = 15$. Let V_t be some node at level t (with $1 \le t \le n$). The descendant nodes of V_t (denoted by V_{t+1}) must satisfy $V_t \subset V_{t+1}$. For example, the descendant nodes of node {C, D} (last node in the second level) are the nodes {A, C, D} and {B, C, D} since {C, D} ⊂ {A, C, D} and {C, D} ⊂ {B, C, D}. This condition ensures that BBPW will produce feasible solutions according to the pathwidth definition. It is important to mention that in the search tree the nodes could have multiple parent nodes. For example, the node {A, B} has two parent nodes: {A} and {B}. This is so because {A, B} ⊃ {A} and {A, B} ⊃ {B}.

In order to obtain a feasible solution, it is required to generate a path $V_1, \ldots, V_n$ from the first level to the last one. For example, the linear ordering (A, B, C, D) must be obtained by consecutively visiting the nodes: V_1 = {D} from the first level, V_2 = {C, D} from the second level, V_3 = {B, C, D} from the third level, and V_4 = {A, B, C, D} from the last level. Notice that all the $n!$ linear orderings can be obtained from this search tree.

In order to visit all the solutions for a given instance, BBPW performs a depth-first search on the search tree by using a *last-in-first-out*(LIFO) stack called L. This stack L is the list of active nodes. Since the levels of the search tree contain nodes with different sizes, the list L is divided into n sub-lists, namely, $L_1, L_2, \ldots, L_n$. Thus, the sub-list L_t contains the active nodes from the level t.

3.2.2 Lower Bound, Upper Bound, and Cost of the Nodes

In the context of BBPW, the *lower bound LB* of some instance is a value smaller than the optimal value. Thus, a natural lower bound for some connected graph is $LB = 1$. Similarly, the *upper bound UB* of some instance is a value larger than or equal to the optimal value. For some connected graph with n vertices, the largest possible objective value is $UB = n - 1$. This objective value is obtained when the graph is completely connected. Finally, the *cost* of some node V_t (denoted by $c(V_t)$) is used to decide whether the algorithm continues the exploration from the current node V_t or not. This cost is obtained by using the cut value from the pathwidth definition, i.e., $c(V_t) = |N(V_t) \backslash V_t|$, where V_t is the subset of vertices corresponding to the node visited at level t and $N(V_t)$ is the set of vertices adjacent to every vertex in V_t.

3.2.3 Branching and Pruning Strategies

Basically, the algorithm BBPW tries to iteratively construct paths whose maximum cost is k (with $LB \le k \le UB$). In order to generate these paths, the algorithm computes the cost of the node visited, $c(V_t)$. If $c(V_t) \le k$, BBPW must continue the exploration of the search tree from V_t. In this case the algorithm must perform the *branching* process. Conversely, if $c(V_t) > k$, this means that it is not possible to discover any complete path whose maximum cost is k.This is so because at least one of the nodes has a cost (cut value) larger than k; therefore, the pruning process must be performed. Since BBPW uses a LIFO stack to store the active nodes, the branching and pruning processes become trivial.

In particular, the branching process consists in entering to the corresponding sub-list L_{t+1} all the descendant nodes from the current node V_t. Recall that these descendant nodes are those nodes in the level $t+1$ such that $V_t \subset V_{t+1}$ the number of these descendant nodes is $n-t$. The pruning process consists in simply not performing the branching process.

3.2.4 Algorithm BBPW

The basic idea of BBPW is iteratively searching for a path whose cost is at most k, that is, solutions whose objective values are less than or equal to a fixed value k. In the context of pathwidth, the values for k range from 1 to $n-1$. In order to find the optimal solution, the algorithm BBPW performs a binary search for the value of k. This implies that the value of k depends on the current lower and upper bounds. More precisely, the value of k is iteratively computed by

$$k = \left\lfloor \frac{LB + UB}{2} \right\rfloor.$$

When the algorithm discovers a path of cost k, this path is not the optimal solution. In fact, the path recently found is the new incumbent solution, and hence, its objective value k becomes the new upper bound UB. The incumbent solution is defined as the best solution found so far. Sometimes, when BBPW is searching for a path of cost k, another path of cost $k' < k$ is discovered instead. In this case the upper bound is updated with k', i.e., $UB \leftarrow k'$ since pathwidth is a minimization problem. In the case that the algorithm BBPW does not find any path of size k (or less), the value k is considered the new lower bound on the instance, i.e., $LB \leftarrow k$. The algorithm finishes when the upper bound is one unit more than the lower bound, that is when $UB = LB + 1$. In this case, the optimal solution is the incumbent solution and the optimal objective value is the current upper bound UB.

4 Theoretical and Experimental Evaluation

In this section, we evaluate the solution methods proposed. Particularly, Sect. 4.1 describes the instances used in the experiment. In Sect. 4.2, the ILP formulations are theoretically analyzed by comparing their number of variables and constraints generated. Finally, in Sect. 4.3, we report the computational experiment conducted to measure the performance of our methods in practice.

4.1 Test Bed Instances

In this chapter, we use 108 different small graphs in order to assess the performance of the methods proposed. In particular, we select graphs with at most 50 vertices since we are evaluating exact methods for solving an NP-hard problem. The description of the graphs used in the experiments is the following:

- GRID: The graphs in this dataset have a well-defined structure that resembles a two-dimensional mesh. This dataset contains only square grids, that is, graphs whose numbers of rows and columns are the same. This dataset originally contains 50 grid graphs [16]. However, we use only 5 graphs whose numbers of vertices and edges range from 9 to 49 and from 12 to 84, respectively. The optimal values for the instances in this dataset are known by construction.
- TREE: The graphs in this dataset have a tree structure, that is, they are connected and do not have cycles. This dataset originally contains 50 trees [16]. From this dataset we use only 15 instances whose numbers of vertices and edges are 22 and 21, respectively. Like the GRID instances, the optimal values for this dataset are known by construction.
- HB: Originally, this dataset contains 62 instances derived from the well-known *Harwell-Boeing Sparse Matrix Collection* [16]. From this dataset, we select only 4 instances whose numbers of vertices and edges range from 24 to 49 and from 46 to 176, respectively. Currently, there is only one instance from this dataset whose optimal value has been found [21].
- SMALL: This dataset consists of 84 small graphs whose optimal value cannot be determined by construction. Nevertheless, Fraire et al. found 81 optimal values for the whole dataset [21]. The numbers of vertices and edges for this dataset range from 16 to 24 and from 18 to 49, respectively.

4.2 Theoretical Analysis of the ILP Formulations

Each integer linear programming formulation generates different numbers of variables and constraints. The importance of these numbers lies in the fact that they determine the efficiency of the formulations in practice. In simple words, the less number of variables and constraints, the less computational effort.

Table 1 presents the equations to compute the total number of variables (#V) and the total number of constraints (#C) generated by each formulation. These equations

Table 1 Formulas to compute the number of variables (#V) and the number of constraints (#C) generated by IPSAGE and IPPW

	# V	# C
IPSAGE	$3n^2+1$	$3n^2+2mn$
IPPW	$2n^2+1$	$2n^2+n$

are in function of the number of vertices n and/or the number of edges m of the graph.

Asymptotically, the numbers of variables generated by both formulations are $\mathcal{O}(n^2)$. Similarly, the number of constraints generated by IPSAGE and by IPPW are, respectively, $\mathcal{O}(n^2+mn)$ and $\mathcal{O}(n^2)$. It is easy to see that the number of variables generated by the formulations is very similar. However, our formulation IPPW generates the smallest number of constraints. This is because the asymptotic function of IPPW only depends on the number of vertices of the graph. Conversely, the asymptotic function of IPSAGE depends on both the number of vertices and the number of edges. In a complete graph, the number of edges is exactly $n(n-1)/2$. Thus, the complexity of IPSAGE becomes $\mathcal{O}(n^3)$ as the number of edges increases.

Figures 2a and 2b depict two plots that, respectively, show the growth of the number of variables and the number of constraints of IPSAGE and IPPW with respect to the size of the graphs. Both plots have the same structure. The x-axis presents the 108 small instances sorted by the number of vertices in ascending order. The y-axis presents the number of variables for Fig. 2a, and the number of constraints for Fig. 2b. This axis is logarithmically scaled in both plots since some values are considerable large.

We can observe that the numbers of variables generated by IPSAGE and IPPW are very close to each other. Conversely, the gap between the numbers of constraints generated by the formulations is not too close. This is because IPPW generates n^2 less variables and $n^2+2mn-n$ less constraints than IPSAGE. It is easy to see that our formulation IPPW saves more constraints than variables with respect to formulation from the literature IPSAGE. Therefore, the theoretical analysis suggests that IPPW is better than the reference formulation IPSAGE since IPPW generates the smallest numbers of variables and constraints.

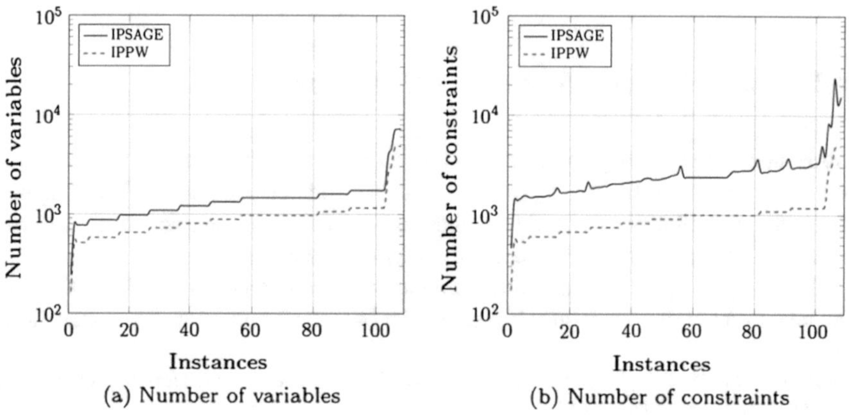

(a) Number of variables (b) Number of constraints

Fig. 2 Number of variables and constraints generated by IPSAGE and IPPW for a set of 108 instances with $9 \leq n \leq 49$

4.3 Computational Experiment

The computational experiment was conducted on a standard computer with an Intel Core i7-5500U processor at 2.4 GHz and 8 GB of RAM. All the algorithms were implemented in Java (JRE 1.8.0_66) and the ILP formulations were solved by the well-known optimization software CPLEX (v12.5) through its Java API.

The experiment was divided into two parts. In both parts, we fix a time limit of 300 CPU seconds for instance. In the first part of the experiment, we compare our algorithm BBPW with EASAGE. We use only 103 instances with $n \leq 32$ since the spatial complexity of EASAGE is $\theta(2^n)$ [19]. The goal of the first part of the experiment is to evaluate the impact of the LIFO stack, that is, the amount of memory saved by BBPW with respect to EASAGE. In the second part of the experiment, we evaluate the methods IPSAGE, IPPW, and BBPW with all the 108 instances. We do not include EASAGE because of its spatial complexity. Specifically, the biggest graph considered in the experiment has $n = 49$ vertices. The algorithm EASAGE requires 512 Terabytes of memory for solving this graph. This amount of memory exceeds by far the storage capacity of the computer where the experiment was conducted (256 GB of hard disk).

Table 2 shows the results of the first part of the experiment. The table has the following structure. The columns show four statistics: the average objective value (O. V.), the average computing time (Time), the number of optimal solutions found (# Opt.), and the average amount of memory required expressed in Megabytes (Mem.). The rows of the table present the results of the algorithms.

The results of the first part of the experiment show that both algorithms optimally solve all the instances evaluated before reaching the time limit of 300 s. This means that EASAGE and BBPW have an effectiveness of 100 % for the set of instances considered. In addition, we can observe that EASAGE is the fastest algorithm since it solves an instance in 2 ms in average. BBPW solves an instance in approximately 74 ms in average. This means that EASAGE was approximately 37 times faster than BBPW. However, as mentioned previously, the computing time of EASAGE is due to the huge amount of memory required. In particular, EASAGE required more than 4 MB in average to carry out the experiment. Conversely, our BBPW algorithm only required 1.7 KB of memory. This means that BBPW only required 0.04 % of the memory used by EASAGE for solving the same instances. Summarizing, EASAGE is 37 times faster than BBPW but requires 2073 times more amount of memory than that required by BBPW.

Table 2 Comparison of EASAGE and BBPW over a set of 103 small graphs with $n \leq 32$

	O. V.	Time	# Opt.	Mem.
EASAGE	3.16	0.002	103	4.146
BBPW	3.16	0.074	103	0.002

Table 3 Comparison of IPSAGE, IPPW, and BBPW over 108 small graphs with $n \leq 50$

	O. V.	Time	# Opt.	# Var.	# Const.
IPSAGE	3.37	20.52	104	1495.03	2850.86
IPPW	3.37	17.46	104	997.02	1017.58
BBPW	3.33	5.70	106	–	–

Table 3 shows the results of the second part of the experiment. This table has five columns that report: the average objective value (O. V.), the average computing time (Time), the number of optimal solutions found (#Opt.), the average number of variables generated by the ILP formulations (#Var.), and the average number of constraints (#Const.). Both #Var. and #Const. were obtained directly from CPLEX. The first two rows of the table present the results of the ILP formulations and the last row the results of BBPW.

From the results of the ILP formulations we can observe the following. Both ILP formulations obtained the same solution quality and the same number of optimal solutions found. This means that IPSAGE and IPPW had an effectiveness of 96.3 %. However, the computing time of IPPW was smaller than that of IPSAGE. Specifically, IPPW saved about 14.9 % of the computing time with respect to IPSAGE. This saving in execution time could be partially explained by the fact that our formulation IPPW generates a smaller number of variables and constraints than IPSAGE. In particular, IPPW generated 33.3 % less variables and 64.3 % less constraints than IPSAGE in average.

The best exact method in the experiment was BBPW since it achieved the best records for all the attributes measured. Specifically, the solution quality obtained by BBPW was better than those obtained by IPSAGE and IPPW by 1.2 %. In addition, the effectiveness of BBPW was about 98.1 %, which is better than those of the ILP formulations by approximately 1.8 %. Perhaps the best attribute of our algorithm was the execution time. BBPW solved an instance in approximately 5.7 CPU seconds in average. This means that BBPW saved about the 67.4 % of the execution time with respect to IPPW, which was the fastest ILP formulation.

5 Conclusions

In this chapter, we face the problem of finding the pathwidth of a simple connected graph. Particularly, we propose two new exact solution methods: one integer linear programming formulation (IPPW) and one branch and bound algorithm (BBPW).

Our formulation IPPW is an improvement of the formulation from the mathematical software project *SageMath* (IPSAGE). The improvement consists in redesigning IPSAGE in such a way that the number of variables and constraints generated is reduced. Specifically, IPPW reduces the number of variables by n^2 and the number of constraints by $n^2 + 2mn - n$ with respect to IPSAGE. Recall that n is the number of vertices and m is the number of edges.

Our branch and bound BBPW is also an improvement of the enumerative algorithm from *SageMath* (EASAGE). In this case, the improvement consists in reducing the spatial complexity of EASAGE by using a *last-in-first-out* stack rather than the array of size 2^n. The use of this LIFO stack allowed our BBPW algorithm to solve bigger instances than those that EASAGE can solve. Furthermore, our algorithm BBPW performs a binary search to find the optimal objective value of the instances rather than the sequential search used by EASAGE.

We conducted a computational experiment to assess the performance of our methods in practice. The experiment was divided into two parts. In the first part, we test the performance of EASAGE and BBPW by solving 103 small graphs with at most 32 vertices. We use graphs of this size because of the spatial complexity of EASAGE. The results from this part of the experiment showed that both algorithms were able to find all the optimal objective values, which gives them an effectiveness of 100 %. The results also showed that our BBPW algorithm was 37 times slower than EASAGE, but it only required 0.04 % of the amount of memory used by EASAGE.

In the second part of the experiment, we evaluated the performance of IPSAGE, IPPW, and BBPW by solving 108 graphs with at most 50 vertices. The results showed the following. The formulations IPPW and IPSAGE achieved the same solution quality and the same number of optimal solutions found (104 out of 108). Moreover, IPPW saved approximately 14.9 % of the computing time spend by IPSAGE and generated 33.3 % less variables and 64.3 % less constraints than IPSAGE. The algorithm BBPW was the best method since it found 106 optimal solutions out of 108 instances. BBPW also achieved the lowest computing time, that is, it solved an instance in 5.7 s in average. This means that BBPW saved about 67.4 % of the computing time of the second fastest method: IPPW.

Considering the results from the computational experiment, we can conclude that our mathematical formulation IPPW outperforms one of the best formulations in the literature: IPSAGE. We also conclude that our exact algorithm BBPW handles the use of memory much more efficiently than EASAGE. Therefore, we believe that the methods proposed in this chapter are good alternatives to efficiently solve the Pathwidth problem.

Acknowledgments This research was partially supported by the Mexican Council of Science and Technology (CONACYT). The second author would like to thank CONACYT for his Ph.D. scholarship. We also wish to thank the IBM Academic Initiative for allowing us to use the optimization software CPLEX.

References

1. Díaz, J., Petit, J., & Serna, M. (2002). A survey of graph layout problems. *ACM Computing Surveys (CSUR)*, *34*(3), 313–356.
2. Lengauer, T. (1981). Black-white pebbles and graph separation. *Acta Informatica*, *16*(4), 465–475.

3. Leiserson, C. E. (1980, October). Area-efficient graph layouts. In *Foundations of Computer Science, 1980, 21st Annual Symposium on* (pp. 270–281). IEEE.
4. Linhares, A., & Yanasse, H. H. (2002). Connections between cutting-pattern sequencing, VLSI design, and flexible machines. *Computers & Operations Research*, *29*(12), 1759–1772.
5. De Oliveira, A., & Lorena, L. A. (2002). A constructive genetic algorithm for gate matrix layout problems. *Computer-Aided Design of Integrated Circuits and Systems, IEEE Transactions on*, *21*(8), 969–974.
6. Bodlaender, H., Gustedt, J., & Telle, J. A. (1998, January). Linear-time register allocation for a fixed number of registers. In *SODA* (Vol. 98, pp. 574–583).
7. Kornai, A., & Tuza, Z. (1992). Narrowness, pathwidth, and their application in natural language processing. *Discrete Applied Mathematics*, *36*(1), 87–92.
8. Lopes, I. C., & Carvalho, J. M. (2010). Minimization of open orders using interval graphs. *International Journal of Applied Mathematics*, *40*.
9. Dinneen, M. J. (1996). *VLSI Layouts and DNA physical mappings*. Technical Report, Los Alamos National Laboratory.
10. Bollobás, B., & Leader, I. (1991). Edge-isoperimetric inequalities in the grid. *Combinatorica*, *11*(4), 299–314.
11. Castillo-García, N., Huacuja, H. J. F., Rangel, R. A. P., Flores, J. A. M., Barbosa, J. J. G., & Valadez, J. M. C. (2014). On the Exact Solution of VSP for General and Structured Graphs: Models and Algorithms. In *Recent Advances on Hybrid Approaches for Designing Intelligent Systems* (pp. 519–532). Springer International Publishing.
12. Ellis, J. A., Sudborough, I. H., & Turner, J. S. (1994). The vertex separation and search number of a graph. *Information and Computation*, *113*(1), 50–79.
13. Skodinis, K. (2000). *Computing optimal linear layouts of trees in linear time* (pp. 403–414). Springer Berlin Heidelberg.
14. Bodlaender, H. L., & Möhring, R. H. (1993). The pathwidth and treewidth of cographs. *SIAM Journal on Discrete Mathematics*, *6*(2), 181–188.
15. Bodlaender, H. L., Kloks, T., & Kratsch, D. (1995). Treewidth and pathwidth of permutation graphs. *SIAM Journal on Discrete Mathematics*, *8*(4), 606–616.
16. Duarte, A., Escudero, L. F., Martí, R., Mladenovic, N., Pantrigo, J. J., & Sánchez-Oro, J. (2012). Variable neighborhood search for the vertex separation problem. *Computers & Operations Research*, *39*(12), 3247–3255.
17. Sánchez-Oro, J., Pantrigo, J. J., & Duarte, A. (2014). Combining intensification and diversification strategies in VNS. An application to the Vertex Separation problem. *Computers & Operations Research*, *52*, 209–219.
18. Kinnersley, N. G. (1992). The vertex separation number of a graph equals its path-width. *Information Processing Letters*, *42*(6), 345–350.
19. Cohen, N., & Coudert, D. (2010). Integer linear programming formulation and enumerative algorithm for computing the vertex separation number. http://sagemanifolds.obspm.fr/preview/reference/graphs/sage/graphs/graph_decompositions/vertex_separation.html.
20. Suchan, K., & Villanger, Y. (2009). Computing pathwidth faster than 2^n. In *Parameterized and Exact Computation* (pp. 324–335). Springer Berlin Heidelberg.
21. Huacuja, H. F., Castillo-García, N., Rangel, R. A. P., Flores, J. A. M., Barbosa, J. J. G., & Valadez, J. M. C. (2015). Two New Exact Methods for the Vertex Separation Problem. *International Journal of Combinatorial Optimization Problems and Informatics*, *6*(1), 31–41.

Iterated VND Versus Hyper-heuristics: Effective and General Approaches to Course Timetabling

Jorge A. Soria-Alcaraz, Gabriela Ochoa, Marco A. Sotelo-Figueroa, Martín Carpio and Hector Puga

Abstract The course timetabling problem is one of the most difficult combinatorial problems, it requires the assignment of a fixed number of subjects into a number of time slots minimizing the number of student conflicts. This article presents a comparison between state-of-the-art hyper-heuristics and a newly proposed iterated variable neighborhood descent algorithm when solving the course timetabling problem. Our formulation can be seen as an adaptive iterated local search algorithm that combines several move operators in the improvement stage. Our improvement stage not only uses several neighborhoods, but it also incorporates state-of-the-art reinforcement learning mechanisms to adaptively select them on the fly. Our approach substitutes the adaptive improvement stage by a variable neighborhood descent (VND) algorithm. VND is an ingredient of the more general variable neighborhood search (VNS), a powerful metaheuristic that systematically exploits the idea of neighborhood change. This leads to a more effective search process according course timetabling benchmark results.

Keywords Course timetabling · Iterated local search · Variable neighborhood descend · Hyper-heuristics

J.A. Soria-Alcaraz (✉) · M.A. Sotelo-Figueroa
Division de Ciencias Economico-Administrativas, Departamento de Estudios Organizacionales, Universidad de Guanajuato, Leon, Mexico
e-mail: jorge.soria@ugto.mx

M.A. Sotelo-Figueroa
e-mail: masotelof@ugto.mx

G. Ochoa
Department of Computer Science and Mathematics,
University of Stirling, Striling, UK
e-mail: gabriela.ochoa@cs.stir.ac.uk

M. Carpio · H. Puga
División de Estudios de Posgrado e Investigacion,
Instituto Tecnológico de León, León Guanajuato, León, México
e-mail: jmcarpio61@hotmail.com

H. Puga
e-mail: pugahector@yahoo.com

P. Melin et al. (eds.), *Nature-Inspired Design of Hybrid Intelligent Systems*,
Studies in Computational Intelligence 667, DOI 10.1007/978-3-319-47054-2_45

1 Introduction

The timetabling problem is a common and recurring problem in many organizations. This paper focuses on a variant of the timetabling problem, named Course Timetabling Problem (CTTP). The CTTP, commonly seen at every university, works with the assignment of a fixed number of events into a number of timeslots. The main objective of this problem is to obtain a timetable that minimizes the number of conflicts for a student [9]. Many possible conflict types exist in CTTP, but the principal conflicts are usually time-related i.e. one student with two or more subjects assigned to the same time period.

Like most timetabling problems, the CTTP is known to be NP-Complete [10, 26]. Due to this complexity and the fact that course timetables are still often constructed by hand, it is necessary to automate timetable construction in order to improve upon the solutions reached by the human expert [25]. Unfortunately, the automation of a timetable construction is not an easy task, it requires a deep knowledge of the problem. This is the principal reason for the high popularity of hyper/metaheuristic solvers for the CTTP. Hyper-heuristics represents a novel direction on the heuristic optimization field, these algorithms aim to provide generic frameworks to solve differents problems. The advantage of this approach is that once we had a hyper-heuristic algorithm with reasonably good performance, that algorithm could be applied to similar problems without more changes that the update of a pool of low level (and problem dependent) operators. Hyper-heuristics can be defined as (meta) heuristics that choose or generate a set of low level heuristics to solve difficult search and optimization problems [4]. Hyper-heuristics aim to replace bespoke approaches by more general methodologies with the goal of reducing the expertise required to solve a problem [22]. In most of the previous studies on hyper-heuristics, low-level heuristics are uniform, i.e. they are either constructive or perturbative (improvement) heuristics [4, 6]. We use perturbative heuristics only.

Hyper-heuristics can be designed to select or construct heuristics when solving a problem. In this work we use hyper-heuristics that select the most promising heuristics to guide the search process. In order to create this kind of hyper-heuristics, a high level heuristic chooser is needed. In this work we use as high level chooser a hybrid ILS-VND algorithm. Iterated local search (ILS) is a simple but successful algorithm [13]. It operates by iteratively alternating between applying an operator to the incumbent solution and restarting local search from a perturbed solution. Variable neighborhood descend (VND) is a powerful component of variable neighborhood search (VNS), this simple strategy applies a set of hierarchically ordered heuristics to an incumbent solution, expecting that less perturbative heuristics are used more frequently than complex operators. The main contribution of this work is a newly proposed iterated variable neighborhood descent algorithm as a high level heuristic chooser when solving the course timetabling problem.

The paper is organized as follows. Section 2 defines the problem and some important concepts. Section 3 presents the solution approach and its justification. Section 4 contains the experimental set-up, results, their analysis and discussion. Finally, Sect. 5 gives the conclusions and describes some future work.

2 Main Concepts

In this section, the generic representation (methodology of design) is briefly explained. We also give a brief description of hyper-heuristic algorithms and high level choosers. Our base algorithms: ILS and VND are also detailed here.

2.1 Problem Definition

The CTTP, part of the Constraint Satisfaction Problems (CSP), considers its main variables as events to be scheduled according to a given curricula and its constraints as time-related restrictions (e.g., specific events to be assigned into specific timeslots). CTTP formulation consist of several sets [9]: a set of events (courses or subjects), $E = \{e1, e2, \ldots, en\}$ are the basic element of a CTTP. In addition, there are a set of time-periods, $T = \{t1, t2, \ldots, ts\}$, a set of places (classrooms) $P = \{p1, p2, \ldots, pm\}$, and a set of agents (students registered in the courses) $A = \{a1, a2, \ldots, ao\}$. Each member $e \in E$ is a unique event that requires the assignment of a period of time $t \in T$, a place $p \in P$ and a set of students $S \subseteq A$, so that an assignment is a quadruple (e, t, p, S). A timetabling solution is a complete set of n assignments, one for each event, which satisfies the set of hard constraints defined by each university or college. This problem is known to be at least NP-complete [10, 19].

The CTTP has been studied intensively, from early solution approaches based on logic programming [2, 5, 14] to metaheuristic schemes such as Tabu-list [16], Genetic Algorithms [28], Ant Colony [17], PSO [12], Variable Neighborhood Search [3] and Bee Algorithms [20]. In the same way, a great number of surveys of metaheuristics solution schemes that have been used to solve the CTTP problem are available [7, 8, 15, 18]. In addition, in recent years Hyper-heuristic frameworks has been applied with encouraging results [22, 23]. Almost all of this research used the ITC2007 [1] important benchmark in order to gather evidence about the efficiency of each proposed approach. This paper takes hyper-heuristics from the state of the art and applies them to ITC2007 instances.

2.2 Methodology of Design for the Course Timetabling Problem

In the literature it can be seen that a problem with the diversity of course timetabling instances exists due different university policies. This situation directly impacts on the reproducibility and comparison between course timetabling algorithms [21]. The state of the art indicates some strategies to avoid this problem; for example, a more formal problem formulation [15], as well as the construction of benchmark instances [1]. These schemes are useful for a deeper understanding of the university timetabling complexity, but the portability and the reproducibility of a timetabling solver in another educational institution is still in discussion [21]. In this sense, we use a context-independent layer for the course timetabling solution representation. This layer, named 'Methodology of Design', has been used previously with encouraging results for CTTP [23]. In addition it has been defined formally on Soria-Alcaraz et al. [24].

When using this layer, a heuristic algorithm only needs to choose a pair (timeslot–classroom) for each event e from a previously defined structure, as seen on Fig. 1. The construction of this structure of valid assignations exceeds the scope of this paper; for further information, please consult [22–24].

The solution representation used by any heuristic algorithm that uses this methodology can therefore be seen as an integer array with length equal to the number of variables (events/subjects). The objective is then to search in this space for a solution that minimizes student conflicts given by Eqs. (1) and (2). One example of this representation can be seen in Fig. 1.

$$\mathrm{Min}(\mathrm{FA}) = \sum_{i=1}^{k} \mathrm{FA}_{V_i} \tag{1}$$

$$\mathrm{FA}_{V_j} = \sum_{s=1}^{(M_{V_j})-1} \sum_{l=1}^{M_{V_i}-s} \left(A_{j,s} \wedge A_{j,s+l}\right) \tag{2}$$

where

FA	Student conflicts of current timetabling.
V_i	Student conflicts from "Time slot" i of the current Timetabling.
$A_{j,s} \wedge A_{j,s+l}$	students that simultaneously demand subjects s and $s+1$ inside the timeslot j.
A	student that demands subject s in timetabling j.

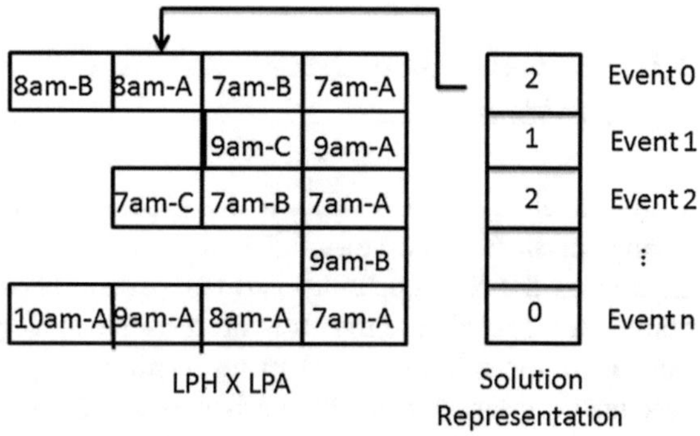

Fig. 1 Representation used

2.3 Hyper-heuristics

The term "hyper-heuristics" is relatively new, having first appeared as a strategy to combine artificial intelligence methods. The un-hyphenated version of the term initially appeared in Burke et al. [4] describing hyper-heuristics as "*heuristics to choose heuristics*" in the context of combinatorial optimization. Hyper-heuristics can be designed in a wide variety of ways, but in general they are constructed to achieve two objectives [4]: *heuristic selection* or *heuristic generation*. This work focuses on heuristic selection. Selection hyper-heuristics generally use a meta-heuristic to choose, from a pool of low-level heuristics, the best possible heuristic/operator to be applied to a problem instance to guide the search effectively. The basic scheme for the selection of hyper-heuristics can be seen in Fig. 2.

The high level heuristic chooser must decide what heuristic is the most promising one to continue the search at a given point. Obviously these algorithms need to have some desired characteristics including a fast and consistent response to changes in the current state of the problem, a mechanism to select heuristics previously not common selected and a mechanism to identify whenever the current operator has more potential to be discover by further applications. The domain barrier prevents the high level chooser from being problem dependent, i.e. having a low level knowledge of the problem. In practice, this means that the high level chooser only knows the last fitness improvement product of the last heuristic applied. This domain barrier ensures the generality of the whole model. Low level heuristics are another important part of the hyper-heuristic framework, this pool of operators receive a signal from the high level chooser whenever they must in order to produce a change in the current state of the problem instance and they return the fitness improvement of its application (if any). Finally, the problem instance state changes dynamically thought the successive application of low level heuristics to an optimal state.

Fig. 2 Selection of hyper-heuristic

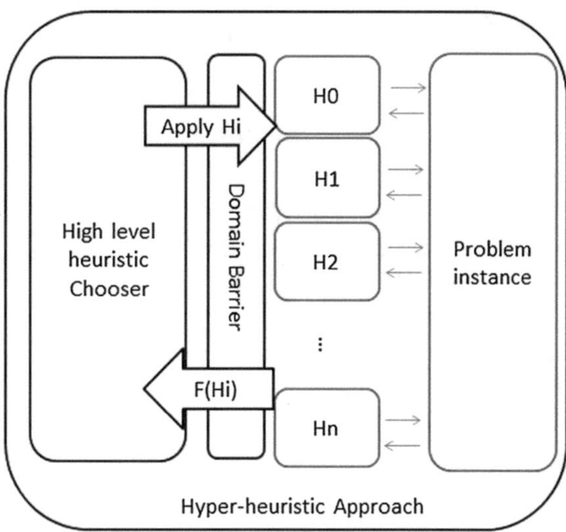

2.4 Iterated Local Search High Level Chooser

Iterated local search is a relatively simple yet powerful strategy. It operates by iteratively alternating between applying a move operator to the incumbent solution (perturbation stage) and restarting the local search from the perturbed solution (improvement stage). The term *iterated local search* (ILS) was proposed in [13].

A number of adaptive variants of multi-neighborhood iterated local search have been recently proposed with encouraging results in other problem domains [27]. If several options are available for conducting perturbation and improvement, a mechanism needs to be provided to choose between them. The idea is to use online learning to adaptively select the operators either at the perturbation stage or the improvement stage, or both. These approaches inspired the algorithm implemented in this article, which can be seen in Algorithm 1 and 2. In this implementation, the perturbation stage (step **4** in Algorithm 1) applies a single fixed move operator to the incumbent solution. This move operator (Simple Random Perturbation) simply selects uniformly at random a single variable and substitutes it for another variable in the range selected uniformly at random. Learning is then applied to the improvement stage (Algorithm 2), in which a perturbation operator is applied to the incumbent solution (steps 4 and 5 in Algorithm 2).

Algorithm 1 Iterated Local Search (ILS)

```
1: S0= GenerateIntialSolution()
2: S*= ImprovementStage(S0)
3: while !StopCriteria() do
4:    S' = SimpleRandomPerturbation(S*)
5:    S*' = ImprovementStage(S')
6:if f(S*')  better than f(S*) then
7:        S* = S*'
8:      end if
9:end while
11: return S*
```

Algorithm 2 Improvement Stage

```
1: ls = IncumbentSolution()
2: S*= ImprovementStage(S0)
3: while !LocalStopCriteria() do
4:    hi = Select Heuristic()
5:    ls* = apply(hi,ls)
6:    if f(ls*) better than f(ls) then
7:        ls = ls*
8:      end if
9: end while
11: return ls
```

2.5 Variable Neighborhood Descent

Variable neighborhood search (VNS), developed by Hansen and Mladenovic [11] in 1996, is a metaheuristic for solving combinatorial and global optimization problems. Unlike many metaheuristics where only a single operator is employed, VNS systematically changes different operators within a local search. The idea is that a local optimum defined by one neighborhood structure is not necessarily the local optimum of another neighborhood structure, thus the search can systematically traverse different search spaces, which are defined by different neighborhood structures. This potentially leads to better solutions, which are difficult to obtain by using single neighborhood based local search algorithms [11]. The basic principles of VNS are simple in execution, since parameters are kept to a minimum. Our high level chooser is based on variable neighborhood descent search (VND), a variant of the VNS algorithm. VND algorithm differs from VNS in that VND needs a hierarchically ordered list of operators; this ordering is commonly made by putting simpler and less perturbative heuristics at the beginning of the ordering, so more complex and perturbative heuristics appear latter in the ordering. The main idea is to choose in higher proportion heuristics that act as local search and when no further improvement can be achieved from this heuristics, to choose more complex operators in order to scape from the local optima. Algorithm 3 details our implementation of this algorithm. Lines 5–9 from Algorithm 3 detail the behavior of our VND-based heuristic chooser.

Algorithm 3 Variable Neighborhood Search (VND)

```
1: ls = IncumbentSolution()
2: H_k = getHierarchicallyOrderedOperators()
3: i=0
4: while !LocalStopCriteria() do
5:    ls* = apply(H_i,ls)
6:    if f(ls*) better than f(ls) then
7:        ls = ls*
8:        i=0
8:    else
9:        i= i+1
10:  end if
11:  end while
12:  return ls
```

2.6 Low Level Heuristic Pool

An important design decision in any hyper-heuristic algorithm is the selection of a pool of neighborhood structures or heuristics. We used as a base the successful operators proposed in [23] extending some of them by multiple applications in order to have larger-scale neighborhoods. The base operators are briefly described below.

- **Simple Random Perturbation** (SRP): uniformly at random chooses a variable and changes its value for another one inside its feasible domain.
- **Swap** (SWP): selects two variables uniformly at random and interchanges their values if possible. Otherwise leaves the solution unchanged.
- **Statistical Dynamic Perturbation** (SDP): chooses a variable following a probability based on the frequency of variable selection in the last k iterations. Variables with lower frequency will have a higher probability of being selected.
- **Double Dynamic Perturbation** (DDP): similar to heuristic SDP, it selects a new variable with a probability inversely proportional to its frequency of selection in the last k iterations. It differs from SDP in that it internally maintains an additional solution (which is a copy of the first initialized solution) and makes random changes to it following the same distribution.
- **Best Single Perturbation** (BSP): chooses a variable following a sequential order and changes its value to that producing the minimum conflict.

Four additional neighborhoods were created by applying SRP two times, SWP three times, SDP five times, and DDP five times. Therefore, a total of nine neighborhood structures were considered in that order.

3 Solution Approach

3.1 Combining Methodology of Design with Hyper-heuristics

As seen in Sect. 2, each high level chooser detailed so far utilizes a pool of low level heuristics, this pool is defined in Sect. 2.6. In this section, we define the codification and parameters used, as well as several details of each hyper-heuristic configuration.

We use the methodology of design approach shown in Sect. 2.2 in order to generalize the implementation of our metaheuristic over different CTTP instances. For each instance, the valid moves structure is built (Fig. 1). The detailed construction process for each of these lists is beyond the scope and purpose of this article, but interested readers are referred to [22, 24]. In the same manner as discussed above, these lists ensure by design that every variable in our solution represents a feasible selection in terms of time-space constraints. The main optimization exercise is to minimize the student conflict by means of the permutation of the events into time slots (i.e. integer values in our representation in Fig. 1). The function to be minimized has been described in Sect. 2.2, Eqs. (1) and (2). Initially our algorithm starts from a random solution and then we use Algorithm 1 to change the initial solution to a more promising one.

3.2 High Level Iterated Variable Neighborhod Decsent

In a hyper-heuristic framework with only ILS as the high level chooser the poll of low lever heuristics are presented without a particular ordering [27]. In our proposed approach, the improvement stage not only uses several neighborhoods, but it also incorporates a hierarchically ordering as seen in Sect. 2.6. We propose an hybridization of high level chooser, this hybridization takes the framework seen in Algorithm 1 but also utilizes a list of ordering heuristics as Algorithm 3. Algorithm 4 details our approach.

Algorithm 4 Iterated Variable Neighborhood Decent (IVND)

```
S0 = GenerateIntialSolution()
S*= VND(S0) {Improvement Stage}
while !StopCriteria() do
    S' = PerturbationStage(S*) {Using SRP Neighborhoods}
    S*' = VND(S') {Improvement Stage}
    if f(S*') better than f(S*) then
        S* = S*'
    end if
end while
return S*
```

The approach proposed here (IVND, Algorithm 4), takes as a base the ILS hyper-heuristic framework, but substitutes the adaptive improvement stage by a variable neighborhood descent (VND) algorithm. if an improvement to S is not possible using the first neighborhood H_1 then it is changed to H_2 *and* so forth with subsequent neighborhoods. As soon as an improvement is found with the current neighborhood, the sequence is restarted and H_1 is used again. This is a desired behavior since less perturbative heuristics are positioned at the beginning of the operator ordering.

4 Experiments and Results

In this section, several experiments are performed in order to find evidence about the performance of our approach against the CTTP state of art. We describe each experiment together with the characteristics of the benchmarks adopted.

4.1 *Test Instances*

The methodology of design allows the solution of diverse problem formulations it is merely necessary that each instance be expressed in terms of the generic structures (MMA, LPH and LPA). A well-known CTTP benchmark from the second international timetabling competition, PATAT ITC2007 Track 2 [1], is used for comparison between sequential and parallel approaches. This benchmark has 24 instances with main characteristics as follows:

Patat 2007

- A set of n events that are to be scheduled into 45 timeslots.
- A set of r rooms, each which has a specific seating capacity.
- A set *features* that are satisfied by rooms and required by events.
- A set of S students who attend various different combination of events.

The hard constraints are:

- No student should be required to attend more that one event at the same time
- In each case the room should be big enough for all the attending students
- Only one event is put into each room in any time slot.
- Events should only be assigned to time slots that are pre-defined as available *
- Where specified, events should be scheduled to occur in the correct order. *

The soft constraints are:

- Students should not attend an event in the last time slot of a day.
- Students should not have to attend three or more events in successive timeslots.
- Student should not be required to attend only one event in particular day.

4.2 *Experimental Design*

Experimental conditions used throughout this section resemble those of the international timetabling competition (ITC) 2007 track 2 (post-enrollment course timetabling). A total of 24 instances are available [1]. The objective function minimizes the sum of hard and soft constraint violations. For the post enrolment track, the number of hard constraint violations is termed the '*distance to feasibility*' metric, and it is defined as the number of students that are affected by unplaced events. In general, the cost of a solution for each timetabling problem is denoted using a value, *sv*, where *sv* and *sv* are the sum of soft constraint violations, In this paper, we only record results with a value of 0 Hard violations. Following the *ITC2007* rules, 10 independent runs per instance were conducted, and results are reported as the average of *sv*. The stopping condition for each run corresponds to a *time limit* of about 10 min, according to the benchmark algorithm provided in the competition website. Finally, these instances range from 400 to 600 events. Experiments were conducted on a CPU with Intel i7, 8 GB Ram using the Java language and the 64 bits JVM.

4.2.1 Results

Table 1 shows the best results (out of 10 runs, as this was the experimental setting used in the competition). The last column correspond to the proposed IVND, the results from the competition winner (Cambazard) are taken from the *ITC-2007*

Table 1 Results experiments ITC2007

Instance	Cambazard	Ceschia	Lewis	Jat and Yang	ILSHH	itVND
1	571	59	1166	501	650	677
2	993	0	1665	342	470	450
3	164	148	251	3770	290	288
4	310	25	424	234	600	570
5	5	0	47	0	35	30
6	0	0	412	0	20	10
7	6	0	6	0	30	15
8	0	0	65	0	0	**0**
9	1560	0	1819	989	630	620
10	2163	3	2091	499	2349	1764
11	178	142	288	246	350	250
12	146	267	474	172	480	450
13	0	1	298	0	46	30
14	1	0	127	0	80	68
15	0	0	108	0	0	30
16	2	0	138	0	0	20
17	0	0	0	0	0	**0**
18	0	0	25	0	20	10
19	1824	0	2146	84	360	299
20	445	543	625	297	150	150
21	0	5	308	0	0	**0**
22	29	5	x	1142	33	15
23	238	1292	3101	963	1007	892
24	21	0	841	274	0	**0**

website and the other columns from recent articles as indicated. The comparison was conducted using the 2007 competition rules and corresponding running time. Since all solutions are feasible, the values in the table correspond to the soft constraint violations, i.e. have zero hard constraint violation (except those marked with an *x*).

5 Conclusions and Future Work

This paper has presented a comparison between state-of-the-art hyper-heuristics and a newly proposed IVND algorithm when solving the course timetabling problem. Our approach substitutes the adaptive improvement stage by a VND algorithm with encouraging results over ITC 2007 track 2 instances.

Results indicate that the conceptually simpler IVND outperforms the recently proposed adaptive hyper-heuristic, and it shows competitive results against more complex state-of-the-art approaches.

Future work will study the extent to which the simpler and deterministic VND mechanism compares against more sophisticated reinforcement learning counterparts within and outside the competition setting. We will also explore whether applying multiple neighborhoods and adaptation to the perturbation stage in addition to (or instead of) the improvement stage provides improved results.

References

1. URL http://www.cs.qub.ac.uk/itc2007/
2. Boizumault, P., Delon, Y., Peridy, L.: Logic programming for examination timetabling. Logic Program **26**, 217–233 (1996)
3. Burke, E., Eckersley, A., McCollum, B., Petrovic, S., Qu, R.: Hybrid variable neighborhood approaches to university exam timetabling. European Journal of Operational Research **206**(1), 46 – 53 (2010)
4. Burke, E.K., Gendreau, M., Hyde, M., Kendall, G., Ochoa, G., Ozcan, E., Qu, R.: Hyper-heuristics: a survey of the state of the art. Journal of the Operational Research Society (JORS) 64(12), 1695–1724 (2013)
5. Cambazard, H., Hebrard, E., OSullivan, B., Papadopoulos, A.: Local search and constraint programming for the post enrolment-based course timetabling problem. Annals of Operations Research **194**, 111–135 (2012)
6. Carter, M.: A survey of practical applications of examination timetabling algorithms. Operations Research **34**, 193–202 (1986)
7. Causmaecker, P.D., Demeester, P., Berghe, G.V.: A decomposed metaheuristic approachfor a real-world university timetabling problem. European Journal of Operational Research **195**(1), 307 – 318 (2009)
8. Colorni, A., Dorigo, M., Maniezzo, V.: Metaheuristics for high-school timetabling. Computational Optimization and Applications **9**, 277–298 (1997)

9. Conant-Pablos, S.E., Magaa-Lozano, D.J., Terashima-Marin, H.: Pipelining memetic algorithms, constraint satisfaction, and local search for course timetabling. MICAI Mexican international conference on artificial intelligence **1**, 408–419 (2009)
10. Cooper, T.B., Kingston, J.H.: The complexity of timetable construction problems. Ph.D. thesis, The University of Sydney (1995)
11. Hansen, P., Mladenovic, N.: Variable neighborhood search. In: Burke, E., Kendall, G. (eds.) Search Methodologies, pp. 211–238. Springer US (2005)
12. Jarboui, B., Damak, N., Siarry, P., Rebai, A.: A combinatorial particle swarm optimization for solving multi-mode resource-constrained project scheduling problems. Applied Mathematics and Computation **195**(1), 299 – 308 (2008)
13. Lourenço, H., Martin, O., Stützle, T.: Iterated local search. In: Glover, F., Kochenberger, G., Hillier, F.S. (eds.) Handbook of Metaheuristics, International Series in Operations Research & Management Science, vol. 57, pp. 320–353. Springer New York (2003)
14. Lajos, G.: Complete university modular timetabling using constraint logic programming. In E Burke and P Ross editors. Practice and Theory of Automated Timetabling (PATAT) I **1153**, 146–161 (1996)
15. Lewis, R.: Metaheuristics for university course timetabling. Ph.D. thesis, University of Nottingham. (August 2006)
16. L, Z., Hao, J.K.: Adaptive tabu search for course timetabling. European Journal of Operational Research **200**(1), 235 – 244 (2010)
17. Mayer, A., Nothegger, C., Chwatal, A., Raidl, G.: Solving the post enrolment course timetabling problem by ant colony optimization. International Timetabling Competition 2007 (2008)
18. Qu, R., Burke, E.K., McCollum, B.: Adaptive automated construction of hybrid heuristics for exam timetabling and graph coloring problems. European Journal of Operational Research **198**(2), 392 – 404 (2009)
19. Rudova, H., Muller, T., Murray, K.: Complex university course timetabling. Journal of Scheduling **14**, 187–207 (2011).
20. Sabar, N.R., Ayob, M., Kendall, G., Qu, R.: A honey-bee mating optimization algorithm for educational timetabling problems. European Journal of Operational Research **216**(3), 533 – 543 (2012)
21. Schaerf, A. & Gaspero, L.Burke, E. K. & Rudová, H. *(Eds.)*Practice and Theory of Automated Timetabling VI: 6th International Conference, PATAT 2006 Brno, Czech Republic, August 30–September 1, 2006 Revised Selected Papers Measurability and Reproducibility in University Timetabling Research: Discussion and Proposals *Springer Berlin Heidelberg,* **2007**, 40-49
22. Soria-Alcaraz, J.A., Terashima-Marin, H., Carpio, M.: Academic timetabling design using hyper-heuristics. Advances in Soft Computing, ITT Springer-Verlag **1**, 158–164 (2010)
23. Soria-Alcaraz, J.A., Ochoa, G., Swan, J., Carpio, M., Puga, H., Burke, E.K.: Effective learning hyper-heuristics for the course timetabling problem. European Journal of Operational Research 238(1), 77 – 86 (2014).
24. Soria-Alcaraz Jorge, A., Carpio, M., Puga, H., Sotelo-Figueroa, M.: Methodology of design: A novel generic approach applied to the course timetabling problem. In: P. Melin, O. Castillo (eds.) Soft Computing Applications in Optimization, Control, and Recognition, *Studies in Fuzziness and Soft Computing*, vol. 294, pp. 287–319. Springer Berlin Heidelberg (2013)
25. de Werra, D.: An introduction to timetabling. European Journal of Operational Research**19**(2), 151 – 162 (1985)

26. Willemen, R.J.: School timetable construction: Algorithms and complexity. Ph.D. thesis, Institute for Programming research and Algorithms (2002)
27. Ochoa, G., Walker, J., Hyde, M., Curtois, T.: Adaptive evolutionary algorithms and extensions to the hyflex hyper-heuristic framework. In: Parallel Problem Solving from Nature - PPSN 2012, Lecture Notes in Computer Science, vol. 7492, pp. 418–427. Springer, Berlin (2012).
28. Yu, E., Sung, K.S.: A genetic algorithm for a university weekly courses timetabling problem. Transactions in Operational Research **9**, 703–717 (2002).

AMOSA with Analytical Tuning Parameters for Heterogeneous Computing Scheduling Problem

Héctor Joaquín Fraire Huacuja, Juan Frausto-Solís, J. David Terán-Villanueva, José Carlos Soto-Monterrubio, J. Javier González Barbosa and Guadalupe Castilla-Valdez

Abstract In this paper, the analytical parameter tuning for the Archive Multi-objective Simulated Annealing (AMOSA) is described. The analytical tuning method yields the initial and final temperature, and the maximum metropolis length. The analytically tuned AMOSA is used to solve the Heterogeneous Computing Scheduling Problem with independent tasks and it is compared versus the AMOSA without parameter tuning. We approach this problem as multi-objective, considering the makespan and the energy consumption. Also, in the last years this problem has gained importance due to the energy awareness in high performance computing centers (HPCC). The hypervolume, generational distance, and spread metrics were used in order to measure the performance of the implemented algorithms.

Keywords Multi-objective simulated annealing · Analytical tuning · Scheduling

H.J. Fraire Huacuja (✉) · J. Frausto-Solís · J.D. Terán-Villanueva · J.C. Soto-Monterrubio · J.J. González Barbosa · G. Castilla-Valdez
Instituto Tecnológico de Ciudad Madero, Ciudad Madero, Mexico
e-mail: automatas2002@yahoo.com.mx

J. Frausto-Solís
e-mail: juan.frausto@gmail.com

J.D. Terán-Villanueva
e-mail: david_teran00@yahoo.com.mx

J.C. Soto-Monterrubio
e-mail: soto190@gmail.com

J.J. González Barbosa
e-mail: jjgonzalezbarbosa@hotmail.com

G. Castilla-Valdez
e-mail: gpe_cas@yahoo.com.mx

P. Melin et al. (eds.), *Nature-Inspired Design of Hybrid Intelligent Systems*,
Studies in Computational Intelligence 667, DOI 10.1007/978-3-319-47054-2_46

1 Introduction

In this section, the Heterogeneous Computing Scheduling Problem (HCSP) is introduced as an important model to save energy in High Performance Computing Centers (HPCC). A heterogeneous computing system is an interconnection of multiple resources, as processors or machines, to provide a high power of computing. HPCC are widely used to solve complex problems and requires a high amount of power [1]. In the last years, the number of HPCC had increased as it is shown in [2], in November 2008, there was a total of 14 HPCC whose energy consumption was higher than 1 MW. Also, in 2006 the consumption of energy by servers and data centers in United States was around 61 billion of kilowatt-hour and the cost of this energy was around 4.5 billion dollars. The energy used to compute applications in the HPCC have implications in the economy cost and the heat generated. We address this problem due to the importance that energy awareness has acquired in HPCC during the last years.

The aim of this paper is to solve the problem of scheduling independent tasks on heterogeneous machines. One important part of this problem is that the machines should have the Dynamic Voltage and Frequency Scaling (DVFS) technology, which is implemented in most of modern machines [3]. This technology allows the user to decrease the energy consumption at the cost of computing power. Therefore, one objective is to minimize the energy consumption, the other objective is to minimize the maximum completion time (*makespan*). As both objectives are in conflict we deal with a multiobjective optimization problem (MOP). To solve this MOP, we propose the use of the Archive Multi-objective Simulated Annealing (AMOSA) algorithm [4] with an analytical tuning. This analytical tuning was implemented for the mono-objective version of Simulated Annealing (SA) in [5]. This method has been used to solve diverse problems [6, 7].

The remaining content of this paper is organized as follows. Section 2 describes the addressed problem. MOP concepts are presented in Sect. 3. Section 4 shows the SA's characteristics and analytical tuning. The AMOSA and the analytical tuning methods are described in Sect. 5. The experimentation and results are shown in Sect. 6. And Sect. 7 presents the conclusions and future works.

2 Problem Definition

The addressed problem is found in heterogeneous computing systems. In these systems the machines are used to execute large loads of works. In this research, the next characteristics are taken into consideration: the machines have the DVFS technology, the machines are heterogeneous with multiple computing capabilities, and tasks without precedence restrictions are received. The DVFS technology means different voltage configurations in machines at different speeds.

Given a set of tasks $T = \{t_1, t_2, \ldots, t_n\}$, a set of heterogeneous machines $M = \{m_1, m_2, \ldots, m_k\}$, the execution times of every task in every machine are $P = \{p_{1,1}, p_{1,2}, \ldots, p_{n,k}\}$. The *makespan* is defined in (1) and should be minimized for AMOSA algorithm which needs to be tuned to obtain a good performance.

$$\mathrm{MAX}_{j=1}^{k}\left(\sum P_{i,j} \forall\, t_i \in m_j\right) \tag{1}$$

Since machines are DVFS capable and heterogeneous, different voltage levels exist for each machine m_j with a relative speed associated. When the highest voltage is selected speed is equal to 1 (the normal execution time in P), with a lower voltage selected the relative speed decrease for example when speed equal to 0.5 (50 % of the normal speed), there is a relative execution time $P'_{i,j}$ calculated as follows.

$$P'_{i,j} = \frac{P_{i,j}}{\text{speed}} \tag{2}$$

Given a set of tasks T, a set of heterogeneous machines M and a set of voltages $V_j = \{v_{i,1}, v_{i,2,\ldots,} v_{j,l}\} \forall\, m_j \in M$ of different l sizes. The minimum energy is produced by the assignation of machine/voltage/task that minimizes

$$\sum_{j=1}^{k} V_{j,p}^{2} P'_{i,j} \forall\, t_i \in m_j \tag{3}$$

where p is the index of the selected voltage in V_j, since DVFS machines capabilities are being used in the energy consumption, the objective function for *makespan* is modified as follows.

$$\mathrm{MAX}_{j=1}^{k}\left(\sum P'_{i,j} \forall\, t_i \in m_j\right) \tag{4}$$

The MOP studied in this work consists in finding the assignation of machine/voltage/task that minimizes (4) and (3).

3 MOP Concepts

The basic concepts of multiobjective optimization are presented in this section. Without loss of generality, we will assume minimization functions only.

Definition 1 *MOP* Given

(a) A vector function $\vec{f}(\vec{x}) = [f_1(\vec{x}), f_2(\vec{x}), \ldots, f_k(\vec{x})]$ and
(b) A feasible solution space Ω,

The MOP consists in finding a vector $\vec{x} \in \Omega$ such that optimizes the vector function $\vec{f}(\vec{x})$.

Definition 2 *Pareto dominance*: A vector $\vec{x}$ dominates $\vec{x}'$(denoted by $\vec{x} \prec \vec{x}'$):

(a) If $f_i(\vec{x}) \leq f_i(\vec{x}')$ for all i functions in $\vec{f}$, and
(b) There is at least one i such that $f_i(\vec{x}) < f_i(\vec{x}')$.

Definition 3 *Pareto optimal*: A vector $\vec{x}^*$ is Pareto optimal if it does not exist a vector $\vec{x} \in \Omega$ such that $\vec{x} \prec \vec{x}^*$.

Definition 4 *Pareto optimal set*: The Pareto optimal set for a MOP is defined as $P^* = \{\vec{x}^* \in \Omega\}$.

Definition 5 *Pareto front*: Given a MOP and its Pareto optimal set P^*, the Pareto front is defined as $\text{PF}^* = \{f(\vec{x}) | \vec{x} \in P\}$.

4 Simulated Annealing

The SA algorithm was proposed by Kirpatrick in 1983 [8]. SA is an analogy with the annealing of solids, whose fundaments came from the physics area known as statistical mechanics. In this area, a set of solutions represents the particles of the solid moving from one position to another because the internal energy and temperature of the system. This energy is the Gibbs energy and the solid reaches the optimal state when this energy is minimal. The internal energy can be modified because the particles change their position, and/or the temperature is changed. The situation is far to be simple, i.e., because for a given temperature the particles change their position thus modifying the Gibbs energy and the temperature. The process continues until a thermal equilibrium is reached. Then, a new temperature is achieved and the process is started again. In this process, everything happens as if a set of rules or an intelligent algorithm were applied to the particles.

In the analogy of solids annealing, the solid is the solution of the problem, the energy is the function objective, and the movement of a solid represents a perturbation over the solution. The temperature is a control parameter and the thermal equilibrium is the solution of the heuristic. The movement of a solid is done by a perturbation function; this perturbation can be a genetic operator or another function that modifies the solution. The changes from one state to another are accepted by the Boltzmann probability function, $p_{x,y}(T) = \min\left\{1, e^{\frac{f(x)-f(y)}{T}}\right\}$.

The principal SA's parameters and characteristics are the following:

- **Initial temperature (T_i)**: Defines the start of the algorithm.
- **Final temperature (T_f)**: Defines the stop criterion.

Algorithm 1 Simulated Annealing

```
1: procedure SIMULATED ANNEALING(T_i, T_f, α)
2:     T_k = T_i
3:     S_i = generateRandomSolution()
4:     while T_k ≥ T_f do
5:         while Metropolis length do
6:             S_j = perturbation(S_i)
7:             if S_j < S_i then
8:                 S_i = S_j
9:             else if e^{-Δ(E(S_j)-E(S_i))/T_k} < random(0, 1) then
10:                S_i = S_j
11:        T_{k+1} = αT_k
```

Fig. 1 Simulated annealing algorithm

- **Cooling scheme**: It is a function which specifies how the temperature is decreased. The most common is the geometric scheme: $T_{k+1} = \alpha T_k$.
- **Markov's chain or metropolis length (L_k)**: This is the number of iterations done in each temperature T_k.
- **Acceptance criterion**: It is a probability function which defines if a new solution is accepted given a previous one.

Figure 1 shows the classic SA and is described as follows: Line 1 of the algorithm receives the initial temperature, the final temperature and the α value. In Line 2 the temperature is initialized. A random solution is generated in Line 3. Line 4 indicates the stop criterion. Then, in Line 5 the metropolis cycle begins. Inside this cycle (Line 6) a neighbor is generated. If the neighbor is better than the current solution then the new solution becomes the current solution, in other case the Boltzmann probability is computed in Line 9. In Line 11 the geometric cooling scheme is applied to reduce the temperature.

The T_i and T_f have an impact in the behavior of the algorithm. If T_i is too high the algorithm could expend a lot of time in the first iterations, and if it is too low the SA could be trapped in a local optimum. The T_f parameter has other impact, if T_f is too high the algorithm could finish too early, and if it is very low SA could take a long time to finish. Also the Markov's chain has an impact in the exploration of the solutions space. In the first iterations L_k needs to be small because the Boltzmann probability would accept almost any solution even if it does not improve the quality of the solution, and in the last iterations it needs to be bigger in order to increase the effort and yield better solutions. With an analytical tuning, these parameters can be optimized to take advantage of each step of the algorithm.

4.1 Analytical Tuning of Parameters in Simulated Annealing

The parameter tuning can be done by an analytical approach [5]. The analytical tuning adjusts the initial temperature, final temperature, and the Markov's chain length. The tuning is done following the subsequent analysis. In high temperatures the probability to accept a new solution is close to one. The initial temperature (T_i) is associated with the maximum deterioration permitted and the acceptance criterion. Let S_i and S_j be the actual and the new solution, respectively, $Z(S_i)$ and $Z(S_j)$ are the associated costs to S_i and S_j. The maximum and minimum deterioration are given by $\Delta Z_{\max}$ and $\Delta Z_{\min}$, then the probability to accept a new solution $P_A(S_j)$ with a maximum deterioration is 1 and T_i can be obtained by

$$e^{-\frac{\Delta Z_{\max}}{T_i}} = P_A(\Delta Z_{\max}) \tag{5}$$

$$T_i = -\frac{\Delta Z_{\max}}{\ln(P_A(\Delta Z_{\max}))} \tag{6}$$

Similarly the final temperature is associated with the probability to accept a new solution with the minimum deterioration $P_A(\Delta Z_{\min})$

$$e^{-\frac{\Delta Z_{\min}}{T_f}} = P_A(\Delta Z_{\min}) \tag{7}$$

$$T_f = -\frac{\Delta Z_{\min}}{\ln(P_A(\Delta Z_{\min}))} \tag{8}$$

The Markov's chain length (L_k) has the following features: a) at high temperatures only a few iterations are required, because the equilibrium stochastic is reached rapidly, b) at low temperatures a more exhaustive exploration is needed, so a bigger L_k is used. Let L_i be equal to L_k at T_i and $L_{\max}$ be the maximum length of the Markov's chain. L_k is decreased by the geometric scheme and L_k is calculated with

$$T_{k+1} = \alpha T_k \tag{9}$$

$$L_{k+1} = \beta L_k \tag{10}$$

where (9) is the geometric cooling scheme, L_k is the Markov's chain length at temperature k and β is the increment coefficient of the Markov's chain. Then T_f and $L_{\max}$ are obtained by

$$T_f = \alpha^n T_i \tag{11}$$

$$L_{\max} = \beta^n L_i \tag{12}$$

where n is the number of steps from T_i to T_f and it is calculated by the following equations.

$$n = \frac{\ln(T_f) - \ln(T_i)}{\ln \alpha} \tag{13}$$

$$\beta = e^{\frac{\ln L_{\max} - \ln L_i}{n}} \tag{14}$$

The probability to obtain the solution S_j from N random samples in the neighborhood V_{S_i} is

$$P(S_j) = 1 - \exp(-(N/|V_{S_i}|)) \tag{15}$$

from (15) N can be obtained by

$$N = -|V_{S_i}| \ln(1 - P(S_j)) = C|V_{S_i}| \tag{16}$$

where C establishes the exploration level, $C = \ln(\mathrm{P_R}(S_j))$. Then the Markov's chain length can be calculated by

$$L_{\max} = N = C|V_{S_i}| \tag{17}$$

In [5] to guarantee a good level of exploration of the neighborhood the values of C are established between $1 \leq C \leq 4.6$.

5 Multiobjective Simulated Annealing

The first adaptation of SA to multi-objective optimization was proposed by Serafini in 1993 [9]. This first adaptation consists in the adjustment of the acceptance criterion to combine all the objective functions using an aggregative function [10]. The principal differences between SA and MOSA are in the evaluation of the energy function and the use of a set of solutions containing the Pareto front. This energy function is required in the acceptance criterion. Fig. 2 shows the MOSA algorithm.

In Fig. 2 we can see the differences between Algorithms 1 and 2, which are in Line 7, 9, and 12–13. In the Algorithm 2, the Pareto dominance is used and three cases can occur in this evaluation. This cases are in Line 7, if the S_j dominates S_i then S_j is accepted and takes a place in the Pareto front. In Line 9, if S_i dominates S_j then in Line 10, we use the Boltzmann probability to decide whether to accept or reject S_j. In line 12, if both solutions are non-dominated, then the new solution is

Algorithm 2 Mullti–Objective Simulated Annealing

procedure MULTI–OBJECTIVE SIMULATED ANNEALING(T_i, T_f, α)
  $T_k = T_i$
  $S_i = generateRandomSolution()$
  while $T_k \geq T_f$ **do**
    while Metropolis length **do**
      $S_j = perturbation(S_i)$
      if $S_j \prec S_i$ **then**
        $S_i = S_j$
      else if $S_i \prec S_j$ **then**
        if $e^{\frac{-\Delta(E(S_j)-E(S_i))}{T_k}} < random(0,1)$ **then**
          $S_i = S_j$
      else ▷ If are non-dominated
        $S_i = S_j$
    $T_{k+1} = \alpha T_k$

Fig. 2 Multi-objective simulated annealing algorithm

accepted. Multiple versions of MOSA have been proposed in the last years [4, 11–14], and the AMOSA is among them [4].

5.1 AMOSA

The AMOSA algorithm is based on Pareto front. The principal features of AMOSA are: (a) the use of an archive to store all the non-dominated solutions [4], this archive has two limits known as hard limit (*HL*) and soft limit (*SL*), (b) the amount-of-dominance concept, which represent the area between two solutions. If the archive reaches *SL* size then a single linkage clustering algorithm is used to reduce their size to *HL* keeping the most diverse solutions [15]. The amount-of-dominance helps to identify the more dominated and less dominated solutions. Also, it is used in the acceptance criterion. AMOSA is capable to solve problems with 15 objective functions. AMOSA considers three cases can occur in the dominance

1. The current solution dominates the new solution; in this case the acceptance criterion is applied.
2. The current solution and the new solution are non-dominated between each other; this case is subdivided in three cases

   (a) If the new solution is dominated by more than one solution in the archive, then the acceptance criterion is applied with the average of amount-of-dominance.
   (b) If the new solution is non-dominated with respect to all the solutions in the archive, then the new solution is added to the archive.

(c) If the new solution dominates more than one solution in the archive, then the new solution is added to the archive and the dominated solutions are removed.

(3) The new solution dominates the current solution; this case considers the three same subcases as the second case. However, for the current case in the subcase (a) the acceptance criterion considers the minimum amount-of-dominance instead of the average amount-of-dominance.

Our proposal, the analytical tuning AMOSA (AMOSA-AT) is done as in SA but an aggregative function is used in the maximum and minimum deterioration. The perturbation function is an integer polynomial mutation. In this mutation the probability distribution is a polynomial function instead of a normal distribution [16].

6 Experimentation

In this experimentation three quality indicators are used to measure the algorithm's performance, hypervolume, generational distance, and spread. The hypervolume calculates the volume of the objective space covered by members of the non-dominated set of solutions. Generational distance is used for measuring the distance between the points in the obtained non-dominated front and the true Pareto front. Spread measures the average distance from a point to its nearest neighbor and the Euclidean distance to the extreme points of the true Pareto front [17].

The experiment consists of 10 independent runs with 14 instances from different authors. AMOSA uses the following parameters: Initial population = 100, $T_i = 4000$, $T_f = 10$, $\alpha = 0.95$, $L = 30$, $HL = 100$, and $SL = 150$. In AMOSA-AT the parameters are: Initial population = 100, $P_A(\Delta Z_{\min}) = 0.10$, $P_A(\Delta Z_{\max}) = 0.90$, $C = 2$, $\alpha = 0.95$, $HL = 100$ and $SL = 150$. The polynomial mutation is used in both algorithms. The algorithm performance is measured with three quality indicators: hypervolume for quality solutions, generational distance for diversity of the front, and spread for the dispersion of the front.

Table 1 shows the experimentation results. The first column contains the algorithm's name. The second column contains the average of hypervolume (HV). The third column shows the average of generational distance (GD). The fourth column contains the average spread (S). The best results are shown in bold font.

HV is focused on the goodness of the solutions, larger values are more desirable. If GD is equal to zero, it indicates that all the generated solutions are in the true Pareto front. In Spread, lower values indicate a better spread of Pareto optimal

Table 1 Results summary

	HV	GD	Spread
AMOSA	0.191236	0.281337	**0.883498**
AMOSA-AT	**0.259468**	**0.164582**	0.955863

Table 2 p-values computed by Wilcoxon test

	HV	GD	Spread
p-value	0.2716	0.1607	0.001187

solution. In this experiment, the AMOSA-AT achieves better results in HV and GD, but AMOSA gets a better spread.

The Wilcoxon non-parametric statistical test is performed to evaluate if the difference in the algorithm performance is statistically significant. The p-values computed by the Wilcoxon test are shown in Table 2. It is important to remember that a significant difference between two algorithms exists, if the p-value is equal or lower to 0.05. From Table 2 we can observe that there is only a significant difference in the Spread indicator.

7 Conclusions and Future Work

The results show that the AMOSA-AT has a best performance in HV and GD. AMOSA only won in the spread indicator. Despite the fact that the spread shows a significant statistical difference, both the HV and GD showed about 73 and 84 % of certainty of a significant statistical difference; which might increase with more instances. This results show that the parameter tuning produces a good performance for two metrics.

As future work we propose: the implementation of a fuzzy logic controller for parameter tuning, the use of variable-length Markov's chain to determine the dynamic equilibrium in real time, and experimentation with diverse functions in the maximum and minimum deterioration. Also, we think that the amount of dominance or the number of dominated solutions should be analyzed. Additionally, the use of a larger set of instances might clarify the statistical results. We consider the implementation of diverse techniques of SA to MOSA, such as the multi-quenching, re-annealing, and chaotic functions. Hybrid AMOSAs with other heuristics like genetic algorithms should be implemented. Finally, parallel programming technology for AMOSA is one of the future works in this research.

Acknowledgments The authors would like to acknowledge with appreciation and gratitude to CONACYT. This work has been partial supported by CONACYT Project.254498 and PRODEP.

References

1. W. C. Feng, "The importance of being low power in high performance computing," *CT Watch Quarterly,* vol. 1, no. 3, pp. 11-20, 2005.
2. Y. Liu and H. Zhu, "A survey of the research on power management techniques for high-performance systems," *Software: Practice and Experience,* p. 943–964, 2010.

3. G. Magklis, G. Semeraro, D. H. Albonesi, S. G. Dropsho, S. Dwarkadas and M. L. Scott, "Dynamic frequency and voltage scaling for a multiple-clock-domain microprocessor," *Micro, IEEE,* vol. 23, no. 6, pp. 62-68, 2003.
4. S. Bandyopadhyay, S. Saha, U. Maulik and K. Deb, "A simulated annealing-based multiobjective optimization algorithm: AMOSA},," *IEEE Transactions on Evolutionary Computation,* vol. 12, no. 3, pp. 269-283, 2008.
5. J. Frausto-Solís, H. Sanvicente-Sánchez and F. Imperial-Valenzuela, "ANDYMARK: an analytical method to establish dynamically the length of the markov chain in simulated annealing for the satisfiability problem," in *Simulated Evolution and Learning*, Springer, 2006, pp. 269-276.
6. J. Frausto-Solis, J. P. Sánchez-Hernández, M. Sánchez-Pérez and E. L. García, "Golden Ratio Simulated Annealing for Protein Folding Problem," *International Journal of Computational Methods,* 2015.
7. J. Frausto-Solis, E. Liñan-García, M. Sánchez-Pérez and J. P. Sánchez-Hernández, "Chaotic Multiquenching Annealing Applied to the Protein Folding Problem," *The Scientific World Journal,* vol. 2014, pp. 1-12, 2014.
8. S. Kirkpatrick, "Optimization by simulated annealing: Quantitative studies," *Journal of statistical physics,* vol. 34, no. 5-6, pp. 975–986, 1984.
9. P. Serafini, "Simulated annealing for multi objective optimization problems," in *Multiple criteria decision making*, Springer, 1993, pp. 283-292.
10. D. Nam and C. H. Park, "Multiobjective simulated annealing: A comparative study to evolutionary algorithms," *International Journal of Fuzzy Systems,* vol. 2, no. 2, pp. 87-97, 2000.
11. M. Alrefaei, A. Diabat, A. Alawneh, R. Al-Aomar and M. N. Faisal, "Simulated annealing for multi objective stochastic optimization," *International Journal of Science and Applied Information Technology,* vol. 2, pp. 18-21, 2013.
12. Y. Xu, R. Qu and R. Li, "A simulated annealing based genetic local search algorithm for multi-objective multicast routing problems," *Annals of Operations Research,* vol. 206, no. 1, pp. 527-555, 2013.
13. A. Zaretalab, V. Hajipour, M. Sharifi and M. R. Shahriari, "A knowledge-based archive multi-objective simulated annealing algorithm to optimize series–parallel system with choice of redundancy strategies," *Computers & Industrial Engineering,* vol. 80, pp. 33-44, 2015.
14. S.-W. Lin and K.-C. Ying, "A multi-point simulated annealing heuristic for solving multiple objective unrelated parallel machine scheduling problems," *International Journal of Production Research,* vol. 53, no. 4, pp. 1065-1076, 2015.
15. A. K. Jain and R. C. Dubes, Algorithms for Clustering Data, Upper Saddle River, NJ, USA: Prentice-Hall, Inc., 1988.
16. K. Deb and M. Goyal, "A combined genetic adaptive search (GeneAS) for engineering design," *Computer Science and Informatics,* pp. 30-45, 1996.
17. J. Wu and S. Azarm, "Metrics for quality assessment of a multiobjective design optimization solution set," *Journal of Mechanical Design,* vol. 123, no. 1, pp. 18-25, 2001.

Increase Methodology of Design of Course Timetabling Problem for Students, Classrooms, and Teachers

Lucero de M. Ortiz-Aguilar, Martín Carpio, Héctor Puga, Jorge A. Soria-Alcaraz, Manuel Ornelas-Rodríguez and Carlos Lino

Abstract The aim of the Course Timetabling problem is to ensure that all the students take their required classes and adhere to resources that are available in the school. The set of constraints those must be considered in the design of timetabling involves students, teachers, and classrooms. In the state of the art are different methodologies of design for Course Timetabling problem, in this paper we extend the proposal from Soria in 2013, in which they consider variables of students and classrooms, with four set of generic structures. This paper uses Soria's methodology to adding two more generic structures considering teacher restriction. We show an application of some different Metaheuristics using this methodology. Finally, we apply nonparametric test Wilcoxon signed-rank with the aim to find which metaheuristic algorithm shows a better performance in terms of quality.

Keywords Course timetabling · Faculty timetabling · Cellular genetic algorithm · Metaheuristics · Iterated local search

L. de M. Ortiz-Aguilar · M. Carpio (✉) · H. Puga · M. Ornelas-Rodríguez · C. Lino
División de Estudios de Posgrado E Investigación, TNM-Instituto Tecnológico de León, León, Mexico
e-mail: juanmartin.carpio@itleon.edu.mx

L. de M. Ortiz-Aguilar
e-mail: Ldm_oa@hotmail.com

H. Puga
e-mail: pugahector@yahoo.com

M. Ornelas-Rodríguez
e-mail: mornelas67@yahoo.com.mx

C. Lino
e-mail: carloslino@itleon.edu.mx

J.A. Soria-Alcaraz
Universidad de Guanajuato, Guanajuato, Mexico
e-mail: jorge.soria@ugtomx.onmicrosoft.com

P. Melin et al. (eds.), *Nature-Inspired Design of Hybrid Intelligent Systems*, Studies in Computational Intelligence 667, DOI 10.1007/978-3-319-47054-2_47

1 Introduction

The timetabling problem is present in different organizations such as schools, universities, hospitals, transport, etc. Universities need timetabling that included students, teachers, and others [1]. Therefore, having a good design of timetable help us optimize resources. And provide to the students necessary tools to finish on time his carrier. Therefore, this type of problem has been considered in the state of the art as NP-complete.

The design of Timetable depends directly over the specific school, university, or curricula; hence there is not universal timetable that can be applied in any case. The set of constraints that must be considered in the design of timetable involves students, teachers, and infrastructure. A key for the timetabling are the test instance, i.e., information regarding about supply and demand of resources, students, and teachers. This work generated instances that were created by artificial simulation techniques.

This work focuses on generating acceptable solution to the problem of Course timetabling, faculty timetabling and classroom assignment, using Metaheuristics. There are a diverse number of approaches that have been used to solve the problem of Course Timetabling as graph coloring [2], IP/LP (Integer programming/Linear programming) [3], genetic algorithms [4–6], Memetic Algorithms [7, 8], Tabu search [9, 10], simulated annealing [11] local search [12] Best-Worst Ant System (BUS), and Best-Worst Ant Colony System [13], in recent years Hyperheuristic approaches [14] and CSP [15] have been raised as generics and good alternatives when solving this problem.

The course timetabling problem has been classified as NP-complete [16]. In state or the art exist a lot of different problems at least class NP, which can be solved with different Metaheuristics, but as we indicated Theorem No Free Lunch [17] there is no Metaheuristics that outperforms all others for all known problems of class NP. Accordingly in this paper we show comparison between algorithm Cellular and Iterated Local Search, using nonparametric statistical tests.

2 Concepts

In this Section, we focus to concepts about University Timetabling, Methodology of design, and Metaheuristics.

2.1 University Timetabling

The university Timetabling is present in the most of universities and schools. This problem was described by Adriaen et al. [18]:

- **Faculty Timetabling**: assign each teacher to courses.
- Class teacher Timetabling: assign courses with minimum conflicts between groups of students.
- **Course timetabling**: assign courses with minimum conflicts between individual students.
- Examination timetabling: assign examinations with minimum conflicts between students.
- **Classroom assignment**: after assigning classes to teachers, assign this class teacher to each classroom.

University timetabling problem can be described as a set events *e*, a set *s* time slots, and set *c* restrictions between events, where we have to set each event *e* in one timeslot *s* [19]. We focused into generate solutions of the mixed problem of: faculty, Course, and Classroom assignment timetabling.

2.1.1 Faculty Timetabling

In [4] consider that faculty timetabling extends from basic model Class Teacher timetabling and defined as follows: given *m* teachers t_1, t_2,…,It_m, *n* class c_1, c_2,…,c_n and the set $\{1, 2, \ldots, t\}$ of periods of time, so each tuple (*t*, *c*) must be set in only one *t* period of time.

This case each teacher must be assigned in only one class in a specific period of time. For example of hard restriction is: one teacher cannot be scheduled in two classes at the same time; and example of soft restriction is: one teacher wants 2 or more consecutive classes.

2.1.2 Course Timetabling

The course timetabling problem can be described as a process of assign classes to resources (timeslots, classrooms, and teachers) while satisfying a set of restrictions [20]. The CTTP was defined by Conant-Pablos [21]: a set of events $E = \{e_1, e_2,\ldots,e_n\}$, a set of periods of time $T = \{t_1, t_2,\ldots,t_s\}$, a set of classrooms $P = \{p_1, p_2,\ldots,p_m\}$ and a set of students $A = \{a_1, a_2,\ldots,a_o\}$, so a quadruple (*e, t, p, S*) is a complete solution which satisfies the set of hard constraints defined by each university.

2.1.3 Classroom Assignment

From the point of the view of University, classroom assignment problem is the most restrictive due to the fact limited infrastructure, so the university cannot enroll more students than its capacity. Moreover in some cases, depending on the methodology of design from the university, the classroom assignment is related to the teachers, students, or subjects.

After forming the timetabling design with minimum conflicts, we should assign the subject to one classroom [22]. Each classroom has capacity, available schedule, etc.

2.2 *Methodology of Design*

The methodology of design proposed by Soria in [23] let us manage different constraints and restrictions of Course timetabling problem through conversion of all restrictions of time and space in a simple kind restriction: student conflicts. This methodology proposed 4 structures: MMA matrix, LPH list, LPH list, and LPS list. The first 3 structures represent the hard restrictions and the last represents the soft restrictions.

In our work we used three structures: MMA, LPH, and LPA. In [23] describe the structures as follows:

MMA matrix contains the number of students in conflict between classes, i.e., the number of conflicts if two classes are assigned in the same timeslots. Also this matrix shows the number of students that demand one class, this information is on the diagonal of the matrix. The Fig. 1 shows an example, for instance we have the subject GEE0925 which 172 students demand it, and it has 91 conflicts with GEF0914. The algorithm 1 shows the procedure to generate this matrix, taken from [24].

Fig. 1 MMA matrix

CLAVE	ACF0901	ACC0906	GEC0906	AEF1074	GEC0913	GEF0914	PDH	AE	ACF0902	AEB1082	GED0904
ACF0901	160	99	91	97	137	109	89	136		66	64
ACC0906		101	88	88	80	96	88	78	2	9	9
GEC0906			91	90	70	88	88	69		2	2
AEF1074				97	76	93	89	75		7	6
GEE0925					172	91	68	168	35	97	96
GEF0914						115	88	93	4	19	19
PDH							1	244	4	1	
AE								1	45	99	94
ACF0902									58	33	32
AEB1082										103	95
GED0904											96

Algorithm 1 MMA Construction

```
Require: int N Students, int[][] LD Students Demands
for i = 0 to N do
  Starr = LD[i]
for j = 0 to size(Starr) do
    for k = j + 1 to size(Starr) do
        MMA[Starr[j]][Starr[k]]+ = 1
        MMA[Starr[k]][Starr[j]]+ = 1
  end for
  end for
end for
return MMA
```

LPH this list contains information about each class and feasible time domain when can be assigned. The Fig. 2 shows an example, for instance we have the subject M4 has only 1 possible time slot where can be assigned. The algorithm 2 shows the procedure to generate this list, taken from [24].

Algorithm 2 LPH Construction

```
Require: int Nm CMaxEvent, int[] SI Initial solution
for i = 0 to size(SI) do
  intCrandom= random(Nm)
int [] lpht = nuevo
  for j = 0 to Crandom do
lpht[j]= randomEntre(Nm)
end for
lpht[Nm]= SI[i]
LPH[i]:add(lpht)
end for
return LPH
```

LPA this list contains information about feasible space domain from each class and where (classroom) can be assigned. The Fig. 3 shows an example, for instance

Fig. 2 LPH List

Event	Timeslots							
M_1	1	2	3	4	5	6	7	8
M_2	1	2	3	4	5		7	8
M_3	1	2	3	4	5	6	7	8
M_4	0							
⋮								
M_{n-1}	1	2	3	4	5	6	7	8
M_n	1	2	3	4	5		7	8

Fig. 3 LPA List

Events	*Classrooms*
M_1	$< A_1, A_2, A_3 >$
M_2	$< Lab_1 >$
M_3	$< A_1, A_2, A_3, A_4, A_5, A_6 >$
M_4	$< A_1, A_2, A_3 >$
⋮	⋮
M_{n-1}	$< A_1, A_2, A_3 >$
M_n	$< A_1, A_2, A_3 >$

we have the subject M4 can be assigned in classroom 1, 2, and 3. The algorithm 3 shows the procedure to generate this list, taken from [24].

Algorithm 3 LPA Construction

Require: *int Nm Subjects, int[][] CA Room features, int[][] DA =Subjects Demands, int[] Rms Room List*

```
for i = 0 to Nm do
for j = 0 to size(CA) do
for k = 0 to size(CA[j]) do
  if DA[i] <= CA[j][k] then
        LPA[i]:add(Rms[k])
  end if
 end for
 end for
end for
return LPA
```

2.2.1 Extending Methodology of Design of Course Timetabling Problem Through Teacher Management

In this paper we used two auxiliary structures LMS (list Subject Semester) and LPT (List Possible of Teachers). In [1] describe the structures as follows:

LMS this list contains each subject with their corresponding semester. The number of semester is given by the curriculum. This structure represents a hard constraint that must be satisfied. Figure 4 shows an example, for instance we have the subject M1 to M3 are semester first.

LPT list contains information related to each subject in terms of its available teacher. This list is built with the preferences from each teacher about schedules and subjects. Each university could have different teacher with defined working hours, academic profile, and minimum hours per week. Once having this information, we can create the LPT to work with the methodology of design. LPT list reports in its rows the possible teachers available to impart the Subject. The Fig. 5 shows an example, in this example M4 can be assigned to timeslot 1 assigning teacher seven.

Fig. 4 LMS List

Subject	*Semester*
M_1	1
M_2	1
M_3	1
M_4	2
M_5	2
$\vdots$	$\vdots$
M_{n-1}	9
M_n	9

Fig. 5 LPT List

Subject	$Teacher - Timeslot$
M_1	$< P_1H_2, P_2H_3, P_3H_3 >$
M_2	$< P_2H_6 >$
M_3	$< P_2H_3, P_1H_3 >$
M_4	$< P_7H_1 >$
$\vdots$	$\vdots$
M_{n-1}	$< P_2H_3, P_1H_3, P_2H_4 >$
M_n	$< P_9H_3 >$

This list should contain at least one teacher for each subject, due to the fact if only one subject do not have it, the timetabling design cannot be formulated, and this means that university does not have enough resources to cover all classes.

2.3 *Metaheuristics*

In this paper we used two Metaheuristics: cellular algorithm and Iterated Local Search. This section shows our adaptations for course timetabling problems.

2.3.1 Parallel Cellular Genetic Algorithm

The parallel cellular algorithm usually creates a conceptual population where each processor contents some individuals. Its main feature is the population structure form of grid, where each individual only relates to its neighbors [25]. In evolutionary algorithms (EA) model most widespread since its origins is called panmictic, where the evolutionary process works in a single population [26] and among other models are evolutionary algorithms Cellular (Fig. 6, taken from [1]).

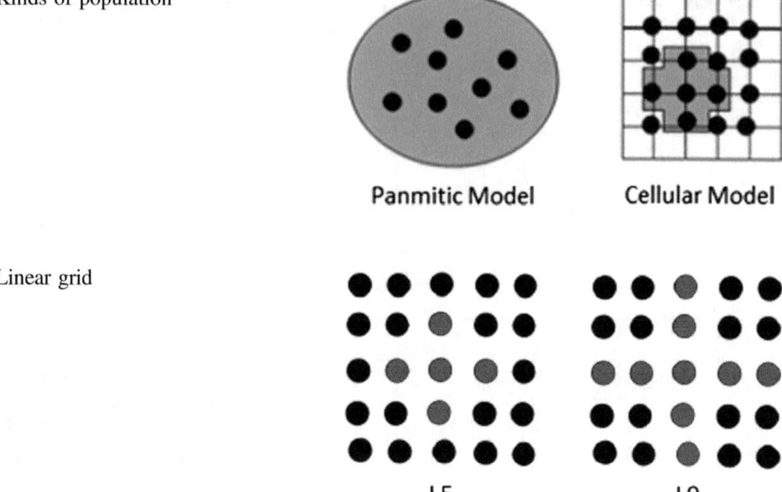

Fig. 6 Kinds of population

Fig. 7 Linear grid

The use of distributed parallel populations in cEAs is based on the idea that the isolation of populations allows maintain a higher genetic diversification. In some cases, these algorithms with decentralized population do a better sampling of the search space and improve behavior and the runtime as in algorithm panmixia [25].

The type of neighborhood used in this work, was linear (LR), where neighbors include $r-1$ structures closest chosen on the horizontal and vertical axes (see Fig. 7, taken from [1]). The algorithm 4 shows the Parallel Algorithm, taken from [8, 27].

Algorithm 4 Pseudo-code of a canonical cGA

```
procEvolve(cga)
GenerateInitialPopulation(cga.pop)
Evaluation(cga.pop)
while !StopCondition() do
for individual ← 1 to cga.popsizedo
neighbors ← CalculateNeighborhood(cga,position(indicidual));
   parents ← Selection(neighbors);
   offspring ← Recombination(cga.Pc,parents);
     offspring ← Mutation(PM);
   Replacement(position(individual),auxiliaryPop,offspring);
   end for
cga.pop← auxiliaryPop;
end while
return Best(cga.pop)
```

2.3.2 Iterated Local Search

The core of the iterated local search metaheuristic focus on initial solution, then iteratively builds a sequence of solutions generated by embedded heuristic [28]. In [29] describe the algorithms 5 and 6. This algorithm has perturbation stage (see step 4 on algorithm 5) where it applies a simple random perturbation to the solution. This simple random Perturbation operator uniformly selects at random value to replace it with another feasible value.

Algorithm 5 High Level Iterated Local Search (ILS)

```
1: s0 = GenerateInitialSolution
2: s* = ImprovementStage(s0)
3: while !StopCondition() do
4:     s' = SimpleRandomPerturbation(s*)
5:     s*' = ImprovementStage(s')
6:     if f(s*') < f(s*) then
7:         s* = s*'
8: end if
9: end while
10: return s*
```

Algorithm 6 Improvement Stage

```
ls ← IncumbentSolution
while !LocalStopCriteria() do
    hi = Perturbate()
    ls* = apply(hi, ls)
    if f(ls*) < f(ls) then
        ls = ls*'
    end if
end while
return ls
```

3 Metaheuristics Adapted to the Methodology of Design

3.1 Matrix Construction

We use the methodology of design proposed by Soria in [23] and we extend this for teachers. Now we will describe the structure of solutions as follows:

Subject	LPH	LPA		LPHxLPA Matrix
1	1	4 2 7 9		H1A4 H1A2 H1A7 H1A9
2	1 2 3	3	→	H1A2 H2A3 H3A3
3	1 6	4 5		H1A4 H1A5 H6A4 H6A5

Fig. 8 Representation LPH × LPA matrix construction

In this work we use artificial instances and the list LPH, list LPA, list LPT, and matrix MMA for the solution construction. However to generate inputs for meta-heuristic, in this case we construct a matrix LPT × LPH × LPA as follows: first have LPH × LPA matrix and then make a Cartesian product between the matrix LPH × LPA and the LPT List.

The Fig. 8 shows an example of LPH × LPA matrix construction. For instance we have 3 subjects and their LPH and LPA list for each one; the subject 1 can be assigned in timeslot 1 and classroom 4, 2, 7 and 9, whereas this information the feasible list for subject 1 is: timeslot 1 and classroom 4, timeslot 1 classroom 2, timeslot 1 classroom 7 and timeslot 1 classroom 9.

Now we do Cartesian product between LPT and LPH × LPA matrix. Figure 9 shows an example of LPT × LPH × LPA construction.

For example subject 2 has teacher 2 and timeslot 4 and 1, matrix LPH × LPA has timeslot 1 classroom 2, timeslot 2 classroom 3 and timeslot 2 and classroom 3. When the Cartesian product is applied the feasible result is teacher 2 timeslot 1 and classroom 2. This matrix has inside all the hard constraints.

3.2 Solution Representation

In this case we had the LPT × LPH × LPA matrix, so for the solution we keep the column position from each subject (row). An illustrative example it shows in Fig. 10 take from [1]. In this example we can see each subject has at least one possible tuple, therefore the final solution is a feasible solution for the current instance. Each semester has each different color, and the highlighted boxes are the option that we choose. For example, the subject 1 has the teacher 10 timeslot 7 and classroom 14, but we keep only the position of this tuple. This let us use an integer codification on the Metaheuristics instead a string codification.

Subject	LPT List	LPHxLPA Matrix		Matrix LPTxLPHxLPA
1	P1H2 P3H4 P1H1	H1A4 H1A2 H1A7 H1A9		P1H1A4 P1H1A2 P1H1A7 P1H1A9
2	P2H4 P2H1	H1A2 H2A3 H3A3	→	P2H1A2
3	P5H6 P2H6 P3H1	H1A4 H1A5 H6A4 H6A5		P5H6A4 P2H6A4 P3H1A4 P3H1A5

Fig. 9 Representation LPT × LPH × LPA matrix construction

Materia	Posición	1	2	3	4	5	6	7	8	9	10	11	12	13	14	15	16	17
Solución		1	2	3	4	5	6	7	8	1	2	3	4	5	6	1	2	3
	1	P10H7A14	P2H7A15	P6H5A7	P3H8A6	P6H6A1	P4H4A10	P3H1A5	P5H8A9	P6H3A14	P1H3A15	P3H4A14	P6H5A8	P9H3A5	P7H6A2	P1H2A15	P8H3A2	P7H4A12
	2	P9H6A11	P7H2A5	P2H6A7	P8H3A9	P6H4A11	P4H1A15	P7H4A6	P4H8A10	P7H8A11	P5H7A11	P2H3A7	P3H4A15	P7H7A6	P9H6A5	P5H6A12	P6H8A9	P8H4A2
	3	P8H2A4	P5H7A14	P2H6A2	P8H8A15	P7H5A9	P5H1A15	P2H2A7	P3H5A14	P4H1A5	P4H8A13	P9H8A10	P2H2A5	P8H4A1	P1H7A1	P10H7A13	P5H1A1	P8H7A14
Matriz (LPHxLPAxLPT)	4	P3H5A14	P10H1A10	P4H5A1	P3H2A2	P3H8A8	P10H1A15	P1H8A9	P1H4A8	P6H3A4	P8H3A12	P3H7A5	P3H2A10	P6H2A12	P9H8A2	P10H1A13	P2H4A9	P5H2A6
	5	P3H7A3	P8H4A5	P2H2A10	P10H3A8	P1H7A7	P7H4A6	P2H1A9	P10H3A15	P4H5A14	P1H8A11	P9H4A12	P4H8A4	P5H6A3	P5H4A3	P8H2A11	P2H1A1	P9H1A11
	6	P6H1A4	P6H3A2	P4H3A12	P2H2A1	P1H2A7	P7H5A6	P7H8A3	P10H8A2	P1H8A2	P4H1A6	P4H5A12	P7H4A4	P9H2A2	P2H7A4	P6H4A3	P2H3A12	P10H8A14
	7	P1H6A11	P3H7A11	P1H1A14	P4H5A9	P7H6A2	P6H7A14	P10H4A14	P6H5A8	P1H1A10	P4H6A2	P5H4A1	P8H8A11	P9H7A1	P4H3A7	P7H4A11	P3H2A11	P8H6A2
	8	P5H8A3	P1H2A7	P7H8A7	P5H8A10	P1H4A3	P8H2A7	P4H5A1	P10H7A1	P2H4A11	P1H2A12	P6H4A7	P10H1A15	P3H1A1	P10H1A15	P5H3A6	P5H4A14	P8H2A10

Materia		18	19	20	21	22	23	24	25	26	27	28	29	30	31	32	33	34
Solución		4	5	6	7	1	2	3	4	5	6	7	8	7	6	5	4	3
	1	P2H4A1	P7H1A2	P9H8A7	P2H7A4	P9H1A10	P1H2A3	P2H1A15	P10H2A12	P8H4A15	P7H1A8	P9H7A1	P1H1A7	P5H5A12	P7H6A10	P2H3A14	P2H2A10	P4H1A4
	2	P9H6A8	P2H1A12	P2H8A12	P4H4A2	P10H3A7	P1H3A11	P10H1A11	P7H1A8	P4H5A10	P8H5A6	P8H5A12	P8H3A8	P2H8A4	P6H2A11	P6H3A6	P1H5A14	P10H2A11
	3	P2H6A8	P3H2A7	P2H2A5	P2H5A14	P9H7A14	P9H7A5	P8H8A1	P1H5A5	P4H5A6	P6H2A14	P5H8A10	P9H3A6	P2H1A6	P7H4A12	P1H8A3	P2H6A9	P5H7A4
Matriz (LPHxLPAxLPT)	4	P10H6A4	P2H3A15	P2H5A13	P8H1A12	P5H1A12	P9H4A11	P9H2A3	P8H5A3	P10H1A9	P1H7A4	P6H3A10	P2H4A3	P9H3A7	P5H7A5	P6H7A15	P5H2A10	P3H2A9
	5	P4H7A13	P6H7A10	P8H1A14	P2H6A4	P3H4A12	P6H6A13	P6H8A13	P10H7A7	P5H8A13	P8H2A14	P4H4A4	P8H5A8	P7H2A8	P8H8A12	P3H1A9	P8H1A8	P10H2A15
	6	P8H2A5	P1H6A14	P10H8A15	P4H7A13	P8H4A13	P2H7A3	P8H2A14	P8H5A9	P2H2A11	P8H6A3	P2H4A14	P9H2A15	P8H6A3	P7H1A5	P7H4A12	P6H3A9	P1H8A13
	7	P3H7A12	P3H6A1	P8H3A14	P1H8A9	P2H7A7	P1H4A7	P8H5A10	P10H1A1	P1H2A15	P5H8A9	P2H5A6	P6H6A4	P4H7A11	P6H6A4	P8H8A6	P10H4A15	P4H3A7
	8	P6H7A9	P6H2A7	P6H6A2	P3H6A15	P10H5A3	P9H3A3	P9H4A1	P9H6A11	P3H3A7	P10H5A8	P8H2A2	P9H7A8	P5H7A13	P5H1A1	P9H5A12	P8H3A14	P5H8A15

Materia		35	36	37	38	39	40	41	42	43	44	45	46	47	48	49	50	51
Solución		2	1	2	3	4	5	6	7	8	7	6	5	4	3	2	1	2
											3							
	1	P10H3A5	P7H2A8	P6H3A5	P5H7A6	P7H7A12	P9H6A2	P8H3A14	P1H1A3	P8H7A7	P5H4A2	P2H5A7	P5H5A7	P5H7A15	P2H8A3	P4H4A12	P5H3A6	P4H1A11
	2	P10H6A14	P9H4A12	P10H6A4	P2H4A13	P10H5A11	P5H3A2	P2H6A7	P2H5A13	P10H6A3	P3H7A11	P10H1A1	P9H2A6	P7H2A12	P1H2A6	P1H3A1	P10H4A14	P5H3A9
	3	P8H5A7	P1H7A2	P10H1A4	P9H5A10	P5H6A2	P6H4A4	P5H2A8	P7H6A6	P1H1A5	P7H4A15	P2H4A8	P6H7A15	P4H4A11	P3H8A9	P1H1A15	P7H1A12	P8H1A7
Matriz (LPHxLPAxLPT)	4	P8H5A5	P10H3A8	P7H3A10	P6H2A11	P9H2A11	P6H1A6	P4H1A13	P5H4A4	P2H8A5	P9H2A12	P4H1A12	P8H2A5	P10H4A15	P2H3A10	P5H8A14	P1H7A10	P9H3A3
	5	P10H6A8	P2H7A15	P1H1A3	P3H5A7	P5H5A13	P4H5A5	P2H6A14	P1H3A14	P9H7A3	P8H1A9	P8H2A3	P10H6A2	P8H2A12	P8H6A10	P4H5A8	P1H4A13	P8H1A6
	6	P9H6A14	P9H3A11	P6H3A10	P9H1A5	P2H5A14	P9H5A11	P5H3A9	P2H6A13	P6H6A4	P6H5A6	P4H8A14	P1H1A10	P10H2A1	P6H2A8	P2H6A4	P4H4A1	P9H4A8
	7	P9H5A13	P5H8A9	P4H1A15	P9H8A2	P8H4A3	P4H2A12	P6H4A13	P1H7A7	P5H1A4	P6H7A5	P1H5A13	P10H7A1	P3H4A4	P5H7A10	P8H3A1	P3H7A5	P6H2A9
	8	P7H6A2	P7H6A12	P7H1A11	P4H5A15	P2H6A5	P3H7A13	P4H2A8	P10H6A9	P2H5A2	P5H2A1	P5H5A5	P4H2A3	P9H3A12	P5H3A13	P10H1A13	P3H7A11	P1H6A6

Fig. 10 Solution Representation

If not exist at least one possibility for the subject it means that not exist a solution for the timetable.

3.3 Fitness Function

In [1] defined fitness function as follows:Given a B matrix, $P = \{p_1, p_l,\ldots,p_r\}$ teachers where $1 \leq l \leq r$, $A = \{a_1, a_m,\ldots,a_s\}$ classrooms where $1 \leq m \leq s$, M a set of $\{m_1, m_h,\ldots,m_n\}$ subjects where $1 \leq h \leq t$. Each element $p\ \varepsilon\ P$ is unique and requires on classroom $a\ \varepsilon\ A$ and one subject $m\ \varepsilon\ M$, so we have tuple denote as (m, p, a). Given $V = \{v_1, v_w,\ldots, v_{u-1}\}$ timeslot, where $0 \leq w \leq u$-1, each $v_v = \{s_1, s_d, \ldots,s_e\}$ and $s_d = (m_h, p_l, a_m)$, therefore our aim is to minimize the following function:

$$f(x) = \sum_{i=0}^{u-1} \sum_{j=0}^{e} \sum_{k=j+1}^{e-1} B_{[m_{jh}][m_{kh}]} + FP(j,\ k) + GA(j.k), \tag{1}$$

where $FP(j,k)$ is:

$$FP(j,k) = \begin{cases} B_{[m_{jh}][m_{kh}]} + B_{[m_{jh}][m_{kh}]} & \text{if } p_{jl} = p_{kl} \\ 0 & else \end{cases}, \tag{2}$$

where $GA(j,k)$ is:

$$GA(j,k) = \begin{cases} B_{[m_{jh}][m_{kh}]} + B_{[m_{jh}][m_{kh}]} & \text{if } a_{jl} = a_{kl} \\ 0 & else \end{cases} \quad (3)$$

$GA(j,k)$ will be zero in case that does not have conflicts between classrooms and in another case we add the number of students that demand the subject. And $FP(j,k)$ will be zero in case does not have conflicts between teachers and in another case we add the number of students that demand the subject. In this work we take Eq. 1 for measure fitness.

4 Experiments and Results

In this section, we show experiments and the performance of Metaheuristics applies to the problem of Course timetabling, faculty timetabling, and classroom assignment. We also describe the most important characteristics of our set instances.

4.1 Test Instances

We generated 35 artificial instances, the Table 1 shows the information about this instances.

The instances are divided in subsets, for example: sets 1–5 have 52 subjects, 8 time slots, 15 classrooms, and 11 teachers. Each instance has different constraints, therefore not exist equals instances. In general we have instances with 52–300 subjects, 8–32 timeslots, 15–60 classrooms, and 11–50 teachers.

4.2 Experimental Design and Results

The configuration initial for Metaheuristics is on Table 2. We use call functions as stop criteria for both algorithms.

Table 1 Artificial instances

Number	Subjects	Timeslots	Classrooms	Teachers
1–5	52	8	15	11
6–10	52	16	15	11
11–15	52	8	15	11
16–20	104	16	30	25
21–30	208	32	60	50
31–35	300	32	60	50

Table 2 Configuration for Cellular Algorithm and Iterate Local Search

	Cellular	ILS
Subpopulations	4	10
Elitism	1 per subpopulation	
Crossover	0.9	
Mutation	0.1	
Individuals	16	
Local Search Iterations	N/A	10
Stop Criteria (functions Call)	5,000,000	

Table 3 has the results in terms of conflicts obtained by both Metaheuristics. We calculate the median, best fitness, worst fitness, and standard deviation for 35 instances.

The first column is the instance number; followed by median, best fitness, worst fitness, and standard deviation from both algorithms. For instance the bold number means the best result between each median, e.g., in instance 1–5 ILS has 0 conflicts, and cellular algorithm has only 0 in instance 2 and 3. The best fitness columns show that cellular algorithm has 0 conflicts in instances 1–20, and ILS has some 0 in instances like 1–5, 8, 11, 13, and 18–20. Now the worst results columns, the Cellular Algorithm has the lowest results except in first two instances. Also we show the standard deviation, because we need an algorithm with low standard deviation, and again the Cellular algorithm has the lowest result except in instances 1, 2 and 3; but this is not a significant answer, so we applied a nonparametric test.

4.3 Wilcoxon Signed-Rank Test

Now we want to know which algorithm has the best performance in terms of conflicts. First we applied Wilcoxon signed-rank test between ILS and Cellular algorithm, which say if this algorithms come from the same distribution.

Applying the Wilcoxon signed-rank test for populations $n > 30$, under the assumption that the ILS and Cellular algorithm come from populations with equal median, and $\propto = 0.05$ and $\propto = 0.01$, and $h_0 =$ There is no difference between the performance of the algorithms and $h_a =$ There are differences between the performance of the algorithms, we got following results:

$$\begin{aligned} W_{-} &= 615 \\ W_{+} &= 12 \\ Z &= -4.96 \\ P_{value} &= 3.472E-07 \end{aligned}$$

Table 3 Results experiments

	Median		Best fitness		Worst Fitness		Std deviation	
	ILS	Cellular	ILS	Cellular	ILS	Cellular	ILS	Cellular
1	**0**	5	**0**	**0**	**0**	50	**0**	5.95
2	**0**	**0**	**0**	**0**	**2**	10	**0.5**	4.4
3	**0**	**0**	**0**	**0**	76	**15**	13	**7.55**
4	**0**	3	**0**	**0**	2	**1**	**0.4**	18.04
5	**0**	2	**0**	**0**	138	**7**	58.2	**10.13**
6	732	**5**	161	**0**	1791	**42**	430.3	**10.5**
7	925	**7**	320	**0**	2526	**45**	509.5	**27.54**
8	703	**0**	**0**	**0**	1837	**8**	495.7	**1.57**
9	504	**0**	103	**0**	3904	**29**	1021.4	**3.88**
10	439	**0**	36	**0**	1993	**30**	555.4	**20.63**
11	341	**0**	**0**	**0**	1565	**21**	303.7	**6.96**
12	943	**3**	135	**0**	2030	**34**	449.8	**16.96**
13	397	**5**	**0**	**0**	1566	**15**	398	**3.08**
14	350	**16**	52	**0**	1925	**10**	519.1	**9.72**
15	1237	**56**	434	**0**	2627	**1**	506.6	**7.78**
16	9609	**12**	2357	**0**	25,846	**39**	7708.6	**12.3**
17	6497	**35**	2	**0**	20,395	**8**	5066.7	**17.14**
18	5018	**12**	**0**	**0**	21,073	**7**	5578.3	**17.17**
19	3414	**35**	**0**	**0**	20,775	**20**	4650.2	**48.5**
20	1872	**21**	**0**	**0**	11,761	**26**	2856.6	**11.83**
21	137,650	**234**	10,3701	**144**	178,027	**847**	16,513	**32.66**
22	140,174	**357**	99,970	**238**	176,874	**404**	21,146.2	**32.96**
23	165,490	**493**	129,316	**454**	196,366	**761**	15,712.9	**68.78**
24	138,172	**390**	108,764	**116**	176,308	**136**	14,397.5	**71.06**
25	157,661	**282**	124,817	**244**	178,698	**252**	13,769.1	**16.23**
26	149,412	**289**	127,638	**60**	174,593	**184**	11,879.5	**28.6**
27	130,602	**423**	105,590	**180**	162,570	**576**	13,296.5	**59.1**
28	117,600	**218**	87,222	**210**	154,111	**606**	16,285.5	**57.23**
29	161,040	**566**	120,276	**383**	202,775	**976**	14,930.2	**86.71**
30	142,257	**598**	114,886	**228**	193,008	**594**	19,459.2	**52.19**
31	539,375	**404**	451,269	**262**	661,147	**780**	45,605.1	**7.33**
32	562,611	**495**	435,852	**402**	656,118	**314**	50,244	**6.11**
33	611,247	**393**	471,273	**353**	695,651	**722**	58,068.4	**55.73**
34	544,623	**495**	453,478	**92**	622,181	**476**	45,566.5	**69.9**
35	551,297	**504**	477,894	**414**	665,014	**987**	44,513.2	**54.4**

So, we have that in case $\propto = 0.05{:}0.05 < 3.47\text{E-07}$, we have enough evidence to reject h_0 and with $\propto = 0.01{:}\ 0.01 < 3.47\text{E-07}$, also we have enough evidence to reject h_0. In other words our two algorithms had different median and the algorithm with best performance is Cellular algorithm.

4.4 Conclusions and Future Work

This paper presents an extension of the methodology of design approach for course timetabling, faculty timetabling, and classroom assignment. We used the Methodology of design proposed by Soria in [23] and extend it adding teacher-related restrictions. This improvement permits to use this approach when solving different universities.

Also we present a comparison between ILS and Cellular algorithm Metaheuristics in the context of the course timetabling, faculty timetabling, and classroom assignment problem. For this experimentation we generate artificial instances to test the methodology proposed, due to the fact in state of the art in which we searched, we did not found real or benchmark instances.

Our results show in this experimentation, that the metaheuristic with best performance was Cellular Algorithm; due to the fact this metaheuristic has neighborhoods operators and work in parallel with different solutions.

As future work we propose to implement difference metaheuristic as Memetic algorithms, Tabu Search, etc. Also we propose to use real instances for test the methodology and Metaheuristics

Acknowledgment Authors thanks the support received from the Consejo Nacional de Ciencia y Tecnologia (CONACYT) México

References

1. Ortiz-Aguilar Lucero de M.: Diseño de horarios de alumnos y maestros mediante técnicas de Soft Computing, para una Institución Educativa. Master's thesis, InstitutoTecnológico de León (2016).
2. De Werra, D.: An introduction to timetabling. European Journal of Operational Research(2), 151 – 162 (1985).
3. Obit, J. H., Landa-Silva, D., Ouelhadj, D., Khan Vun, T., & Alfred, R.: Designing a multi-agent approach system for distributed course timetabling. IEEE. (2011).
4. Asratian, A. S., de Werra, D., Luleå.: A generalized class–teacher model for some timetabling problems. University of Technology, Department of Engineering Sciences and Mathematics, Mathematical Science, & Mathematics. European Journal of Operational Re-search, **143**(3), 531-542. (2002). doi:10.1016/S0377-2217(01)00342-3.
5. Deng, X., Zhang, Y., Kang, B., Wu, J., Sun, X., Deng, Y.: An application of genetic algorithm for university course timetabling problem. 2119-2122 (2011). doi:10.1109/CCDC.2011.5968555.
6. Mahiba, A. A., Durai, C. A. D.: Genetic algorithm with search bank strategies for university course timetabling problem. Procedia Engineering, 38, 253-263. (2012). doi:10.1016/j.proeng.2012.06.033.
7. Soria-Alcaraz, J. A., Carpio, J. M.; Puga, Hé., Melin, P.; Terashima-Marn, H., Reyes, L. C., Sotelo-Figueroa, M. A. Castillo, O., Melin, P., Pedrycz, W. &Kacprzyk, J.: Generic Memetic Algorithm for Course Timetabling ITC2007 Recent Advances on Hybrid Approaches for Designing Intelligent Systems, Springer, 547, 481-492. (2014).
8. Nguyen K., Lu T., Le T., Tran N.: Memetic algorithm for a university course timetabling problem, **132**(1) 67-71. (2011). doi:10.1007/978-3-642-25899-2_10.

9. Aladag C., Hocaoglu G. A tabu search algorithm to solve a course timetabling problem. hacettepe journal of mathematics and statistics, **36**(1), 53-64. (2007).
10. Moscato, P.: "On Evolution, Search, Optimization, Genetic Algorithms and Martial Arts: Towards Memetic Algorithms". Caltech Concurrent Computation Program (report 826). (1989).
11. Zheng, S., Wang, L., Liu, Y., & Zhang, R.: A simulated annealing algorithm for university course timetabling considering travelling distances. International Journal of Computing Science and Mathematics, **6**(2), 139-151. (2015). doi:10.1504/IJCSM.2015.069461.
12. Joudaki M., Imani M., Mazhari N. Using improved Memetic algorithm and local search to solve University Course Timetabling Problem (UCTTP). Doroud, Iran: Islamic Azad University. (2010).
13. Thepphakorn T., Pongcharoen P., Hicks C.: An ant colony based timetabling tool. International Journal of Production Economics, 149, 131-144. (2014). doi:10.1016/j.ijpe.2013.04.026.
14. Soria-Alcaraz J., Ochoa G., Swan J., Carpio M., Puga H., Burke E.: Effective learning hyper-heuristics for the course timetabling problem. European Journal of Operational Research, **238**(1), 77-86. (2014) doi:10.1016/j.ejor.2014.03.046.
15. Lewis, M. R. R.: Metaheuristics for university course timetabling. Ph.D. Thesis, Napier University. (2006).
16. Cooper, T. B. & Kingston, J. H.: The Complexity of Timetable Construction Problems. PhD thesis, The University of Sydney. (1995).
17. Wolpert, H., Macready, G.: No free lunch Theorems for Search. Technical report The Santa Fe Institute, 1 (1996).
18. Adriaen M., De Causmaecker P., Demeester P., VandenBerghe G.: Tackling the university course timetabling problem with an aggregation approach. In Proceedings of the 6th International Conference on the Practice and Theory of Automated Timetabling, Brno.,**1**, 330-335. (2006).
19. Soria-Alcaraz, J.A.: Diseño de horarios con respecto al alumno mediante tecnicas de computo evolutivo. Master's thesis, Instituto Tecnologico de Leon (2010).
20. McCollum., B. University timetabling: Bridging the gap between research and practice. In Proceedings of the 6th International Conference on the Practice and Theory of Automated Timetabling, Brno, 1, 15-35. (2006).
21. Conant-Pablos, S.E., Magaa-Lozano, D.J., Terashima-Marin, H.: Pipelining memetic algorithms, constraint satisfaction, and local search for course timetabling. MICAI Mexican international conference on artificial intelligence **1**, 408–419 (2009).
22. DammakAbdelaziz, ElloumiAbdelkarim, K. H. Classroom assignment for exam timetabling. Advances in Engineering Software, **37**(10), 659-666. (2006).
23. Soria-Alcaraz Jorge, A., Carpio, M., Puga, H., Sotelo-Figueroa, M.: Method-ology of design: A novel generic approach applied to the course timetabling problem. In: P. Melin, O. Castillo (eds.) Soft Computing Applications in Op-timization, Control, and Recognition, Studies in Fuzziness and Soft Compu-ting, vol. 294, pp. 287–319. Springer Berlin Heidelberg (2013).
24. Alcaraz J. A. S.: Integración de un esquema de Diseño en Algoritmos de Dos fases con técnicas CSP para calendarización de eventos. PhD thesis, Instituto Tecnologico de Leon. (2014).
25. Alba, E.: Parallel Metaheuristics: A New Class of Algorithms. Wiley-Interscience. (2005).
26. Goldberg, D. E.: Genetic Algorithms in Search, Optimization and Machine Learning. Boston, MA, USA: Addison-Wesley Longman Publishing Co., Inc., 1st edition. (1989).
27. Alba, E. & Dorronsoro, B.: The state of the art in cellular evolutionary algorithms. Cellular Genetic Algorithms, **1**, 21-34. (2008).
28. Lourenco, H., Martin, O., &Stützle, T. Iterated local search. In F. Glover, G. Kochenberger, & F. S. Hillier (Eds.), Handbook of Metaheuristics, volume 57 of International Series in Operations Research & Management Science, 320-353. Springer New York. (2003).
29. Talbi, E.-G. Metaheuristics: From Design to Implementation. Wiley Publishing. (2009).

Solving the Cut Width Optimization Problem with a Genetic Algorithm Approach

Hector Joaquín Fraire-Huacuja, Mario César López-Locés, Norberto Castillo García, Johnatan E. Pecero and Rodolfo Pazos Rangel

Abstract The Cut width Minimization Problem is a NP-Hard problem that is found in the VLSI design, graph drawing, design of compilers and linguistics. Developing solutions that could solve it efficiently is important due to its impact in areas that are critical for society. It consists in finding the linear array of an undirected graph that minimizes the maximum number of edges that are cut. In this paper we propose a genetic algorithm applied to the Cut width Minimization Problem. As the configuration of a metaheuristic has a great impact on the performance, we also propose a Fuzzy Logic controller that is used to adjust the parameters of the GA during execution time to guide it during the exploration process.

Keywords Genetic algorithm · Cut width · Optimization · Combinatorial · Metaheuristic · Fuzzy logic

1 Introduction

The Cut width Minimization Problem (CMP) is a combinatorial optimization problem that belong to the NP-Problem class, as demonstrated by Garey and Johnson [1].

In Korach et al. [2] is shown that the Path width Decomposition Problem is closely related to the CMP, and that the solution for a given graph is also less than or equal to the solution of the same graph to CMP, so it works as a lower bound for CMP.

The CMP is usually found in the design of VLSIs, as a method to partition circuits into smaller subsystems, with a small number of components on the boundary between those, as described by Ohtsuki [3].

H.J. Fraire-Huacuja · M.C. López-Locés (✉) · N.C. García · R.P. Rangel
Tecnologico Nacional de Mexico, Instituto Tecnológico de Ciudad Madero, Ciudad Madero, Mexico
e-mail: mariocesar@lopezloc.es

J.E. Pecero
Computer Science and Communications Research Unit, University of Luxembourg, Esch-sur-Alzette, Luxembourg

P. Melin et al. (eds.), *Nature-Inspired Design of Hybrid Intelligent Systems*, Studies in Computational Intelligence 667, DOI 10.1007/978-3-319-47054-2_48

In Suderman et al. [4] the problem of graph drawing is defined as the number of parallel lines on which the vertices of a tree can be drawn with no edge crossings and is equivalent to the pathwidth of the tree.

The CMP is also present in the compilation of programming languages, in the reordering sequences of code without control flow branches or loops, in such a way that all the values computed in the code can be placed in machine registers and not in the relatively slower main memory, according to Bodlaender et al. [5].

The CMP can also be found in the linguistics field, specifically in the natural language processing where sentences are modeled as graphs, in which the vertices represent words and the edges represent relationships between words [6].

In this paper we proposed a solution for the CMP based on a Genetic Algorithm (GA), which takes inspiration on the process that drives the evolution by natural selection, and uses a Fuzzy Inference System (FIS) to adjust the mutation parameter during the execution of the algorithm.

2 Problem Definition

The CMP is an NP-Hard optimization problem. This statement was proven by generating a polynomial transformation from the Partition Problem to the CMP [1]. The CMP is formally defined by Pardo et al. [7] as follows:

Given $G = (V, E)$ with $n = |\mathrm{V}|$ y $m = |E|$, see Fig. 1, a labeling π that designate a different number from the set $\{1, 2, \ldots, n\}$ for every element $v \in V$, as shown in Fig. 2.

The cut width (CW) of a node v in regard to π $(CW_\pi(v))$, is the number of edges $(u, w) \in E$ that satisfies $\pi(u) \leq \pi(v) < \pi(w)$, as shown in Fig. 3, where the cut width value of the node labeled as 4 is equal to 4, π $(CW_\pi(4)) = 4$.

The CW of the graph G in regard to the permutation π $(CW_\pi(G))$ is the maximum CW of all its nodes. In Fig. 4 it is shown the value of the cut width of the graph, which is equal to 5.

$$CW_\pi(G) = max_{v \in V}\, CW_\pi(v)$$

The objective of CMP is to find, among all the possible labelings, the one that minimizes the CW of the graph.

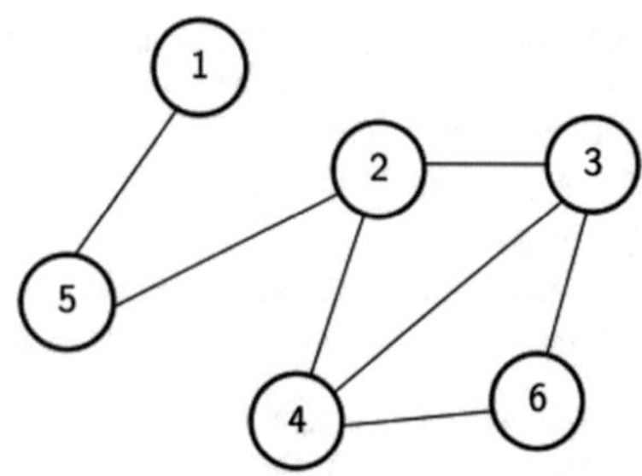

Fig. 1 Non directed graph

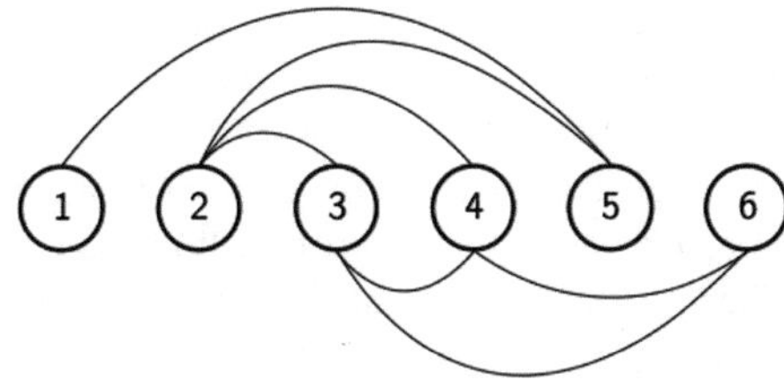

Fig. 2 Labeling of non directed graph

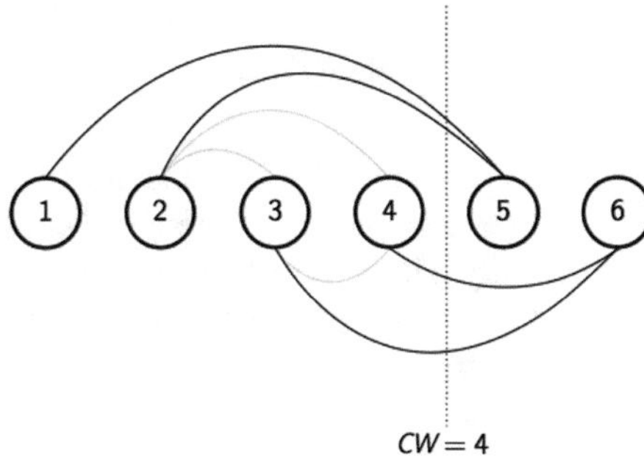

Fig. 3 Cut width value of node with label 4 in linear array

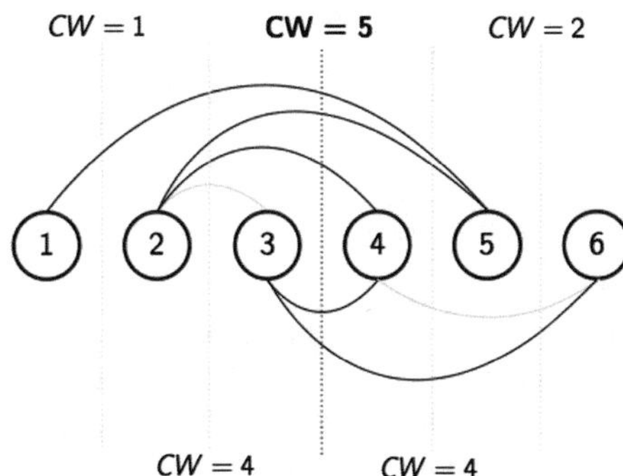

Fig. 4 The cut width value of G

3 Related Works

In Andrade et al. [8, 9] is described the design of a GRASP metaheuristic combined with Path Relinking to the Network Migration Scheduling Problem, which is equivalent to CMP. The local search algorithm uses the swap neighborhood, evaluating the elements and performing the insertion if the move results in a real improvement on the objective value of the whole graph.

In Pantrigo et al. [10] is described the design of a Scatter Search Algorithm applied to CMP. The local search mechanism implemented stops at the moment of the first real improvement on the objective value of the solution, and uses the insertion neighborhood. Then, it uses a Path Relinking algorithm from the current solution to the one improved by the local search method.

In Campos et al. [11] is described the development of a hybrid metaheuristic of Tabu Search and Scatter Search as a solution for the Matrix Band with Minimization Problem, equivalent to CMP. The proposed algorithm, implements a local search that stops on the first improvement and uses the swap neighborhood, selecting the elements to be reallocated from a list of critic and near critic elements. The list of critic elements is formed with those who have an objective value equal or near to the objective value of the whole graph.

4 Genetic Algorithm

Originally proposed by Holland in 1975 in its current form by Holland [12]. The GA is a population based metaheuristic that is inspired by the theory of evolution by natural selection, proposed by Charles Darwin in 1859.

The idea behind this metaheuristic is that an unguided process could yield fitting individuals by following relatively simple rules. In the context of computer science, the individuals represent solutions for a particular problem, and the environment is determined by the objective function of the problem. High quality individuals are considered more fitting for a given environment. Thus, having more probabilities of surviving and to reproduce, causing the whole population to converge to a local optimum in the search space.

4.1 Static GA

Algorithm 1 shows the structure and the parameters of the GA proposed. Such parameters are: the size of the population; the maximum number of evaluations per individual; the crossover probability, this value defines the probability of producing offspring from two parents; and the mutation probability, which defines the probability of change of a specific gene of one individual.

The process begins by generating a population with random individuals; each one is a possible solution for the CMP. Then, in the main loop, two candidate solutions are selected by a binary tournament function to produce an offspring solution. The new solution is produced using the two point crossover method to combine both parents. The offspring solution is then mutated to move it to another region in the search space. Then, the algorithm evaluates the fitness value of the offspring solution for the CMP. And if its quality is better than the worst solution in the population, then the newly generated solution replaces that worst solution and the main loop continues until the termination condition is met. Finally, the best solution in the population is returned and the GA process finishes.

Algorithm 1: ***StaticGA***

Description: Metaheuristic inspired by the evolution process by natural selection

Input:
Instance: *instance*
Integer: *populationSize* := 20
Integer: *maxEvaluations* := 1000
Double: *crossoverProbability* := 0.95
Double: *mutationProbability* := 0.2

Output:
Solution: *bestSolution*

Initialize *population* with *populationSize* random candidate solutions
While*currentEvaluation*<*maxEvaluations*:
(Selection)
Select *parent1* and *parent2* from *population* by *binaryTournament()*

(Crossover)
Obtain *offspring* by applying *crossover()* to *parent1* and *parent2*

(Mutation)
Apply *bitFlipMutation()* to *offspring*

Evaluate *fitness()* of *offspring*
If*offspring*fitness is better than *worstSolution* in *population***Then**
Remove *worstSolution* from *population*
Add *offspring* to *population*
End If
End While

Return *bestSolution* from *population*

4.2 *FuzzyGA*

Fuzzy Logic was initially proposed by Zadeh in 1965 [13]. It is a form of logic that can take different amounts of truth, ranging from 0 to 1, in contrast with traditional Boolean logic; that is limited to two values, truth or false.

The use of fuzzy logic is mainly applied when some elements are not precisely defined by a classification; e.g., are 33° considered as hot for a cup of tea? It might be too drastic to say that 33° is not hot for a cup of tea. However, it might more precise to say that 33° are considered 85 % hot for a cup of tea.

The use of Fuzzy Inference System (FIS) to adjust of the parameters of evolutionary metaheuristic algorithms is a technique that has been used for a large variety of problems, such as Michael and Takagi [14], Valdez et al. [15] and Valdez and

Melin [16]. This approach provides flexibility and is general, in the sense that we can apply a set of rules for all possible values of a specific variable; e.g., if the variable is high then change the value of other variable to low; and a value of 8 for one variable might be consider as 90 % high and also a 30 % medium and 0 % low, which allows the use of such general rules for all possible values of the former variable.

The Fuzzy GA, shown in Algorithm 2, differs from Static GA on the use of a FIS to adjust dynamically the percentage of the mutation parameter applied to the solutions being generated.

Algorithm 2: *FuzzyGA*

Description: Metaheuristic inspired by the evolution process by natural selection that includes a FIS to control the mutation parameter.

Input:
Instance: *instance*
Integer: *populationSize* := 20
Integer: *maxEvaluations* := 1000
Double: *crossoverProbability* := 0.95
Double: *mutationProbability* := ∅

Output:
Solution: *bestSolution*

```
Initialize population with populationSize random candidate solutions
Load FIS and adapt it to instance
While currentEvaluation<maxEvaluations:
    (Selection)
    Select parent1 and parent2 from population by binaryTournament()

    (Crossover)
    Obtain offspring by applying crossover() to parent1 and parent2

    (Mutation)
    Apply bitFlipMutation() to offspring

    Evaluate fitness() of offspring
    If offspring fitness is better than worstSolution in population Then
        Remove worstSolution from population
        Add offspring to population
    End If
    Evaluate out = FIS(fitness(best))
    Set probability = out
End While

Return bestSolution from population
```

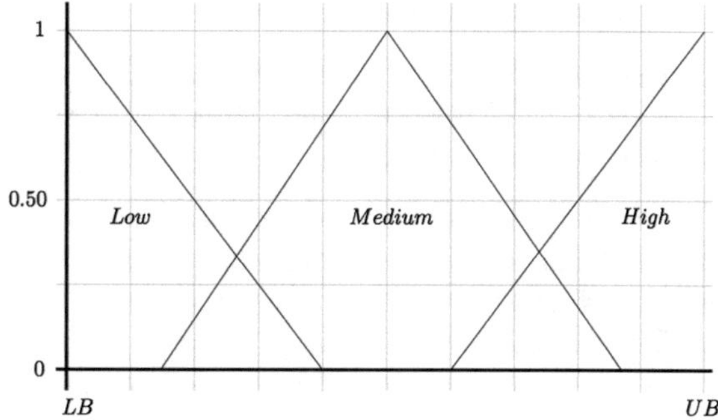

Fig. 5 Triangular membership functions of the *objective value* variable

The input variable used in the FIS was the *objective value* of the solution being evaluated.

The membership functions used for this variable are shown in Fig. 5, where the lower bound and upper bound were calculated as follows:

- The lower bound was calculated with the method described in [17], this approach constructs an ideal graph, which is not the graph of the instance being solved, but a graph with the same number of vertices and edges, where the connectivity does not matter, just that the vertices are added to a linear array such that it increases the cut width value by one on each pass.
- The upper bound was determined by the number of edges of the instance, which is the maximum possible cut width value for any linear array for the graph of the instance.

The *Mutation* variable determines the percentage of mutation that will be applied to the offspring solution, after the crossover operation, which is the probability of swapping two elements of the offspring solution. The membership functions used for this variable are shown in Fig. 6, and as the *objective value* each membership function range in value from 0.0–1.0.

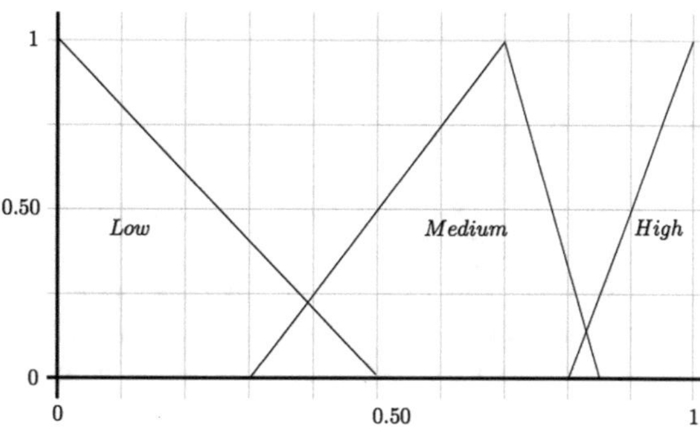

Fig. 6 Triangular membership function of the *Mutation* variable

Next we show the rules applied to the FIS:

If *objective value* is *Low* then *Mutation* is *Low*
If *objective value* is *Medium* then *Mutation* is *Medium*
If *objective value* is *High* then *Mutation* is *High*

The motivation for the previous rules is that: when a solution is in a good region of the search space, then is desirable to continue to explore closer to that point; when a solution is regular, a good strategy is to balance the exploration and exploitation near to that region; and when the quality of the solution is closer to the upper bound, then is a good idea to explore other regions of the search space.

The inference method used was the one proposed by Mamdani, in which consequents and antecedents are related by the min operator or a t-norm, according to Rutkowski [18].

5 Computational Experimentation

To evaluate the performance of the proposed GAs we used the sets of instances Small and Grid, that can be found in the Optsicom Project [19]. Those sets were selected because all the instances have known optimal values.

In the experimentation we tested the two variants of the GA: the *Static GA*, with a small sized population of 30 individuals and a mutation value of 0.05; and the *Fuzzy GA*, with 30 individuals and a dynamic mutation value which is updated by the previously described FIS. Both variants were tested with a maximum number of 3000 iterations and a crossover probability of 0.95.

The two GA variants were implemented in Java. The characteristics of the hardware on which the experimentation was carried out are the following: a CPU Intel i5 at 2.3 GHz, 8 GB of DD3 RAM at 1333 MHz, and a SSD drive of 256 GB. Running on Mac OS X 10.11.1.

The FIS was implemented in the fuzzy logic module from MATLAB 8.3.0, and was connected to the GA using the Matlab control API 4.1.0.

6 Results

The results of the computational experimentation described in the previous section are presented in Table 1.

Table 1 show that the Fuzzy GA obtains the best results on both of the sets of instances evaluated, but with a significant increase in computation time due to the use of MATLAB for calculating the output value of the FIS.

Table 1 Results for instance group

		Small	Grids
StaticGA	Avg. Obj. Val	6.25	23.83
	Time	1.42	3.15
	% Error	0.71	4.19
FuzzyGA	Avg. Obj. Val	5.35	23
	Time	444.05	397.76
	% Error	0.46	3.99

7 Conclusions

In this paper we presented an approximated approach to solve the CMP, using the GA metaheuristic. We observed that with a FIS applied to the control and tuning of the mutation parameter, the metaheuristic was able to outperform in quality the other method, which established the value of the mutation before the execution of the algorithm.

Although the improvement might seem modest, the CMP is a problem that tends to have a flat landscape and finding better solutions is difficult because when moving to a neighborhood solution from the current one usually does not reflect any change in the objective value of the graph.

Future work will include the exploration of alternative membership function for the variables used in the Fuzzy Inference System (FIS), the design and implementation of a FIS on the same programming language, to reduce the computational time during the solution of the instances of the CMP, and extend the proposed FIS to control additional parameters of the metaheuristic.

References

1. M. R. Garey, D. S. Johnson, and others, *Computers and Intractability: A Guide to the Theory of NP-completeness*. WH freeman San Francisco, 1979.
2. E. Korach and N. Solel, "Tree-width, path-width, and cutwidth," *Discret. Appl. Math.*, vol. 43, no. 1, pp. 97–101, 1993.
3. T. Ohtsuki, H. Mori, E. Kuh, T. Kashiwabara, and T. Fujisawa, "One-dimensional logic gate assignment and interval graphs," *Circuits Syst. IEEE Trans.*, vol. 26, no. 9, pp. 675–684, 1979.
4. M. Suderman, "Pathwidth and layered drawings of trees," *Int. J. Comput. Geom. Appl.*, vol. 14, no. 03, pp. 203–225, 2004.
5. H. Bodlaender, J. Gustedt, and J. A. Telle, "Linear-time register allocation for a fixed number of registers," in *Proceedings of the ninth annual ACM-SIAM symposium on Discrete algorithms*, 1998, pp. 574–583.
6. Z. Tuza, "Narrowness, Path-width, and their Application in Natural Language Processing."
7. E. Pardo, N. Mladenović, J. Pantrigo, and A. Duarte, "Variable formulation search for the cutwidth minimization problem," *Appl. Soft Comput.*, vol. 13, pp. 2242–2252, 2013.
8. D. Andrade and M. Resende, "GRASP with path-relinking for network migration scheduling," ... *Int. Netw.* ..., pp. 1–7, 2007.

9. D. Andrade and M. Resende, "GRASP with evolutionary path-relinking," *Proc. Seventh Metaheuristics ...*, pp. 6–9, 2007.
10. J. J. Pantrigo, R. Martí, A. Duarte, and E. G. Pardo, "Scatter search for the cutwidth minimization problem," *Ann. Oper. Res.*, vol. 199, no. 1, pp. 285–304, 2012.
11. V. Campos, E. Piñana, and R. Martí, "Adaptive memory programming for matrix bandwidth minimization," *Ann. Oper. Res.*, pp. 1–17, 2006.
12. J. H. Holland, *Adaptation in Natural and Artificial Systems*, vol. Ann Arbor. 1975.
13. L. A. Zadeh, "Fuzzy sets," *Inf. Control*, vol. 8, no. 3, pp. 338–353, Jun. 1965.
14. A. Michael and H. Takagi, "Dynamic control of genetic algorithms using fuzzy logic techniques," in *Proceedings of the Fifth International Conference on Genetic Algorithms*, 1993, pp. 76–83.
15. F. Valdez, P. Melin, and O. Castillo, "Fuzzy control of parameters to dynamically adapt the {PSO} and {GA} Algorithms," in *{FUZZ-IEEE} 2010, {IEEE} International Conference on Fuzzy Systems, Barcelona, Spain, 18-23 July, 2010, Proceedings*, 2010, pp. 1–8.
16. F. Valdez and P. Melin, "A New Evolutionary Method with Particle Swarm Optimization and Genetic Algorithms Using Fuzzy Systems to Dynamically Parameter Adaptation," in *Soft Computing for Recognition Based on Biometrics*, 2010, pp. 225–243.
17. R. Martí, J. Pantrigo, A. Duarte, and E. Pardo, "Branch and bound for the cutwidth minimization problem," *Comput. Oper. ...*, vol. 40, pp. 137–149, 2013.
18. L. Rutkowski, *Flexible neuro-fuzzy systems: structures, learning and performance evaluation*, vol. 771. Springer Science & Business Media, 2006.
19. D. A. Martí R. Pantrigo J.J. and P. E.G., "Optsicom Project." 2010.

Part VII
Hybrid Intelligent Systems

A Dialogue Interaction Module for a Decision Support System Based on Argumentation Schemes to Public Project Portfolio

Laura Cruz-Reyes, César Medina-Trejo, María Lucila Morales-Rodríguez, Claudia Guadalupe Gómez-Santillan, Teodoro Eduardo Macias-Escobar, César Alejandro Guerrero-Nava and Mercedes Pérez-Villafuerte

Abstract Organizations are facing the problem of having more projects than resources to implement them. In this paper, we present a dialogue interaction module of a framework for a Decision Support System (DSS) to aid in the selection of public project portfolios. The Interaction module of this DSS is based on multiple argumentation schemes and dialogue games that not only allow the system to generate and justify a recommendation. This module is also able to obtain new information during the dialogue that allows changing the recommendation according to the Decision Maker's preferences. Researchers have commonly addressed the public portfolio selection problem with multicriteria algorithms.

L. Cruz-Reyes (✉) · M.L. Morales-Rodríguez · C.G. Gómez-Santillan · T.E. Macias-Escobar · C.A. Guerrero-Nava
Tecnológico Nacional de México, Instituto Tecnológico de Ciudad Madero, 1o. de Mayo y Sor Juana I. de la Cruz S/N C.P. 89440 Cd, Ciudad Madero, Mexico
e-mail: lauracruzreyes@itcm.edu.mx

M.L. Morales-Rodríguez
e-mail: lmoralesrdz@gmail.com

C.G. Gómez-Santillan
e-mail: cggs71@hotmail.com

T.E. Macias-Escobar
e-mail: teodoro_macias@hotmail.com

C.A. Guerrero-Nava
e-mail: cesaragn1990@hotmail.com

C. Medina-Trejo · M. Pérez-Villafuerte
Tecnológico Nacional de México, Instituto Tecnológico de Tijuana, Tijuana, B.C, Mexico
e-mail: cesarmedinatrejo@gmail.com

M. Pérez-Villafuerte
e-mail: pvmercedes@gmail.com

P. Melin et al. (eds.), *Nature-Inspired Design of Hybrid Intelligent Systems*, Studies in Computational Intelligence 667, DOI 10.1007/978-3-319-47054-2_49

However, in the real life the final selection of the solution depends on the decision maker (DM). We modeled the reasoning of DM by a Dialogue Corpus. This corpus is a database, supported by an argument tree that validates the system's recommendations with the preferences of the DM.

Keywords Argumentation schemes · Dialogue corpus · Recommendation system

1 Introduction

The public project portfolio selection is a primary task in institutions or organizations because decisions impact directly on the society. Commonly the organizations have more projects than resources to support them; they have to choose the portfolio that provides more benefits based on the criteria of the organization.

This problem is mainly addressed through multicriteria algorithms, which do not generate a solution, but a set of solutions on the Pareto front; recent studies have focused on seeking a privileged region of this front, where the preferences of the decision maker are reflected. However, the generated recommendations are presented without any justification or explanation of why certain projects are or are not within the portfolio. The selection of a portfolio depends on the decision maker (DM); a justification of the recommendation would help to make this decision in a less demanding way.

This paper presents a framework of decision support, in order to provide explanations and justifications to a recommended project portfolio generated by a multicriteria algorithm. The proposed approach makes a hybrid combination of an approximate algorithm with an exact method.

Using the information generated we can enter into a dialogue game with the DM based on the argumentation theory and the proposed interaction module to state valid arguments of why the presented recommended portfolio satisfies the DM preferences, also, we can explain why certain projects are or are not in the portfolio and how the construction of the portfolio was made.

2 DSS to Aid in the Selection of Public Project Portfolios

The decision support systems (DSS) are important tools because they allow users to have a support on why they reached a decision. According to [1] a DSS is defined as "an interactive computer-based system or subsystem intended to help decision makers use communications technologies, data, documents, knowledge and/or models to identify and solve problems, complete decision process tasks, and make decisions". It is a general term for any computer application that improves the ability of a person or group of persons to make decisions. A DSS helps to retrieve, summarize and analyze data relevant to the decision.

A DSS for public project selection have different subsystems that work together to support a decision maker, this article focuses mainly on the subsystem to generate recommendations in an interactive way.

The definition of the term "*interactive*", taken from the Oxford Dictionary is: "permit a flow of two-way information between a computer and a computer user." Furthermore, in the context of computer science, it is defined as the interaction with a human user, usually in a conversational way to obtain data or commands and provide immediate results or updates [2].

In this section, the proposed methodology to obtain explanations in public project portfolios is detailed. Figure 1 shows the complete methodology, ranging from obtaining information to the calculus of reduced decision rules. An overview of the process is presented below:

1. First, we need to obtain the preferential information of the organization, these preferences are necessary for the selection of the adequate multicriteria algorithm.
2. With the correct selection of the multicriteria algorithm, using the preferential information, we obtain the set of portfolios to be presented, which is reduced to a subset in the region of interest of the DM.

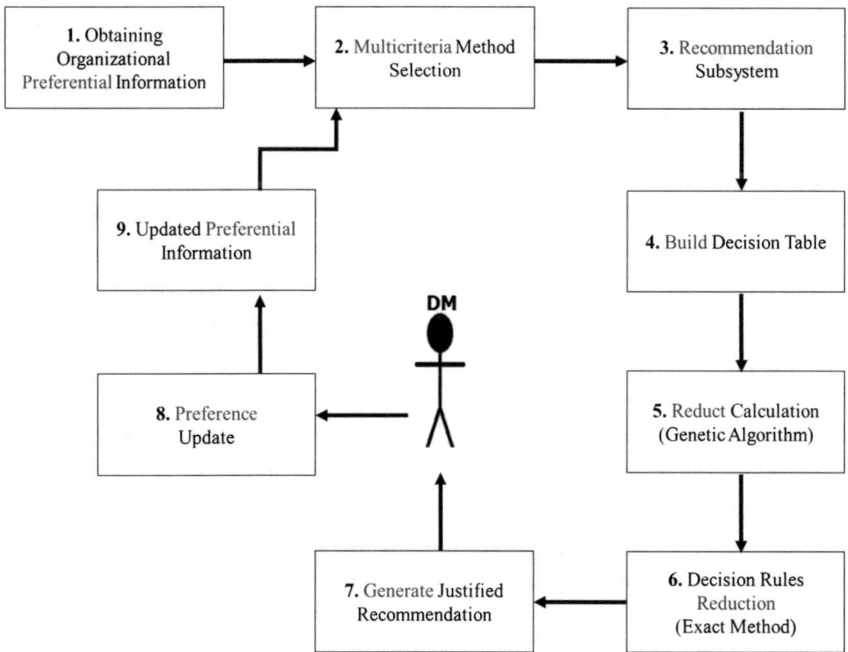

Fig. 1 Proposed methodology for recommendation of project portfolio with reduced decision rules

3. In the recommendation subsystem there are stored the possible portfolios for recommendation to the DM, the selection of one of these portfolios is used to generate reducts.
4. A transformation of the public portfolio instance to a decision table is made. The solution generated from the multicriteria method is added at the end of the original instance to act as a decision attribute, the original attributes will be the condition attributes of the decision table.
5. This tabular representation is necessary to function with the rough set approach. The reduct calculation is performed starting with the decision table with the purpose of generation a reduced decision table using the reducts generated with the genetic algorithm.
6. By means of an exact method, we work with the newly generated reduced decision table to get the attributes of the rows decrement (value reduction). The number of attributes is diminished, rule by rule (row by row), reducing the quantity of clauses on each rule.
7. Using a small subset of rules on the projects in the portfolio (which projects are supported and which ones are not), we can summarize the information about the construction of the portfolio and verify how far is from the organization policy. This decision is a critical problem the DM is faced with.
8. The process of recommending a justified portfolio can end when such a recommendation is presented to the DM. At this stage, he can make an introspection when he sees the summary of the characteristics of the projects conforming the recommended portfolio, and—based on such information-there is a possibility that the DM is not satisfied with this portfolio. The DM can update his preferences and make a revaluation of the attributes values of some projects. The option to perform this updates is a work in progress, this process must be repeated if the DM changes his preferences, an—hereby—a new justification (on how the new portfolio is constructed) is obtained.
9. The updated preferential information can be used to restart the process to re-run the multicriteria algorithm.

The main contribution of this methodology is the reduction of attributes and the simplification of decision rules by means of a hybrid algorithm, which consist of two stages:

1. A genetic algorithm to generate reducts, and
2. An exhaustive calculation of the decision rules with an exact method, using the reduced decision table generated by the genetic algorithm.

The argumentation theory comes into play when the final recommendation is presented, within a dialog manager the DM and the framework can enter in an interaction to justify the solution presented to him, if this solution satisfies his base of preferences.

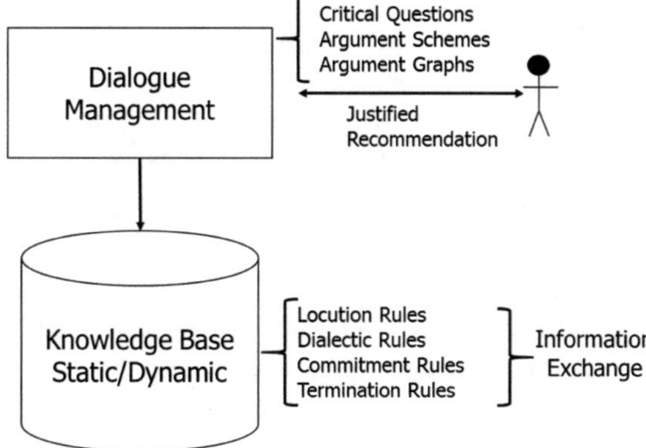

Fig. 2 Argumentation theory dialogue management interactivity

The argumentation theory will be sustained with a knowledge database, as shown in Fig. 2, which consists in a set of rules (locution, dialectic, commitment and termination rules) necessary to make a valid interchange of arguments and explain the recommendation. This type of dialog must be supported by a static and dynamic knowledge base; the static part is composed by elements such as rules, argument schemes and initial preferences; on the other hand, the dynamic part consists on argument graphs and updated preferences.

3 Argumentation Theory for Decision Making

The decision making problems usually contain a huge amount of alternatives, which makes it difficult to solve using only the human capabilities. Normally, to reduce the complexity of the problem, recommendation systems are used to allow the DM to see the best alternatives that suit his preferences are. These systems attempt to create a user model and apply heuristics to anticipate what information could be of interest [3].

We believe that a DSS will benefit from using the *argumentation theory* to justify its recommendations. It is a growing field of artificial intelligence, related to the process of constructing and evaluating arguments in order to justify conclusions, providing a non-monotonic reasoning mechanism, which means that conclusions might change as more information is obtained [4].

3.1 Argumentation Schemes

The *argumentation schemes*, capture the stereotypical patterns of human reasoning, especially the defeasible ones [5], the arguments are shown as general inference rules, when given a set of premises a conclusion can be reached [6]. However, because of the defeasible nature of the arguments, said schemes are not strictly deductive. The schemes allow the arguments to be represented within a certain context and take in consideration that the presented reasoning can be modified in the light of new proof or rule exceptions.

The argumentation schemes are composed of the following elements:

- Premises: Arguments that works as either support or opposition of the conclusion.
- Conclusion: Statement that is reached from analyzing the premises.
- Critical Questions: Attacks or challenges that weak the argument within the scheme if they are not answered.

Another important point to consider within the argumentation theory is how the statements are going to be evaluated, using a *proof standard*, allows comparing the arguments made in favor and against a certain statement to evaluate if it holds true or not [7]. However, every proof standard works in a different way, which means that while some might consider a statement true others might not; it is important to be careful when choosing which proof standard to use when evaluating a statement.

3.2 Dialogue System

The dialogue games (or dialogue systems) essentially define the principle of consistent dialogue and conditions under which a statement made by an individual is adequate. There are several formal dialogues, taking into account various information such as the participants, the communication language, roles of participants, the aim of the dialogue, etc. These types of dialogues system are generally sustained by a set of rules:

- *Locution Rules (speech acts, movements).* The rules indicate which expressions are allowed. Generally, legal phrases enable the participants to affirm propositions, allowing others to question or challenge the above statements, and allow those affirmations that can be claimed, questioned or challenged to justify the statement. The justifications may involve submitting a proof of the proposition or an argument for it.
- *Commitments Rules*. Rules defining the effects of movements in the "commitments"; associated with each player is a compromise to maintain the statements that the players have made and the challenges they have issued; so there are rules that define how the commitments are updated.
- *Dialogue Rules*. Rules to regulate movements. It specifies, for example, all the acts of speech allowed in a dialogue and types of responses allowed at a certain state. Various dialogue protocols can be found in the literature, especially for persuasion [8] and negotiation [9, 10]

- *Termination rules*. The rules that define the circumstances in which the dialogue ends.
- *Acceptability*. In a process of argumentation, it is important to define (or evaluate) the status of arguments based on all the ways in which they interact. Thus, the best or acceptable arguments must be identified at the end of the process of argumentation.

Most arguments systems are based on the notion of acceptability as identified by Dung [11], he has proposed an abstract framework for the argument which only focuses on the definition of the status of the arguments. In this framework, the acceptability of an argument depends on its membership to some sets, called acceptable sets or extensions. In other words, the acceptability of the arguments is defined without considering the internal structure of the arguments.

4 Dialogue Interaction Module for a Decision Support System

Based on Ouerdane [4] proposal, the recommendation system explained in this document will work using a set of modules, these modules will allow the system to implement the decision support, using a dialog game and argumentation schemes to reach a conclusion that satisfies the DM. The four main modules of the interaction module of the DSS can be seen in Fig. 3.

4.1 Proposed Framework Supported by the Argumentation Theory and a Dialogue Game

In this section the modular diagram shown in Fig. 3 will be described in detail.

Load Instance: The system reads a file which provides the system with the necessary information to generate a recommendation and start a dialogue game between it and the user, the data that this module requires to permit the system to continue its process are the following:

- Number of alternatives
- Number of criteria
- Weight of each criterion

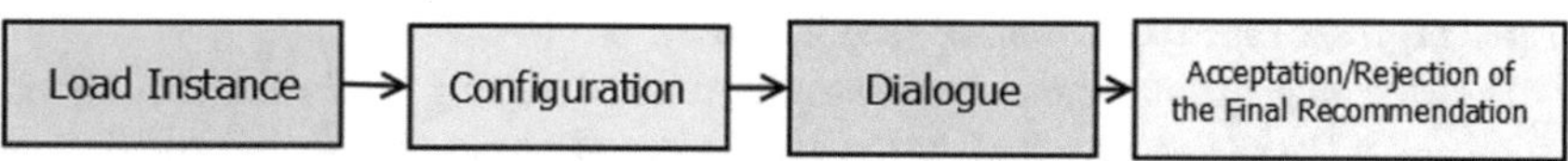

Fig. 3 Modular diagram of the interaction module of the DSS

- Lexicographic order
- Performance table
- Number of solutions
- Solution matrix

Configuration: The system analyses the information obtained from the previous process to determinate the initial configuration of elements such as locution rules, the state transition diagram and which proof standard to use. This will allow the system to generate an initial recommendation, that lets both it and the user start a dialogue game.

Dialogue: In this process, the user and the system initialize an exchange of arguments, supported by the argumentation schemes, in which the system will look to convince the user to accept the recommendation established by it. However, the user is also capable of rejecting the current recommendation or to manipulate the initial configuration to force the system to generate a new recommendation, all according to the user's preferences.

Acceptation/Rejection of the Final Recommendation: The user can conclude if the recommendation received is satisfactory or not to his needs, even if it is the best possible solution available. This rejection option is established in consideration of the human factor (the user) that will be in contact with the system. Since he doesn't follow a strict set of rules, the user could simply ignore the dialogue following the initial recommendation and reject it as soon as it is shown, instead of searching for a better solution.

4.2 Argumentation Schemes for the Interaction Module

Based on the work of Walton et al. [12], and Ouerdane [4], some argumentation schemes were found to be useful for the system for its interaction process with the user, the schemes presented in this section are formally described in [12].

The system requires defining a proof standard to create an initial recommendation. It also needs to be capable of changing the proof standard on use based on the information obtained in the interaction with the user. The abductive reasoning is a process that allows the system to select from the set of proof standards the one that is closer to the active properties. Said properties are defined both in the configuration process and in the dialogue game. The argumentation scheme for the abductive reasoning is shown in Table 1.

After choosing a proof standard, the system must generate a recommendation. Two argumentation schemes have been identified that put the system as a capable entity for this action. The *argument from position to know* (see Table 2) and the *argument from an expert opinion* (see Table 3).

The *argument from position to know* is utilized during the initial recommendation and a few states after that, as the system doesn't have enough information yet to be considered an expert for the instance that is being analyzed. The *argument*

Table 1 Argumentation scheme for the abductive reasoning argument

Premises	*F* is a finding or given set of facts
	E is a satisfactory explanation of *F*
	No alternative explanation *E*' given so far is as satisfactory as *E*
Conclusion	Therefore, *F* is plausible, as a hypothesis
Critical Questions	How satisfactory is *E* as an explanation of *F*, apart from the alternative explanations available so far in the dialogue?
	How much better an explanation is *E* than the alternative explanations available so far in the dialogue
	How far has the dialogue progressed? If the dialogue is an inquiry, how thorough has the investigation of the case been?
	Would it be better to continue the dialogue further, instead of drawing a conclusion at this point?

Table 2 Argumentation scheme for the argument from position to know

Premises	Major premise: Source *E* is an expert in subject domain *F* containing proposition *A*
	Minor Premise: *E* asserts that proposition *A* is true (false)
Conclusion	*A* is true (false)
Critical Questions	How credible is *E* as an expert source?
	Is *E* an expert in the field that *A* is in?
	What did *E* assert that implies *A*?
	Is *E* personally reliable as a source?
	Is *A* consistent with what other expert assert?
	Is *E*'s assertion based on evidence?

Table 3 Argumentation scheme for the argument from an expert opinion

Premises	Major Premise: Source *a* is in position to know about things in a certain subject domain *S* containing proposition *A*
	Minor Premise: *a* asserts that *A* is true (false)
Conclusion	*A* is true (false)
Critical Questions	Is *a* in position to know whether *A* is true (false)?
	Is *a* an honest (trustworthy, reliable) source?
	Did *a* assert that *A* is true (false)?

from an expert opinion defines the system not only as capable of making a recommendation of high quality, but also confirms it as an expert of the current instance as it has obtained enough information.

Several argumentation schemes were found to be useful for the system during the interaction process that could be used to establish a coherent dialogue with the user. These schemes allow the system to defend its recommendation or obtain new information. The use of each of these schemes will be explained below.

Table 4 Argumentation scheme for the muticriteria pairwise evaluation

Premises	Action *a*
	Action *b*
	A set of criteria *h*
	There are enough supporting reasons *SR*
	There are no sufficiently strong opposing reasons *OR*
Conclusion	*a* is at least as good as *b*
Critical Question	Are the reasons in *SR* strong enough to overcome the reasons in *OR*?
	Is the difference between *SR* and *OR* big enough to accept the conclusion?

Table 5 Argumentation scheme for the practical argument from analogy

Premises	The right thing to do in *S1* was to carry out *A*
	S2 is similar to *S1*
Conclusion	Therefore, the right thing to do in *S2* is carry out *A*
Critical Questions	*A* really applies on *S2*?
	Is there any other action for *A* that works better than *S2*?
	Have *S1* y *S2* enough similarities?

Table 6 Argumentation scheme for the ad ignorantiam argument

Premises	Major premise: If *A* were true, then *A* would be known to be true
	Minor Premise: It is not the case that *A* is known to be true
Conclusion	Therefore, *A* is not true
Critical Questions	Is *A* known?
	Are there proofs that *A* is false?

Multicriteria pairwise evaluation: The system must be able to compare the recommendation with the other alternatives and explain why it is better than them (see Table 4).

Practical argument from analogy: Since it is possible that two or more solutions could have a certain degree of similitude, a set of actions that were used in the previous alternatives could be used for new recommendations that are almost similar (see Table 5).

Ad Ignorantiam: The system cannot generate inferences; therefore, it must only consider the known information as true. Everything that is unknown for the system must be considered false, even if this is considered a fallacy (see Table 6).

Cause-Effect: A change in the value of a criterion or property can lead to a change in a process within the system; whether it is the proof standard or the recommendation shown to the user (see Table 7).

Argument from bias: This fallacy must be considered because the user is not always unbiased as he could have a preference over a certain criterion or alternative (see Table 8).

Table 7 Argumentation scheme for the cause-effect argument

Premises	Major Premise: Generally, if *A* occurs, then *B* will (might) occur
	Minor Premise: In this case, *A* occurs (might occur)
Conclusion	Therefore, in this case, B will (might) occur
Critical Question	How strong is the causal generalization?
	Is the evidence cited (if there is any) strong enough to warrant the causal generalization?
	Are there other causal factors that could interfere with the production of the effect in the given case?

Table 8 Argumentation scheme for the argument from bias

Premises	Major premise: If x is biased, then x is less likely to have taken the evidence on both sides into account in arriving at conclusion *A*
	Minor Premise: Arguer a is biased
Conclusion	Arguer a is less likely to have taken the evidence on both sides into account in arriving at conclusion *A*
Critical Question	What type of dialogue are the speaker and hearer supposed to be engaged in?
	What evidence has been given to prove that the speaker is biased?

4.3 Argumentative Dialog Corpus

A dialogue corpus will be located in the dialogue module, it will permit the characterization of the decision maker, allowing emulating him to make experiments relevant to decision making.

In Fig. 4 the each of the modules for the interaction module can be seen in greater detail; in the configuration module the premises, locution rules, state transition diagram and the proof standards are set.

Furthermore, in the dialog module the argumentation schemes and the corpus of argumentative dialogue interact each other; criteria may be updated and the proofs standard, according to different conditions that may occur in the dialogue. This process is performed with the acceptance or rejection of the recommendation.

The architecture shown in Fig. 4 is based on the work of Querdane [4], she presents a theoretical architecture for a recommendation system, but the implementation was not realized. The architecture of the recommendation system of this work will change according to the context of this particular problem.

We present in Fig. 5, the architecture that show the interaction between the different elements that are involved in the corpus dialogue; the DM preferences will

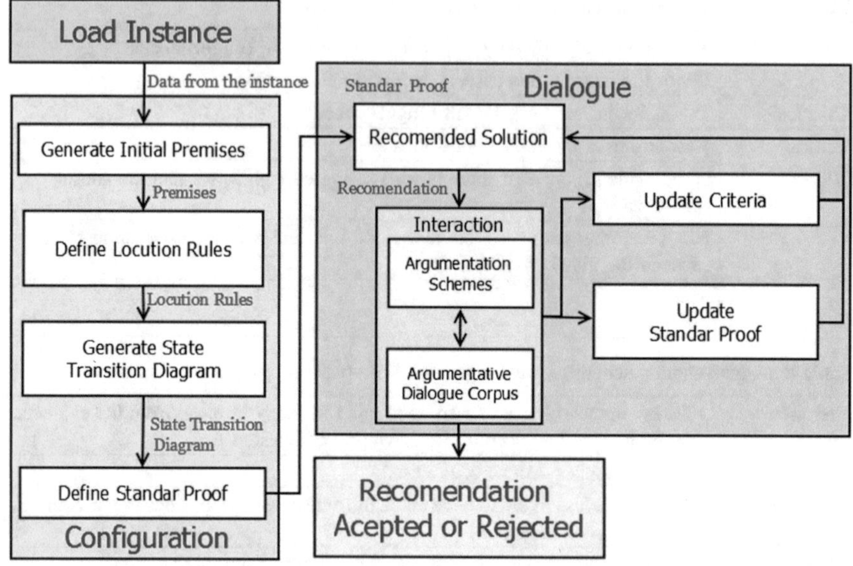

Fig. 4 Second layer of the modular diagram for the interaction module

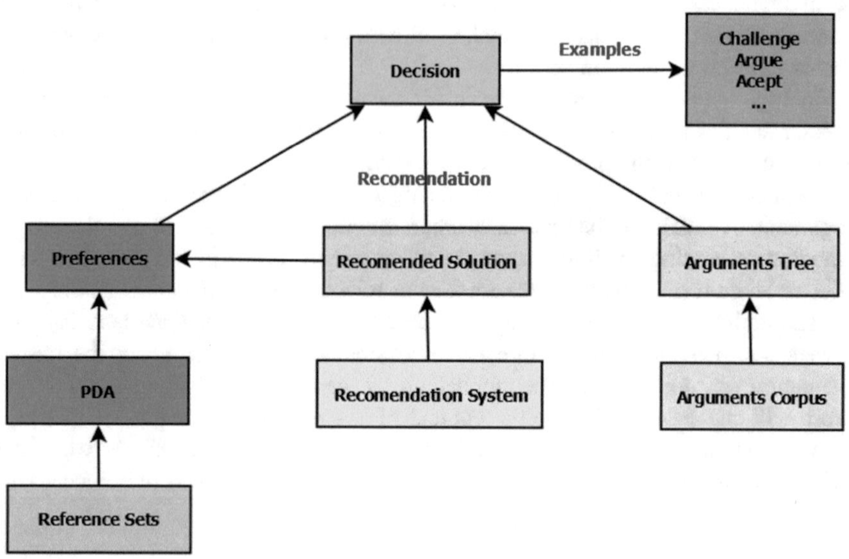

Fig. 5 Architecture for the interaction in the recommendation system

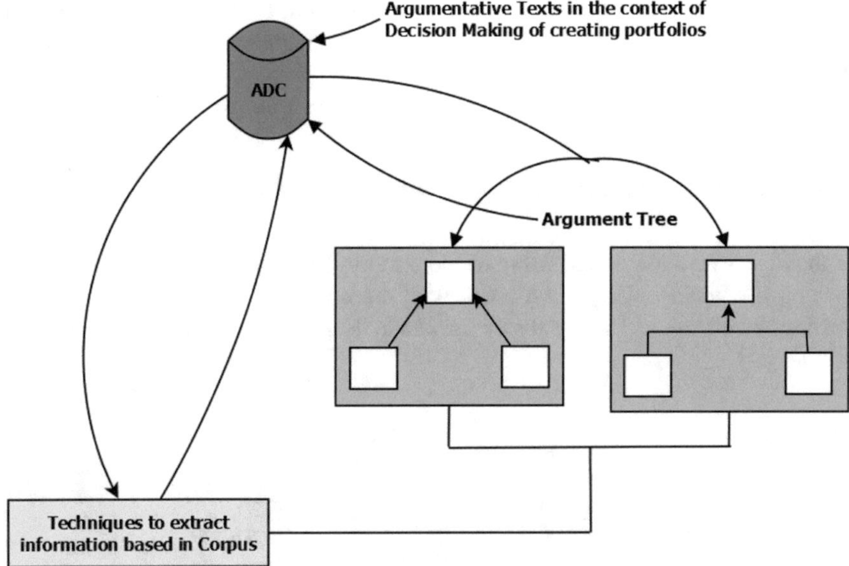

Fig. 6 Task diagram of the Interaction with the Corpus

play a very important role in the final decision; there are other elements that interact such as PDA, arguments tree, the dialogue corpus, reference sets, work and so on, they each other in the recommendation system, in order to help in the decision making.

The proposal in this article covers the generation of argument trees, we are building this generation based in the *Araucaria software* [13], and this software has the distinction of generating an arguments tree with premises and conclusions

The recommendation system will generate an initial solution, this solution will be verified against the preferences of the DM (these will be based on a PDA, linked to a reference set), and the *arguments corpus*, which is a repository of dialogues that characterize a DM, in order to construct an argument tree.

The recommended solution will be evaluated by the preference and argument tree to provide an argument as conclusion, which will be determined by the actions established in the state transition diagram (the user); from here, the interaction will continue in the dialogue established between the system and the user.

A proposal for the argumentation corpus is shown on Fig. 6, it will be filled with argumentative texts, then, it generates different argument trees [14]. These argument trees will characterize the DM, and interact with techniques to extract information in the corpus, we can obtain a conclusion and generate new data to construct arguments in the corpus.

Figure 7 shows a resume of the different tasks of the interaction process of the expert and the simulated DM; the expert will generate a recommendation (solution), if this solution matches the one with the arguments tree, to conclude if the solution is within the data in the corpus, with that comparison, the artificial DM will generate an answer according to the state transition diagram and continue with the interaction.

4.4 Development Status of the Interaction Module

In the Fig. 8 we show a screenshot of the prototype version of the recommendation system, it contains a dropdown menu with the option of loading an instance and starting the dialogue. The list shows the available actions in each phase of the state

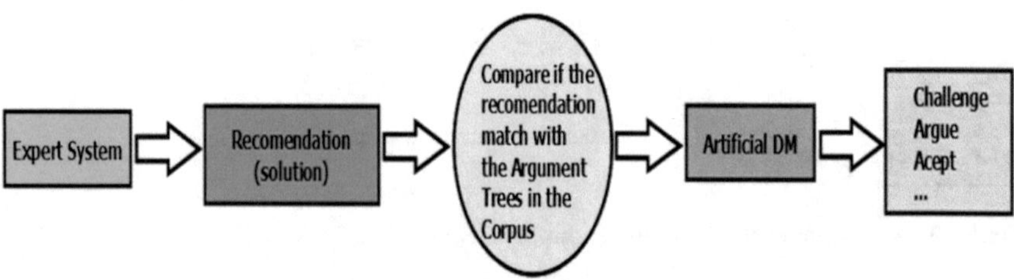

Fig. 7 Process of interaction between the expert and the artificial DM

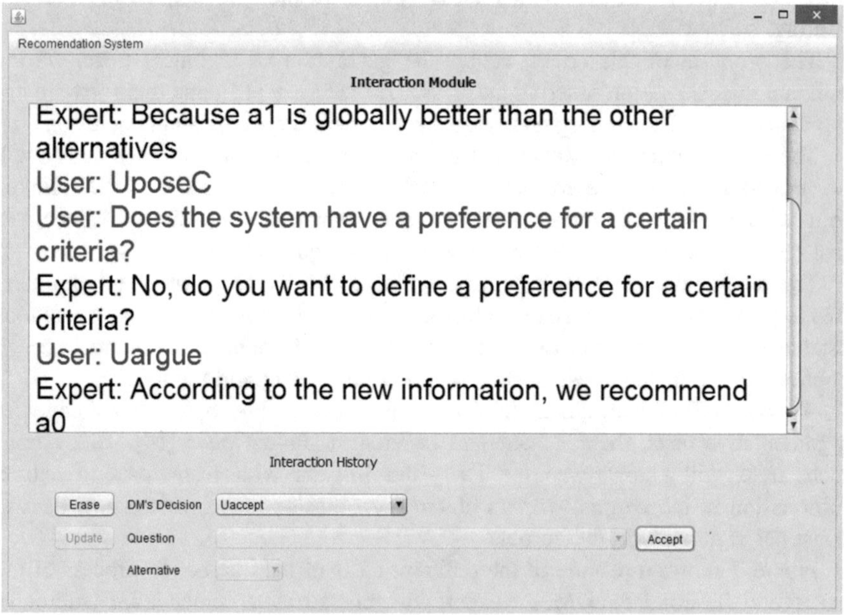

Recomendation System

Interaction Module

Expert: Because a1 is globally better than the other alternatives
User: UposeC
User: Does the system have a preference for a certain criteria?
Expert: No, do you want to define a preference for a certain criteria?
User: Uargue
Expert: According to the new information, we recommend a0

Interaction History

Erase DM's Decision Uaccept
Update Question Accept
Alternative

Fig. 8 Screenshot of the recommendation system

transition diagram. Finally, the text area located in the central part of the application shows the dialog between the DM and the Expert.

5 Analysis and Future Work

The architecture introduced has been implemented in the Java programming language. The current development has been focused on the initial configuration setup, recommendation process, proof standard selection and acceptation or rejection of the final recommendation. We have also worked on the update modules, for properties and criterion values.

Two of the four main modules have been completed (Load instance and acceptation/rejection of the final recommendation). However, the remaining modules (Configuration and Dialogue) are the most code-heavy processes of the system and require more time to be developed.

Although some of the argumentation schemes and part of the interaction have been already developed, it is necessary to keep working on those sections to have a full working system, capable of generating high quality recommendations.

The argumentative dialogue corpus is in the final design, when completed, it will be incorporated to interact with the argumentation schemes and continue with the experimentation.

6 Conclusions

Under the conditions of uncertainty and vagueness, it is required an effective model for justification that fits the nature of the problem and users.

The argumentation theory has a solid base for justifying arguments, serves as a complement for a portfolio obtained with a multicriteria method, allowing explaining why this option fits the DM model of preferences. This theory may help with the cognitive capacity of the mind in relation to non-monotonic reasoning which is commonly used by humans to make decisions.

The multicriteria methods for the generation of the initial recommendations has been tested thoroughly, due this, we have confidence that the complementation of the argumentation theory for justify this recommendation will be most adequate, since it begins with a recommendation with a base of preferences.

Once completed the development, the DSS will be a tool to help Decision Making of organizations to gain benefits such as money, time, by choosing the portfolio with the most impact.

Acknowledgments We express our gratitude to CONACYT for partially financing this work.

References

1. Power, D.: Decision Support Systems Resources – DSSResources.COM. [online] Dssresources.com. Available at: http://dssresources.com/ (Accessed 18 Jul. 2015).
2. Mousseau, V., & Stewart, T.: Progressive methods in multiple criteria decision analysis (Doctoral dissertation, PhD Thesis, Université Du Luxemburg) (2007).
3. Resnick, P., & Varian, H. R.: Recommender systems. Communications of the ACM, 40(3), 56-58 (1997).
4. Ouerdane, W.: Multiple criteria decision aiding: a dialectical perspective (Doctoral dissertation, Université Paris-Dauphine) (2009).
5. Walton, D.: Argumentation methods for artificial intelligence in law. Springer Science & Business Media (2005).
6. Walton, D.: Argumentation schemes for Presumptive Reasoning. Mahwah, N. J., Erlbaum (1996).
7. T. Gordon, H. Prakken, and D. Walton.: The Carneades model of argument and burden of proof. Artificial Intelligence, 171(4):875–896 (2007).
8. Prakken,H.: Relating protocols for dynamic dispute with logics for defeasible argumentation. Synthese, 127:187–219 (2001).
9. Parsons S., Sierra C., Jennings, N.: Agents that reason and negotiate by arguing. Journal of Logic and Computation, 8(3):261–292 (1998).
10. Amgoud L., Parsons S., Maudet N.: Arguments, dialogue, and negotiation. In W. Horn, editor, Proceedings of the European Conference on Artificial Intelligence (ECAI'00), pages 338–342. IOS Press. (2000)
11. Dung, P. M.: On the Acceptability of Arguments and its Fundamental Role in Nonmonotonic Reasoning, Logic Programming and n-person games. Artificial Intelligence, 77(2):321–358 (1995).
12. Walton, D., Reed, C., & Macagno, F.: Argumentation Schemes. Cambridge University Press (2008).
13. Reed,C.: Araucaria 3.1 User Manual, http://araucaria.computing.dundee.ac.uk/, (2004).
14. Botley, S. P., & Hakim, F. (2014). ARGUMENT STRUCTURE IN LEARNER WRITING: A CORPUS-BASED ANALYSIS USING ARGUMENT MAPPING. Kajian Malaysia, 32(1), 45 (2014).

Implementation of an Information Retrieval System Using the Soft Cosine Measure

Juan Javier González Barbosa, Juan Frausto Solís, J. David Terán-Villanueva, Guadalupe Castilla Valdés, Rogelio Florencia-Juárez, Lucía Janeth Hernández González and Martha B. Mojica Mata

Abstract The retrieval information models have been of important study since 1992. These models are based on comparing a user query and a collection of documents taking into account the concurrency of the terms, with the objective to classify a set of relevant documents and retrieve them to the user in accordance with the evaluations criterion. There are metrics to classify a set of documents according to the grade of similarity, such as cosine similarity and soft cosine measure. In this paper, we perform a comparative study of these similarity metrics. The Vector Space Model (VSM) was implemented for retrieving information. A sample of the Collection of the Association for Computing Machinery (CACM) in the domain of Computer Science was used in the evaluation. The experiment results show that the recall is of 96 % in both metrics, but the soft cosine achieves 2 % more in mean average precision.

J.J.G. Barbosa (✉) · J.F. Solís · J. David Terán-Villanueva · G.C. Valdés
R. Florencia-Juárez · L.J.H. González · M.B. Mojica Mata
Instituto Tecnológico de Ciudad Madero, México, Tecnológico Nacional de México,
1o. de Mayo y Sor Juana I. de la Cruz S/N, Ciudad Madero, Mexico
e-mail: jjgonzalezbarbosa@hotmail.com

J.F. Solís
e-mail: juan.frausto@gmail.com

J. David Terán-Villanueva
e-mail: david_teran00@yahoo.com.mx

G.C. Valdés
e-mail: gpe_cas@yahoo.com.mx

R. Florencia-Juárez
e-mail: rogelio.florencia@live.com.mx

L.J.H. González
e-mail: luciajaneth.hernandez@gmail.com

M.B. Mojica Mata
e-mail: marthamm_mx@hotmail.com

P. Melin et al. (eds.), *Nature-Inspired Design of Hybrid Intelligent Systems*,
Studies in Computational Intelligence 667, DOI 10.1007/978-3-319-47054-2_50

Keywords Vector space model · Similarity cosine · Soft cosine measure · CACM · Recall · Precision

1 Introduction

The web search engine is used to retrieve information from the Internet. Frequently, the retrieved information to the user is not accurate with the user request. This occurs due to the big amount of information available on the Internet. For this reason, models or systems that permit to obtain accurate and relevant information are today a real need that distress to the Internet users and researchers. Given a user query, these search engines uses models to retrieve and show information with a certain grade of priority. These models are based on the classification of the information taking into account the concurrency of the term in the query and in the documents collection. These models are well known as Information Retrieval Model (IRM).

There are diverse IRM that helps to search and classify the information, such as VSM. With the objective of retrieving the required documents by the user this VSM uses matrix, vectors, and normalization techniques of the terms. This model usually implements the similarity cosine measure to show to the user a list of ranked documents that are related to the user request.

Similarity cosine measure is one of the most used metrics in the IRM. It is based on vector calculus; also, the VSM's representation is a data structure of vectors. These characteristics made this metric appropriate for this model. Although this metric is efficient in the VSM, Sidorov in [1] proposes a new version of this metric, called it soft cosine measure. This metric considers the base vectors as part of the solution.

2 Problem Description

As it was mentioned before, the issue of search in the Information Retrieval Systems (IRS) has grown because of the colossal information increment. These systems, often related with web searchers, allow to the user view or get information classified according to the requested query. This procedure, by using natural language processing, extracts keywords of the query. The set of documents with higher concurrency of the keywords are indexed with the rest of the collection.

In general, the process of a query in the IRM rarely shows the information that the user request. Usually, there are two issues in the classification: false positive and false negative. A false positive is when the system retrieves a lot of insignificant information to the user. A false negative is when the retrieved information does not include the documents that should be in the solution.

With the purpose of minimizing these issues, we pretend to use a similarity measure that improves the classification of the information and retrieves a set of ranked documents with the higher relevancy to the query.

3 Justification

With the evolution of the technology and sciences, knowledge and information in the Internet have grown in an exponential way. This growth has increased the difficulty of the task of finding the information related to our interests. The IRMs appeared with the purpose of improve the ranking of the information in the Internet. These IRMs retrieve the relevancy of the documents in a collection using statistical or probabilistic methods.

However, these models are limited because they are based on the lexical of the words in the query and in the documents. This means the classified documents as relevant are only those, which contain the higher frequency of the words in the query. Due to this, the similarity measure is important in the classification mechanism. If the quality of the similarity measure is higher than the precision index will be higher.

The purpose of this work is to implement an IRM to improve the precision index in the classification of the documents for a given user query. This will allow improving the limitation that the IRMs have, also improving with this the quality and the precision of the classification mechanism.

4 VSM Description

VSM is one of the most used models in IR. It has an efficiency and utility proved by numerous works [3–6]. Taking the IRM definition from [2], it is characterized as:

Definition *An information retrieval model is a quadruple* $\left[D,\ Q,\ \mathcal{F},\ R\left(q_i,\ d_j\right)\right]$. *Where D is a set composed of logical views (or representations) for the documents in the collection; Q is a set composed of logical views (or representations) for the user information needs, such representations are called queries;* $\mathcal{F}$ *is a framework for modeling document representations, queries, and their relationship;* $R\left(q_i, d_j\right)$ *is a ranking function which associates a real number with a query* $q_i \in Q$ *and a document representation* $d_j \in D$*, such ranking defines an ordering among the document with regard to the query* q_i.

Considering this definition on the implemented VSM, D is a set of vectors used to characterize the essence of the vocabulary in each document from the collection. Q is a vector used to store the main terms from the query, that is, those words that contribute with significant information to the model. $\mathcal{F}$ is a matrix which represents the weights of the query terms of each document where these terms concur at least one time. Finally, $R(q_i,\ d_j)$ is the metric to evaluate the similarity of each document vector with the query vector.

This model requires certain structures to present to the user an index of all the documents that concur with their query. The (a) from Fig. 1 presents the *inverted index*, this is one of the principal functions. It obtains a compendium from the documents collection vocabulary. Likewise, it is required a structure to retrieve the information from a particular document, this structure is called *document* (Fig. 1b). Both structures are needed to generate a matrix to view the concurrency of the query terms with the document terms, this structure is called *term frequency matrix* (Fig. 1c).

Each position i of the inverted index contains two important elements for the VSM, a term t_{ic} and a list of documents d_{ij} where t_{ic} concurs. This structure in the first column contains the vocabulary obtained from the collection. The second column contains a vector that contains in each position j a reference of the documents where t_{ic} concurs.

Each index d_{ij} of Fig. 1b is of the type *document*. This structure contains the characteristics from a document such as their identifier, document path, and the matrix that contains all the document terms with their respective frequency in each term. Both structures provide the necessary elements to generate the matrix (Fig. 1c), which reflects each term from the collection.

The values of the data structure shown in the Fig. 2 are obtained from the structures in the Fig. 1. the Eq. (1) is used to generate the weight for each term from the weight query vector (Fig. 2a). the Eq. 2 is applied to obtain the inverse frequency from the document shown in Fig. 2b.

$$P_{q_i} = \left(\frac{tf_{q_i}}{\max\left(tf_{q_i}\right)}\right)(idf_t) \tag{1}$$

$$idf_t = \log\left(\frac{N}{df_t}\right) \tag{2}$$

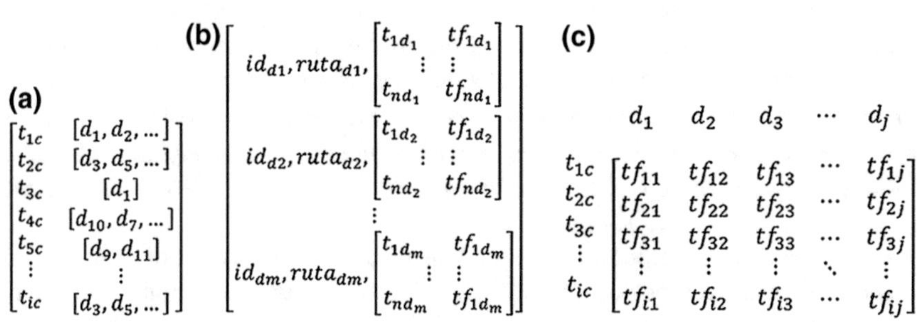

(a)

$$\begin{bmatrix} t_{1c} & [d_1, d_2, \ldots] \\ t_{2c} & [d_3, d_5, \ldots] \\ t_{3c} & [d_1] \\ t_{4c} & [d_{10}, d_7, \ldots] \\ t_{5c} & [d_9, d_{11}] \\ \vdots & \vdots \\ t_{ic} & [d_3, d_5, \ldots] \end{bmatrix}$$

(b)

$$\begin{bmatrix} id_{d1}, ruta_{d1}, \begin{bmatrix} t_{1d_1} & tf_{1d_1} \\ \vdots & \vdots \\ t_{nd_1} & tf_{nd_1} \end{bmatrix} \\ id_{d2}, ruta_{d2}, \begin{bmatrix} t_{1d_2} & tf_{1d_2} \\ \vdots & \vdots \\ t_{nd_2} & tf_{nd_2} \end{bmatrix} \\ \vdots \\ id_{dm}, ruta_{dm}, \begin{bmatrix} t_{1d_m} & tf_{1d_m} \\ \vdots & \vdots \\ t_{nd_m} & tf_{1d_m} \end{bmatrix} \end{bmatrix}$$

(c)

$$\begin{array}{c} \\ t_{1c} \\ t_{2c} \\ t_{3c} \\ \vdots \\ t_{ic} \end{array} \begin{array}{c} \begin{matrix} d_1 & d_2 & d_3 & \cdots & d_j \end{matrix} \\ \begin{bmatrix} tf_{11} & tf_{12} & tf_{13} & \cdots & tf_{1j} \\ tf_{21} & tf_{22} & tf_{23} & \cdots & tf_{2j} \\ tf_{31} & tf_{32} & tf_{33} & \cdots & tf_{3j} \\ \vdots & \vdots & \vdots & \ddots & \vdots \\ tf_{i1} & tf_{i2} & tf_{i3} & \cdots & tf_{ij} \end{bmatrix} \end{array}$$

Fig. 1 **a** Inverted Index, **b** Document, and **c** frequency term matrix

$$\textbf{(a)}\ \begin{matrix} t_{1c} \\ t_{2c} \\ t_{3c} \\ t_{4c} \\ \vdots \\ t_{ic} \end{matrix} \begin{bmatrix} w_{1q} \\ w_{2q} \\ w_{3q} \\ w_{4q} \\ \vdots \\ w_{iq} \end{bmatrix} \quad \textbf{(b)} \begin{bmatrix} idf_{1c} \\ idf_{2c} \\ idf_{3c} \\ idf_{4c} \\ idf_{5c} \\ \vdots \\ idf_{nc} \end{bmatrix} \quad \textbf{(c)}\ \begin{matrix} & d_1 & d_2 & d_3 & \cdots & d_j \\ t_{1c} & w_{11} & w_{12} & w_{13} & \cdots & w_{1j} \\ t_{2c} & w_{21} & w_{22} & w_{23} & \cdots & w_{2j} \\ t_{3c} & w_{31} & w_{32} & w_{33} & \cdots & w_{3j} \\ \vdots & \vdots & \vdots & \vdots & \ddots & \vdots \\ t_{ic} & w_{i1} & w_{i2} & w_{i3} & \cdots & w_{ij} \end{matrix}$$

Fig. 2 **a** Weight query, **b** inverted document frequency, and **c** weight matrix

Where tf_{q_i} represents the term frequency q_i in the query vector, $\max(tf_{q_i})$ denotes the maximum frequency value in the query. Finally, idf_t is the inverse frequency value given by the logarithm from the total terms and the quantity of the document in which the term appears. Immediately the construction of the weight matrix shown in Fig. 2c is started with the Eq. 3.

$$w_{t,d} = tf\_idt_{t,d} = tf_{t,d} \times idf_t \tag{3}$$

The weight of each term can be getting by the product of the term frequency and the inverse frequency of the same term. This structure and the weight vector of the query are the most important elements because those are replaced in the similarity measure (Eqs. (4) and (5)). This metrics allows measuring the relevance grade of each document for a given user query.

$$ssim(q, d_j) = \frac{\sum_{i=1}^{n} P_i \cdot W_{ij}}{\sqrt{\sum_{i=1}^{n} P_i^2}\sqrt{\sum_{i=1}^{n} W_{ij}^2}} \tag{4}$$

$$\text{soft\_cosine}_1(a, b) = \frac{\sum\sum_{i,j}^{N} S_{ij} a_i b_j}{\sqrt{\sum\sum_{i,j}^{N} S_{ij} a_i a_j}\ \sqrt{\sum\sum_{i,j}^{N} S_{ij} b_i b_j}} \tag{5}$$

The difference between soft cosine measure and similarity cosine is in the base vector presented as S_{ij}.

5 CACM Benchmark

The CACM benchmark [7] is one of the most used in the IRM researches. This is a collection that contains a set of papers in the computer science domain from 1958 to 1979.

In this paper, we worked with 126 documents and 5 queries. Natural language processing techniques were applied in order to identify keywords in the user query.

Table 1 Queries in the original format and without stopwords

Query	Original	Processed
1	What articles exist which deal with TSS (Time Sharing System), an operating system for IBM computers?	What, articles, exist, deal, TSS, Time, Sharing, System, operating, system, IBM, computers
2	I am interested in articles written either by Prieve or Udo Pooch; Prieve, B.; Pooch, U	I, interested, articles, written, either, Prieve, Udo, Pooch, Prieve, B, Pooch, U
3	Intermediate languages used in construction of multi-targeted compilers; TCOLL	Intermediate, languages, used, construction, multi, targeted, compilers, TCOLL
4	I'm interested in mechanisms for communicating between disjoint processes, possibly, but not exclusively, in a distributed environment. I would rather see descriptions of complete mechanisms, with or without implementations, as opposed to theoretical work on the abstract problem. Remote procedure calls and message-passing are examples of my interests	I'm, interested, mechanisms, communicating, disjoint, processes, possibly, exclusively, distributed, environment, I, would, rather, see, descriptions, complete, mechanisms, without, implementations, opposed, theoretical, work, abstract, problem, Remote, procedure, calls, message, passing, examples, interests
5	I'd like papers on design and implementation of editing interfaces, window-managers, command interpreters, etc. The essential issues are human interface design, with views on improvements to user efficiency, effectiveness and satisfaction	I'd, like, papers, design, implementation, editing, interfaces, window, managers, command, interpreters, etc., The, essential, issues, human, interface, design, views, improvements, user, efficiency, effectiveness, satisfaction

The *stopword* technique was applied to the queries. The results from this process are shown in the Table 1. The first column contains the query's number, the second column the original query and the third column shows the processed query.

CACM includes references to all the documents in the collection. Also, it specifies the relevant documents to each query, and the more frequent words in the collection. All these with the aim of facilitate the evaluation of IRMs.

6 Similarity Measure and Metrics

The ranking Precision, Recall, Average Precision by Query, and Mean Average Precision (6–10) were applied to measure the performance in each query.

$$\text{Precision} = \frac{\text{DRR}}{TD_{\text{Recovered}}} \qquad (6)$$

$$Recall = \frac{\text{DRR}}{TD_{\text{Relevant}}} \tag{7}$$

$$\text{Avg}P = \frac{\sum_{k=1}^{n} P(k) \times \text{rel}(k)}{\text{NDR}} \tag{8}$$

$$P(k) = \frac{k}{\text{pos}} \tag{9}$$

$$\text{MAP} = \frac{\sum_{q=1}^{Q} \text{Avg}P(q)}{Q} \tag{10}$$

Where DDR represents the total of recovered relevant documents. $TD_{\text{Recovered}}$ is the total of recovered documents. TD_{Relevant} is the total of relevant documents for the given query. In the *averageprecision*, $P(k)$ is the document precision of the k document given by $P(k) = \frac{k}{\text{pos}}$, where pos is the position that k has in the recovered documents, $\text{rel}(k)$ indicates the relevancy of the document k for the given query, 1 if it is relevant and 0 elsewhere. NDR is the total of relevant documents. Finally, Q is the total of queries.

7 Experimental Results

Table 2 shows the results of similarity measures applied to the queries from the CACM collection with the VSM. The column 1 is the query number (Query). Column 2 is the total of retrieved documents (NumDocRec). The column 3 has the relevant documents (ID_Doc). Column 4 has the position of the relevant document in the retrieved documents for the soft cosine measure. The column 5 has the same

Table 2 Position and metrics values

			Soft Cosine measure	Cosine measure
Query	NumDocRec	ID_Doc	Pos	Pos
1	113	116	5	5
		114	2	2
		99	57	57
		43	34	34
		124	–	–
2	125	123	2	**1**
		112	**27**	34
		118	**1**	3

(continued)

Table 2 (continued)

			Soft Cosine measure	Cosine measure
Query	NumDocRec	ID_Doc	Pos	Pos
3	112	106	1	1
		110	3	3
		96	17	17
		105	8	8
		126	73	73
		101	110	110
4	126	102	1	1
		115	2	2
		122	4	**3**
		113	**6**	7
		125	7	**6**
		109	**15**	18
		119	17	17
		121	21	**20**
		104	26	**25**
		108	**25**	26
		97	86	**85**
5	124	103	1	1
		100	3	3
		111	**8**	9
		107	7	**6**
		117	**17**	18
		120	21	21
		98	121	121

as column 4 but for the cosine measure. The number in bold indicates the measure that has a better performance than the other.

As it can be seen, both metrics work with the same collection of retrieved documents by VSM, thus these metrics are only focused on ranking these documents by using their ranking criterions.

Analyzing the results, we can observe that the query one does not retrieve all the relevant documents. The relevant document with the *id* 124 was not retrieved.

Although both metrics obtain the same precision value, a difference can be observed in the Table 3. They share the same *recall* values because they used the same retrieved documents.

In Table 3, the first column contains the query's number. The second and third columns show the average precision value of the relevant documents for soft cosine measure and for cosine measure, respectively. The last column denotes the recall value. The number in last row indicates the MAP's value for each measure of similarity.

Table 3 Metric position results

Query	AVG precision by position		Recall
	Soft cosine measure	Cosine measure	
1	**0.21168215**	**0.21168215**	0.8
2	*0.7037037*	0.58496732	1
3	**0.3999999**	**0.3999999**	1
4	0.56329017	0.58152815	1
5	0.4618459	0.4617792	1
MAP	**0.46810436**	**0.44799134**	**0.96**

Table 4 Experimentation results of cosine measure obtained by the recovery documents

Query	AVG Precision in general		Recall
	Soft cosine measure	Cosine measure	
1	**0.083650319**	**0.083650319**	0.8
2	**0.09099322**	**0.085118192**	1
3	**0.126727617**	**0.126727617**	1
4	**0.231569164**	**0.233212877**	1
5	**0.152584471**	**0.152446109**	1
MAP	**0.137104958**	**0.136231023**	**0.96**

From Table 3, we can observe that soft cosine measure achieves 2 % more precision than cosine measure. Also, in the queries 1 and 3, we can observe that both metrics get the same precision. In query 2, soft cosine measure obtains a 70 % of precision. These results show the percentage of precision based on the position of the relevant document in the retrieved documents (8), if this value is close to one it indicates that the documents were ranked in the first positions. These positions were obtained by the similarity metrics.

In Table 4, the first column contains the query's number. The second and third columns show the average precision value for soft cosine measure and cosine measure over all the recovered documents, respectively. The last column denotes the recall value. The number in last row indicates the MAP's value for each measure of similarity.

Table 4 shows the percentage that the relevant documents have over the total retrieved documents. We can note that the queries 1 and 2 get more irrelevant documents than relevant because the values are more close to zero, and that means that the relevant documents have a less impact in the retrieved documents. Also, the soft cosine measure achieves a better result in the precision measure than cosine measure.

8 Conclusions and Future Work

One of the issues in the development of this work was the dimensionality that the vectors and matrix acquires. This is due to the large number of terms in the pre-processing of the data collection. With the aim of reducing the size of this set, it is essential to apply other techniques as the stemming when the set of terms is built.

With respect to the comparative of the similarity metrics, we can observe a significant difference. But this does not minimizes that the size of the collection and the number of queries used in the experimentation could not be the adequate. We should do more experiments with a larger sample size of documents from the CACM.

However, the results obtained in the Precision and Recall metrics show that the soft cosine measure can be improved. Other conclusion is that the percentage of retrieved documents in the VSM. It achieves 96 % of relevant documents.

As future work, we consider to focus in the study of the query processing. We plan to use an expansion of queries using ontologies as an alternative to improve the precision and the coverage of the classification of documents in IRMs.

References

1. Grigori Sidorov, Alexander Gelbukh, Helena Gómez-Adorno, and David Pinto. Soft Similarity and Soft Cosine Measure: Similarity of Features in Vector Space Model. Computación y Sistemas, Vol. 18, No. 3, 2014, pp. 491–504, DOI:10.13053/CyS-18-3-2043.
2. Baeza - Yates, R., & Ribeiro - Neto, B. (1999). Modern Information Retrieval (Vol. 463). New York, The United States of America: Addison Wesly, ACM press.
3. Farrús, M., & R. Costa-jussá, M. (2013). Presencia de IRRODL evaluación automática del aprendizaje electrónico utilizando el análisis semántico latente: caso de uso. Revista mexicana de bachillerato a distancia., 153 - 165.
4. La Serna Palomino, N., Pró Concepción, L., & Román Concha, U. (2013). Diseño de un sistema de recuperación de imágenes de individuos malhechores para seguridad ciudadana. Revista de Investigación de Sistemas e Informática, 25 - 32.
5. Monsalve, L. S. (2012). Experimento de Recuperación de Información usando las medidas de similitud coseno, Jaccard and Dice. Revista de Investigación: TECCIENCIA, 14 - 24.
6. La Serna Palomino, N., Román Concha, U., & Osorio, N. (2009). Implementación de un Sistema de Recuperación de Información. Revista de Ingeniería de Sistemas e Informática, 57 - 64.
7. The JNT Association. (01 de October de 2015). *CACM collection*. Obtenido de CACM collection: http://ir.dcs.gla.ac.uk/resources/test_collections/cacm

TOPSIS-Grey Method Applied to Project Portfolio Problem

Fausto Balderas, Eduardo Fernandez, Claudia Gomez, Laura Cruz-Reyes and Nelson Rangel V

Abstract Project portfolio selection is one of the most difficult, yet most important decision-making problems faced by many organizations in government and business sectors. The grey system theory proposed by Deng in 1982 is based on the assumption that a system is uncertain and that the information regarding the system is insufficient to build a relational analysis or to construct a model to characterize the system. The aim of this chapter is to compare a multi-attribute decision-making (MADM) that incorporates a system of preferences TOPSIS (Technique for Order Performance by Similarity to Ideal Solution) by Hwang and Yoon in (Multiple attribute decision making: methods and applications. Springer, Berlin, 1981) with TOPSIS-Grey by Lin et al. in (Expert Syst Appl 35:1638–1644, 2008).

Keywords TOPSIS-Grey · Multi-attribute decision making · Project portfolio problem · Grey system theory

---

F. Balderas (✉) · C. Gomez · L. Cruz-Reyes · N. Rangel V
Instituto Tecnologico de Ciudad Madero, Tecnologico Nacional de Mexico, Tamaulipas, Mexico
e-mail: fausto.balderas@itcm.edu.mx

C. Gomez
e-mail: cggs71@hotmail.com

L. Cruz-Reyes
e-mail: lauracruzreyes@itcm.edu.mx

N. Rangel V
e-mail: nrangelva@conacyt.mx

E. Fernandez
Universidad Autonoma de Sinaloa, Sinaloa, Mexico
e-mail: eddyf@uas.edu.mx

P. Melin et al. (eds.), *Nature-Inspired Design of Hybrid Intelligent Systems*,
Studies in Computational Intelligence 667, DOI 10.1007/978-3-319-47054-2_51

1 Introduction

TOPSIS (technique for order performance by similarity to ideal solution) is a useful technique in dealing with multi-attribute or multi-criteria decision making (MADM/MCDM) problems in the real world [1].

TOPSIS helps the DM organize the alternative solutions that has to resolve to perform an analysis, comparisons and ordering of the possible alternatives. According to TOPSIS, the selection of the most suitable alternative will take place. This technique is based on the idea that the optimal solution must have the shortest distance to the ideal alternative, and the farthest distance to the alternative non-ideal. A solution is determined as ideal if it maximizes the benefit of the criteria. TOPSIS simultaneously considers these distances to classify the solutions in a preference order, using the relative closeness, which is obtained with both distances (positive and negative ideal alternatives). The alternative with the highest value is selected as the best alternative.

In this work, we carry out a comparative study between the TOPSIS and TOPSIS-Grey method. The purpose of the study is to use a multi-attribute technique applied to project portfolio problem. The comparison was done in the project portfolio problem by using an instance with two objectives.

2 Background

In this section, some basic definitions about grey system theory and TOPSIS method is given. Then, the original NSGA-II is described.

2.1 Grey System Theory

Definition 1 A grey system is defined as a system containing uncertain information presented by a grey number and grey variables.

Definition 2 Such a number instead of its range whose exact value is unknown is referred to as a grey number. In applications, a grey number in fact stands for an indeterminate number that take its possible value within an interval or a general set of numbers. This grey number is generally written using the symbol "$\otimes$". There are several types of grey numbers.

Definition 2.1 *Grey numbers with only a lower bound*: This kind of grey number $\otimes$ is written as $\otimes \in [\underline{a}, \infty)$ or $\otimes(\underline{a})$, where $\underline{a}$ stands for the definite, known lower bound of the grey number $\otimes$. The interval $[\underline{a}, \infty]$ is referred to as the field of $\otimes$.

Definition 2.2 *Grey numbers with only an upper bound*: This kind of grey number $\otimes$ is written as $\otimes \in (-\infty, \bar{a}]$ or $\otimes\ (\bar{a})$, where a stands for the definite, known upper bound of $\otimes$.

Definition 2.3 *Interval grey numbers*: This kind of grey number $\otimes$ has both a lower a and an upper bound a, written $\otimes \in [\underline{a}, \bar{a}]$

Definition 2.4 *Continuous and discrete grey numbers*: A grey number that only takes a finite number or a countable number of potential values is known as discrete. If a grey number can potentially take any value within an interval, then it is known as continuous.

Definition 2.5 *Black and white numbers*: When $\otimes \in (-\infty, +\infty)$, that is, when $\otimes$ has neither any upper nor lower bound, then $\otimes$ is known as a black number. When $\otimes \in [\underline{a}, \bar{a}]$ and $\underline{a} = \bar{a}$, $\otimes$ is known as a white number.

Definition 2.6 *Essential and non-essential grey numbers*: The former stands for such a grey number that temporarily cannot be represented by a white number; and the latter such grey numbers each of which can be represented by a white number obtained either through experience or certain method. The definite white number is referred to as the whitenization (value) of the grey number, denoted $\tilde{\otimes}$. Also, we use $\otimes(a)$ to represent the grey number(s) with a as its whitenization.

Definition 3 Grey number operation is an operation defined on sets of intervals, rather than real numbers. The modern development of interval operation began with R.E. Moore's dissertation [2]. We cite literatures [3, 4] to define the basic operation laws of grey numbers $\otimes a_1 = [\underline{a_1}, \overline{a_1}]$ and $\otimes a_2 = [\underline{a_2}, \overline{a_2}]$, on intervals where the four basic grey number operations on the interval are the exact range of the corresponding real operation.

$$\begin{aligned}
\otimes a_1 + \otimes a_2 &= [\underline{a_1} + \underline{a_2},\ + \overline{a_1} + \overline{a_2}] \\
\otimes a_1 - \otimes a_2 &= [\underline{a_1} - \overline{a_2}, \overline{a_1} - \underline{a_2}] \\
\otimes a_1 \times \otimes a_2 &= [\min(\underline{a_1 a_2}, \underline{a_1}\overline{a_2}, \overline{a_1}\underline{a_2}, \overline{a_1 a_2}), \max(\underline{a_1 a_2}, \underline{a_1}\overline{a_2}, \overline{a_1}\underline{a_2}, \overline{a_1 a_2})] \\
\otimes a_1 \div \otimes a_2 &= [\underline{a_1}, \overline{a_1}] \times \left[\frac{1}{\underline{a_2}}, \frac{1}{\overline{a_2}}\right]
\end{aligned}$$

Definition 4 The length of grey number $\otimes G$ is defined as

$$L(\otimes a) = [\overline{a_1}, \underline{a_2}]$$

Definition 5 Comparison of grey numbers

Shi et al. [5] proposed a degree of grey possibility to compare the ranking of grey numbers.

For two grey numbers $\otimes a_1 = [\underline{a_1}, \overline{a_1}]$ and $\otimes a_2 = [\underline{a_2}, \overline{a_2}]$, the possibility degree of $\otimes a_1 \leq \otimes a_2$ can be expressed as follows:

$$P(\otimes a_1 \leq \otimes a_2) = \frac{\max(0, L^* - \max(0, \overline{a_1} - \underline{a_2})}{L^*}$$

For the position relationship between $\otimes a_1$ and $\otimes a_2$, there exist four possible cases on the real number axis. The relationship between $\otimes a_1$ and $\otimes a_2$ is determined as follows:

1. If $\underline{a_1} = \underline{a_2}$ and $\overline{a_1} = \overline{a_2}$, we say that $\otimes a_1$ is equal to $\otimes a_2$, denoted as $\otimes a_1 = \otimes a_2$. Then $P(\otimes a_1 \leq \otimes a_2) = 0.5$
2. If $\underline{a_2} > \overline{a_1}$, we say that $\otimes a_2$ is larger than $\otimes a_1$, denoted as $\otimes a_2 > \otimes a_1$. Then $P(\otimes a_1 \leq \otimes a_2) = 1$.
3. If $\overline{a_2} < \underline{a_1}$, we say that $\otimes a_2$ is smaller than $\otimes a_1$, denoted as $\otimes a_2 > \otimes a_1$. Then $P(\otimes a_1 \leq \otimes a_2) = 0$.
4. If there is an intercrossing part in them, when $P(\otimes a_1 \leq \otimes a_2) = 0.5$, we say that $\otimes a_2$ is larger than $\otimes a_1$, denoted as $\otimes a_2 > \otimes a_1$. When $P(\otimes a_1 \leq \otimes a_2) < 0.5$, we say that $\otimes a_2$ is smaller than $\otimes a_1$, denoted as $\otimes a_2 < \otimes a_1$.

2.2 *TOPSIS-Grey Method*

Lin et al. [6], define the following procedure to integrate the TOPSIS method with the grey philosophy:

Step 1: Constructing the grey decision-making matrix $\otimes D$. The matrix $\otimes D$ is defined in Eq. (1):

$$\otimes MD = \begin{bmatrix} \otimes y_{11} & \cdots & \otimes y_{1m} \\ \vdots & \ddots & \vdots \\ \otimes y_{n1} & \cdots & \otimes y_{nm} \end{bmatrix}; i = 1, \ldots, n; j = 1 \ldots, m, \quad (1)$$

Step 2: The normalized grey decision matrix is established, using the Eq. (2), giving as result $\otimes r_{ij}$

$$\otimes r_{ij} = \frac{\otimes y_{ij}}{max_i(\bar{y}_{ij})} = \left(\frac{\underline{y}_{ij}}{max_i(\bar{y}_{ij})}; \frac{\bar{y}_{ij}}{max_i(\bar{y}_{ij})}\right), \quad (2)$$

Step 3: The positive ideal A^+ and negative ideal A^- alternatives for each objective are determined, using the Eqs. (3) and (4).

$$A_j^+ = \left\{(\max_i \bar{r}_{ij} | j \in J), | i = 1, 2, \ldots, n\right\}, \quad (3)$$

$$A_j^- = \left\{\left(\min_i \underline{r}_{ij} | j \in J\right), | i = 1, 2, \ldots, n\right\}, \quad (4)$$

Step 4: Calculate the *distance of separation to the ideal and negative ideal alternatives*, using the m-dimensional Euclidean distance. The distance that separates the ith alternative from the ideal A^+ is obtained through Eq. (5):

$$d_i^+ = \sqrt{\frac{1}{2}\Sigma_{j=1}^m w_j}\left[\left|r_j^+ - \underline{r}_{ij}\right|^2 + \left|r_j^+ - \bar{r}_{ij}\right|^2\right], \tag{5}$$

where w_j is the weight of the criterion j. Likewise, the separation distance to the *negative ideal alternative* A^-, is obtained through Eq. 6:

$$d_i^- = \sqrt{\frac{1}{2}\Sigma_{j=1}^m w_j}\left[\left|r_j^- - \underline{r}_{ij}\right|^2 + \left|r_j^- - \bar{r}_{ij}\right|^2\right], \tag{6}$$

Step 5: Calculate the *relative closeness* to the ideal solution. The relative closeness of the alternative i, C_i^+, with respect to the ideal alternative is defined by Eq. (7):

$$C_i^+ = \frac{d_i^-}{d_i^+ + d_i^-}, \tag{7}$$

where $0 \leq C_i^+ \leq 1$. The closer is the value of relative closeness to 1; the better will be the evaluation of the alternative.

2.3 Non-dominated Sorting Genetic Algorithm II (NSGA-II)

The NSGA-II [7] is considered one of the benchmarks of the multi-objective optimization to solve problems, preferably of two or three objectives. The NSGA-II (Algorithm 1) is based on the creation of non-dominated fronts, establishing elitism on the first front; it also includes an indicator of diversity called crowding distance.

Algorithm 1. NSGA – II [13]

$R_T = P_T \cup Q_T$	combine parent and children population
$F = fast - non - dominated - sort\ (R_T)$	$F = (F_1, F_2, \ldots)$, all non-dominated fronts ofR_T
$P_{T+1} = \emptyset\ or\ i = 1$	
while $\lvert P_{T+1}\rvert + \lvert F_i\rvert \leq N$ do	till the parent population is filled
crowding – distance – $assignment(F_i)$	calculate crowding distance in F_i
$P_{T+1} = P_{T+1} \cup F_i$	include i-th non-dominated front in the parent pop
$i = i + 1$	
end while	
$SORT(F_{i,} \prec_i)$	sort in descending order using $\prec_i$
$P_{T+1} = P_{T+1} \cup F_i[1:(N - \lvert P_{T+1}\rvert)]$	choose the first N elements of P_{T+1}
$Q_{T+1} = make - new - pop(P_{T+1})$	use selection, crossover and mutation to create a new population Q_{T+1}
$t = t + 1$	

3 Experimentation and Results

In this section, we present the case of study and the results of the experimentation carried out.

3.1 *Case of Study: The Project Portfolio Problem*

In the project portfolio problem, we have a set A of N projects, the DM should describe the contributions of the objectives $f_j(i)$, which refer to the number of persons benefited by each project; the costs of those projects c_i must also be described. As for the general information of the problem, there is a budget B to distribute that had to be analyzed, evaluated and calculated by the DM previously. The proposals are classified into three areas by their nature a_i, and by their impact location into two regions. The DM requires a certain level of balance in the supported portfolio. Each project i is represented as a p-dimensional vector $f(i) = \langle f_1(i), f_2(i), f_3(i), \ldots, f_p(i)\rangle$, where each $f_j(i)$ represents the benefit of project i to objective j into a problem of p objectives.

Based on this, a portfolio is feasible when it satisfies the constraint of the total budget (Eq. 8) and the constraint for each area j (Eq. 9).

$$\left(\sum_{i=1}^{N} x_i c_i\right) \leq B \tag{8}$$

$$L_j \leq \sum_{i=1}^{N} x_i g_i(j) c_i \leq U_j \tag{9}$$

where g may be defined as:

$$g_i(j) = \begin{cases} 1 & \text{if} \quad a_i = j, \\ 0 & \text{otherwise} \end{cases} \tag{10}$$

The union of the benefits of each of the projects that compose to a portfolio determine its quality. This may be expressed as:

$$z(x) = \langle z_1(x), z_2(x), z_3(x), \ldots, z_p(x)\rangle \tag{11}$$

where $z_j(x)$ in its simplest form, is determined by:

$$z_j(x) = \sum_{i=1}^{N} x_i f_i(i) \tag{12}$$

Therefore, considering R_F as the feasible region, the problem of project portfolio is to determine one or more portfolios that solve:

$$\max_{x \in R_F}\{z(x)\} \quad (13)$$

3.2 Results

In this section, we provide the experimental results. We compared the results obtained from NSGA-II and them we applied the TOPSIS method and TOPSIS-Grey method. We experimented with one instance addressing the project portfolio problem.

The parameters of the evolutionary search were as follows, crossover probability = 1; mutation probability = 0.05.

To the results obtained from NSGA-II and for simplicity and without loss of generality, we will give a random number between 1 and 4 % to the cost of the projects, and that it is symmetrically distributed around the value. For the imprecision in the benefit of projects is a random amount between 0.15 and 1.5 % of its value. This is for the purpose to work with grey numbers and the TOPSIS-Grey method.

The NSGA-II was run thirty times with the instance. Subsequently to the results, we apply the TOPSIS method and the TOPSIS-Grey method for the grey numbers. The results are illustrated in Tables 1 and 2.

Table 1 Five best solutions obtained with TOPSIS

Id	C_l	o_{i1}	o_{i2}	*Card	C_i^+
1	$244,545	1,330,915	316,400	38	0.7090257
2	$244,310	1,336,730	314,230	38	0.7037742
3	$244,550	1,336,550	314,000	38	0.7016510
4	$244,840	1,334,070	314,540	38	0.7011177
5	$244,905	1,353,470	310,045	38	0.6989283

Table 2 Five best solutions obtained with TOPSIS-Grey

Id	$\otimes C_l$	$\otimes o_{i1}$	$\otimes o_{i2}$	*Card	C_i^+
1	244010, 254810	1367065, 1376265	322490, 331990	39	0.729962157
2	244760, 255160	1386180, 1395080	316640, 326240	39	0.729604841
3	244445, 255245	1384885, 1394185	316975, 326375	39	0.729376421
4	244295, 254895	1394410, 1403210	314630, 324030	39	0.728997948
5	244545, 254945	1388175, 1396875	315920, 325620	38	0.728522268

We can see in the results shows in Tables 1 and 2, that when we transform the results of NSGA-II to grey numbers and applied a technique that deal with multi-attribute like TOPSIS, and TOPSIS-Grey, those provided by TOPSIS-Grey are clearly better than those obtained by TOPSIS.

4 Conclusions and Future Work

We have presented a comparative study between TOPSIS and TOPSIS Grey method. The Application of TOPSIS and TOPSIS-Grey methods to the project portfolio problem is presented. As immediate work we are going to work with three and more objectives and applied a different preferences system to find the best solution.

Acknowledgments This work was partially financed by CONACyT, TECNM and ITCM.

References

1. C.L. Hwang, K. Yoon. (1981). Multiple Attribute Decision Making: Methods and Applications, Springer, Berlin, Heidelberg, New York.
2. Moore, R.E.: Interval Analysis. Prentice-Hall, Englewood Cliffs (1966)
3. Wang, Q., Wu, H.: The concept of grey number and its property. In: Proceedings of NAFIPS, USA, pp. 45–49 (1998)
4. Wu, Q., Zhou, W., Li, S., Wu, X.: Application of grey numerical model to groundwater resource evaluation. Environ. Geol. 47, 991–999 (2005)
5. Shi, J.R., Liu, S.Y., Xiong, W.T.: A new solution for interval number linear programming. J. Syst. Eng. Theor. Pract. 2, 101–106 (2005) (in Chinese).
6. Lin, Y. H., Lee, P. C., & Ting, H. I. (2008). Dynamic multi-attribute decision making model with grey number evaluations. *Expert Systems with Applications*, *35*(4), 1638-1644.
7. Deb, K., Pratap, A., Agarwal, S., & Meyarivan, T. A. M. T. (2002). A fast and elitist multiobjective genetic algorithm: NSGA-II. *Evolutionary Computation, IEEE Transactions on*, *6*(2), 182-197.
8. Deng, J.L. (1982). Control problems of grey systems. Systems & Control Letters, 1(5), 288-94.

Comparing Grammatical Evolution's Mapping Processes on Feature Generation for Pattern Recognition Problems

Valentín Calzada-Ledesma, Héctor José Puga-Soberanes, Alfonso Rojas-Domínguez, Manuel Ornelas-Rodríguez, Juan Martín Carpio-Valadez and Claudia Guadalupe Gómez-Santillán

Abstract Grammatical Evolution (GE) is a grammar-based form of Genetic Programming. In GE, a Mapping Process (MP) and a Backus–Naur Form grammar (defined in the problem context) are used to transform each individual's genotype into its phenotype form (functional representation). There are several MPs proposed in the state-of-the-art, each of them defines how the individual's genes are used to build its phenotype form. This paper compares two MPs: the Depth-First standard map and the Position Independent Grammatical Evolution (πGE). The comparison was performed using as use case the problem of the selection and generation of features for pattern recognition problems. A Wilcoxon Rank-Sum test was used to compare and validate the results of the different approaches.

Keywords Grammatical evolution · Mapping process · Feature generation · Pattern recognition

1 Introduction

Grammatical Evolution (GE) is a grammar-based form of Genetic Programming (GP), which includes at its core a Mapping Process (MP) and a Bacus–Naur Form (BNF) grammar, which can be incorporated regardless of the problem. The GE has a modular design, which enables the use of different search strategies, either evolutionary or some other kind of heuristic [1]. A BNF grammar, describes the structures generated by the GE, these can be modified by simply changing the grammar.

V. Calzada-Ledesma (✉) · H.J. Puga-Soberanes · A. Rojas-Domínguez · M. Ornelas-Rodríguez · J.M. Carpio-Valadez
Tecnológico Nacional de México, Instituto Tecnológico de León, León, Gto, Mexico
e-mail: valecalzada@hotmail.com

C.G. Gómez-Santillán
Tecnológico Nacional de México, Instituto Tecnológico de Ciudad Madero, Ciudad Madero, Tamaulipas, Mexico

P. Melin et al. (eds.), *Nature-Inspired Design of Hybrid Intelligent Systems*, Studies in Computational Intelligence 667, DOI 10.1007/978-3-319-47054-2_52

The MP is an important aspect of representation in evolutionary computing, allowing disengage the search space in two parts, the genotype and the phenotype space [2]. A genotype represents a potential solution, it consists of a chain of components called genes, which pass to their phenotypic-form (fitness-based selection is determined by the phenotypes) by decoding in the components associated with the problem under study [2].

In the state-of-the-art have been reported different approaches of MPs for GE, including the Depth-First standard map, Position Independent Grammatical Evolution (or πGE), and others [3–6].

A performance analysis of different MPs on GE is reported in [3], using as use cases four benchmark problems: Even-5-parity, Symbolic Regression, Santa Fe ant trail, and Max. These problems are classic and commonly used in GE by its simplicity [3, 4, 7–9]. In this article, a comparison is performed between two MPs: the Depth-First standard map and πGE; using as use case a problem outside of the benchmark in GE. The problem used is the selection and generation of features, which can be extended to solve real-world instances. A methodology to solve this problem was proposed in [10], by using the canonical GE to improve the accuracy in classification problems, which adopts Depth-first as MP.

The performance was measured in terms of the number of successful solutions found to each problem instance, (in this case, classification datasets from UCI [11]) and by examining the average best fitness, which was used as comparison criterion. The results can provide a starting point for deciding which MP to use for different instances of the problem, used to compare both approaches. In order to provide a statistical support of the results, Wilcoxon Rank-Sum test was performed.

The structure of this paper is as follows. Section 2 introduces some background of the canonical Grammatical Evolution, the GE's MPs, and the Differential Grammatical Evolution. In Sect. 3 the benchmark problem is described. In Sect. 4, the methodology, the proposed Bacus–Naur Form, the evaluation process using both MP approaches and the datasets are described. In Sect. 5, the experimental results given by Depth-First and πGE, and the results of Wilcoxon test are presented. In Sect. 6, a discussion of the results is presented. In the last section, the conclusions are presented.

2 Grammatical Evolution

Grammatical Evolution is a grammar-based form of Genetic Programming, which can be used to solve problems, by means of the automatic generation of solutions adapted to the problem domain [1].

In GE, a population of individuals is initialized; the chromosome of each one contains genes with binary or integer values. An individual is considered as a potential solution, which has to be transformed from its genotypic-form to its phenotypic-form by means of a MP and a BNF grammar, which provides the language and the grammatical rules in the context of the problem. Each phenotype

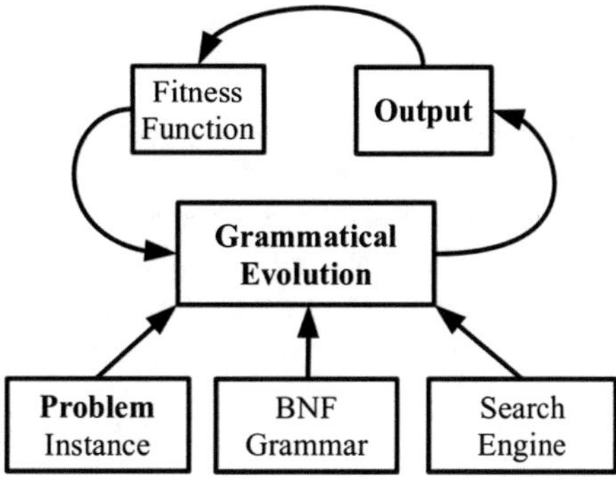

Fig. 1 Grammatical Evolution's Modular Design [12]

is then evaluated in the fitness function, to verify the quality of the solution [1]. Figure 1 shows the modular design of the GE [12].

2.1 Grammatical Evolution's Mapping Process

The MP decouples the search space in two parts, search space (within the genotype) and solution space. The MP is governed by a BNF grammar, which must be defined according to the context of the problem. Generally, grammar defines the syntax of phenotypic programs that will be created by the GE [13].

A BNF grammar is made up of the 4-tuple {***N***, ***T***, ***P***, ***S***}, where:

- ***N*** is the set of all Nonterminal symbols.
- ***T*** is the set of Terminals.
- ***P*** is the set of Production rules that map ***N*** to ***T***.
- ***S*** is the initial Start symbol and a member of ***N***.

The number of production rules can be applied to a nonterminal symbol, a "|" (or) symbol separates each production rule.

2.2 Depth-First

This mapper starts by identifying the start symbol in the grammar and using Eq. 1 (proposed in [13]) the production rule replaces the start symbol and the process begins always replacing the leftmost nonterminal symbol on the grammar, chosen by the production rules based on the current genotype, until no more *NT*s remain.

$$Rule = c\%r. \tag{1}$$

where c is the codon value and r is the number of production rules available for the current nonterminal.

2.3 πGrammatical Evolution

The Depth-first standard map consumes nonterminal symbols from left to right, whereby there is a positional dependence. With πGE this dependency does not exist, because the mapper chooses which nonterminal symbol will be consumed [6]. The codon is composed by a pair of genes from the chromosome, the first value is the ***nont*** and the second one is the ***rule***. The ***nont*** value contains the coding to select the nonterminal symbol to consume, it is done by Eq. 2 (proposed in [13]).

$$NT = nont\%count. \tag{2}$$

where:

- ***NT*** is the nonterminal to be consumed (counted *0, 1,...,n* from left to right of the remaining nonterminals).
- ***nont*** is the value from the individual.
- ***count*** is the number of nonterminals remaining.

The ***rule*** part of the codon is applied as in the Depth-First.

2.4 Grammatical Differential Evolution

The GE adopts Differential Evolution (ED) coupled with the MP. ED maintains a population of potential solutions, which are mutated and recombined to produce new individuals; these are selected according to its performance. The ED uses trial solutions, which compete with individuals of the current population to survive [13]. The ED algorithm (taken from [14]) is detailed below.

Algorithm:

1. Initialize a random population of *N* with (*j*) individuals, *d*-dimensional vectors:
$$X_j = (x_{i1}, x_{i2}, \ldots, x_{id}), \quad j = 1, 2, \ldots, n. \tag{3}$$
2. Evaluate initial population on fitness function *fitness*.
3. Under mutation operator, for each vector $X_j(t)$, a variant solution $V_j(t+1)$ is obtained using Eq. 4:
$$V_j(t+1) = X_m(t) + F \cdot (X_k(t) - X_l(t)). \tag{4}$$

where:

- $(k, l, m) \in 1, \ldots, N$: are mutually different, randomly selected indices and $(k, l, m) \neq j$.
- X_m: is the base vector.
- $X_k(t) - X_l(t)$: is the difference vector.
- $F \in (1, 2)$: is the scaling factor.

4. Calculate a trial solution $U_j(t+1) = \left(u_{j1}, u_{j2}, \ldots, u_{jd}\right)$ using Eq. 5:

$$U_{jn}(t+1) = \begin{cases} V_{jn}, & \text{if } (rand \leq CR) \text{ or } (j = rnbr\,(i)); \\ X_{jn}, & \text{if } (rand > CR) \text{ and } (j \neq rnbr\,(i)). \end{cases} \tag{5}$$

where:

- $n = 1, 2, \ldots, d$
- $rand \in (0, 1)$: is a uniform random number.
- $CR \in (0, 1)$: is the user-specified crossover constant.
- $rnbr\,(i)$: is a randomly chosen index selected from the range $(1, 2, \ldots, n)$.

5. Selection: The resulting trial solution replaces its predecessor, if it has higher fitness, otherwise the predecessor survives unchanged into the next iteration (Eq. 6), then a new population is created, and the process is repeated.

$$U_{jn}(t+1) = \begin{cases} U_i(t+1), & \text{if fitness}\,(U_i(t+1)) < \text{fitness}\,(X_i(t)); \\ X_i(t+1), & \text{otherwise.} \end{cases} \tag{6}$$

3 Benchmark Problem

In classification theory, a pattern p describes an entity or object through a set of features:

$$p = \{x_1, x_2, \ldots, x_n\}. \tag{7}$$

Features are measurements, attributes, or primitives derived from the patterns that may be useful for their characterization [15]. In classification problems, the number and type of features in the pattern are critical to the classification accuracy. However, as the number of features grows, the amount of data needed to generalize accurately grows exponentially. This phenomenon, generally called "the curse of dimensionality" introduces two major challenges when a limited number of data are available: the selection of a subset from the original set of features (primitives) that preserves the classification scheme and the generation of a set of new features from nonlinear transformations of the primitives [16–18].

4 Methodology

We want to test the null hypothesis that there is no difference in performance of the grammatical evolution's MPs Depth-First and πGE, on feature generation for classification datasets (shown in Table 1). A Grammatical Evolution-based method (FSC method) for selecting and generating features was proposed in [10]. In order to compare the performance of Depth-first and πGE, the idea of the FCS method was taken, but replacing the canonical GE for the Grammatical Differential Evolution. The best MP will be the one that offers the best fitness for each dataset, in this case de minimal fitness value.

The method steps are the following:

1. The dataset is split in two independent sets, train and test set.
2. The Differential Evolution's parameters are defined, shown in the following Table 2.
3. Chromosome initialization: each gene is initialized randomly in the range [0, 255].
4. The BNF grammar is defined:

$$\begin{aligned}
\mathbf{S} &::= \ <\text{expr}> (\mathbf{0}) \\
<\mathbf{expr}> &::= (<\text{expr}> <\text{op}> <\text{expr}>)(\mathbf{0}) \\
&\quad |<\text{func}> (<\text{expr}>)(\mathbf{1}) \\
&\quad |<\text{terminal}> (\mathbf{2}) \\
<\mathbf{op}> &::= +(\mathbf{0})|-(\mathbf{1})| * (\mathbf{2})|/(\mathbf{3}) \\
\mathbf{func}> &::= \sin(\mathbf{0})|\cos(\mathbf{1})|\exp(\mathbf{2})|\log(\mathbf{3}) \\
<\mathbf{terminal}> &::= <\text{xlist}> (\mathbf{0})|<\text{digitlist}> . <\text{digitlist}> (\mathbf{1}) \\
<\mathbf{xlist}> &::= x_1(\mathbf{0})|x_2(\mathbf{1})\ldots\ |x_n(\mathbf{n}-\mathbf{1}) \\
<\mathbf{digitlist}> &::= <\text{digit}> (\mathbf{0})|<\text{digit}> <\text{digitlist}> (\mathbf{1}) \\
<\mathbf{digit}> &::= 0(\mathbf{0})|\ 1(\mathbf{1})|2(\mathbf{2})\ldots|9(\mathbf{9})
\end{aligned}$$

Grammar 1. BNF Grammar proposed in [10] for feature generation.

5. Fitness evaluation. Each chromosome is evaluated as follows:

   (a) The chromosome is split into Requested Features (*RF*) equal parts. Each sub-chromosome is used to generate a new feature using a given MP. An

Table 1 Classification datasets selected from UCI machine learning repository [11]

Dataset	Instances	Attributes	Number of class
Wine	178	13	3
Glass	214	10	6
Liver disorders	345	6	2
Ionosphere	351	30	2
WD breast cancer	569	30	2
PIMA indians diabetes	768	8	2

Table 2 Differential Evolution's Parameters

Generations	Population	Chrom. length	RF	Mutation rate	Selection rate
200	500	120	3	0.05	0.5

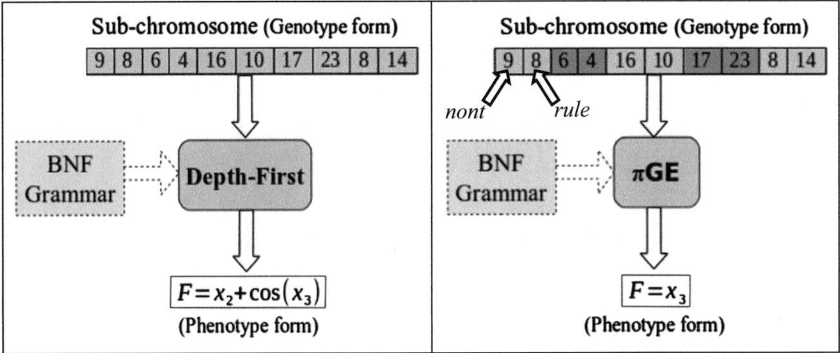

Fig. 2 An example of Genotype–Phenotype map using Depth-First and πGE Mapping Processes

example is shown in Fig. 2. As can be seen, the resulting phenotype using Depth-first and πGE MPs are different between approaches.

(b) The training and test datasets are transformed, according to the generated features proposed by GE. New sets (training and test) are generated.

(c) The new sets are used to train and test the classification system. The fitness function to minimize is the Classification Error:

$$E_r = \left(100 - \left(\frac{\sum_{i=1}^{Nfolds} \text{Classifier}\left[\text{acc}_i\right]}{Nfolds}\right)\right). \tag{8}$$

where:

- E_r: Is the error rate of classification on 0–100 scale.
- *Nfolds*: Is the cross-validation parameter, in this case equal to 10.
- Classifier $[\text{acc}_i]$: Is the accuracy computed for each $fold_i$. In this case we have used the *k*-Nearest Neighbor classifier, with *k* varying from 1 to 10.

5 Experimental Results

Table 3 shows the minimal classification error (the best fitness) for each dataset, obtained by the GE algorithm, using the different MP approaches. Each experiment was repeated 31 times.

Table 3 The best fitness for each dataset, using the different MPs

	Wine		Glass		Liver Disorders		Ionosphere		WDBC		PIMA	
k	DF	πGE	DF	πGE	DF	πGE	DF	πGE	DF	πGE	DF	πGE
1	**0**	1.39	14.95	14.95	**20.81**	22.54	4.55	**3.41**	1.40	1.40	20.83	20.83
2	**0**	2.78	**15.89**	16.82	24.86	**23.70**	4.55	**3.98**	1.75	**1.40**	20.83	20.83
3	**0**	1.39	15.89	**14.95**	20.81	**18.50**	3.41	3.41	1.40	1.40	19.53	19.53
4	**0**	2.08	18.69	**17.76**	24.28	**22.54**	**3.41**	3.98	1.75	**1.40**	**18.23**	19.53
5	**0**	2.08	18.69	**16.82**	21.39	**20.23**	4.55	**3.98**	1.40	1.40	17.45	**17.19**
6	**0**	2.78	19.63	**18.69**	21.97	**18.50**	4.55	**3.98**	1.75	**1.40**	**17.45**	18.49
7	**0**	2.08	19.63	**17.76**	20.81	**19.65**	3.98	**3.41**	1.40	**1.05**	18.49	**18.23**
8	**0**	2.78	**18.69**	19.63	21.39	**20.23**	5.11	**3.98**	1.75	**1.40**	**17.45**	17.71
9	**0**	2.08	21.50	**19.63**	21.39	**19.65**	4.55	**2.84**	1.40	**1.05**	18.23	**17.71**
10	**0**	1.39	**18.69**	20.56	21.97	**20.81**	4.55	**3.98**	1.75	**1.40**	18.49	**18.23**

Table 4 The *W-value* and *Critical_value* computed for the Wilcoxon tests

Dataset	W-value	Critical value
Wine	**0**	**8**
Glass	11	5
Liver disorders	**7**	**8**
Ionosphere	**1.5**	**5**
WDBC	**0**	**2**
PIMA	12	2

The values in bold typeface are the minimal results for each dataset and ***k*** value.

5.1 Statistical Significance of the Results

The Wilcoxon Rank-Sum test is used to determine if Depth-first and πGE are drawn from the same distribution [19]. We want to test the null hypothesis that there is no difference in performance (Eq. 9) when different approaches of MP are adopted by the GE to solve the problem presented in previous sections.

$$\begin{aligned} H_0 &: \text{Median}_{\text{differences}} = 0 \\ H_1 &: \text{Median}_{\text{differences}} \neq 0 \end{aligned} \tag{9}$$

The null hypothesis is that the samples are the same, and the test returns a ***W_value*** and a ***Critical_value***. For this article if $W\_value < Critical\_value$, will be taken to indicate that the samples are drawn from different distributions [19].

The Wilcoxon Rank-Sum test was applied to each dataset to compare the performance of both approaches of MP, with a significance level of 0.05 and a sample size of 10 (for each ***k*** value in the ***k-N*** *N*classifier), the results are shown in Table 4.

The values in bold typeface show that the corresponding differences are significant at $p \leq 0.05$.

6 Discussion

As can be seen, Table 3 shows that sometimes one approach is better than the other depending on the value of the parameter *k*. The results of the Wilcoxon Rank-Sum test (Table 4) shows that the ***W_value*** is lower than the ***Critical_value*** in the case of Wine, Liver Disorders, Ionosphere, and WDBC, it means that the result is significant at $p \leq 0.05$, in other words, there is statistical evidence to say that the results of the MPs are different to each other.

By comparing the fitness value obtained by the GE for each dataset and each ***k*** value, it can be said that for the Wine dataset the best MP was Depth-First; in the case of Liver Disorders, Ionosphere, and WDBC datasets, πGE was better.

7 Conclusions and Future Work

In this work, we presented a statistical comparison of the performance between two MP approaches in GE: Depth-First and πGE. The use case was the problem of the selection and generation of features applied to improve the classification performance over different datasets.

The results show that for different values of *k*, the performance with both approaches is similar. Sometimes one is better than the other and vice versa. Therefore, a strategy was sought to determine statistically if the approaches are drawn from different distributions. We have found significative evidence, by means of Wilcoxon Rank-Sum tests to say that, for these datasets and parameters, the performance of the GE is different when different MP approaches are used. In the case of the Wine dataset, Depth-first was the best; on other hand, in the case of Liver Disorders, Ionosphere, and WDBC datasets, πGE was better. For Glass and PIMA datasets, both approaches showed the same performance.

Future work will include a comparison with other approaches of MP such as Breadth-First and Random-Map. Also, we want to compare the results using different search engines.

References

1. Michael Q'Neill and Conor Ryan, *GRAMMATICAL EVOLUTION Evolutionary Automatic Programming in an Arbitrary Language*, 1st edition, Springer, 2003.
2. Peter F. Stadler and Bärbel M. R. Stadler, *Genotype-Phenotype Maps*, Biological Theory, Vol. 1, pp. 268-279, 2006.
3. David Fagan, Michael O'Neill Edgar Galvan-Lopez, Anthony Brabazon and Sean McGarraghy, *An Analysis of Genotype-Phenotype Maps in Grammatical Evolution*, Genetic Programming Volume 6021 of the series Lecture Notes in Computer Science, pp 62-73, Springer, 2010.
4. Eoin Murphy, Michael O'Neill, Edgar Galván-López and Anthony Brabazon, *Tree-Adjunct Grammatical Evolution*, 2010 IEEE Congress on Evolutionary Computation (CEC), 2010.
5. Anthony Brabazon, Michael O'Neill and Seán McGarraghy, *Natural Computing Algorithms*, Natural Computing Series, Springer, 2015.
6. Michael O'Neill, Anthony Brabazon, Miguel Nicolau, Sean Mc Garraghy, and Peter Keenan, *πGrammatical Evolution*, Genetic and Evolutionary Computation — GECCO 2004, Springer, 2004.
7. Paulo Urbano and Loukas Georgiou, *Improving Grammatical Evolution in Santa Fe Trail using Novelty Search*, Home advances in artificial life, ECAL 2013.
8. Loukas Georgiou and W. J. Teahan, *Grammatical Evolution and the Santa Fe TrailProblem*, ICEC 2010 - Proceedings of the International Conference on Evolutionary Computation, 2010.
9. Takuya Kuroda, Hiroto Iwasawa, Tewodros Awgichew and Eisuke Kita, *Application of Improved Grammatical Evolution to Santa Fe Trail Problems*, Natural Computing Volume 2 of the series Proceedings in Information and Communications Technology, pp. 218-225, 2010.

10. Dimitris Gavrilis, Ioannis G. Tsoulos and Evangelos Dermatas, *Selecting and constructing features using grammatical evolution*, Pattern Recognition Letters 29, 1358–1365, Elsevier, 2008.
11. M. Lichman, *UCI Machine Learning Repository* [http://archive.ics.uci.edu/ml]. Irvine, CA: University of California, School ofInformation and Computer Science, 2013.
12. Marco Aurelio Sotelo-Figueroa, Héctor José Puga Soberanes, Juan Martín Carpio, Héctor J. Fraire Huacuja, Laura Cruz Reyes and Jorge Alberto Soria-Alcaraz, *Improving the Bin Packing Heuristic through Grammatical Evolution Based on Swarm Intelligence*, Hindawi Publishing Corporation Mathematical Problems in Engineering Volume 2014, 2014.
13. I. Dempsey, M. O'Neill, and A. Brabazon, *Foundations in Grammatical Evolution for Dynamic Environments*, vol. 194, Springer, 2009.
14. M. O'Neill and A. Brabazon, *Grammatical differential evolution*, in Proceedings of the International Conference on Artificial Intelligence (ICAI'06), CSEA Press, Las Vegas, Nev, USA, 2006.
15. P. Devijver and J.Kittler, *Pattern recognition: A statistical approach*, Prentice/Hall International, 448 p, 1982.
16. Marques de Sá, *Pattern Recognition Concepts*, Methods and Applications, J.P, Springer, 2001.
17. Menahem Friedman and Abraham Kandel, *Introduction to pattern recognition: statistical, structural, neural, and fuzzy logic approaches*, volume 32 of Machine perception and artificial intelligence. Singapore River Edge, N.J. World Scientific, 1999.
18. Tatjana Pavlenko, *On feature selection, curse-of-dimensionality and error probability in discriminant analysis*, Journal of Statistical Planning and Inference Volume 115, Issue 2, pp. 565–584, Elsevier, 2003.
19. D.J. Sheskin, *Handbook of Parametric and Nonparametric Statistical Procedures*, 2nd ed, CRC, 2000.

Hyper-Parameter Tuning for Support Vector Machines by Estimation of Distribution Algorithms

Luis Carlos Padierna, Martín Carpio, Alfonso Rojas, Héctor Puga, Rosario Baltazar and Héctor Fraire

Abstract Hyper-parameter tuning for support vector machines has been widely studied in the past decade. A variety of metaheuristics, such as Genetic Algorithms and Particle Swarm Optimization have been considered to accomplish this task. Notably, exhaustive strategies such as Grid Search or Random Search continue to be implemented for hyper-parameter tuning and have recently shown results comparable to sophisticated metaheuristics. The main reason for the success of exhaustive techniques is due to the fact that only two or three parameters need to be adjusted when working with support vector machines. In this chapter, we analyze two Estimation Distribution Algorithms, the Univariate Marginal Distribution Algorithm and the Boltzmann Univariate Marginal Distribution Algorithm, to verify if these algorithms preserve the effectiveness of Random Search and at the same time make more efficient the process of finding the optimal hyper-parameters without increasing the complexity of Random Search.

Keywords Parameter tuning · Support vector machines · Hyper-parameters · Estimation of distribution algorithms · Pattern classification

L.C. Padierna · M. Carpio (✉) · A. Rojas · H. Puga · R. Baltazar
Tecnológico Nacional de México, Instituto Tecnológico de León, León, Mexico
e-mail: juanmartin.carpio@itleon.edu.mx

L.C. Padierna
e-mail: luiscarlos.padierna@itleon.edu.mx

A. Rojas
e-mail: alfonso.rojas@gmail.com

H. Puga
e-mail: pugahector@yahoo.com

R. Baltazar
e-mail: r.baltazar@ieee.org

H. Fraire
Tecnológico Nacional de México, Instituto Tecnológico de Cd. Madero,
Ciudad Madero, Mexico
e-mail: automatas2002@yahoo.com.mx

P. Melin et al. (eds.), *Nature-Inspired Design of Hybrid Intelligent Systems*,
Studies in Computational Intelligence 667, DOI 10.1007/978-3-319-47054-2_53

1 Introduction

Support Vector Machines (SVMs) are supervised learning models that can be formulated to solve problems from areas such as classification, regression, feature selection, density estimation, and so on [1].

SVMs were initially designed to handle classification problems with two classes under the assumption that these classes could be linearly separated; as this case is unusual in real problems two mechanisms were added to the initial SVM design in order to deal with nonlinearly separable classes: a kernel function and a penalty factor that led to the development of the standard version of an SVM named C-SVM [2] (*cf.* Sect. 2.1 for a brief explanation).

The penalty factor C and the kernel parameters are called hyper-parameters in the field of machine learning because they are in the highest level of the SVM implementation; their main characteristic is that hyper-parameters define the set of available decision functions [3]. From this dependency, we can observe that obtaining the optimal hyper-parameters leads to the best decision function of an SVM model.

The problem of hyper-parameter tuning consists in finding a set of values that maximize some performance criteria such as the accuracy index and the proportion of support vectors in the case of Support Vector Classification (SVC). Hyper-parameter tuning for SVC has been widely studied in the past decade from diverse approaches.

Chapelle, Vapnik, and others developed a gradient descent technique to optimize hyper-parameters considering a different kernel parameter for each attribute in the input vector [4]. Friedrich and Igel applied the Covariance Matrix Adaptation Evolution Strategy (CMA-ES) to optimize non-differentiable kernel functions considering the reduction of the number of support vectors [5].

With the objective of avoiding local optima in the optimization of hyper-parameters, many nondeterministic and stochastic methods have been employed. Among these methods there can be found a Bayesian approach [6], Genetic Algorithms (GA) [7] and Particle Swarm Optimization (PSO) [8].

One interesting study ([9]) argues that the Random Search (RS) technique is as effective as more sophisticated methods when optimizing hyper-parameters; experiments in this study were carried out on Neural Networks. A comparison among Grid Search, RS, PSO, and a particular EDA called copula-EDA to optimize hyper-parameters in SVC tested the effectiveness of RS on several datasets concluding that there was no statistical difference among these methods [10].

In this chapter, an analysis of two EDAs is presented to verify if these algorithms preserve the effectiveness of RS and at the same time make more efficient the process of finding the optimal hyper-parameters for SVC without significantly increasing the complexity of RS.

The rest of this chapter is structured as follows. Section 2 provides theoretical background about SVMs and EDAs. The test instances and experimental methodology are explained in Sect. 3. Section 4 presents the results of the experiments and offers relevant findings. Finally, conclusions are drawn and future work is proposed in Sect. 5.

2 Theoretical Background

This section provides elemental definitions of SVMs and EDAs, a brief description of each technique used in this work and references to further material are recommended to the reader interested in getting a deeper understanding on the basics of SVMs and EDAs.

2.1 Support Vector Machines

Given a set of training data: $\{x_i,\ y_i\}_{i=1}^{m}$, where $x_i \in R^d$ is the ith input vector and $y_i \in \{+1,\ -1\}$ are the class labels, an SVM attempts to separate the data by finding the optimal hyper plane, defined as

$$D\mathbf{x} = \mathbf{w}^T\mathbf{x} + b \tag{1}$$

where the weight vector w^T is given by a linear combination of a relatively few data points called support vectors.

When the data is not linearly separable, the hyper plane is obtained by introducing a set of slack variables, $\{\xi_i\}_{i=1}^{m}$, a penalty factor C and by solving the following quadratic programming problem:

$$\begin{gathered}\text{Min}\left(\frac{1}{2}\mathbf{w}^T\mathbf{w} + C\sum\nolimits_{i=1}^{m}\xi_i\right) \\ s.t. \quad y_i(\mathbf{w}^T\mathbf{x}_i + \mathrm{b}) \geq 1 - \xi_i,\ \xi_i > 0 \quad \text{for} \quad i = 1,\ldots,m\end{gathered} \tag{2}$$

Introducing the nonnegative Lagrange multipliers α and following the Karush–Kuhn–Tucker conditions, the problem (2) can be proved to be equivalent to the following dual problem [11]:

$$\begin{gathered}\text{Max } L(\alpha) = \sum_{i=1}^{m}\alpha_{\mathrm{i}} - \frac{1}{2}\sum_{i=1}^{m}\sum_{j=1}^{m} y_i y_j \alpha_i \alpha_j K(\mathbf{x}_i, \mathbf{x}_j) \\ s.t. \quad C \geq \alpha_i \geq 0\ \forall i = 1,\ldots,m \quad \text{and} \quad \sum_{\mathrm{i=1}}^{\mathrm{m}}\alpha_i y_i = 0\end{gathered} \tag{3}$$

where the function $K(x, z)$ defined on $R^d \times R^d$ is called a kernel if there exists a map ϕ from the space R^d to the Hilbert space, $\phi : R^d \rightarrow \mathcal{H}$ such that $K(x, z) = \langle \phi(x), \phi(z) \rangle$, [12]. Kernel functions that have been widely used since the beginning of the SVM theory are: the Radial Basis Function (RBF or Gaussian) kernel, $K(\boldsymbol{x}, z) = e^{-\gamma\|\boldsymbol{x}-z\|^2}$ and the Linear kernel, $K(x, z) = x^T z$.

In Eq. (3) it can be observed the role that hyper-parameters C and γ play in the SVM optimization problem when the function $K(x, z)$ is the RBF. The next subsection describes the algorithms to tune these hyper-parameters.

2.2 *Estimation Distribution Algorithms*

Estimation distribution algorithms (EDAs) are stochastic optimization techniques that explore the space of potential solutions by building and sampling explicit probabilistic models of promising candidate solutions; the general procedure of an EDA is the following [13]:

```
Algorithm 1 General EDA
1   Initialize a generation counter g ← 0
2   Generate initial population P(0)
3   WHILE (stopping criteria is not fulfilled) DO
4           Select population of promising solutions S(g) from P(g)
5           Build a probabilistic model M(g) from S(g)
6           Sample M(g) to generate new candidate solutions O(g)
7           Incorporate O(g) into P(g)
8       g ← g + 1
9   END WHILE
```

Depending on the specific probabilistic model $M(g)$ in line number 5, the EDA algorithm takes a particular name. Two of these particular cases of probabilistic models are studied in this work, the UMDA and BUMDA.

2.2.1 Univariate Marginal Distribution Algorithm (UMDA)

This EDA uses Gaussian distributions to create the next generation and is denoted as UMDA_c^G where G stands for Gaussian and C for continuous; this may be the simplest EDA and its algorithm is the following [14]:

Algorithm 2 $UMDA_c^G$

Initialize parameters: dimension of the individual (d), population size (N), range of the search space $[min, max]$
Initialize a population of candidate solutions $C = \{x \in \mathbb{R}^d\}_{i=1}^N$, $x \sim U[min, max]$
WHILE NOT(termination criteria is fulfilled) DO
Select M individuals from C according to fitness, where $M < N$.
Index the M selected individuals as $\{x_i\}$, $i \in [1, M]$
Build a Gaussian model by taking:
  $\mu_k \leftarrow \frac{1}{M}\sum_{j=1}^{M} x_j(k)$, where k is the k-th component in vector x_j
  $\sigma_k \leftarrow \left[\frac{1}{M-1}\sum_{j=1}^{M}\left(x_j(k) - \mu_k\right)^2\right]^{\frac{1}{2}}$
  FOR $i = 1$ to N
    FOR $k = 1$ to d
      $x_i(k) \leftarrow Normal(\mu_k, \sigma_k)$
    NEXT variable
  NEXT individual
END WHILE

The above algorithm basically uses two user-defined parameters, the population size (*N*) and the proportion of individuals with higher fitness (*M*) from which the Gaussian model is built. A Random Search algorithm requires just the parameter *N* and can be seen as a special case of the UMDA_C^G where the Gaussian model is replaced with a random function.

The reason to employ techniques with the minimum number of parameters (one or two) is fundamental to maintaining the calibration of the optimization strategy simpler than the SVM hyper-parameter tuning problem (where two, three, or four parameters are required). As an immediate alternative to reduce the number of user-defined parameters, the BUMDA algorithm is considered.

2.2.2 Boltzmann Univariate Marginal Distribution Algorithm (BUMDA)

BUMDA [15] requires just the same parameter as Random Search, the population size. As additional advantages to UMDA one can point out the fact that BUMDA guarantees convergence by adding elitism and a truncation method that automatically chooses the proportion of best individuals to build the probabilistic model. Furthermore, the BUMDA makes a Normal-Gaussian approximation to the Boltzmann PDF by computing the Kullback–Leibler divergence.

These slight modifications have shown significant improvement in convergence when applied to benchmark optimization problems. The BUMDA algorithms are the following (a MATLAB implementation of the Algorithms 3 and 4 can be found their author's website[1]).

Algorithm 3 $BUMDA$

1. Initialize parameters: dimension of the individual (d), population size (N), range of the search space $[min, max]$, minimum variance allowed $minvar$.
2. Initialize a population of candidates $C = \{x \in \mathbb{R}^d\}_{i=1}^{N}, x \sim U[min, max]$.
3. Set $t \leftarrow 0$
3. WHILE $v > minvar$ AND NOT(other termination criteria is fulfilled) DO
4. $t \leftarrow t+1$
5. Evaluate and truncate the population according to Algorithm 4.
6. Compute the approximation to μ and v (for all dimensions) by using the selected set of size M as follows:
7. $$\mu \approx \frac{\sum_{i=1}^{M} x_i \bar{g}(x_i)}{\sum_{i=1}^{M} \bar{g}(x_i)} \qquad v \approx \frac{\sum_{i=1}^{M} \bar{g}(x_i)(x_i - \mu)^2}{1 + \sum_{i=1}^{M} \bar{g}(x_i)}$$
8. Where $\bar{g}(x_i) = g(x) - g(x_M) + \epsilon$, and ϵ is a small number to avoid division by 0.
9. The individuals can be sorted to simplify the computation, and $g(x_M)$ is the minimum objective value of the selected individuals.
10 Generate $M-1$ individuals from the new model $Normal(\mu, v)$ and insert the elite individual.
11 END WHILE
12 Return the elite individual as the best approximation to the optimum.

Algorithm 4 Truncation Method for BUMDA

1. For the initial generation $t \leftarrow 0$, let be $g(x_i, 0)$ for $i = 1, \ldots, N$ the objective values of the initial population. Define: $\theta_0 = \min g(x_i, 0)$.
2. FOR $t > 0$, set
   $\theta_t = \max\left(\theta_{t-1}, \min(g(x_i, t) \mid g(x_i, t) \geq \theta_{t-1}\right)$
4. If for the decreasing sorted individuals $g(x_{N/2}) \geq \theta_t$, set $\theta_t = g(x_{N/2})$.
5. Truncate the population such that $g(x_s, t) \geq \theta_t$. Where x_s are all the individuals whose objective values are equal or greater than θ_t.
6. END FOR
7. RETURN the indexes of the M individuals that remain after truncation.

---

[1] www.cimat.mx/~ivvan/bumda.m.

3 Experimental Methodology

This section describes the test instances and the configuration defined to evaluate the performance of EDAs for tuning the hyper-parameters of SVC.

A total of 11 datasets were employed to verify the convenience of using EDAs instead of Random Search. These datasets correspond to classification problems of two classes. Table 1 summarizes relevant information about these datasets.

For each of these datasets the process depicted in Fig. 1 was followed. In this process the first step consists in dividing the dataset into two parts, one for training and the other one for testing; the proportion of elements in each part was 80 and

Table 1 Datasets containing classification problems with two classes

No	Problem	Instances	Attributes
1	Breast	683	10
2	Diabetes	768	8
3	Fourclass	862	2
4	Haberman	306	3
5	Heart	270	13
6	Ionosphere	351	34
7	Liver	345	6
8	Monks-1	124	6
9	Monks-2	169	6
10	Sonar	208	60
11	Wpbc	194	33

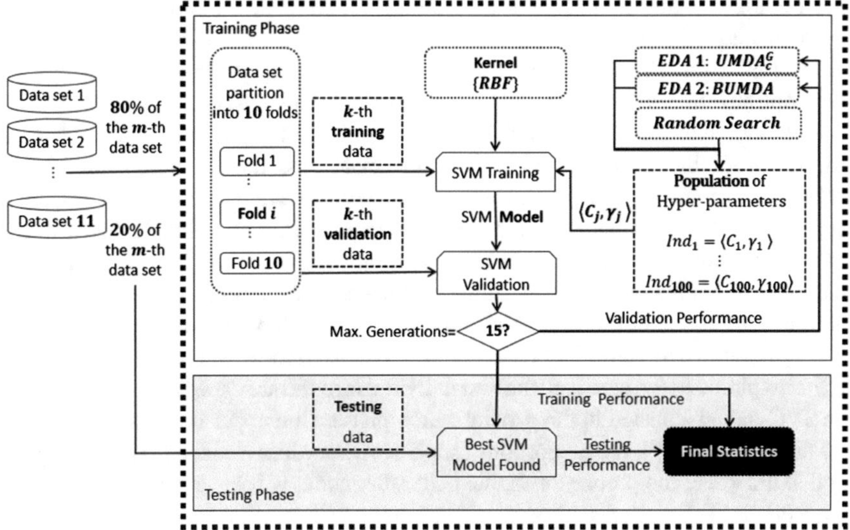

Fig. 1 Methodology followed for hyper-parameter tuning of SVC

20 % respectively. A tenfold cross validation was then applied to the training set to get a validation proportion of around 8 % of the original data.

The Radial Basis Function was the kernel selected for our experiments. The two EDAs UMDA_C^G and BUMDA and the Random Search method were initialized with the same population of 100 individuals. Each individual consists of two hyper-parameters, the penalty factor C and the parameter γ for the RBF kernel. An SVM is trained using the kth training data, and the jth individual to build an SVM model. The fitness function is calculated as the accuracy obtained by the SVM model on the kth validation data

The training phase of the hyper-parameter tuning process finishes when one of the two stopping criteria is met: the maximum number of generations is equal to 15 or the minimum variance on the fitness in one generation is reached.

The outcome from the training phase is the best model found in the search process. Thus a model is selected for each one of the algorithms UMDA_C^G, BUMDA, and RS. These models are then evaluated on the data that was not used for training, called testing data. In the end, each algorithm will produce three performance measures, a training accuracy index, a testing accuracy index, and the proportion of support vectors that were used to build the best model. This methodology was replicated 35 times in order to get the final statistics summarized in Tables 2 and 3

4 Results

This section summarizes the findings in the comparison of EDAs against Random Search. Table 2 presents the statistics on 35 replications of the experimental methodology, taking into account the training accuracy index and testing accuracy index. In this table it can be noticed that in 7 out of 11 datasets the statistics are quite similar; fourclass and Haberman datasets are in this group and show how the three optimization algorithms were capable to find equivalent individuals.

Notably, BUMDA method reached higher indexes in 4 out of 11 datasets considering both the training and the testing measures; these datasets are: heart, ionosphere, monks-1, and sonar. Computing a general average from these statistics, again BUMDA obtained superior indexes. This finding is valuable because the three algorithms started from exactly the same configuration (initial population, training and testing sets, and the same pseudo-random number generator) leaving the only difference in the outcome to the natural development of the algorithms.

Results shown in Table 2 were favorable to BUMDA; however, these results can be further explored to see how well the best individuals found are expected to behave when the SVC model is tested in the generalization phase. One upper bound for the generalization expectancy is the proportion of support vectors that were used to build the SVC model; the lower the proportion is, the better the model will behave in generalization phase [4]. Table 3 presents the proportion of support vectors that were employed for the models built with the best individuals found in the optimization process.

Table 2 Performance on training and testing indexes

No	Dataset	Random Search				UMDA				BUMDA			
		Train		Test		Train		Test		Train		Test	
		Avg	Std. Dev.	Avg	Std. Dev.	Avg	Std. Dev.	Avg	Std. Dev.	Avg	Std. Dev.	Avg	Std. Dev.
1	Breast	95.1	0.16	97	0.44	95.1	0.18	97.1	0.49	95.6	1.67	97.2	1.56
2	Diabetes	72	0.99	71.9	3.46	71.1	1.31	70.5	2.52	76.7	3.18	73.6	4.01
3	Fourclass	100	0	100	0	100	0.07	100	0	100	0	100	0
4	Haberman	76.5	0.21	64.5	0	76.3	0.12	64.5	0	76.3	0.07	64.5	0
5	**Heart**	73.5	6.66	71	3	67.9	8.23	68.8	4.86	**81.9**	**7.79**	**85**	**8.54**
6	**Ionosphere**	86.6	10.6	86.8	8.92	79.3	13	79.5	12.7	**92.1**	**8.44**	**91.4**	**8.24**
7	Liver	71.6	3.21	70.2	2.44	71	2.66	70.2	2.49	70.2	2.77	68.3	2.32
8	**Monks-1**	61.4	7.7	37.1	15.7	60.7	7.42	34.5	13.4	**68.4**	**14.8**	**48.2**	**22.4**
9	Monks-2	65.7	5	65	10.7	65.4	4.66	64.9	10.6	67	6.88	66.4	13
10	**Sonar**	58.9	10.2	56.9	9.66	58.6	9.97	56.5	8.99	**64.4**	**15.4**	**66.3**	**19.5**
11	Wpbc	77.8	0.45	69	0.96	77.8	0.34	69.2	0.43	78.2	1.23	69.2	0.43
Global Avg		76.3	4.1	71.8	5.03	74.8	4.36	70.5	5.13	**79.2**	**5.66**	**75.5**	**7.27**

Table 3 Proportion of support vectors from the training data

No	Dataset	Random Search		UMDA		BUMDA	
		Avg	Std. Dev.	Avg	Std. Dev.	Avg	Std. Dev.
1	Breast	0.43	0.02	0.41	0.06	0.24	0.18
2	Diabetes	0.56	0.13	0.63	0.16	0.57	0.13
3	Fourclass	0.11	0.02	0.11	0.05	0.11	0.02
4	Haberman	0.96	0.21	0.96	0	0.96	0
5	**Heart**	0.88	0.11	0.92	0.13	**0.49**	**0.19**
6	**Ionosphere**	0.65	0.27	0.83	0.15	**0.53**	**0.16**
7	Liver	0.72	0.16	0.69	0.15	0.84	0.17
8	**Monks-1**	0.95	0.12	0.96	0.11	**0.75**	**0.28**
9	Monks-2	0.96	0.09	0.96	0.09	0.92	0.14
10	**Sonar**	0.99	0.04	0.99	0.02	**0.89**	**0.14**
11	Wpbc	0.96	0.09	0.97	0.09	0.95	0.13
General average		0.74	0.11	0.77	0.09	**0.66**	**0.14**

Results in Table 3 are again favorable to BUMDA method. With the only exception of liver, the rest of the datasets were modeled with lower or equal number of support vectors. Consistent with results in Table 2, the four datasets that reached higher training and testing accuracies also got lower proportion of support vectors which is a strong evidence that the individuals explored by the BUMDA are preferable to those explored by Random Search.

Finally, in order to get a closer insight in the evolution of the optimization algorithms, the distribution of all the explored individuals are shown in Figs. 2a, b, and 3a, b. One representative of each group of datasets was selected for analysis. The first group consists of those datasets that reached similar or even equal performance and its representative is shown in Fig. 2a, b. The second group consists of datasets for which BUMDA got better individuals; its representative is depicted in Fig. 3a, b.

The fourclass is the representative dataset of the group where the three optimization algorithms reached equivalent results. Figure 2a shows the pattern found on this group where it can be observed that EDAs make a more efficient search because they found a higher proportion of better individuals using less function calls. This result is very important since the algorithmic complexity of training an SVM is of $O(n^3)$, where n is the size of the dataset.

Figure 2b complements the information in Fig. 2a by showing while the accuracy index increases, the proportion of support vector decreases simultaneously; which is a desirable property for SVC models. In the particular case of BUMDA, the highest frequency of individuals corresponds to the better individuals.

The group of datasets that includes: heart, ionosphere, monks-1, and sonar, were better optimized by the BUMDA algorithm. Figure 3a illustrates the fact that this method managed to escape from local optima and explored a bigger amount of individuals. It is relevant to emphasize that the three methods were initialized with

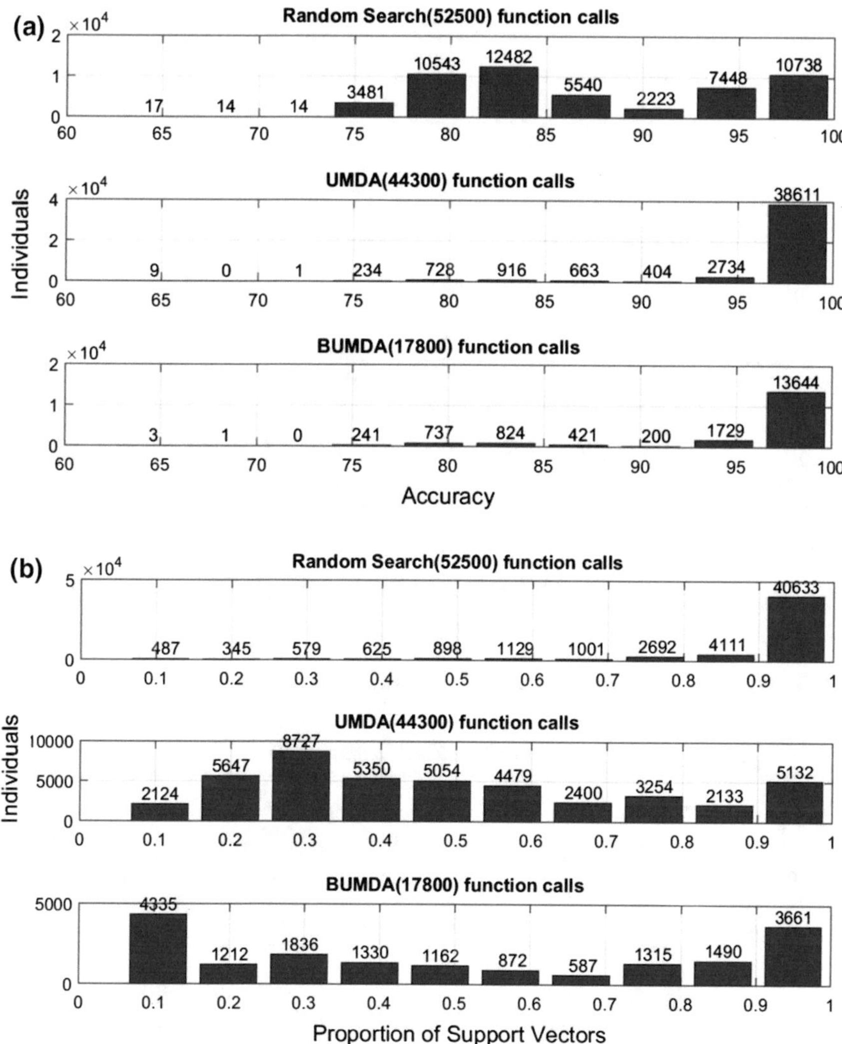

Fig. 2 **a** Distribution of accuracies reached by the best hyper-parameters on the fourclass dataset, **b** Distribution of the proportion of support vectors reached by the best hyper-parameters on the fourclass dataset

the same population and were provided the same data, leaving the algorithm as the only variable in the experimental methodology.

Figure 3b complements the Fig. 3a illustrating how BUMDA achieved the highest frequency of individuals with the better proportion of support vectors. It also can be noticed that the number of function calls is bigger in BUMDA than in the other algorithms because the former avoids premature convergence to local optima. The same patter was found on the ionosphere, monks-1 and sonar datasets.

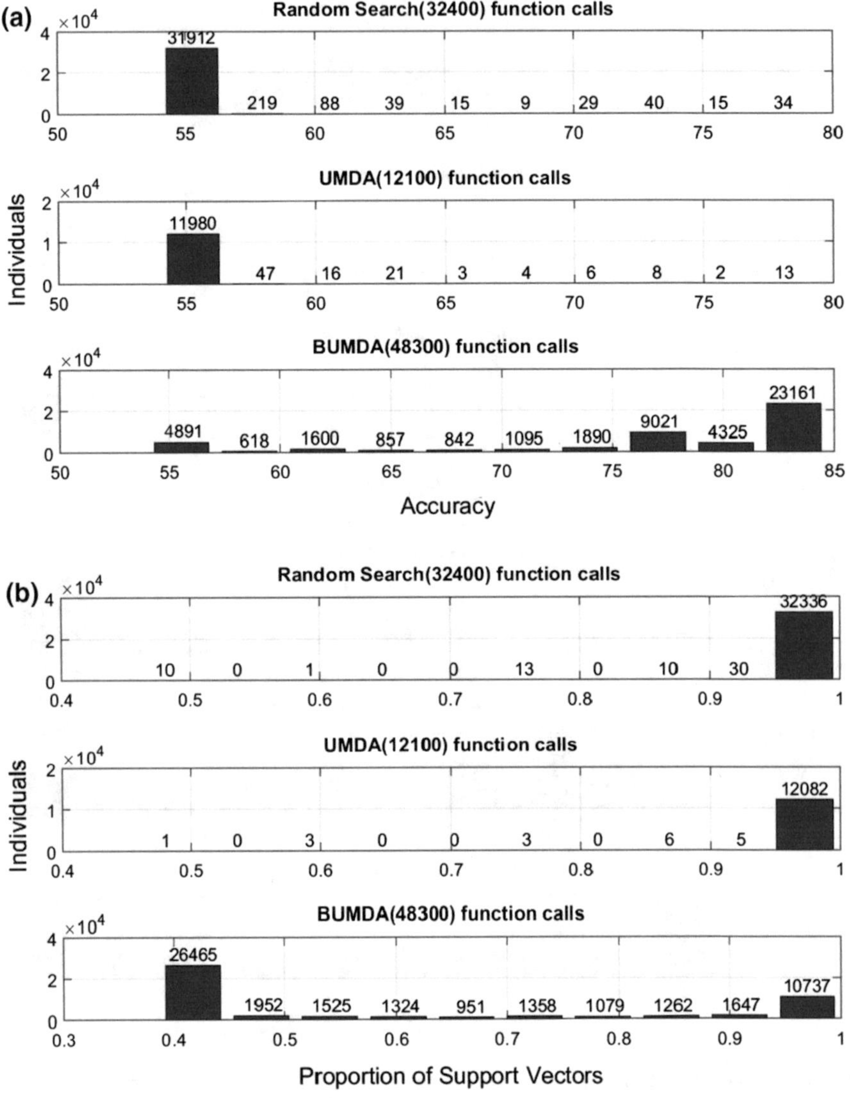

Fig. 3 **a** Distribution of accuracies reached by the best hyper-parameters on the heart dataset, **b** Distribution of the proportion of support vectors reached by the best hyper-parameters on the heart dataset

5 Conclusions and Future Work

The main conclusion is that EDAs, and particularly BUMDA, was capable of reaching better results than Random Search with no more added algorithmic complexity than a sorting and the computation of simple statistics.

In four out of eleven problems, BUMDA offers the advantage of exploring better individuals. In these cases BUMDA managed to find good regions; meanwhile, UMDA_C^G remained stuck into and prematurely converged to local optima. On the other hand, in six out of eleven datasets BUMDA reached similar or even equivalent results than the other algorithms; however, BUMDA explored a richer variety of individuals and converged faster to the global optima using considerably fewer number of function calls than RS and UMDA_C^G. For all these reasons the authors propose the use of BUMDA as a better option for hyper-parameter tuning of SVC than Random Search.

One exception is the case of the liver dataset, that departed from the pattern marked by the other results and we currently do not have a clear insight to why this happened, so further analysis of this dataset is left for future work. Although our results provide a strong evidence of the benefits that EDAs offer over the traditional RS, a more comprehensive analysis (considering test hypothesis) is proposed as the natural next step to the results presented so far. In the near future BUMDA will be proposed as an efficient alternative to solve the hyper-parameter tuning in SVC and will be compared with more sophisticated metaheuristics like PSO and GA.

Acknowledgments Luis Carlos Padierna and Alfonso Rojas wish to acknowledge the financial support of the Consejo Nacional de Ciencia y Tecnología (CONACYT grants 375524 and CATEDRAS-2598). The authors also thank Dr. Ivann Valdez from the Center of Research in Mathematics for his assistance and sharing his BUMDA-code.

References

1. N. Christianini and J. Shawe-Taylor. An Introduction to SVM and other Kernel Based Methods. Cambridge, U.K.: Cambridge University Press, 2000.
2. V. Vapnik, Statistical Learning Theory, New York: John Wiley and Sons, 1998.
3. M. Kanevski, V. Timonin and A. Pozdnukhov, Machine Learning for Spatial Environmental Data: theory, applications and software, CRC Press, 2009.
4. O. Chapelle, V. Vapnik, O. Bousquet and S. Mukherjee, *Choosing Multiple Parameters for Support Vector Machines* Machine Learning, vol. 46 (1-3), pp. 131-159, 2002.
5. F. Friedrichs and C. Igel. Evolutionary tuning of multiple SVM parameters. Neurocomputing, vol 64 pp. 107-117, 2005.
6. C. Gold, A. Holub y P. Sollich. Bayesian approach to feature selection and parameter tuning for support vector machine classifiers. Neural Networks, vol 18 (5) pp. 693-701, 2005.
7. C.-L. Huang and W. Chieh-Jen. A GA-based feature selection and parameters optimization for support vector machines. Expert System with Applications, vol 31(2) pp. 231-240, 2006.
8. S.-W. Lin, K.-C. Ying y S.-C. L. Z.-J. Chen. Particle Swarm Optimization for Parameter Determination and Feature Selection of Support Vector Machines. Expert System with Applications, vol 35(4) pp. 1817-1824, 2008.
9. J. Bergstra and Y. Bengio. Random Search for Hyper-Parameter Optimization. Journal of Machine Learning Research, 13(1) pp. 281-305, 2012.
10. R. Mantovani, A. Rossi, J. Vanschoren and B. d.-C. A. Bischl. Effectiveness of Random Search in SVM hyper-parameter tuning. The International Joint Conference on Neural Networks (IJCNN), 2015.
11. A. Shigeo, Support Vector Machines for Pattern Classification, New York: Springer, 2010.

12. N. Deng, Y. Tian and C. Zhang, Support Vector Machines, Boca Raton: CRC Press, 2013.
13. M. Hauschild and M. Pelikan. An introduction and survey of estimation of distribution algorithms. Swarm and Evolutionary Computation, vol 1(3) pp. 111-128, 2011.
14. D. Simon, Evolutionary Optimization Algorithms: Biologically Inspired and Population-Based Approaches to Computer Intelligence, Hoboken: John Wiley and Sons, 2013.
15. S. I. Valdez, A. Hernández and S. Botello. A Boltzmann based estimation of distribution algorithm. Information Sciences, vol 236 pp. 126-137, 2013.

Viral Analysis on Virtual Communities: A Comparative of Tweet Measurement Systems

Daniel Azpeitia, Alberto Ochoa-Zezzatti and Judith Cavazos

Abstract This study shows the results of a comparison of different measurement systems that help measure tweets virality within virtual communities. Likewise, the history of this type of virtual social networks in the context of marketing are essential to creating effective proposals for the study of computer systems, software developers and marketing professionals and advertising are presented. Ultimately, a proposal for a graphic tweets measurement system is presented.

Keywords Viral marketing · Digital social networks · Social networking

1 Introduction

Virtual social networks were named as such when compared with the mathematical graph theory initially developed by Leonhard Euler in 1739 and used in social studies by Cartwright and Zander in 1953 in the field of sociology. Network theory originated from Kutt Lewin, who in 1944 set a field of relations in which the perception and behavior of a group are part of a social space formed by the group [1].

Cartwright, Zander, Harary, Norman, Bavelas and Festinger studied these structures using graph theory, working in small groups, analyzing their social structure, and establishing how individual behavior was affected in these structures (Fig. 1). Then at Harvard in the 1930s and 40s, anthropological functionalism was

D. Azpeitia · A. Ochoa-Zezzatti (✉)
Juarez City University, Ciudad Juarez, Mexico
e-mail: alberto.ochoa@uacj.mx

J. Cavazos
Universidad Popular Autónoma del Estado de Puebla, Puebla, Mexico

P. Melin et al. (eds.), *Nature-Inspired Design of Hybrid Intelligent Systems*,
Studies in Computational Intelligence 667, DOI 10.1007/978-3-319-47054-2_54

Fig. 1 Social structures

developed, in which Warner and Mayo researched the structure of the subgroups in their work in an electrical factory in the city of Chicago [2].

From 1997 to 2001, the communities that allowed users to share information, create sections called profiles containing personal information, professional or just dummy information which could detect tastes and preferences began to significantly grow [3]. Thus, new sites such as AsianAvenue, BlackPlanet, MiGente, LiveJournal and Cyworld, which usually catered to specific geographic niches.

In subsequent years, an accelerated pace continued perfecting these networks, mainly based on unexplored areas meet virtually; Ryze.com helped connect people with businesses, Friendster allow customization and showed a more professional interface attracting consumer niches early [4].

2 Virtual Social Networks Worldwide

Globally, new era for social networks began. With the start of Orkut, a Google creation, began a massive invasion of the country of Brazil. Windows Live Spaces, better known as MSN Spaces came into its own in the USA and the creation of MySpace in Santa Monica, California, began a globalization of the use of social networks and innovation to attract users to the [5] network strategies. Following the global phenomenon, the Dutch adopted Hyves, Polish adopted Grono, Hi5 saw conquests in Latin America, South America and Europe, and Bebo dominated New Zealand and Australia (Fig. 2).

Meanwhile in 2004, Facebook began with a very specific niche to connect students from Harvard University in the USA [6]. From that moment, a stampede Facebook growth began by connecting several universities worldwide to finally open their services free of charge to anyone who wishes to open an account.

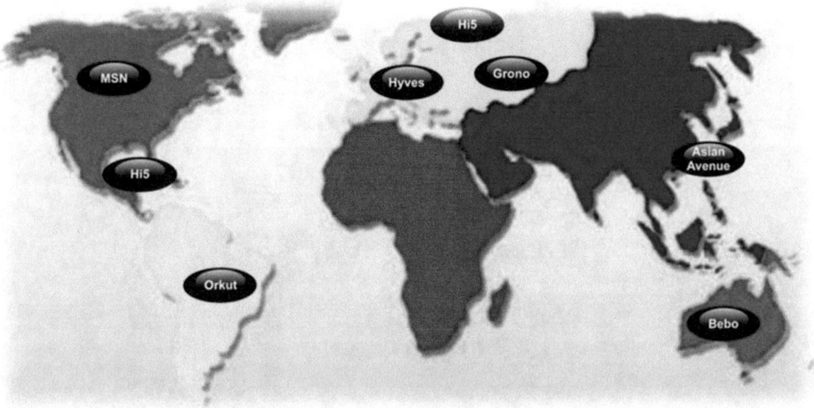

Fig. 2 Growth of virtual social networks in the world (2004) [Prepared by the authors]

3 Comparison of Measurement Systems

Knowing the growth and importance of online social networks, four different measurement systems are presented within virtual social networks, which then start with a proposal for a system of measurement using a graphic basis of the above systems and databases. Including those presented in the following study are: SocialRank, Klout, databases Google Trends and Google Alerts, and finally, Sea of Tweets.

First, SocialRank allows monitoring of precisely who are the followers of a given user within the social network Tweeter, resulting in accurate statistics on the number of followers, the number of people following the user in different time ranges, either by day, week or month. Furthermore, the application SocialRank helps to know the number of retweets, to know whether there has been a response to these and to observe retweets that have been marked as favorites (Fig. 3).

On the other hand, a Klout score is estimated according to the growth that has been shown within the social network with the use of different tools that help decision-making according to the influence obtained through tweets. Likewise, features and custom options allow detailed observation of best results by time and day (Fig. 4).

On the other hand, Klout, by inserting code into the site, allows for real-time critical data, which is known as a user arrives at a site. Information can be obtained if users arrive via a virtual social network, how long they lasted into the site before returning to the social network and provide some tools that can be applied to obtain more visitors. Klout also provides data from earlier dates, date ranges, keywords, and studies how you are affected by keywords positioned within the search engines. It also provides data on competition in positioning of key words.

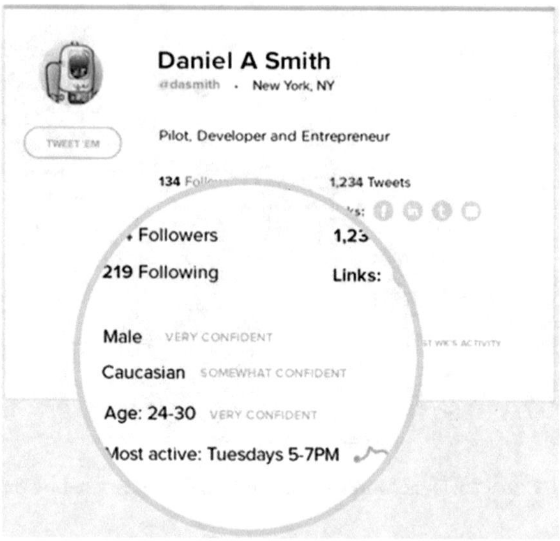

Fig. 3 Screenshot of SocialRank

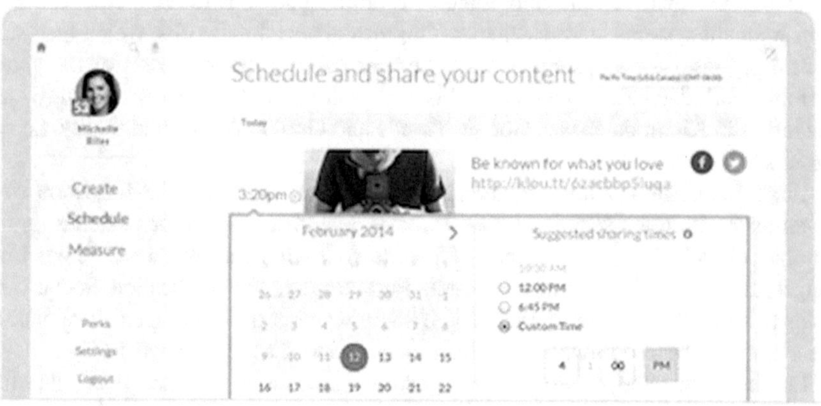

Fig. 4 Screen measurement system Klout

Twitter accounts and Facebook graphics showing the scope of each of the messages. It also allows us to observe the behavior of the messages for hours and the number of times the message appeared on screen. These data can be weekly, monthly, global and personal. Specifically, to track the messages within virtual social networks, there are tools that provide information with a high degree of reliability for measuring performance. To observe the behavior of a message and to follow through user profiles, Postling provides information, providing comparative charts on a message and the response that generated the message, so it is useful in

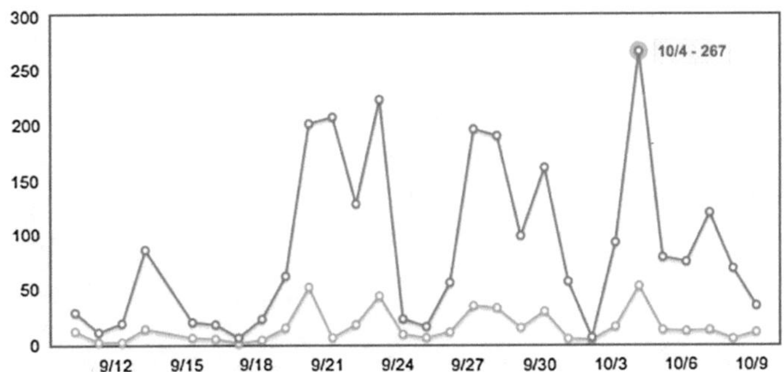

Fig. 5 Comparison between a message and commentary generated

determining the success of an advertising campaign on social networking. Figure 5 shows the comparison between a message (in blue) within the social network and the number of comments generated in this message (in red).

On the other hand, allows obtaining relevant statistics YouTube playback of the videos into your social network. Allows chronological identification of visitors to a particular video, provides the date that video was exhibited for the first time, as well as behavior in terms of visits over time (Fig. 9). Likewise, Youtube displays the number of reproductions and a list of keywords for the video, which was citadel, number of comments generated, the geographical areas of reproduction, as well as general user data including age, sex, origin, and another.

4 Proposed Measurement System

Finally, a measuring system tweets list showing the performance of five different users, performs a comparative graph simulating said users intends wave manner. The graph is displayed together (Fig. 6) or separately (Fig. 7) showing by a mouse over generating comments in the retweets high.

The system, through waves, allows the user to monitor five different Twitter users and so it makes a comparative graph of tweets that generate virality, offering the user a nice perspective that allows decision-making in different marketing campaigns. On the other hand, flexible shell allows the user to change the interface quickly in order to return back to the individual interface.

To make the system proposed, JavaScript is used. With the use of databases, this can utilize user-friendly programming to help to obtain user-friendly graphics. Figure 8 presents the JavaScript code for this graphics system proposed for measuring the virality of tweets through their intensity on social networking.

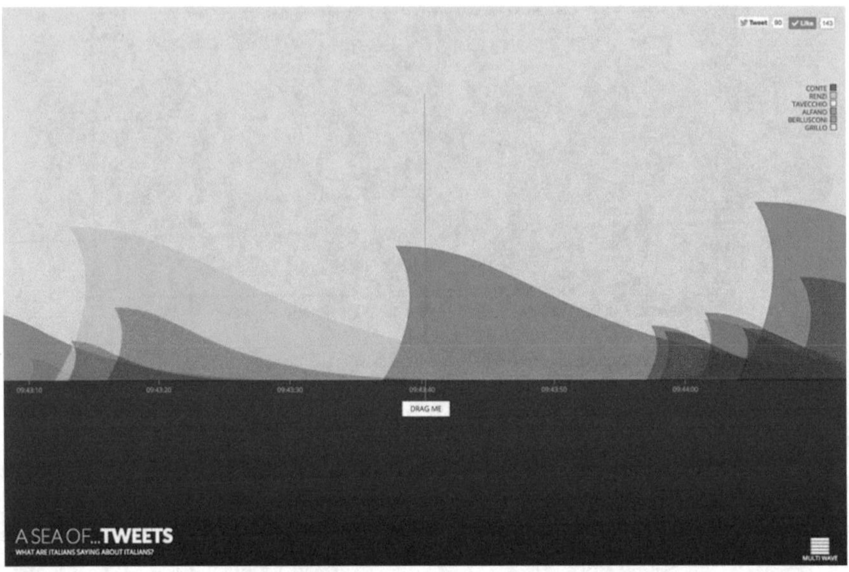

Fig. 6 Measurement system virality through wave

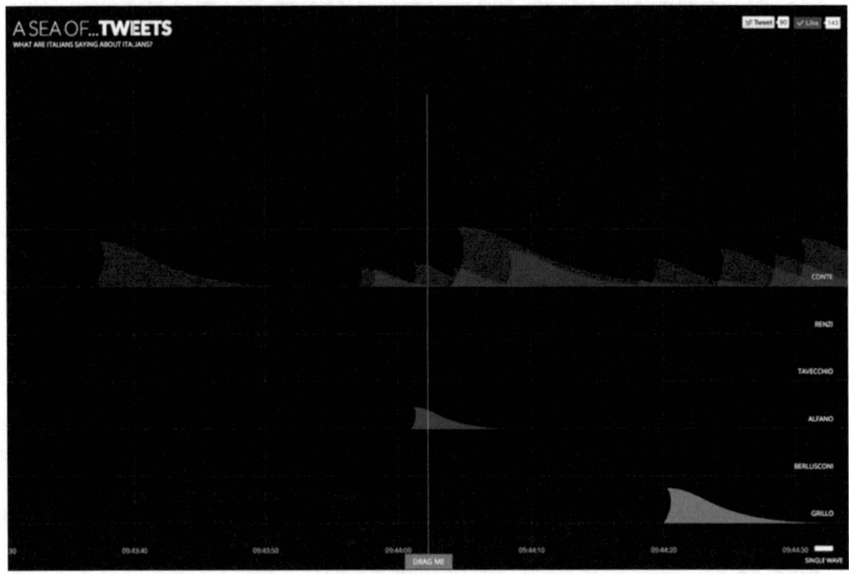

Fig. 7 Simultaneous measurement system virality

```
}

function showTweets(tweets){
        var same=true,
                ids=tweets.map(function(d){
                        return d.id;
                });

        var tmp_tweets={};


        same=statuses.selectAll("div.status").data().some(function(d){
                return (ids.indexOf(d.id) == -1)
        });

        var __data=statuses.selectAll("div.status").data();
        if(__data.length == ids.length) {
                var same=true;
                for(var i=0;i<__data.length;i++) {
                        same=(ids.indexOf(__data[i].id)>-1);
                        if(!same)
                                break;
                }
                if(same) {
                        return;
                }
        }
```

Fig. 8 Java code of waves measuring system

5 Conclusions

After the comparison of the different measurement systems of tweets, it is concluded that there are several proposals that, on the one hand, are of high cost to users in the market. To perform cost–benefit analysis, it is concluded that the measurement system by means wave is sufficient for decision making. For future research, it is suggested that the measurement system waves allows the user to enter the accounts that the user wants to monitor, in order to perform the corresponding measurements themselves to assist in decision making for marketing professionals. In addition, this will help in assisting with making the right decisions to help effective performance of the proposed marketing and advertising as well as monitoring.

References

1. Lozares, C. (1996). La Teoria de Redes Sociales. *Papers No. 48* , 103-126.
2. Campos, F. (2008). *Revista Latina de Comunicación Social*. (L. d. Social, Producer, & Facultad de Ciencias de la Informacion de la Universidad de la Laguna) Retrieved Febrero 20, 2012, from www.ull.es: www.ull.es/publicaciones/latina/_2008/23_34_Santiago/forma.html

3. Wattanasupachoke, T. (2010). Success Factors of Online Social Networks. *The Journal of Global Business Issues* , *5* (2), 11-22.
4. Boyd, D. M., & Ellison, N. B. (2008). Social Network Sites: Definition, History, and Scholarship . *Journal of Computer-Mediated Communication* , *13*, 210-130.
5. Wiedemann, D. G. (2009). Exploring the Concept of Mobile Viral Marketing through Case Study Research. *Chair of Business Informatics and Systems Engineering* (pp. 49-60). Augsburg: University of Augsburg.
6. Trusov, M., Bodapati, A. V., & Bucklin, R. E. (2010). Determining Influential Users in Internet Social Networks. *Journal of Marketing Research* , *XLVII*, 643-658.
7. Woerndl, M., Papagiannidis, S., Mourlakis, M., & Li, F. (2008). Internet-induced marketing techniques: Critical factors in viral marketing campaigns . *International Journal Of Business Science and Applied Management* , *3* (1).
8. Akar, E., & Topçu, B. (2011). An Examination of the Factors Influencing Consumers' Attitudes Toward Social Media Marketing. *Journal of Internet Commerce* , 35-67.

Improving Decision-Making in a Business Simulator Using TOPSIS Methodology for the Establishment of Reactive Stratagems

Alberto Ochoa, Saúl González, Emmanuel Moriel, Julio Arreola and Fernando García

Abstract Nowadays using a robust simulator is very important to support an organization in its first step to consolidate in the market, in this research we make a challenging organizational tool based on different components to make an adequate strategic planning methodology, unlike current applications, it is focused on an environment that goes beyond simple numerical forecasts and statistical processes. It is based on advanced components to optimize the strategies and stratagems to be followed within the company, helping businesses to achieve competitive advantage in the market. A business simulator is flexible, adaptive, has learning ability, is robust and fault tolerant. Our intelligent tool uses different methodologies to provide optimal strategies to improve competitiveness of a company, the ability of the model can provide strategies that are not obvious because they can find no obvious relationship among variables that can help the manager or leader of an organization to realize a better decision. This tool is an aid in the process of improving competitiveness because it supports the strategic decisions made in an organizational level.

Keywords Strategic planning · Decision support system · Business simulator

1 Introduction

The use of technology in business in a way helps a lot in making decisions, i.e., approaches to reality, figuring that we can have different scenarios, identifying the complexity of daily work that occurs in organizations within which can be found: industrial organizations, trade organizations for goods and services, public organizations, educational organizations, the nonprofit organizations. It is important and helpful to rely on a certain part in an organized and generally focused on the events that occur in business. Thus, the use of simulation as a system created to streamline

A. Ochoa (✉) · S. González · E. Moriel · J. Arreola · F. García
Juarez City University, Ciudad Juárez, Mexico
e-mail: alberto.ochoa@uacj.mx

P. Melin et al. (eds.), *Nature-Inspired Design of Hybrid Intelligent Systems*,
Studies in Computational Intelligence 667, DOI 10.1007/978-3-319-47054-2_55

decision-making and delineate the scenarios we can test and company information is certainly an advantage for employers who do business and turn all seeking profitability. A business simulator is a decisive intelligent tool, usually using Artificial Intelligence, which allows playback and feedback of a system. The simulators reproduce sensations and experiences that in reality may come to pass. A simulator is intended to play both physical sensations (speed, acceleration, perception of the environment) and the behavior of the machine equipment that is meant to simulate.

2 TOPSIS Methodology to Improve a Business Simulator

Business Simulator is a learning tool and modeling which allows the entrepreneurial experience of creating and managing own business in an environment that does not risk money, a special situation which do not occur in the real life. This will help you to acquire the necessary experience to learn to distinguish the important from the unimportant when it comes to managing a business.

Business Simulator can learn without risk—an important and transcendental factor in the real situations, but empirically, it should not be done and what cannot be left to do. This is a systemic game in which the entrepreneur enters a simulator that behaves autonomously. Our actions affect others and those of others influence us. The entrepreneur must meet all the needs of your business: finance, marketing, sales, production, human resources, tax, competition, marketing positional, and quality of services or products. An entrepreneur must learn to make decisions in a fog of uncertainty, should know to take advantage of the moments when all men doubt to gain competitive advantage. An entrepreneur must learn to calibrate each decision to understand the effect of short, medium, and long term as in our Business simulator see Fig. 1. The chosen Alternative should have the shortest distance from the ideal solution and the farthest from the negative-ideal solution. It is very difficult to justify the selection of A1 or A2. Each Attribute in the Decision Matrix takes either monotonically increasing or monotonically decreasing utility, a Set of Weights for the Attributes is required to Any Outcome which is expressed in a nonnumerical way, should be quantified through the appropriate scaling technique and rank, the preference order specify a set of alternatives that can now be preference ranked according to the descending order of C_i^*.

Learn that ill-considered decisions are priced and that the right decisions with calculated risk and it can lead to success. In the game simulation, the entrepreneur will get to know how hard it is to get funding. Even so, many resources have so longed for necessary capital to implement their project. Each has its advantages and disadvantages. Inside should address both manage their employees, equipment, processes purchase, production, in another—as sales and advertising campaigns. You should pay attention to the fees, investment in Research & Development, quality and even overall competition because this simulator is not playing alone.

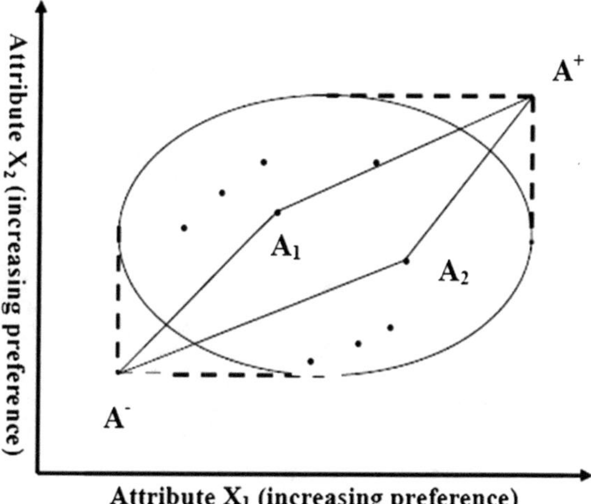

Fig. 1 Intelligent selection of stratagems using TOPSIS methodology

And the actions of others affect us with their decisions. In the uncertain world, for making a large and specific decision naturally requires a team. It is expected that a virtual company to begin operations within the simulator is managed by one person. For this reason, importance begins with the formation of real teams of entrepreneurs who are virtually your company and begin to make their business operations, technology, production, and finance, in competition with other teams.

The simulator is an experience not individually but collectively. No one individual plays against a preset machine but real teams play against other real equipment. So there is a specific date that begins and ends the game. In the simulated environment, each quarter equals 1 week in the real environment. Therefore, for each company to compete for a minimum period of 2 years (where introduce new products on the market, will expand its workforce will require new investments, media advertisings, and market implications). The business simulator requires a minimum of 8 weeks of real time. This is the time we spend on each new edition of the simulator. Go ahead, if you feel the entrepreneurial vein, no issues and learn how to run a business preserving your money for when you are really ready. Participate in an exciting take of decisions to a real world where success is measured by profit and failure goes straight to oblivion.

We propose the development of a model according to specifications of a small company, we propose that the final users can adequate the actual equipment to our necessities using the same software development on the real equipment but making a computer run more conventionally (and therefore cheaper).

The latter option is known as “Rehosted Software.” The more complex simulators are evaluated and qualified by the competent authorities using a Likert scale and a grand model prix based on evaluation of components. This simulator business so far consists of four modules: Visual Marketing Module, Marketing under uncertainty, Decisions under uncertainty, Detail Marketing Module, and Financial Module (Fig. 2).

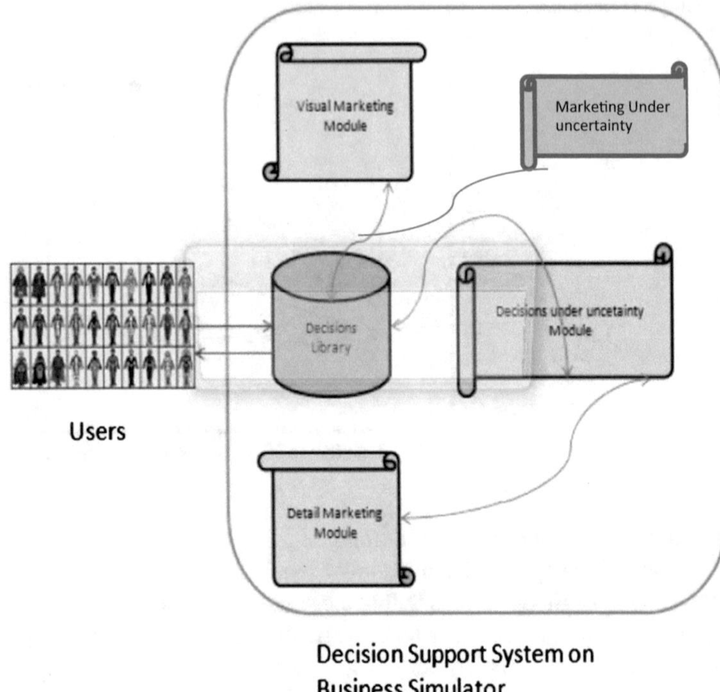

Fig. 2 Prototype of our business simulator

3 Implementation of an Intelligent Application

The simulator provides this data to the simulator, so users will have information that allows them to have some knowledge of the market they will face and make decisions for the current stage. Once all this is well defined, we can start with the dynamic part. They must decide which products to implement in the shelves on the first week of the simulation. After that, during the course of the simulation, shall be presented, week after week, different situations related to marketing variables and in these situations, the user must take several decisions that can impact both positively and negatively in the business.

The simulator will give the user to know the event or situation itself and three different possible options to choose from, each of which would cause a different result, affecting either sales or income of the consumers directly. Each answer leads to three possible scenarios: one positive, one negative, and one neutral. The positive scenario is the result of having made the right choice of the three that the simulator gives a choice to the user, this leads to an increase in sales or profits earned. The negative scenario is the result of poor decision-making of the situation. The third scenario, the neutral, when will choose the remaining option, which does not impact or a negative or positive result, leaving the sales or profits of a similar way as they were in the previous week, although this is the least risky option, the user can represent a loss because they may overlook potential growth opportunities stagnating business development and ending with a poor result at the end of the simulation. In addition to weekly decisions, it must decide which products will replenish the stock to avoid the lack of these in case of high demand from consumers, since the lack of stock of a product represents dissatisfied customers, resulting in a lower turnover. Based on the decisions taken either increase the simulator input to business customers or demand for certain products, forcing the user to adapt their future decisions on the possible scenarios that could happen. At the end of each period the business simulator, based on the information you have about the products on sale, price, promotions, stock there and each assessment or demand that consumers have of each, throw a turnover, which minus the cost of sales will result in net profits, which, along with the decisions taken at each stage, determine user performance in the simulation. The variables used in these five modules are enlisted below: (Tables 1 and 2).

Table 1 Input variables

Variable	Nomenclature	Definition
Advertising	A_{D}^{*}	Spending on advertising media to reach more consumers and to position our company in a wider market
Quantity	C^{*}	Is the number of products purchased for sale. This decision requires planning because they buy products to be used in the next period and the present
Functional strategies and differentiation strategies	E_{D}^{*}	These variables contain the strategic evaluations, i.e., strengths and opportunities presented, as well as the weaknesses and threats, within the internal approach will have what are the strengths and weaknesses, and external focus within the opportunities and threats, this SWOT-based model. The results will enable functional and differentiation strategies
Competitive strengths	F^{*}	This group of variables contains a group of competitive strength measurements. Evaluation of how relevant is the company's differentiation in the market
Retained earnings	G_{A}^{*}	It is the sum of net earnings over periods
Inventory management	G_{I}^{*}	Is measured existing physical units and sales in the period
Uncertainty in marketing decision	G_{P}^{*}	Also assess the economic performance of the company in the period
Income	I^{*}	Regardless of the market share of the company will assess the income that they generate

(continued)

Table 1 (continued)

Variable	Nomenclature	Definition
Uncertainty support decisions	I^{*}_{M}	These variables measure the scenarios in which the answers are given to the uncertainty that is generated when making decisions, making the only alternative approach to marketing
Marketing	M^{*}	The monetary value allocated to expenditure on advertising the product
Market and competitors position	M^{*}_{C}	This set of variables contains all the parameters evaluating the position which deals in the market, as well as the position of competitors.
Target marketing	M^{*}_{M}	These variables are related to the market which requires reaching, i.e., the focus and function of the business simulator is based on the market that is contemplated. Which gives results of different scenarios of possible markets
Market share	M^{*}_{P}	The percentage occupied by the sales joint venture in relation to other companies and the overall market
Financial impact	N^{*}	How will translate the monetary costs and benefits of this option in the final results by means net present value and what will be the opportunity of this result
Prices and discounts	P^{*}_{D}	The selling price of each product and the discounts can encourage increased consumption
Product mix	P^{*}_{M}	The products will be available for sale on the shelves during simulation
Place	P^{*}_{L}	The visual improvement of point of sale and special markings may have certain products
Budgets and projections	P^{*}_{P}	This group of variables contains all projections, budgets and objectives to be used In the strategy
Price	P^{*}_{R}	Is the value at which the product will be marketed has an acceptance range is defined in part by the system and in part by the administrator
Human Resources	R^{*}_{H}	Personal optimal to operate the company
Gifts and promotions	R^{*}_{P}	Spurs made to motivate consumers may be discounts on products or a gift attached to a product for sale as a gift
Suppliers	S^{*}_{P}	Managing the various suppliers, the difference in product quality, price and credit management between them
Return time	T^{*}_{I}	The time that It takes for the return of the investment

(continued)

Table 1 (continued)

Variable	Nomenclature	Definition
Time	T_P^*	These variables refer to the time that the business targeted, this in order that business life is given in medium and long term
Units in stock	U_S^*	The amounts of stock we have of each product and the same weekly refill.
Evaluation of the company's current strategy	V_e^*	This group of variables contains all the information for the indicators of current strategy fro company in terms of visual communication and advertising. (budget applied, means used creative type, frequency and period post exposure, existence and use of logo and corporate identity)
Competitive forces evaluation	V_F^*	This group of variables is composed by competitive forces facing the market Knowing how the competition is conformed and how it behaves
		Within this group can be found evaluating substitute products, buyers, degree of influence on each of them. Considering the actions of competitors regarding measures within commercial visual communication. Relative distance from the market leader
Expansion	X^*	This group contains all the parameters that evaluate the position it occupies in the market as well as the position of competitors. Based on a forward-looking approach (leadership)
Visual advertising budget	X_C^*	In this group of variables includes today's budget spent for the purposes of advertising and visual communication company's business SMEs
Budget	X_T^*	These variables are related to the investment you have in an SME (small budget) and prices that exist in the different marketing strategies that can be implemented, taking into account what can be used economically on investment SME (as intended for this module)
Identification of strengths, weaknesses, opportunities, and threats	Y^*	This group of variables contains the evaluation of the strengths and weaknesses of the resources of a company, its business opportunities and external threats to their future welfare. An example of this group of variables are SWOT indicators

Table 2 Output variables

Variable	Nomenclature	Definition
Implementation of corrective actions	A'	In the simulator has the advantage of having alternatives, i.e., you can play with the different decisions before implementing an SME, therefore, to correct, to get the expected result (optimal)
Competition	C'_{P}	Strategies and behaviors implemented by competition for greater market share and response to take to avoid losing our consumers
Effectiveness of strategy applied	E'_{A}	Sales generation: incremental sales as a result of advertising stimuli • Remembrance of advertising • Willingness to purchase • Generate leads or leads
Economic environment	E'_{E}	Changes in the general economic environment where the business simulation and the possible actions and reactions to face these changes with the least possible reduction in the profit margin
Proposed strategy	E'_{P}	This output shall make recommendations to the administrative leader through various strategies proposed to be included in the projections about advertising goals. Each strategy suggest advertising media, public choice you target, ad frequency, period and amount of investment publication
Risks	R'	What kind of risks associated this alternative? e.g., Could cause loss of profits or competitive advantage? How competition respond? Since the risk and uncertainty are essentially the same, what information would reduce this uncertainty?
Follow-up marketing strategies	S'_{E}	This variable grant support marketing strategies, i.e., does the implementation of these are contributing? Does it add value?
Monitoring marketing results	S'_{M}	In this variable is intended to grant a balance of the results obtained, i.e., really worked as intended? It has made the right decisions in this module? In short, you get a feedback
Costs	T'	How much will cost the alternative? Will result in cost savings now or long term? Can additional costs arise on the way? Is the alternative in the budget?
Viability	W'	Can be implemented alternative really? Can be an obstacle to be overcome? If the alternative is implemented, what resistance could be from inside to outside the organization?

4 Design of Experiments

We determine and evaluate using TOPSIS Methodology with four different scenarios:

Weight	0.1	0.4	0.3	0.2
	Style business reliability financial economics cost of operation			
Scenario A	7	9	9	8
Scenario B	8	7	8	7
Scenario C	9	6	8	9
Scenario D	6	7	8	6
After calculate $(\Sigma x_{ij}^2)^{1/2}$ for each column and divide each column by that to get r_{ij}				
Scenario A	0.46	0.61	0.54	0.53
Scenario B	0.53	0.48	0.48	0.46
Scenario C	0.59	0.41	0.48	0.59
Scenario D	0.40	0.48	0.48	0.40
After multiply each column by w_j to get v_{ij}				
Scenario A	0.046	0.244	0.162	0.106
Scenario B	0.053	0.192	0.144	0.092
Scenario C	0.059	0.164	0.144	0.118
Scenario D	0.040	0.192	0.144	0.080
Determine Ideal Solution A^*				
$A^* = \{0.059, 0.244, 0.162, 0.080\}$				
Scenario A	0.046	*0.244*	*0.162*	0.106
Scenario B	0.053	0.192	0.144	0.092
Scenario C	*0.059*	0.164	0.144	0.118
Scenario D	0.040	0.192	0.144	*0.080*
Find negative ideal solution $A-$. $A- = \{0.040, 0.164, 0.144, 0.118\}$				
Scenario A	0.046	0.244	0.162	0.106
Scenario B	0.053	0.192	*0.144*	0.092
Scenario C	0.059	*0.164*	*0.144*	*0.118*
Scenario D	*0.040*	0.192	*0.144*	0.080
Determine separation from ideal solution				
$A^* = \{0.059, 0.244, 0.162, 0.080\}$ $S_i^* = [(v_j^* - v_{ij})^2]^{1/2}$ for each row j				
Scenario A	$(\mathbf{0.046} - 0.059)^2$	$(\mathbf{0.244} - 0.244)^2$	$\mathbf{(0)^2}$	$\mathbf{(0.026)^2}$
Scenario B	$(\mathbf{0.053} - 0.059)^2$	$(\mathbf{0.192} - 0.244)^2$	$\mathbf{(-0.018)^2}$	$\mathbf{(0.012)^2}$
Scenario C	$(\mathbf{0.053} - 0.059)^2$	$(\mathbf{0.164} - 0.244)^2$	$\mathbf{(-0.018)^2}$	$\mathbf{(0.038)^2}$
Scenario D	$(\mathbf{0.053} - 0.059)^2$	$(\mathbf{0.192} - 0.244)^2$	$\mathbf{(-0.018)^2}$	$\mathbf{(0.0)^2}$
Determine separation from ideal solution $\mathbf{S_i^*}$				
$\sum(v_j^* - v_{ij})^2$ $S_i^* = [\sum(v_j^* - v_{ij})^2]^{1/2}$				

(continued)

(continued)

Scenario A	**0.000845**	0.029		
Scenario B	**0.003208**	0.057		
Scenario C	**0.008186**	0.090		
Scenario D	**0.003389**	0.058		
Determine separation from negative-ideal solution S−				
$\sum(v_j - v_{ij})^2$ Si− $= [\sum(v_j - v_{ij})^2]^{1/2}$				
Scenario A	**0.006904**	0.083		
Scenario B	**0.001629**	0.040		
Scenario C	**0.000361**	0.019		
Scenario D	**0.002228**	0.047		
Calculate the relative closeness to the ideal solution $C_i^* = S_i/(S_i^* + S_i)$				
	$S_i/(S_i^* + S_i)$	C_i^*		
Scenario A	**0.083/0.112**	0.74—BEST		
Scenario B	**0.040/0.097**	0.41		
Scenario C	**0.019/0.109**	0.17—WORSE		
Scenario D	**0.047/0.105**	0.45		

Finally we generate a "narrative guide" to explain each relevant aspect to support a specific scenario and will be include our reactive stratagems.

5 Conclusions and Future Work

Each business simulation modules comprising the customization process of filling information or input variable data, analysis, decision-making, and finally the simulation of such decisions. As a first step it is necessary to enter into the module the company's information that will be assessed in this way to create real conditions today and to make more accurate prognosis and create visual advertising strategy that best suits the needs of the user and business goals. Through the design of this tool can find a support, which is of paramount importance because, when making decisions under uncertainty speaking specifically about marketing strategies, gives SMEs an advance and improve the development of strategies to be implemented, taking into account the limited budget that account. It is very important to have the display functions such strategies in the medium and long term, and make corrections at the right time, because the simulator is achieved through trial and error decisions. The results of the periods set offer a competitive advantage to SMEs that are within the same niche. Concluding about the decision-making under uncertainty module, can be said that the decision-making as an experience is a key element, and that decisions should be taken on a reality that in many cases is complex because there are many variables involved, both within the organization and in the outside. Accumulating experience is over (for the time that you learn) and expensive (make mistakes), considering that the more you gain experience is the consequence of errors. Therefore, to achieve a high level of experience in the workplace can have very high costs. The role of a simulator business, specifically in the form of decision-making, is that the immediate consequences of all the experience you can gain without the effects that might result from a wrong decision or simply a nonoptimal decision, will be welcomed and cheaper, whatever the cost. Precisely the risk is that we do not know if it has made a good decision, not knowing whether an idea will work or not, and exactly a bad decision taken what leads to success or failure of a company. The marketing module will place the user in detail simulator business in an environment where they can become familiar with the development and design strategies, with the variables to consider in any election relating to products and services offered by a company using a safe method where not risk economic resources and increasing experience in business management. Utilities that can be given to this module are varied, ranging from an educational perspective to prepare for a future business, in a dynamic and fun way, preparing the user to deal with different situations that occur in the world of sales detail.

References

1. Barnes, J. (1984). Cognitive Biases and Their Impact on Strategic Planning. Strategic Management Journal, 5(2), 129.
2. Kanooni, A. (2009). Organizational factors affecting business and information technology alignment: A structural equation modeling analysis. Ph.D. dissertation, Capella University, United States – Minnesota.
3. Klayman, J. & Schoemaker, P. (1993). Thinking about the future: A cognitive perspective. Journal of Forecasting, 12(2), 161.
4. Laudon K. & Laudon J. (2002). Management Information Systems Managing the Digital Firm. (7ª Ed). E.U.A. Pearson Prentice-Hall. (pp 401- 465).
5. Peyrefitte J., Golden P., & Brice J. Jr. (2002). Vertical integration and economic performance: A managerial capability framework. Management Decision, 40(3), 217-226.
6. Philip, G. (2007). IS Strategic Planning for Operational Efficiency. Information Systems Management, 24(3), 247-264.
7. Porter, M. (1998). Estrategia Competitiva: Técnicas para el Análisis de los Sectores Industriales y de la Competencia. México: Continental.
8. Pretorius, M. (2008). When Porter's generic strategies are not enough: complementary strategies for turnaround situations. The Journal of Business Strategy, 29(6), 19-28.
9. Salem, M. (2005). The Use of Strategic Planning Tools and Techniques in Saudi Arabia: An Empirical study. International Journal of Management, 22(3), 376-395,507.
10. Sundin S., & Braban-Ledoux C. (2001). Artificial Intelligence–Based Decision Support Technologies in Pavement Management. Computer-Aided Civil & Infrastructure Engineering, 16(2), 143.

Non-singleton Interval Type-2 Fuzzy Systems as Integration Methods in Modular Neural Networks Used Genetic Algorithms to Design

Denisse Hidalgo, Patricia Melin and Juan R. Castro

Abstract In this paper, we propose the use of Non-Singleton Interval Type-2 Fuzzy Systems (NSIT2FI) automatically designed through genetic algorithms as integration method of modular neural networks (MNN's) for multimodal biometrics. The goal is to obtain such fuzzy systems as integrators, better recognition rate, and best mean square error in MNN. The results shown comparison between interval type-2 fuzzy systems and Non-singleton Type-2 Fuzzy Systems, where we can observe showing a significant difference that we can get higher recognition rate using non-singleton type-2 fuzzy logic.

Keywords Non-singleton Type-2 fuzzy logic · Genetic algorithms · Neural networks

1 Introduction

Optimization is the process of maximizing a desired objective function while satisfying the prevailing constraint. Limited material or labor resources must be utilized to maximize profit. Often, optimization of a design process saves money for a company by simply reducing the developmental time. This is because there is considerable effort needed to apply optimization techniques on practical problems to achieve an improvement. This effort invariably requires tuning algorithmic parameters, scaling, and even modifying the techniques for the specific application.

D. Hidalgo (✉) · P. Melin
Division of Graduate Studies and Research, Tijuana Institute of Technology, Tijuana, Mexico
e-mail: dra.denisse.hidalgo@hotmail.com

P. Melin
e-mail: pmelin@tectijuana.mx

J.R. Castro
School of Engineering, UABC University, Tijuana, Mexico
e-mail: jrcastror@uabc.edu.mx

P. Melin et al. (eds.), *Nature-Inspired Design of Hybrid Intelligent Systems*, Studies in Computational Intelligence 667, DOI 10.1007/978-3-319-47054-2_56

Moreover, the user may have to try several optimization methods to find one that can be successfully applied. To date, optimization has been used more as a design or decision aid, rather than for concept generation or detailed design [1]. For many years, interest in algorithms that are based on the analogies of natural processes has been increasing. For optimization tasks functions of real variables, the strategic evolution has emerged as a major contender to various traditional solution methods. Among the most remarkable techniques of this group, include genetic algorithms (GA), evolution strategies (ES), evolutionary programming (EP), and genetic programming (GP).

To manually fuzzy systems designing is known that it takes considerable time and a great knowledge of the expert for solving the problem, as we do this trial and error; that is why we use the optimization of such systems through of different optimization methods and get optimal results to the analyzed problem.

The main goal of this paper is design of non-singleton type-2 fuzzy integration in modular neural network, where we use a genetic algorithm for the automatic design. On this occasion we used the modular neural networks for multimodal biometrics as face, fingerprint, and voice so whit expect the type-2 non-singleton fuzzy integrators obtain better recognition rate in MNN.

This paper is organized as follows: Sect. 2 presents the theoretical. Section 3 introduces the problem statement. Section 4 it provides the explanation of design of interval type-2 non-singleton fuzzy integration in modular neural network, where we use a genetic algorithm for the automatic design. Simulation results for integration to Modular Neural Networks for Multimodal biometry described in Sect. 5. Finally, Sect. 6 presents the conclusions.

2 Theoretical Basis

This section describes the some basis about modular neural networks, non singleton type-2 fuzzy systems, and genetic algorithms.

2.1 Modular Neural Networks

Modularity can be defined as subdivision of a complex object into simpler objects. The subdivision is determined either by the structure or function of the object and its subparts. Modularity can be found everywhere; in living creatures as well as in inanimate objects. The subdivision in less complex objects is often not obvious.

Artificial neural networks (ANN) are information processing systems that have certain computational properties analogous to those which have been postulated for biological neural networks. The origins of the concept of artificial neural networks can be traced back more than a century as a consequence of man's desire for understanding the brain and emulating its behavior.

The emulation of the principles governing the organization of human brain forms the basis for their structural design and learning algorithms. Artificial neural networks exhibit the ability to learn from the environment in an interactive fashion and show remarkable abilities of learning, recall, generalization, and adaptation in the wake of changing operating environments.

Neural networks are attractive for classification problems because they are capable to learn from noisy data and to generalize. The first neural network model (perceptron) was developed by Rosenblatt in the late 1950s. Since then, several other models have been proposed. Examples are: generalized feedforward networks, radial basis function networks, the Hopfield model, the multilayer perceptron, modular networks, etc. [1]. These models differ in their architectures and in the way they learn and behave, so they are suitable for different types of problems. In this case, Modular Neural Networks will be considered.

An artificial neural network is comprised of a set highly interconnected nodes. According to the interconnection scheme used for a neural network, it can be categorized as a feedforward or a recurrent neural network. Learning in artificial neural networks refers to a search process, which aims to find an optimal network topology and an appropriate set of neural network weights to accomplish a given task. A monolithic artificial neural network can be viewed as an unstructured black box in which the only states, which are observable are the input and output states. Considering the shortcomings of monolithic artificial neural networks, modularization of artificial neural network design and learning seems to be an attractive alternative to the existing artificial neural network design and learning algorithms. Therefore, modular neural network architectures are a natural way of introducing a structure to the otherwise unstructured learning of neural networks.

It is important to mention that the use of the modular neural networks to solve a problem in particular, requires ample knowledge of the problem to be able to make the subdivision of the problem, and to build the suitable modular architecture to solve it, in such a way that it is possible to train each of the modules independently, and later to integrate the knowledge learned by each module, in the global architecture [2].

2.2 Genetic Algorithm

Genetic Algorithms (GAs) were invented by John Holland and developed by him and his students and colleagues. This lead to Holland's book "*Adaption in Natural and Artificial Systems*" published in 1975.

Genetic algorithms (GAs) are adaptive heuristic search algorithm based on the evolutionary ideas of natural selection and genetics. As such they represent an intelligent exploitation of a random search used to solve optimization problems. Although randomized, GAs are by no means random, instead they exploit historical information to direct the search into the region of better performance within the search space. The basic techniques of the GAs are designed to simulate processes in natural systems necessary for evolution; especially those that follow the principles first laid down by Charles Darwin of "survival of the fittest." Since in nature, competition among individuals for scanty resources results in the fittest individuals dominating over the weaker ones.

Algorithm is started with a set of solutions (represented by chromosomes) called population. Solutions from one population are taken and used to form a new population. This is motivated by a hope that the new population will be better than the old one. Solutions that are selected to form new solutions (offspring) are selected according to their fitness—the more suitable they are the more chances they have to reproduce. This is repeated until some condition (for example, number of populations or improvement of the best solution) is satisfied.

2.3 Singleton Type-2 Fuzzy Logic Systems

When something is uncertain (example, a measure), we have problems to determine their exact value, and if necessary use type-1 fuzzy sets makes more sense to use classical ensembles. But then, even in type-1 fuzzy sets, membership function specify exactly. If we cannot determine the exact value of the amount of uncertainty then use type-2 fuzzy sets [2]. In Fig. 1 we can see a membership function type-1 (a) and membership function type-1 with uncertainty (b) respectively.

Type-2 FLS are the extension of the type-1 FLS. In type-2 FLSs, the antecedent and/or consequent membership functions of type-2 FLS are type-2 fuzzy sets [3]. A type-2 fuzzy set A is characterized by a type-2 membership function $0 \leq \mu_{\tilde{A}}(x, u) \leq 1$, where $x \in X$ and $u \in Jx \subseteq [0,1]$, as [1]:

$$\tilde{A} = \left\{ \left((\mathrm{x,u}), \mu\, \tilde{A}(\mathrm{x,u}) \right) | \forall \mathrm{x} \in \mathrm{X}, \quad \forall \mathrm{u} \in \mathrm{Jx} \subseteq [0,1] \right\} \tag{1}$$

Jx is called primary membership of x, where $Jx \subseteq [0,1]$ for $\forall x \in X$. The uncertainty in the primary membership grades of a type-2 fuzzy membership function consist of a bounded region, that we call the Footprint of Uncertainty (FOU) of a type-2 membership function. In Fig. 2 we can see the structure of singleton type-2 FLS where crisp inputs are first fuzzified into input type-2 fuzzy

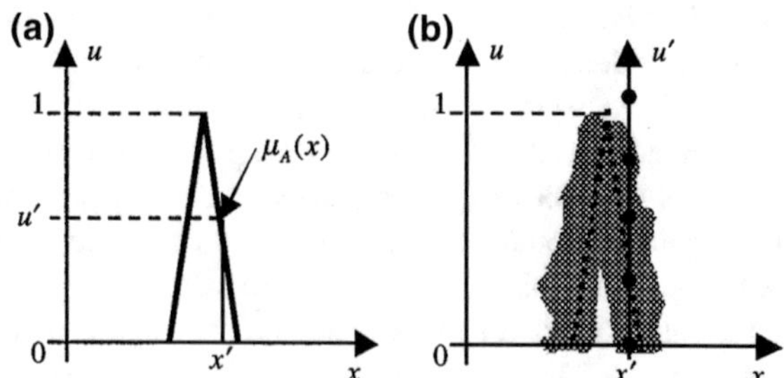

Fig. 1 Membership functions. **a** Membership function of type-1 and **b** membership function of type-1 with uncertainty

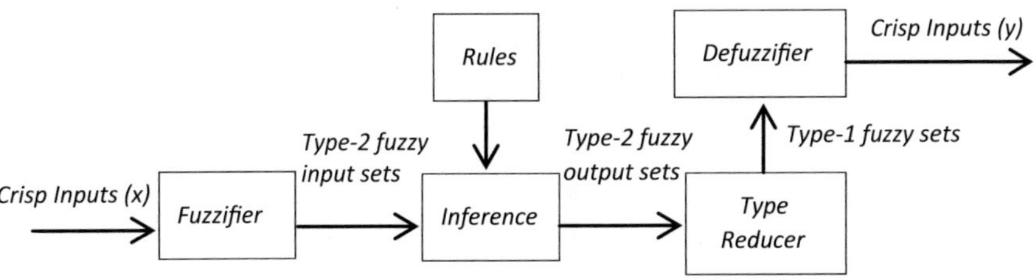

Fig. 2 Structure of singleton interval type-2 FLS

sets (in singleton fuzzification) which then activate the inference engine and the rule base to produce output type-2 fuzzy sets. These output type-2 fuzzy sets are then processed by the type-reducer, which combines the output sets and then performs a centroid calculation, which leads to type-1 fuzzy sets called the type-reduced sets [1, 4].

The structure of the rules in a type-1 FLS and a type-2 FLS is the same, but in the latter the antecedents and the consequents will be represented by type-2 fuzzy sets. So for a type-2 FLS with p inputs $x_1 \in X_1, \ldots, x_p \in X_p$ and one output $y \in Y$, which is a multiple input single output (MISO) system, if we assume there are M rules, the lth rule in the type-2 FLS can be written as follows [5]:

$$R^l\text{:IF}\, x_1 \text{ is } \widetilde{F}_1^l \text{ and} \ldots \text{and}\, x_p \text{ is } \widetilde{F}_p^l,\ \text{THEN}\, y \text{ is } \widetilde{G}^l \quad l = 1, \ldots, \text{M} \tag{1}$$

2.3.1 Singleton Interval Type-2 Fuzzy Inference System (SIT2FIS)

The Mamdani IT2FIS, is designed with **n** inputs, **m** outputs and **r** rules [6]. The kth rule with interval type-2 fuzzy antecedents $\overset{\approx}{A}_{k,j} \in \left\{\mu_{i,l_{k,i}}\right\}$, interval type-2 fuzzy consequent $\overset{\approx}{C}_{k,j} \in \left\{\sigma_{j,l_{k,j}}\right\}$ and interval type-2 fuzzy facts $\overset{\approx'}{A}_i$ are inferred as a direct reasoning.

$$R^k\text{:IF } \mathrm{x}_1 \text{ is } \overset{\approx}{\mathrm{A}}_{\mathrm{k},1} \text{ and} \ldots \text{and } \mathrm{x}_\mathrm{n} \text{ is } \overset{\approx}{\mathrm{A}}_{\mathrm{k,n}} \text{ THEN } \mathrm{y}_1 \text{ is } \overset{\approx}{\mathrm{C}}_{\mathrm{k},1} \text{ and} \ldots \text{and } \mathrm{y}_\mathrm{m} \text{ is } \overset{\approx}{\mathrm{C}}_{\mathrm{k,m}}$$
$$H\text{:IF}\, x_1 \text{ is } \hat{x}_1 \text{ and} \ldots \text{and}\, x_n \text{ is } \hat{x}_n$$

$$C\text{:}y_1 \text{ is } \overset{\approx'}{C}_1 \text{ and} \ldots \text{and}\, y_m \text{ is } \overset{\approx'}{C}_m$$

$$\begin{aligned}\mu_{\overset{\approx''}{C}_{k,j}}(y_j) &= \left[\sqcap_{i=1}^{n}\left(\mu_{\overset{\approx''}{A}_{k,i}}(\hat{x}_i)\right)\right] \sqcap \mu_{\overset{\approx}{C}_{k,j}}(y_j) \\ &= \left\{\int_Y \left[\begin{array}{c}\int 1/\alpha \\ \alpha \in \left[\underline{\mu}_{\overset{\approx''}{C}_{k,j}}(y_j), \bar{\mu}_{\overset{\approx''}{C}_{k,j}}(y_j)\right] \subseteq [0,1]\end{array}\right] \Big/ y_j\right\}\end{aligned}$$

$$\underline{\mu}_{\underset{C_{k,j}}{\approx''}}(y_j) = \left[\underset{i=1}{\overset{n}{\tilde{*}}} \left(\underline{\mu}_{\underset{A_{k,i}}{\approx}}(\hat{x}_i) \right) \right] \tilde{*} \, \underline{\mu}_{\underset{C_{k,j}}{\approx}}(y_j)$$

$$\bar{\mu}_{\underset{C_{k,j}}{\approx''}}(y_j) = \left[\underset{i=1}{\overset{n}{\tilde{*}}} \left(\bar{\mu}_{\underset{A_{k,i}}{\approx}}(\hat{x}_i) \right) \right] \tilde{*} \, \bar{\mu}_{\underset{C_{k,j}}{\approx}}(y_j)$$

$$\mu_{\underset{C_j}{\approx'}}(y_j) = \sqcup_{k=1}^{r} \left[\sqcap_{i=1}^{n} \left(\mu_{\underset{A_{k,i}}{\approx}}(\hat{x}_i) \right) \sqcap \mu_{\underset{C_{k,j}}{\approx}}(y_j) \right]$$

$$= \left\{ \int_Y \left[\begin{array}{c} \int 1/\alpha \\ \alpha \in \left[\underline{\mu}_{\underset{C_j}{\approx'}}(y_j), \bar{\mu}_{\underset{C_j}{\approx'}}(y_j) \right] \subseteq [0,1] \end{array} \right] \Big/ y_j \right\}$$

$$\underline{\mu}_{\underset{C_j}{\approx'}}(y_j) = \bigvee_{k=1}^{r} \left(\underline{\mu}_{\underset{C_{k,j}}{\approx''}}(y_j) \right) = \bigvee_{k=1}^{r} \left(\underset{i=1}{\overset{n}{\tilde{*}}} \left[\underline{\mu}_{\underset{A_{k,i}}{\approx}}(\hat{x}_i) \right] \tilde{*} \, \underline{\mu}_{\underset{C_{k,j}}{\approx}}(y_j) \right)$$

$$\bar{\mu}_{\underset{C_j}{\approx'}}(y_j) = \bigvee_{k=1}^{r} \left(\bar{\mu}_{\underset{C_{k,j}}{\approx''}}(y_j) \right) = \bigvee_{k=1}^{r} \left(\underset{i=1}{\overset{n}{\tilde{*}}} \left[\bar{\mu}_{\underset{A_{k,i}}{\approx}}(\hat{x}_i) \right] \tilde{*} \, \bar{\mu}_{\underset{C_{k,j}}{\approx}}(y_j) \right)$$

The defuzzification of the interval type-2 fuzzy aggregated output set $\overset{\approx'}{C_j}$ is: $\hat{y}_j = \text{idefuzztype2}\left(\mu_{\underset{C_j}{\approx'}}(y_j), 'type' \right)$ where type is the name of the defuzzification technique (see Fig. 3).

2.4 Non-singleton Type-2 Fuzzy Logic Systems

Type-2 FLS are very useful in circumstances, where it is difficult to determine an exact membership function, and there are measurement uncertainties [3]. As a justification for the use of type-2 fuzzy sets, in [4] at least four sources of uncertainties not considered in type-1 FLS are mentioned:

1. The meaning of the words that are used in the antecedents and consequents of rules can be uncertain (words mean different things to different people).
2. Consequents may have histogram of values associated with them, especially when knowledge is extracted from a group of experts who do not all agree.
3. Measurements that activate a type-1 FLS may be noisy and therefore uncertain.
4. The data used to tune the parameters of a type-1 FLS may also be noisy.

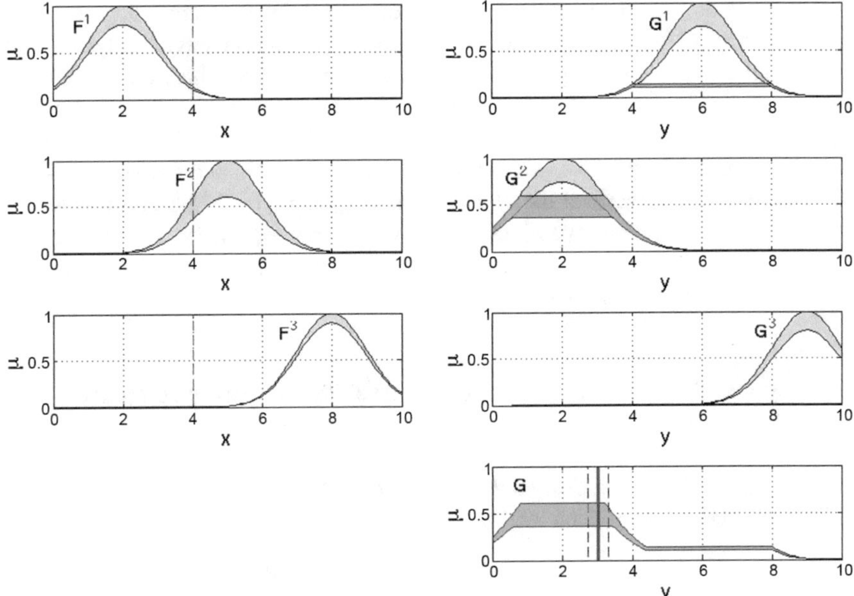

Fig. 3 The defuzzification of *SIT2FIS*

All of these uncertainties translate into uncertainties about fuzzy set membership functions. Type-2 fuzzy sets are able to model such uncertainties because their membership functions are themselves fuzzy.

A type-2 FLS is again characterized by IF-THEN rules, but its antecedent or consequent sets are now of type-2. Type-2 FLS can be used when the circumstances are too uncertain to determine exact membership grades, such as when the training data is corrupted by noise. Similar to type-1 FLS, a type-2 FLS includes a fuzzifier, a rule base, fuzzy inference engine, and an output processor [7], as we can see in Fig. 1. The output processor includes a type-reducer and defuzzifier; it generates a type-1 fuzzy set output (from the type-reducer) or a crisp number (from the defuzzifier) [8, 9].

Now a type-2 FLS whose inputs are modeled as type-2 fuzzy numbers is referred to as a type-2- non-singleton FLS. The major difference between a non-singleton FLS and a singleton FLS is in the fuzzification part [5].

A non-singleton FLS is one whose inputs are modeled as fuzzy numbers. A type-2 FLS whose inputs are modeled as type-1 fuzzy numbers is referred to as "type-1 non-singleton type-2 FLS." This kind of a fuzzy system not only accounts for uncertainties about either the antecedents or consequents in rules, but also accounts for input measurement uncertainties [10].

A type-2 non-singleton type-2 FLS is described by the same diagram as in singleton type-2 FLS are the same. The rules of a type-2 non-singleton type-2 FLS are the same as those for a type-1 non-singleton type-2 FLS, which are the same as those for a singleton type-2 FLS. What is different is the fuzzifier, which treats the inputs as type-2 fuzzy sets, and the effect of this on the inference block. The output of the inference block will again be a type-2 fuzzy set; so, the type-reducers and defuzzifier that we described for a type-1 non-singleton type-2 FLS apply as well to a type-2 non-singleton type-2 FLS [10]. In this research, we use type-2 non-singleton type-2 fuzzy systems as integration methods in modular neural networks and a comparison with type-2 fuzzy systems.

2.4.1 Interval Type-2 Non-singleton Interval Type-2 Fuzzy Inference System (NSIT2FIS)

The Mamdani IT2FIS, is designed with **n** inputs, **m** outputs and **r** rules [6]. The IT2FIS Takagi-Sugeno-Kang we can see [11].The kth rule with interval type-2 fuzzy antecedents $\overset{\approx}{A}_{k,j} \in \left\{\mu_{i,l_{k,i}}\right\}$, interval type-2 fuzzy consequent $\overset{\approx}{C}_{k,j} \in \left\{\sigma_{j,l_{k,j}}\right\}$, and interval type-2 fuzzy facts $\overset{\approx'}{A}_i$ are inferred as a direct reasoning.

$$R^k\text{:IF } x_1 \text{ is } \overset{\approx}{A}_{k,1} \text{ and} \ldots \text{and } x_n \text{ is } \overset{\approx}{A}_{k,n} \text{ THEN } y_1 \text{ is } \overset{\approx}{C}_{k,n} \text{ and} \ldots \text{and } y_m \text{ is } \overset{\approx}{C}_{k,m}$$

$$H\text{:IF } x_1 \text{ is } \overset{\approx'}{A}_1 \text{ and} \ldots \text{and } x_n \text{ is } \overset{\approx'}{A}_{k,n}$$

$$C\text{:}y_1 \text{ is } \overset{\approx'}{C}_m \text{ and} \ldots \text{and } y_m \text{ is } \overset{\approx'}{C}_m$$

The evaluation of this reasoning is:

$$\overset{\approx}{R}_{k,j} = \overset{\approx}{A}_{k,1} \times \cdots \times \overset{\approx}{A}_{k,n} \to \overset{\approx}{C}_{k,j} = \left(\overset{\approx}{A}_{k,1} \to \overset{\approx}{C}_{k,j}\right)\left(\overset{\approx}{A}_{k,n} \to \overset{\approx}{C}_{k,j}\right), k\text{th rule}$$

$$\begin{aligned}\mu_{\overset{\approx}{R}_{k,j}}(x,y) &= \mu_{\overset{\approx}{A}_{k,1}}(x_1) \sqcap \cdots \sqcap \mu_{\overset{\approx}{A}_{k,n}}(x_n) \sqcap \mu_{\overset{\approx}{C}_{k,j}}(y_j) \\ &= \mu_{(\overset{\approx}{A}_{k,1} \to \overset{\approx}{C}_{k,j})}(x_1, y_j) \sqcap \cdots \sqcap \mu_{(\overset{\approx}{A}_{k,n} \to \overset{\approx}{C}_{k,j})}(x_n, y_j)\end{aligned}$$

$$\begin{aligned}\mu_{\overset{\approx}{R}_{k,j}}(x,y) &= \mu_{\overset{\approx}{A}_{k,1}}(x_1) \sqcap \cdots \sqcap \mu_{\overset{\approx}{A}_{k,n}}(x_n) \sqcap \mu_{\overset{\approx}{C}_{k,j}}(y_j) \\ &= \left[\sqcap_{i=1}^{n} \mu_{\overset{\approx}{A}_{k,i}}(x_i)\right] \sqcap \mu_{\overset{\approx}{C}_{k,j}}(y_j)\end{aligned}$$

$$\tilde{\tilde{H}} = \tilde{\tilde{A}}'_1 \times \cdots \times \tilde{\tilde{A}}'_n, \text{facts}$$

$$\mu_{\tilde{\tilde{H}}_{k,j}}(x) = \mu_{\tilde{\tilde{A}}'_1}(x_1) \sqcap \cdots \sqcap \mu_{\tilde{\tilde{A}}'_n}(x_n) = \sqcap_{i=1}^{n} \mu_{\tilde{\tilde{A}}'_i}(x_i)$$

$$\begin{aligned} \tilde{\tilde{C}}''_{k,j} &= \tilde{\tilde{H}}\, o\, \tilde{\tilde{R}}_{k,j} = \left[\tilde{\tilde{A}}'_i o (\tilde{\tilde{A}}_{k,i} \rightarrow \tilde{\tilde{C}}_{k,j})\right] \\ &= \left\{ \int_Y \left[\begin{matrix} \int 1/\alpha \\ \alpha \in \left[\underline{\mu}_{\tilde{\tilde{C}}''_{k,j}(y_j)}, \bar{\mu}_{\tilde{\tilde{C}}''_{k,j}(y_j)}\right] \subseteq [0,1] \end{matrix} \right] / y_j \right\} \end{aligned}$$

$$\mu_{\tilde{\tilde{C}}''_{k,j}}(y_j) = \mu_{\tilde{\tilde{H}}\, o \tilde{\tilde{R}}_{k,j}}(y_j) = \sqcap_{i=1}^{n} \left(\sqcup_{x \in X} \left\{ \mu_{\tilde{\tilde{A}}'_i}(x_i) \sqcap \mu_{\tilde{\tilde{A}}_{k,i} \rightarrow \tilde{\tilde{C}}_{k,j}}(x_i, y_j) \right\} \right)$$

$$\begin{aligned} \mu_{\tilde{\tilde{C}}''_{k,j}}(y_j) &= \mu_{\tilde{\tilde{H}}\, o \tilde{\tilde{R}}_{k,j}}(y_j) \\ &= \sqcap_{i=1}^{n} \left(\sqcup_{x \in X} \left\{ \left[\mu_{\tilde{\tilde{A}}'_i}(x_i) \sqcap \mu_{\tilde{\tilde{A}}_{k,i}}(x_i) \right] \sqcap \mu_{\tilde{\tilde{C}}_{k,j}}(y_j) \right\} \right) \end{aligned}$$

$$\underline{\mu}_{\tilde{\tilde{C}}''_{k,j}}(y_j) = \left[\mathop{\tilde{*}}_{i=1}^{n} \left(\sup_{x_i \in X_i} \underline{\mu}_{\tilde{\tilde{A}}'_i}(x_i) \tilde{*} \underline{\mu}_{\tilde{\tilde{A}}_{k,i}}(x_i) \right) \right] \tilde{*} \underline{\mu}_{\tilde{\tilde{C}}_{k,j}}(y_j)$$

$$\bar{\mu}_{\tilde{\tilde{C}}''_{k,j}}(y_j) = \left[\mathop{\tilde{*}}_{i=1}^{n} \left(\sup_{x_i \in X_i} \bar{\mu}_{\tilde{\tilde{A}}'_i}(x_i) \tilde{*} \bar{\mu}_{\tilde{\tilde{A}}_{k,i}}(x_i) \right) \right] \tilde{*} \bar{\mu}_{\tilde{\tilde{C}}_{k,j}}(y_j)$$

$$\begin{aligned} \tilde{\tilde{C}}'_j &= \bigcup_{k=1}^{r} \tilde{\tilde{C}}''_{k,j} = \bigcup_{k=1}^{r} \left(\mathop{\times}_{i=1}^{n} \left[\tilde{\tilde{A}}'_i o \left(\tilde{\tilde{A}}_{k,i} \rightarrow \tilde{\tilde{C}}_{k,i} \right) \right] \right) \\ &= \left\{ \int_Y \left[\begin{matrix} \int 1/\alpha \\ \alpha \in \left[\underline{\mu}_{\tilde{\tilde{C}}'_j}(y_j), \bar{\mu}_{\tilde{\tilde{C}}'_j}(y_j)\right] \subseteq [0,1] \end{matrix} \right] / y_j \right\} \end{aligned}$$

$$\tilde{\tilde{C}}'_j = \sqcup_{k=1}^{r} \tilde{\tilde{C}}''_{k,j} = \sqcup_{k=1}^{r} \left[\sqcap_{i=1}^{n} \left\{ \sqcup_{x \in X} \left[\mu_{\tilde{\tilde{A}}'_i}(x_i) \right] \sqcap \mu_{\tilde{\tilde{C}}_{k,j}}(y_j) \right\} \right]$$

$$\underline{\mu}_{\underset{\sim}{\tilde{C}}'_j}(y_j) = \overset{r}{\underset{k=1}{\vee}} \left(\underline{\mu}_{\underset{\sim}{\tilde{C}}''_{k,j}}(y_j) \right)$$

$$= \overset{r}{\underset{k=1}{\vee}} \left(\left[\overset{n}{\underset{i=1}{\tilde{*}}} \left(\sup_{x_i \in X_i} \underline{\mu}_{\underset{\sim}{\tilde{A}}'_i}(x_i) \,\tilde{*}\, \underline{\mu}_{\underset{\sim}{\tilde{A}}_{k,i}}(x_i) \right) \right] \tilde{*}\, \underline{\mu}_{\underset{\sim}{\tilde{C}}_{k,j}}(y_i) \right)$$

$$\bar{\mu}_{\underset{\sim}{\tilde{C}}'_j}(y_j) = \overset{r}{\underset{k=1}{\vee}} \left(\bar{\mu}_{\underset{\sim}{\tilde{C}}''_{k,j}}(y_j) \right)$$

$$= \overset{r}{\underset{k=1}{\vee}} \left(\left[\overset{n}{\underset{i=1}{\tilde{*}}} \left(\sup_{x_i \in X_i} \bar{\mu}_{\underset{\sim}{\tilde{A}}'_i}(x_i) \,\tilde{*}\, \bar{\mu}_{\underset{\sim}{\tilde{A}}_{k,i}}(x_i) \right) \right] \tilde{*}\, \bar{\mu}_{\underset{\sim}{\tilde{C}}_{k,j}}(y_j) \right)$$

The defuzzification of the interval type-2 fuzzy aggregated output set $\overset{\approx'}{C_j}$ is: $\hat{y}_j = \text{idefuzztype2}\left(\mu_{\underset{\sim}{\tilde{C}}'_j}(y_j)', \text{type}' \right)$.Where type is the name of the defuzzification technique (see Fig. 4).

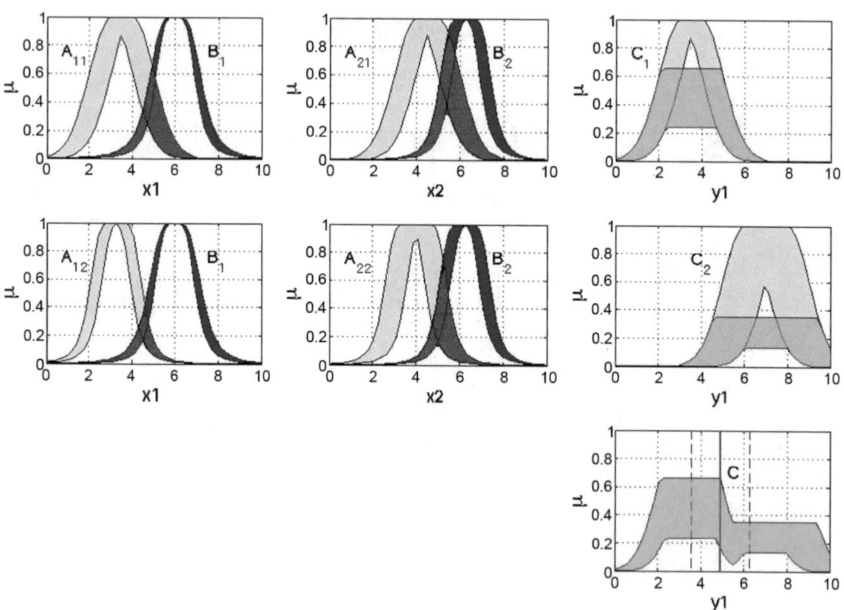

Fig. 4 The defuzzification of *NSIT2FIS*

3 Problem Statement

The main goal of this work is to create a non-singleton interval type-2 fuzzy systems automatically designed through genetic algorithms as an integration method for modular neural networks (MNN's) for multimodal biometrics; for that reason, it has been proposed to make a comparison between the method of integration results in a modular neural network; the comparison made with the method of fuzzy systems integration where they used technical type-2 and non- singleton interval type-2 Fuzzy Logic. These integration methods areimplemented in modular neural networks, that is, which were trained to perform the recognition of people through his face, fingerprint, and voice that we were used in previous works [12].

4 Non-singleton Interval Type-2 Fuzzy Systems Design

The principal objective of this research is design of non-singleton interval type-2 fuzzy integration in modular neural network, where we use a genetic algorithm for the automatic design. On this occasion we used the modular neural networks for multimodal biometrics as face, fingerprint, and voice so whit expect the type-2 non-singleton fuzzy integrators obtain better recognition rate in MNN. In this section, we show the design of a non-singleton interval type-2 fuzzy systems using the interval type-2 fuzzy logic toolbox [13].

A non-singleton type-2 fuzzy logic integrator is designed, basing on the optimal type-2 integrator; we set the parameters of the membership functions of a type-2 non-singleton fuzzy logic using a genetic algorithm. We set the parameters of the GA using a chromosome with 24 genes of real values that represent the three inputs, face, fingerprint, and voice activations; we used different values in the genetic operator's; real-value mutation and single point crossover. In Table 1 we can see the results of execution to the GA for fuzzy integrators obtained, which they were tested in different modular neural networks.

5 Simulation Results

In this section, we only present simulations results of the non-singleton type-2 fuzzy integrator; we can find in [14] the previous results.

We organized this section as follows: first, we present the simulations results for the training the neural networks (see Table 2) [15], integrations methods using type-2 Non-singleton fuzzy logic showing the behavior of each integrator.

Table 1 Results of executions to the genetic algorithms to non-singleton type-2 fuzzy integrators

GA	Min error	Time execution	Individuals	Crossover	Mutation	GGAP	Max_gen
sw1	0.000171	00:23:00	80	0.15	1.00E−04	0.85	100
sw2	3.09E−10	00:22:18	80	0.15	1.00E−04	0.85	100
sw3	4.53E−13	00:56:19	200	0.30	0.001	0.80	100
sw4	0.000494	00:40:42	100	0.25	1.00E−04	0.85	150
sw5	3.26E−14	01:05:30	120	0.30	0.01	0.85	200
sw6	6.97E−13	00:44:05	120	0.3	0.01	0.85	200
sw7	4.63E−11	00:24:32	50	0.4	0.01	0.85	180
sw8	3.09E−12	00:08:20	20	0.4	0.1	0.85	150
sw9	1.44E−10	00:08:26	20	0.4	0.1	0.85	150
sw10	9.94E−13	00:20:29	50	0.3	0.1	0.85	150
sw11	2.02E−10	00:14:59	35	0.35	0.01	0.85	150
sw12	6.08E−05	00:26:49	50	0.3	0.01	0.85	200
sw13	9.98E−13	00:24:21	60	0.25	0.01	0.85	150
sw14	5.30E−12	00:30:19	75	0.3	0.01	0.9	150
sw15	1.39E−05	00:08:10	15	0.4	0.01	0.85	200
sw16	5.46E−09	00:16:27	30	0.4	0.01	0.85	200
sw17	4.60E−12	00:23:13	55	0.2	0.01	0.9	150
sw18	8.38E−04	00:28:30	65	0.25	0.01	0.9	150
sw19	3.52E−12	00:31:00	70	0.4	0.01	0.9	150
sw20	9.37E−11	00:43:44	100	0.4	0.01	0.9	150

Table 2 Results of training to the modular neural network for multimodal biometric no fuzzy integration

Training neural Networks	Submodule layers			Persons recognized by submodule	Recognition rate (%)	Execution time
	Submodule 1	Submodule 2	Submodule 3			
1	65, 50	110, 90	30, 35	12	26.67	01:14:16
	75, 73	115, 95	35, 43	1		
	80, 85	120, 100	40, 32	59		
2	112, 98	121, 93	52, 36	11	27.78	01:50:48
	112, 98	121, 93	52, 36	1		
	112, 98	121, 93	52, 36	63		
3	96, 78	113, 95	56, 44	13	31.11	00:47:26
	96, 78	113, 95	56, 44	0		
	96, 78	113, 95	56, 44	71		
4	102, 88	111, 83	62, 46	8	30.37	01:06:37
	102, 88	111, 83	62, 46	0		
	102, 88	111, 83	62, 46	74		

(continued)

Table 2 (continued)

Training neural Networks	Submodule layers			Persons recognized by submodule	Recognition rate (%)	Execution time
	Submodule 1	Submodule 2	Submodule 3			
5	99, 87	118, 96	78, 53	10	32.59	00:47:33
	99, 87	118, 96	78, 53	0		
	99, 87	118, 96	78, 53	78		
6	85, 70	90, 80	60, 70	14	33.33	01:10:06
	85, 70	90, 80	60, 70	0		
	85, 70	90, 80	60, 70	76		
7	60, 40	150, 100	50, 25	10	12.59	01:07:10
	95, 80	150, 100	50, 25	0		
	90, 75	150, 100	50, 25	24		
8	64, 32	120, 60	80, 40	13	27.41	00:54:56
	64, 32	120, 60	80, 40	0		
	64, 32	120, 60	80, 40	61		
9	100, 80	100, 100	65, 50	11	31.85	01:11:56
	100, 80	100, 100	65, 50	0		
	100, 80	100, 100	65, 50	75		
10	50, 70	90, 70	75, 35	7	21.15	01:49:43
	50, 70	90, 70	75, 35	0		
	50, 70	90, 70	75, 35	51		
11	100, 90	50, 25	60, 60	15	31.48	00:48:04
	100, 90	50, 25	60, 60	1		
	100, 90	50, 25	60, 60	69		
12	110, 80	40, 60	80,70	19	35.93	00:47:26
	110, 80	40, 60	80,70	1		
	110, 80	40, 60	80,70	77		
13	120, 85	20, 30	60, 50	10	30.74	00:35:34
	120, 85	20, 30	60, 50	1		
	120, 85	20, 30	60, 50	72		
14	130, 90	50, 15	68, 32	17	26.67	01:22:33
	130, 90	60, 20	68, 33	1		
	130, 90	65, 25	68, 34	54		
15	125, 90	100, 150	70, 65	21	35.93	00:50:17
	125, 90	100, 150	70, 65	0		
	125, 90	100, 150	70, 65	76		
16	120, 85	70, 40	70, 30	15	34.81	01:10:40
	110, 80	60, 35	65, 45	2		
	125, 90	65, 45	75, 35	77		
17	135, 100	200, 100	80, 50	16	32.59	00:48:45
	135, 100	200, 100	80, 50	0		
	135, 100	200, 100	80, 50	72		

(continued)

Table 2 (continued)

Training neural Networks	Submodule layers			Persons recognized by submodule	Recognition rate (%)	Execution time
	Submodule 1	Submodule 2	Submodule 3			
18	150, 100	200, 150	80, 50	16	33.33	00:39:13
	150, 100	200, 150	80, 50	0		
	150, 100	200, 150	80, 50	74		
19	128, 72	95, 100	55, 43	21	23.70	01:17:49
	128, 85	100, 85	55, 43	0		
	128, 92	90, 78	55, 43	43		
20	165, 100	10, 15	75, 30	12	31.48	00:43:57
	165, 100	30, 50	75, 30	3		
	165, 100	50, 85	75, 30	70		

Our objective using the optimal integrator is to obtain the best mean square error and the highest percentage of recognition.

Table 2 shows the percentage of recognition obtained for 20 different training of the modular neural network; where we can see that without integration for network performance, we obtain low recognition rate of people. Table 3 shows the results of the non-singleton fuzzy integrators obtained and tested in the different modular neural networks; we can see that the recognition rate is improved.

Table 3 Non-singleton fuzzy integrators obtained with the GA tested in the modular neural networks, where the optimal integrator NS2FI_13 obtains the best error and achieving better recognition rate in the tested modular neural networks

No-singleton type-2 fuzzy integrator	Best error	Time (min)	Persons recognized	Recognition rate (%)	MNN tested
NS2FI_8	3.09E−12	00:08:20	30	100	1
NS2FI_11	2.02E−10	00:14:59	30	100	1
NS2FI_12	6.08E−05	00:26:49	22	73.33	1
NS2FI_13	9.98E−13	00:24:21	30	100	1
NS2FI_16	5.46E−09	00:16:27	1	3.33	1
NS2FI_8	3.09E−12	00:08:20	29	96.67	2
NS2FI_11	2.02E−10	00:14:59	27	90.00	2
NS2FI_12	6.08E−05	00:26:49	23	76.67	2
NS2FI_13	9.98E−13	00:24:21	29	96.67	2
NS2FI_16	5.46E−09	00:16:27	9	30.00	2
NS2FI_8	3.09E−12	00:08:20	30	100	4
NS2FI_11	2.02E−10	00:14:59	30	100	4
NS2FI_12	6.08E−05	00:26:49	30	100	4
NS2FI_13	9.98E−13	00:24:21	30	100	4

(continued)

Table 3 (continued)

No-singleton type-2 fuzzy integrator	Best error	Time (min)	Persons recognized	Recognition rate (%)	MNN tested
NS2FI_16	5.46E−09	00:16:27	28	93.33	4
NS2FI_8	3.09E−12	00:08:20	22	73.33	7
NS2FI_11	2.02E−10	00:14:59	25	83.33	7
NS2FI_12	6.08E−05	00:26:49	21	70	7
NS2FI_13	9.98E−13	00:24:21	30	100	7
NS2FI_16	5.46E−09	00:16:27	20	66.67	7
NS2FI_8	3.09E−12	00:08:20	30	100	10
NS2FI_11	2.02E−10	00:14:59	30	100	10
NS2FI_12	6.08E−05	00:26:49	29	96.67	10
NS2FI_13	9.98E−13	00:24:21	30	100	10
NS2FI_16	5.46E−09	00:16:27	28	93.33	10
NS2FI_8	3.09E−12	00:08:20	30	100	11
NS2FI_11	2.02E−10	00:14:59	30	100	11
NS2FI_12	6.08E−05	00:26:49	24	80.00	11
NS2FI_13	9.98E−13	00:24:21	30	100	11
NS2FI_16	5.46E−09	00:16:27	2	6.67	11
NS2FI_8	3.09E−12	00:08:20	30	100	15
NS2FI_11	2.02E−10	00:14:59	30	100	15
NS2FI_12	6.08E−05	00:26:49	30	100	15
NS2FI_13	9.98E−13	00:24:21	30	100	15
NS2FI_16	5.46E−09	00:16:27	30	100	15
NS2FI_8	3.09E−12	00:08:20	30	100	17
NS2FI_11	2.02E−10	00:14:59	30	100	17
NS2FI_12	6.08E−05	00:26:49	30	100	17
NS2FI_13	9.98E−13	00:24:21	30	100	17
NS2FI_16	5.46E−09	00:16:27	30	100	17

With the results presented, we can see that the NS2FI 13 provides good results; because when the modular neural network has low recognition rate by poor training, this fuzzy integrator performs better.

In previous sections, we presented the automatic design of non-singleton fuzzy integrators using a genetic algorithm to generate. To test the integrators we used a 20 training of modular neural networks for multimodal biometry, in which we can see that when using non-singleton type-2 fuzzy integrators, significantly increased the percentage of recognition. Table 4 is a comparison of the results obtained with interval type-2 fuzzy integrators and non-singleton interval type-2 fuzzy integrators presents.

Table 4 Comparison result using the interval IT2FI and NST2FI

FIS	Error achieved	% Recognition with interval type-2 fuzzy	Error achieved	% Recognition non-singleton type-2 fuzzy integration
1	0.003115353	100	0.00017105	100
2	0.001610318	100	3.09E−10	100
3	0.00294546	100	4.53E−13	100
4	7.87919E−05	100	0.000493723	100
5	0.00115187	100	3.26E−14	100
6	0.0001781	100	6.97E−13	96.67
7	0.0000218	96.67	4.63E−11	100
8	0.00298097	96.67	3.09E−12	100
9	0.00239668	100	1.44E−10	93.33
10	0.00681549	100	9.94E−13	100
11	0.00031891	73.33	2.02E−10	100
12	0.01700403	86.67	6.08E−05	100
13	0.00047306	93.33	9.98E−13	100
14	9.15214E−05	100	5.30E−12	100
15	0.00089849	100	1.39E−05	100
16	1.03302E−05	43.33	5.46E−09	93.33
17	0.00135865	100	4.60E−12	100.00
18	0.001623653	100	8.38E−04	93.33
19	0.00373201	100	3.52E−12	93.33
20	0.00031598	100	9.37E−11	93.33
Total percentage		94.50		98

We can observe how the type-2 non-singleton fuzzy integrators improved the mean square error and the total percentage of recognition obtaining a 98 % recognition.

6 Conclusions

This paper describesthe use of non-singleton interval type-2 fuzzy systems (NSIT2FS) automatically designed through genetic algorithms as integration method of modular neural networks (MNN's) for multimodal biometrics. Also a comparison was made with results of interval type-2 fuzzy systems and non-singleton type-2 fuzzy systems as integrators; the comparison between these methods of integration was made with the simulation results for Pattern Recognition; where it is noted that Type-2 Fuzzy Logic is superior to integration response in Modular Neural Networks for Multimodal Biometrics. Using Fuzzy Logic as method of integration, we can see the advantages we get to use it, and to make a comparative study between the use of fuzzy systems interval type-2 and non-singleton interval type-2 was found that non-singleton interval type-2 fuzzy logic allows us to equate or improve the results obtained with Interval Type-2 Fuzzy Logic.

In summary, it is concluded that the fuzzy integration response in modular neural networks applied to biometrics, gives good results, especially in singleton interval type-2 Fuzzy Systems, as it was noted in this case that better results in less time are achieved. We have obtained satisfactory results with the GA to design non-singleton interval type-2 fuzzy systems as integrators in modular neural networks for multimodal biometry.

References

1. J.-S.R. Jang, C.-T. Sun, E. Mizutani, Neuro-Fuzzy and Soft Computing, A Computational Approach to Learning and Machine Intelligence, Prentice Hall, 1997.
2. J. Mendel, UNCERTAIN Rule-Based Fuzzy Logic Systems, Introduction and New Directions, Prentice Hall, 2001.
3. J. Mendel, Uncertain Rule-based Fuzzy Logic Systems, Prentice Hall, 2001.
4. Mendel, J., John, R., «Type-2 fuzzy sets made simple,» *IEEE Transactions on Fuzzy Systems*, nº 10, pp. 117-127, 2002.
5. Sahab, Nazanin and Hagras, Hani, «Adaptive Non-singleton Type-2 Fuzzy Logic Systems: A Way Forward for Handling Numerical Uncertainties in Real World Applications.» *International Journal of Computers, Communications and Control,* vol. 3, nº 5, pp. 503-529, 2011.
6. T. S. M. Takagi, «Fuzzy identification of systems and its application to modeling and control,» *IEEE Transactions on Systems, Man, and Cybernetics,* vol. 1, nº 15, 1985.
7. Qilian Liang, Jerry M. Mendel, «Interval Type-2 Fuzzy Logic Systems:,» *IEEE TRANSACTIONS ON FUZZY SYSTEMS,* vol. 8, nº 5, pp. 535-550, 2000.
8. Liang, Q., Mendel, J. , «Interval type-2 fuzzy logic systems: theory and design,» *IEEE Transactions on Fuzzy Systems,* vol. 5, nº 8, pp. 535-550, 2000.
9. Mendel, J., Mouzouris, George C., «Type-2 fuzzy logic systems,» *IEEE Transactions on Fuzzy Systems,* nº 7, pp. 643-658, 1999.
10. O. Castillo, P. Melin, Studies in Fuzziness and Soft Computing, Type-2 Fuzzy Logic: Theory and Applications, Tijuana , Baja California: Springer-Verlag Berlin Heidelberg, 2008.
11. Ricardo Martínez-Soto, Oscar Castillo, Juan R. Castro, «Genetic Algorithm Optimization for Type-2 Non-singleton Fuzzy Logic Controllers. : 3-18,» de *Recent Advances on Hybrid Approaches for Designing Intelligent Systems*, 2014, pp. 3-18.
12. Denisse Hidalgo, Patricia Melin, Oscar Castillo, «Type-1 and Type-2 Fuzzy Inference Systems as Integration Methods in Modular Neural Networks for Multimodal Biometry and its Optimization with Genetic Algorithms,» *Journal of Automation, Mobil Robotics & Intelligent Systems,* vol. 2, nº 1, pp. 53-73, 2008.
13. Juan R. Castro, Oscar Castillo, Luis G. Martínez, «Interval Type-2 Fuzzy Logic Toolbox.,» *Engineering Letters,* vol. 1, nº 15, pp. 89-98, 2007.

14. Karnik, N.N., Mendel, J., «Centroid of a type-2 fuzzy set,» *Information Sciences,* Vols. %1 de %2(1-4), nº 132, pp. 195-220, 2001.
15. Denisse Hidalgo, Oscar Castillo, Patricia Melin, «Type-1 and type-2 fuzzy inference systems as integration methods in modular neural networks for multimodal biometry and its optimization with genetic algorithms,» *Information Sciences,* vol. 179, nº Issue 13, p. 2123–2145, 13 June 2009.